GEOLOGY

OF

U.S. PARKLANDS

GEOLOGY
OF
U.S. PARKLANDS
Fifth Edition

Eugene P. Kiver
Eastern Washington University

and

David V. Harris
(Deceased)
Colorado State University

John Wiley & Sons, Inc.
New York • Chichester • Weinheim • Brisbane • Singapore • Toronto

Copyright © 1999 by John Wiley & Sons, Inc. All rights reserved.

Published simultaneously in Canada.

This publication is designed to provide accurate and authoritative information in regard to the subject matter covered. It is sold with the understanding that the publisher is not engaged in rendering professional services. If professional advise or other expert assistance is required, the services of a competent professional person should be sought.

Library of Congress Cataloging-in-Publication Data:

Kiver, Eugene P.
    Geology of U.S. parklands / Eugene P. Kiver and David V. Harris.—
5th ed.
        p.   cm.
    Rev. ed. of: The geologic study of the national parks and monuments / David V.
Harris, Eugene P. Kiver : graphics by Gregory C. Nelson. 4th ed. c1985.
    Includes bibliographical references and index.
    ISBN 0-471-33218-6 (cloth)
    1. Geology—United States.   2. National parks and reserves—United
States.   3. Natural monuments—United States.   I. Harris, David V.
II. Harris, David V. Geologic story of the national parks and
monuments.   III. Title.
QE77.K59   1999
557.3—dc21   98-31447

Printed in the United States of America.

10 9 8 7 6 5 4 3 2

Contents

Preface

Anew concept was formalized in 1872—one that was to have a profound effect on a fledgling nation and the world. In that year Yellowstone National Park in the wilds of the Wyoming Territory was established as the world's first national park. Here was a reserve to be used for recreation not only by the wealthy and the privileged, but one that all Americans could visit and share equally as our national heritage. The desire to set aside areas of special natural and later cultural significance grew as national pride helped assemble the world's finest system of national parks.

Each of the 376 units under the jurisdiction of the National Park Service is a special place and each has a story to tell. Here in the fifth edition the geologic stories of 78 of these units as well as those of two areas administered by the U.S. Forest Service (Mount St. Helens and Newberry Volcanic Area) and one by the Bureau of Reclamation (Grand Staircase–Escalante) are covered. The need for more in-depth coverage of park areas by students and inquisitive visitors has resulted in a complete revision and a longer book, which in turn required that fewer areas be covered. Thus the emphasis in this edition is on more complete coverage of those parks in the "lower 48" and Hawaii, and requires the omission of the spectacular but less visited parks in Alaska.

The more in-depth coverage of the fifth edition enables those students and park visitors who wish to delve deeper into the workings of the earth to do so without excessive outside reading. Recognition that the earth is a heat-driven dynamo that responds to internal heat by movement of rigid surface plates floating on a more fluid zone at depth (plate tectonics) has revolutionized the geological sciences. Tentative working models of how these plate processes influenced specific areas are presented with full realization that many of these explanations will be modified or replaced in the future.

As in previous editions, the fine areas belonging to the hundreds of state parks and national forest and Bureau of Land Management lands across the nation are excluded. The U.S. Forest Service and Bureau of Land Management lands in particular contain not only lands of special scientific and scenic interest, but also the last remnants of the expanses of undeveloped land that inspired and continues to inspire the free spirit of a nation and its people.

All of these public lands are precious to the psychological and spiritual health of our people and provide an opportunity for us to pass on to future inhabitants a land worth living in. If we as individuals could somehow return in 100 years and look our successors in the eye, could we honestly say to them that we are proud of what has happened to the land and the environment?

This edition follows the same organizational scheme established by Dave Harris in the first edition in 1976. A regional approach based on geomorphic provinces enables one to better develop the "big picture" of how the restless earth has changed through geologic time and how similar sequences of events have created large geographic areas with similar scenery and geologic characteristics.

Also included are short historical introductions that briefly record human influences, including those that eventually resulted in the establishment of a park. The individual(s) and groups who devoted their efforts and often much of their lives and energies to the establishment of parks for future generations to enjoy were noble people indeed. Such role models should inspire others to recognize that similar work still needs to be done and that such work must be done now before the window of opportunity closes!

It was with great sadness that this revision had to be undertaken without the able leadership, enthusiasm, and humor of Dave Harris who passed away in 1989. Dave's long career of passing on his knowledge and enthusiasm to generations of students and those who seek to understand the science behind our national parks will be sorely missed. It was Dave who put together the first edition of this textbook in 1976—the first textbook to cover not only the geology of the national parks but some of the equally significant geology in the national monuments as well. Although the title *The Geologic Story of the National Parks and Monuments* is being changed to *Geology of U.S. Parklands* in this new edition, the coverage remains essentially unchanged except for the necessity of dropping the Alaskan parks in order to keep book costs down. New additions include Great Basin National Park, Grand Staircase–Escalante, Hagerman Fossil Beds, Newberry Volcanic, and El Malpais National Monuments.

Numerous people have generously contributed photographs, figures, and information included in the book. Many of those people are given credit in the figure captions and in acknowledgements included with earlier editions. A number of Park Service personnel and others graciously furnished information that made the revision work much easier. Special mention should be given to Sid Ash, Wally Bothner, Carl Bowman, Larry Chitwood, Bob Jensen, Bruce Kaye, Allyson Mathis, Greg McDonald, Tom Miller, Felix Mutschler, and Mimi McColloch.

Special thanks go to Cleo Harris who read and corrected the entire manuscript in an effort to make it more intelligible. I now realize the truth of Dave Harris's 1985 comment that "Cleo Harris disentangled the truth from many otherwise bewildering paragraphs." Although most of the grammatical problems were skillfully recognized and solved by Cleo, any mistakes in interpreting the science are entirely my responsibility.

Eugene P. Kiver

ONE

Introduction

Our national parks are a unique American heritage. Unlike many of our institutions that have their roots in European soil, the national park concept is our own. Preservation of land and open space as hunting preserves for the benefit of a few is quite old—preserving an area for all people for all time is a unique accomplishment that requires people with vision and an unselfish attitude.

The concept of a park for all the people was born in 1870 in the wilderness that is now Yellowstone National Park. Prior to this time, little was known of this remote area. Indians roamed the Yellowstone country, but apparently they were tight-lipped about what they had seen. Historical records indicate that the first white man to see the wonders of Yellowstone was John Colter, one of Lewis and Clark's men, in 1807. His account of what he saw was too incredible for most people. Several years later, mountain-man Jim Bridger went into the Yellowstone country and came out with stories similar to Colter's, and similarly, they were ridiculed. According to Chittenden (1895), Bridger resented the fact that no one took him seriously and retaliated with fabrications so preposterous that they would give him some sort of renown. It is likely that his tales will stand tall for some time: very few people accept as fact his claim that the glass in Obsidian Cliff acts as a telescope lens to bring within shooting distance an elk 25 miles (42 km) away!

Eventually, curiosity as to whether there was any truth in Colter's and Bridger's stories led to the exploration of the Yellowstone country. In 1870 General Henry D. Washburn led a party of Montanans into the area to find out what was actually there. At the end of the summer they realized that here was something truly unique and that once it was publicized, people would eventually flock in to see it. General Washburn and his group could stake their claims and make a good thing out of it for themselves.

Sitting around a campfire, watching the flames lift into the darkness, is likely to bring out the best in a person, and finally Judge Cornelius Hedges startled his companions by saying that this area with all its awesome and beautiful features should belong to all the people to see and enjoy. In due time they agreed, and the

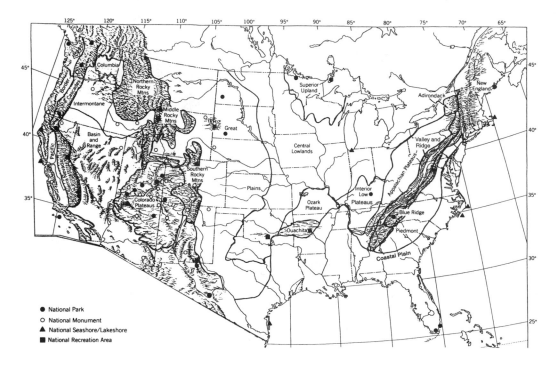

PLATE 1   Geomorphic provinces of conterminous United States. For names of parklands shown here, refer to the province maps in various chapters.

idea of national parks took shape—parks to be used as "pleasuring grounds for all of the people."

Although Judge Hedges may not have been aware of it, almost 40 years earlier George Catlin had proposed to the government that a large tract of land around the Yellowstone geysers be preserved as the "Nation's Park" (Chittenden, 1895). Catlin is better known for his work of preserving records of the American Indian in his writings and in his hundreds of paintings, many of which are in the Smithsonian Institution in Washington. Later, in 1864, the idea of a large public park was furthered when Abraham Lincoln ceded federal land in the Yosemite area to the state of California for public use and recreation.

Although Catlin must be given credit for the concept and California pioneered the state park idea, it was Judge Hedges, N.P. Langford, and others who, by their tireless efforts, saw to it that this time the seed would grow and bring forth fruit. More evidence was needed to convince the skeptics in Washington, and in 1871 the highly respected government geologist, F.V. Hayden, and a party of scientists were sent to explore the area (Fig. 1–1). Fortunately, their official report was convincing and on March 1, 1872, Congress established Yellowstone as a national park, the first in the world.

Sensitive people recognized that the human spirit needs more than "creature comforts" and material possessions and that to reestablish our connection with the

FIGURE 1–1   Camp of F.V. Hayden's geological survey party on Yellowstone Lake in 1871. (Photo by W.H. Jackson, U.S. Geological Survey)

earth and its beauties would help satisfy that need. So the national park idea spread rapidly even into areas that were then remote. Australia was the first to follow our lead; then the New Zealanders indicated their approval by creating their first national park, later increasing the number to 10. Our neighbors the Canadians are, with good reason, proud of their Glacier, Banff, and Jasper national parks, along with many others. Now, more than 100 nations have national parks or reserves; unquestionably the world's highest national park is at Mt. Everest. But the idea took shape as the flames rose above that campfire in Yellowstone.

All these areas are monuments to our unselfish predecessors who have made important contributions and sacrifices for the benefit of all. Let us hope that these natural monuments will inspire the present occupants of the land to be good stewards and do the same for future generations. If we were to return in 100 years, would we be proud of what has happened to the land, water, air, plants, and animals in our lifetimes?

NATIONAL PARK SYSTEM

As pride in our nation and our heritage has grown, the original national park concept has been expanded. Now certain park areas preserve sites of significant historic interest, such as Valley Forge; others preserve archeological sites, such as Mesa Verde. There are also national seashores, national lakeshores, national parkways, national memorials, national preserves, and national recreation areas. Essentially all of the national parks and many of the monuments are noted primarily for their classic geologic features and their spectacular scenery. However, all of the parks and monuments have a larger story to tell that includes not only the history of the land but also of its plants and animals and of the interactions of humans with all of these elements. The complete park experience is to see the "big picture," not usually possible in one visit, especially if it is hurried.

The parks and other areas administered by the National Park Service are among the most common vacation destinations for U.S. citizens as well as many visitors from other countries. It is particularly affordable for families and promotes a closeness that strengthens the human bonds and helps develop the respect for the land and the values that make better citizens and individuals.

In general, parks are larger and contain a wider variety of natural features than monuments. The main distinction between the two is the way they are established. An act of Congress is necessary for the creation of a national park, whereas the President can establish a monument. Thus a threatened area can be quickly set aside as a monument and, if there is justification, reclassified later as a national park. Inasmuch as the difference is mainly administrative, the term *park,* as used in this book, refers also to monuments, except where otherwise specified.

By no means are all of the country's beautiful areas within the National Park System. Particularly significant are the vast areas that are administered by the U.S. Forest Service and the Bureau of Land Management. Then there are the state parks, which serve much the same purpose as the national parks, namely, to preserve natural and historic features (Fig. 1–2).

FIGURE 1–2   Reflected in a quiet pool on Sugar Creek is the bluff of Mansfield Sandstone (locally called "millstone grits") in Turkey Run State Park, western Indiana. Created in 1916 to preserve hardwood trees, the park also protects relict plants hidden in the cool recesses of the deep ravines. (Photo by D. Harris)

NATIONAL PARK SERVICE

When Yellowstone was established as a national park, much of the West was still untamed and so were many of the people who roamed these remote areas. They were rugged individuals who were in the habit of doing as they pleased and taking what was there. In general they were not at all impressed with the national park concept, and park boundaries were commonly ignored.

N.P. Langford, one of the Washburn party, was named as the first superintendent of Yellowstone National Park. Langford was given the title of superintendent—and the nickname "National Park Langford"—but for 5 years, Congress neglected to appropriate any money for his salary or for the park. Consequently, he was powerless to carry out his duties. Poachers decimated the bison herd, and serious damage was done to the formations in Yellowstone's thermal areas. Similar problems were occurring in Yosemite as well. Finally, as a stop-gap measure, the U.S. Army was called in to provide protection for the fledgling parks until a more permanent solution could be found. Other inroads that threatened the park's integrity were made, especially in Yosemite National Park where, despite John Muir's valiant efforts, the power interests prevailed and Hetch Hetchy Dam destroyed for all time one of the truly magnificent canyons in Yosemite. Instead of two incredible valleys at Yosemite, only one remains; the other is flooded by a reservoir—the triumph of economics (small scale at that) over irreplaceable scenery.

Stephen T. Mather, a prominent Chicago businessman, was appalled at this sorry treatment of our national treasures, and he wasted no words in expressing his feelings on the matter. The result was that he was given the job of managing all the parks and monuments. Mather was very persuasive. In less than 2 years, on August 25, 1916, Woodrow Wilson signed the act that established the National Park Service within the Department of Interior. As director, Mather devoted his life to his job and when federal funds were lacking, he used his own money for the operation and improvement of the parks. At last, the way was paved for the preservation of the parks as prescribed by law more than 40 years earlier.

The National Park Service is headed by a director, appointed by the President and approved by Congress, who is responsible to the Secretary of the Interior. In each of the parks and monuments a superintendent is in charge of operations; with a staff of assistants, he or she sees to it that the natural, historic, and cultural features are properly interpreted for park visitors. Whether you go into the visitor center down at Oconaluftee or up at Ohanapecosh, the park people will do their best to make your visit both enjoyable and educational. They also do their best to protect the park from unthinking people who could unwittingly damage the environment.

Through the years, the National Park Service has evolved into an organization of well-trained men and women. Unfortunately, government cutbacks have made it difficult to hire and hold a sufficiently large cadre of dedicated professionals. Lack of funds often means that, understandably, law enforcement and park protection is funded better than the interpretation staff. Personal sacrifice for the rangers and their families is the rule. The fact that there are park rangers who qualify for government food stamps does not bode well for the long-term professional management of our precious parklands. Many of the ranger staff are employed only during the summer, and they too are dedicated to sharing the park's mission "to conserve the scenery and the national and historic objects and the wildlife therein and to provide for the enjoyment of the same in such a manner and by such means as will leave them unimpaired for the enjoyment of future generations," as set forth in the original National Park Organic Act.

This key phrase in the Organic Act is the basis for many philosophical and park management conflicts—how do you allow people to use an area and yet preserve it in an unimpaired condition? In the early days development was emphasized to ensure public and congressional support of the park system. The railroads also had a stake in park development with passenger trains and operation of tourist facilities at Yellowstone, Glacier, and other parks. Their powerful influence helped the park system succeed. The hard times of the Great Depression gave rise in 1933 to a federal organization called the Civilian Conservation Corps (CCC). Here was a work force that enabled "jump start" development of physical facilities in our park system. Campgrounds, roads, trails, and many other necessary improvements that would have otherwise taken decades to complete were completed in a few years by this hard-working group.

Sometime later, after the CCC efforts, things got out of hand—golf courses, barber shops, an airport, and banks are not facilities appropriate to a national park. However, there are some who believe that all sorts of accommodations and recreational facilities should be constructed; others are quite certain that no development should be permitted. Theodore Roosevelt's comment concerning the Grand Canyon is just as applicable today to our parks as it was in the early 1900s, especially to those parks that emphasize natural features: "Leave it as it is. The ages have been at work upon it, and man can only mar it. What you can do is to keep it for your children, your children's children, and for all who come after you."

With the overwhelming majority of Americans strongly supportive of our national parks, it is amazing to learn that certain economic interests and some members of Congress are promoting closure, privatization, and even elimination of some of our parklands. Those who subscribe to economic determinism, some of whom are in high government offices, would give away public lands to the states, including some of our parklands. The states would retain the economically profitable land and sell the rest of our national heritage to the highest bidder—speeding up the process of making the United States more like our European cousins with almost every square centimeter being used for an economic purpose. Local control of public lands would further reduce the national pride that binds the widely spaced areas of our country into one nation—a goal that some are pursuing.

Even if the economic and political pressures on our parklands can be brought under control, we have yet another problem to contend with—population increases that result in more visitors and more development and loss of the "buffer zones" around the park edges. With 250 million visitors to our parks in 1992 and an expected 500 million by 2010, serious problems in park management lie ahead. With over 6600 people added to the United States population every day and that number increasing each year, the country will double its population in about 40 years. This pressure by itself will place immense stress on the environment and will reduce the open spaces that have given Americans a special spirit of freedom and self-reliance. As the cartoon strip character Pogo once said, "We have met the enemy and he is us."

Given these circumstances, it is alarming to consider what can take place if people in high office have but little understanding of or concern for the environment or for the future of our parks and wilderness areas. A strong effort toward education of the general public and public officials about the long-term outlook for our parks, our nation, and our planet is apparently necessary. Without constant vigilance on the part of the general public, such officials can exercise their power to reinterpret the law or change the law. Although it is true that a substantial amount of geothermal energy could be developed in Yellowstone National Park, the overwhelming majority of Americans would not be willing to trade Old Faithful for a whole nest of hissing steam wells.

PARKS AND PEOPLE

National parks mean different things to different people. To one person a park may be an escape from a hectic life; to another it is a great place to ski. As mentioned previously, family vacations often involve stays in national parks—many of these families camp and hike together—activities that promote family values, self-sufficiency, and respect for our humble beginnings and our continued survival (Fig. 1–3).

Scientists are particularly interested in the park areas. In these relatively undisturbed natural environments biologists find plants and animals that have been largely eradicated elsewhere. Over eons of geologic time chemical compounds that have evolved in some rare and endangered species in our parks may hold the key to control cancer and other diseases. Discovery of the microorganism *Thermopolis aquillus* in the thermal pools at Yellowstone has enabled researchers and industry to speed up the DNA replication process and has helped open up new frontiers in science. A geologist can see on full display, in our parklands, classic examples of rocks and structures that reveal the history of the earth.

Most people visit the parks because of the beautiful scenery, still mostly unspoiled. They are likely to be more knowledgeable than visitors of earlier days. They are interested in the total environment, both physical and biological, and they want to know what is back of the scenery—what caused it to be as it is.

Behind the scenery and beneath the soil growing the plants that feed the animals are the rocks that made this all possible. The story of the rocks and the landscapes and how they came to be is the geologic story of our national parks. As we shall shortly see, behind spectacular scenery there is spectacular geology.

FIGURE 1–3   Park naturalist points out glacial features in the mountains above Nymph Lake in Rocky Mountain National Park. Note the sample of park visitors—retired people, families with young children, and others. (Photo by National Park Service)

GEOLOGIC PRINCIPLES AND PROCESSES

This section is for the benefit of those who have had no previous training in geology. A number of geologic concepts are capsulized here and elaborated on later in the discussions of various parks. Somewhere in the National Park System examples of practically every geologic principle can be found, often among the earth's finest. Although this introductory treatment will probably be sufficient for most readers, some may be interested in delving more deeply, and for them several geology textbooks are listed at the end of the chapter.

Measuring Geologic Time

Geology is the study of the earth and its history, as revealed in its rocks, sediments, and landforms. What is revealed in the national parks and monuments, when pieced together, gives us a remarkably complete history of the earth.

The earth is constantly changing, and it has been changing throughout its 4.5-billion-year history. In some cases, as when Mexico's Paricutin Volcano was born in 1943 or when Iceland's Surtsey Volcano boiled up out of the sea in the 1970s, the changes are spectacular because they are, geologically speaking, instantaneous. In most cases, though, the changes are so gradual that many people assume that a mountain or a canyon "has always been there." But the mountain was not always there; Grand Canyon was no canyon at all 10 million years ago.

The concept of geologic time is one of the hardest to grasp. It is difficult for some to imagine back 5000 years ago when the Egyptians were constructing the pyramids or 30,000 years ago when our ancestors were drawing on cave walls. To imagine a time 4 million years ago when no humans existed or a time tens or hundreds of millions of years ago when the rock in which the caves formed was nonexistent requires a real intellectual jump! Most people are in the habit of thinking of time in terms of human existence. In the early days, especially in western societies, people believed that the earth was only a few thousand years old and that mountain ranges were the result of "colossal convulsions" within the earth that formed almost overnight. Later, when the true meaning of fossils became apparent and biological evolution was finally accepted, the immensity of geologic time was gradually recognized.

Much later, when the radiometric method of age determination was perfected in the mid 1900s, we had for the first time the means for assigning more accurate ages in terms of years or millions of years. Although the method is highly technical, the principles are given here in simplified form. Certain forms of several chemical elements are radioactive; that is, their nuclei decay spontaneously and transmute them from a "parent" to a "daughter" element (or isotope) at a known rate. Newly formed minerals exclude daughter elements; thus the number of radioactive parent elements decreases with time while the number of daughters increases. For example, one form of the element uranium decays at an extremely slow rate, so that after approximately 4.5 billion years, one-half of the uranium is left. After another 4.5 billion years, one-half of that half, or one-fourth of the original amount, is left. Therefore, this form of uranium has a *half-life* of 4.5 billion years.

Radioactive carbon has a much shorter half-life, only about 5700 years. As long as an organism is alive it absorbs small amounts of naturally produced carbon 14 from its environment. Upon death the carbon 14 is no longer replaced and the amount remaining follows a predictable path of decrease with time. Measurement of small quantities of radioactive material after nine or ten half-lives becomes difficult and limits the upper range of radiometric dating, especially for those isotopes such as carbon 14 that have very short half-lives. Radiocarbon dating is therefore used for age determinations of extremely young rocks, usually those that are less than 50,000 years old. However, some newer instruments can now extend the useful radiocarbon dating range back to 125,000 years, still a drop in the geologic bucket of time.

Unfortunately, many rocks do not contain minerals that have radioactive elements suitable for dating purposes. And even though a mineral in a sedimentary rock contains a radioactive element, the age determined is likely not the age of the sedimentary rock but, rather, the age of the original igneous rock in which the radioactive mineral crystallized. An exception is a very young rock or sediment, less than 125,000 years old, in which radioactive carbon from living organisms was incorporated. For these reasons, and others, many formations are assigned only approximate ages based on extrapolations of data. Nevertheless, the radiometric method represents significant progress, a breakthrough in geologic dating. Age dates given in this book will often be given as millions of years ago and will be abbreviated as *Ma*.

How does a geologist go about deciphering the history of the earth, the complex structures exposed in the mountains, and the surface features we see everywhere? Prior to 1785, there was no systematic approach; instead, there was mere speculation. In 1785, a Scotsman by the name of James Hutton published a paper in which he stated that the geologic processes now operating are the same processes that have, throughout geologic time, changed the earth and formed it as we see it today. Thus, "the present is the key to the past." This simple concept, which became known as the uniformitarian principle, is fundamental in geologic thinking. In fact, geology became a science when the uniformitarian principle was found to be sound and acceptable. For the first time, geologists had a rational, logical approach to use in interpreting the structures and features of the earth.

Although Hutton was the first to publish, others also deserve much credit; John Playfair undertook the task of restating in more understandable form many of Hutton's ideas. And later on, in 1830, Charles Lyell published his three-volume *Principles of Geology* in which the uniformitarian principle was firmly interwoven. It was a monumental work and was largely responsible for the acceptance of the uniformitarian principle.

Minerals: Building Blocks of Rocks

The solid part of the earth is made up of rocks that differ in composition, structure, and color; with a few exceptions each of these rocks is composed of several minerals. For example, granite is made up of the *essential minerals* quartz and orthoclase feldspar and may contain one or more of the following *accessory minerals:* muscovite mica, biotite (black mica), hornblende, or garnet. A few rocks, however, are composed of one mineral; for example, pure limestone contains millions of tiny crystals of the mineral calcite.

Concentrations of certain minerals can be found in fractures and other openings in the rocks. When such minerals have economic value and the quantity is sufficient, it is called an *ore deposit*. Frequently, two or more minerals containing lead, zinc, copper, iron, tungsten, or other valuable metals occur in these concentrations. During the mining boom in the 1800s and early 1900s, large quantities of gold, silver, copper, lead, and zinc were extracted from huge mines in Colorado, Montana, Idaho, Arizona, California, and elsewhere. In the United States, natural resources of all kinds were once abundant—inexhaustible in the minds of those promoting mining. But now an alarming number of our nonrenewable resources have been exhausted, or nearly so. Clearly, conservative use of what we have left is indicated to postpone the day when we become one of the "have-not" nations—when we will be forced to mine our landfills. Efficient use, substitution of materials, and conscientious recycling by individuals and corporations will lead us away from a "throw-away" mindset and will postpone and perhaps eliminate the have-not nation status toward which we are heading.

The term *mineral* is used in both a general and a specific sense. When we speak of mineral resources we include oil, gas, and coal, none of which is a true mineral. To mineralogists, minerals are naturally occurring, inorganic, solid substances, each with its own specific physical properties (hardness, color, crystal form, cleavage, etc.) and its own special chemical composition and internal arrangement of atoms. Consequently, by making certain simple tests and observations, we can identify in the field most of the common minerals that we find. For the less common ones, the mineralogist uses the petrographic microscope, which measures optical properties, or X-ray analysis.

Minerals are formed by different processes, the simplest of which is by crystallization from a water solution, as a result of evaporation. You can observe this process by dissolving a teaspoon of table salt (NaCl) in a half-cup of hot water; then, using a flat pie pan, evaporate it slowly over a low flame. With evaporation, the solution becomes saturated with sodium (Na) and chlorine (Cl), which will combine to form crystals of artificial halite (NaCl). When crystallization begins, allow the solution to cool very slowly and observe the tiny cubic crystals, using a magnifying glass if necessary. In a few hours, you observe what has taken place many times during geologic history, where over thousands of years seawater evaporated in enclosed basins, thus forming layers or beds of salt.

Minerals also form when *magmas* or *lavas* cool. Should you time your trip to Hawaii fortuitously, you could observe minerals being formed as lava cools and crystallizes. Lavas are *extrusive igneous rocks* ("fire-formed" magma erupted onto the earth's surface) that cool rapidly, too rapidly for the crystals to grow to appreciable size; therefore, most or all of the minerals in lava rocks are too small to see with the naked eye. A few glassy, green crystals of *olivine*—crystals that grew in the slowly cooling magma before it reached the surface—might be visible. When large masses of magma crystallize far beneath the surface as *intrusive igneous rocks* all of the crystals grow to considerable size, large enough to be more easily recognized and identified. Granite, the most common coarse-grained igneous rock, is formed under such conditions.

Another way minerals form is by crystallization directly from the vapor state. *Hematite* (iron oxide), *sulfur,* and other minerals have been observed in the process of crystallization around tiny volcanic vents where extremely hot gases are pouring forth. You could have observed this process in the Valley of Ten Thousand Smokes in Alaska if you had been up there when the vents (*fumaroles*) were especially active from 1912 to about 1940. Possibly only two or three vents are still emitting gases today from the nearly cool ash sheet deposited during the giant 1912 eruption.

The world of minerals is large; more than 2500 minerals have been described and cataloged. Of this vast array, however, almost all of the minerals that you will likely find can be identified by searching through the 20 described in Table 1–1.

Most minerals are relatively stable in the geologic environment where they were formed. Even so, very gradual changes are brought about on the surface by such processes as *weathering* and *erosion*. And if the rocks containing the minerals

TABLE 1–1   Minerals

MINERAL	PROPERTIES[a,b]	MISCELLANEOUS
AMPHIBOLE A group of complex Ca, Mg, Fe, Al silicates	H: 5–6 Cl: 2 planes at about 60°	Generally dark-colored to black. Hornblende is a common variety.
BIOTITE Complex silicate of K, Fe, Al, and Mg	H: 2.5–3.0 Cl: 1 plane, perfect cleavage	Black mica.
CALCITE Ca carbonate	H: 3.0 Cl: 3 planes, not at right angles	Generally white. Transparent variety—Iceland Spar.
CHALCEDONY (Cryptocrystalline quartz) Si dioxide	H: 7 Cl: none; conchoidal fracture	Commonly white to gray; translucent.
CHLORITE Complex hydrous Mg, Fe, Al silicates	H: 1–2.5 Cl: 1 plane (like mica)	Color: grass-green. Common in metamorphic rocks.
EPIDOTE Complex Ca, Fe, Al silicates	H: 6–7 Cl: 1 plane	Pistachio green.
GARNET A complex silicate mainly of Ca, Fe, and Al	H: 6.5–7.5	Commonly brown to red; some varieties are semiprecious stone.
GYPSUM Hydrous Ca sulfate	H: 2.0 Cl: Perfect in 1 plane; imperfect in 2 other directions	Colorless to white. Fibrous variety—satinspar.
HALITE Na chloride	H: 2–2.5 Cl: 3 perfect at right angles	Rock salt; salty taste.
HEMATITE Fe oxide	H: 5–6.5 (earthy varieties softer)	Generally red to reddish-brown; may be black. Most important iron ore.
KAOLINITE Hydrous Al silicate	H: 1–2	White when pure. One of the important clay minerals.
LIMONITE Hydrous Fe oxide	H: 3–5.5	Yellow, brown to black.
MAGNETITE Magnetic Fe oxide	H: 5.5–6.5	Strongly attracted by a magnet.
MUSCOVITE K, Al silicate	H: 2–3 Cl: perfect in 1 plane	White mica; may be transparent.

TABLE 1–1   (*Continued*)

MINERAL	PROPERTIES[a,b]	MISCELLANEOUS
OLIVINE		
Mg, Fe silicate	H: 6.5–7	Yellowish-green to brownish-green; olive green.
ORTHOCLASE FELDSPAR		
K, Al silicate	H: 6 Cl: 2 planes at 90°	Typically pink; important constituent of granite.
PLAGIOCLASE FELDSPAR		
Na, Ca, Al silicate	H: 6 Cl: 2 planes not at 90°	Generally white to gray.
PYRITE		
Fe sulfide	H: 6–6.5 Cl: none	Fool's Gold; brassy-yellow color. Source of sulfuric acid.
PYROXENE		
Ca, Mg, Fe silicate	H: 5–6 Cl: 2 planes nearly 90° apart	Augite is a common variety; greenish black.
QUARTZ		
Si dioxide	H: 7 Cl: none; uneven to conchoidal fracture	Hexagonal (6-sided) crystals common. Important constituent of granite and rhyolite.

[a]*Hardness* is the resistance to scratching or abrasion. A copper penny has a hardness of 3; a fingernail about 2.5; the steel in a good pocketknife about 5.5.

[b]*Cleavage:* When broken, most minerals will split along one or more planes of weakness. Quartz (and a few other minerals) has no planes of weakness in the crystal structure and therefore has no cleavage.

More than 2000 minerals are found in the rocks of the earth's crust; the 20 given in this table are the most common. The physical properties of hardness (H) and cleavage (Cl) are most helpful in the identification of minerals. A few minerals have a characteristic color, but for most minerals the color varies greatly due to impurities. CHEMICAL SYMBOLS: Al-aluminum; Ca-calcium; Fe-iron; K-potassium; Mg-magnesium; Na-sodium; Si-silicon.

are moved into a distinctly different geologic environment—where temperatures and pressures are extremely high—they will be profoundly changed by the process known as *metamorphism* and may ultimately melt and become magma (liquid rock) again. By these and other geologic processes, rocks are broken down or changed, thus destroying their identities; but other rocks are constructed from their remains, in a never-ending cycle—the rock cycle.

Rock Cycle

In visualizing the rock cycle (Fig. 1–4), magma is a good starting point. The early earth and solar system likely began with large accumulations of cosmic debris that

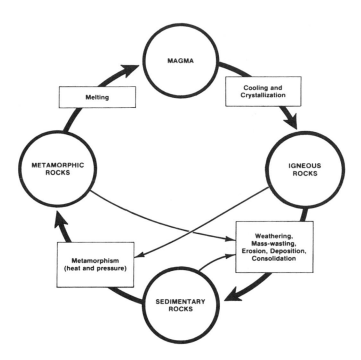

FIGURE 1–4   The rock cycle, showing the main geologic processes involved in the change from one genetic rock group to another.

gravitationally compressed into the sun and planets. Heat generated by this process caused melting of the earth with solidification and formation of a solid crust about 4.5 billion years ago! Thus all of our rock materials, except for newer infalls of cosmic sediment, is derived from igneous rock. Magma is molten rock material that is beneath the surface; the term lava is applied to the same material that has reached the surface. When magma (or lava) cools sufficiently, minerals begin to crystallize, and when all of the liquid has become solid, it is called igneous rock. If such rock is exposed at or near the earth's surface, perhaps by erosion, it eventually becomes weathered; that is, it is broken down, both physically and chemically, and is ready to be eroded away. The loosened materials that are transported away by some eroding agent (water, wind, ice, etc.) and then deposited are sedimentary rocks in the making. The loose materials must be cemented together or compacted sufficiently to form a solid rock. If these *sedimentary* rocks are subjected to enough pressure and heat, they become unstable and are transformed into *metamorphic* rocks. Then, if the metamorphic rocks are subjected to sufficiently high temperatures, they melt, thus forming magma. In this way the rock cycle is completed (indicated by the heavy line in Fig. 1–4). This is the ultimate in recycling—material from Mount St. Helens or other volcanoes may be making its appearance at the earth's surface for the *n*th time.

There are also other possible changes. Sedimentary rocks may be weathered and reconstructed into younger sedimentary rocks, or they may be metamorphosed and eventually melted into a magma. Igneous rocks may be metamorphosed or weathered. And metamorphic rocks may be weathered and reformed into sedimentary rocks (indicated by the light lines in Fig. 1–4).

The rock cycle can be used as a skeletal outline in the study of geology. Here, a synthesis of some of the more important aspects of geology is presented. It is not complete, but for those without geologic training it may provide sufficient background for the understanding of this book, with occasional reference to any good geology textbook, a few of which are listed in the references at end of chapter.

Origin of Magma

Before proceeding further, perhaps we should consider the problem of the origin of magma, lest someone assume (as the ancients did) that magma is present everywhere, beneath a thin "crust." Actually, almost all of the earth is solid, except for a liquid envelope around the small inner core at the earth's center, and for masses of magma that underlie specific limited areas such as parts of the Hawaiian Islands, Yellowstone, Mount St. Helens, and other areas where mountain building continues today.

We know that temperatures increase with depth below the surface, but we must also remember that pressures increase in the same manner and that pressure tends to prevent rocks from expanding and melting. Consequently, we have an extremely complex system on our hands. Exhaustive discussion of the problem would require more pages than we have space for, or need, in this book. Brief treatment of the principles, however, may be helpful—without inflicting any permanent psychological damage.

At a depth of several miles the temperature is above the melting point of rocks under surface pressure, but they remain solid because of the enormous pressure of the overlying rocks. Yet, if for any reason the temperature is increased sufficiently to overcome the effect of pressure, partial melting will take place. If radioactive minerals are concentrated in certain deep-seated rocks, the heat generated by their disintegration may provide the necessary increase in temperature. Or, in an active fault zone where rocks are being thrust over other rocks, the heat produced by friction might be sufficient. Heat from this mechanism would be particularly significant deep in a subduction zone, a topic discussed later in this section. On the other hand, should widespread, rapid erosion remove huge quantities of rock material from a mountain range, the resulting reduction of pressure would perhaps permit melting, directly or in combination with temperature increases. Slow upward movement of the viscous rock located deep along a subduction zone can also bring hotter rock closer to the surface where pressures are less. Thus, release of pressure would again permit partial melting of a rock body—the resulting fluid would be less dense than the surrounding rock and would tend to rise buoyantly toward the surface.

Because of the difficulties involved in determining with accuracy just what conditions exist where magmas form, the problem of the origin of magma, like certain other geologic problems, may not be completely solved for some time. But with new additions to our knowledge and with new techniques, we are much closer to the solution than we were a century ago.

Igneous Rocks

Magmas differ in chemical composition, mainly because of the differences in the mineral composition of the particular rocks that were originally melted. Because silicon (Si) and oxygen (O) are the two most abundant chemical elements in rocks of the earth's crust (the outer layer averaging about 25 miles, or 40 km, in thickness), all magmas contain silica (SiO_2) in moderate to large quantities. For this reason, the silica content of igneous rocks serves as one of the two major bases of igneous rock classification (see Table 1–2).

Partial melting of the layer below the crust, the mantle, produces a fluid of relatively low silica content whose chemical composition is that of the dark-colored lava called *basalt*. If the melt moves rather rapidly through the high silica rocks characteristic of the overlying crust, contamination is minimized and basalt lavas, like those in Craters of the Moon, Lava Beds, Sunset Crater, and other areas, reach the earth's surface practically unchanged chemically. However, if the basaltic magma resides at the base of the crust or in the high silica crustal rocks for a significant period of time, complete melting of the silica-rich crustal rocks can produce a very high silica magma possessing the composition of a rock called rhyolite. Partial melting of the crustal rock or mixing of magmas of different composition can produce extrusive igneous rock of intermediate silica composition, such as andesite or dacite.

The second criterion used as the basis for the igneous rock classification shown in Table 1–2 is the texture of the rock—are the mineral grains large (seen

TABLE 1–2   Simplified Igneous Rock Classification

| | SILICA (SiO_2) CONTENT | | |
TEXTURE	HIGH (75–62% SiO_2)	MEDIUM (62–54%)	LOW (54–45%)
Coarse-grained (intrusive)	Granite Granodiorite	Diorite	Gabbro
Fine-grained (extrusive)	Rhyolite Dacite	Andesite	Basalt
Glassy	Obsidian		Basalt Glass (rare)
Vesicular	Pumice		Scoria

Volcanic tuff (fine-grained) and breccia (coarse, angular fragments) are common rocks formed in areas of explosive volcanic activity. Note: as silica (Si), sodium (Na) and potassium (K) increase; calcium (Ca), iron (Fe), and magnesium (Mg) decrease.

with the naked eye), small (visible with a microscope), or absent (glassy or obsidian-like)? Does the rock have a frothy appearance due to gas bubbles emanating from the hot lava and being trapped by the hardening rock?

High-silica rocks are composed mainly of light-colored minerals such as quartz (pure SiO_2) and orthoclase feldspar (a silicate, $KAlSi_3O_8$). If the magma cooled slowly as an intrusive igneous rock and the rock is therefore coarse-grained, it is called granite; if it is fine-grained, its name is rhyolite; and if it was chilled too rapidly or the magma is extremely low in water content, crystals do not form and a volcanic glass called *obsidian* forms. Obsidian, like granite and rhyolite, should be light colored; instead, it appears to be black! It is actually light-colored, as you will observe if you break off a very thin flake; you can read through it. There are enough tiny particles of black magnetite crystals dispersed throughout the essentially colorless glass to make it opaque and appear black in thicker specimens.

Low-silica magmas are those in which the silica content is too low for the formation of quartz; instead, the resulting rocks are composed dominantly of *pyroxene* and other dark-colored minerals, in some cases olivine. *Gabbro* is formed from the slow cooling of such magmas, and basalt is the rock derived from the lava equivalent. More will be said about basalt when we get to Hawaii and later when we look into the Craters of the Moon and certain other park areas. Intermediate magmas produce either diorite or andesite or dacite or granodiorite, as indicated in Table 1–2. Some rocks are highly porous; pumice, nature's equivalent of popcorn, is actually rock froth (not edible!) created by expanding volcanic gases. Often it is light enough to float on water. Basalt scoria contains fewer gas-bubble holes and is heavier than pumice. When *tephra* (rock fragments or magma clots) is blown out of a volcano, it forms a pyroclastic sediment that, after consolidation, forms rocks such as *tuff* if fine-grained or *volcanic breccia* if the fragments are mainly large.

Types of Volcanoes

Volcanoes build cones that differ in configuration, based on the explosiveness of the eruptions, which is in turn a function of temperature, gas pressure, and chemical composition (especially silica content) of the magma. If the magma is relatively low in silica content and extremely hot, it will be highly fluid and will ordinarily be forced out of the vent without a violent explosion.[1] During a relatively quiet eruption, highly mobile, low-silica basaltic lava pours from the vent and flows rapidly out away from the center, often for many miles. With successive eruptions, flow upon flow, a flattish, circular cone is eventually built up (Fig. 1–5). These *lava shields,* or *shield cones,* so named because they resemble the round shields of ancient warriors, are the dominant type in the Hawaiian Islands. Figure 1–5 illustrates the connection between silica content of magma and type of volcanic cone produced.

[1]Note that the use of the words "violent" and "quiet" are only relative terms used by geologists and volcanologists. Most people witnessing a quiet eruption might use stronger words that are proportional to the flow of adrenalin in their bodies!

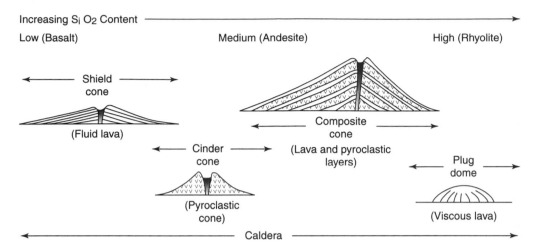

FIGURE 1–5   Relation of silica content and type of volcanic cone. Note that more than one type of cone can be produced by magma with the same silica content—thus gas content, temperature, and other factors must also play a role.

As conditions change, particularly with an increase in silica content, the lava becomes more viscous, just as 30-weight motor oil is much "stiffer" than 10-weight oil, especially in cold weather. The viscous lava inhibits the escape of dissolved gases, and, when gas pressure in a volcanic vent concentrates sufficiently, a violent explosion occurs, one that blasts through the solid rock, lifting large blocks and finer tephra materials high into the air. Gases trapped in airborne clots of still liquid lava expand, much like a giant popcorn machine, forming the porous texture of *cinder*. Larger clots, from baseball to beach-ball size are called *volcanic bombs*. Accumulations of cinder and other coarse *pyroclastic* debris around the vent forms a steep-sided cinder cone (Fig. 1–5). Finer materials, mainly volcanic ash and dust, are carried for varying distances downwind.

Volcanoes fed by magma with an intermediate silica content often shift from violent explosions to quiet lava-producing eruptions; thus a layered cone is constructed, as shown in Figure 1–5. The cone is built higher during pyroclastic events and extends outward during lava flow eruptions. A large number of the world's majestic volcanoes are *composite*, or *stratovolcanoes*—such as Fujiyama, Vesuvius, Pinnatubo, and many in the Andes and in our Cascade Mountains.

An even greater increase in silica content of the magma could result in the extrusion of a pasty, viscous lava that forms a *plug dome volcano* (Fig. 1–5) like that at Lassen Volcanic National Park and the newly formed one nestled in the crater at Mount St. Helens. This stiff lava has a consistency reminiscent of toothpaste or honey that was refrigerated! However, if gas pressures are sufficiently high in the rising magma, then the magma body can instead turn into a time bomb when confining pressures are reduced closer to the earth's surface. The result could be a Yellowstone-type eruption, which produced one of the most violent explosions known to have ever occurred on the face of the earth. As one might surmise, this latter type of eruption is not the type in which a "front-row" seat is desirable.

An igneous process that rapidly depletes part of the underlying magma chamber, like that accompanying the huge volcanic explosions at Yellowstone National Park 600,000 years ago or at Crater Lake a mere 6800 years ago, removes support for the overlying rock and allows a large, circular section of crustal rock to collapse downward into the void producing a *caldera* (surface depression many times larger in diameter than the underlying volcanic vent). Such collapses are not all catastrophic events only related to high-silica magmas—some are slow, piecemeal events involving low-silica magma like that in Hawaii.

Intrusive Igneous Rock Bodies

Magma that cools beneath the surface forms igneous bodies of different sizes and shapes. By far the largest is the *batholith* in which most of the earth's granite is found (Fig. 1–6). Granitic magmas are formed by the partial melting of deep-seated rocks whose fluids rise into the crust where they reside for a few million years. The partial melting of the surrounding silica-rich crustal rocks forms a new magma or perhaps changes the chemistry of the original magma body toward the granite or rhyolite end of the igneous spectrum. Eventually, while the top of the magma chamber is still miles below the surface, melting ceases and later on crystallization begins. Because of the extremely large volume of the magma and the great thickness of overlying rocks, the heat escapes at a very slow rate, thus

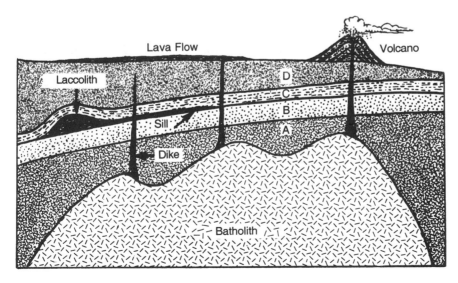

FIGURE 1–6   Generalized cross section showing intrusive igneous bodies, a volcano, and a lava flow. Rock unit A is the oldest, igneous rocks are younger than rock layer D. (Illustration by Gregory Nelson)

providing time for the growth of large mineral grains—hence the rocks are invariably coarse-grained.

As the magma cools, it shrinks, and eventually the reduction of volume of this huge mass is sufficient to produce severe stresses on the overlying rocks, which are broken, or "fractured," thus providing potential avenues for the escape of some of the still molten material. During crystallization water and other volatiles that do not fit into the crystal structure of minerals forming at these high temperatures become concentrated. This increased gas pressure forces liquids into the fractures and in some cases even lifts the surface of the earth above. As shown in the cross section[2] in Figure 1–6 this magma either solidifies in much smaller bodies such as dikes, sills, or laccoliths or it escapes to the surface to form volcanoes and lava flows. Being much smaller and much closer to the surface, these intrusive bodies cool more rapidly than their deep-seated cousins and are composed of finer-grained rock, particularly those that are formed at very shallow depths. Consequently, the latter are fine-grained or medium-grained; often grain sizes are similar to those found in lava flows.

In your travels you are likely to observe in the wall of a canyon something similar to what is shown in Figure 1–6. It is worth a stop to ponder what occurred and in what sequence. Which is the oldest rock and which is the youngest? The age of a rock refers to the time that it was formed, not where it was formed or where it is now. Where layers of sedimentary rock are exposed, the first of these rocks to be deposited is the bottom layer (layer A in Fig. 1–6). In our cross section we observe that the dikes cut through all of the sedimentary layers; therefore, the dikes are younger than the youngest layer, bed D. Since the rocks in the batholith and in the other igneous bodies are the same age, they are the youngest rocks. Formation A must have been there when the magma intruded; therefore, formation A is the oldest rock exposed in the canyon.

When an intrusive igneous body is exposed, like the dike in Figure 1–7, it means that erosion has been taking place over a long period of time. Imagine how long it must have taken to remove thousands of feet of overlying rock and expose the huge Sierra Nevada batholith in California. The incredibly slow attack by *chemical* and *physical weathering* processes ultimately frees individual mineral grains and permits them to be picked up, grain by grain, and transported to new sites by the processes of erosion. Ultimately, this leads to another category of rocks called sedimentary rocks.

Surface Processes and Sedimentary Rocks

As indicated earlier, sedimentary rocks form from all kinds of pre-existing rocks, either by the consolidation of rock fragments or by chemical precipitation of material derived by chemical decomposition of pre-existing rocks. Ordinarily, those

---

[2]Cross sections are vertical slices down through the earth's crust showing structures and the arrangement of rock layers; plan-views show surface features as viewed from the air.

FIGURE 1–7　Dike exposed by stream erosion near Willow Creek Pass in Middle Park, Colorado. (Photo by E. Kiver)

materials are transported a considerable distance from the source; in any case, before they can be transported they must first be loosened from the bedrock.

Weathering is the loosening process. When water penetrates into cracks or fractures and then freezes, the rocks are pried apart. Tree roots grow into the fractures and enlarge them. Expansion and contraction of the rocks as they are heated and cooled assists in the crumbling or weathering process. In such cases, physical forces are involved; the term physical weathering is applied to these processes. When water combines with carbon dioxide they form a weak acid, carbonic acid, that attacks the minerals and decomposes them; this is an example of chemical weathering. When orthoclase feldspar is slowly decomposed by chemical weathering an entirely different mineral, namely *kaolinite clay,* is formed. *Calcite,* the main mineral in *limestone,* is taken into solution and carried away.

Physical and chemical weathering work hand-in-hand, each assisting the other, in the breakdown and separation of materials from the solid bedrock. The resulting soils vary widely, depending on the *parent material,* climate, time, and other factors. Parent material derived from granite is distinctly different from basalt parent material; therefore the soils are different. A warm, humid climate is conducive to rapid weathering and soil development; therefore, *residual soils,* those formed in place, are generally deeper (thicker) than residual soil formed in arid regions.

Plants are selective in their habitats. Climatic factors—temperature and precipitation—provide the basis for climatic zones, each with its own assemblage of plants; within each zone the soil determines the distribution of the plants, reinforcing what many people forget—everything in this world is tied to everything else. Change one thing and there is a chain reaction of adjustments; nowadays, such changes are usually human-caused and the results are mostly negative.

The soil and other unconsolidated material do not accumulate indefinitely. They may move slowly downslope when freezing and thawing occur. On steeper slopes, a mass of the loose material, perhaps saturated with water, may slide down to the base of the hill, creating a *landslide.* These and other downslope movements are forms of *mass-wasting,* an important process in the leveling of the landscape. Note that there is no transporting agent involved; the force of gravity is sufficient.

Erosion is the removal and transportation of rock materials on the earth's surface. The face of the earth is gradually changed by erosion; high areas, even entire mountain ranges, are eventually eroded away, and the erosional debris is deposited in low areas, much of it in the sea.

The main eroding agents are water (streams, waves, and currents), wind, gravity, and glacier ice. Although all play important roles, a significant part of the landscape bears the unmistakable imprint of stream erosion, quantitatively the most important agent of erosion.

The evolution of landscapes, particularly those developed by stream erosion, is a fascinating subject. In mountainous areas where there are steep-walled canyons and fast-flowing streams with waterfalls and rapids abundant, we see a youthful landscape. One day this spectacular scenery will be gone: the sides of the canyons will be much less steep, stream gradients (slope of channel) will be flatter and more uniform, and most or all of the waterfalls will have vanished. It will become a *mature landscape.* But neither you nor I will be here to see it; it requires too long a period—millions of years. To see what the country will be like when it reaches the mature stage, we must go to a place where the mountains have been eroded for a longer period, perhaps the Ozarks or the Catskills. In another several million years they too will be distinctly different; they will be worn down—in fact, they will be erased, and undulating plains will be there instead. In this *old-age stage* of the *erosion cycle,* or *fluvial cycle,* the streams will be sluggish and unable to lower the land surface. Many mountain ranges have suffered this fate, but here or elsewhere, new mountain ranges have risen, thus preventing the land areas of the earth from becoming monotonous plains—*peneplains.*

At one time some geologists and *geomorphologists* (scientists specializing in landscape features and processes) thought that a few 100,000 or a few million years were all that was needed to complete the fluvial cycle. However, realization that the elevation of the earth's surface is a reflection of the delicate balance between the

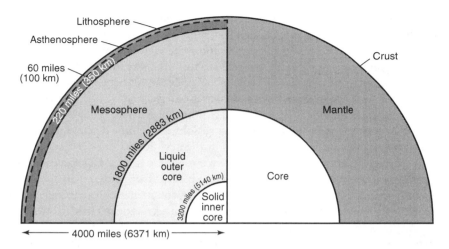

FIGURE 1–8   Layering of the earth: (Right side) boundaries of compositional differences between the crust, mantle, and core. (Left side) transitional boundary zones where physical properties change from rigid lithosphere to more fluid asthenosphere and then to rigid mesosphere, liquid outer core, and solid inner core. (From Skinner and Porter, 1987. Reprinted by permission of John Wiley & Sons, Inc.)

thickness and mass of the earth's crust with the underlying dense mantle (Fig. 1–8), a balance condition known as *isostasy* accounts for the long time required to level a mountain range—a few 100 million years!

The less dense crust (density about 2.7) literally "floats" in the denser mantle below (density about 3.3 at top of mantle; Fig. 1–8), much like a marshmallow in a cup of hot chocolate. A thinner marshmallow does not float as high. So too for the earth's crust; areas of thicker crust tend to ride higher (higher elevations) than areas with thinner crust. Crustal areas underlain by denser rock, such as the basalt (density 3.0) that underlies all of the ocean basins, tend to float at lower elevations. Thus, when erosion removes a thin layer of crustal material, eventually the crust will bob back up; but not quite to the original elevation. Thus, ranges like the Appalachians that last experienced major mountain building over 245 Ma (millions of years ago) are highly eroded and worn down, whereas the Rockies were only built "yesterday," about 50–70 Ma, and are topographically more rugged and are located at much higher elevations. Fortunately both mountain ranges have their own special beauty and grandeur that qualify many areas for national park status.

What becomes of the vast amount of material eroded away during one of these erosion cycles? The streams, large and small, pick up the material loosened by weathering and carry it away. These materials, particularly sand and larger particles, act as abrasive tools to grind away at the bedrock. The impact of boulders against the sides and bottom of the channel breaks off chips or fragments of the bedrock. Abrasion and impact are particularly effective during floods when the stream is capable of moving coarse material along its bed, as *bedload*. *Potholes* are formed when, by the swirling action of the stream, boulders grind against the walls and floor and drill large wells in the bedrock floor, especially at the base of waterfalls. Here, the boulders are rounded and shaped, some of them into near-perfect spheres.

Other remarkable things happen to the materials as they are swept along, banging against each other in the violence of a turbulent stream. The feldspar minerals have good cleavage; that is, they break or split easily along planes of weakness. Consequently, they are readily broken down into smaller and smaller particles. Soft minerals such as the micas are ground down into tiny flakes. Small particles are formed by physical and chemical weathering along the channel and are thus made ready to be transported as *suspended load* by the next flood.

Quartz is less affected by weathering as it is carried to the sea. Quartz is the most durable of all the common minerals; it has no planes of weakness and can be broken only with difficulty. Consequently, there is an abundance of quartz sand along the seashore while many of the other minerals have been reduced to silt and clay.

At the edge of the sea, the coarsest materials are deposited first, near shore. Farther out, beyond the sand zone, silts and clays slowly settle out. In the right environment—warm, shallow water which is the habitat of lime-secreting marine animals—the material carried in chemical solution, mainly calcium bicarbonate, precipitates out in the form of the mineral calcite ($CaCO_3$), and from these limy muds limestones are formed.

Although most sedimentary rocks are formed in the sea, as *marine* rocks, some form from deposits laid down in lakes (*sublacustrine*) or on land (*subaerial*). Gypsum and rock salt are nonmarine rocks that are formed in large land-locked basins or shallow bays barely connected to the sea where, by evaporation, the solutions become sufficiently concentrated to cause chemical precipitation. Coal is another special case; it is formed in huge swamps where over a long period the plant materials that accumulate are buried and slowly converted to peat and later to bituminous coal (soft coal).

Consolidation, or *lithification,* must take place to convert loose sediment into hard sedimentary rocks (Table 1–3). Pebbles and sand grains are cemented together by a cementing agent, usually silica or calcium carbonate, to form *conglom-*

TABLE 1–3   Sedimentary Rocks

SEDIMENT	SEDIMENTARY ROCK
Gravel, cobbles, and boulders	Conglomerate
Sand	Sandstone
Silt and Clay	Shale
Calcium carbonate[a]	Limestone[c]
Hydrous calcium sulfate[b]	Gypsum
Sodium chloride[b]	Rock salt

[a]Chemical and biochemical precipitate.
[b]Chemical precipitate.
[c]Dolomite, calcium-magnesium carbonate.

erate and *sandstone,* respectively. The source of the cementing material is the dissolved chemicals contained in the water trapped between the sediment particles at the time of deposition or in groundwater that later flows slowly through the loose sediment. Silts, clays, and calcareous muds are compacted by the weight of the materials deposited on top, helping to form shale and limestone.

Fossils are rarely found in igneous and metamorphic rocks; almost all fossils occur in sedimentary rocks. However, there are tree molds in the lavas of Hawaii, Craters of the Moon National Monument, and Newberry Volcanic Monument; insects, leaves, and tree trunks are preserved in the volcanic ash of Florissant National Monument.

Sandstones may contain fossils—the bones of the "Terrible Reptiles" in Dinosaur National Monument, for instance—but most fossils are invertebrates and occur in limestones and shales. The limestones in Grand Canyon and in Big Bend contain excellent records of past life for *paleontologists* to decipher. By identifying the fossils and fitting them into the evolutionary scale, the geologic age of the rock formation containing the fossils is established. Because each episode of time has a distinct assemblage of plants and animals, fossils are extremely useful to establish ages of rocks.

Environments of the past can be reconstructed by using the evidence found in sedimentary rocks. Here again, fossils are invaluable keys to the past. Shales that contain marine fossils were obviously laid down in the sea, where the descendants of these animals now live. Dinosaur skeletons indicate a *terrestrial* (land) environment, in most cases low-lying swampy conditions. Sedimentary structures such as ripple marks, mud cracks, and raindrop impressions suggest deposition in areas that were covered by shallow water for part of the time.

Sandstone with *eolian* (wind) *cross-bedding* is usually indicative of a dry climate, perhaps a desert where there was little vegetation to prevent active wind erosion and deposition, as in the Sahara today. As our travels take us into park areas where sedimentary rocks occur, these bits of evidence will help us interpret environments that existed millions of years, even hundreds of millions of years, ago.

Metamorphic Rocks

The development of metamorphic rocks is the next step in the rock cycle. Metamorphism means change in form. The changes are brought about by unusually high heat and pressure, although on occasion, solutions derived from magmas play a significant role. While weathering also brings about change, weathering is not a type of metamorphism because it occurs on or near the surface where the rocks are at low temperatures and pressures.

Because metamorphism takes place at depth, no one has observed metamorphic rocks until, after a long period, they have been exposed by erosion. Their formation has been established by carefully and logically applying the principles of physics and chemistry to the deep earth environment. Those processes are too complex to be outlined here in more than the simplest way.

Most metamorphic rocks are formed by *regional metamorphism* far below the surface, in an environment where the rocks are subjected to enormous pressures, mainly *shearing* and *load pressure.* At a depth of several miles where regional metamorphism occurs, temperatures are also high. Minerals that were stable in a near-surface environment cannot withstand the pressures and heat of the metamorphic environment. Blocky minerals such as orthoclase recrystallize and are changed to platy minerals, mainly muscovite mica. The platy minerals, and those that have elongate forms, tend to orient perpendicular to the direction of applied pressure, forming a texture called foliation. Slates, phyllites, mica schists, and chlorite schists are examples. There are also banded rocks called gneisses and massive (nonfoliated) quartzite and marble (see Table 1–4).

On a lesser scale, thermal or *contact metamorphism* affects rocks that are in contact with igneous intrusions. The heat from the magma, in some cases aided by hot solutions that penetrate into the adjacent rocks, bakes and hardens the *country rock.* A good place to observe the effects of contact metamorphism is in a road cut where a dike is exposed in contact with sedimentary rocks (Fig. 1–9).

Although much remains to be learned about regional metamorphism, it probably occurs mainly in places where the earth's crust is being subjected to severe deformation by almost unbelievable forces. An area of active mountain building would be such a place, where rocks near the surface are being folded and broken by faults. Below this so-called *zone of fracture,* the zone where shearing and load pressures dominate, is the *zone of plastic flow,* or the zone of metamorphism.

Mountain Building and Plate Tectonics

Mountain building takes place in many ways, but most major mountain ranges are formed by lateral (horizontal) forces, the origin of which is not completely under-

TABLE 1–4   Simplified Metamorphic Rock Classification

ORIGINAL ROCK	METAMORPHIC ROCK LOW GRADE—MEDIUM GRADE—HIGH GRADE
Sandstone	Quartzite[a] ⟶
Limestone	Marble[a] ⟶
Shale[b]	Slate—Phyllite—Schist[c,d]—Gneiss[c,e]
Granite	Granite Gneiss[c,e]
Basalt	Greenstone[c]—Amphibolite[c]—Granulite

[a]Nonfoliated.
[b]Shale may be baked by the heat of intruding magma, the resulting rock is called Hornfels.
[c]Foliated.
[d]Slate is composed of microscopic plates of muscovite; phyllite and schist are composed of larger plates of mica, the product of prolonged metamorphism.
[e]Gneiss is made up of bands of light and dark minerals.

FIGURE 1–9   Dike cutting through sandstone and shale, on Interstate 25 near the Colorado–New Mexico boundary. The shale in contact with the dike is now hornfels; the sandstone is now quartzite. (Photo by D. Harris)

stood. At one time these forces were assumed to be the horizontal components of the downward force of gravity. It was believed that the earth was cooling and shrinking and that, as the interior continued to shrink, the crust was crumpled into mountains. But is the earth actually getting smaller? True, heat is flowing from the interior to the surface, but heat is concurrently being generated by the decay of radioactive elements deep in the earth. Probably our planet is not changing materially in size. Then what causes the crust to be bent, broken, and thrust up in the form of mountains?

Until a few decades ago, geologic interpretations were based almost entirely on the rocks and structures exposed on the land masses; essentially no consideration was given to the evidence hidden in the vast ocean basins that cover more than 70 percent of the earth's surface. When geologists acquired a significant amount of information about the ocean floors, it became evident to a few imaginative minds that it was time to reexamine our time-honored geologic concepts. As is invariably the case when "sacred" ideas are challenged, the new concepts met with considerable opposition. Even so, a major upheaval in geologic thinking was in progress, and by the late 1960s the *plate tectonics theory* was widely accepted.

Geologists and especially geophysicists had already concluded through the study of earthquakes and the determination of the mass of the earth that concentric layers comprise our planet. The *crust* is composed of lower density, silica-rich rocks, and the underlying *mantle* contains much denser rock with lower amounts of silica (right side of Fig. 1–9). Below this is the dense core, likely rich in iron and nickel. Moreover, it was recently recognized that the physical behavior of materials at depth does not coincide with the compositional boundaries as indicated on the two sides of Figure 1–8. The upper part of the earth (crust and upper mantle) is called the *lithosphere* because it is rigid and seems to float on a weak, plastic zone in the upper mantle called the *asthenosphere*. The large quantity of heat trapped and being generated in the earth by the decay of radioactive materials cannot escape rapidly enough to the earth's surface by conduction—thus part of this heat energy is trapped and converted to mechanical movements (Fig. 1–10) of the dozen or so independent lithospheric plates that make up the upper 60 miles (100 km) of the earth.

Continent-size chunks of real estate pull and push against each other like slabs of ice jostling about in a river during the spring thaw. Movements originate in the asthenosphere, causing the rigid plates of the lithosphere above to either ride up on top of others, grind past one another, or separate and leave a space between.

By referring to the cross section in Figure 1–10, you will soon grasp the essentials of the plate tectonics concept. Note first that in places the ocean floor is spreading apart as the plastic material below the lithosphere spreads outward. Seafloor spreading is taking place along the Mid-Atlantic Ridge and along similar ridges in the other oceans. Basaltic magmas that rise to fill the gap between the separating plates pour out onto the ocean floor, and the resulting rocks are added to the plates on both sides of the ridge. Where volcanoes develop and build their cones above the sea, islands are formed. Iceland was once a tiny island like the island of Surtsey a few miles to the south; now Iceland is about 70 miles (115 km) across and still growing. Can you picture an Atlantic Ocean so narrow that you could almost jump across it—a few hundred million years ago? Next, turn the geologic clock back one more tick and see that there was one gigantic continent about 200 million years ago, Pangaea, a composite of all of the continents in existence today.

As the rising heat concentrated beneath this supercontinent, this huge, rigid plate began to break apart as asthenosphere material flowed laterally away, dividing the landmass of Pangaea into smaller continent-size plates. The continents moved apart, their leading edges colliding with oceanic crust. In these collision zones, the less dense continental rocks overrode the oceanic crust and were crumpled, broken, and faulted, thus forming long ranges of coastal mountains. The zone where the heavier oceanic crust "dives" slowly under the adjacent plate is called a *subduction zone* (Fig. 1–10). Here, mantle materials are mixed as the plates slide past one an-

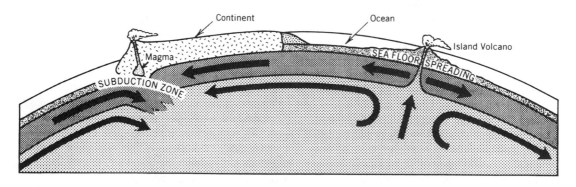

FIGURE 1–10   Schematic cross section showing possible mechanics involved in plate tectonics. (Illustration by Gregory Nelson)

other—one result is to bring hotter mantle material closer to the surface. The lessened pressure enables decompression melting of the rock to occur, which in turn sends magma upward toward the surface where it breaks through in places as volcanoes. Thus, the giant heat engine inside the earth is the driving force of plate tectonics and ultimately mountain building. The Andes Mountains, a classic example, result from the oceanic plate diving under the western coast of South America.

The earth is three-dimensional and thus plates may move in any direction. Along the west coast of North America the oceanic plate is moving generally northward and is carrying a slice of California with it—the part that is west of the San Andreas Fault. And all of what is now Alaska was apparently rafted northward over long distances, perhaps from as far south as the equator! In fact, much of western North America is composed of these plate slices (*microcontinents* or *exotic terranes*) that have been assembled along the west edge of the North American plate like a giant log or ice jam in a river (Fig. 1–11). Each slice or microcontinent has its own distinct geology and is separated from its neighbors by faults. What a nightmare for the early geologists trying to make sense of these isolated land areas without having a workable, overall model into which to fit the pieces of information.

Supporting evidence for plate movements is impressive. It comes from many branches of geology and is now regarded as overwhelming. For example, similar kinds of rocks and fossils at the abrupt north end of the Appalachian Mountains in New Foundland are found in the British Isles some 1700 miles (2800 km) away. Other continents show similar associations in which the continents are believed to have once been connected. The shapes of the continent edges in the Atlantic basin in particular resemble pieces of a child's picture puzzle that scream out—"Connect me, connect me!"

When newly erupted basalt along an oceanic ridge cools, the direction of the earth's magnetic field is preserved in iron-rich minerals such as magnetite that form in the basalt. The direction of the earth's magnetic field (either north or south) is determined by movements in the liquid iron–nickel core (Fig. 1–8). Occasional changes in flow patterns in the liquid core can produce a magnetic field completely reversed from the present, "normal" field. During times of magnetic reversal, the north end of a compass would actually point to the south! The numerous flip-flops of magnetic field polarity over the past few tens of millions of years are recorded as broad magnetic "stripes" in the oceanic floor basalts (Fig. 1–12). The stripes are oriented parallel to the ocean ridge and form a mirror image across the spreading ridge, best explained by the seafloor-spreading process accompanied by magnetic reversals.

Other lines of evidence for moving plates are discussed in relation to appropriate parks throughout the book. Precise measurement of present-day geographic locations using satellites verifies that the positions of areas on separate plates is changing by a few inches each year. For example, in the Pacific Basin, Hawaii and Japan are approaching each other at the rate of about 3 inches each year (8 cm/yr)—ongoing separations in the Atlantic are much less, only 0.07 inch (17 mm) each year. Most of the rates measured by satellites and geologic means

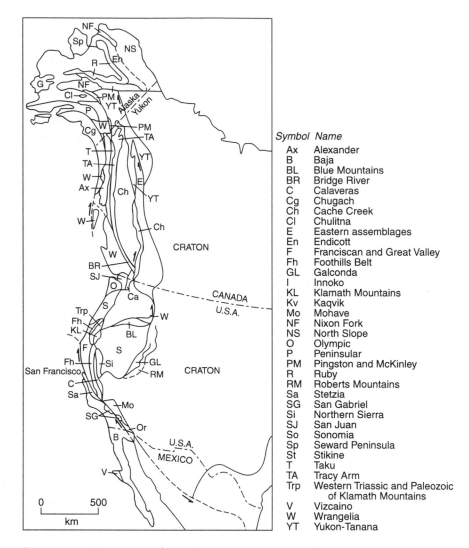

Symbol	Name
Ax	Alexander
B	Baja
BL	Blue Mountains
BR	Bridge River
C	Calaveras
Cg	Chugach
Ch	Cache Creek
Cl	Chulitna
E	Eastern assemblages
En	Endicott
F	Franciscan and Great Valley
Fh	Foothills Belt
GL	Galconda
I	Innoko
KL	Klamath Mountains
Kv	Kaqvik
Mo	Mohave
NF	Nixon Fork
NS	North Slope
O	Olympic
P	Peninsular
PM	Pingston and McKinley
R	Ruby
RM	Roberts Mountains
Sa	Stetzia
SG	San Gabriel
Si	Northern Sierra
SJ	San Juan
So	Sonomia
Sp	Seward Peninsula
St	Stikine
T	Taku
TA	Tracy Arm
Trp	Western Triassic and Paleozoic of Klamath Mountains
V	Vizcaino
W	Wrangelia
YT	Yukon-Tanana

FIGURE 1–11   Western North America tectonic terranes. Each terrane was added as a "sliver" of land to plate edge during the Mesozoic and Cenozoic. Each terrane is bounded by faults and has rocks, structure, and a geologic history different from its neighbors. (From Skinner and Porter, 1987. Reprinted by permission of John Wiley & Sons, Inc.)

produce rates of movement about as rapid as the growth of a human fingernail. This may not seem significant until one considers what your fingernails would look like after a hundred years, and especially after a few million years if they were never cut!

Here we can only glimpse this earth-shaking concept; later, when discussing the geomorphic provinces where plate tectonics are active today, particularly in

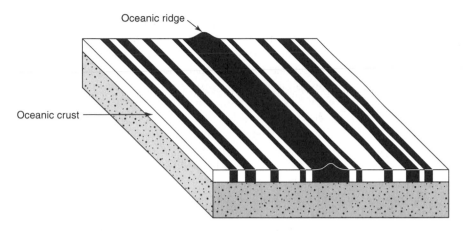

FIGURE 1–12   Generalized diagram showing the record of normally polarized oceanic crust (dark areas) and reversely magnetized crust (white areas). Note symmetry of pattern around an oceanic ridge. (From Skinner and Porter, 1987. Reprinted by permission of John Wiley & Sons, Inc.)

those bordering the Pacific, we will be in a better position to visualize what is taking place.

Perhaps before accepting the "new" geology as being the final, the everlasting theory, we should pause and reflect on the fact that more than a century ago Charles Lyell was being congratulated for *his* new geology. From this we may conclude that what we have now is merely the newest new geology and that revolutionary concepts will continue to be brought to light in the future. Such is the way of science—science uses the best answer that fits the facts and observations available at the time. Given new information, scientists modify their ideas, and if necessary replace old concepts with new ones, and the science undergoes a revolution in thought. What an exciting time for a geologist to witness and be part of the changes! Without question the plate tectonics theory is one of the most important developments of geology in our time and will provide generations of geologists with many new problems to ponder and solve.

Folds and Faults

In our discussion of the origin of sedimentary rocks, we left them as they were deposited, essentially flat-lying. Should you drive westward across the Great Plains, you will go through occasional road cuts where horizontal sedimentary rocks are exposed. When you get to the Black Hills in South Dakota, however, you will note that the rocks are not flat-lying but tilted and that on the east side they are tilted (*dipping*) to the east, away from the mountains. Later, in the center of the Black Hills near Mt. Rushmore, you will see much older metamorphic and igneous rocks that form the higher mountains. Still later, when you have crossed the Black Hills, you suddenly realize that here on the west side you are seeing the same sedimentary rocks, but they are dipping in the opposite direction, toward the west. What happened? How did they get that way? That night, sitting around your campfire, suddenly the lights begin to flash and the answer is there. The Black Hills were domed up and the sedimentary rocks on top were eroded away, leaving the "stubs" dipping away from the core on all sides.

Then, if you swing down through Wyoming into Colorado and head west into the Front Range, again you will see sedimentary rocks dipping away from the mountains. Could the Rockies have been formed in about the same way as the Black Hills, only on a much larger scale?

By now you may have decided that after rocks are formed, they are deformed —at least in places. Many of the rocks have been bent, twisted, wrinkled (folded), and/or broken, depending on the type and intensity of the forces applied. Two of the many geologic structures formed are *anticlines* (upfolds) and *synclines* (downfolds); they are discussed together because they frequently occur together (Fig. 1–13).

You can simulate folding by pushing against a throw rug on a smooth floor. The rug represents thick layers of rock. When you push against one side, the rug buckles up and forms an anticline; continue pushing and another anticline develops, with a downfold, a syncline, between the two. Of course, rugs are limber and fold easily; rocks are rigid and brittle, and they would surely break when pressure is

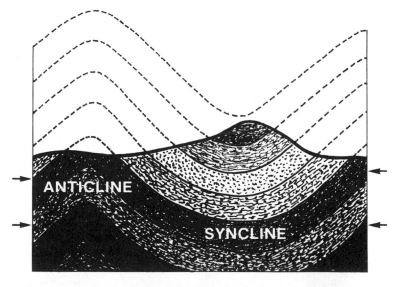

FIGURE 1–13   Cross section of folded sedimentary beds after prolonged erosion; restoration shown by dashed lines. Arrows indicate the direction of forces involved. Note that depending on the level of erosion and the resistance to erosion of different beds, anticlines are not necessarily topographically higher than a syncline in a folded area. (Illustration by Gregory Nelson)

applied. But under normal conditions, the pressures build up gradually, allowing time for the rocks to adjust internally; consequently, although some fracturing does occur, it is of minor importance. Also, except for those on the surface, the rocks are under great confining pressure, enough to prevent more than tiny cracks from forming. Thus, instead of breaking, the rocks are folded into broad, essentially symmetrical anticlines and synclines (Fig. 1–13). Note that anticlines and synclines are structures below the surface and that surface topography may not correspond to those structures, especially after a prolonged interval of erosion has occurred. Here in our cross section the synclinal structure is represented by a ridge, a common condition in parts of the Appalachian Mountains in the eastern United States.

If compressional forces are long-continued, the anticline will become asymmetrical, as shown in Figure 1–14. Structures such as Sand Creek Anticline occur in many places along the base of the Rockies and elsewhere. Some of these anticlines contain oil and gas—those in which a porous formation, usually sandstone, is capped by an essentially impervious shale, which traps the upward-migrating oil and gas.

In some cases, however, the pressures continue for a longer period than at Sand Creek; then the rocks are broken and *thrust faulted,* as shown in Figure 1–15. Refer again to Sand Creek Anticline (Fig. 1–14), noting the rocks shown in the distance. At the time folding occurred, the rocks now exposed in the anticline were buried deep beneath the rocks you see in the background. Sand Creek and its tributaries stripped off all of those rocks, exposing the hard rocks of the anticline; then Sand Creek cut a small canyon into the hard rocks, thus exposing the "innards" of the anticline. Such exposures afford a good opportunity for geologists to learn more of the details of the process of folding.

Thrust faults (Fig. 1–15) may develop to enormous size. Later, when we travel to Glacier National Park, we will see a thrust fault that is more than 100 miles (60 km) long, in which the *displacement* (amount of offset along the fault) is more than 30 miles (48 km)!

FIGURE 1–14   Sand Creek Anticline in northern Larimer County, Colorado, looking north to the Gangplank in southern Wyoming. (Airview by J.A. Campbell)

FIGURE 1–15   Cross section in which an asymmetrical anticline has been thrust faulted. Note how beds line up if the upper plate is moved to the right in this illustration. (Art work by Gregory Nelson)

Folds and thrust faults shorten the earth's crust and are the result of compressional forces. But there are places, such as in the Basin and Range Province, where the earth's crust is being pulled apart. This causes faulting to occur, but the faults are distinctly different; one block simply slides down the sloping fault plane, responding to the force of gravity. There are many of these *gravity faults* and a few are extremely large. The Sierra Nevada is a giant *fault block;* a system of mostly gravity faults, also called *normal faults,* lies at the base of the precipitous east face of the Sierra Nevada. Figure 1–16 shows in simplified form what the Sierra Nevada looked like before it was dissected by streams and later by glaciers.

There are many more faults than most people realize because in many places they are not well exposed. Many unusually straight mountain valleys are straight because the streams cut deeply into the broken, easily eroded material in fault zones; mountain passes are frequently at the head of two such valleys. Many faults are covered by overwash material, glacial deposits, and soil. Road cuts, particularly the huge ones along some of the interstate highways, are likely places to observe faults. Note the small gravity faults in Figure 1–17, where a 6-inch (15-cm) layer of white volcanic ash serves as a key bed in measuring the displacements. The displacement along the fault at the right cannot be measured because the ash bed is at an unknown depth below the roadbed.

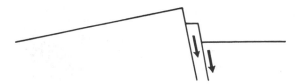

FIGURE 1–16   Schematic cross section of a normal fault. (Illustration by Gregory Nelson)

(a)

FIGURE 1–17(*a*)   Small normal faults offsetting beds of volcanic ash in a highway cut on U.S. 180 near Alma, southwestern New Mexico. (Photo by D. Harris)

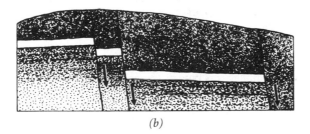

(b)

FIGURE 1–17(*b*)   Cross section sketch of Figure 1–17a. (Illustration by Gregory Nelson)

Of far greater magnitude are the faults or fault systems along which huge plates of the earth's crust grind past one another. The San Andreas Fault system in California is our best known example; here, the Pacific plate is moving northward past the continental plate. The displacements are almost entirely in a horizontal direction in contrast to both gravity and thrust faults, which have mostly vertical components of movement. The San Andreas and faults with similar horizontal movements are generally called *transcurrent* or *strike-slip faults*.

Faulting consists of sudden adjustments interspersed with long periods of quiet, during which stresses gradually build up—to the breaking point. Then, without warning, displacement occurs that can be measured in inches or sometimes in tens of feet. Regardless of the amount of movement, these abrupt displacements generate earthquake waves that radiate out from the *focus,* the area along the fault where movement occurred. The *epicenter,* the area on the earth's surface directly above the focus, is the location usually given in news reports. A major earthquake in August of 1959 had its epicenter in the West Yellowstone area and was caused by gravity faulting; maximum displacement was 19 feet (6 m), almost all vertical movement. There are a number of possible causes of earthquakes and earth tremors, including volcanic activity, but most of the devastating quakes are generated by faulting produced as the plates shift restlessly about.

Unconformities

Folds and faults generally document single events in the geologic history of an area, perhaps of a mountain range; *unconformities* record a series of events, in some cases involving many millions of years. An unconformity is a buried erosion surface, one that represents a gap in the geologic record—a time when erosion exceeded deposition of new material. Figure 1–18*a* shows a continuous sequence, a *conformable sequence,* in which the shale bed B was laid down on top of the sandstone bed A. Then the limestone bed C was deposited on the shale, without any interruption. In Figure 1–18*b,* perhaps a hundred miles away, we find the same sandstone and the same limestone, but the shale bed B is missing. Assuming that all of these rocks were deposited beneath the sea, the following sequence is indicated for the second area: the sandstone was deposited and then covered by part of the shale, as in the first area; then the area was uplifted above the sea and the shale was eroded away; later, with lowering beneath the sea again, limestone was deposited on the erosion surface, converting it into an unconformity. Because the beds above and below the unconformity are parallel, this type of unconformity is called a *disconformity,* the simplest of the three types.

Suppose that several beds were deposited and then folding and uplifting began. As the folded beds appeared above the sea, erosion by streams would begin to plane them down to sea level. Eventually they would be truncated, as shown in the lower part of Figure 1–19. Then with the submergence of the area, deposition of beds Y and then Z (Fig. 1–19) would take place. This unconformity is called an *angular unconformity* for obvious reasons. Although exceptions are possible, generally much

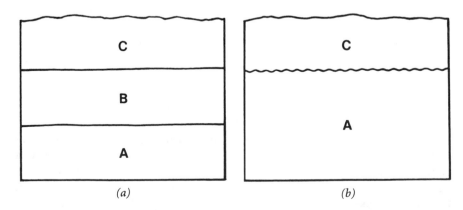

(a)   (b)

FIGURE 1–18   (*a*) Conformable sequence of sedimentary rocks A, B, and C laid down in order—layer A is the oldest. (*b*) Parallel unconformity or disconformity; bed B was eroded away before C was laid down. (Illustrations by Gregory Nelson)

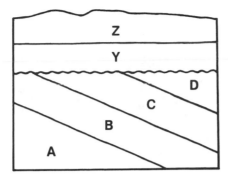

FIGURE 1–19   Angular unconformity; the rocks were tilted after bed D was deposited. Then, after the area was leveled by erosion, beds Y and Z were laid down. (Illustration by Gregory Nelson)

more time is involved here than in the case of the disconformity, inasmuch as folding is an extremely slow process. And, frequently, the folding is of widespread significance, as in the building of mountains—folded mountains.

Actually, Figure 1–19 is a simplified cross section of the rocks exposed in northernmost Colorado and southern Wyoming where the Front Range meets the Great Plains. The presently tilted rocks were originally deposited in a nearly horizontal position in a large depositional area called a *geosyncline* before the Rockies were born. Subsequent folding and uplift during the *Laramide* (Rocky Mountain) *Orogeny* (mountain building) brought the now-tilted layers above the sea where these rocks were eroded by streams, finally creating an erosion surface across the tilted beds. Still later, streams deposited huge quantities of material on the erosion surface, thus forming the angular unconformity. To complete the sequence, the Rockies were uplifted with the result that the streams cut down and exposed the unconformity and Sand Creek Anticline as we see it today. By determining the geologic age of the rocks above the unconformity and of the youngest tilted formation, geologists were able to establish the time when the Rocky Mountains were born, roughly 50–70 Ma.

The third type of unconformity is called a *nonconformity*. Here, the rocks below the unconformity are deep-seated rocks, either metamorphics or granite igneous rocks or a combination of the two. Many of these rocks, formed miles beneath the surface in the core of a developing mountain range, were originally sediments deposited into a geosyncline. Later they were consolidated into sedimentary rocks, subjected to high temperatures and pressures (metamorphism) as the mountain range was forming, and later unroofed and then deeply eroded during the formation of the erosion surface; thus the length of time involved is simply mind-boggling. Then, as with the other two types of unconformities, sedimentary rocks or lavas were deposited on the old erosion surface (Fig. 1–20). Thus we see that unconformities contribute much in the deciphering of really ancient history—geologic history. They provide essential information in localized areas that applies to entire mountain ranges as well.

FIGURE 1–20   Unconformity (nonconformity) in the Grand Coulee area, Washington. Eocene (about 50 Ma) granite batholith was unroofed by erosion before layers of Miocene basalt (about 15 Ma) were laid down on top. Therefore, missing time represented by the unconformity is about 35 million years! (Photo by E. Kiver)

Types of Mountains

The geologist looks at a mountain or mountain range in terms of the dominant geologic processes, especially structural processes, involved. Volcanic activity was the process that formed the Hawaiian Archipelago, a gigantic mountain chain. Folding was mostly responsible for the Black Hills deformation, as discussed, and the Uinta Mountains of Utah, a huge anticline. Block faulting has formed many small ranges in Nevada and the majestic Sierra Nevada in California, so-called fault-block mountains. Thrust faulting was the main structural process involved in the creation of the mountains in the Glacier National Park area. In some cases, as in the Southern Rockies, two processes are of about the same importance. Folding and thrust faulting, both the result of lateral compressive forces, are the primary structural features in the Southern Rockies and in many other large mountain systems.

Erosion: The Great Leveler

Mountains are products of internal geologic processes; such processes are constructive and increase the elevation of the land. External processes such as running water, glaciers, mass movement, wind, and waves and currents are generally destructive processes that overall produce a net loss or reduction in elevation. Stream erosion is quantitatively the most important and ubiquitous of the so-called *geomorphic processes;* its effects dominate landscapes except those covered by glaciers. High in the mountains and in the polar areas where more snow falls than melts each year, mighty rivers and sheets of ice bury entire landscapes. Our most spectacular mountain scenery, the sharp spines, or *horns,* were formed by three or more glaciers enlarging their *cirques.* Grinding away the mountain on all sides produces these spectacular Matterhorn-like features.

As the glaciers grind their way down the valleys, they deepen and widen them, and the V-shaped valleys are remodeled so that they are parabolic or U-shaped. As you drive up a canyon that has been only partially glaciated, you will immediately recognize the glaciated section when you reach it, even though the end moraine may be absent, having been eroded away. In the U-shaped glaciated portion you will see glacial polish on the rocks along the sides, and close up you will find *glacial grooves* and scratches, or *striations.* Thus, the glaciers have left their own special imprints on the landscape, features both bold and delicate.

Glaciers can be classified by geographic location; those confined to glacially excavated, bowl-shaped depressions (cirques) high in the mountains are called cirque glaciers. Those that expand and flow out of the cirque into the valley below are called *valley, mountain,* or *alpine glaciers* (Fig. 1–21). Glaciers that coalesce high in the mountains are called *ice caps* and those that flow out of their valleys onto the flat terrain in front of the mountain (the *piedmont*) are called *piedmont glaciers.*

Not all glaciers originate in mountains. *Continental glaciers,* or *ice sheets,* develop on relatively flat terrain, generally in the high latitudes. Those that covered northern United States originated in central Canada; they extended southward at least four times during the *Pleistocene Epoch* and overran essentially all areas north of the Missouri and the Ohio rivers in the Midwest and all of New England and much of New York State. Unlike the mountain or *alpine glaciers,* which sharpen topographic features, ice sheets tend to smooth the topography by shaving the hilltops and filling the valleys. Thus the areas covered by these glaciers were also profoundly changed, but in a very different way.

FIGURE 1–21   Alpine or valley glacier in Glacier Bay, Alaska. Note medial moraine (black ridge). McBride Glacier flows into Muir inlet (a fjord) in foreground. (Photo by D. Harris)

We have seen that the topography in certain areas is clearly the work of streams, in others the work of glaciers. In still other places the landscapes were unmistakably fashioned by the wind, particularly in dry regions where the sparse vegetation is unable to hold the soil in place. Large dune areas occur in some deserts around the world; our western deserts, however, contain only localized dunefields, in addition to a variety of other fascinating desert landforms. But sand dunes can also form in other environments—wherever loose sand is abundant—such as seashores and lakeshores. Dunes are major features at Great Sand Dunes, White Sands, Death Valley, Padre Island National Seashore, and Indiana Dunes National Lakeshore. The dunes of ancient days were even more widespread; the Navajo Sandstone in Zion and other Colorado Plateau parks show the same eolian *cross-bedding* that is found in modern dunes. By observing how the modern dunes are being formed, we can re-enact, in our mind's eye, the *Jurassic* scene.[3]

Along seashores, and to a lesser extent along lakeshores, the work of the waves is clearly marked. In the bold wave-cut cliffs in Acadia National Park and on the opposite side of the continent in Olympic National Park, the force of the endless waves is evident. *Sea caves, sea arches,* and *stacks* are other landforms produced in the ever-changing oceanscape.

Thus far, we have concerned ourselves mainly with surface features. In some places, what has been developing in the underworld for centuries of centuries is for some people even more fascinating. Caverns, some small and some mammoth, are found in most of the states; in the park system there are several including Carlsbad Caverns (New Mexico), Wind Cave (South Dakota), Lehman Cave in Great Basin National Park (Nevada), Russell Cave (Alabama), and Mammoth Cave (Kentucky), the largest cavern system in the world.

GEOLOGIC TIME

People are in the habit of thinking of time in terms of years, months, and days; consequently, when geologists talk casually about millions, even billions, of years, some people become ill at ease. But when we stand on the rim of Grand Canyon (Fig. 1–22), we see many layers of sedimentary rock, each of which required thousands or even millions of years to form. And when we look deeper, down into the inner gorge, and see metamorphic rocks that must be incredibly old, we suddenly realize that we must develop a "feel" for the immensity of geologic time if we are to appreciate the ancient history so vividly written in the rocks.

Our present conception of geologic time was developed slowly and with great difficulty. It was a long road from Archbishop Ussher's 1650 decree that the earth was created in 4004 B.C. to the generally accepted view that the earth is about 4.5 billion years old. Acceptance of uniformitarianism required that vast amounts of time be available to accomplish the huge changes recorded in the rocks and fossils.

---

[3]The Jurassic is a geologic period during the Mesozoic Era; refer to the geologic time scale in Figure 1–23.

FIGURE 1–22   Grand Canyon, John Wesley Powell's "Book of Geology," where over 2 billion years of earth history are recorded in the rock record. (Photo by D. Harris)

Through the years, many different methods were used in attempting to determine the age of the earth; however, without exception, each method involved assumptions that ultimately caused it to be discredited as grossly inaccurate. Finally, as discussed earlier, the radiometric method was developed and is now accepted as reliable. The oldest rock that has been dated so far is in Canada and it is about 4 billion years old. It, in turn, is surrounded by rocks that are somewhat older. Because the earth is believed to have experienced a stage of melting that terminated about 4.5 billion years ago when the crust solidified, we do not expect to find rocks much more than 4.2 billion (4200 million) years old.

Long ago, geologists saw the need for divisions and subdivisions of geologic time, and a standard, worldwide geologic time scale was eventually agreed upon (Fig. 1–23). The eras, or intervals, of longest time span, were set up to represent the time of one worldwide orogeny (mountain-building event) to the next. We know now that orogenies occur more frequently but in restricted areas—not as worldwide events. However, the names of eras do reflect the fossil forms found in each. *Paleozoic* ("ancient life") was an appropriate name. Lifeforms from this era are considerably different from modern forms and are mostly extinct. *Mesozoic* ("middle life") and *Cenozoic* ("recent life") contain lifeforms progressively more modern with fewer extinct organisms as one examines younger and younger fossils. The sequential progression of lifeforms from older to younger rocks is undeniable and records the change, or evolution, of organisms through geologic time (Fig. 1–24).

Eras older than the Paleozoic were believed at the time not to contain any evidence of life. These *Precambrian* Eras (time before the Cambrian, the oldest period of the Paleozoic Era) were rather arbitrarily divided into two eras, the older Archeozoic ("beginning of life") and the younger Proterozoic ("fore life"), without the realization that each of these two eras was of longer duration than all three

ERAS	PERIODS	EPOCHS	DOMINANT LIFE
CENOZOIC	Quaternary	Holocene 0.1 Ma Pleistocene 1.6 Ma	"Age of Mammals"
	Tertiary	Pliocene	
		Miocene	
		Oliogocene	
		Eocene	
		Paleocene 66 Ma	
MESOZOIC	Cretaceous		"Age of Reptiles"
	Jurassic		
	Triassic	245 Ma	
PALEZOIC	Permian		"Age of Invertebrates"
	Pennsylvanian		
	Mississippian		
	Devonian		
	Silurian		
	Ordovician		
	Cambrian	545 Ma	

PRECAMBRIAN ERAS   Approx. 7/8 of Geologic time
   PROTEROZOIC
 	
   ARCHEZOIC   2500 Ma   (Fossil Record
                                                                                      Very Incomplete)
(Other divisions of the Precambrian are used in certain areas)
                                                        4500 Ma			

Ma—millions of years ago.

FIGURE 1–23   Geologic time scale.

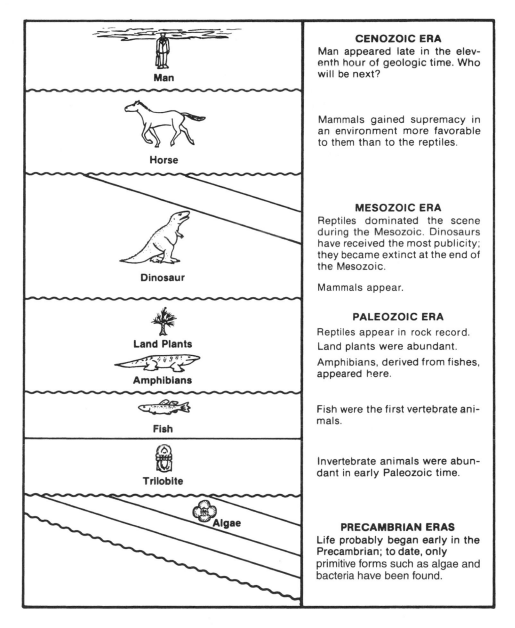

CENOZOIC ERA

Man appeared late in the eleventh hour of geologic time. Who will be next?

Mammals gained supremacy in an environment more favorable to them than to the reptiles.

MESOZOIC ERA

Reptiles dominated the scene during the Mesozoic. Dinosaurs have received the most publicity; they became extinct at the end of the Mesozoic.

Mammals appear.

PALEOZOIC ERA

Reptiles appear in rock record.

Land plants were abundant.

Amphibians, derived from fishes, appeared here.

Fish were the first vertebrate animals.

Invertebrate animals were abundant in early Paleozoic time.

PRECAMBRIAN ERAS

Life probably began early in the Precambrian; to date, only primitive forms such as algae and bacteria have been found.

FIGURE 1–24 Glimpses of the evolution of life beginning with primitive forms in the ancient Precambrian rocks. For proper sequence read from the bottom to the top. (Illustration by Gregory Nelson)

of the latest eras put together! Now we know from radiometric dating that geologic time prior to the Paleozoic amounts to almost 88 percent of geologic time; in addition we know that each successive era is significantly shorter than the one before it. Although we now have a different concept of geologic time, the original system, imperfect though it is, serves its intended purpose reasonably well.

The Paleozoic and younger eras were divided into shorter time-intervals called *periods;* the periods of the Cenozoic Era were divided into even smaller time units called *epochs,* as indicated in the geologic time scale (Fig. 1–23). Note that the oldest era, the Archeozoic, is at the bottom and that the youngest era, the Cenozoic is at the top. This convention is followed in most geologic reports and legends for geologic maps.

The Pleistocene or *Glacial Epoch,* which began about 1.6 Ma, was the period during which the glaciers advanced over the land several times, only to melt away (retreat) during warm intervals called *interglacials.* The retreat of the last great ice sheets began about 11,000–12,000 years ago. As shown by the temperature curve in Figure 1–25, the warming of the climate continued until about 6000 years ago when it was considerably warmer than now; the period from about 8000 years ago to 5500 years ago is called the *Thermal Maximum* or *Hypsithermal Interval.* Again

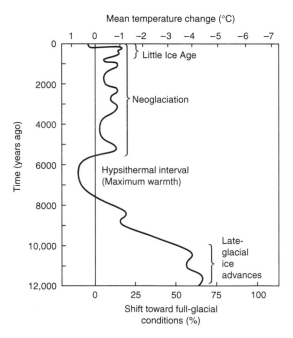

FIGURE 1–25   Temperature changes from the late Wisconsin (Pleistocene) to present inferred from the study of glacial deposits, vegetation, historical records, and archeological studies. (From Skinner and Porter, 1987. Reprinted by permission of John Wiley & Sons, Inc.)

the climate cooled, beginning about 6000 years ago, leading the world into the present cooler climate episode known as the Neoglacial. Existing glaciers expanded, and many cirque glaciers were reborn and later extended far downvalley, many to their greatest extent since the end of the Pleistocene.

The latest significant period of cooler climate is called the *Little Ice Age* and began in the middle thirteenth century and ended in the middle 1800s. This worldwide climate fluctuation is well recorded by historic records. Crop failures, a decrease in the fish harvest, and a significant loss of the human and livestock population occurred in Iceland. Life became more difficult in Greenland and the Norse colonies there were abandoned by 1410 (Skinner and Porter, 1987). During the 1600s the glaciers moved down the valleys in the Alps, and one village, St. Jean de Perthuis, was overwhelmed by ice. The helpless villagers could do nothing but stand and watch the monster demolish and overrun one after another of their homes. (There goes the neighborhood!)

The warming since the late 1800s has continued with short excursions to slightly cooler conditions in the 1940s and 1960s. By the 1980s world temperatures swung in the direction of warmer climate with year after year of increasing average world temperatures. Changes like these are partially the result of natural changes in the complex solar–atmosphere–ocean system that controls our climate and partially, perhaps mostly, the result of human activities.

With increasing world population and more people demanding more food, energy, and manufactured goods, the amount of carbon dioxide, methane, particulate matter, and other climate-influencing substances in the atmosphere is increasing dramatically. Massive cutting of tropical rainforests and other forests, extensive irrigation projects, paving over of huge areas in our cities, as well as other human activities also influence world climates. Where this great unplanned experiment will lead us is uncertain—most of the possible consequences do not bode well for humans, glaciers, or other inhabitants of the globe. Moreover, the condition of the world's glaciers is not just a scientific curiosity but a sensitive gauge that integrates yearly climate trends. It can well serve as a warning to us to consider the cost to human society as well as other earthly inhabitants if climate change, in either direction, persists. Modern society and in particular, agriculture, is finely adjusted to average temperature and precipitation conditions. Even a small change will have profound and costly effects on the volume and distribution of the world's food supplies and on the stability of governments as well as international harmony.

Before leaving the subject of glaciation, we should take note that although the Pleistocene is frequently referred to as *the* glacial period, actually it is merely the latest of several; glaciers were widespread during a number of geologic periods, such as the Permian and several times during the Precambrian. However, in comparison to the total of geologic time, ice ages are very rare events in the geologic record—more moderate conditions are the rule. The latest glaciation is uppermost in our minds because essentially all of the glacial features we see today were formed during the Pleistocene. An interesting coincidence(?) is that the rise of humanity corresponds to the retreat of the last ice sheets and the warmer climates of the Holocene.

GEOMORPHIC PROVINCE

The regional concept involving geomorphic provinces is used in our treatment of the national parks. A *geomorphic province* (Plate 1) is a region with definable borders in which the rocks, geologic structure, geologic and geomorphic history, and landforms are similar. However, within a province, smaller areas may have detailed features that distinguish them from surrounding areas in the same province. Such an area is referred to as a "section." For example, the Colorado Piedmont and the High Plains are 2 of the 10 or so sections of the Great Plains Geomorphic Province. Some provinces, like the Columbia Intermontane, have a dozen sections, others like the Blue Ridge Province have only 2.

The parks within a geomorphic province are related geologically, although each has its unique features. For example, specifics about the four volcanic national park areas in the Cascade Mountain Province are all different, but much of their geological development is the same. The geology of the park areas will unfold with, first, a summary of the geomorphic province and then a discussion of the distinguishing features of each of its national parks and monuments. This sequence will make the relationships between parks within each province more apparent than an alphabetical or geologic-process approach. The geomorphic province approach will also assist in developing an appreciation of some of the regional geology that, in combination with some of the fine Roadside Geology guides to many of our states (a few are listed in reference section), will make traveling an even more interesting experience. Also listed in the references are a number of other books that provide more detailed information about geologic subjects for those who wish to delve deeper into the mysteries of our planet.

REFERENCES

Blackburn, William H., and Denner, W.H., 1988, Principles of Mineralogy: Wm. C. Brown, Dubuque, Iowa, 413 p.

Chesterman, Charles W., 1995, Audubon field guide to North American rocks and minerals: Alfred A. Knopf, New York, 850 p.

Chittenden, H.M., 1895, The Yellowstone National Park: Historical and descriptive: R. Clarke Co., Cincinnati, Ohio.

Mason, Brian, and Berry, L.G., 1968, Elements of mineralogy: W.H. Freeman, San Francisco, 550 p.

Roadside Geology Series, excellent books available for various states including Alaska, Arizona, California, Colorado, Hawaii, Idaho, Louisiana, Montana, New Mexico, New York, Oregon, South Dakota, Texas, Utah, Vermont and New Hampshire, Virginia, and Washington; check with the publisher for additional titles: Mountain Press, Missoula, MT.

Skinner, Brian J., and Porter, Stephen C., 1987, Physical Geology: Wiley, New York, 750 pages.

Skinner, Brian J., and Porter, Stephen C., 1995, The dynamic Earth: An introduction to physical geology: Wiley, New York, 567 p.

Wicander, Reed, and Monroe, J.S., 1993, Historical Geology: West Publishing, Minneapolis, 640 p.

TWO

Hawaiian Archipelago

The Hawaiian Archipelago is composed of a linear group of islands (including a special type called *atolls,* which are rings of coral islands with a central lagoon) and *seamounts* (submarine mountains) that extend from the island of Hawaii northwestward beyond Midway to near Daikakuji where it makes a sharp bend and continues northward as the Emperor Seamount Chain—ultimately ending in the Aleutian Subduction Trench (Fig. 2–1), a total distance of about 2200 miles (3520 km). The southern group of eight islands are known collectively as the Hawaiian Islands with Haleakala and Hawaii Volcanoes National Parks located on the southernmost islands—Maui and Hawaii. The "Big Island" (Hawaii) is less than 19° north of the equator and rises to a spectacular 13,796 feet (4206 m) above sea level on Mauna Kea (north of Hawaii Volcanoes National Park) and slopes down some 19,700 feet (6000 m) from the beach to the sea floor. With a total relief of nearly 33,500 feet (9756 m), this is the highest mountain on the globe. Mount Everest, the tallest land-based mountain in the world, is 29,002 feet (8842 m) high.

A common origin—submarine volcanism—is envisioned for the entire chain. A Polynesian legend describes the red-headed goddess of volcanoes, Pele, being pursued by her older sister from island to island, leaving a trail of volcanoes behind. It is also acknowledged that red-haired Pele seems to have a fiery disposition! The first significant geologic study of the area was by James Dwight Dana in 1840–1841 who recognized that the islands to the northwest were more eroded, and therefore older, and that indeed volcanism did progress from north to south as told in Hawaiian legend! More recent evidence from radiometric dates of the major lava flows on each island confirms that islands decrease in age from the north to south along the entire chain (Fig. 2–1). Indeed, in the Hawaiian group the same trend prevails: Kauai is about 5 million years old; Oahu is 3 million years old; Molokai is 1.9 million years old; Maui is 1.3 million years old; and Hawaii is less than 700,000 years old and still going strong (Tilling and co-workers, 1987).

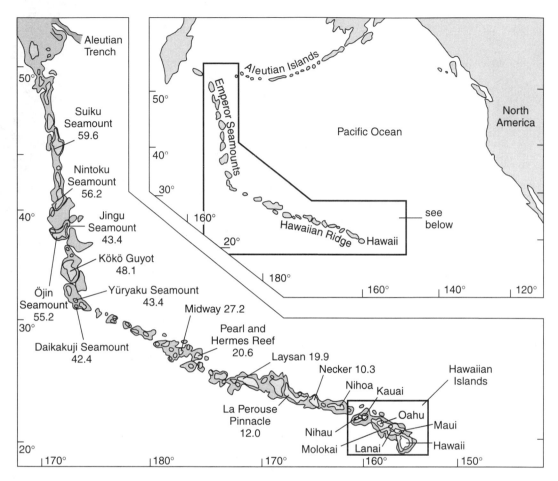

FIGURE 2–1   Hawaiian-Emperor chain showing location of Hawaiian Islands and the oldest reliable dates in millions of years of selected islands and seamounts. (From Skinner and Porter, 1987. Reprinted by permission of John Wiley & Sons, Inc.)

Initially it was reasoned that a fissure progressively opened from north to south, allowing release of pressure, melting, and volcanic activity to follow. Acceptance of plate tectonics caused geologists to rethink the simple model where tectonic and volcanic activity is limited to the plate edges. Questions were asked—why is the world's most active volcano located in the *middle* of a plate rather than at its edge? The Emperor and Hawaiian ridges are connected and obviously have the same origin. Why should the ridge bend sharply where the two ridges join?

Hotspot Origin

The new hypothesis first suggested by Wilson (1963) requires that the deep-seated "fire," or hotspot, remain stationary and that the plate move slowly over the fire. This conveyor belt system is illustrated in Tilling and co-workers (1987) and in Figure 2–2. When the plate is carried too far from the rising heat source, the connection between new magma and the feeder pipes is severed and the volcano becomes extinct—slowly eroding and subsiding to sea level where it may become an atoll or subside below sea level to become one of the thousands of seamounts in the Pacific basin (Fig. 2–3). As the northwestward plate movement continues at a few inches each year for the next few tens of millions of years, the present Hawaiian Islands are heading for much colder climates before being dumped into and swallowed up by the Aleutian subduction zone (Fig. 2–1). Thus, yet another reason is provided for visiting the national park as soon as possible! The "bend" in the island chain is bracketed by ages of rocks on nearby islands at about 43 Ma (millions of years ago)—a time when plate motions shifted from north to northwest in the Pacific.

Origin of Hawaiian Rock

The heat source for the hotspot is quite deep, perhaps at the core–mantle boundary. Rising hot, but solid, mantle material causes partial melting of the upper mantle, generating a magma whose composition is that of basalt. As discussed in Chapter 1, basalt has a relatively low content of silica and is high in iron, magnesium, and calcium compared to other common volcanic rocks. Much to the delight of those who want to simplify rock identification, almost all of the archipelago is

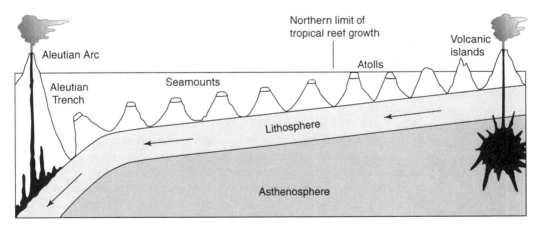

FIGURE 2–2   Diagram depicting island formation above the Hawaiian hotspot and lateral transport and evolution into atolls and seamounts before disappearing into the Aleutian Trench. (From Skinner and Porter, 1987. Reprinted by permission of John Wiley & Sons, Inc.)

composed of basalt with small amounts of limestone formed by coralline animals living in the sea. Pick up a rock and you can be more than 99 percent sure it is basalt! If it is in the national park put it back down—it belongs to Pele and the national park!

Many of the basalts display large crystals of the greenish mineral olivine surrounded by smaller crystals. The small crystals of calcium-rich plagioclase feldspar, augite, and other minerals that cannot be seen with the naked eye formed rapidly in the cooling lava, and large, easily visible crystals (called *phenocrysts*) formed more slowly and grew larger in a magma chamber before being carried to the surface during an eruption. In places olivine phenocrysts are large and flawless and are used as semiprecious stones generally marketed under the name "peridot." Any igneous rock with two distinct sizes of crystals is called a *porphory*.

Most of the Hawaiian lavas are a type called *tholeiitic basalt,* which differs chemically from a much less abundant type called *alkalic basalt.* The alkalic basalt often appears very late in the history of a Hawaiian type of volcano. Tholeiitic basalts are extremely fluid and typically form gently sloping, broad-base landforms known as *shield volcanoes* because of their resemblance to a Roman soldier's shield. The alkalic basalts are more viscous, tend to produce shorter, steeper flows, and generate an *alkalic cap* on the shield and small, steep-sided cinder cones on the volcano's flanks (Fig. 2–3). In the late stages of activity some Hawaiian volcanoes will erupt both types of lava at different times, suggesting that different magma chambers, stratified chambers, or compartmentalized chambers exist beneath the surface. The gradual compositional change away from tholeiitic basalt during the late stages of activity can, at the extreme, lead to generation of magmas that produce andesitic lavas—a lava much richer in silica than basalt.

Island Formation

Islands composed of more than one shield volcano, like Maui and Hawaii (Fig. 2–4), are called *volcanic shield clusters*. The islands are riding "piggy-back" on the Pacific plate, and shield building will continue as long as the volcano is positioned near the stationary hotspot below.

The sequence of island formation supported by radiometric dating was first summarized by McDougall (1964). An excellent discussion of the development of the Big Island shield cluster is given by Moore and Clague (1992), and a generalized series of sketches is given in Figure 2–3. A new volcano begins on the underwater slopes created by its predecessor rather than beginning anew from the deep ocean floor (Fig. 2–3). Submarine eruptions build the growing cone to sea level if the volume of lava emitted exceeds the rate of subsidence due to *isostatic adjustment*. The heavy load of the shield cone slowly adjusts its elevation to establish a balance with the dense mantle material that it "floats" on below, much like a ship will ride lower or higher in the water as cargo is loaded or unloaded. The adjustment of crustal thickness, mass, and surface elevation to the denser mantle mate-

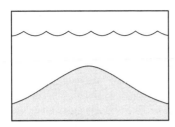

Stage 1.

Submarine shield builds; pillow lava forms.

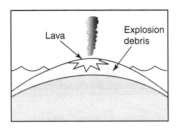

Stage 2.

Explosion debris forms as water shallows, lava in subaerial eruptions builds higher shield.

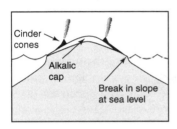

Stage 3.

More silica-rich magma forms alkalic cap, waves erode island edges.

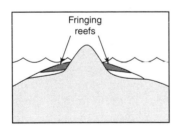

Stage 4.

Eruptions slow or end, erosion and reef formation dominate.

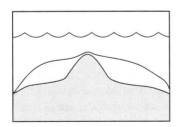

Stage 5.

Submergence exceeds reef building, seamount forms.

FIGURE 2–3   Generalized evolution of a Hawaiian volcano island. Based on diagrams by Dalrymple and others, 1973.

rial below is called *isostasy*. Whereas a ship's draft, or vertical position in the water, is achieved rapidly, earth materials respond much more slowly and require intervals of thousands or hundreds of thousands of years to achieve equilibrium. Vertical growth on the active volcanoes in Hawaii greatly exceeds the measured subsidence rate of 0.1 inch (2.6 mm)/year.

Once above sea level the shield grows in area and height as the shield-building stage continues (Fig. 2–3). Eruption after eruption of fluid basalt adds layer after layer to the growing volcanic edifice. Lavas flow smoothly across the surface but slow when they are cooled more rapidly as they enter the sea (Fig. 2–5)—resulting in a marked steepening at sea level of the volcano slope. As the volcano is carried off the hotspot by the moving plate, the shield-forming stage slows and ends, and the prominent break in slope that marks former sea level subsides along with the rest of the island. Depth charts show these former levels, and deposits of coralline limestone that formed as fringing reefs also mark former sea levels. Radiometric dating of the basalt and the reef rock thus establishes the time of the end of the shield-building stage. Moore and Clague (1992) estimate that it requires about 300,000 years to grow from the sea floor to sea level and another 300,000 years to reach the end of shield building. Further, if the activity span is 600,000 years and the plate is moving at about 5 inches/year (13 cm/year), the hotspot must be at least 50 miles (80 km) to perhaps as much as 200 miles (322 km) in diameter.

In the later stages of shield development when the more viscous alkalic lavas emerge, a distinctive alkalic cap and steeper volcano slopes develop. Moore and Mark (1992) report that slope inclinations on the Big Island systematically increase from 3.3° at Kilauea (youngest shield) to 5.4° at Mauna Loa, 6.4° at Hualalai, 7.0° at Mauna Kea, and 11.3° at Kohala (oldest shield) on the north end of the island (Fig. 2–4). In addition, alkalic eruptions are more explosive and also produce cinder cones that are seen as "bumps" on the skyline as one scans across the smooth profile of a shield.

Somewhere in the late stages as the addition of volcanic material to the shield slows, erosion and subsidence gain the upper hand. Erosion is a downhill process, and streams cut incredible canyons, such as Waimea Canyon on Kauai, and those on Kohala Volcano on the north end of the Big Island. Without fresh lava to heal these erosional "wounds," the scars become deeper and more plentiful.

Stream Erosion

Drainage on volcanic cones typically radiates outward from a central high like spokes on a wheel, forming a *radial drainage pattern*. The water concentrates in ready-made "valleys" between flows and begins its downcutting and valley development there. These are *consequent* streams, streams whose routes are the consequences of the uneven, sloping surface. Valley development continues only until the next eruption fills the valley with lava or other debris. The water seeks out another route of the same kind, and this stream is also consequent.

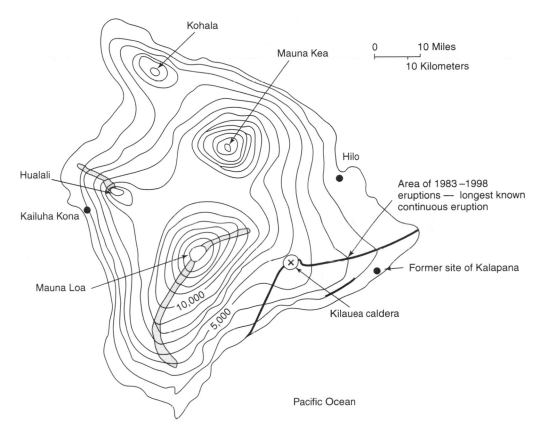

FIGURE 2–4   Contour map of the "Big Island" of Hawaii showing the five peaks of the shield cluster and rift zones that were active in historic times. (Modified from Heliker, 1991)

Wave Erosion

There is, by human standards, a slow-motion battle between the waves that seek to destroy the island edges and the volcanoes that periodically "roll back the waves" and push the shorelines farther seaward. When the volcanoes become inactive, only the destructive processes continue; consequently, one by one the Hawaiian Islands will be planed off at sea level and the fringing reefs of today will become offshore barrier reefs and, with subsidence, atolls. Highly eroded Midway Island is a snapshot of what the Hawaiian Islands will eventually look like.

Wave-cut platforms and wave-cut cliffs are common features around the islands, particularly on their northeast (windward) sides. Waterfalls have developed where hanging valleys empty their waters over the cliffs. The streams, many of them relatively short, have been unable to cut down fast enough to keep pace with cliff recession.

FIGURE 2–5   Lava cascading down sea cliff into Pacific Ocean from Mauna Ulu eruption in 1972. Pillow lavas form at and beneath the sea surface. (Photo by W.H. Parsons)

Glaciation in the Hawaiian Islands

Glaciation, the geologic process least expected in the Hawaiian Islands, is clearly recorded on 13,784-foot-high (4202-m) Mauna Kea, north of Hawaii Volcanoes National Park. Glacial features were first recognized by Daly (1910); recent studies by Porter (1986) indicate that a 27-square-mile (70-km^2) ice cap covered the top of Mauna Kea at least four times during the colder climatic episodes of the last 280,000 years. Older events may have occurred, but their record is buried or removed by erosion. At times, glaciers and lava competed for the same space—not a very compatible couple! Radiometric dates of lava flows provide excellent time marks that help geologists reconstruct events. Not surprisingly, the oldest glacial sediments contain fragments mostly of tholeiitic basalt, and younger sediments contain fragments of mostly alkalic composition. The youngest interval of glaciation began some time after about 69,500 years ago and ended perhaps 10,000 years ago.

The name Mauna Kea ("White Mountain") describes its frequent appearance during the winter months after a storm. The stratigraphy on Mauna Kea involves tholeiitic basalt overlain by alkalic basalts, in turn overlain by glacial sediments, and finally overlain by a large group of telescopes operated by universities and government groups. The cold, clear, unpolluted air and the absence of large cities with light pollution of the night sky are commodities that are rare in today's crowded world.

HAWAII VOLCANOES NATIONAL PARK (HAWAII)

For many thousands of years the tropical island group of Hawaii was one of the most remote areas of the world. Chance arrival of seeds, plants, insects, snails, and birds combined with nearly complete isolation enabled a unique ecosystem to develop that contained many biological species found nowhere else in the world. No mammals were there to exploit the island until about 2000 years ago when Polynesian settlers carrying pigs, plants, and seeds from distant islands began the process of change.

The Polynesian culture and the unique environment were drastically changed after Captain Cook's arrival in 1778 and the later influx of American whaling fleets and missionaries that began in the early 1800s. Numerous species of plants and birds became extinct as forests were cut and exotic plants and animals were introduced.

Fascination with the spectacular volcanoes on the Big Island came early—the first scientific study was by J.D. Dana, a prominent American geologist, during the 1838–1842 U.S. Exploring Expedition. The islands became a U.S. territory in 1900 and through the efforts of local newspaper editor L.A. Thurston, an effort to include parts of the Big Island and nearby Maui as national parks came to fruition in 1916—forty-three years before Hawaiian statehood. Thurston also played a key role in helping to establish the world-famous Hawaiian Volcano Observatory (HVO) in 1911. Professor T.A. Jaggar chose the rim of Kilauea's caldera as the location for the modest observatory facilities and conducted research on one of the world's most interesting, and by volcano standards, one of the safer places to study an erupting volcano. Much of our basic understanding of volcanoes and eruption-monitoring techniques developed under the auspices of the Hawaiian Volcano Observatory, which has been operated by the U.S. Geological Survey since 1948. An excellent compilation of the results of over 75 years of volcanological study in Hawaii is available in Professional Paper 1350 (Decker and others, 1987).

Geographic and Geomorphic Setting

As indicated earlier, the Big Island is the southernmost island along the archipelago and is a cluster of five volcanic shields (Fig. 2–4). Except for the 1800–1801 eruption at Hualalai Volcano, all historic eruptions have occurred at Mauna Loa and Kilauea on the south end of the island (Heliker, 1991). The vent areas and parts of the flanks of these two giant volcanoes form the nucleus of Hawaii Volcanoes National Park (Fig. 2–6). Giant Mauna Loa ("Long Mountain") is 13,680 feet (4170 m) tall and Kilauea ("source of great spreading or spewing"; elevation 4077 feet, 1243 m) is nestled against its southeast edge.

Because the island is growing toward the south, that edge of the island has no older island or seamount to support it. Thus, large masses of the submarine shield

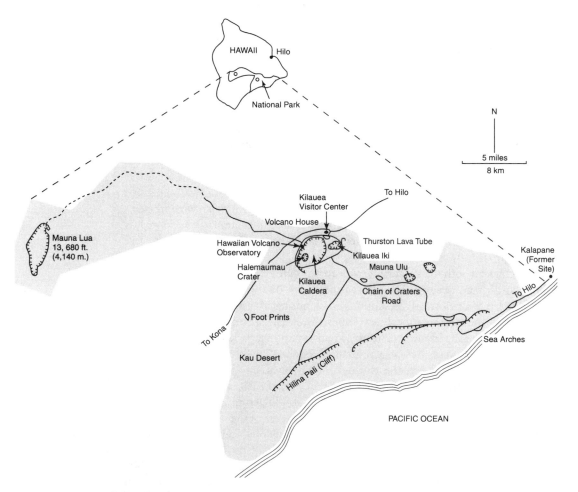

FIGURE 2–6   Map of Hawaii Volcanoes National Park.

occasionally slide seaward along faults with curved planes, much like that of a landslide, but on a much larger scale. The November 1975 fault movement dropped some areas 11 feet (3.4 m) and shifted areas laterally as much as 24 feet (7.3 m) to the south. The abrupt *palis* (Hawaiian for cliffs) on the south flank of the island are surface expressions of *fault scarps*. The 1975 fault movement dropped a section of palm-covered coast and a group of camping boy scouts into the Pacific Ocean (two lives were lost in the national park) and generated a substantial 7.2 magnitude earthquake—not a small event, even by California standards!

Recent studies indicate that Kilauea volcano is now sliding southeastward toward the sea at about 4 inches a year (10 cm/yr) along a gently inclined fault. The seafloor edge of the fault is moving slower or is locked and an ominous bulge has formed where stresses are building. When this locked section breaks loose, another earthquake will occur, and possibly one of the incredibly large submarine avalanches that can carry debris over 120 miles (200 km) offshore will result. Earthquakes accompanying larger events could have magnitudes of 8 or more—very major events! Besides considerable local damage from the earthquake, underwater movements will generate water disturbances that manifest themselves as ocean waves called *tsunamis*. Such waves can move at velocities of 300–435 miles per hour (500–700 km/hr) across the ocean and would devastate Pacific Rim coastal areas, as they did after the 1975 and previous large earthquakes.

Both volcanoes have large collapsed areas at their summits called *calderas*. A caldera is much larger than the underlying feeder vent whereas a volcanic *crater* is about the same size as the feeder vent—usually less than a mile (1.6 km) in diameter. Cracking produced by inflation and deflation of the volcano combined with draining of magma encourages foundering or sinking of a large block of rock into the magma chamber below. Collapsing of immediately adjacent blocks enlarges the depression, thus forming a caldera.

All large shield volcanoes have rift or fissure zones that allow the volcano to adjust mechanically to inflation and deflation as magma rises or drains from the shield. If a shield were to form on a flat plain away from rugged topographic features, it would likely develop three rift zones about 120° apart. Here the close spacing of the volcanoes produces a buttressing or supporting effect that allowed Mauna Loa and Kilauea each to develop only two rifts—one to the southwest and one to the northeast. As we will see, the rift zones are extremely important because they form natural conduits where lava can move laterally great distances to erupt at the surface as a *flank eruption*. Such eruptions within the park boundaries are part of what makes this park unique; those that occur farther down the rift where villages, farms, and housing developments are located are not quite so welcome.

Mechanical movements along the rift zones as well as pumping and draining of magma weaken these zones and allow large, circular collapse features called *pit craters* to form. Pit craters are accessible along the Chain of Craters Road, and impressive Kilauea Iki ("little Kilauea") is accessible by trail from Crater Rim Drive (Fig. 2–7). It was here in 1959 that a major eruption began and the highest (1900-foot, 579-m) recorded *lava fountain* played. Gaze down the 400-foot-deep (122-m)

FIGURE 2–7   Kilauea Iki caldera in 1965. A lava lake during the 1959 eruption. (Photo by D. Harris)

vertical walls of Kilauea Iki and across to the now defunct vent and imagine the events that transpired here. The flat floor was the surface of a 350-foot-deep *lava lake* that was trapped here during the eruption. Hawaiian Volcano Observatory scientists moved drill rigs onto its crusted surface (always a tricky business a few days after an eruption) and have monitored the rate of crust formation and properties of the liquid pool beneath. The last liquid solidified in 1987, twenty-eight years after the lake formed. Other pit craters occur near Mauna Loa's summit and can be reached by a long and difficult trail. Any hiking to Mauna Loa should not be taken lightly—wind and hypothermia are serious hazards—check with a ranger before attempting such a trip.

The geographic isolation of the Hawaiian Islands and their tropical setting have enabled a diverse and unique flora and fauna to develop. The islands have been a "biological test tube" in which accidental arrivals of seeds, plants, birds, and insects have colonized the islands. Geographic isolation has allowed unique species, found nowhere else in the world, to develop. Although still a biological wonder, the presence of humans and the unfortunate introduction of numerous plant and animal species have greatly disrupted the original environment. Feral pigs and goats consume Hawaii's exotic plants, and house cats gone wild feast on its native birds.

The islands are in the zone of trade winds where the dominant air masses move from the northeast. Trade-wind-generated wave erosion is intense at the northern end of the islands where spectacular sea cliffs abound. The same moisture-charged winds, combined with high topography, produce high rainfall and a jungle environment on the windward side but a rain-shadow desert (the Kau Desert on the southwest slope of Kilauea) that even contains large cacti on the leeward side!

Volcanoes in Action

Both Mauna Loa and Kilauea are still in the shield-building stage with no sign of slowing down. The alkalic lavas that mark the end of shield building on large Hawaiian volcanoes have yet to appear, and the volume of lava production continues to be very large. Lockwood and Lipman (1987) report that 90 percent of the surface of Mauna Loa is covered by tholeiitic basalt that is less than 4000 years old, and Holcomb (1987) notes that 90 percent of Kilauea's surface is younger than 1100 years.

A written history of eruptions began with the arrival of American missionaires about 1820. The eruption patterns alternate between the two volcanoes with only very rare simultaneous eruptions occurring at both—suggesting that both volcanoes share the same magma source. Eruptions, on the average, occur every 2–3 years. The Pu'u 'O'o eruption at Kilauea that began in January 1983 is the longest and largest on record—lasting more than 15 years and still going strong as this chapter is written!

Anatomy of an Eruption

Studies begun in 1912 when HVO was built have led to recognition that certain physical and chemical phenomena occur before, during, and after an eruption. Conceptual models attempt to tie in these changes with what is likely occurring at depth and provide scientists with predictive tools that work here and on volcanoes elsewhere in the world. As a result of these efforts and those of other volcanologists around the world, countless lives and property have been saved. Numerous scientific and summary articles are available on the Hawaiian volcanoes and a few are listed at the end of this park discussion.

Occasional deep earthquakes occur some 30 miles (48 km) or more below the surface where rising pockets of magma generated from the underlying hotspot push rock from their path, producing the breaking and grinding that is recorded as an earthquake. Shallow earthquakes of 1–4 miles (1.6–6.4 km) depth indicate that magma is accumulating as liquid pockets in a reservoir zone within the volcano. As more magma enters the reservoir, the shield swells much like a balloon that is being inflated. Small vertical and horizontal displacements at the summit and on the slopes are detectable by extremely sensitive instruments such as laser-measuring distance equipment and tiltmeters. Continuing inflation and an increase in shallow, short-period earthquakes indicate that an eruption is in the offing, usually within days or weeks.

The beginning of an eruption is preceded by rapid inflation and *harmonic tremor* as moving fluids create a continuous vibration—similar to that produced when a fluid moves energetically through a pipe or firehose. Shallow, short-period earthquakes (single bursts of energy) often occur close to where magma will soon break through to the surface, allowing a welcoming party of scientists to be on hand to observe the initial phases of the eruption—no doubt reminding themselves that Hawaiian eruptions are relatively quiet and safe. Other precursors to an eruption include changes in concentration of sulfur dioxide, carbon dioxide, and other gases from volcanic vents (*fumaroles*); measurable changes in the magnetic and gravitational fields; and other geoelectrical properties as well.

The growing magma pressure breaks through rock that has solidified during previous rift eruptions, and magma leaves the confines of the earth to become lava. Just as an automobile tire will fail at its weakest part, so will a volcano break through at a weak point. The eruption often begins at the summit but usually shifts to one of the two rift zones that radiate from the summit. Summit deflation occurs during intrusion of magma in the rift zone or during a flank eruption. When the flank eruption or intrusion ends, summit reinflation proceeds as magma continues to rise from below. The volcano is considered to be in an eruptive state as long as inflation persists, even if no lava is flowing at the surface. The eruption "officially" ends when the volcano deflates to its original shape.

The initial outburst is spectacular; lava fountains hundreds of feet high form a continuous line along the newly opened rift producing a curtain of fire for the first few hours until the excess magma pressure is reduced. The eruption localizes at a single vent and may form a small shield cone such as 400-foot-high (122-m) Mauna Ulu during the 1969–1974 eruption or 830-foot-high (253-m) Pu'u 'O'o that formed along Kilauea's East Rift from 1983–1986. In 1986 activity at Pu'u 'O'o shifted 1.8 miles (2.9 km) eastward to a new vent location called Kupaianana. Slowly but relentlessly the lava flowed from the vent, through a system of lava conduits called *lava tubes,* and emerged at the surface. It crept through housing subdivisions and the historic town of Kalapana, and covered the famous Black Sands Beach on its way to the ocean. The park service poured truckloads of water on the flow front—a technique that slows its advance and sometimes protects property—but this time to no avail as the beautiful Wahaula Visitors Center was engulfed in flames in June of 1989.

Most eruptions have numerous episodes of activity that begin with fountaining events followed by flow material that may be voluminous enough to produce a river of fire. The record for the highest lava fountain measured thus far is at Kilauea Iki with 1900 feet (579 m), 523 feet (160 m) taller than the World Trade Center Building in New York City!

As explained in Chapter 1, lower silica lava tends to be more fluid than higher silica lavas. Thus gases escape more readily, and less explosive eruptions occur with lower silica magma. However, also as mentioned earlier, other factors such as gas pressure can significantly change the eruptive behavior. For example, lava fountaining often occurs at the beginning of a new eruptive episode. Magma confined underground contains large quantities of dissolved gas. Release of pressure when a vent or rift opens or when the lava reaches the surface enables much of the dissolved gas to escape rapidly. A similar phenomenon occurs when a can of warm, carbonated soda pop is uncapped. The demonstration is even more spectacular if the can is shaken violently before uncapping (kids, don't try this at home)!

The lava fountains often fragment the magma and form *pyroclastic* (airborne lava or rock fragments) debris. Larger fragments are called *volcanic bombs,* smaller masses with openings (*vesicules*) created by escaping gases are called cinder, or if very light in weight, *pumice* or *reticulite.* Drops of lava that acquire a streamlined form as they sail through the air are called *Pele's tears,* and fine threads of volcanic glass spun in the fire fountains are called *Pele's hair.*

Unusual amounts of gas pressure can turn eruptions to the deadly side as occurred in 1790 and again in 1924. Magma draining from the summit area occasionally allows groundwater to flow in, mix, and violently explode sending forth ash clouds and hurtling large blocks, some weighing several tons, around the caldera area. The 1924 steam explosions at Kilauea blasted dust clouds more than 20,000 feet (6098 m) into the air, and large blocks of rock were thrown a half-mile (0.8 km) from the vent. There was one fatality, the only one in the park's history caused by volcanism. Definitely recorded, although details are lacking, is another explosive eruption which in 1790 killed a group of perhaps 80 Hawaiian warriors trekking across the Kau Desert on the southwest flank of Kilauea. Likely, the pyroclastic cloud seared their lungs causing them to suffocate. The survivors walked through the wet ash deposited by the blast wave and left their footprints (Fig. 2–8) along the Mauna Iki (Footprints) trail in the Kau Desert.

Lava Flows and Related Features

The low-silica content, high temperature (up to 2200°F; 1200°C), gas content, and other factors contribute to the high fluidity (low viscosity) and relatively quiet eruptions that characterize Hawaiian volcanoes. From the standpoint of topography, volcanism is a constructive process that, in Hawaii, mostly forms shield cones by piling up layer upon layer of lava.

The more fluid flows form a thin, smooth lava skin that may wrinkle as the interior continues to move, forming a *pahoehoe* (pronounced "pay-hoy-hoy") type of lava (Figs. 2–9 and 2–10). Most flows begin as pahoehoe but often change downslope into a slow moving, clinkery-surfaced *aa* (pronounced "ah ah") lava flow (Fig. 2–11). A stroll over the cooled surfaces of these two flow types without

FIGURE 2–8   Footprints left by Hawaiians crossing ash erupted from Kilauea in 1790. A number of warriors were killed by the ash cloud. (Photo by E. Kiver)

your shoes will make it clear why it was important for native Hawaiians traveling about to have names for flows with different surface characteristics.

Where pahoehoe flows move through a forest, around highway sign posts, and around stop signs (usually without even slowing down!), a protective smear of lava wraps around the obstruction. In the case of *lava trees* the tree eventually catches fire and is destroyed, leaving a hole where the trunk was formerly located. Sometimes a ghost forest of cylindrical lava casts (molds) projecting eerily above the cooling lava surface replaces what was once a section of tropical forest.

Areas covered by recent flows are essentially barren of vegetation. There are "islands," however, where there is a lush growth of trees and bushes. These are *kipukas* (pronounced "kipookas"), areas that were surrounded but not covered by recent lava flows (Fig. 2–12).

Fluid pahoehoe in long-lasting eruptions often becomes channeled into narrow lava streams that build *lava levees* at their edges that allow the lava level to rise. Floating rafts of solid crust may attach to the levee edges and the channel may roof over forming a lava tube. Lava can then flow within these rock-insulated tubes without

FIGURE 2–9 Slow moving toe of pahoehoe lava flow with a temperature close to 1700°F (1000°C). Lava is about 3 feet from camera. (Photo by E.H. Gilmour in 1995)

FIGURE 2–10 Pahoehoe lava flow a few months old in 1995 buries a section of Chain of Craters road. (Photo by E.H. Gilmour)

FIGURE 2–11   Aa flow blocks highway in Hawaii Volcanoes National Park. (Photo by National Park Service)

FIGURE 2–12   Kipuka standing above and surrounded by 1935 pahoehoe lava flow on north flank of Mauna Loa. (Photo by E. Kiver)

appreciable heat loss and reappear as surface flows many miles downslope—enabling the lava to reach the sea and to once again make the maps of Hawaii obsolete. Where the end of a lava tube is below sea level, the pounding of the waves within the tube may develop a vertical chimney along fractures. These *blow-holes* erupt as "cold-water geysers" with each major wave. Blowholes may also develop from sea caves formed by erosional processes in many different types of rocks.

Flows entering water develop a special form known as *pillow lava* (Fig. 2–5) whose process of formation was not well understood until divers observed and photographed them forming as lava poured into the sea during the 1969–1974 Mauna Ulu eruption. Tubes, a few feet in diameter, form underwater and lava inside flows like toothpaste from a tube. The front edge of the advancing tongue of lava cools rapidly when it contacts the ocean and forms discrete, pillow-shaped lava blobs that resemble a pile of pillows or sacks of grain. The foundation of each of the islands is built on a platform of rocks exhibiting these underwater flow features.

Under the right conditions lava entering the sea will cause steam explosions or, *littoral explosions* (the littoral zone is the area between high and low tides), and spray small glassy fragments of basalt about. Small accumulations are quickly reworked by the sea into *black sand beaches,* and larger deposits may be more permanent and form a *littoral cone* such as Diamond Head (Fig. 2–13) on the island of Oahu. The littoral cone, Pu'u O Mahana at South Point outside of the national park, is at the seaward edge of the Southwest Rift Zone of Mauna Loa and contains numerous sand-size crystals of olivine that were reworked by the sea into one of the famous *green sand beaches* of Hawaii. Erosion of coral deposits elsewhere in the islands forms white sand beaches to complete the technicolor scheme of beaches.

FIGURE 2–13   Looking north over Diamond Head, a tuff cone on the island of Oahu, southeast of Honolulu. (Photo by D. Harris)

The Future

Using what we have learned, we can predict what events to expect in the future. Mauna Loa and Kilauea are young cones and will be located above the underlying hotspot for many tens of thousands, or at most a few hundred thousand, years. Eventually their connection to the hotspot will be severed and alkalic eruptions, with progressively longer noneruptive intervals, will follow. By then, geologists will have something new to watch. As the plate tectonics and hotspot theories predict, the moving plate will place new ocean floor in reach of the rising plume of heat and new seamounts and islands will form. Submarine surveys using instruments and deep-diving research submarines indicate that the next island is well on its way to the surface. The seamount Loihi is 20 miles (32 km) south of the Big Island and is now within 3100 feet (945 m) of the water surface—ETA (estimated time of arrival) is 200,000 years in the future (Moore and Clague, 1992).

Hopefully, when Loihi is much closer to sea level, caring people will be here to establish it initially as an underwater national park and later as a tropical island park with an active volcano. Whoever is in charge at that time will have an opportunity to "start from scratch" and do the "best" thing for this new piece of real estate. Because of expanding human population and monetary motives, pressure to use public lands, even national parks, for commercial purposes is being pushed by small but powerful groups. Fewer areas are being added to the park system, and some are seriously suggesting that some of the parks be closed or given to the states. However, in Hawaii, Pele is still a strong park supporter and continues to add land to our park system!

The Civil Defense Agency and HVO scientists are very concerned about the earthquake and associated tsunamis (seismic sea wave) hazard, especially to the south end of the island where a large resort was planned, a major geothermal power plant was proposed, and developers began to construct large housing developments. According to Harry Kim, head of the Civil Defense Agency for the state of Hawaii, the death and destruction potential is now greatly reduced because "nature—others would say Pele—has solved a problem for us by removing the concentration of people." The large 1975 earthquake squelched resort development plans, the 1983 lava flows covered the proposed geothermal site, and lava flows in the late 1980s and early 1990s destroyed the town of Kalapana and over 200 houses in the Kalapana Gardens subdivision!

Park Address

Hawaii Volcanoes National Park
HI 96718

REFERENCES

Dalrymple, G.B., Silver, E.I., and Jackson, E.D., 1973, Origin of the Hawaiian Islands: American Scientist, v. 61, p. 294–308.

Daly, R.A., 1910, Pleistocene glaciation and the coral reef problem: American Journal of Science, v. 180, p. 297–308.

Decker, Robert W., Wright, T.L., and Stauffer, P.H., 1987, editors, Volcanism in Hawaii: U.S. Geological Survey Professional Paper 1350, 1667 p.

Heliker, Christina, 1991, Volcanic and seismic hazards on the island of Hawaii: U.S. Geological Survey General Interest Publication, 48 p.

Holcomb, R.T., 1987, Eruptive history and long-term behavior of Kilauea volcano: U.S. Geological Survey Professional Paper 1350, p. 261–350.

Lockwood, J.P., and Lipman, P.W., 1987, Holocene eruptive history of Mauna Loa volcano, Hawaii: U.S. Geological Survey Professional Paper 1350, p. 509–536.

McDougall, Ian, 1964, Potassium-argon ages from lavas of the Hawaiian Islands: Geological Society of America Bulletin, v. 75, p. 107–120.

Moore, James G., and Clague, D.A., 1992, Volcano growth and evolution of Hawaii: Geological Society of America Bulletin, v. 104, p. 1471–1484.

Moore, James G., and Mark, R.K., 1992, Morphology of the island of Hawaii: GSA Today, Geological Society of America, v. 2, no. 12, p. 257–262.

Porter, Stephen C., 1986, Glaciation of Mauna Kea, Hawaii. In Sibrova, V., Bowen, D.Q., and Richmond, G.M., Eds.: Quaternary glaciations in the northern hemisphere, Quaternary Science Reviews, v. 5, p. 181–182.

Skinner, Brian, and Porter, S.C., 1987, Physical Geology: Wiley, 750 p.

Tilling, Robert I., Heliker, C., and Wright, T.L., 1987, Eruptions of Hawaiian volcanoes: Past, present, and future: U.S. Geological Survey, General Interest Publication, 55 p.

Wilson, J.T., 1963, A possible origin of the Hawaiian Islands: Canadian Journal of Physics, v. 41, p. 863–870.

HALEAKALA NATIONAL PARK (HAWAII)

The name Haleakala (pronounced hah-lay-ah-kah-lah), which means "house of the sun," is related, according to legend, to a project by the Polynesian god Maui to capture the sun and persuade it to move at a slower pace across the sky—so his mother could get all of her work done! Although this early attempt at daylight saving and mother-appeasement failed, the Haleakala name prevailed. A sunrise picture across the top of the mountain is convincing evidence that the name is appropriate.

Established first in 1916 as a section of Hawaii National Park, Haleakala gained full status as a national park in 1961. Originally the park included only the upper part of the 10,023-foot-high (3056 m) volcanic cone; in 1969 it was extended down the Kipahulu Valley on the east side of the mountain to the sea, adding a beach environment, a tropical rain forest (over 300 inches, 762 cm of rain each year!), and the Seven Pools area (Fig. 2–14).

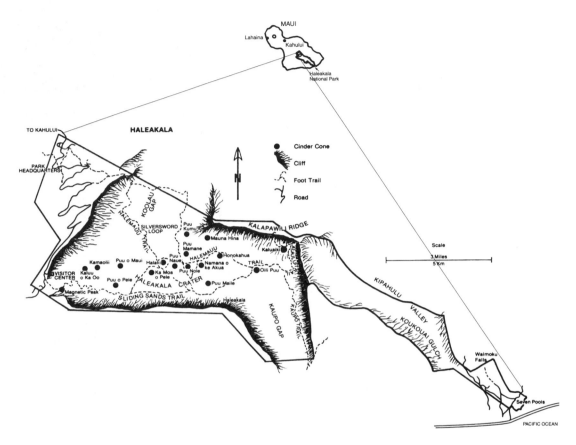

FIGURE 2–14   Map of Haleakala National Park.

The new addition contains a number of rare and endangered species and the upper slopes are not open to the public. In 1980 the park was declared a biosphere reserve, in recognition of its unique biological environment. The beach environment is accessible to the public and spectacular Waimoku Falls, at the head of one of the valleys, can be reached by hiking up through the tropical forest. Above 6000 feet (1829 m) on the northeast side and on much of the southwest side, semiarid conditions prevail and vegetation is less dense.

In order to reach the Seven Pools area, you drive a slow road from the town of Kahuli around to the southeastern part of the island; to reach the crater rim you follow a tortuous road up the west side of the cone. First view is from Kalahaku Overlook, one you will not forget.

Haleakala is distinctly different in appearance from Mauna Loa and Kilauea (Fig. 2–15). Haleakala is highly colored whereas the others are mainly black. Gases from fumaroles and degassing lava have deposited colorful crusts of mineral pre-

FIGURE 2–15   Erosional caldera of Haleakala Volcano. Cinder cones, including Puu O Maui, rise from the caldera floor. (Photo by National Park Service)

cipitates and sublimates. Also, the greater age of Haleakala has allowed many of the iron-bearing minerals in the rocks to alter to brighter yellow and red iron-oxides.

The crater, or caldera, at the summit is about 7 miles (11 km) long and 3 miles (5 km) wide, and more than half a mile (about 1.0 km) deep. Actually, it is neither a crater nor a caldera in the true sense because the present topography has resulted mainly from stream erosion—not directly from volcanic activity. Thus the term *erosional caldera* is appropriate. Here, headward erosion by two streams created two large amphitheaters side by side to form the calderalike feature. At its highest, the mountain was about 12,000 feet (3659 m) high. Now, the highest point is 10,023 feet (3056 m) above the sea.

Geologic Sequence

The islands of Maui, Kahoolawe, Lanai, and Molokai are part of a volcanic shield cluster of at least six separate volcanoes whose tops project above the sea. About 20,000 years ago, as a result of glaciers around the world being more plentiful and very large, sea level was hundreds of feet lower. A view of the shield cluster at that time would show a connected group of peaks making up a single island. Maui now consists of two peaks; Haleakala on the east and a highly eroded, older shield cone in west Maui. A low isthmus connecting the two was submerged during higher stands of Pleistocene sea level creating, at times, two islands rather than one.

As would be anticipated from the discussion of the hotspot theory at the beginning of this chapter, the volcanoes on Maui are older than their Big Island (Hawaii) cousins. As active shield building occurs, a sharp increase in slope angle is maintained where lava flows from the land into the sea (Fig. 2–3, stage 3). When shield building stops or slows, the break in slope subsides along with the rest of the volcano. Radiometric dates of the last flows across the slope break establish when cone building stopped.

Moore and Clague (1992) report that when the flow of tholeiitic basalt and active shield building came to an end about 950,000–1,100,000 years ago, a cap composed of alkalic basalt (richer in sodium and potassium) and andesite (higher in silica) over 2500 feet (762 m) thick formed on the summit (see Fig. 2–3, stage 3). Erosion attacked the volcano and streams flowing out Koolau Gap on the north and Kaupo Gap on the southeast changed the summit into an erosional caldera. Occasional eruptions, or episodes of eruptions, characterize the very late stage of a Hawaiian volcano's history. Rifts reopened following the long episode of erosion and added fresh-appearing lava flows and cinder cones such as 600-foot-high (183 m) Puu O Maui (Fig. 2–15) and Red Hill (elevation 10,023 feet, 3056 m), the high point on the summit rim. Some of these later eruptions were highly explosive because cinders and ash were distributed widely, particularly on the leeward flank of the cone. The latest major eruption probably occurred about 1750 (Macdonald and Hubbard, 1989); however, two small flank eruptions sent lava flows down into the sea around 1790.

By no means should Haleakala be regarded as an extinct volcano; it is a quiescent or dormant volcano that might erupt again at any time. The connection with the hotspot is likely broken, but rising magma pockets and trapped magma pockets still remain. Periodic earth tremors serve to remind us that although calm prevails at the surface, there is still some activity below.

Many of the rock materials that you see are pyroclastics of andesitic composition, but the bulk of the cone is made up of basalt flows. Some of the flows are porphyritic, and large olivine phenocrysts are common. Blanketing the basalts in many places are the andesitic materials, mainly cinders and ash.

Haleakala's Other Features

Within the caldera, there are many interesting features for energetic hikers or for those on horseback to enjoy. Cabins, available by reservation, are situated at strategic locations for those who wish to stay 2 or 3 days. Lava tubes, Bubbler Cave, and Pele's Paintpot are among the attractions that can be visited. Vegetation is sparse or lacking in much of the caldera due to the low precipitation and the high permeability of the cinder floor. Growing in some of the high, dry areas within the caldera and around the rim is a remarkable silvery, yuccalike plant, the silversword (Fig. 2–16). When it matures, perhaps in 20 years, its tall stalk produces purple flowers once; then it dies. The caldera floor was carpeted with silversword in the 1800s, but by 1920 it was nearing extinction, having been devastated by wild goats and dug up for sale by dealers in ornamental plants. Under the protection of the National Park Service, silversword is now making a comeback. A few plants are found in similarly high places on the island of Hawaii. Also unique is the nene

FIGURE 2–16   A mature silversword plant on the flank of Haleakala. (Photo by D. Harris)

(pronounced nay-nay), or Hawaiian goose, which has renounced its aquatic heritage and lives only on land; it is even found within the caldera.

The eastern part of the park is an oasis where trees, grasses, and ferns are abundant. Rare plants grow here, including the apeape with leaves up to 5 feet (1.6 m) across. Birds of many kinds decorate the jungle, among them two endangered species—the Maui parrotbill and the nukupuu, formerly believed extinct. Before about 1900, birds were far more numerous in the area than now, and ornithologists had reported seeing about 25 kinds of birds that are no longer found. Scientists found that they had succumbed to disease carried by mosquitoes that were introduced to the island when sailors emptied their near-empty water barrels. Mosquito larvae (brought from Mexico) dumped into the stream during barrel refilling spelled doom for these strikingly beautiful birds.

Haleakala is small compared to Hawaii Volcanoes National Park. Although still possible, it is unlikely that it will stage a spectacular eruption for you. However, it is different; it represents an older stage in the volcanic cycle than either Mauna Loa or Kilauea. Shield building was followed by caldera development; then there was a long period of erosion, after which there was the final (?) spasm of volcanism—the cinder cones (Fig. 2–3). Perhaps we have here a prophetic picture of the future Mauna Loa and Kilauea.

REFERENCES

Decker, R.W., Wright, T.L., and Stauffer, P.H., editors, 1987, Volcanism in Hawaii: U.S. Geological Survey Professional Paper 1350, 1667 p.

Macdonald, G.A., and Hubbard, D.H., (revised by Jon W. Erickson), 1989, Volcanoes of the national parks in Hawaii: Hawaii Natural History Association, 65 p.

Moore, James G., and Clague, D.A., 1992, Volcano growth and evolution of the island of Hawaii: Geological Society of America Bulletin, v. 104, p. 1471–1484.

Park Address

Haleakala National Park
Box 369
Makawao, HI 96768

THREE

Pacific Border Provinces

The combined Pacific Border Provinces are lean and long—almost 5000 miles (8050 km) long. They extend from Attu Island at the tip of the Aleutian Islands near Russia to the Kenai Peninsula and the coastal mountains around the Gulf of Alaska, and southeastward through Alaska's panhandle and British Columbia, all the way south through Washington, Oregon, and California (see Plates 1 and 3).

The Aleutian Range includes an island arc of active volcanoes that extends to the Gulf of Alaska where lofty mountains rise abruptly above the sea—mountains deeply indented by huge *fjords,* or *fiords.*[1] Here, countless glaciers actively relandscape the mountains and act as silent reminders that this area is still in the Ice Age. They also serve as living examples of what much of the mountainous West and the northeastern United States looked like during the glacial episodes of the Pleistocene.

South of Olympic National Park in Washington, in the Southern Pacific Border Province, the coastal ranges are at lower elevations and topographically subdued. However, rugged wave-cut cliffs, stacks, and sea arches (Fig. 3–1) present picturesque oceanscapes; elsewhere along the seacoast there are sandy beaches and coastal dunes. East of the coastal mountains, structural troughs can be traced southward from Puget Sound in Washington through the Willamette Valley in Oregon, the Great Valley of California, and the Gulf of California. A very generalized geologic overview of the entire Pacific Border is given here, but details of the Alaskan parklands in the Northern Pacific Border are not presented in this edition.

Tectonic History

Because continents grow by adding new land to their margins, especially along the leading edge of a moving continental plate, we should expect that active, dynamic forces are operating along the West Coast of North America. As expected, the heat

[1]Fjords are deep valleys cut by glaciers and now partially submerged by seawater. Fiord is an anglicized spelling.

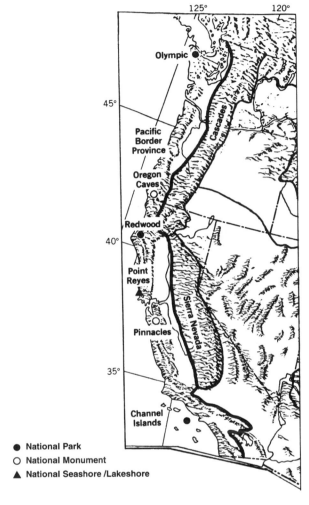

PLATE 3   Southern Pacific Border Province. (Base map copyright Hammond Inc.)

engine that drives plate movement has indeed created an incredible array of complex folds and faults that will provide generations of geologists many years of study and "head scratching" as the geologic puzzle is slowly assembled. As residents of Los Angeles were rudely reminded in January of 1994 by a magnitude 6.6 earthquake, and as the entire West Coast was reminded by Mount St. Helens in 1980, plate movements are not just ancient history; somewhere in our future are more big earthquakes, more volcanic eruptions, and more mountains to be born.

The Pacific Border Provinces have four major tectonic components: the Aleutian Arc–Gulf of Alaska subduction zone, the coastal Alaska–British Columbia transform fault system, the Washington–Oregon–northern California spreading ridge subduction system, and the California San Andreas transform fault

(a)

(b)

FIGURE 3–1   Sea arches near Santa Cruz, California, before and after a January storm in 1980. (Photos by R. Scott Creely)

system. The Pacific plate has migrated north for millions of years—in some places jamming sediments under the continent edge along subduction zones, in other areas the Pacific plate slides horizontally past the North American plate along huge *strike-slip,* or *transform, faults*. In even older times blocks and slices of continental crust moved huge distances along faults and assembled much of western North America into a mosaic of distinct terranes (see Fig. 1–11). But we must turn the clock back farther yet if we are to understand why the continent was pieced together in this fashion.

Mountain building is directly related to plate movements. For example, during the Paleozoic Era the North American Plate was moving eastward and deforming the leading, or Appalachian, edge of the plate. Later, during the Mesozoic Era, the direction of plate movement reversed, and mountain building began along a subduction zone along the western edge of the continent. The eastern margin then became the continent's tectonically inactive trailing edge.

Another major change in tectonic stress along the continent's west edge occurred when the continental plate collided with the spreading ridge associated with the Pacific–Farallon plates (Fig. 3–2) about 20–30 Ma (millions of years ago). As the ridge is even today being overrun and eliminated, large *transcurrent*, or strike-slip, *faults* such as the Queen Charlotte in British Columbia and the San Andreas in California are still increasing in length. Such vertical faults, along which major plates slide horizontally past one another, are a special type of strike-slip fault called a transform fault. Because the Pacific Plate is moving northward relative to the North American Plate, movement along these faults is *right lateral* (if one faces the fault from either side, the apparent motion is such that the block across the fault appears to be displaced to the right). At the ends of a transform fault the plate motion changes, or transforms, to a converging, subduction movement or a diverging, spreading-ridge type of movement.

On the ends of the San Andreas Fault and the south end of the Queen Charlotte Fault are unique sites where three plates join in one area—a *triple junction*. The Mendocino triple junction lies at the north end of the San Andreas Fault just off the northern California coast where the North American, Pacific, and a remnant of the Farallon (now called the Juan de Fuca) plates join (Fig. 3–2). The Riveria

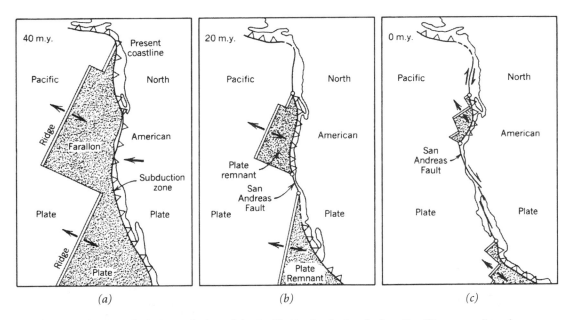

FIGURE 3–2   Diagrams depicting evolution of the Pacific Border during the last 40 million years (based on diagrams in Dickinson, 1979). (*a*) Westward-moving North American Plate overrides Farallon Plate and begins to interact with the oceanic ridge sometime after 40 million years ago. (*b*) Note the early development of the San Andreas Fault and the change to right-lateral transverse faulting about 20 Ma as the oceanic ridge is overrun by the North American Plate. (*c*) Present-day plate elements along the Pacific Border. Small remnants of the original Farallon Plate remain, but right-lateral transverse faults have appeared where ridges are overrun. (Illustration by Gregory Nelson)

triple junction lies on the south end of the San Andreas Fault just off the Mexican coast.

The stresses and, therefore, the types of geologic activity and features found north and south of a triple junction are profoundly different and easily justify dividing the Pacific Border into different sections or even separate provinces as have Thornbury (1965) and Hunt (1974). In this edition of the book, only the parks in the Southern Pacific Border Province of Washington, Oregon, and California are covered.

The northeasterly movement of the Pacific Plate and the westward movement of the North American Plate have been the dominant tectonic influences in the Pacific Border Provinces since mid-Cenozoic time—the unrelenting, but much "slower than a speeding snail," pace of the plates continues today. Presented above are mere glimpses of what has been discussed in detail by many workers, among them Maxwell (1974), Dickinson (1979), and Oldow and co-workers (1989).

To visualize how western North America has been broken into plates that have been shoved about and formed into mountains, the following analogy may be helpful: Picture a wide river that is flowing northward toward the Arctic Ocean several hundred miles away. The river is frozen over, but in its southerly reaches, water is flowing beneath the ice. When the spring thaw begins, the ice cracks into blocks, or "plates," some large, others small, some mere slivers. Soon the ice plates begin to move, carried slowly northward by the flowing water under the ice. But in the cold north, the ice is still frozen solid, and here is the collision zone where the plates pile up in an ice jam. At first they create merely a ridge of ice blocks, but soon they become an elongate "mountain range," much like the mountains bordering the Gulf of Alaska. Some of the ice plates that have traveled hundreds of miles are now on top of, underneath, or beside blocks of "resident" ice—like some of the blocks of rock in the Chugach, Wrangell, and St. Elias mountains of southern Alaska. Thus the location of a plate of rock completely different from its neighbors in the Wrangell Mountains, yet identical to rocks in southern British Columbia, can now be reasonably explained using the concepts of plate tectonics. These slivers of transported crustal blocks are called *microplates, exotic terranes,* or *suspect terranes.* Their importance in constructing western North America can be better appreciated by noting their extent in Figure 1–12. Oldow and co-workers (1989) estimate that about 200 microplates make up the western Cordillera of North America.

Admittedly, ice blocks or ice plates on a river are distinctly different from 20- to 30-mile-thick (32–48-km) plates of crustal rocks, and the plastic, slow-moving subcrust only remotely resembles water flowing beneath ice blocks. However, this analogy may help in visualizing what took place in the Pacific Border Provinces, particularly in the north where essentially all of Alaska has been assembled by wayward pieces of crustal rock derived from southern locations.

Farther south, in Washington and Oregon, subduction zone deformation dominated the later scene. Rocks that were carried along on the northeastward-moving oceanic plate were crumpled and thrust up into the coastal ranges, including the Olympic Mountains. Rocks that were to become the Coast Ranges began to accumulate early in the Tertiary about 65 Ma. However, the uplift that raised the mountains to their present height began much later, late in the Pliocene. For more detail on the tectonics of the Pacific Northwest see Alt and Hyndman (1978, 1984), Drake (1982), Oldow and co-workers (1989), Orr and co-workers (1992), and Orr and Orr (1996).

Without question, tectonic movements continue in the Pacific Border Provinces today. The San Andreas Fault system in California is of particular concern because of the millions of people living near it and the vivid recollections of devastating historic earthquakes: San Francisco in 1906, the Imperial Valley in 1940, San Fernando in 1971, and Los Angeles in 1994. The damaging earthquakes experienced in the Willamette Trough in Oregon and in Puget Sound in Washington are believed to be related to subduction activity. Earthquakes here are less frequent and historically less destructive than some of those in California and Alaska. However, geologists have recently discovered coastal forests that were suddenly lowered below sea level and covered by marine sediments. Such events accompany large coastal earthquakes. These "ghost" forests on the Washington and Oregon coasts form about every 300–500 years and result from large earthquakes—perhaps some with magnitudes comparable to the devastating 1964 Alaska earthquake (magnitude, M = 9.2). According to Atwater (1992), the last major coastal earthquake occurred about 300 years ago—thus a similar large earthquake could occur at any time. One of the consequences of living on the "edge" is that residents of the Pacific Border Province are in the most tectonically active province on the continent and, ultimately, they will bear witness to more of the earth's "growing pains."

Geomorphic Processes and Features

Landforms are the products of the interaction of internal and external processes. Often the external processes such as running water, glaciers, waves, currents, gravity, and wind dominate because of the slowness or inactivity of the internal processes. However, when internal forces operate vigorously, distinct landform expressions result. For example, Pleistocene changes in sea level, both higher and lower, can have either tectonic (internal process) or climatic (external) causes. Periods of worldwide glaciation correspond to times of lower sea levels—periods lacking extensive glaciation are associated with episodes of higher sea level. Sea level lowering due to the volume of water locked up in glacial ice amounts to, at most, 400 feet (122 m). Increases above current sea level during glacial meltdowns larger than that of the present were perhaps as much as 100 feet (30 m). Yet, *wave-cut terraces* (gently sloping surfaces cut by waves at sea level) of Pleistocene age occur as much as 1600 feet (488 m) above sea level in Oregon and over 1700 feet (518 m) in California—hardly the result of changes in glacier size. Thus, vertical land movements due to tectonic processes must be at work. Further, many of these marine terraces are warped and bent—more evidence that tectonic forces continue to reshape the landscape in the Pacific Border Provinces.

As previously noted, shore processes are operating vigorously in the coastal zones and are major elements in the geologic story of Olympic, Redwood, and Channel Islands National Parks and Point Reyes National Monument in the Southern Pacific Border Province. In the northern areas where climate is favorable, glacial and cold climate phenomena dominate. Glacial activity is also encouraged by high elevation and high winter precipitation. The lower elevation coastal mountains south of the Olympics lacked extensive glaciation—even during the height of Pleistocene glaciation. High topography in the path of moisture-charged air masses moving inland from the Pacific produces high precipitation. Air masses cool as they rise and consequently they release large amounts of precipitation—over 140 inches (356 m) each year in the Olympics, a significant amount of which occurs as snow. Chemical weathering is intensified by high precipitation—particularly in the lower, warmer, southern localities. The high rainfall and thicker soils result in dense vegetation—a delight to biologists and foresters—but not so good for "rock people."

REFERENCES

Alt, David, and Hyndman, D.W., 1978, Roadside Geology of Oregon: Mountain Press, Missoula, Montana, 272 p.

Alt, David, and Hyndman, D.W., 1984, Roadside Geology of Washington: Mountain Press, Missoula, Montana, 282 p.

Atwater, Brian F., 1992, Geologic evidence for earthquakes during the past 2000 years along the Copalis River, southern coastal Washington: Journal of Geophysical Research, v. 97, p. 1901–1919.

Dickinson, W.R., 1979, Cenozoic plate tectonic setting of the Cordilleran region in the United States: in Cenozoic paleogeography of the western United States, Society of Economic Paleontologists and Mineralogists, p. 1–13, Anaheim, California.

Drake, E.T., 1982, Tectonic evolution of the Oregon continental margin: Oregon Geology, v. 44, no. 2, p. 15–21.

Hunt, Charles B., 1974, Natural regions of the United States and Canada: W.H. Freeman, San Francisco, 725 p.

Maxwell, J.C., 1974, Anatomy of an orogen: Geological Society of America Bulletin, v. 85, p. 1195–1204.

Oldow, John S., Bally, A.W., Ave'Lallemant, H.G., and Leeman, W.P., 1989, Phanerozoic evolution of the North American Cordillera; United States and Canada: in The Geology of North America—An overview, Geological Society of America, The Geology of North America, v. A, p. 139–232.

Orr, Elizabeth L., and Orr, W.N., 1996, Geology of the Pacific Northwest: McGraw-Hill, New York, 408 p.

Orr, E.L., Orr, W.N., and Baldwin, E.M., 1992, Geology of Oregon: Kendall-Hunt, Dubuque, Iowa, 254 p.

Thornbury, William D., 1965, Regional geomorphology of the United States: Wiley, New York, 750 p.

OLYMPIC NATIONAL PARK (WASHINGTON)

Within the park system there is no area of greater contrasts than those of Olympic National Park. Water plays a major role in the uniqueness of this area—not only in its streams and lakes out also in its vaporous state as large fog and cloud banks, in its salty form in the water bodies surrounding the peninsula, and in its frozen state in the 60 or so glaciers shrouding Mt. Olympus (7965 feet, 2428 m) and the other high peaks in the core of the range. With the highest precipitation in the conterminous United States, the glaciers are well nourished; so is the rain forest region on the Pacific slope. A visit to the Hoh rain forest clearly shows what large quantities of water can do for plant growth. Although Hoh is an Indian word, some might think that it represents the chemical notation for water! The situation is distinctly different on the eastern side. Moisture-bearing air masses moving inland from the Pacific Ocean shed much of their moisture on the temperate rain forest and on the glaciers on the west side, leaving little to drop on the eastern, "rain shadow" area. In further contrast is the 57-mile-long (92-km) section of the Pacific coast where there are wave-cut cliffs, sea arches, and other spectacular features of the oceanscape (Fig. 3–3).

The Olympic Peninsula is in the upper-left-hand corner of the conterminous United States and is the crown jewel of the coastal ranges in Washington, Oregon,

FIGURE 3–3   Wave-cut bench at low tide in Olympic National Park. Note the upturned edges of the mid-Tertiary marine sedimentary rocks and the abundant sea stacks. (Photo by E. Kiver)

and California. To the east lies the broad, Puget Sound structural depression that contains Seattle and other large cities and beyond lies the rugged North Cascades (Plate 3). Park headquarters and the main visitor center are located in Port Angeles (Fig. 3–4). The magnificent view of glacier-clad peaks from the Hurricane Ridge Road south of Port Angeles is a "must" for visitors. Roads and trails in the Hoh and Quinault River valleys provide access into the rain forest on the western slope, where the moss-draped giants of trees blot out the sun. On the east side, the dirt road up the Dosewallips River from the Hood Canal Highway leads into the drier eastern-slope section. Access to the coastline environment is from short trails leading from U.S. Highway 101 north of Kalaloch (Fig. 3–4) and by hiking along the "low maintenance" trails—the beaches—both here and along the wilderness coast farther north. To see a significant portion of the park up close, the rugged and truly exciting parts, one must use the trails, the most rewarding way to experience the park. About 600 miles (968 km) of trails penetrate into the back country along which there are a number of camping shelters.

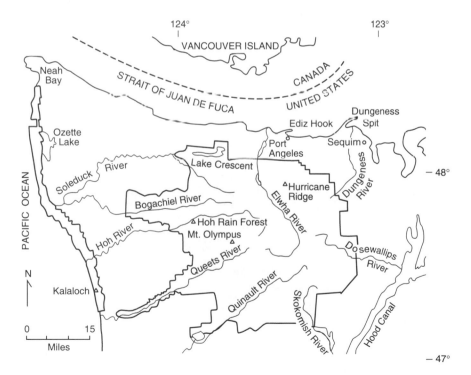

FIGURE 3–4   Olympic National Park and surrounding areas. Heavy lines are park boundaries.

Olympic Area History

Little is known about the first human inhabitants of the Olympic Peninsula except that they moved in after the large alpine glaciers had greatly receded and the most recent mile-thick (1.6-km) Puget Sound ice lobe had thinned and retreated. We do know that the Indians were hunting bison and mammoth 12,000 years ago on the east side of the Olympics near the present town of Sequim and that forest-dwelling mastodons were also present. We know a good deal more about the Native American inhabitants during the past 2000 years—not only from oral histories but from archeological sites, especially the one at Cape Alava where an entire village and its contents were buried by mudslides. Perfectly preserved and skillfully constructed and decorated wooden and cloth artifacts reveal an elaborate culture. Many of these cultural tools are also art objects. Displays of the actual artifacts at the Makah tribal museum at Neah Bay, just north of the park, are outstanding. A side trip to Cape Flattery on the reservation, the westernmost point in the conterminous United States, is well worth the time.

Europeans came in ships to explore the coast for various reasons, one of which was the search for the Northwest Passage, the imagined shortcut from Europe to Asia. Spaniards Juan Perez (1774) and Bruno Heceta (1775) led the way and were followed by Englishmen James Cook (1778), Charles Barkley (1786–1787), John Meares (1788), and George Vancouver (1792). American Captain Robert Gray in his ship *Columbia* plied the nearby waters from 1787 to 1793 and named the only large river on the west coast of what is now the conterminous United States after his vessel. At least two small landing parties on the Olympic coast, one of Heceta's and one of Barkley's, were never seen again after making landfall.

Captain Meares noted a "lofty mountain" rising above the others that he named Mt. Olympus. Vancouver in 1792 extended the name to the entire mountain group and called it the Olympic Mountains. The gradual settlement and exploration of the area occurred in the 1840s—except for the impenetrable wilderness in the core of the mountains.

Lieutenant Joseph O'Neill convinced his U.S. Army superiors to support an expedition into the unknown interior in 1885. With great difficulty a mule trail was cut through the dense forest near what was to become Hurricane Ridge Road. Hurricane Ridge and some of the surrounding ridges were explored; the rest of the vast interior still awaited exploration.

Washington became a state in 1889 and in the same year the *Seattle Press* challenged someone to explore the unknown wilderness. James Christie came forward to lead the famous Press expedition. Ironically, this was the first area to be named in what is now Washington state, and after 100 years it was the only unexplored area in the state! Rumors of a huge central lake draining subterraneously and guarded by hostile cannibals inspired explorers to determine what the Olympic interior was really like. The Press expedition with six men, two mules, and four dogs started up the Elwha River on the north end of the peninsula in December, 1889, one of the snowiest winters on record. Six months later in tattered clothes and shoes falling apart, they exited on the coast along the Quinault River.

Significant explorations were conducted in 1890 by Lieutenant O'Neill again and by Tacoma judge James Wickersham and Charles and Samuel Gilman. All were enthralled by the rugged beauty of the interior, and reports by O'Neill and Wickersham included a recommendation to establish a national park. A *National Geographic* article by the Gilmans helped the cause, but commercial interests compromised the proposal enabling only an Olympic Forest Preserve (later a national forest) to be established in 1897. Further attempts to establish a park were stymied until Theodore Roosevelt, by presidential proclamation, established Mount Olympus National Monument in 1909.

Preservation of the diminishing herds of Roosevelt elk were the major concern to park advocates, but commercial interests again defeated park proposals in 1911, 1912, 1926, 1935, and 1937. As the recreational use of the area boomed, pressure mounted—finally in 1937 another Roosevelt, Franklin Delano, visited the Olympics and became a strong supporter of a national park, essentially ensuring its designation. Pressure mounted for park status following the Great Elk Hunt boondoggle of 1937 where an open season on the protected elk turned the peninsula into a "war zone." The elk had lost much of their fear of man and over 5000 hunters butchered and crippled over 800 elk in a few days. The hunters also did in one of their own, a pack horse, cow, and a dog. Ten elk near Schmidt's Crossing were killed by 160 shots! The public was incensed and the park was established in 1938, again over the opposition of then governor Clarence Martin, the U.S. Forest Service, and the logging industry. The coastal strip was added in 1953 bringing the total park area up to 1420 square miles (3692 km^2).

Geologic Story—Old

Just as the exploration of the interior required a herculean effort, so have the Olympics only grudgingly given up the secrets of its hidden geologic past. The rugged topography is covered by a temperate rainforest and glaciers. Rock exposures occur mostly along rivers, in alpine areas, and the coastal zone. Where the rocks are exposed, they too are not overly friendly. Recrystallization, especially in the core of the range, has destroyed most of the fossils; rapid lateral changes in rock characteristics make long-distance correlation difficult; and, last but not least, breaking rocks apart in the subduction "eggbeater" presents many difficult problems to those who try to read nature's hidden diary.

A rather uncomplicated structure was envisioned by Danner (1955) and other workers—the Olympics are a simple dome or anticlinal structure. An outer ring of basalt (the Crescent Formation), the so-called basaltic horseshoe (Fig. 3–5), was early identified as Eocene in age. If this were truly an anticline whose summit rocks were eroded off, the anticline core should contain older rocks, perhaps Cretaceous

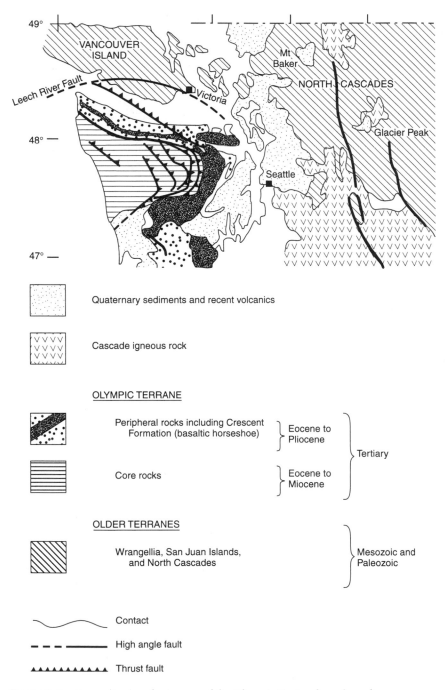

FIGURE 3–5   Generalized geologic map of the Olympic Peninsula and nearby areas. (Modified from Tabor, 1987b)

in age. Careful, detailed field work led to the discovery of a few microscopic fossil "hangers on"—Foraminifera—in the core rocks. Rather than Cretaceous, these fossils and the rock enclosing them were Eocene and *younger*. In fact, rocks become progressively younger toward the Pacific Ocean. Thus the anticline or dome theory was "shot down." Some new explanation was needed—enter plate tectonics.

Geologic Story—New

Recognition that the earth's surface is composed of a dozen or so major plates that move either away from, toward, or laterally past an adjacent plate provides an explanation for the mechanical forces needed to create mountains and other tectonic features. Knowing that large blocks can sometimes move great distances and be incorporated as a microplate or suspect terrane onto a larger plate also greatly helps in reconstructing the history of our planet.

As the North American plate began its westward movement about 200 Ma (early Mesozoic), an oceanic plate overlain with sedimentary rocks and embedded with islandlike blocks moved eastward. Like a giant conveyor belt, the oceanic plate delivered material to a collision or subduction zone where it was welded onto the edge of the North American plate (Fig. 3–6). The North Cascades and Wrangellia blocks containing mostly Paleozoic and Mesozoic rocks were added to the west edge of the continent (Fig. 1–11) about 50 Ma. As these embedded blocks, or microplates, were docking onto North America, plate movements slowed and a group of Hawaii-like seamounts began to form (Fig. 3–7*a*) off the Washington and Oregon coasts. The resulting shield cones are built of flattened tubelike forms called *pillow lava* and fragmented lava material from underwater explosions that form *volcanic breccia*.

A subduction zone on the west side of the newly docked North Cascade microplate carried the seamounts and associated sediments (the Crescent Formation) eastward where, about 30 Ma, they too began to be squeezed against the continent

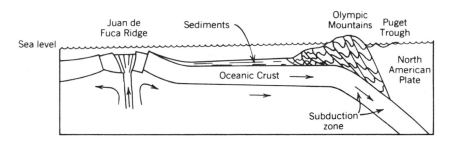

FIGURE 3–6   Collision of the Juan de Fuca and North American plates "scrapes off" and deforms sedimentary rocks, forming the Olympic Mountains (refer to Fig. 3–2). (Modified from Rau, 1980; illustration by Gregory Nelson)

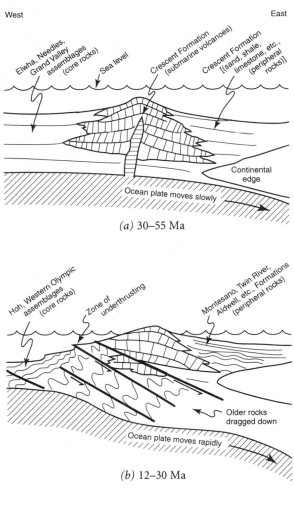

West   East

Elwha, Needles,
Grand Valley
assemblages
(core rocks)

Sea level

Crescent Formation
(submarine volcanoes)

Crescent Formation
[(sand, shale, etc.,
limestone, etc.,
peripheral
rocks)]

Continental
edge

Ocean plate moves slowly

(a) 30–55 Ma

Hoh, Western Olympic
assemblages
(core rocks)

Zone of
underthrusting

Montesano, Twin River,
Aldwell, etc., Formations
(peripheral rocks)

Older rocks
dragged down

Ocean plate moves rapidly

(b) 12–30 Ma

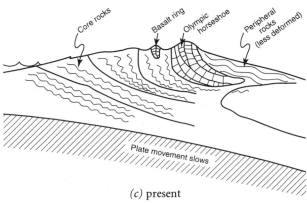

Core rocks

Basalt ring

Olympic
horseshoe

Peripheral
rocks
(less deformed)

Plate movement slows

(c) present

FIGURE 3–7 Evolution of the Olympic
Mountains from Eocene to present.
[Generalized diagrams from Tabor (1987a);
generalized stratigraphy from Tabor and Cady
(1978b).] *(a)* Submarine volcanic rocks and
sediments accumulate (30–55 Ma), *(b)*
moving plate jams sedimentary rock *under*
basalt buttress, rapid plate movement holds
crust down (12–30 Ma), *(c)* slower plate
movement allows thickened section of low-
density rock to rise vertically (present).

edge. The seamount was too much for the subduction zone to "swallow," and it acted as a buttress or logjam against which sediments riding "piggy back" on the oceanic plate were scraped off as slices of rock and stuffed under the Crescent Formation (Fig. 3–7b). Just like a massive freeway accident, the sandstones, siltstones, and shale layers continued to ride the giant conveyor belt to the "scene of the accident." Some of the rock slices were shoved beneath others, some were bent like folds in a pleated skirt, still others were shattered into fragments of all sizes called a *melange*.

The slices of rock were jammed under each other—like slipping playing cards under a deck of cards and slowly rotating the deck into a nearly vertical position (Figs. 3–6, 3–7c, and 3–8). Just as a highway patrol officer reconstructs the events represented by the twisted wreckage in the freeway junkyard, so too must the geologist interpret the story of the rock wreckage. Figure 3–7 is based on Tabor's (1987a,b) and Tabor and Cady's (1978a,b) studies and summarizes our present understanding of what the accident report should look like.

The packaged sedimentary and volcanic rocks were shoved into a pocket against the older microplates and bent into the horseshoe shape reflected by the Crescent Formation and the core rocks in the Olympic interior. Metamorphic conditions were greatest in the eastern core where slates formed and less so to the west in the Hoh Rock Assemblage (Fig. 3–7) and younger rocks. The rate of plate movement slowed or stopped about 12 Ma, allowing this thick wedge of relatively

FIGURE 3–8 Layers of ocean floor sediments on Mount Angeles thrust under the Eocene Crescent Formation and rotated to a nearly vertical position. Subsequently the rocks were uplifted many thousands of feet above sea level. See Figure 3–7c. (Photo by E. Kiver)

light sedimentary strata to rise vertically, much like a bobber floats higher in the water when the fish stops pulling at the bait below. The rocks were thickest near the center around the Mount Olympus area so that this area was lifted higher, thus creating a topographic dome. Small streams formed as uplift began and grew to rivers that cut valleys and canyons. Because the streams flow away from a central high, they radiate like spokes on a wheel forming a *radial drainage pattern*.

The presence of oil seeps along the coast has intrigued oil prospectors for over a century. Organic materials trapped in the sedimentary strata at the time of deposition were subjected to pressure and heat and were converted to oil and gas. Early oil explorers either landed drilling equipment by barge or, with great difficulty, built roads through the temperate rainforest. After nearly 1000 wells were drilled, commercial quantities have yet to be discovered.

Glaciation

The worldwide cooling of the Pleistocene age brought about by the changes in the earth's climatic machinery, plus the higher elevation of the rising Olympic Mountains, caused alpine glaciers to form and to extend down their valleys. The glacial amphitheaters called *cirques* form best near the *snowline*—the elevation where last winter's snow remains through the summer melt season—especially when average snowline persists at this elevation for a long period. Cirques at lower elevations and currently unoccupied by glaciers tell a story of former lower snowlines and colder climate. Similarly, pollen grains in the park's bog deposits record the presence of a former extensive tundra, much like that found today on Alaska's North Slope. Analysis of the pollen record and cirque elevations suggests that summer temperatures during the Pleistocene were 3–11°F cooler. A similar cooling would again put the world into another ice age.

The extensive Pleistocene valley glaciers and the existence of small glaciers today associated with a relatively low mountain range (less than 8000 feet; 2440 m) at this latitude at first seems anomalous until one considers precipitation. Moisture-charged Pacific air masses moving onshore are cooled as they rise over their first topographic obstacle, the Olympics, and produce heavy precipitation. Eighty inches (203 cm) per year near the coast is common, 140 inches or 11.7 feet (3.6 m) occurs in the temperature rainforest higher up, and over 200 feet (61 m) of snow falls on Mount Olympus during snowier years! Because most of the precipitation occurs during the winter months, most falls as snow at higher elevations and nourishes the 60 existing cirque and valley glaciers. The large Pleistocene alpine glaciers sculptured the mountain interior into steep-walled cirques, glacial valleys, aretes, and horns. End moraines and lateral moraines occur around Quinault Lake and in other low-elevation areas as well as in the major valleys. Large areas downstream and along the coasts are covered by sand and gravel deposits called *outwash*, which is derived from debris washed along by meltwater streams. Within these deposits all the rock types of the drainage basin can be found.

The presence of granite boulders seemed anomalous to the pioneer geologists because there is no granite bedrock in all of the Olympic Mountains! Some of these boulders on the north side are at elevations of 3000 feet (915 m). As with most geologic problems, careful observations by persistent researchers produced a reasonable solution. The granite boulders had originally been a part of the Coast Range Batholith, which is in British Columbia. Glaciers originating in the Canadian mountains pushed down into the lowland areas and coalesced to form a broad piedmont glacier that then headed southward and crossed the boundary into the Puget Sound area. Here it divided into lobes; one continued south in the Puget Lowland into what is now the area of Washington State's capital, Olympia. The other extended westward through the Strait of Juan de Fuca along the north end of the Olympics. The ice pushed down into the Puget Sound area at least six times (Easterbrook, 1986). At least once this ice lobe was greatly thickened and extended up onto the flanks of the mountains as high as 3000 feet (915 m), to deposit the granite boulders as *glacial erratics.* The presence of glacial lake sediments in many of the valleys on the north indicates that alpine glaciers had receded by the time the Puget Lobe arrived.

The Canadian glacier was also responsible for a number of lake basins in lower areas. Near the northern boundary of the park, the ice scoured out the irregular, elongate basin now occupied by Lake Crescent and Lake Sutherland. After the ice withdrew, a large landslide divided the lake into two lakes. This explanation differs only slightly from the Indian legend in which Storm King Mountain hurled a big rock into the lake!

The rigors of the Ice Age were hard on plant and animal life. Migrations and elimination of species occurred in response to the presence of large glaciers and the changing climate. Some, like the porcupine, pika, grizzly, and golden-mantled ground squirrel, now occupy the Cascades to the east but were unable to migrate to the Olympics in the 12,000 years since the ice retreated.

Today's Glaciers

Perhaps as many as 60 glaciers, most of them small, are clustered around the higher peaks (Fig. 3–9). The largest mass of ice, about 10 square miles, clusters around Mount Olympus. The well-studied Blue Glacier is 2.0 miles (3.2 km) long and the Hoh Glacier, the longest in the park, is 3.3 miles (5.3 km). As in other mountainous areas of the western United States, most of the alpine glaciers completely disappeared during the warm interval from about 5000 to 8000 years ago and reformed with the coming of cooler climates during the Neoglacial interval, probably within the last 5000 years (see Fig. 1–25). Larger glaciers, like the Hoh, Blue, and White may have survived warmer times and are relics of the Ice Age. Using *dendrochronology* (age-dating trees by use of growth rings) researchers have established that the time of maximum extent of Holocene ice in the Olympics was during the Little Ice Age in about 1820 AD (Heusser, 1957) (Fig. 3–9).

FIGURE 3–9   The Hoh River valley near White Glacier on the north flank of Mt. Olympus. Note Neoglacial moraines and trimlines as well as other glacial landforms. (Photo by E. Kiver)

Coastal Processes

The thin line that separates the land from the sea is a very special place—unique because of the shore processes affecting this narrow band and unique because of the fascinating adaptations of organisms to a high-energy wave environment complicated by the rise and fall of tides twice during every 25-hour lunar cycle.

The appearance of a given coast depends on the interaction of wave energy, offshore currents, the nature of the bedrock or unconsolidated materials edging the continent, sediment supply, and the stability of sea level and land elevations.

Wind blowing across water surfaces creates disturbances called *swells*. Water in these disturbances moves only vertically in small circles and transfers the waveform laterally until it reaches shallow water. Here the wave front is slowed allowing the swells to organize into linear waves that grow higher and narrower—eventually the tops fall forward as *breakers* into the turbulent area known as the *surf zone*.

Tons of water, especially in storm waves, beat against the shore materials. This "fire-hose process" attacks shore material by moving and grinding loose debris and breaking bedrock by trapping and compressing air in rock fractures. Waves armed with rock particles hurtle them against each other and the exposed bedrock causing further grinding and fracturing. Sand and other debris is carried up the beach by breaking waves (*swash*), and the returning water flow, *backwash* (*swish*), rolls debris back toward the sea. A single sand particle might be bounced and rolled 10 feet up and 10 feet down the beach by a single wave. When multiplied by the number of waves per day, movements equivalent to many miles of grinding can affect a single particle.

Coastal Landforms

As the sea wears away the land's edge and the shoreline advances landward, an erosional "bite" consisting of a *wave-cut bench* and a *wave-cut cliff* forms. In places a *wave-cut notch* undercuts the wave-cut cliff encouraging landslides to drop debris into the sea or onto the beach below. Waves begin their slow but untiring process of redistributing material into beaches covering parts of the wave-cut platform (Fig. 3–3). Sediments carried by streams toward the open sea are pushed back by waves forming a sand ridge called a *baymouth bar.* Sand moved laterally along the coast by *longshore drift* sometimes forms an elongate ridge called a *spit.* Dungeness and Ediz Hook along the Strait of Juan de Fuca north of the park are good examples.

If rock characteristics were identical everywhere and land elevations and sea level remained constant, the shoreline would eventually be straightened. Such is not the case along the Olympic coast. Indications on the Olympic coast, and the West Coast in general, including the sparsity of good harbors, the presence of ancient wave-cut benches 100 or more feet above present sea level, and precise measurements of changing land elevations indicate that an *emergent coastline,* one that is being actively uplifted by tectonic or isostatic forces, dominates. Also contributing to the rugged shoreline is the variety of rocks and the complicated fracturing accompanying their journey through the subduction zone pressure mill. Just as running water selectively erodes along rock weaknesses, so too do waves erode areas or zones of weaker rock more rapidly, creating an infinite variety of shapes along this rocky coast. Patches of sandy beach promise an easy path for hikers, but a glance ahead foretells of more challenging route finding.

Indentations of the shoreline are called *coves* or *bays* and projections of land jutting seaward are called *heads* or *points.* Hoh Head and Taylor Point are examples. Wave energy focuses on these resistant heads, eventually separating them from the mainland to become, if large enough, islands, or smaller isolated monoliths called *sea stacks* (Fig. 3–3). Joints, faults, and weaker bedding planes are often excavated by the untiring waves into *sea caves* that ultimately collapse and also help to form some of the thousands of sea stacks along the Olympic coast.

The degree of development of the tectonic melange greatly influences shoreline appearance. During the scraping and stacking of sedimentary layers described earlier (or perhaps as a result of submarine landslides on the former continental slope), the previously formed brittle sandstone, siltstone, metamorphic, and volcanic blocks broke apart in the grinding mill, some into boulders the size of a house or apartment building, and others into millions of small fragments. The more intensely fractured zones erode rapidly and form more gently sloping sea cliffs. Nonfractured or less fractured sections of rock form heads, spectacular rock cliffs, and islands and large stacks. The spaces between the blocks were filled in by a relatively soft clay-rich matrix rock (Rau, 1987). Some of the melange is like a thick crunchy-style peanut butter. Removal of the "creamy" matrix by shore processes leaves the individual "peanut chunks" as isolated blocks or stacks.

The wilderness character of the Olympics was one of the main "selling points" in establishing the national park. Designation as a Biosphere Reserve in 1976 and a World Heritage Site in 1981 was based partly on its relatively intact assemblage of plants and animals in a large roadless area. However, as rapid population growth in the Puget Sound area and elsewhere has increased, new pressures are being exerted on the Olympic environment. Unchecked growth cannot be tolerated indefinitely. The twenty-first century will show us where our future lies—hopefully we will pursue a course that protects the Olympics and other special places in perpetuity. If changes in the environment are minimal, then the threat of extinction of organisms, including *Homo sapiens,* is lessened.

REFERENCES

Danner, W.R., 1955, Geology of Olympic National Park, University of Washington Press, Seattle.

Easterbrook, Don J., 1986, Stratigraphy and chronology of Quaternary deposits of the Puget Lowland and Olympic Mountains of Washington and the Cascade Mountains of Washington and Oregon: in Šibrava, V., Bowen, D.Q., and Richmond, G.M., Eds., Quaternary glaciations in the northern hemisphere: Quaternary Science Reviews, v. 5, Pergamon Press, New York, p. 145–159.

Heusser, C.J., 1957, Variations in Blue, Hoh, and White glaciers during recent centuries: Arctic, v. 10, p. 139–150.

Rau, Weldon W., 1980, Washington coastal geology between the Hoh and Quillayute rivers: Washington Division of Geology and Earth Resources Bulletin 72, 57 p.

Rau, W.W., 1987, Melange rocks of Washington's Olympic coast: in Hill, M.L., ed., Cordilleran Section of the Geological Society of America, Centennial field guide, v. 1, p. 373–376.

Tabor, Rowland W., 1987a, Geology of Olympic National Park: University of Washington Press, Seattle, 144 p.

Tabor, R.W., 1987b, Tertiary accreted terrane: Oceanic basalt and sedimentary rocks in the Olympic Mountains, Washington: in Hill, M.L., ed., Cordilleran Section of the Geological Society of America Centennial Field Guide, v. 1, p. 377–382.

Tabor, R.W., and Cady, W.M., 1978a, The structure of the Olympic Mountains, Washington—Analysis of a subduction zone: U.S. Geological Survey Professional Paper 1033, 38 p.

Tabor, R.W., and Cady, W.M., 1978b, Geologic map of the Olympic Peninsula, Washington: U.S. Geological Survey Miscellaneous Investigations, Map I-994.

Park Address

Olympic National Park
600 East Park Avenue
Port Angeles, WA 98362

OREGON CAVES NATIONAL MONUMENT (OREGON)

In the Siskiyou Mountains in southwestern Oregon is a 1600-foot-long (488-m) cave that was discovered by a bear, a dog, and Elijah J. Davidson, in that order. Davidson is credited with the discovery in 1874. The bear's reward for his or her part in the discovery was not so glorious after man, dog, and bear groped in the darkness for some exciting moments. After the discovery visitors came, most to see and to admire the "Marble Halls of Oregon," a few to break off the stalactites to take home as souvenirs. One not so knowledgeable visitor suggested that broken stalactites would "grow back again next year!" To protect the cave from vandalism the area was set aside, under U.S. Forest Service administration, as Oregon Caves National Monument in 1909. In 1933, the monument, consisting of 480 acres, was transferred to the National Park Service.

Most true caverns—those formed by solution by subsurface water—are in limestone. Oregon Caves are unusual because they were formed in marble—the metamorphic equivalent of limestone.

Geologic Development

The Siskiyou Mountains are part of the geologic complex located in northwestern California and southwestern Oregon called the Klamath Mountains. Like much of the coastal region, rocks originated elsewhere and were transported by moving plates and "jammed under" or accreted to the west edge of the growing North American continent (Orr and co-workers, 1992). The slab called the Rattlesnake Creek subterrane contains the Oregon Cave marble (Applegate Formation). The limey muds that were to eventually contain Oregon's largest limestone cave were first deposited on the seafloor during the Triassic Period, about 200 Ma. Subduction began shortly after as the North American plate reversed its motion and began to move westward. The colliding continental and oceanic plates initiated a long interval of continuous mountain building that began as the Nevadian (Sevier) Orogeny about 150 Ma along the southwestern edge of North America. Deformation migrated to the northeast where, in the Rocky Mountain area, it is known as the Laramide Orogeny.

The limestone was folded and recrystallized into marble from heat and pressure developed as slice after slice of rock was scraped from the ocean plate as it slid beneath the advancing continent edge. Sandstones were converted to quartzite and shale to slate as the subduction vise in western North America tightened during the mid-Mesozoic orogeny. The thick slices of rock in the Klamath Mountains today resemble a series of toppled books on a shelf with each volume leaning westward, supported by its neighbor. Metamorphism was helped along as the slices were further welded together when granitic magma intruded during the Nevadian Orogeny. Uplift followed by extensive erosion during the mid-Tertiary helped create today's subdued mountain topography. By late Tertiary or early Quaternary (1–2 Ma) time the marble was closer to the earth's surface where groundwater began the slow but persistent process that would form Oregon's finest limestone cave.

Cave Development

The conversion of limestone to marble results in the calcite crystals reorganizing and growing larger—but still vulnerable to solution when subjected to acidic groundwater—especially if fractures are present. The combination of water with carbon dioxide in the atmosphere, and especially in the soil zone, produces a small amount of carbonic acid. Cavern development begins deep in the zone of saturation where small solutional openings form along tiny fractures in the rock. Enlargement and integration of these openings occur much later when the rock is located just below the water table where maximum groundwater flow and more chemically aggressive water dissolve the rock material faster.

When a fragment of calcite is put in a strong acid, such as hydrochloric acid, a violent reaction takes place and soon the calcite disappears. The material is still there, but as calcium bicarbonate in solution, not as solid calcium carbonate. With a weak acid such as carbonic acid, the reaction is extremely slow—much too slow for human observation. The missing key to cavern formation is time—hundreds of thousands of years were required for the development of Oregon Caves.

As erosion further reduces the landscape and the water table lowers, air enters the draining passages. Closeness to the land surface also permitted water to flow through the cave and, like a surface stream, River Styx eroded its bed and banks, further enlarging the cave passages.

Speleothem Formation

If only solution and underground stream erosion had taken place, the caverns with their bare walls would be relatively unexciting. But these caverns are adorned with many and varied formations that resulted from another process, chemical precipitation. When water saturated with calcium bicarbonate drips into an air-filled cave, some of its carbon dioxide is lost to the cave atmosphere making the water less acidic. The calcium bicarbonate in solution is no longer stable under these conditions and calcite is deposited as breath-taking mineral decorations (*speleothems*) on the roof, floor, and walls of the cave (Figs. 3–10 and 3–11). Evaporation could play a role in calcium deposition, but the nearly 100 percent humidity of most caves minimizes this process—loss of carbon dioxide is the main cause of precipitation. Water droplets on the cave roof can form hollow, tubelike deposits, called *soda-straw stalactites,* that may eventually become plugged and form iciclelike mineral deposits, *stalactites,* that hang from the ceiling. Stalactites may also form without experiencing a soda-straw stage. Other speleothems include the more blunt upward mineral projections called *stalagmites* and the sheetlike forms on the walls and floors called *flowstone* (Fig. 3–12). When stalactite and stalagmite join, a *column* is formed (Fig. 3–10). An infinite variety of forms are created as small irregularities of deposition occur. Irregular, knobby deposits of *cave popcorn* (also called *globulites*) cover many cave surfaces and record at least one interval of a higher water table and resubmergence of the cave, probably in response to some past Ice Age climate change.

FIGURE 3–10   Columns and stalactites in Oregon Caves. (Photo by National Park Service)

FIGURE 3–11   Unusual stalactite and stalagmite developed on a limestone ledge in Oregon Caves. (Photo by E. Kiver)

FIGURE 3–12   Calcium-rich water flowing down the cave walls leaves a coating of calcite known as flowstone. (Photo by E. Kiver)

These processes continue today. River Styx is still enlarging the caverns, the stalactites are growing minutely longer, and the stalagmites, although not detectable by the human eye, will be very slightly taller next year. Therefore, Oregon Caves are "living caves." Eventually, however, they will be destroyed by collapse of the cavern roofs and by erosion of surface streams. But that will not take place for a long, long time.

Many other fascinating features—draperies in the Ghost Room, flowstone in Joaquin Miller's Chapel, the soda-straw stalactites and *helictites*—are well illustrated and discussed by Contor (1963) in his informative booklet, *The Underworld of Oregon Caves*. Contor also describes the plants and animals that live in or frequent the caves. Bats, one of the world's more important mammal groups, are there in small numbers; most visitors do not see any of these elusive, beneficial creatures.

REFERENCES

Contor, R.J., 1963, The underworld of Oregon Caves National Monument: Crater Lake Natural History Association.
Orr, Elizabeth L., Orr, W.N., and Baldwin, E.M., 1992, Geology of Oregon: Kendall-Hunt, Dubuque, Iowa, 254 p.

Park Address

Oregon Caves National Monument
19000 Caves Highway
Cave Junction, Oregon 97523

REDWOOD NATIONAL PARK (CALIFORNIA)

Redwood National Park was created to preserve some of the earth's oldest, and most majestic living organisms—the coast redwood trees (*Sequoia sempervirens*) (Fig. 3–13). Although most redwood trees live from 500 to 700 years, some of these trees were already growing when Christ was born—about 2000 years ago. The

FIGURE 3–13   Trail through grove of redwoods at Redwood National Park. (Photo by E. Kiver)

coast redwoods are not the oldest redwoods, however, their cousins in Sequoia and Yosemite National Parks, *Sequoia gigantea,* are as much as 3500 years old. These magnificent trees, including the world's tallest (367.8 feet; 112 m) in the Tall Trees Grove at Redwood National Park, were appropriately named sequoia in honor of an early American who also stood tall; Sequoyah was chief of the Cherokee.

Redwood National Park is located in the northwest corner of California in Humbolt and Del Norte counties where sea cliffs and occasional narrow beaches form its western border, and the eastern, arbitrary border is among the steep-sided ridges of the Coast Range. The Klamath River begins its journey at Klamath Marsh in Oregon and flows through the Klamath and Coast Range mountains entering the Pacific Ocean near the town of Klamath near the middle of the park. Both state and federal parklands are interconnected here (Fig. 3–14) and managed cooperatively as a single park unit.

The animal life of the park is highly varied because of the distinctly different environments—from the mountains to the ocean. In the redwood forests you may see beaver, raccoon, bobcat, deer, mountain lion, and black bear. Birds are numerous, but many species spend their time in the treetops—difficult to observe from the forest floor. A herd of about 300 Roosevelt elk inhabits the park and are often seen in the Gold Bluffs Beach area and elsewhere. The elk are sometimes unfriendly and are most safely observed from a distance or from inside a car. Marine life is abundant and includes seals, sea lions, and occasional gray and other whales as they are "passing by."

History

The Klamath and other tribes used the redwoods to make dugout canoes and planks for construction. The trees have incredibly thick bark and a natural chemistry that protects them from fire, disease, and insects. However, the trees are no match for nature's number one destructive enemy—profit-motivated humans. The redwoods on the 500-mile-long (806-km) coastal strip were vigorously attacked by the newly arriving Americans and rapidly decimated.

Theodore Roosevelt set aside Muir Woods National Monument in 1908 on private lands purchased and donated to the federal government by Congressman William Kent. The Save-the-Redwoods League was established in 1918 to help preserve redwoods and create a Redwood National Park. Numerous studies and legislative proposals were put forth but to no avail. A 1963–1964 National Park Service–National Geographic Society study revealed that only 15 percent of the original 2 million acres of original redwood forest remained, and only 2 percent of the surviving forest was protected in state parks! The rest, including the world's tallest tree, was located on private, unprotected land.

Incredible as it may seem, it took *50* years of effort to finally establish Redwood National Park. The park was approved in 1968 and enlarged in 1978 to 106,000 acres. Similarities to the present debate about old-growth forests are re-

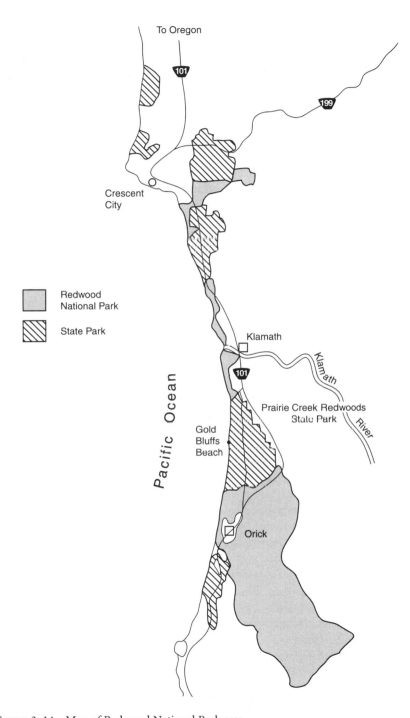

To Oregon

101

199

Crescent
City

Redwood
National Park

State Park

Klamath

Pacific Ocean

101

Klamath

Prairie Creek Redwoods
State Park

River

Gold
Bluffs
Beach

Orick

FIGURE 3–14 Map of Redwood National Park area.

markable. Only 7 percent of the original western old-growth trees of all species remain, yet those supporting "economic development" want to cut as many of these as possible. A "smokescreen" debate about the spotted owl that inhabits these forests and jobs for lumber workers is an attempt to disguise the motives of those few who would receive short-term financial gain from unlimited cutting on public land in our national forests.

Without question the big trees are the main attraction. The mature trees have branchless trunks for the lower first hundred feet or more; thus branches are well out of reach of flames from forest fires. The bark of mature trees is about a foot thick, further protecting the tree from most fires. The tree tops form a canopy that "creates the illusion of a great natural cathedral which visitors find both inspiring and humbling" (Sunset, 1965).

But Redwood National Park has more for park visitors than just its trees. There are the sea cliffs and the stacks offshore where sea cliffs stood at an earlier time. The beaches along the 50 miles (80 km) of ocean front are preserved in their natural state where one can see the results of the eternal pounding of the waves that have been shaping the shoreline. In places the wind has picked up beach sand and piled it up as dunes in nearby areas. Waves strike the shore from the north, creating a southward movement of water and sediment called *longshore drift*. Sand and other sediment from the sea cliffs and that spilling out from the mouth of the Klamath River and other streams replenish beach sediment washed away by waves and currents. Locally, sediments accumulate to form long, partially submerged ridges of sand called *spits*.

In the sea cliffs and in fresh road cuts along Highway 101 and other roads are rocks that hold some of the keys to the geologic past. The mild temperatures and rainy winters with 25–122 inches (64–310 cm) of rain have helped produce a thick soil and dense cover of vegetation that hides the bedrock in most of the park. Lack of good exposures, in combination with a group of "unfriendly" rocks, have helped create what Alt and Hyndman (1975) call, the "Coast Range nightmare." Similar difficulties were encountered in trying to understand the geology of the Olympic Mountains in northwestern Washington state. Many of the geologic explanations developed to explain the structure and history of the Olympics are applicable to the Redwood Park and other Coast Range areas as well.

The Geologic Past

The dominant rock along the northern California coast is a dark, somber-appearing rock unit of Jurassic-Cretaceous age belonging to the Franciscan Series (about 100–150 million years old). The silt, sand, and gravel-size sediment and submarine basalt flows accumulated in a deep oceanic trench associated with a subduction zone, much like the Peru–Chile Trench off the west coast of South America or the Middle America Trench off Mexico today. The sand is mostly composed of small rock fragments rather than single mineral grains, a special type of sediment that ultimately forms a *graywacke* sandstone. Slurries of sediment that cascaded down the steep slopes of the continental shelf forming *turbidity currents* account for the abundance of graywacke sand in the Franciscan Series. A combination of factors make the Franciscan difficult to work with: (1) the lack of distinct beds that can be followed long distances, (2) metamorphism that has destroyed most of its fossils, and (3) intense folding and faulting that have greatly disrupted the original bedding. Metamorphism has turned the sedimentary rocks into *metasediments* and the basalt into a *metavolcanic* rock called a *greenstone*. The abundance of the green-colored mineral called *chlorite* accounts for the grayish-green to green color of many Franciscan rocks. A few rocks were recrystallized sufficiently to form "higher grade" metamorphic rocks like *phyllite* and *schist*.

Just as in the Olympic Mountains and other Coast Range areas, rock layers were "scraped off" and stuffed under the North American plate edge in the Redwood Park area as the oceanic plate was overridden. The highly fractured melange (mixture of rock fragments of diverse size) was metamorphosed and so intensely contorted that it looks as if it were "stirred with a stick" (Alt and Hyndman, 1975). Outcrops of an interesting rock called *serpentinite* in the sea cliffs in the northern part of the park may be slices of ocean crust or mantle caught in the subduction zone. Again, as was true in the Olympic Mountains, a few fossils did survive the grinding mill. At Redwood Park and at other Franciscan rock areas the fossils and the rocks that contain them are Mesozoic, much older than rocks in the Olympics and elsewhere in the Coast Range.

Early Tertiary uplift and erosion were followed by middle Tertiary (Miocene) subsidence and later (Pliocene) folding and uplift into the present rugged mountain topography. Seaward shifting of the subduction zone and changing stresses as the San Andreas transform fault began to form may account for the "on again–off again" tectonic activity. As more of the remnant of the once large Farallon Plate (Fig. 3–2a) is consumed, the San Andreas, and its northern counterpart, Queen Charlotte transform fault, in British Columbia, will grow longer. Currently the San Andreas swings westward out to sea along the Mendocino Ridge and fracture system about 50 miles (80 km) south of the park. The small plate remnant off the northern California coast, the Gorda Plate, continues to be consumed as the relentless movement of plates continues (Fig. 3–2c). At the rate of movement of a few inches (centimeters) every year, in a few tens of millions of years the entire tectonic setting of western North America will change as the last trace of the spreading ridge disappears.

The Pliocene-Pleistocene mountain building here energized streams and generated an episode of valley filling. Coarse gravels overlying the Franciscan rocks contain clasts (fragments) of granite and metamorphic rocks (Fig. 3–15) that are completely foreign to any rock found in the Coast Range. The mystery of their source is solved by following the Klamath River to its headwaters in Oregon. Here, in the Klamath Mountains, is a whole mountain range composed of the same rock types. Sediments flushed down the Klamath River as a result of the late Cenozoic uplift formed a delta where it entered the Pacific Ocean. Remnants of the Pliocene-

FIGURE 3–15   Conglomerate at Gold Bluffs contains granite and metamorphic clasts derived from the Klamath Mountains many miles upstream. (Photo by E. Kiver)

Pleistocene delta sediments are located about 8 miles (12 km) south of the present river mouth at Gold Bluffs Beach. Through time the river has shifted northward, pushed by its own pile of sediment. Fossil clams and plants found in the sediments suggest an age of approximately 2 million years.

The yellow cliffs at Gold Bluffs Beach are a pleasant change from the somber cliffs of Franciscan rock found elsewhere along the northern California coast. The color is caused by oxidized iron, in the form of the mineral limonite. The obvious color is not the only reason for the Gold Bluffs name. The Klamath Mountains contain gold and were the site of intensive gold mining in the 1800s. In the 1850s enterprising miners, unaware of why the sediments were there, found flakes of gold in the loose sediments below the cliffs. Waves concentrated denser, heavier minerals in pockets that could be worked by miners only at low tide. The mineral concentrates contained mostly black-colored, heavy minerals like magnetite, but enough gold was present initially to keep miners busy.

By 2.0 Ma (near the end of the Pliocene) the entire area was above sea level. Climate in the southern Coast Range was warmer than in its northern counterpart, and low elevations of only a few thousand feet above sea level also helped prevent glaciers from forming here. Erosion prevailed in Quaternary times, especially along the coastal areas where the sea has been winning its battle with mountain building. Numerous stacks and small islands off the coast tell of a long episode of retreating shorelines. Wave-cut platforms were eroded into bedrock and geologically recently uplifted into the wave-cut terraces that are present in the north end of the park near Crescent City and at other places along the California coast. Thus, the building of the Coast Ranges here still continues.

The Tall Trees

Although the rocks and soils are suitable for redwood growth in a large area in the California and Oregon Coast Range, they are greatly restricted in their geographic distribution to a narrow band within 30 miles (50 km) of the ocean and below 3000 feet (915 m) in elevation. Thus, the dominant factor must be the climate. High rainfall during the winters and dry summers with an abundance of fog and a moderate annual temperature range must be the ideal climate for redwoods. As some have noted, the trees "follow the fog." Without fog, the excessive loss of water stresses the trees or even kills them. Weakened trees, or sometimes apparently healthy mature trees, fall during high winds.

Today's redwoods are relics of the past—so-called living fossils. They shared the earth with the dinosaurs during the Cretaceous (about 100 Ma) and formed vast forests across Asia, Europe, and North America during the early to mid-Tertiary (12–65 Ma) (Chronic, 1986). Tertiary-age fossil specimens are found in Eocene rocks at Yellowstone National Park, in Oligocene sediments at Florissant National Monument in Colorado, and at many other sites in the world. About 12 species had evolved—only three remain today. In addition to the coast redwoods and the giant redwoods in Yosemite and other areas in the Sierra Nevada Mountains, a primitive species, the dawn redwood, exists in a remote part of China today. Significant elevation increases of land masses and the cooler climates of the Pleistocene dramatically reduced the distribution of redwoods, confining the coastal species to a 30-mile-wide (48-km) band that stretches for 500 miles (800 km) from northern California just south of Monterey to Curry County in extreme southern Oregon. Explorers in the early 1800s report traveling only about one mile a day through this dense forest; in less than 100 years human greed had decimated the pristine forest so that only small groves of redwood now remain intact.

Preservation of even these protected areas can be difficult. In the surrounding privately owned lands, forest practices, especially those of clear-cutting, logging road construction, and movement of heavy equipment across the forest soil greatly disturb the land surface (Fig. 3–16). The soil's ability to absorb water is reduced allowing larger volumes of water to run down the forest slopes and into nearby streams. Also, the churned-up soils are more vulnerable to erosion, thus large quantities of sediment are carried down slopes and clog nearby stream channels.

In the 1960s logging companies were ruthlessly clear-cutting old-growth redwoods, in some cases right up to the national park boundary. The winter rains in 1964 sent a spectacular flood, intensified by the logging activity, down Redwood Creek and into the national park. The grove containing the world's tallest tree narrowly missed destruction. Sediment washed into the channel from upstream caused the floodwater to reach even greater depths and the stream to start under-

FIGURE 3–16   Devastated area, the result of improper logging practices, is the source of the excessive floods and erosion downstream in the Tall Trees Grove. (Photo by W.E. Weaver, National Park Service)

cutting its banks. Also, the rise of the water table because of the sediment-filled channel threatened to drown the tree roots and kill the trees.

As a consequence of this frightening event, we were again reminded that a national park is not a tranquil "island" isolated from its surroundings, but a small piece of the total environment. What happens immediately outside of a park boundary, and sometimes half a continent away due to air pollution or political decisions, can significantly affect a park.

The park area was increased by 80% in 1978 as private lands were added. The "park protection zone" added at the south end of the park was a "no-man's-land" of logging roads and mostly burnt, redwood stumps (Fig. 3–16). Rehabilitation of this devastated area was critical to control erosion and protect the park forests along lower Redwood Creek. Given time and care, this area will be the most extensive redwood forest in the world—what a wonderful, thoughtful gift to those who will visit this area 500 years from now!

The Future

Other subtle geologic changes are occurring at Redwood National Park. In the 1850s no beach existed at Gold Bluffs. Today there is a narrow but substantial beach. What happened? Recall that longshore drift is south from the mouth of the Klamath River along this section of the California coast. In the 1800s and early 1900s hydraulic mining (now illegal) was practiced in the Klamath Mountains area. Huge firehoses were turned onto loose sediment to free the gold and catch it with sluice boxes as the water and sediment flowed toward the river. Large volumes of sediment, further enriched by additional sediment from logging operations, were carried to the ocean where they were redistributed down the coast. New beaches formed and old ones were enlarged as the sediment supply increased.

The highly fractured Franciscan rock is extremely susceptible to landslides. Many natural slides occur (Fig. 3–17), but others are generated by humans when

FIGURE 3–17   Landslide along redwood coast in highly fractured Franciscan rocks. (Photo by E. Kiver)

the land is disturbed by logging, road building, and other construction activities. Periods of heavy rain add tons of weight to slopes and lubricate particles, making slope failure more likely. Highway 101, the scenic road that follows the coast through California, Oregon, and Washington is often blocked by landslides, especially in the area where Franciscan rock occurs at the surface.

The tranquil landscape appears to be at peace and frozen in time, especially in the majestic groves of tall trees. Glimpses of change are found along the beaches and sea cliffs; not so apparent is the slow vertical rise of about 0.01 inches (3 mm) per year measured along the coast. Frequent earthquakes rattle the area as the subducting plates and their associated faults occasionally move. The November 1980 earthquake that occurred west of the Park beneath the ocean floor was particularly large, a magnitude 7.0. At Redwood National Park we see a "snapshot in time," a blink of the geologic eye, as the slow processes of change continue.

REFERENCES

Alt, David D., and Hyndman, D.W., 1975, Roadside geology of northern California: Mountain Press, Missoula, Montana, 244 p.

Chronic, Halka, 1986, Pages of stone: Geology of western national parks and monuments, vol. 2. The Mountaineers, Seattle, WA, 170 p.

Park Address

Redwood National Park
1111 Second Street
Crescent City, CA 95531

POINT REYES NATIONAL SEASHORE (CALIFORNIA)

Sir Francis Drake was the first European to visit Point Reyes. When his ship, the *Golden Hinde,* sailed into the bay in 1579 to make repairs, he was greeted by the friendly Miwok Indians. Commemorating his visit are Drakes Bay, Drakes Estero, Drakes Beach, and Sir Francis Drake Highway, which crosses the Seashore from Tomales Bay southward to Point Reyes (Fig. 3–18). Spaniard Sebastian Vizcaino named the area on January 6, 1603, in honor of the Day of the Three Kings of the Nativity. Pacific Punta de los Reyes translates as the Point of the Kings. Oddly, this area, which is only a few minutes' drive from San Francisco, has not changed greatly since the time of the Miwoks.

In our travels down the Pacific Coast, we make our first direct contact with the famous San Andreas Fault at Point Reyes. The Coast Ranges south of here are west of the fault and are geologically better understood than areas to the north. More

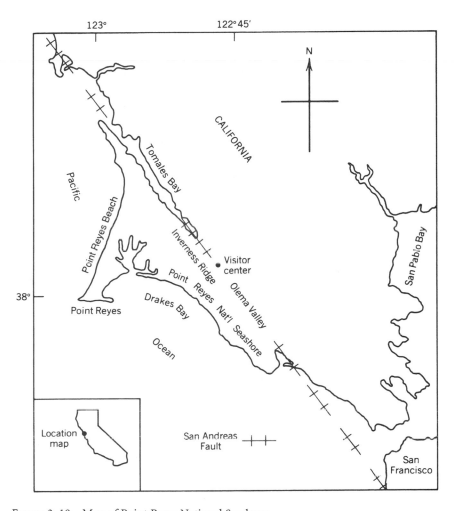

FIGURE 3–18   Map of Point Reyes National Seashore.

rock exposures due to drier climate, more land development, and more oil drilling provide considerable geologic information from areas to the south. About 225 miles (360 km) north of Point Reyes the San Andreas Fault swings westward out to sea along the Mendocino Ridge and fracture system (Fig. 3–2). The Coast Ranges are located east of the San Andreas north of Point Reyes. Wetter climate and denser vegetation produce fewer bedrock exposures along the northern California coast. However, even where more information is available in the Coast Ranges, because of their complexity many major questions about their geologic development remain unanswered.

Like the Fairweather Fault in Glacier Bay and the Queen Charlotte Fault in western British Columbia, the San Andreas Fault is an important zone where movement between the Pacific and North American plates is accommodated. The San Andreas and all of its branch faults are the cause of most of California's earthquakes.

Point Reyes National Seashore, established in 1962, is included with the national parks and monuments because of its geologic significance. With later additions it now includes 71,049 acres of rugged coasts, beaches, lagoons, and forested ridges. It is a triangular-shaped peninsula about 40 miles (66 km) north of San Francisco that looks out of place on maps, as if it had perhaps been added on (Fig. 3–18). Indeed, that is exactly what happened. Several million years ago it was far south of San Francisco, closer to Los Angeles; a few million years from now it will be a triangular island northwest of its present location as it is carried out to sea along the San Andreas Fault. In other words it is a peninsula merely "in passing." On April 18, 1906, as reported by Molenaar (1982), Point Reyes suddenly jumped 21 feet (7 m) northwestward; this sudden displacement along a 270-mile-long (435 km) segment of the San Andreas Fault produced the catastrophic earthquake that destroyed San Francisco. The maximum movement along the fault occurred not in San Francisco, but near the small town of Point Reyes Station on the peninsula (Galloway, 1977).

Plate Tectonic Setting

The San Andreas Fault, which essentially parallels the northeastern boundary of the seashore, is actually a broad fault zone locally about 1.5 miles (2.4 km) wide composed of badly crushed and sheared Jurassic and Cretaceous marine sedimentary rocks that are called the *Franciscan Series,* the same rocks that we encountered in Redwood National Park, some 275 miles (445 km) north. The rocks are easily eroded—especially where the already highly fractured rocks are caught up and further "squished" about along one of the world's most famous and most active faults. The north section of the fault at Point Reyes lies beneath Tomales Bay; the south part underlies Olema Valley (Fig. 3–18). The broad valleys topographically delineate the line along which Point Reyes and the Pacific plate are jammed tightly against the North American plate. West of the fault zone the entire peninsula is part of a long sliver of granitic and metamorphic rocks belonging to the *Salinian Block* (Galloway, 1977; Norris and Webb, 1990). Only on the inner edge along Inverness Ridge (Fig. 3–18) and the outer tip of the peninsula around Point Reyes is the granite well exposed; elsewhere it is buried by gently folded marine sedimentary rocks of Tertiary (mostly Miocene and Pliocene) age.

The Point Reyes peninsula is part of the Pacific Plate that has been moving north-northwestward along the San Andreas Fault for at least 30 million years. The 84-million-year-old granite on the peninsula was once continuous with the granite in the Tehachapi Mountains, some 370 miles (592 km) to the south. The Tehachapi Mountain granite is part of the Mesozoic-age Sierra Nevada granite and has remained close to where it originally formed. When the San Andreas Fault formed about 30 Ma, it cut through the west edge of the Tehachapi granite. In the intervening years this wayward chunk of granite has "jerked" its way to the northwest. At 5-, 10-, or perhaps 20-foot (6-m) "jerks" during a single earthquake, it doesn't require a mathematical genius to realize that hundreds of thousands or more of earthquake-producing displacements have occurred along this fault during the past 30 million years. Evidence for displacements of hundreds of miles along the San Andreas Fault also occurs elsewhere. Pinnacles National Monument, the next park area on our trip through the Coast Ranges, has similar evidence for large-scale movements.

Most visitors will be intrigued to learn that they have stepped off the North American Plate at Point Reyes and that they are on the next major plate to the west. Point Reyes and the Pacific Plate of which it is a part are only temporarily attached at this site. Visitors are also on an "endless belt"—one that moves by jerks—and one that could suddenly be jerked out from under their feet. Even if surface displacement was only a fraction of 20 feet (6 m), a location so close to a moving fault would make standing virtually impossible. Minor faulting has recently occurred in several places along the fault; stresses are building up, and the "big one" will soon occur. That it will come and that it will generate a severe earthquake are not guesswork; it is merely the moment when the rocks will snap that is in question.

Coastal Processes and Landforms

The rocks of Point Reyes vary greatly in type and in response to weathering and erosion. Rocks that are highly resistant to wave erosion form headlands extending out into the ocean; there, arches and stacks are common features. Where the rocks are alternately hard and soft, as near the Coast Campground, sea caves and tunnels are carved out of the soft rocks. Thus, the variety of features is truly remarkable.

The prevailing winds from Alaska and Siberia drive across thousands of miles of open ocean to the California coast. Swells on the open sea organize into lines of waves as the water shallows. The water piles up and forms breakers that crash with incredible fury, especially onto the Point Reyes Beach on the northwest side of the triangular-shaped peninsula. Water returning from the beach area concentrates into narrow bands called *rip currents* that are extremely dangerous to swimmers. Waves farther west bend around the Point Reyes headland, concentrating their fury on the granite bedrock exposed there. Lighthouses act as warning sentinels for traffic at sea (Fig. 3–19).

Waves striking the shore at an angle produce *longshore currents* that move sand laterally along the outer edge of the beaches. The south coast along Drakes Bay (Fig. 3–18) is better protected from storm waves than the northwest coast; consequently, the longshore currents build long *spits* along the south coast, like the 2.5-mile-long (4-km) Limantour Spit near Drakes Bay. Bars built across the mouths of small streams act as dams to impound water in small lakes. Estuaries such as Drakes Estero form large indentations in the coastline. For a discussion of the

FIGURE 3–19   Lighthouse at Point Reyes National Seashore.

complexities of life in an estuarine ecosystem or in the tidepools, see Russell Sackett's book, *Edge of the Sea* (1983).

Swash (landward movement of breaking waves on the beach) moves sand high up onto the beach during high tide where it may dry before the next high tide. Winds blow the sand about and form low beach dunes that are out of reach of the next high tide. Some are free-moving—others are "locked down" by vegetation growing in the back beach area.

The seas have not pounded forever at this location—recall that Pleistocene (Ice Age) glaciers temporarily locked up large quantities of water and caused sea-level changes of hundreds of feet during the past 1.6 million years. The last rapid rise in sea level began relatively recently—about 15,000 years ago as glaciers of the last ice advance reduced in volume due to an ameliorating climate. The edge of the sea was many miles to the west at that time and must have produced many shoreline features that are presently submerged beneath the ocean. The slow rise of sea level today will greatly increase as global warming continues and will further change the location and appearance of shoreline features the world over.

Visiting Point Reyes

Point Reyes has many miles of trails to explore, each with its own fascinating features. Molenaar's map (1982) is informative and highly recommended for hikers and other visitors. Elk, deer, sea lions, harbor seals, and gray whales can be observed from viewpoints on the sand dunes and from trails on top of the sea cliffs. At the end of the Bear Valley Trail is Arch Rock, a sea arch tunneled through the rock by the endless pounding of the waves. Crawl through the arch at low tide but be sure to crawl back through before the tide comes in or you will be marooned until the next low tide! Some of the trails are steep, such as those leading up Mt. Wittenberg (1407 feet; 407 m), high point on Inverness Ridge. Carry a good supply of water because the water in the streams is not fit to drink. And watch out for poison oak and stinging nettles. There are places for swimming and places where the hammering of the surf is extremely hazardous; check your seashore brochure, and be safe.

REFERENCES

Galloway, Alan J., 1977, Geology of the Point Reyes peninsula, Marin County, California: Bulletin 202, California Division of Mines and Geology, 72 p.
Molenaar, D., 1982, Pictorial landform map, Point Reyes National Seashore and the San Andreas Fault, California: Wilderness Press, Berkeley, CA.
Norris, Robert M., and Webb, R.W., 1990, Geology of California: Wiley, New York, 541.
Sackett, Russell, 1983, Edge of the sea: Time-Life Books, Alexandria, Virginia.

Park Address

Point Reyes National Seashore
Point Reyes, CA 94956

PINNACLES NATIONAL MONUMENT (CALIFORNIA)

Pinnacles National Monument is small—slightly more than 25 square miles (65 km^2) including the additions made since it became a national monument in 1908. This unusual area of craggy columns and pinnacles (Fig. 3–20), some up to 1200 feet (366 m) high, dramatically stands out from the rest of the rolling, grassy hills of the Gabilan and other ranges in this part of the Coast Range section of the Southern Pacific Border Province. Something is very different about the geology at the Pinnacles compared to surrounding areas. As we shall soon see, pinnacles here form only on rocks from an extinct volcano of Miocene age (about 23.5 Ma). The giant San Andreas Fault passes through the monument and also played a major role in the area's development. The area has been a popular attraction since the late eighteenth century when even George Vancouver, the famous coastal explorer, took a mule trip to the Pinnacles in 1794. David Starr Jordan, a naturalist who was president of Stanford University in the early 1900s, was instrumental in helping convince Theodore Roosevelt to declare the area a national monument.

FIGURE 3–20   Sharp pinnacles in volcanic rocks in Pinnacles National Monument. (Photo by V. Matthews III)

The only entrance to Pinnacles is on the east side, about 35 miles (56 km) southeast of Holister, which is east of California's Monterey Bay. Only a small part of the 16,265-acre monument can be seen from a car; about 13,000 acres are designated wilderness. Several foot trails lead through this fabulous geological area, including the Condor Gulch Trail in the central section, which took early visitors through the breeding grounds of the California condor. Now the condors are gone—yet another animal falling victim to excessive habitat modification by you know who. You are likely to see black-tailed deer and raccoons but not the nocturnal bobcat and gray fox. The cougar is king of the mountain but usually keeps out of sight. Keep on the lookout for rattlesnakes and poison oak, both of which could take the edge off your visit. Carry a canteen on the trails; some of the streams are polluted. Parts of the trails are steep, and summer days are likely to be hot.

The long, dry summers are unfavorable for many plants but are evidently ideal for chaparral, the "pygmy forest." Greasewood (chamise) with lesser amounts of manzanita, buckbrush, hollyleaf cherry, and toyon make up the chaparral community within the monument. In the moister areas, sycamore, live oak, willows, and alders abound; yellow pine and digger pine are dominant in the higher areas.

Geologic Development

The Gabilan Range is bounded on the east and west by north-by-northwest-trending faults that are part of the San Andreas Fault zone. The Gabilan Range contains metasedimentary rocks of the Sur Series of probable late Paleozoic age—

some of the oldest rocks identified in the Coast Range. Next in the rock record are widespread intrusions of Cretaceous-age (about 80–90 Ma) granitic rocks that were part of the huge Sierra Nevada batholith complex. Extensive erosion during the early Tertiary (30–60 Ma) Coast Range Orogeny unroofed the granite batholith and formed an erosion surface near sea level that was soon to serve as a platform for the Pinnacles Volcanic Formation, the rocks in which the spectacular pinnacles and other landforms would later form.

The Pinnacle Volcano erupted from at least five vents during the Miocene, about 23–24 Ma, extruding a wide variety of eruptive products, mostly of rhyolitic composition (Norris and Webb, 1990). Dikes and domes mark the vent positions in the estimated one half to one third of the volcano that has not been removed by erosion. Rhyolitic tuffs, flow-banded lavas, obsidian, and some andesite and basalt lavas occur in the east part of the monument. A stratovolcano stage built a lofty peak, perhaps 8000 feet (2440 m) tall (Chronic, 1986), and slightly smaller than the present-day Mount St. Helens volcano in the Cascade Mountains. Volcanic sediment derived from the Pinnacles Volcano grades into nearshore Miocene-age marine deposits. Thus the volcano was located close to the Miocene sea covering this part of southern California and was synchronous with deposition of the nearby fossiliferous marine sediments of the Monterey Shale. Steam explosions near the end of the eruptive cycle blasted out rock fragments and blobs of lava. Avalanching and mixing of the debris on the steep western slopes of the volcano formed a *volcanic breccia* in which most of the more interesting landforms of Pinnacles would later form.

An estimated 6–10 cubic miles (25–40 km³) of volcano graced the landscape 23 Ma, but only 3 cubic miles (12.5 km³) remain (Norris and Webb, 1990). What happened to the rest of the volcano? Certainly much of the missing mountain can be accounted for by erosion. However, movements along the San Andreas Fault zone, which extends across the area, have moved the Pinnacles Volcano a considerable distance to the north from its point of origin and dropped part of the volcano downward into the underlying granitic rocks of the Gabilan Range. Dropping part of the volcano downward into a small fault trough, or *graben,* protected the volcanic rocks from more extensive erosion. Areas remaining at higher elevations are more likely to be ravaged and destroyed by erosion.

Another part of the answer to the question of the missing volcanic rocks is found farther south in Los Angeles County, where identical rocks of exactly the same age were mapped earlier. The Neenach Volcanic Formation (Dibblee, 1967) is located 180 miles (290 km) southeast, on the opposite side or North American Plate side of the San Andreas Fault (Fig. 3–21)! The Neenach rock layers are the same and in the same sequence; moreover, major structures would again line up if the two areas could somehow be joined again. The obvious conclusion is that after the Pinnacle–Neenach Volcano formed in the Los Angeles area, the San Andreas Fault cut through the area and carried the west part of the volcano, by a series of jerks, northwestward nearly 200 miles (320 km) to its present location (Ryder and

FIGURE 3–21   Sketch map showing displacement of Pinnacles volcanics along the San Andreas Fault and the location of Point Reyes and Channel Islands. (Modified from National Park Service brochure)

Thomson, 1989). The narrow sliver of lithosphere moving northwest is part of the Salinian Block, the same moving block found at Point Reyes National Seashore farther north near San Francisco.

The evidence from the Pinnacle and Neenach volcanics provides an answer to the frequently asked geologic question, "How much displacement has taken place along the San Andreas Fault?" Previous lines of evidence submitted to answer this question were subject to alternative interpretations. With the Pinnacle–Neenach discovery the problem was solved—at least for the period since the early Miocene. Ryder and Thomson (1989) believe that movement along the San Andreas through the Pinnacles volcano may not have started until perhaps 12 or 13 Ma (middle Miocene). Similar evidence described in the section on Point Reyes National Seashore, further supports displacements of hundreds of miles along the San Andreas Fault. When Hill and Dibblee suggested in 1953 that displacement has been as much as 350 miles (565 km), a certain amount of skepticism greeted them. With convincing data from the Pinnacles area, however, most geologists now accept their 350-mile (565-km) displacement without difficulty, and others suspect that Hill and Dibblee's estimate of total offset along the San Andreas Fault may be too conservative—perhaps displacement was as much as 620 miles (1000 km) (Oldow and others, 1989)!

Landforms

Numerous cracks or joints developed as the hot avalanche debris that forms the rhyolite breccia and other volcanic rocks cooled and contracted. Movements along the Gabilan Range boundary faults also twisted and stressed the volcanic pile creating additional fractures. Weathering, particularly in the rhyolite breccia, widened the cracks and formed broad columns that were later narrowed into *pinnacles, balanced rocks,* and other bizarre forms (Fig. 3–20). Unusual, weirdly shaped landforms like many of those found at Pinnacles are also called *hoodoos.* The streams that developed had sinuous courses and undercut their banks; where the rocks were least resistant, the undercutting was at its maximum, and remarkable overhangs or "caves" were formed. Other caves were formed by rocks tumbling or sliding down into the narrow canyons and wedging between the walls. You will need a flashlight when you explore Bear Gulch Cave on the Moses Spring Trail south of the visitors center. Weathering, mass-wasting, and stream erosion worked hand in hand in the fashioning of the fascinating, sometimes grotesque features that entice people into the Pinnacles.

Many people will visit Pinnacles National Monument and marvel at the columns, spires, balanced rocks, and caves without particular concern for the fact that the monument is "on the move;" others will want the whole story. We have offered only a few glimpses of the vast amount of work involved in putting this puzzle together.

REFERENCES

Chronic, Halka, 1986, Pages of Stone, v. 2: The Mountaineers, Seattle, 170 p.
Dibblee, Jr., T.W., 1967, Areal geology, western Mojave Desert, California: U.S. Geological Survey Professional Paper 522.
Hill, M.L., and Dibblee, Jr., T.W., 1953, San Andreas, Garlock, and Big Pine faults, California: Geological Society America Bulletin, v. 64: p. 443–458.
Norris, Robert M., and Webb, R.W., 1990, Geology of California: Wiley, New York, 541 p.
Oldow, John S., Bally, A.W., Ave'Lallemant, H.G., and Leeman, W.P., 1989, Phanerozoic evolution of the North American Cordillera; United States and Canada: in The Geology of North America—An overview: Geological Society of America, The Geology of North America, v. A, p. 139–232.
Ryder, Robert T., and Thomson, A., 1989, Tectonically controlled fan delta and submarine fan sedimentation of late Miocene age, southern Temblor Range, California: U.S. Geological Survey Professional Paper 1442, 59 p.

Park Address

Pinnacles National Monument
Paicines, CA 95043

CHANNEL ISLANDS NATIONAL PARK (CALIFORNIA)

A unique experience awaits those who would explore one or more of the fabulous islands in Channel Islands National Park. As close as 11 miles (18 km) from the California mainland and about 50 miles (80 km) from Los Angeles, one of the most populous areas in the United States, lies a virtual wilderness controlled by the whims of the sea and the weather. A visit to an island wilderness would seem to be an experience reserved for only the wealthy—not so thanks to the foresight of those who came before us and established a national monument in 1938 and later enlarged the people's holdings and designated this area a national park in 1980. Although the National Park Service administers the park, land ownership also involves the U.S. Navy, some private holdings, and the Nature Conservancy. This latter non-profit organization purchases unique areas and ensures that they are protected and managed properly, in this case in cooperation with the National Park Service.

The eight Channel Islands are separated from the California mainland on the north by the Santa Barbara Channel (Fig. 3–21)—hence the name Channel Islands. The four in the Northern Channel Islands are, from west to east, San Miguel, Santa Rosa, Santa Cruz, and closest to the mainland and the most visited, Anacapa. The islands are the tops of a submerged mountain range—an extension of the Santa Monica Mountains on the mainland. Of the four islands in the southern group of Channel Islands, only Santa Barbara, the smallest in the Southern Channel Islands shown in Figure 3–21, is part of the park.

The Santa Monica and other mountain ranges in this part of California are all part of the Transverse Ranges—a group of east–west trending mountains that are "transverse" to the northwest–southeast trend of topography (Plate 1) and geologic structures in adjacent areas. The San Andreas Fault bisects the Transverse Ranges farther east, placing all of California west of the fault, including the Channel Islands, on the northwest-moving Pacific Plate (Fig. 3–21).

The sea dominates here. Prevailing northwest winds cause cold, nutrient-rich bottom currents to upwell along the north, and the warm California Current brings warm waters from the south. Sea life is abundant and spectacular—especially the seal and sea lion populations. Steep, formidable *sea cliffs* on the island perimeters, especially on the "weather" side where the prevailing northwest winds create the highest waves, make landings difficult except in a few protected coves. Commercial boat service is available from the mainland to Anacapa and Santa Barbara where camping is permitted. Because of sensitive environmental conditions and some private ownership, permission is required to visit San Miguel, Santa Cruz, and Santa Rosa islands.

Human History

Numerous archeological sites indicate that seafaring Indian groups lived in harmony with their environment here for many thousands of years. When the first

European explorer, Juan Rodriquez Cabrillo visited the islands in 1542, the Chumash Indians inhabited the islands. The productive environment supported one of the highest human population densities in the New World. Cabrillo returned to San Miguel in 1543 to winter over, where he died from a gangrene infection. Cabrillo is reportedly buried on San Miguel although his grave has never been found.

The "fur frenzy" of the 1800s brought greedy Yankees, Russians, and local Spaniards to the islands where they mercilessly annihilated the sea otter population in order to make money by shipping furs to Europe. As the sea otter numbers plummeted, the hunters began to slaughter the northern fur seals, Guadalupe fur seals, and later the elephant seals. By the late 1800s all of these magnificent creatures were extinct in the Channel Islands and presumably everywhere along the Pacific coast (Howorth, 1982). Amazingly, a few small colonies of what had been considered extinct species were later found on some isolated Mexican islands, and a remnant colony of about 125 sea otters, the last of the world's sea otter population, was discovered on the central California coast. Today thousands of California sea lions inhabit the Channel Islands and small numbers of other species, including the sea otters, are slowly re-establishing their presence after being nearly decimated by humans.

The Chumash Indians fared as badly as their seal neighbors. Villages were raided by Aleut Indian hunters brought down by the Russian fur traders, and European diseases took their toll. The last of the Chumash were removed to mainland missions in 1813.

In the 1850s sheep and cattle were brought to the islands where overgrazing and other human activities greatly reduced the plant and associated animal communities. Some rare plant species, found nowhere else in the world, were lost or greatly diminished, particularly in years of drought when sheep would dig down 2 feet (60 cm) into the soil to eat plant roots. The strong, drying summer winds turned the western two-thirds of San Miguel Island into a sandy desert. Shifting sands have been producing the "ghost forests" of San Miguel and Santa Rosa for thousands of years as calcium carbonate impregnated dead tree trunks and roots were uncovered by moving sand, but not on the scale of the past century. Amazingly, many endangered species have survived man's poor stewardship and still remain on these isolated islands.

A number of maritime disasters occurred on the islands as ships ran aground in fog or during storms. Initially it was considered too difficult to construct navigation aids on these remote islands. The most notable wreck occurred in 1853 when the paddle-wheel steamer *Winfield Scott* ran aground on Anacapa in the fog. The 250 passengers spent a few terrifying days waiting to be rescued. The U.S. Lighthouse Service constructed a beacon on East Anacapa Island a few years later. A lighthouse was built in 1932 and U.S. coastguardsmen lived on the island until 1969 when the light and foghorn were automated.

The U.S. Navy acquired several of the Channel Islands and used them as early-warning outposts during World War II. Ranchers had been removed from San

Miguel in the 1930s so that the island could be used for bombing practice! The National Park Service took over management of San Miguel in 1963.

The Rock Record

Mostly marine sedimentary and submarine volcanic rocks of Mesozoic through Pleistocene age make up the Northern Channel Islands and their land counterpart, the Santa Monica Mountains. Cenozoic volcanic rocks are abundant here as well as on Santa Barbara in the Southern Channel Islands. Only parts of the geologic story are preserved on individual islands; but, taken together with the record in other parts of the Transverse Ranges, they help outline the geologic development of the entire area.

The Santa Cruz Island schist, the oldest rock exposed in the Channel Islands, was initially a section of Jurassic(?) volcanic and sedimentary rock that was deposited in a subduction zone and later intruded by Mesozoic plutons and intensely deformed during the late Mesozoic. Locally thick marine sediments accumulated in basins created as Cenozoic faulting and folding began to form the Transverse Range mountains beginning some 25 Ma. Over 15,000 feet of sediments and submarine *pillow lavas* were deposited in a deep marine trough of Miocene age and are extensively exposed on the Channel Islands.

Movements along the San Andreas and related faults in late Miocene time (about 12 Ma) caused the rotation of the Transverse Ranges block and added further complications to the already complicated arrangement of rocks. Block rotation and strike-slip faults with large horizontal displacements have moved pieces of landscape many miles from their point of origin. East–west strike-slip faults cut through Santa Cruz and Santa Rosa islands and place very dissimilar rocks next to one another, some of which began their journey in the Los Angeles Basin some 50 miles to the east! A central fertile valley with deep soil follows the trace of the strike-slip fault that extends the length of Santa Cruz Island. Diablo Peak towers 2400 feet (732 m) above sea level north of the fault and the more subdued, but still rugged, topography on the south rises some 1500 feet (457 m). When added to the 3000 foot (915 m) depth of the Santa Barbara Channel to the north, one can better appreciate the significant mountain building that occurred here.

Uplift of the Transverse Ranges in late Pliocene to early Pleistocene (about 2.0 Ma) formed an island mountain system as the west extension of the Santa Monica Mountains. The northern island chain was thrust above the sea while the Santa Barbara Channel, a continuation of the Ventura Basin on the mainland, subsided as a long trough. Continued, but episodic, uplift formed a series of flat, wave-cut platforms as the landmass continued to rise. For example, Anacapa displays at least six elevated marine terraces. Each terrace was initially cut by the pounding surf and later uplifted to form a marine terrace. An episode of stability followed and yet another wave-cut platform formed, in turn being uplifted as the island continued to rise from the sea. Marine terraces of the same age are at different levels on different islands. These in turn are at different levels from those on the mainland. Thus, different areas are uplifting at different rates along the California coast.

Geologic Enigma—The Transverse Ranges

The east–west trending Santa Monica Mountains and their Channel Island extension, as well as adjacent, similarly oriented mountains in southern California, are part of a puzzling group of mountains called the Transverse Ranges. Rather than being oriented north–south or northwest–southeast as are most of the mountains in California, and North America for that matter, these ridges and their associated folds and faults are oriented east–west, the obvious product of north–south compression. Why these mountains defy the "rules" of how we think mountains should "behave" has not been adequately explained.

Perhaps there is an east–west fracture zone in the Pacific Plate controlling the mountain orientation, or as Luyendyk and others (1980) suggest, a small plate was detached and caught between the North American plate and the northwesterly moving Pacific Plate and rotated clockwise 60–80 degrees from its original northwest orientation to an east–west trend to form the Transverse Ranges. Norris and Webb (1990) summarize these and other ideas relating to this puzzling geologic problem.

Even the monstrous San Andreas Fault is not immune from the forces that created the Transverse Ranges. The San Andreas Fault bends to the west and back to the northwest where it forms the northern margin of the Transverse Ranges on the mainland (Fig. 3–21). Whatever its origin, the "big bend" causes the San Andreas Fault to bind and produce the north–south compression that accounts for the east–west-oriented synclinal basins, anticlines, thrust faults, and strike-slip faults that have created the most complicated geology in California and the entire Coast Range.

Why should so many geologists study and puzzle over this complex geology when more simple, and perhaps more interesting problems could be solved elsewhere? The coincidence of three factors makes the answer apparent: (1) the presence of sedimentary rocks such as the Monterey Formation that is very rich in organic compounds, (2) permeable rock that can hold substantial amounts of fluids and gases derived from the organic-rich beds, and (3) ideal structures that permit fluids and gases to migrate upward where they encounter an impermeable material that causes the fluids to collect in pockets as oil and gas reservoirs. Much of California's lucrative oil industry is built around extracting petroleum and natural gas from structures in the Transverse Ranges, including reservoirs beneath the Santa Barbara Channel.

The impermeable seals, or *oil traps,* are not always perfect. Numerous *oil seeps* exist in the Santa Barbara Channel area and nearby areas including the famous Rancho La Brea tar pits near Los Angeles. As crude oil loses its more volatile components, it converts to a viscous tar or asphaltlike material. A thin sheet of water covering the tar at Rancho La Brea turned into a death trap for thousands of Pleistocene animals that waded out and became mired in the sticky tar. Oil seeps and associated tar deposits occur on San Miguel, and an oil seep on Santa Cruz Island has burned continuously for hundreds of years. Energy companies find such oil seeps interesting but seldom drill into them because the breached or imperfect seal prevents commercial quantities of oil from accumulating.

Pygmy Mammoths

A fascinating paleontological find in the Northern Channel Islands is the discovery of the bones of a diminutive species of prehistoric elephant that occurred nowhere else in the world. As Howorth (1982) points out, the "pygmy mammoth" or "dwarf mammoth" is an oxymoron—an amusing contradiction of terms. The significance of these 4-foot-high (1.2 m) mammoths was instrumental in helping to establish Channel Islands Park. Understanding that a subspecies of the mammoth could develop because of geographic isolation is not a problem—however, how the animals initially came to occupy the islands is a source of debate. The evolution of the explanations, or hypotheses, as new information or ideas are brought forth is an interesting study in scientific methodology. Without a time machine the answer cannot be known with absolute certainty, but reasonable hypotheses that fit all or most of the facts have evolved over the years.

The initial assumption was that these North American elephants had to walk across a land bridge to reach what was later to become a chain of islands. Sea level is known to have been much lower during episodes of intense Pleistocene glaciation; thus it was suggested that the necessary land connection existed during these times. Unquestionably the islands at that time were all united into one "super island," but if the Pleistocene topography was the same as at present, then a 4–5 mile (6 km) water gap separated the mainland and Anacapa Island during the most drastic known conditions of sea level lowering. Could recent tectonic processes drop the section of land between Anacapa and the mainland—after the elephants had migrated along this theoretical peninsula? What if the original assumption that the animals had to walk on a land bridge is incorrect? Elephants have been observed to swim a few miles out to sea, in one case repeatedly to visit a small island off the African coast (Norris and Webb, 1990). If this is what indeed happened in the Channel Islands, then other observations, like the incomplete representation of mainland species noted by biologists, would be adequately explained. A land bridge should enable deer, bear, rabbits, rattlesnakes, coyotes, and many other mainland species to populate the hypothetical peninsula. Yet the expected abundance of mainland species is not present; only unique subspecies of the mainland deermouse, a housecat-sized island fox, and a spotted skunk are the only native mammals present.

Another intriguing question is why are the mammoths no longer there? They likely shared the islands with early human inhabitants, just as their mainland cousins did. Charred bones of pygmy mammoth in a fire pit on Santa Rosa Island has fueled speculation that the demise of the mammoths is due to the increasing presence of *Homo sapiens!* A similar explanation is favored by many to explain the dramatic extinctions of Pleistocene mammals following the retreat of the last great ice sheets about 15,000 years ago. The demise of so many Pleistocene mammals and the simultaneous rise in human populations and better hunting technologies may be causally connected.

Ocean Processes

The prevailing northwest winds not only bring high surf and storms to the Northern Channel Islands, but they also encourage strong upwelling of the deep, cold, nutrient-rich Pacific waters that encourages the growth of kelp beds and the amazing abundance of sea life, including the sea lions and elephant seals for which this area, especially San Miguel Island, are famous.

The coastal processes here are similar to those previously discussed at Point Reyes National Seashore and Redwood and Olympic parks. The conversion of wind energy to ocean disturbances called swells, and ultimately to surf when shallow water is encountered, delivers energy to the shoreline—enough energy to tear away at the shore materials and form distinctive landforms such as the spectacular sea cliffs that ring the Channel Islands and make boat landings treacherous except in a few natural harbors. Wave erosion is king, forming the wave-cut platforms, stacks, sea caves (Fig. 3–22), and arches that give the area its "other world" appearance. The 40-foot-high (12-m) Arch Rock on the south end of Anacapa is considered to be the "trademark" view in the Channel Islands. Surf flowing up the surge channel near Cat Rock on the south end of Anacapa compresses air in rock cavities causing seawater to erupt spectacularly from a feature called a *blowhole*.

Noteworthy among the over 380 sea caves in the Channel Islands is Painted Cave on Santa Cruz (Fig. 3–22). The impressive 130-foot-high (40-m) entrance leads back some 1215 feet (373 m), probably qualifying it as the world's longest sea cave according to Bunnell and Vesely (1983) who were part of the first team to carefully explore this most unusual cave. The cave is even longer according to an early report by a seal hunter who chased a herd of seals into a nearby sea cave only to have them emerge from Painted Cave. Apparently a submarine passage connects the two caves. Determined divers have found underwater passages and air-filled, dry passages (with seals!) in the back of Seal's Secret Cave near Painted Cave. The rest of the seal's secret passage still remains undiscovered by humans (Bunnell, 1993).

FIGURE 3–22   Painted Cave on Santa Cruz Island, Channel Islands National Park. (Photo by D. Bunnell)

Island Biology

As glaciers rapidly melted around the world about 15,000 years ago, the volume of water in the oceans increased and sea level began to rise. As the singular "super island" in the Channel Islands was gradually inundated by rising water, low areas between the higher peaks were flooded, and the individual islands gradually took on their present configuration. Land animals, such as the mammoth, deermouse, fox, and skunk that had accidentally found their way to the island by swimming or clinging to floating debris, had at least 15,000 years of separation from their mainland ancestors and developed the genetic characteristics of a distinct subspecies. A similar change affected many plant species after many millennia of geographic isolation. Human introduction of mainland plants, weeds, cats, and other animals forever changed the once pristine islands.

Of particular interest is the large population of pinniped ("fin-footed") mammals such as fur seals, elephant seals, and sea lions that breed and give birth to their pups along the rugged island shores. They have figuratively "returned from the dead" after becoming nearly extinct as a result of the shameful slaughter of the 1800s. On San Miguel five different species of seal and sea lion have rookeries—making this the most diverse seal rookery in the world (Howorth, 1982). The sea lions arrive by the thousands in the spring to calve and breed. Porpoise, dolphin, shark, and occasional gray, humpback, and blue whales also frequent or migrate past the islands.

Another of the few success stories of humans reversing some of our destructive ways is that of saving the California brown pelican from extinction. Their major breeding ground and rookery is on West Anacapa Island. Direct interference here was not the problem. The culprit was the DDT poison that was spread widely across the world landscapes in the 1950s and 1960s as a pesticide. The poison found its way into marine invertebrates and fish—the food source of the brown pelican. The resulting thin eggshells were crushed under the weight of the nesting female and very few chicks were born alive. The pelican population plummeted and they were on their way to extinction when the United States banned the use of DDT in 1972, enabling this species, as well as many other bird and bat species throughout the world, to slowly recover. The increased concentration in human tissue of DDT poison was also an ominous sign of imminent danger to other members of the top of the food chain.

The thick kelp beds and the upwelling currents provide ideal conditions for one of the world's richest ecosystems to exist. The 1980 legislation establishing the park also established a 6-nautical-mile (11-km) National Marine Sanctuary around each island where commercial exploitation is not permitted.

Visitation—Geologic Future

Camping is permitted on East Anacapa near the lighthouse (associated historic Coast Guard buildings now used as a ranger station) and on Santa Barbara Island. Commercial boat service is available, particularly during the summer months. One should write or visit the visitor center and park headquarters in Ventura for current information and schedules.

The other islands require special permission to visit. Special regulations are necessary to protect the sensitive landscape and bird and mammal populations. In spite of human mismanagement of the land, the islands still provide a habitat for many endangered native species, who find shelter in one of the most important marine wildlife sanctuaries in the United States.

At present the islands continue to shrink as the sea and its power are directed onto the edge of these lonely, isolated land masses. However, the power of the moving plates over the long run can be even more significant. Energy generated by the moving plates is transferred to the myriad of faults and folds in the southern California area in a complex way that, as yet, is not completely understood. Whether uplift of the partially submerged mountain ridge will exceed the rate of destruction by the sea is a drama that will require many millennia to play out before a definitive answer is available. Recent movements have unquestionably occurred. Pleistocene marine terraces, cut by wave action near or below present sea level, are now located well above our current high sea level. Island stream channels are sliced by and offset by strike-slip faults—in a geologic sense these features must have appeared "yesterday!" Rapid change here is inevitable and, because of its national park status, generations of Americans will bear witness to these natural changes.

REFERENCES

Bunnell, David, 1993, Sea caving in the Channel islands—A decade of intertidal adventure: National Speleological Society News, v. 51, no. 6, p. 150–159.

Bunnell, David, and Vesely, C., 1983, The amazing caves of Santa Cruz Island: National Speleological Society News, v. 41, no. 2, p. 86–91.

Howorth, Peter C., 1982, Channel Islands, the story behind the scenery: K.C. Publications, Las Vegas, Nevada, 48 p.

Luyendyk, Bruce P., Kammerling, M.J., and Terres, R., 1980, Geometric model for Neogene crustal rotations in southern California: Geological Society of America Bulletin, v. 91, p. 211–217.

Norris, Robert M., and Webb, R.W., 1990, Geology of California: Wiley, New York, 541 p.

Park Address

Channel Islands National Park
1901 Spinnaker Drive
Ventura, CA 93001

FOUR

Cascade Mountain Province

The quiet of Sunday morning, May 18, 1980, was broken by a thunderous explosion that shook southwestern Washington. It was Mount St. Helens blowing its top, after being essentially quiet for more than a century. Minor eruptions had begun on March 27 (Fig. 4–1), but this was the "big one" that blew away the mountaintop, killed 57 people, and devastated a large area.

Fortunately, geologists Dwight (Rocky) Crandell and Don Mullineaux had recently published U.S. Geological Survey Bulletin 1383-C in 1978 in which they warned that a major eruption was due in the near future. The Cascades, in the short period of accurate record keeping from about 1800 on, have managed to provide each generation of Americans with at least one volcanic eruption. Steve Harris (1988) further points out that a cluster of eruptions occurred from 1800 to 1857 and wonders if we are not due for another episode of accelerated activity. Humans are unlikely to devise a means of capping an ebullient volcano; however, if sufficient warning is given—as it was at St. Helens—the loss of life and property can be kept to a minimum. Most of the 300,000 or more human deaths attributable to volcanoes in historic times were caused by stratovolcanoes, the prominent type of volcano in the Cascades.

Mount St. Helens is the youngest and most active of the dozen or so major, active stratovolcanoes that grace the crest of the Cascade Range. The range extends from southern British Columbia, through Washington and Oregon, and terminates a short distance south of Lassen Peak in northern California (Plate 4; Fig. 4–2). All of these volcanoes are recent additions to the landscape. The oldest events on these peaks are no more than a few hundred thousand years old with most displaying youthful volcanic landforms hardly scratched by the rigors of erosion.

Geomorphic Province Description

Topographically the Cascades are continuous except where cut by the spectacular Columbia River Gorge (Plate 4; Fig. 4–3), a scenic feature in its own right and worthy of protection. The gorge separates the South and Middle Cascade sections as

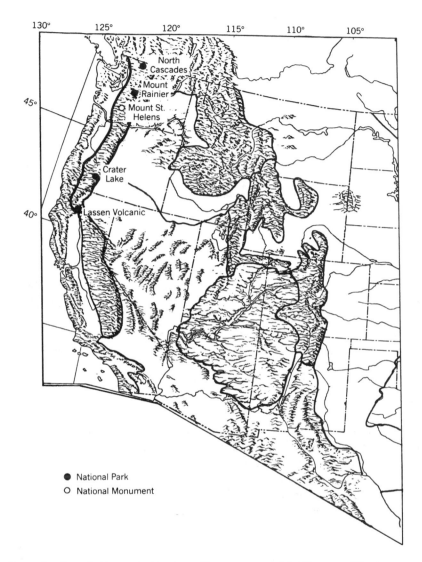

PLATE 4   Cascade Mountain Province. (Base map copyright Hammond Inc.)

well as the states of Oregon and Washington. Farther north, Interstate 90 follows a topographic corridor that crosses the Washington Cascades along a major northwest–southeast trending fault zone (Olympic–Wallowa lineament, or OWL) that separates the North and Middle Cascade sections (Kiver and Stradling, 1994). The OWL continues southeast beneath the Columbia Intermontane Province, through the Hanford Nuclear Reservation (where over 760 million gallons of high-level radioactive waste are stored!), and into the Northern Rocky Mountains of Idaho.

FIGURE 4–1   Air view of Mount St. Helens on April 7, 1980. The crater was formed by volcanic explosions that began on March 27. (Photo by Wes Guderian of the Portland *Oregonian*)

The Northern Rocky Mountains, Columbia Intermontane, and Basin and Range provinces form distinct topographic boundaries east of the Cascade Range, and Puget Sound and the Willamette Valley Sections of the Pacific Border Province make sharp topographic boundaries to the west (Plate 4). The Province boundaries are indistinct in British Columbia where the Cascades merge into the Coast Range and in southern Oregon and northern California where they join the Klamath Mountains. The southern boundary is a short distance south of Lassen Volcanic National Park, where the Cenozoic volcanics of the Cascades give way to the Mesozoic granites and metamorphic rocks of the Sierra Nevada.

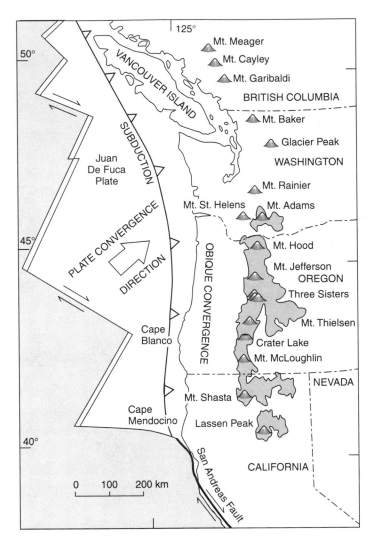

FIGURE 4–2   Cascade stratovolcanoes active during Pleistocene time and location of important plate tectonic features. (Modified from Skinner and Porter, 1987)

Excluding the high volcanic peaks, elevations range from about 1700 feet (520 m) to 5800 feet (1770 m) in the Older, or Western, Cascades and 4000 feet (1220 m) to 8000 feet (2440 m) in the Eastern, or High, Cascades of Oregon and southern Washington. The North Cascades are exceptionally high, with elevations commonly up to 8000 feet (2440 m). The isolated Pleistocene-age volcanic peaks in the Eastern Cascades are regularly spaced about 50–75 miles (80–120 km) apart and tower above the lower mountain peaks and ridges. At times they seem to "float,"

FIGURE 4–3   The Columbia River cuts through the Cascade Range in a spectacular gorge with walls averaging 1500–3000 feet (460–900 m) high and locally (Mt. Defiance) almost 5000 feet (1525 m)! View is from near Crown Point on the Oregon side of the gorge. (Photo by E. Kiver)

like an apparition, high above the surrounding mountains—making them appear almost unreal. In our opinion, a sweeping view of three or more of these strato-cones on a clear day is one of the top ten spectacular sights available in the United States. All areas surrounding them are "low country" in comparison (Fig. 4–4). For

FIGURE 4–4   Photo of Mt. Shasta and parasitic cone, Shastina, rising boldly above surrounding Cascade peaks. (Photo by E. Kiver)

those who live near high mountains and love the outdoors, seeing high country with snow-capped peaks, even some distance away, acts as an irresistible magnet that draws them back to the mountains again and again.

Mt. Rainier is the topographic champion at 14,411 feet (4393 m) with Mt. Shasta (14,162 feet; 4318 m) in northern California running a close second (Fig. 4–4). However, the future will no doubt bring profound elevation changes. The slow processes of erosion will take their toll as is apparent on some of the older volcanic peaks (Fig. 4–5) but more rapid, instantaneous changes like those at Mount St. Helens in 1980 will also occur. However, lest we forget that volcanism is primarily a topographically constructive force—elevations and volumes of the cones during eruptions will in most cases increase.

The Columbia River Gorge, the 75-mile (120-km)-long "Grand Canyon of the Columbia" is a spectacular canyon (Fig. 4–3) whose unique rock walls record the geologic history of a mountain range. John Allen's *The Magnificent Gateway* (1976) is an excellent geologic reference. The archeological and historical records of the gorge are also of national significance, and in 1986 the area was declared a National Scenic Area in hopes of limiting the environmentally and scenically destructive development that was proceeding at an ever-increasing rate. The attempt has some successes but overall is disappointing. The longer areas are left unprotected the more private lands become involved making protection more economically costly. Insufficient funding for land acquisition in the original legislation combined with state and local government entities unable or unwilling to take strong protection-

FIGURE 4–5   Mt. Thielsen just north of Crater Lake is the "lightning rod of the Cascades." This extinct stratovolcano last erupted during the early Pleistocene and has been ravaged by erosion through three or more major intervals of glaciation. (Photo by E. Kiver)

ist stands have permitted continued scenic deterioration. Clear-cutting, luxury recreational house building, and other construction are some of the problems.

Local control or local influence often means short-term benefit to the local economy (mostly to a few individuals) rather than long-range benefits to the community or country-at-large. It is ironic that public land sold for a few dollars an acre or given away in the "old days" should now have to be purchased back by the general public at greatly inflated prices. It seems that we do not learn the lessons of history. Today there are still politicians who advocate repeating these same mistakes by giving away federal lands owned by all citizens to individual states where they might be developed or sold to the highest bidder. Following this approach we can "look forward to" turning the country into "America, land of the no trespassing signs." The lessons are clear both here and in other seemingly "wild" areas that still remain. If we hope to boast an "America the Beautiful" in the future, we need to act in a strong way to protect special areas, and we need to act sooner rather than later.

Geologic Evolution

The Cascades is an area where active plate tectonics processes, with some interesting complications, best explain this "land of volcanism." The geologic story of the Cascade Range was originally considered to be mostly the story of a simple plate tectonic model in which the dense basaltic rock of the oceanic plate was subducted beneath the less dense rocks of the westward-moving North American plate during Cenozoic time (Fig. 4–6). However, although this explanation is basically cor-

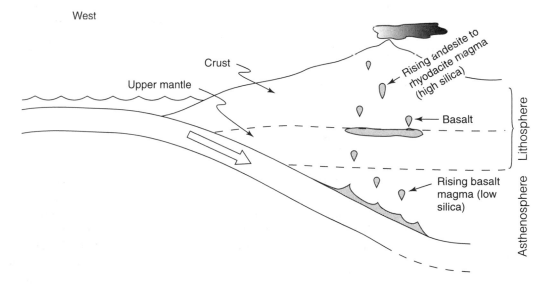

FIGURE 4–6   Generalized subduction model showing (*a*) generation of basalt magma by partially melting the mantle, and (*b*) generation of andesite to rhyodacite magma by partially melting lower crustal rocks.

rect, studies indicate that Mesozoic events, especially in the North Cascades, affected the development of the Cascades and that the simple plate collision explanation involves further complications. Although proposed tectonic models differ in detail, all include subduction, addition of exotic terranes, and oblique plate movements as important components in creating today's Cascade Range.

The long-term picture now emerging indicates that earlier North American plate edges were farther east and shifted westward as new material was added during Paleozoic and Mesozoic orogenies and as exotic plate fragments from afar arrived. The North American plate edge in early Mesozoic time was located in Nevada, western Idaho, and eastern Washington. Along some edges subduction scraped sediments from the descending oceanic plate causing the plate to widen in a westward direction (Alt and Hyndman, 1995). However, difficulty in sometimes matching the geology in one area with adjacent areas has led to the realization that larger masses of continental materials, islandlike masses called *microcontinents, microplates,* or *exotic terranes,* are sometimes imbedded in these oceanic plates and can add, or *accrete,* sizable "chunks" of real estate to a plate edge (see Fig. 1–11). Such additions during the Mesozoic added land to eastern Oregon and Washington and shifted the plate edge westward. Boundaries between these individual puzzle pieces are marked by faults—weak zones that localize tectonic adjustments and igneous activity. One block or microplate containing rocks as old as Precambrian and perhaps derived from somewhere in Mexico "docked" against North America about 100–90 Ma—this exotic terrane contains North Cascades National Park. Although much still needs to be discovered and some ideas and ages of events will be substantially revised as new information comes to light, the plate tectonics concept places us much closer to understanding the complex Cascade Mountain geology.

Table 4–1 presents a highly generalized sequence of major events to affect the Cascade Mountains and may be helpful in understanding the following discussion. For those seeking additional general information and summaries, the *Geology of the Pacific Northwest* by Orr and Orr (1996), the *Geology of Oregon* (Orr and others, 1992), *Fire Mountains of the West* (Harris, 1988), *Northwest Exposures* (Alt and Hyndman, 1995), and the *Roadside Geology Series* (Alt and Hyndman, 1981a, 1981b, 1984) are strongly recommended.

The first of the Cascade volcanoes erupted about 42 Ma (late Eocene) beginning the formation of the older Western Cascades. Partial melting of rock in the subduction "furnace" produced the raw materials to fuel the magma chambers that were destined to build this remarkable volcanic range. The range was oriented northwesterly at this time and would later rotate clockwise along with the microplates with which it was associated to achieve today's north–south orientation. Subduction continued and formed rows of volcanic vents that become younger to the east. This eastward-migrating trend is particularly prominent in the Oregon and southern Washington Cascades where the Western and Eastern Cascade division is well developed. During the middle Miocene (10–20 Ma) folding and tilting of the Cascade rocks, the mountains still remained quite low. About 17 Ma the voluminous eruptions of the Columbia River Basalt occurred in the Columbia

TABLE 4–1   Time Chart Listing Significant Geologic Events to Affect the Cascades

C	HOLOCENE	Modern global warming begins Cooler climate generates small glaciers (Neoglaciation) Huge eruption of Mt. Mazama (Crater Lake) 6800 years ago
I	PLEISTOCENE	Ice age ends about 10,000–12,000 years ago, climate warms Mount St. Helens forms about 40,000 years ago Modern stratovolcanoes begin (about 400,000 years ago), basaltic volcanism continues in Middle and Southern Cascades Ice Age begins (about 1.6 Ma)
O	PLIOCENE	Localized silica-rich (andesite and dacite) volcanic centers form Mostly basaltic eruptions form plateau in Middle and Southern Cascades
N	MIOCENE	Grabens form in Middle and Central Cascades, Eastern or High Cascades begin (about 10 Ma) Mountains still topographically low, partially covered by Columbia River Basalt (about 15–17 Ma) Folding and tilting of Western Cascades Volcanism continues to migrate eastward
O	OLIGOCENE	Oblique subduction begins to rotate Cascades to north-south orientation (25–30 Ma) Volcanism continues, tropical to subtropical climate
Z	EOCENE	Volcanism starts about 42 Ma in Western or Older Cascades, volcanism migrates eastward, range oriented northwest-southeast North Cascades microcontinent docks (about 50 Ma)
O	PALEOCENE	No record in Cascades
E		
C	MESOZOIC	Microplates dock onto edge of North American Plate Subduction zone located east of present Cascades

Intermontane Province to the east, sending massive flows against and locally over the low mountains, especially in the area along the Columbia River Gorge where lava flowed completely through the Cascades and into the Pacific Ocean beyond.

An interesting complication occurred in the plate tectonics picture as the spreading ridge off the Pacific coast was beginning to be overridden and destroyed by the westward-moving North American plate about 20–25 Ma (see Fig. 3–2). The direction of convergence between the oceanic and continental plate changed from east to northeast creating lateral (strike-slip) rotational forces (Orr and others, 1992). Extensive fissures opened up in the Columbia Intermontane Province to the east during the Miocene phase of plate rotation producing huge outpourings of Columbia River Basalt that covered vast areas of the Pacific Northwest.

Subduction and accompanying volcanism continued for some 25 million years (ending about 17–9 Ma) in the older Cascade Range with rows of volcanic vents continuing their eastward march (Alt and Hyndman, 1995; Orr and Orr, 1996). The Eastern, or High, Cascades began to arch upward about 10–7 Ma (late Miocene time) as magma rose and the underlying rocks were heated and began to expand. Large lithospheric blocks, like giant keystones, dropped downward along north–south oriented faults along the arch crest as a series of *grabens* (linear depressions bounded by parallel faults). The faults opened pathways that enabled large volumes of magma to rise. The grabens not only filled but overflowed as a series of overlapping shield and cinder cones built atop one another. The resulting volcanic plateau is called the Eastern, or High, Cascades. Basaltic eruptions from hundreds of short-lived vents have dominated up to the present day (last basaltic eruption in Oregon Cascades about 400 years ago) with some localized long-lived volcanic centers erupting more silica-rich andesites and dacites (Fig. 4–2). Although the silica-rich magmas are in the minority, the spectacular stratocones that they produce are scenically irresistible and form the nucleii of our Cascade parks, except for the North Cascades where spectacular alpine topography alone justifies park designation.

Studies of the eruptive record of the present generation of volcanoes indicates that most of the dozen or so stratovolcanoes considered to be active are Quaternary in age and no more than 400,000–600,000 years old. Radiometric dating of volcanic rocks and pyroclastic debris using the potassium–argon method and *paleomagnetic* studies are the bases of this conclusion. The paleomagnetic method measures the earth's magnetic field locked into minerals like magnetite at the time that the hot rock cools below about 500°F (260°C). The present-day magnetic field with north toward Canada and south toward Antarctica has existed for the last 730,000 years. An interval of reversed magnetic polarity existed before this time. Because all of the materials deposited from the peaks currently considered active have a "normal" magnetic signature, we can confidently conclude that they are all younger than 730,000 years old. Mount St. Helens is apparently much younger— the "new kid on the block" is a mere 40,000 years old! No wonder this "teen-age" volcano is the most active in the Cascade Range!

Geologic Trends

Some would believe that eruptions in the Cascade Range are decreasing and that they are in their "final" stages of activity. Again, an appreciation of the vastness of geologic time helps put things in perspective. By cranking our time machine vision forward, there is no doubt that Cascade volcanism will terminate. Because current volcanism will only continue for a short time after the tiny Juan de Fuca oceanic plate is completely overridden by the westward-drifting North American plate, a glance at Figure 3–2 or 4–2 indicates that there is not much plate and therefore not much time left to subduct! Since plates move only a few inches each year, there may only be a mere 10 million years of Cascade eruptions left! This provides yet another reason to plan a visit soon to this remarkable area.

As we crank our time machine backward by using the timing of past eruptive events recorded in the lavas, the pyroclastics, and the mudflow deposits around these volcanoes, it becomes apparent that for most of their history each volcano looked much as it does today—a quiet, serene landscape with a thick forest on the lower slopes and snow and glaciers coating its alpine heights. The glaciers were, of course, much more extensive during intervals of Pleistocene glaciation. Wispy clouds of steam might be seen over some of the dormant peaks, the only outward indicator of the potential turbulence below. Eruptive intervals are few and brief— mere hiccups in the total picture. However, if one is in the wrong place during a hiccup—look out!

An interesting and ominous trend is that the silica content of many of the Cascade volcanoes tends to increase with age. Many volcanoes begin with a shield stage (basalt) and progress to stratocone (andesite) and even plug dome (dacite) stages. As discussed in Chapter 1, silica content and eruption violence are often related. Because the parent magma from the mantle is basaltic, some process or processes must account for the silica increase that produces the andesites that characterize most of the Cascade stratovolcano eruptions. Some of this chemical change might be accounted for when a magma chamber becomes compositionally zoned with denser, heavier fluids and early formed crystals in the lower part (e.g., andesitic in composition) and lighter colored, less dense fluids (e.g., dacitic or rhyodacitic in composition) concentrating in the upper part. Depending on which part of the magma chamber is "tapped" during an eruption, the type of lava produced can be highly variable.

Another possibility for the wide range of chemical composition may be due to an increased *residence time* of magma in silica-rich crustal rocks. Once magma forms by partial melting in the upper mantle, the fluid is less dense than the surrounding rocks and rises buoyantly toward the surface, usually stopping at the base of the crust or somewhere higher in the crust where neutral buoyancy (i.e. *fluid density* equals density of surrounding rock) is achieved (Fig. 4–6). Partial melting of the surrounding crustal rock will "contaminate" the magma with more silica-rich solutions, thereby changing the magma chemistry. To reach the surface either tectonic movements "pump" or squeeze the magma or gas pressure increases and pushes the magma to an area of lower pressure—the earth's surface.

Recognizing that the highest silica content eruptions are confined to areas south of central Oregon (south of the towns of Bend and Sisters), some geologists speculate that the accreted terrain beneath the Cascades to the north contains mostly oceanic crust (basalt), and areas to the south are underlain by more silica-rich, continental crust rocks. A major fault zone, the Brothers Fault Zone, extends northwest into the Cascades near Bend, Oregon and may represent the buried junction of these two very different microplates.

Geomorphology

In earliest Pliocene time there probably was very little elevation difference between the Cascades, Columbia Intermontane, and Basin and Range provinces. The topographically high Cascades as we know them began to take shape during the Pliocene. Broad upwarping and westward tilting of the Cascades, in places several thousand feet, produced an abrupt separation between the Cascades and the provinces to the east.

During the Pliocene upwarping, the west-flowing streams were, one by one, blocked off—unable to keep pace with the uplift athwart their courses. The Columbia River was an exception; perhaps already the largest of the streams, it increased its flow and eroding power by gathering the waters of the streams that were blocked off. Moreover, the uplift across its course was less than that to the north; thus the Columbia maintained its course across the Cascade Range and cut a deeper gorge. By so doing, it became an *antecedent stream* (a river that predates the topographic or structural feature that it cuts across), one of comparatively few whose antecedence cannot be successfully questioned. Other exciting moments in the gorge history include the flow of lava rivers over a hundred miles long from the Columbia Intermontane Province to the Pacific Ocean; diversion of the western segment of the river northward when volcanic eruptions, including those of giant Mt. Hood, one of the Pleistocene stratovolcanoes, buried a segment of the Columbia River valley; and when an ice dam collapse in the Northern Rockies sent 1000-foot-deep (305-m) floodwaters from glacial Lake Missoula through the gorge.

Glaciers and streams have had a profound effect on the volcanic cones, gouging out deep valleys on their flanks. Although a volcano's appearance is not a precise way to measure the amount of time lapsed since a volcano stopped erupting and repairing the wounds of erosion, the degree of preservation of the original topography does provide a broad age indicator. None of what were likely impressive Eocene, Oligocene, or Miocene volcanic landforms in the Western Cascades remain today.[1] Volcanoes, mostly of Pliocene or early Pleistocene age, retain some

[1]Mt. Aix east of Mt. Rainier is an exception. Although greatly eroded, it maintains a broad, but subdued, cone form.

of their original profiles but are deeply furrowed by streams and glaciers. Some, such as Mt. Thielsen (the "lightning rod of the Cascades"), have their more resistant *vent plugs* projecting upward as giant spines (Fig. 4–5), indicating a very long time since the last eruption—so long, in fact, that a volcano like this can safely be regarded as *extinct*. Recently active peaks can repair much of the erosional damage by filling in valleys with lava and pyroclastic debris and reconstructing their craters. The smooth slopes and perfect volcano shape of Mount St. Helens before the 1980 eruption and the fresh-appearing craters such as those that adorn Mt. Rainier's summit are examples—such volcanoes, even without further evidence, should be considered *active volcanoes* and are likely in a *dormant* (a "sleeping" volcano) stage.

Glaciers formed on each volcanic cone when it reached sufficient height, in most cases late in the Pleistocene. Most of the sculpturing now in evidence occurred during the Wisconsin glaciation, which reached its maximum only 15,000–20,000 years ago. At the peak of glaciation, a continuous icecap buried the upper Cascades from Canada to northern California, broken only in the Columbia Gorge area where elevations are close to sea level. In the North Cascades the alpine glaciers merged with the huge *Cordilleran ice sheet* flowing south from Canada. The ice sheet inundated the Okanogan Valley on the east side of the Cascades and dammed the Columbia River in the Grand Coulee area. Ice flowing on the west side of the Cascades covered what is now Seattle and terminated just south of Olympia in the Puget Lowlands section of the Pacific Border Province. Alpine or valley glaciers spawned from the large Pleistocene icefields on Mt. Rainier flowed as far as 65 miles (105 km) from the mountain. Valley glaciers farther south near Mt. Mazama (now Crater Lake) extended a lesser, but still impressive, distance of 17 miles (27 km) from the summit.

Numerous *cirque* and valley glaciers are present today on the high Cascade peaks with, as expected, fewer or no glaciers on the more southern peaks. The glacial complex on Mt. Rainier is the largest and most accessible in the United States outside of Alaska. The North Cascades section contains an incredible 750 individual glaciers, the largest concentration in the conterminous United States. Over half of these were expanding from 1944 to 1976. By 1977 *all* were in retreat as winter precipitation decreased and global warming took hold. If these trends continue, Pelto (1991) predicts that 690 of these 750 glaciers will disappear during the next century! If these changes result in part from human influence on the climate, as many scientists suspect, we have reason to ponder on what else we may be destroying of the earthly environment.

Thermal features such as fumaroles and hot mud pots are surface indications of the sleeping "fires" below. The Bumpass Hell area near Lassen Peak is perhaps the most spectacular. Hazard Stevens and P.B. Van Trump made the first documented climb of Mt. Rainier in 1870. It was bitterly cold, but they survived a long night huddled over one of the *fumaroles* inside a geothermal ice cave in the summit crater. The release of even small quantities of heat and gas serve as reminders that beneath each of these active volcanoes lies a restless magma chamber. That future eruptions will occur is unquestionable—the "when" is the big unknown.

An extension of James Hutton's pronouncement that "the present is the key to the past"—one of the basic principles in geology—is that the past is the key to the future. This was the basis for a study by Kiver (1982) in which he determined that the number of significant eruptions of Cascade volcanoes thus far in the twentieth century is considerably less than during the nineteenth century. This should not be interpreted to mean that volcanic activity is slowing down. If your traveling companions think that you are taking too many pictures of Mt. Rainier or some of the other volcanoes, remind them that these pictures may prove to be valuable "before" pictures, and that the "after" may come sooner than most people are aware.

At present, there are four national parks and one volcanic monument in the Cascades Province. Because of its recent activity, Mount St. Helens will be discussed first, then Lassen, Crater Lake, Mount Rainier, and finally the North Cascades. These park areas, as well as many other Cascade Mountain areas currently lacking park status, contain some of our finest outdoor museums, displaying world-class examples of volcanic and glacial features.

REFERENCES

Allen, J.E., 1979, The magnificent gateway: Timber Press, Forest Grove, Oregon, 144 p.

Alt, D.D., and Hyndman, D.W., 1981a, Roadside geology of northern California: Mountain Press, Missoula, Montana, 243 p.

Alt, D.D., and Hyndman, D.W., 1981b, Roadside geology of Oregon: Mountain Press, Missoula, Montana, 272 p.

Alt, D.D., and Hyndman, D.W., 1984, Roadside geology of Washington: Mountain Press, Missoula, Montana, 282 p.

Alt, D.D., and Hyndman, D.W., 1995, Northwest exposures: Mountain Press, Missoula, Montana, 443 p.

Crandell, D.R., and Mullineaux, D.R., 1978, Potential hazards from future eruptions of Mount St. Helens volcano: U.S. Geological Survey Bulletin no. 1383-C, 26 p.

Harris, S.L., 1988, Fire mountains of the west; the Cascades and Mono Lake volcanoes: Mountain Press, Missoula, Montana, 379 p.

Kiver, E.P., 1982, The Cascade volcanoes—comparison of geologic and historic records, in Keller, S.A.C., ed., Proceedings of the Mount St. Helens One Year Later Symposium: Eastern Washington University Press, Cheney, WA, p. 3–12.

Kiver, E.P., and Stradling, D.F., 1994, Landforms, in Ashbaugh, J.G., ed., The Pacific Northwest: Geographical perspectives: Kendall/Hunt, Dubuque, Iowa, p. 41–75.

Orr, E.L., and Orr, W.N., 1996, Geology of the Pacific Northwest: McGraw-Hill, New York, 409 p.

Orr, E.L., Orr, W.N., and Baldwin, E.M., 1992, Geology of Oregon: Kendall/Hunt, Dubuque, Iowa, 254 p.

Pelto, M.S., 1991, North Cascade glaciers; their recent behavior: Dept. of Environmental Science, Nichols College, Maine, 17 p.

Skinner, B.J., and Porter, S.C., 1987, Physical geology: Wiley, New York, 750 p.

MOUNT ST. HELENS NATIONAL VOLCANIC MONUMENT (WASHINGTON)

Mount St. Helens was, until 1980, one of the truly beautiful mountains of the world—symmetrical, with glistening glaciers streaming down its flanks—the "Fujiyama of America" (Fig. 4–7). Now it is only a vestige of its former self—an empty shell, the shattered remains of the colossal explosion of May 18, 1980.

The explosions blew away about 99 billion cubic feet (2.8 billion m³) of rock material and roughly 3.5 billion cubic feet (0.1 billion m³) of ice (Brugman and Post, 1980). Before the May 18 eruption the mountain was 9671 feet (2931 m) high. In less than one minute, Washington's fifth highest mountain lost 1300 feet (396 m) of its summit elevation and became its thirty-seventh highest!

Is Mount St. Helens destined to remain a jagged stump or will it rise again? Geologists point out that Mount St. Helens is one of the youngest of the Cascade volcanoes, and that the records of the events involved in the building of Mt. Rainier and the other high cones indicate that each had one or more major setbacks before reaching its present height. Therefore, it is almost a certainty that Mount St. Helens will sometime in the future take her place among the "high ones." Sadly, we will not be here to admire her majesty!

Meantime, the paroxysms of the 1980s eruptions have provided a wealth of information that will assist us in predicting what will likely take place during the next

FIGURE 4–7   Mount St. Helens, the "Fujiyama of America," as it appeared prior to 1980. (Photo by E. Kiver)

Cascade volcano eruption. We must pay tribute to the geologists and geophysicists who risked or gave their lives assembling every conceivable type of information about active volcanoes.

The Indians of the several local tribes were well aware that Mount St. Helens was an active and dangerous volcano long before the "new Americans" arrived. Their names for the mountain—for example, Tah-one-lat-clah—means "Fire Mountain" and Louwalaclough means "Smoking Mountain." Their legends are included, along with a beautiful pictorial record, in Williams' *Mount St. Helens: A Changing Landscape* (1980). Fascinating accounts of the eruptions during the 1800s by personnel at military forts in the area, missionaries, and early settlers and explorers are nicely summarized in Harris' *Fire Mountains of the West* (1988). The mountain was named for Lord St. Helens (Alleyne Fitzherbert) by his friend Captain George Vancouver during his exploration of the Pacific coast in 1792 for the British government.

Spirit Lake, at the northern base of the volcano (Fig. 4–8), was for many years a favorite retreat for thousands of people; to the Indians, it was "evil spirit lake." A more recent legend, dating back only to 1924, got its start when some miners claimed that they had killed a "large, hairy ape" on the rim of a canyon—now Ape Canyon—on the east flank of the mountain. The body reportedly tumbled into the canyon and they were unable to produce the evidence. Nevertheless, it was enough to give birth to "Bigfoot," the elusive giant who will probably never be apprehended!

The hairy figment of imagination also gave its name to Ape Glacier and Ape Cave, both on the southern flank of the mountain. The cave is actually a lava tube similar to Thurston Lava Tube in Hawaii Volcanoes National Park. Here also, the tube is in a basalt flow, and it was similarly formed when lava drained out from under the solidified crust, about 1900 years ago. Ape Cave has the distinction of being the second longest single passage lava tube in North America—about 12,500 feet (3788 m) long. Take reliable flashlights (preferably three sources of light) and a warm jacket for the 42°F (5.6°C) year-round temperature and the long underground trip.

The 1980 eruptions brought Mount St. Helens into the limelight and onto television screens throughout the country. Many articles and books have been written about its behavior, both recent and ancient; for an excellent summary, read Stephen Harris' *Fire Mountains of the West* (1988) and Pat Pringle's *Roadside Geology of Mount St. Helens National Volcanic Monument and Vicinity* (1993). Lipman and Mullineaux (1981) edited an impressive volume that contains significant studies by scientists in several fields. The article by Rosenbaum and Waitt on "Eyewitness accounts of the May 18 eruption" is particularly interesting—and frightening. These and other publications that describe the hazards of volcanic outbursts contribute much to human welfare. But without question, the greatest contribution was the 26-page booklet that appeared in 1978, in which Crandell and Mullineaux issued a warning that Mount St. Helens was soon to erupt. Without that warning, and the resulting preparedness plans, the loss of life would without doubt have been much greater.

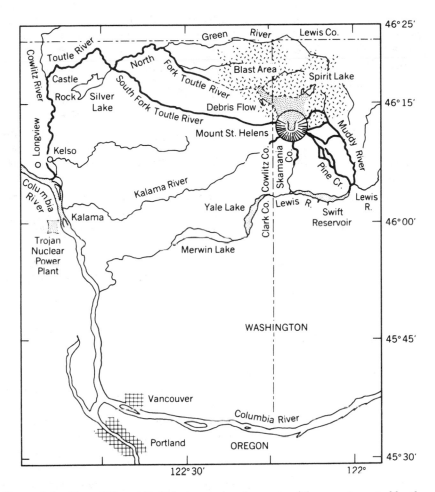

FIGURE 4–8 Map of Mount St. Helens area showing some of the areas impacted by the directed blast, areas covered by the debris avalanche (stipple pattern), and some important geographic features. Heavy lines indicate areas of mudflows and flooding in stream valleys. (Adopted from Brugman and Post, 1980)

Geologic Events (40,000–350 Years Ago)

Mount St. Helens is geologically very young, first making its appearance a mere 40,000 years ago. Dates of these older events come from radiocarbon dates of tree fragments and charcoal found in mudflow and pyroclastic deposits around the mountain flanks. Equivalent-age pyroclastic debris and lava are likely present near the vent, but are buried deep inside the mountain. The volcanic history at Mount St. Helens is divided into four major eruptive intervals or stages (each stage likely includes dozens of specific eruptions) separated by dormant intervals that range from 5000 to as much as 15,000 years! Thus, a Cascade volcano that has not erupted for thousands of years should, for safety reasons, be considered a dormant volcano and not an extinct volcano. Only after many tens of thousands of years should we consider the volcano as "safe."

The most recent eruptive stage, the Spirit Lake, includes the last 4000 years and at least 16 individual eruptions. Dormant intervals within the Spirit Lake Stage range from a few years to perhaps 200 years. Ash volumes produced during some of these eruptions greatly surpassed the 1980 eruption and were therefore larger eruptions. The eruption 3500 years ago was at least 13 times larger than the 1980 event and the eruption of 1480 was about six times larger!

Of the visible part of Mount St. Helen's cone, over 90 percent was added recently, within the past 400 years. The oldest visible rock on the volcano is a mere 2500 years old. Eruptions produce ash, coarser pyroclastic debris, and plug domes of mostly dacitic composition (intermediate to high silica content). In addition, lava flows also occurred and varied from andesite to basalt in composition. The 1900-year-old Cave Basalt flow is of particular interest because it contains significant lava tubes such as Ape Cave, Ole's Cave, and others.

Geologic Events (350 B.P.[2] to May 17, 1980)

Of particular significance to later events is the emplacement of the summit plug dome about 350 years ago that temporarily "corked" the top of the mountain and the Goat Rock dome that sealed a lateral vent on the north flank of the mountain between 1800 and 1857. The summit dome also transformed the jagged peak into the beautiful symmetric mountain that was Mount St. Helens prior to 1980 (Fig. 4–7). With the exception of a few minor steam events in 1898 and perhaps 1920, the stage was now set for a 123-year interval of quiet until the next pulse of new magma moved upward and the gas pressures in the shallow magma chamber exceeded the confining pressure of the many thousands of tons of overlying rock.

Precisely when the cone reached sufficient height for glaciers to form has not been determined but was undoubtedly during the very late Pleistocene. The area was extensively glaciated as large ice caps formed when alpine glaciers from highland areas merged. It is probable that glaciers disappeared or nearly disappeared from the flanks of Mount St. Helens in the warm interval following the Pleistocene and during the major cone-building, post-Pleistocene activity. Glaciers enlarged or re-formed sometime during the cold period known as the Neoglacial (Fig. 1–25), which began about 5500 years ago. An ice cap glacier covered the peak and 13 glaciers streamed down the valleys on all sides prior to the latest eruption. Almost 2 square miles (5.2 km²) of the mountain were covered with about 5 billion cubic feet (177 billion m³) of ice (Brugman and Post, 1980). The largest glaciers—Wish-

---

[2]B.P. = before present.

bone, Loowit, Lesch, and Forsyth—were on the north and northeast flanks where the direct effect of the sun was least. The two longest glaciers, Forsyth and Shoestring, extended down to about 4500 feet (1500 m). The large glaciers on the north flank, containing about 70 percent of the total ice volume on the mountain, were soon to meet a violent and sudden end when they became part of the world's largest historic landslide.

To the local people, and to many from outside, the Cascade volcanoes are there to admire and enjoy. The powder kegs beneath the volcanoes were generally ignored until 1969 when Dwight (Rocky) Crandell addressed an emergency preparedness conference in San Francisco, urging immediate action in assessing potential dangers. That same year a cooperative study involving the U.S. Geological Survey and the University of Washington was initiated. Then in 1978, Crandell and his Geological Survey co-worker Don Mullineaux issued a specific warning that Mount St. Helens was about due to erupt and that it was time to prepare for it. Thus the groundwork was laid for what soon became necessary—perhaps sooner than even they anticipated.

The monitoring system was in operation when, on March 20, 1980, the seismographs suddenly went wild, registering an earthquake of 4.2 magnitude on the Richter scale. By March 25 earthquakes were so numerous that the end of one seismic event could not be separated from the beginning of the next event! University of Washington seismologists alerted the U.S. Geological Survey Volcanic Hazards Team and the U.S. Forest Service in Vancouver, Washington.

On March 27, a thunderous explosion was heard; when the clouds lifted there was a large hole in the mountaintop and volcanic ash blackened the cone; the volcano was operational once again and its 123-year resting period was over. Numerous steam explosions, produced by water contacting hot rock or magma occurred in the month of April sending clouds of ash and debris skyward (Fig. 4–1). Scientists swooped down in helicopters following each of these explosions to sample the pyroclastic debris in hopes of learning the composition of the new magma below. If it were andesitic (lower silica content) then a less explosive eruption is likely— if dacitic or even richer in silica, then a more explosive eruption could be anticipated. The mountain was not yet going to give up its secrets. Only fragments of the summit dome, pulverized by the steam explosions, were produced in these preliminary events. The world would have to wait until May 18 to find out that gas-charged dacite awaited just below the surface in a *cryptodome,* a dome-shaped magma body hidden beneath the earth's surface (Fig. 4–9a).

Besides the usual short pulses of earthquake energy produced by rock fracturing, seismographs on March 31 began to record a rhythmic ground shaking known as *harmonic tremor,* a type of disturbance produced when fluids or gases flow in the subsurface—definitely not a good sign. This type of "vibration" is similar to what one would feel by grasping a fire hose or garden hose when water was flowing rapidly through it.

Another ominous sign occurred as magma continued to be forced into shallow locations beneath the north flank of the volcano. The north flank began to

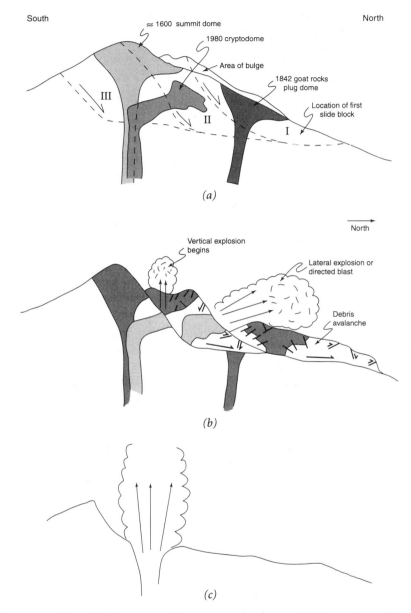

FIGURE 4–9   Generalized sequence of events on May 18, 1980. (*a*) Mountain profile and cross section immediately before the eruption; (*b*) approximately 52 seconds after the magnitude 5.1 earthquake; slide block I has exposed the cryptodome allowing the directed blast to form; a few seconds later the explosion cloud catches and passes the debris avalanche; vertical eruption cloud begins; (*c*) debris avalanche clears volcano flank, some of slide block blown apart as it passes over vent; vertical eruption cloud continues for the next 9 hours. (Sketches based on illustration in Tilling, 1985, and photographs taken by Gary Rosenquist)

swell or bulge (Fig. 4–9*a*) and by mid-April new cracks or *crevasses* appeared on the surface of the Forsyth Glacier on the north flank. The "bulge" continued to enlarge to an alarming 5 feet (1.5 m)/day until some parts were displaced as much as 450 feet (137 m) from their pre-eruption activity positions! The mountain at this stage was like a tire with a bulge that goes thump thump on the road—head for the nearest tire store—or in the case of a volcano—head for cover!

The preparedness plan was now in effect, and only scientists and certain officials were allowed in the red zone—the zone of highest danger. Local residents were evacuated, some with reluctance; one long-time resident, Harry Truman, a feisty 84-year-old operator of a lodge at Spirit Lake, decided to stay with "his" mountain. A group of people with cabins at Spirit Lake were escorted in on Saturday, May 17, by state highway patrolmen and sheriff's deputies to retrieve some of their belongings. A similar caravan was planned for Sunday as well as a local demonstration protesting government restrictions on access to the area. Some logging companies and businesses were irate concerning travel and access limitations. Had the eruption waited for 24 hours, until Monday morning, over a thousand people (mostly loggers) would have been in areas where survival was unlikely—or impossible. Fortunately, the mountain interceded before more people could place themselves in dangerous locations.

Geologic Events (May 18, 1980, 8:32 a.m.—present)

The Sunday morning was clear, ideal for picture-taking. Geologists Keith and Dorothy Stoffel were up above Mount St. Helens in a plane (celebrating Dorothy's birthday!), with cameras shuttering. Of particular interest was the huge bulge that was hanging precariously out over the north flank. They decided to make one last pass over the top of the mountain and then head east back to the airport at Yakima. Suddenly, at 8:32 a.m. (they learned the time later) they were transfixed in horror; directly below them the mountain began to shake like a bowl of Jello. The bulge shuddered violently back and forth, and about 10 seconds later the first of three huge chunks of mountain began to slide downward. The slide blocks disintegrated and quickly merged into a gigantic *debris avalanche* that gained momentum while rushing down the mountainside. Almost immediately, a huge mass of superheated steam and ash exploded from the wound (Fig. 4–10). The hot, searing *blast cloud* roared down the north flank, overtaking and then racing ahead of the slower-moving debris avalanche.

The sequence of events for the first minute is shown in the sketches in Figure 4–9. At exactly 8:32 a.m., a 5.1 magnitude earthquake shook the mountain, including the precariously balanced north flank bulge. The north flank collapsed and turned into a debris avalanche that was accelerated down the mountain flank by the force of gravity. As the north flank slid, it "uncorked" the cryptodome inside the mountain (Fig. 4–9*b*). The dissolved gases in the cryptodome magma were no longer confined, and they expanded instantaneously, similar to what would happen if the lid on a hot pressure cooker on a stove could suddenly be removed. If

FIGURE 4–10   Eyewitness photo of Mount St. Helens, taken from a small aircraft less than one minute after eruption began on May 18, 1980. (Photo by Keith and Dorothy Stoffel)

the confining pressure on water is instantaneously removed, it quickly flashes to steam, expanding its volume about 1100 times. A similar process, on a much smaller scale, occurs every time Old Faithful or some other geyser erupts.

The energy of the explosion cloud initially escaped out the volcano side creating a *directed blast* containing ash and pumice formed by the exploding cryptodome magma (Figs. 4–9b and 4–10). In Figure 4–9b, the debris avalanche is moving at an estimated 180 miles/hour (290 km/hr) and the directed blast is overtaking and passing the avalanche—eventually reaching a velocity of 670 miles/hour (108 km/hr) and locally supersonic speeds (greater than 735 miles/hour; 1183 km/hr)! Momentum of the ground-hugging death cloud was sufficient to carry it up and over high ridges and down into valleys. As the hot, searing gas cloud (temperatures as high as 680°F; 360°C) lost momentum, the cloud lifted, leaving a 230-square-mile (596-km²) area north of the mountain completely devastated. Trees were snapped off or uprooted as far as 16 miles (26 km) away, and fell with crowns away from and roots toward the peak. About 3.2 billion board feet of lumber lay scattered about like so many "pick-up-sticks" or spilled boxes of wooden matches (Fig. 4–11).

The "slower" debris avalanche (about 150 mph; 242 kph) followed behind, one lobe slamming into Spirit Lake at the base of the mountain, a second riding up to the top of Coldwater Ridge to the north, and a third veering westward down the North Fork of the Toutle River valley. The debris contained not only fragments of old rock (up to house-size boulders) but also significant amounts of fresh magma, steam, large chunks of glacial ice, and, after it passed through Spirit Lake, water incorporated from the lake. This is the earth's largest known historic landslide, covering 24 square miles (62 km²) and extending 15 miles (24 km) down valley. Debris about one mile downvalley from Spirit Lake is about 600 feet (183 m) thick and about 150 feet (46 m) thick near the western edge of the avalanche. The entire avalanche event was over in ten blinks of the eye—less than 10 minutes from the start of the initiating earthquake!

Avalanche debris entering Spirit Lake sent the lake water some 600 feet (183 m) high up the Harmony Valley to the north and deposited 295 feet (90 m) of debris in the lake bottom near the lake outlet. The new Spirit Lake dam is about 200 feet (60 m) higher than the previous debris dam that was formed by a similar process about 3500 years ago. The displaced lake rushed back into the new Spirit Lake basin carrying thousands of logs with it that had, just seconds before, been snapped off by the blast wave. Branches and bark were stripped from the trunks

FIGURE 4–11   Trees flattened by the directed blast. (Photo by E. Kiver)

FIGURE 4–12   View from Norway Pass showing the devastated north flank of Mount St. Helens; note the plug dome visible in crater and logs floating in Spirit Lake at base of mountain. (Photo by E. Kiver)

during their tumultuous trip into the Spirit Lake basin where today they form a "floating forest graveyard"—casualties of the big eruption (Fig. 4–12). Whims of the wind move this massive log mat from one part of the lake to another—like a ghost ship forever doomed to sail endlessly about. However, the log mat will not last forever; slowly the logs become water logged and sink to their watery graves. Even after nearly two decades and loss of over one-quarter of the floating forest, it is still an amazing sight—especially for those who get closer views by hiking the Harmony trail to the lake edge (Fig. 4–13).

The explosion was heard as far away as Missoula, Montana (375 miles; 600 km), but not in Portland, Oregon, 50 miles (80 km) away! Apparently, because sound waves were directed upward and were "bounced" (refracted) off the ionosphere in a complex way, rings of sound and quiet were created. Hikers on Mt. Rainier and Mt. Adams less than 50 miles (80 km) away watched the eruption but

FIGURE 4–13　Closeup view of "floating log graveyard" in Spirit Lake. (Photo by E. Kiver)

heard nothing while people on the Washington and Oregon coasts as much as 250 miles (400 km) away heard it loud and clear!

Very quickly after the directed blast the vent area was cleared and a vertical plume shot upward above the volcano. After 10 minutes the mixture of ash, gas, and fragmented rock from the cone rose 12 miles (19 km) above the volcano; in less than 15 minutes the mushroom cloud was 16 miles (25 km) high! The ash was carried north and then northeast, heavily blanketing areas in eastern Washington and with much thinner deposits reported as far east as Minnesota. Some downwind areas were pitch black by 10 a.m. activating switches that turned on streetlights that remained on for the rest of the day. The ash cloud reached Spokane, at the east edge of Washington State, by 2:00 p.m. and Yellowstone National Park by 10:15 p.m. (Harris, 1988). Those areas with even as little as 0.1 inch (0.25 cm) of ash were paralyzed for a week or so until the ash was cleaned from roofs and streets. The abrasive particles of volcanic glass wore at machinery causing at least one geologist to have a car engine overhauled!

The ash plume continued for 9 hours sending over 0.2 cubic miles (1 km^3) skyward. *Pyroclastic flows,* mixtures of hot gas and ash, "boiled" from the crater and collapsed from the ash plume during the eruption, spreading thick layers of frothy dacite ash on top of the debris avalanche deposits north of the mountain.

Potentially the most dangerous hazards associated with an erupting Cascade volcano are *mudflows,* mixtures of water and sediment whose consistency varies from that of a thick soup to concrete as it pours out of the cement mixer. These mixtures are denser and more forceful and destructive than a normal flood and eventually, like water, will concentrate in the valleys. Human settlements also tend to concentrate in the valleys. Even though mudflows are only "passing through," they are not welcome visitors! Some mudflows were generated immediately on the mountain flanks when water, freed by rapid melting of snow and ice, became mixed with ash and loose sediment. The destruction of about a dozen bridges in those first few minutes, created a situation that greatly hindered later search and rescue efforts. Lahar Viewpoint on the southeast flank of the mountain is an impressive stop for visitors who want to see and imagine the destructive power of a "small" mudflow.

The largest mudflow, however, emerged about 1:00 p.m. from the debris avalanche. Parts of the avalanche were saturated with water acquired when Spirit Lake was overrun and when additional water was generated by melting pieces of snow and glacial ice incorporated into the huge avalanche. The release of water was also aided by ground vibrations produced by the harmonic tremor that continued for nine hours on May 18. The mudflows destroyed houses, bridges, and logging operations as far as 45 miles (75 km) downstream along the Toutle and Cowlitz rivers. Water and mud temperatures along the lower Toutle River were still 85°F on May 19, reflecting the heat from magma incorporated into the debris avalanche.

The avalanche dammed tributary valleys along the Toutle valley and produced a number of debris-dammed lakes. As these lakes filled their basins during the next few months, the loose, highly-erodible material in the debris dams would occasionally cause the debris dams to fail and release more mudflows. The Corps of Engineers constructed small dams downstream to stop the smaller floods and mudflows and constructed outflow channels on some of the larger lakes. Failure of the debris dams on larger lakes would generate mudflows even larger than those on May 18. Special concern for the rising level of the largest lake, Spirit Lake, and its precarious avalanche dam, led to bringing a barge and large pumps in to keep water levels down. A tunnel was completed in 1985 that diverts water through a bedrock ridge to South Coldwater Creek and back into the Toutle River valley a few miles downstream.

The full extent of the loss of life will never be known. Countless trees, shrubs, flowers, songbirds, deer, elk, and bear were destroyed. Thirty-five people are known dead; 22 people who never returned are presumed dead. Most of the 57 deaths were from asphyxiation from the blast cloud—mercifully almost instantaneously. Some had climbed the beautiful mountain many times through the years and had returned to pay their last respects. Harry Truman had said, "That mountain is part of me"; now he is a part of the mountain, buried under the volcanic debris, which is perhaps the way he wanted it.

Geologist Dave Johnston was at his post about 5 miles (8 km) north of the ominous bulge on the mountain. At 8:32 a.m. he saw it all and radioed his last words: "Vancouver, Vancouver, this is it!" Unlike many who lost their lives, he was entirely aware of the danger, but he was determined to do his share in obtaining information that would improve our understanding of explosive volcanoes—perhaps it would help save lives in the future.

Dome Building Events

Since May 18, 1980, six smaller eruptions have occurred, numerous steam eruptions, and 17 episodes of dome building. Dacite magma low in dissolved gases forms a thick, pasty fluid that flows only with difficulty; hence it remains close to the vent and forms a steep-sided, dome-shaped landform called a *plug dome*.[3]

The few domes to build in the summer of 1980 temporarily "corked the bottle." However, gas pressure was still too high in the magma below, and these early domes were blown to bits. The dome that formed in October of 1980 persisted and grew in size with a new lobe added with each pulse of new dacite magma that rose into the crater. The last dome-building event occurred from October 21 to 25 of 1986—only steam eruptions from groundwater being heated by hot rock and magma just below the crater floor have occurred since. A spectacular view of the 810-foot (247 m) high, 2800 foot (854-m) diameter dome (Fig. 4–14) is available to those who like the long, steep trail to the crater rim from the south side of the mountain. The newly formed dome has accomplished what its predecessors did in the 1600s and 1800s—sealed the vent and set up conditions that could lead to another gigantic explosion.

FIGURE 4–14   The over 800-feet-high (245-m) plug dome towering over a group of geologists standing on the floor of Mount S. Helens' crater. (Photo by E. Kiver)

[3]A similar feature, in miniature, can be created by emptying a tube of toothpaste on a bathroom counter (kids—don't try this at home!)

The Future

The remarkable ability of nature to heal if left alone is vividly apparent as each year passes. The volcanic wounds are less apparent today as new plants and even forests are taking hold. Biologists have learned important lessons about how reestablishment of plants and animals occurs in natural settings.

Geologic lessons learned from the Mount St. Helens experience are also many and valuable. The need for strong science-based personnel in government agencies and our universities was apparent from day one when the first warning signs, the earthquakes, occurred. The expertise and rapid response of the U.S. Geological Survey was particularly valuable in efficient planning for contingencies and ultimately in greatly limiting the loss of lives and property.

Techniques for volcano monitoring such as the use of seismographs and ground deformation instrumentation were perfected, and valuable experience was acquired in dealing with eruptions near highly populated areas. This knowledge and experience will, of course, be applied to other volcanoes both in the Cascades and elsewhere in the world. Eyewitness accounts and examination of deposits shortly after their formation has provided new insights into volcanic processes.

Unlocking the secrets of a "lady with a past" as Crandell, Mullineaux, and other geologists have done over many years, enables us to anticipate the range of events that the future might bring. Previous eruptions, like the one 3500 years ago, was several times more violent than the 1980 event. Others, like the eruptions in the 1840s, were much less in intensity. Although dormant intervals can last for thousands of years, the recent geologic record indicates that since about 1400, eruptions are relatively frequent, occurring every 100 years or so. Likely, this recent pattern will continue.

By studying the location of earthquakes scientists have determined that the top of the magma body is located 4.3 miles (7 km) and the bottom is 6.8 miles (11 km) beneath the mountain. Further, the diameter of the magma chamber is nearly 1 mile (1.5 km) giving a total magma volume of about 1.7 cubic miles (7 km^3). Only about 0.048 cubic miles (0.2 km^3) of magma was erupted in 1980. Thus, about 35 times as much magma still remains in the now slumbering volcano!

If an eruption should occur on the south flank, the three large reservoirs along the Lewis River (Fig. 4–8) could be destroyed, resulting in colossal flooding in the Columbia River and the Portland, Oregon (a major population center) area. Or, a blast on the west flank would head toward the Trojan Nuclear Power Plant, which is located about 35 miles (56 km) west of Mount St. Helens (Fig. 4–8).

The remaining glaciers are small vestiges of their former selves. Loss of their nourishment areas on the missing mountain top has caused them to slow and stagnate, turning some into icefields rather than glaciers.

However, the profound topographic changes of 1980 have created new conditions for glacier development. The instantaneous formation of a 2000-foot (610-m)-high, north-facing crater wall at these high elevations causes more snow to accumulate than is melted during the summer season. Thus, a new glacier is forming as visitors and scientists look on.

The future of Mount St. Helens is impossible to predict and definitely beyond our control, but the future of the Mount St. Helens area is ours to determine. Unquestionably, the 1980 eruption was the most significant and unique natural event to take place in the United States in recent times. At present, it is a National Volcanic Area administered by the U.S. Forest Service rather than the National Park Service. Both agencies share the same problem—lack of adequate funding and staffing. Our parklands and other federal lands are a major source of national pride and deserve to be properly cared for and interpreted. Individuals, through their political representatives, can make this happen.

REFERENCES

Brugman, M.M., and Post, A., 1980, Effects of volcanism on the glaciers of Mount St. Helens. U.S. Geological Survey Circular 0850-D.

Crandell, D.R., and Mullineaux, D.R., 1978, Potential hazards from future eruptions of Mount St. Helens volcano, Washington: U.S. Geological Survey Bulletin 1383-C.

Harris, S.L., 1988, Fire mountains of the west: Mountain Press, Missoula, Montana, 379 p.

Lipman, P.W., and Mullineaux, D.R., eds., 1981, The 1980 eruptions of Mount St. Helens, Washington: U.S. Geological Survey Professional Paper 1250.

Pringle, P.T., 1993, Roadside geology of Mount St. Helens National Volcanic Monument and vicinity: Washington Division of Geology and Earth Resources, Information Circular 88, 120 p.

Tilling, R.I., 1985, Eruptions of Mount St. Helens: Past, present, and future: U.S. Geological Survey general interest publication, 47 p.

Williams, C., 1980, Mount St. Helens: A changing landscape: Graphic Arts Center Publishing, Portland, Oregon.

Park Address

Gifford Pinchot National Forest
500 W. 12th street
Vancouver, Washington 98660

LASSEN VOLCANIC NATIONAL PARK (CALIFORNIA)

Although geographically two states away from Mount St. Helens, it is appropriate to cover Lassen Peak next because it shares the distinction of being one of the only two Cascade volcanoes to erupt in the twentieth century. In hindsight, it is readily apparent that the warning signs of future eruptions were there, if only in the early 1900s scientists had known how to interpret them. The landforms, the geothermal features, nearby historic eruptions of steam and cinder—these were all clues to what the future held. The abundance of similar features and geologically recent events in many other parts of the Cascades are clues that exciting times lie ahead for volcano watchers. One or more of the Cascade volcanoes will erupt in the twenty-first century—which one(s) is (are) unknown.

Peter Lassen settled in the area in the 1830s and guided people on the 156Emigrant-Noble trail through the future park area on their way to the Sacramento Valley and the California goldfields. Some of the "49'ers" and settlers in August, 1850, report "smoke" (pyroclastic particles) and a persistent glow in the night sky for many nights from Cinder Cone in the northeast part of the park (Fig. 4–15). An excellent, but steep, trail leads to the double crater at the summit of the 600-foot-high (183-m) cone. Also in view from the summit are the 250-year-old basaltic-andesite flows that dammed the valley to form Snag and Butte lakes.

The main focus of the park, however, is the 10,457-foot-high (3188-m) Lassen Peak in the northwestern corner of the park (Fig. 4–15). The area was established as Lassen Peak Forest Preserve in 1905 and as two separate, unattached monuments (Lassen Peak and Cinder Cone) in 1907 by President Theodore Roosevelt. Strong support for national park status was generated when, on May 30, 1914, the quiescent volcano came to life with a violent explosion, similar to the one shown in Figure 4–16. Lassen erupted repeatedly in 1914, but it did not really blow its stack until May of 1915, when a 5-mile-high (8-km) column of volcanic ash and dust shot up into the sky.

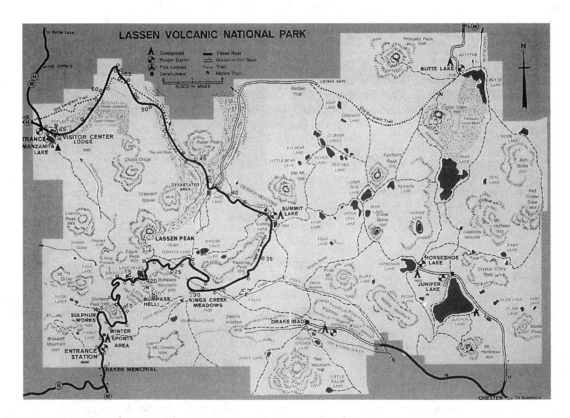

FIGURE 4–15   Map of Lassen Volcanic National Park. (National Park Service)

FIGURE 4–16   Lassen Peak in eruption, May 22, 1915. (Photo by National Park Service)

Regional Geology

Lassen is the southernmost of the series of large volcanoes along the Cascade crest and is located a short distance north of the massive granitic block that makes up the Sierra Nevada Province. Plate tectonics again proves to be a powerful tool to help answer major questions: (1) Why has magma generation been so prolific beneath the Cascade Mountains? (2) What geologic factors control volcano location there?

Magma generation, as elsewhere in the Cascade Mountains, is a consequence of subduction processes. Water is incorporated into the sediments and dragged deep into the lithosphere, where it lowers the melting point of the hot rock. Churning of the upper mantle as plates grind by one another in the subduction zone and localized tension, crustal thinning, and graben formation in the Eastern Cascades brings hot mantle rock closer to the surface where lowered pressure allows partial melting and formation of basaltic magma to occur. *Underplating* of the crust by basalt magma, which in turn partially melts the overlying crustal rocks, generates more silica-rich magma and produces the andesites, dacites, and rhyodacites at Lassen and elsewhere (Fig. 4–6). The squeezing of the subduction "vise" and associated Cascade volcanism terminates a short distance to the south where the San Andreas transform (strike-slip) fault swings seaward along the Mendocino fracture zone.

The location of major faults strongly influences volcano location. Again plate tectonics provides some reasonable ideas. According to Argus and Gordon (1991), the Sierra Nevada is on a small subplate that is rotating northwestward relative to North America at about 0.03 inches (9 mm) per year. As a result of this lateral slippage of plates, a broad zone of *normal,* or *pull-apart,* faults (the Walker Lane Fault Zone) occurs between the Sierra Nevada and Basin and Range Province to the east. A branch of this pull-apart zone extends northward through the Lassen Peak area and along the High Cascades in Oregon and southern Washington. Release of pressure associated with this zone of extension contributes to the magma production beneath the Cascades.

Geologic Sequence—Formation of Mt. Tehama (Brokeoff) Volcano

Activation of plate movements important to the present geology began in the late Cenozoic, first affecting the Lassen Peak area about 7 Ma (late Miocene) (Guffanti and others, 1990) when faults opened up and lava buried Mesozoic rocks closely related to the nearby Sierra Nevada Province. Regional volcanism in the Cascades is mostly characterized by thousands of short-lived basalt-to-andesite eruptions with development of a few long-lived volcanic centers like that found at Lassen. An extensive basalt plateau built up during the Pliocene. Pyroclastic debris first appears in the rock record about 3 Ma, signaling the appearance of the first of four stratovolcano complexes that would build in the Lassen area (Clynne, 1990). A series of normal faults within the Walker Lane pull-apart zone caused large crustal blocks to drop and form a series of grabens along its length, including the Hat Creek Graben that provided the local pathways for magma to flow to the surface. The Hat Creek Graben extends southward from the Pit River, beneath Lassen Volcano, and reappears south of the peak in the Lake Almanor area. Earthquakes and very late Pleistocene or Holocene age lava and fault scarps are abundant within the graben.

Lassen Peak's immediate predecessor was a stratovolcano called Mt. Tehama or Brokeoff Volcano. According to Clynne (1990) and other researchers, Mt. Tehama began to form about 600,000 years ago (Table 4–2). Its development, as well as that of the three other nearby extinct stratocones, is characterized by a trend toward increasing silica content of magma with time and a dramatic change of eruptive behavior—from the relatively quiet eruptions characteristic of lava shields to more explosive stratocone eruptions, and finally to dozens of pyroclastic flows and plug dome eruptions on the north flank of Mt. Tehama. Based on fieldwork and radiometric dating of volcanic units, Clynne (1990) divided the volcanism into three

TABLE 4–2   Geologic Development of Mt. Tehama (Brokeoff)-Lassen Peak Volcanoes

STAGE	AGE (× 1000 YEARS)	VOLCANIC PRODUCTS	VOLCANIC EVENTS
	(1914–1921 A.D.)	Dacite, ash	Explosive eruptions, mudflow
	1.2	Dacite	Choas Crags Domes emplaced
	29	Dacite	Lassen dome emplaced
III	400–30	Dacite, rhyodacite	Reading Peak, Loomis Peak, Bumpass Mt, numerous others
			Erosional caldera forms
	400	Rhyolite, ash flows	Small collapse caldera forms
II	470–400	Andesite, silicic andesite	Thick, cone-building lava on Tehama-Brokeoff cone
I	600–470	Basaltic-andesite, andesite, pyroclastic layers	Mt. Tehama-Brokeoff stratovolcano forms on top of lava shield

stages. Stage I basalt and basaltic-andesite lavas built the Mt. Tehama stratocone on top of a lava shield from 600,000 to 470,000 years ago; thick stage II andesites and siliceous andesites continued to build Mt. Tehama from 470,000 to 400,000 years ago; and the eruption of rhyolitic magma about 400,000 years ago signaled a major change in the mode of eruption as the more silica-rich magmas (dacite and rhyolite) of the present stage III eruptions began (Table 4–2). Reading Peak, Loomis Peak, Eagle Peak, Bumpass Mountain, Mt. Helen, Chaos Crags, and Lassen Peak are all stage III plug domes. Clynne (1990) reports that silica content of the lava increased from about 55 percent during stage I to 75 percent during stage III.

Although Mt. Tehama dominated the local volcanic scene from 600,000 to 400,000 years ago, elsewhere in the region and in the park Pleistocene-age eruptions of basalt, from magma chambers separate from Mt. Tehama, built a number of prominent shield volcanoes. Raker Peak, Prospect Peak, Red Mountain, and Mt. Harkness are examples. A trail to the lookout station on Prospect Peak leads up the shield flanks to the small cinder cone at its top, where excellent views of Lassen and the area devastated on the north flank by its 1915 eruptions are visible.

Mt. Tehama's Missing Cone

It has long been recognized that much of Mt. Tehama is gone and that only remnants of its outer flanks remain. Surrounding peaks such as Mt. Diller, Brokeoff Mountain, and Vulcan's Castle have layers of stage I and stage II lava, breccia, and ash flows that slope westward. Diamond Peak and Mount Conard have equivalent layers sloping eastward. The layers project upward to the sky where an 11,000-foot-tall (3350-m) peak once stood (Williams, 1932). Initially it was assumed that a large eruption followed by inward collapse of the summit into the partially vacated magma chamber below removed much of the cone, similar to the events recorded at Mt. Mazama (Crater Lake). However, remnants of the thick ash layer that would be produced by such an eruption have yet to be found. The thickness and extent of the ash produced by the 400,000-year-old eruption is indicative of a small caldera collapse event—not the massive Mt. Mazama type of collapse. Clynne suggests a small 3.7-mile-diameter (6-km) collapse caldera.

A more likely explanation for most of Tehama's missing cone is that the long-lived hydrothermal activity in the vent area made the cone more vulnerable to erosion by greatly weakening the volcanic rocks by changing their mineralogy to clay, opal, and other minerals more susceptible to erosion. Streams and glaciers attacked these weaker minerals vigorously and created a low area ringed by mountain peaks—an *erosional caldera* rather than a *collapse caldera*. A visit to Sulphur Works, where Brokeoff Volcano's highly eroded vent is located, is a visual and odoriferous experience. It provides a first-hand opportunity to observe what sulfur-rich, acid gases and fluids can do to once solid rock.

Lassen Peak Formation

One of the more important events to affect the present-day park occurred about 29,000 years ago—the emplacement of the Lassen Peak dome. A series of earthquakes no doubt shook the area as the rising dacite magma cracked and pushed the older rocks apart. The thick lava emerged and piled up around the vent, unable to flow far from its source because of its high viscosity. The outer solidified carapace cracked as the dome inflated and red-hot blocks of dacite tumbled and ricocheted down its flanks to form rubbly talus slopes at its base, much as was observed on the carefully studied dacite plug dome that formed in Mount St. Helen's crater from 1980 to 1986. At night, the cascading blocks and freshly exposed lava in the plug dome would display an eerie red glow.

A series of eruptions likely occurred over a period of years as the dome grew in size. At one point the dome reached a height of 1000 feet (305 m), as tall as most plug domes grow. However, Lassen is different—it continued its growth to 1800 feet (550 m), making it the world's tallest plug dome! Also different from most plug domes is the vent system that penetrated the dome and formed a small, quarter-mile-diameter crater at the summit—a feature that would later be obliterated during the 1914 and 1915 eruptions.

Lassen's Glaciers

The abundance of talus around the base of the mountain today and the lack of large cirques cut into the plug dome led early geologists to suggest that Lassen Peak was quite young, perhaps postglacial in age. However, the discovery of pieces of Lassen's distinctive dacite (Gerstal, 1989) in moraines of latest Pleistocene age (Lassen Peak episode of glaciation) in the surrounding valleys indicates that the dome was in place about 25,000 years ago when glaciers were building these moraines. Radiometric dates indicate that a dacite flow that underlies Lassen Peak is about 31,000 years old; thus Lassen Peak dome was emplaced sometime between these events, perhaps about 29,000 years ago (Gerstal and Clynne, 1995). Glaciers removed much of the talus formed during dome emplacement but postglacial frost action at these high elevations has pried numerous blocks from the highly fractured, steep-sided dome forming new talus slopes below.

Glacier "fingerprints" are almost everywhere around Lassen Peak. Striking examples of *glacial striations* (Fig. 4–17) and *glacial grooves* are abundant in the "high country." The directions of ice flow indicated by abundant striations on bedrock at the base of the mountain suggest an ice cap source for the radiating system of glaciers. High on various peaks, Loomis, Reading, and others, there are many cirques and cirque lakes (*tarns*). Ice from these many sources coalesced to form the Lassen ice cap. Valley glaciers from the ice cap spilled down surrounding valleys and built *terminal moraines* as much as 5 miles (8 km) away. Look for the excellent scratched and polished dacite surfaces at the summit trailhead and along

FIGURE 4–17   Glacially polished and striated bedrock surface on south flank of Lassen plug dome. (Photo by E. Kiver)

the trail to Bumpass Hell (Fig. 4–17). A low sun angle enhances their appearance and improves photographs. It is clear that glaciers were once widespread; although there are many small icefields, no true glaciers are here at the present time.

Lassen's Latest Eruptions

The evidence discovered thus far suggests that after the initial eruptions, the Lassen Peak dome remained quiet for 29,000 years. Most plug domes experience a single eruptive episode, the massive plug usually sealing the vent, causing later activity to shift to a new location where the rift is weaker. Indeed, the five closely spaced plug domes that make up nearby Chaos Crags were emplaced during the past 1100 years along a rift trending northward from Lassen Peak.

In the very recent geologic past, a series of rockfall avalanches from Chaos Crags, perhaps triggered by a steam explosion or earthquake, roared downslope and traveled over 3 miles (4.8 km) into the Manzanita Creek valley where it formed the dam that holds in Manzanita Lake near the north entrance. Submerged trees in the lake are 300 years old, effectively dating the age of the spectacular avalanche. The descending debris from Chaos Crags likely trapped air beneath and rode the nearly frictionless "air cushion" at speeds up to 100 miles/hour (160 km/hr). The avalanche rode partly up Table Mountain on the opposite valley wall and was deflected westward down the Manzanita Creek valley. The impressive mass of loose dacite boulders and finer pulverized rock along the road from the north entrance is descriptively named the Chaos Jumbles. Large quantities of steam were still rising from Chaos Crags as recently as 1857 (Harris, 1988), a reminder that the "fires" are still there. More eruptions and devastating avalanches are likely in the future, as first recognized by Crandell and others (1974).

On the east side of the park in the Central Plateau lies Cinder Cone, an obvious recent addition to the landscape (Fig. 4–18). The cinder slopes are nearly bar-

FIGURE 4–18   Cinder Cone, about 10 miles (16 km) northeast of Lassen Peak. Eruptions occurred as late as 1851. (Photo by National Park Service)

ren of vegetation as are the lava flows that have blocked drainages to create Butte and Snag lakes. Much of the activity occurred about 250 years ago, but the most recent "fireworks" occurred in 1850–1851 as described by a number of observers. The basaltic andesite at Cinder Cone is unusual in that the mineral quartz occurs in the lava as distinct mineral grains visible to the unaided eye. Such a mixture is chemically incompatible in a single magma chamber and may reflect mixing of magmas of two widely different compositions or contamination of magma as it rose through quartz-rich materials on its way to the surface.

Lassen Eruptions, 1914–1921

History was soon to be made as an infusion of magma beneath Lassen Peak began to punch its way through the old vent system. On May 30, 1914, local resident Bert McKenzie was looking directly at the peak when a dark cloud of steam and ash belched from the summit (see Fig. 4–15 for a similar-appearing eruption). A 7-year-long spectacle had begun.

Government and university geologists were few in number at the time, and no professional geologist was able to observe all of the eruptive events that followed. Fortunately, occasional observations of geologists and excellent records kept by local residents and U.S. Forest Service personnel allow the events of those 7 years to be satisfactorily interpreted. Summaries of eyewitness accounts are skillfully presented by Harris (1988), Hill (1970), and Loomis (1971). The outstanding photographs taken by local resident and amateur photographer Benjamin Loomis provide invaluable pictorial documentation. Some of his important photographs are published in a small pamphlet (Loomis, 1971) available at the visitors center.

Amazingly, no deaths or serious injuries resulted from the eruptions although local millworker Lance Graham received a broken collar bone when bowling-ball size fragments were blown from the crater as he stood on its rim! Curiosity had compelled three men to have a look down into the crater on June 14, 1915. Ominous noises caused two of the three to withdraw rather hurriedly, but Lance was apparently transfixed by the unearthly sounds. The blast occurred as he started off the rim, and it was then that volcanic bombs rained down upon him.

J.S. Diller, geologist with the U.S. Geological Survey, first climbed to the crater rim in 1883. His *Lassen Peak Folio*, published in 1895, includes a geologic map that serves as the basis for "before and after" pictures of what happened during the 1914–1921 eruptions. By then Diller was in his sixties, but on hearing that Lassen was on a rampage, he lost little time in getting to the scene. Again he climbed to the crater rim, but it was not the same one; Lassen had blown out through a new vent this time. His studies during the next several years were published in part, but his long report was still unfinished at the time of his death in 1928. Partly for this reason, Howel Williams, pioneer Cascade volcanologist, undertook his investigations which were published in 1932. Later, Macdonald (1966) prepared a more detailed geologic map as a part of his regional studies of the Cascades. Most recently,

Michael Clynne has improved greatly on our detailed knowledge of the volcanic history of the Lassen Peak area.

The May 30, 1914, eruption was the first of 170 relatively small but significant documented outbursts during the first year. On May 19, 1915, lava welling up into the crater reached the rim and oozed down the southwest and northeast flanks of the mountain (Fig. 4–19). Nighttime eyewitnesses describe a deep-red glow reflected on the steam cloud over the crater and a dark mass that resembled a "titanic slag pot" that glowed red when the thin lava crust cracked and red hot lava chunks rolled down the mountain exposing the hot interior of the thick, pasty lava.

Deep snowbanks on the northeast side of the mountain were almost instantaneously melted as lava and hot debris contacted the snow. The resulting flush of water mixed rapidly with the combined eruption debris, loose rock and soil, and formed a mudflow that rushed down Lassen's northeast flank. Trees were bowled over and pickup-truck-size and larger blocks of hot dacite lava were carried along in and on top of the powerful flow of water and debris that likely reached speeds of 35 miles per hour (56 km/hr) on steeper slopes. One lava boulder—later named Hot Rock—weighing about 30 tons, was rafted 3 miles (4.8 km) from the crater.

FIGURE 4–19   Looking down into Lassen's newest craters. Note the lava-filled 1915 crater with a lava tongue on the southwest flank of the mountain and the smaller 1917 crater in the foreground. (Photo by National Park Service)

About 40 hours after the mudflow when people investigated, water contacting the lava blocks would still sizzle and boil from the high rock temperatures. The roar of the mudflow down the Lost Creek and Hat Creek drainages woke ranchers, and in two cases dog owner's lives were saved when their pets warned them of the impending disaster (Hill, 1970).

With the vent partially sealed by lava, gas pressure increased and 3 days later the most violent of Lassen's outbursts sent a "mushroom cloud" about 5 miles (8 km) above the vent and showered volcanic ash over a large area of northern California and Nevada. This eruption generated a *nuee ardente* (fiery cloud of hot ash and gas) that roared down the northeast flank across the mudflow destruction area snapping more trees off (some with 6-foot-diameter; 1.8-m trunks!) and leaving them with their tops pointing away from the mountain (Fig. 4–20), a scenario that was repeated at a much larger scale at Mount St. Helens during the 1980 eruption. Timing is everything—Benjamin Loomis had the good fortune to photograph the mudflow damage and leave before the "great hot blast" rolled across the area where he had stood a few hours earlier. He also had the good fortune to stop along the trail at a site where he could witness the giant explosion and not be in danger. One misfortune that he had to endure is similar to what every photographer has had happen on more than one occasion, he was out of film!

The Devastated Area is there for all to see. Revegetation at high elevation is very slow due to the rigorous alpine climate and poor soils. Decomposed trees facing away from the peak are still recognizable (Fig. 4–20) and an even-age stand of trees (trees are same age) covers the mudflow-impacted area at lower elevations.

FIGURE 4–20   Devastated area on north flank of Lassen Peak. Partially decomposed tree in foreground was blown over by the great hot blast on May 22, 1915. (Photo by E. Kiver)

Take the interpretive trail through the Devastated Area—gaze up at the mountain and imagine the consequences, if you dare, of a similar sequence of events occurring while you are standing there! Most of the lessons of what an erupting Cascade volcano can do were there for all to see—their reenactment, on a much larger scale, would again play out some 65 years later at Mount St. Helens.

Lassen's Thermal Areas

Momentarily, the thermal areas and abundant earthquakes provide the only direct evidence that the Lassen volcanic area is not dead. Surface water from rain and snow infiltrates through the soil and rock and recharges the groundwater reservoir below. Hot rock and perhaps magma at depth heat the groundwater and drive steam and sulfide gases toward the surface. The vapor emerges in fumaroles or condenses near the surface and then emerges as the acid-sulfate springs at Sulphur Works, Bumpass Hell, Devil's Kitchen, Boiling Springs, and at other less accessible areas in the park. These "mini-Yellowstone" areas contain such features as *solfataras* (vents or fumaroles that emit sulfur fumes), *mudpots, mud volcanoes,* and *hot springs.* Bumpass Hell, at the end of a "mile-long" trail, shows most of the above features in an impressive manner, and the appropriateness of the terms "Big Boiler" and "Steam Engine" is immediately apparent. Kendall Vanhook Bumpass discovered the area in the 1860s and scalded a foot when he broke through a thin crust. When asked where he had been his reply was "In hell." The name caught on. On a return trip he was not as fortunate—again he broke through the thin mud crust—this time it was necessary to amputate his leg! Visitors who stay on the marked trail will be perfectly safe. Those who stray from the trail could suffer a similar or worse fate!

Hydrothermal alteration is widespread in and around the thermal areas. In addition to the yellow and orange colors produced by sulfur, the whites, grays, and other colors that are associated with opal, tridymite, kaolinite, pyrite, and clay minerals are also produced in this highly acid environment. Algae and bacteria colonies also lend color to these unique thermal areas. Geochemical studies indicate that where the acidity of the water is relatively high, nearly pure opal is formed; lower acidity and lower temperatures produce mainly kaolinite. The obvious weakening of the rock by hydrothermal activity, and the location of long-lived thermal areas at or near the center of old Brokeoff Volcano, support the hypothesis discussed earlier that an erosional caldera accounts for the missing summit and core of Brokeoff Volcano.

Lassen's Future

Plate tectonic movements continue unabated. Subduction produces the partial melting to form a new supply of magma, and shifting of microplates continues to stretch the area along the Walker Lane fault zone providing pathways for magma to reach the earth's surface. The rate of magma production and rate of its upward movement is unknown. Also unknown is which of the numerous vent systems will be next—Lassen, Chaos Crags, Cinder Cone, or someplace brand new?

From past experience with active volcanoes we can be reasonably assured that seismic and other monitoring techniques utilized by the U.S. Geological Survey and state agencies will provide ample warning of an impending eruption. Less predictable are the occasional earthquakes and steam explosions that occur as groundwater and hot subsurface rocks interact. Combined with the steep, unstable slopes of Lassen, and especially Chaos Crags, destructive rock avalanches could occur at any time. Numerous small earthquakes occur annually and occasional earthquakes "rattle" the countryside, like the magnitude 5.5 Lassen Peak earthquake in 1950.

As always it is a good idea to visit these unstable areas to see what the "before" looks like. Take pictures, hike the trails, enjoy the gifts of nature, and contemplate your surroundings. Most important, do what you can to ensure that *every* generation can have these same experiences.

REFERENCES

Argus, D.F., and Gordon, R.G., 1991, Current Sierra Nevada–North American motion from very long baseline interferometry: implications for the kinematics of the western United States: Geology, v. 19, p. 1085–1088.

Clynne, M.A., 1990, Stratigraphic, lithologic, and major element geochemical constraints on magmatic evolution at Lassen volcanic center, California: Journal of Geophysical Research, v. 95, no. B12, p. 19,651–19,669.

Crandell, D.R., Mullineaux, D.R., Sigafoos, R.S., and Rubin, M., 1974, Chaos Crags eruptions and rockfall-avalanches, Lassen Volcanic National Park, California: Journal of Research, U.S. Geological Survey, v. 2, no. 1, p. 49–59.

Gerstal, W.J., 1989, Glacial chronology and the relationship to volcanic stratigraphy in the Hat and Lost Creek drainages, Lassen Volcanic National Park, California: MSc thesis, Humboldt State University, 89 p.

Gerstal, W.J., and Clynne, M.A., 1995, Volcanic and glacial stratigraphy in the Hat and Lost Creek drainages and the age of Lassen Peak: in, Quaternary geology along the boundary between the Modoc Plateau, southern Cascade Mountains, and northern Sierra Nevada: Friends of the Pleistocene, 1995 Pacific Cell field trip, Appendix 2-3a, San Francisco, CA.

Guffanti, M., Clynne, M.A., Smith, J.G., Muffler, L.J.P., and Bullen, T.D., 1990, Late Cenozoic volcanism, subduction, and extension in the Lassen region of California, southern Cascade Range: Journal of Geophysical Research, v. 95, no. B12, p. 19,453–19,464.

Harris, S.L., 1988, Fire mountains of the west: Mountain Press, Missoula, 379 p.

Hill, M.R., 1970, "Mount Lassen is in eruption and there is no mistake about that": Mineral Information Service, California Division of Mines and Geology, v. 23, no. 11, p. 211–224.

Loomis, B.F., 1971, Eruptions of Lassen Peak: Loomis Museum Association, Mineral California, third revised edition, 96 p.

Macdonald, G.A., 1966, Geology of the Cascade Range and Modoc Plateau: in, Geology of northern California, California Division of Mines and Geology Bulletin 190, p. 65–96.

Williams, H., 1932, Geology of the Lassen Volcanic National Park, California, University of California Publications in Geological Science, v. 21, no. 8, p. 195–385.

Park Address

Lassen Volcanic National Park
Mineral, CA 96063

CRATER LAKE NATIONAL PARK (OREGON)

History

All who encounter the mountain called Mazama with its huge crystal-clear lake at its summit agree this is indeed a very special place. The magnificent lake and surrounding area make up the 183,224-acre Crater Lake National Park in Oregon's southern Cascade Range (Fig. 4–21). Geologists and native Americans further agree that such a special place must have a special origin. However, although both groups broadly agree that Mazama's top collapsed inward, the details and reasons given are quite different.

According to the legends of the Makalak tribe of the Klamath Indians, the "Lake of the Blue Waters" was a battleground between Llao, Chief of the Below World, who resides deep in Mazama's interior, and Skell, Chief of the Above World, who resides in Mt. Shasta in northern California. The battle produced huge volumes of ash and many days of darkness before Llao was defeated. Skell pushed the mountain top down on Llao, presumably imprisoning him forever.

Gold prospector John Wesley Hillman followed his mule up a gentle ridge in 1853 and was amazed to find himself standing on the edge of a 2000-foot-deep (610-m) hole in the mountain top with a 5-by-6-mile (8-by-10 km)-diameter lake of indescribable beauty gracing its interior (Fig. 4–22). Hillman called it Deep Blue Lake. However, Crater Lake, suggested in 1869 by local newspaperman John Sutton, became its official name. The extreme depth and clarity of the lake enables the water to absorb all colors of light—only the deepest blue color is reflected back to the surface.

A Kansas schoolboy, William Gladstone Steel, read about this fabulous lake in a newspaper used to wrap his lunch. His fascination with the area continued and, a few years later, after moving to Oregon, he spent 13 years looking for the extraordinary lost lake. He devoted 17 years of effort finally convincing Congress in 1902 to establish the nation's fifth national park at Crater Lake. Although the lake had a name, no one had bothered to name what was left of the former 12,000-foot-high (3660-m) mountain until 1896 when a Portland-based mountaineering club, the Mazamas, christened the mountain with their name. Fish were stocked beginning in 1888, forever changing the original population of microorganisms that in-

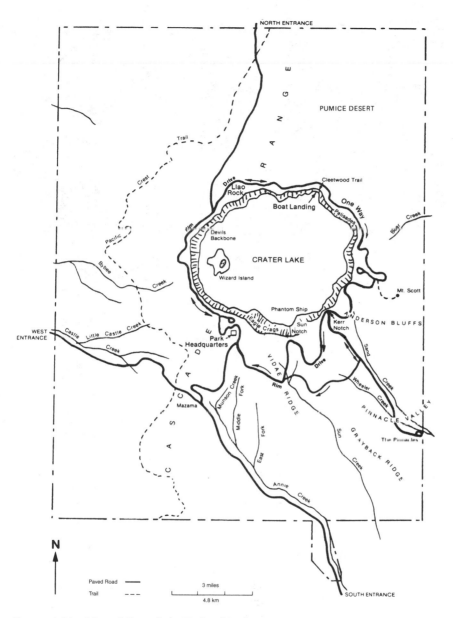

F<small>IGURE</small> 4–21   Map of Crater Lake National Park.

FIGURE 4–22   Beheaded Mt. Mazama Volcano with Crater Lake partly filling the collapse caldera. Wizard Island, the cinder cone rising above the lake waters, formed shortly after the catastrophic eruptions that occurred 6800 years ago. (Photo by E. Kiver)

habited the lake. Today land-locked salmon (kokonee) and rainbow trout, some as long as 27 inches (69 cm) swim in its cold waters.

Scientific investigations began in 1886 when Clarence Dutton of the newly formed U.S. Geological Survey led an exploration party, accompanied by William Steel, to Mt. Mazama. Carrying a 26-foot-long (8-m), one-half-ton boat named the *Cleetwood* over and down the steep walls was not an easy task! Using rolls of piano wire, they determined that the lake was nearly 2000 feet (610 m) deep! More sophisticated instruments later determined its deepest point as 1932 feet (589 m), showing it to be the second deepest lake in North America, second only to the Great Slave Lake in Canada.

Howel Williams's studies of Crater Lake in the 1930s and 1940s enabled him to introduce the concept that volcanic depressions many times larger than their associated feeder vent systems should be called *calderas* (caldera, Spanish for large kettle) rather than craters (Williams, 1942). Additionally, his studies of the area also demonstrated that the missing top of the mountain was not blown off the mountain to form an *explosion caldera* as most geologists believed, but rather the top collapsed inward as the magma chamber below was partially evacuated during the violent eruption that occurred about 6800 years ago. More recently, studies by Charles Bacon (1983, 1988) have modified some of Williams's less important interpretations, but the "big picture" of a mountain top collapsing into a partially emptied magma chamber below still stands. Additional studies of the lake sediments and features on the caldera floor using deep-diving research submarines, sediment cores, and geophysical techniques such as *seismic reflection* (shock waves bounce off buried layers, faults, and erosion surfaces) have proven invaluable in unravelling the recent history of the volcano (Nelson and others, 1995).

Geomorphic Setting

What is left of Mt. Mazama extends from an elevation of 4700 feet (1432 m) at the base of the cone to 7000–8000 feet (2134–2439 m) at the caldera rim. The lake elevation is 6176 feet (1883 m) and it occupies some 27 square miles (70 km²). Mt. Scott, a parasitic cone formed early in Mt. Mazama's history, rises to 8926 feet (2721 m) into the alpine zone and provides a commanding view for those who climb the steep but well-maintained trail to its summit (Fig. 4–21).

The lake has no outlet and its level is controlled by the balance between evaporation and precipitation. Lake levels only vary from 1 to 3 feet (30–90 cm) during a year. Nearly 50 feet (15 m) of snow each winter closes the scenic rim drive from mid-October to early July, but the west entrance to the Rim Village (Fig. 4–21) remains open all year. The average surface temperature is 38°F (3.3°C) and the lake seldom freezes. Drowned hot springs on the caldera floor also contribute to maintaining its ice-free condition during winter months.

Mt. Mazama, Building Mode

Mt. Mazama is located along the eastern border of the Cascade Mountains in what is known as the High, or Eastern, Cascades where late Tertiary and Pleistocene volcanic activity over the past 7 million years has filled a large graben (Orr and others, 1992). Subduction continues as plates grind away beneath the Cascades, generating yet more magma that continually rises slowly toward the surface (Fig. 4–6). Geologically recent eruptions account for the magnificent line of volcanoes that stretch from northern California to southern British Columbia along the Cascade crest (Harris, 1988). Eruptions are infrequent by human time standards but are persistent and expectable when viewed through geologic time.

A long-lived volcanic vent system developed in the Mt. Mazama area about 400,000 years ago (Fig. 4–23) building a complex of stratovolcanoes atop older shield cones and basalt flows. Mt. Scott (420,000 years old) and the Phantom Ship vents (400,000 years old) are mostly andesitic in composition and contain some of the oldest exposed rocks on Mt. Mazama. A series of overlapping stratocones likely gave an impressive but rounded profile to Mt. Mazama. At its peak of building, perhaps 70,000 years ago, glacier-clad Mazama likely stood over 12,000 feet (3660 m) in elevation, making it Oregon's tallest and most impressive mountain.

Glacier and mudflow deposits sandwiched between layers of andesite and pyroclastic deposits stand as mute evidence of the chaotic fire-and-ice battles that

6800 YEARS AGO — Top of Mt. Mazama collapses into magma chamber below, massive landslides enlarge caldera, post-cataclysmic eruption lava flows and cinder cones like Wizard Island form before lake reaches present level.

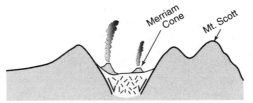

6800 YEARS AGO — Cataclysmic eruption begins, tremendous outpouring of magma as airborne ash and pyroclastic flows partially evacuates magma chamber below.

HOLOCENE — Llao Rock eruption about 7000 years ago breaks 20,000-year-long dormant interval.

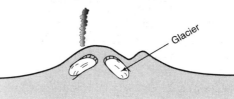

PLEISTOCENE (400,000 to 30,000 years ago) — Mt. Mazama builds by a series of lava flows and plug domes, mostly on flanks; dacite and rhyodacite near summit, basalt on flanks.

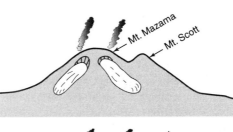

PLEISTOCENE (~ 400,000 YEARS AGO) — Andesitic Phantom Cone and Mr. Scott form.

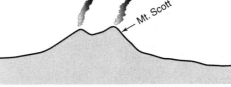

PLEISTOCENE (EARLY PLEISTOCENE) – Basaltic eruptions, shield volcanoes form.

FIGURE 4–23   Generalized evolution of Mt. Mazama Volcano

raged during the relatively short history of this remarkable mountain. Soil profiles further record long intervals of weathering—episodes during which forests covered the mountain's flanks, only to be obliterated during succeeding eruptions.

Emplacement of silica-rich lava (dacite and rhyodacite) as plug domes peripheral to the summit and eruptions of silica-poor basalt and basaltic-andesite magma as cinder and shield cones lower on the mountain flank ended about 30,000 years ago. A 20,000-year-long dormant interval followed. However, the ring of earlier formed plug domes followed a major zone of weakness that would later localize the mountain collapse when the "mother-of-all-Cascade eruptions" occurred.

Mt. Mazama, Self-Destruct Mode

Signs of restlessness began about 7000 years ago when an explosion crater formed in the Llao Rock area on the northwest side of Mazama's cone (Fig. 4–24). High-

FIGURE 4–24 Llao Rock on the north rim of Crater Lake caldera; explosion crater formed on flank of old Mt. Mazama was filled with lava about 7000 years ago, and its cross section was exposed when the mountain top collapsed into the caldera now occupied by "Crater" Lake. (Photo by E. Kiver)

silica rhyodacite lava (a bad sign) welled up into the explosion crater and flowed down slope. About 200 years later the Llao Rock area would be sliced in two when the "big one" on Mt. Mazama occurred.

About 6850 radiocarbon years ago[4] the largest documented eruption ever to occur in the Cascade Mountains began as huge ash clouds darkened the sky and sent plumes of ash as far away as Yellowstone, the Dakotas, and north into Alberta and British Columbia. Figure 9–4 in the Yellowstone section provides an excellent visual comparison of the volume of eruptive materials released during prominent eruptions from selected volcanoes around the world. The 1980 Mount St. Helens event pales in comparison to Mazama's fiery eruption—the equivalent of at least 75 Mount St. Helens eruptions!

Huge ash flows rolled down Mazama's flanks filling valleys with gas-choked deposits of pumice, volcanic glass, and crystals that were belched from the magma chamber. The heat trapped in the ashy deposit melted some of the pyroclastic debris into *welded tuff*. Hard, chimney or pipelike features formed where hot, trapped gases escaped through the tuff to the surface. For years after the eruption there must have been "ten thousand smokes" venting from the ashflow surfaces depositing minerals and case-hardening the vent pipes. These chimney forms are resistant and erode slower than the surrounding ash. Thus, a bizarre landscape of *pinnacles,* some 30 feet (9 m) high, are locally exposed in the Annie, Sand, and Wheeler Creek valleys (Fig. 4–25).

The huge volume of magma ejected in such a short period of time vacated the upper part of the magma chamber causing the summit cone to shatter and plunge downward like a piston into the magma chamber below—which in turn pushed yet more magma to the surface. The steep caldera walls collapsed inward and were blown to bits by the venting volcano. Giant landslides continued to enlarge the caldera. The main eruption likely lasted for only a few days or weeks but the steep unstable caldera walls continued to slide occasionally for years afterward, much as the oversteepened crater walls at Mount St. Helens continue to slide today. When the cataclysm was over the once-tall mountain lay in smoldering ruin—transformed into a 4000-foot-deep (1220 m) depression that was soon to become one of the beauty spots of the world.

At one time it was believed that a gigantic explosion was solely responsible for the caldera. To have a huge mountain literally blown to bits certainly lacks nothing of the sensational, particularly when we become aware that about 10–13 cubic miles (40–52 km³) of missing mountain must be accounted for (Bacon, 1983). The surrounding area is buried with pyroclastic material and a large quantity of dust was carried long distances by the wind. The volume of fragmented rock (cone) material contained in this pyroclastic material (1.4 cubic miles; 6 km³) is woefully inadequate

---

[4]Corrections for past variations in atmospheric abundance of radiocarbon indicate that the cataclysmic eruption occurred 7700 *calendar* years ago. Dates used in this book are usually given in *radiocarbon* years.

FIGURE 4–25   Delightful pinnacles in the Sand Creek–Wheeler Creek area on southeast flank of Mt. Mazama. (Photo by E. Kiver)

to account for Mazama's missing summit. However, the volume of *liquid* magma (not volume of frothy pumice) represented by the ejected debris (13–14 cubic miles; 51–59 km^3) would produce a space in the magma chamber large enough to accommodate the missing summit. Therefore, the collapse or subsidence theory is favored over the explosion hypothesis to account for the missing upper cone.

The height and shape of the original cone has long been debated. Evidence used to reconstruct the pre-eruption Mazama Volcano includes the diameter of the base of the mountain (about the same as Mt. Rainier), the extent of Pleistocene glaciers from the mountain base (greater extent implies a loftier peak), and projecting the remaining slopes upward to estimate the former summit elevation. Elevation estimates place Mazama as high as 16,000 feet (4880 m). However, because the profile of the summit area was likely rounded rather than "pointy," Williams (1942) and Bacon (1983) favor pre-eruption elevations closer to 12,000 feet (3660 m).

Eruptions about 300 years later produced two cinder cones on the floor of the caldera—spectacular andesitic Wizard Island (Figs. 4–22 and 4–26) stands some 764 feet (233 m) above present lake level and, hidden from view beneath 600 feet (200 m) of water, lies Merriam Cone on the north edge of the caldera floor. Wizard Island lava flows found 230 feet (70 m) and more below the present lake level were erupted into water as indicated by the presence of *pillow lavas,* much like those forming off the Hawaiian coast today. Lava flows above the 230 foot (70 m) level were erupted subaerially—thus Crater Lake was near its present level a mere 300 years after the caldera formed (Nelson and others, 1995). A lava platform formed east of Wizard Island, and a small rhyodacite plug dome, produced by the last known volcanic event there, squeezed upward just east of the Wizard Island cone only about 4240 years ago (Nelson and others, 1995).

FIGURE 4–26   Crater Lake, a caldera lake, with Wizard Island cinder cone rising high above the lake's surface. (Photo by D. Harris)

Special Features

The lake and its setting unquestionably form one of the most spectacular views in North America. A short hike to Discovery Point may help one to imagine Hillman's reaction in 1853 when his mule abruptly stopped a few steps short of the caldera edge. Hillman must have gazed in astonishment at the unexpected scene below him. Also at Discovery Point note the tell-tale scratch marks left by glaciers that streamed down from the vanished higher parts of the mountain. Look carefully at the south rim of the caldera from more northerly vantage points or from the launch that takes visitors around Crater Lake (more accurately Caldera Lake). Kerr Notch and Sun Notch are broad glacial valleys truncated by the caldera walls—the cirques and source areas for the Pleistocene glaciers that occupied these valleys are now part of the debris layer hidden by the lake waters and lake sediments at the bottom of Crater Lake.

The extreme clarity of the lake is due to a lack of suspended sediment (no streams flow into the lake) and a sparse population of microscopic algae inhabiting the lake. In the late 1960s researchers measured a visibility of 325 feet (100 m) through the crystal clear waters. An alarming 25 percent decrease in visibility measured 10 years later may have been caused by nutrients discharging from a spring fed by the sewage treatment facility at Rim Village. It was hoped that by changing the location of the sewage system and scaling back an extensive remodeling and expansive construction project on the caldera rim that this frightening trend would stop. However, by 1994 visibility was further reduced to 134 feet (41 m). Although still ranking as one of the clearest bodies of water in the world, its long-term future is uncertain. To best photograph the incredible blue color of the lake waters, shoot down from a high point on the rim with the sun in a clear blue sky behind you.

The Pinnacles, as already discussed, are hardened pyroclastic material surrounding former gas vents. Some are still hollow where the escaping gases created a "ten-thousand smoke" appearance shortly after the eruption. The 300-foot-thick (91-m) ash flow deposit has an interesting characteristic—the lower part is light colored and the top is dark (Fig. 4–25). Chemically the light-colored zone is rhyodacitic (high silica) in composition that grades upward to dark material of dacitic to andesitic (medium-silica content) composition. A reasonable explanation is that the magma chamber was segregated into less dense rhyodacite near its top and denser, heavier andesitic fluids below. Uncapping such a magma chamber lowers the pressure and permits dissolved gases in the magma to be rapidly released— much like a can of soda that has been shaken and uncapped. The lighter rhyodacite blew out of the magma chamber first and progressively deeper, heavier, darker-colored andesitic magma followed.

Long trails are limited in the park or, like the Pacific Crest Trail along the west rim, are situated in close proximity to the automobile road. However, short trails to Garfield Peak, Hillman Peak, and the Watchman provide excellent views that are exhilarating. A more challenging trail takes one up to the historic fire lookout atop of Mt. Scott where an excellent overview of the caldera is available (Fig. 4–22). The lighting in the morning is particularly good for "camera bugs." A trip on one of the tour boats with a ranger-naturalist is a "must" for those wishing a closer look at a volcano's "innards." Close-up views of stratocone layers, feeder dikes, the Phantom Ship vent remnant, and Wizard Island are spectacular.

Mazama's Future

What lies in the future for this now seemingly peaceful mountain? Long, time-based studies provide geologists with the capacity to peer centuries, millennia, or even farther in the future. A number of scenarios are possible—which will occur and when is uncertain. However, that something *will change* is certain!

As already discussed, Mazama, as well as other Cascade volcanoes, have experienced thousands or tens of thousands of years of dormancy only to spring once again into violent action. It would seem that two eruptions of large magnitude from the same volcano are unlikely—however, geologists do not yet know all the "rules" by which these volcanoes "play." Postcaldera eruptions have occurred—another one is possible and perhaps a certainty. Deep-diving research submarines have discovered floating mats of bacteria living in the lake and small amounts of fluorine and sodium—all suggestive of active hydrothermal vents on the lake bottom and ominous signs that the "fires" may only be sleeping. The cause of the mysterious "burps" of blue-gray clouds of smoke or gas in 1945 is unknown. The clouds appeared near the lake center, mushroomed, and drifted away before measurements could be made. Activity persisted from September to December and has not been observed since.

Even if no further volcanic activity occurs, a lake containing over 6 cubic miles (25 km^3) of water held in by steep rock walls that are actively eroding could produce a flood of incredible size on the mountain flanks. The presence of that volume of water during an eruption will unquestionably have serious consequences.

REFERENCES

Bacon, C.R., 1983, Eruptive history of Mount Mazama and Crater Lake caldera, Cascade Range, USA: Journal of Volcanology and Geothermal Research, v. 18, p. 57–115.

Bacon, C.R., 1988, Crater Lake National Park, Oregon; map and text on back of map: U.S. Geological Survey, Washington, DC.

Nelson, H., Bacon, C.R., and others, 1995, The volcanic, sedimentologic, and paleolimnologic history of the Crater Lake caldera floor, Oregon: Evidence for small caldera evolution: Geological Society of America Bulletin, v. 106, p. 684–704.

Orr, E.L., Orr, W.N., and Baldwin, E.M., 1992, Geology of Oregon: Kendall/Hunt, Dubuque, IA, 254 p.

Williams, H., 1942, The geology of Crater Lake National Park, Oregon: Washington, DC, Carnegie Institution of Washington, Publication 540.

Park Address

Crater Lake National Park
P.O. Box 7
Crater Lake, Oregon 97604

MT. RAINIER NATIONAL PARK (WASHINGTON)

Mt. Rainier is high (14,410 feet; 4393 m)—the second highest peak in the "lower 48" (Fig. 4–27). From afar, on a clear day, the glistening white mountain with its graceful shape appears pristine and almost unreal. However, up close the deeply carved glacial valleys and the craggy outcrops (Fig. 4–28) reveal that the mountain is a mere shadow of its former, much larger self. Although much reduced from pre-

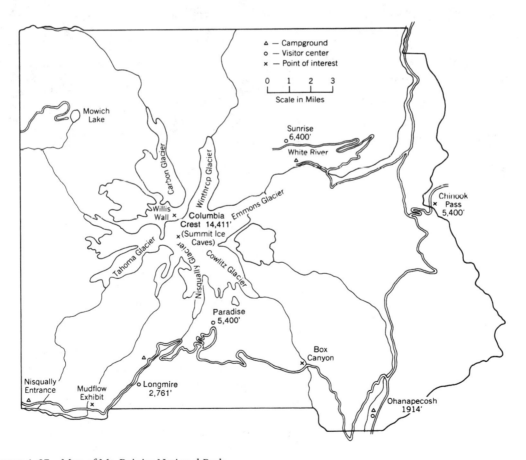

FIGURE 4–27   Map of Mt. Rainier National Park.

FIGURE 4–28   Mount Rainier from Reflection Lake. (Photo by E. Kiver)

vious, more glorious times, what a spectacular "shadow" it is today! Rainier tow-ers above the surrounding mountains, some of which contain significant 6000-foot-high (1830-m) peaks, dwarfed by their massive neighbor. Although its volume is less than those of Mt. Shasta and Mt. Adams, its great height and per-petual snow and ice cover make it a sight to behold, especially when seen in the set-ting sun from Puget Sound. It was from here that Captain Vancouver saw it in 1792. Lest anyone who has visited Mt. Rainier during a rainy period get the im-pression that it is rainier there, Captain Vancouver named the mountain after his friend and superior officer, Rear Admiral Peter Rainier.

To the Indians it was Tahoma—"Snow Mountain" or the "Mountain that was God." The "restless spirit" that inhabits the mountain made itself known not only to the Indians but to the early settlers and explorers as well. Sixteen minor eruptions between 1820 and 1894 were reported. "Fire, noise," and "brown, billowy clouds" issuing from the summit crater are recorded in early reports. One of these events produced a light dusting of ash on the northeast flank of the mountain sometime between 1820 and 1840. For reasons unknown, most of the Cascade volcanoes, ex-cept for Lassen Peak and Mount St. Helens, have not erupted during the twentieth century even though they experienced many minor events during the nineteenth century. To some volcano watchers, this suggests that these volcanoes are overdue and that there may be more than one Cascade volcanic eruption in our immediate future. For more details on the history and geology of Mt. Rainier and other Cascade volcanoes, Steve Harris's *Fire Mountains of the West* (1988), is the volcano watcher's "quick-reference bible" to Cascade volcanoes. For those interested in the climbing and exploration history of the Pacific Northwest's greatest mountain, Dee Molenaar's *The Challenge of Rainier* (1971) makes fascinating reading.

The highest elevation is on a ridge separating the two overlapping craters at the summit. The summit cone, barely touched by erosion, and the perfectly preserved craters (Fig. 4–29) with their hissing fumaroles continuously emitting steam, speak to the recency of events to affect the slumbering giant. Geologists Scott, Vallance, and Pringle (1995) note that Mt. Rainier is "considered to be potentially the most dangerous volcano in the Cascade Range because of its great height, frequent earthquakes, active hydrothermal system, and extensive glacial mantle." A few hundred earthquakes occur near the mountain and over 20 earthquakes occur annually beneath the restless mountain making Rainier second only to Mount St. Helens as the most seismically active volcano in the Cascades. The location of an active volcano close to a mega-metropolis, with all of its valleys leading to densely populated areas, is a disaster waiting to happen.

FIGURE 4–29   Airview showing 2000-year-old summit cone and associated craters. Cone is built on top of an older cone decapitated by a huge debris avalanche about 5000 years ago. Note the truncated andesite layers on Liberty Cap (right side) pointing toward the sky and the former much higher summit. (Photo by National Park Service)

Older Geologic Sequence

Early reconnaissance by S.F. Emmons (1879) and others established that Mt. Rainier is a composite (stratovolcano) cone perched on a foundation of Tertiary-age volcanic and granitic rocks. Detailed geologic analysis was slow in coming, however, because of the physical environment. Below 5000 feet (1524 m) most areas are covered by dense forest and underbrush, and the area above 5000 feet (1524 m) is mainly glacier covered. There are extensive outcrops in the higher areas, but many are on steep slopes in unstable rocks. The Willis Wall, the headwall of the Carbon Glacier cirque, is an example. At the top of the headwall, 3600 feet (1100 m) high, are *shelf glaciers* that push out over the edge and occasionally break off and tumble down the precipitous cliff. The resulting ice-block avalanches and rockfalls are not conducive to prolonged and detailed geologic examination. Thus, credit is due Fiske, Hopson, and Waters (1963) for their work under difficult conditions. Much of the following summary of bedrock geology is based on their U.S. Geological Survey study and other summaries presented by Crandell (1969), Harris (1988), and Orr and Orr (1996). A generalized geologic record is presented in Table 4–3 and a cross section in Figure 4–30 to assist in understanding the following discussion.

TABLE 4–3   Generalized Geologic Record, Mt. Rainier National Park

Time				Geologic Materials	Corresponding Geologic Events
ERA	PERIOD	EPOCH	YEARS		
CENOZOIC	QUATERNARY	Holocene		Neoglacial moraines Andesite lava, pyroclastics Osceola Mudflow Volcanic ashes	Alpine glaciers fluctuate Modern summit cone forms Summit collapses, forms caldera about 5600 B.P. At least 11 post-glacial eruptions Cone steepened by eruptions and erosion
		Pleistocene	–10,000–	Short andesite flows Glacial and interglacial deposits Andesite lava, pyroclastics	Volcanic activity decreases Intense glacial erosion, 10,000–65,000 yrs** B.P. Rainier reaches maximum height, 16,000 feet Rainier and other High Cascade volcanoes begin
		Pliocene	–1.6 Ma–	No local deposits	Uplift, erosion uncovers Tatoosh Granodiorite New Cascade eruptions elsewhere
	TERTIARY	Miocene	–5.0 Ma–	Tatoosh Granodiorite	New Cascade eruptions begin about 7 Ma Granite-like intrusion in park, about 13 Ma
				Stevens Ridge Formation / Fifes Peak Formation	Fifes Peak volcano to east, some dikes and sills in park Stevens Ridge – ash flows, welded tuffs
		Oligocene	–24 Ma–	Ohanepecosh Formation (andesite and breccia)	Unconformity – area uplifted, erosion, gentle folding, tropical to sub-tropical climate "Old Cascades" begin to form, rapid plate subduction, extensive volcanism in Cascades
		Eocene	–36 Ma– –58 Ma–	Puget Group – sand, clay, coal, only exposed west of park	Shallow marine environment, deltas, swamps

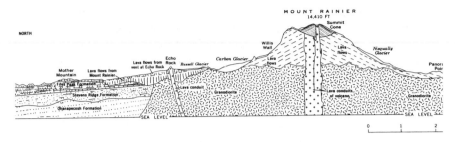

FIGURE 4–30   Cross section of Mt. Rainier. (From Crandell, 1969)

The plateau upon which the Rainier volcano is built is composed of igneous and sedimentary rocks formed as part of the Older, or Western, Cascades. Sands, clay, and coal beds of the *Puget Group* formed a large marine delta complex in western Washington from about 40 to 58 Ma during the Eocene Epoch. Plate movement along the western subduction zone was especially rapid beginning in Oligocene time producing a less steep descent of the oceanic plate and abundant magma. Undersea volcanoes produced the lavas and debris flows of the *Ohanapecosh Formation* in the Mt. Rainier and surrounding areas during the Oligocene, about 28–35 Ma. These andesites, rhyolites, and volcanic breccias are locally over 10,000 feet (3050 m) thick and are well-exposed in the eastern part of the park. Gentle folding and faulting of the older strata were accompanied by uplift of the area above sea level. Exposure of the rocks to surface processes triggered an episode of erosion and formation of deep soils in the tropical to subtropical climate that prevailed at the time.

Renewed volcanic activity from about 27 to 16 Ma (late Oligocene to early Miocene) showered the park area with ash and ash flows of the *Stevens Ridge Formation* and produced the lava, dikes, and sills in the northern part of the park belonging to the *Fifes Peak Formation* (Table 4–3). The Fifes Peak volcano was located east of the park, and the Stevens Ridge volcano was also outside of the present park boundaries. Renewed folding and uplift followed, and compression accentuated the earlier-formed folds and produced some faults with thousands of feet of displacement. Magma pushed upward during the late Miocene (about 12 Ma) through the older Tertiary-age rocks forming the batholith and associated dikes and sills of the *Tatoosh Granodiorite* (granodiorite is a close cousin of granite). Again vertical uplift occurred and the forces of erosion laid bare these older rock layers that permit geologists to examine and reconstruct parts of the area's turbulent past. The deeply eroded plateau of the Older Cascades would soon experience yet another profound change—the development of a new generation of volcanoes.

Birth of a Mountain

As the moving plates slowed about 9 Ma, the angle of descent of the subducting plate steepened. Rather than the broad zone of volcanism that characterized the

Western, or Older, Cascades, volcanism narrowed as the New Cascades formed. Evidence for very early eruptions at Mt. Rainier is mostly buried by younger lava, but the initial eruptions are at least a few hundred thousand years old and may date back over a million years. Extensive voluminous andesite flows filled surrounding valleys and built a broad, 30-mile-diameter (48-km) lava shield. Later, erosion by streams and glaciers attacked weaker rocks alongside the lava-filled valleys producing an *inversion of topography*. Ironically, the long, linear ridges radiating from the mountain, like Burroughs Mountain and Rampart and Klapatche Ridges, are former valleys that are now topographic high points!

Following the period of large, extensive flows, eruptions began to produce short, thin flows that piled up closer to the vent and increased Rainier's elevation to perhaps as much as 16,000 feet (4880 m). By 75,000 years ago Mt. Rainier was at its prime. Its nearly perfect symmetry and extraordinarily high elevation will likely not be reached again. At least three episodes of glaciation during the last 65,000 years have taken their toll in the destruction and steepening of the mountain—especially in combination with declining volcanism that makes repair of the deep erosional wounds difficult or impossible. Little Tahoma Peak, Gibraltar Rock, and other rock projections on the mountain flank are remnants of the former volcano surface that is now mostly eroded away. For example, the layers of andesite, breccia, and pyroclastic material exposed in the spectacular cliffs of Little Tahoma Peak give mute testimony that at least 2000 feet (610 m) of Rainier's side has been eroded away.

Mt. Rainier's Glaciers

The myriad of glacial landforms visible today such as the cirques, glacial valleys, horns, and aretes were mostly excavated during the Pleistocene, particularly from about 25,000 to 10,000 years ago. Glaciers, especially those flowing down steep mountain slopes, are extremely effective erosional agents. Although glaciers were larger during the older episodes of glaciation, their effects were mostly obliterated by the more recent, very late Pleistocene glaciers. During glacial episodes, the mountain, and the entire park, was a gleaming white wilderness with scarcely a bare-rock surface visible. Valley glaciers radiated like spokes on a wheel from the volcano. The Cowlitz Glacier on the southeast side of the mountain (Fig. 4–27) was the longest, extending some 65 miles (105 km) from the mountain and about 8 miles (13 km) west of the present town of Randle.

Many mountain glaciers in the midlatitudes did not survive Holocene (the last 10,000 years of earth history) warming and especially the higher temperatures of the Thermal Maximum (Fig. 1–25). However, most of Mt. Rainier's glaciers merely retreated upslope into the "arctic" of the mountaintop. The extreme height and massive character of the mountain intercepts the moisture-laden westerly winds from the Pacific, producing an annual snow cover that normally exceeds 100 feet (30 m) at midlevels on the mountain! During the winter of 1971–1972 slightly more than 102 feet of snow (31 m) fell—the world's record at an official weather station. Cold temperatures prevail at higher elevations, even during the summer melt season, further encouraging the preservation of last winter's snowpack. Numerous advances and retreats occurred in response to Neoglacial climatic fluctuations during the past few thousand years. However, the most extensive post-Pleistocene advances occurred during the cooler climates of the past 600 years (the Little Ice Age). Burbank (1981) used the extremely slow growth rate of certain lichens on glacier boulders (*lichenometry*) and the growth rings of trees (*dendrochronology*) to date the fluctuations of Rainier's glaciers during the past few centuries.

Changes in glacier extent at Mt. Rainier and elsewhere around the world are mostly synchronous and reflect both natural perturbations and human influences on climate. The Cowlitz and other glaciers have lost as much as 35 percent of their surface area since the mid-1800s. The retreat of the Nisqually Glacier, one of the most readily accessible and photographed glaciers in the world, is pictorially recorded in Veatch's (1969) study. In 1900 the glacier terminus was just upstream from the old highway bridge; by 1960 it had retreated to near its present position, nearly a mile (1.5 km) upstream. Nisqually Glacier was not alone in its extensive retreat during the past 150 years. Mountain glaciers in the Alps and elsewhere behaved similarly. One local victim of warming climate was the gradual disappearance of the Paradise Ice Caves, a major attraction in the park until their final collapse in the fall of 1991. Most scientists studying climate change agree that worldwide *global warming* produced by modifications of the atmospheric and land environment by human activity is real. The disruptions that will be produced by global warming are not fully realized at present, but the costs of climate modifications to society will be enormous.

The 26 named and the numerous smaller, unnamed glaciers covering Mt. Rainier occupy over 36 square miles (93 km^2) of the park, the largest glacier system in the conterminous United States. Rainier's high elevations ensure cool-to-cold temperatures and interception of moisture-laden winds from the Pacific that produce a perpetual blanket of snow. Some of the glaciers, such as Emmons, Nisqually, and Ingraham, originate from the summit ice cap; others such as Carbon, South Tahoma, Wilson, and Russell stream down the mountain flanks from cirques located mostly in the high-precipitation zone located at midelevations. Separating neighboring glaciers are long, linear rock ridges called *cleavers*. Carbon Glacier cirque is huge, with a 3600-foot-high (1100-m), nearly vertical headwall (the Willis Wall); well over a mile wide, this cirque is by far the largest in the Cascade Mountains.

Driedger and Kennard (1986) measured the thickness of the glaciers on Mt. Rainier and other Cascade volcanoes by using an ingenious portable radar device that fits into a backpack. From this they determined that Carbon Glacier, the largest in the park, is locally over 700-feet (215-m) thick and that over one cubic mile of snow and ice perpetually covers the mountain. This volume of ice, if melted rapidly during an eruption, would cause huge floods and mudflows that would easily reach the densely populated areas in the Puget lowland.

Nisqually Glacier is the most accessible in the park and has been photographed millions of times. Excellent views are available at the Paradise Visitor Center—a must for all visitors (Fig. 4–27). Take the trails through the alpine tundra to various vistas to enjoy the magnificent mountain and its glaciers. Observe the color of meltwater emanating from the glacier—the finely ground *rock flour* derived from the mechanical disintegration of rock fragments rubbing against one another and the underlying bedrock gives the water its milky color. Look for the lateral moraines that record ice margins and levels when the glaciers were much thicker in the 1800s. Remind yourself that these huge ice masses are not "frozen in time" as stagnant ice masses but are active, dynamic bodies that are moving forward at the same rate as the ice melts back from the glacier snout.

Postglacial Rainier

With the disappearance of the huge Pleistocene glaciers and the absence of large, extensive lava flows such as those that characterized earlier eruptions, one might suppose that Holocene (the last 10,000 years) events, except for fluctuations of the greatly reduced valley and cirque glaciers, were relatively uneventful. Not so, as we shall soon learn. At least 22 air-fall ash layers blanket the mountain's flanks—11 of which were derived from Rainier's own volcanic activity. A thick yellow ash layer is from the cataclysmic eruption of Mt. Mazama 6850 years ago, and the remaining non-Rainier ashes are from Mount St. Helens—the most recent layer added in 1980. Ten of the Holocene ash layers derived from Rainier lie above the Mazama ash and are, therefore, less than 6850 years old! No wonder the Native Americans had great respect for the power of the mountain.

The prolonged attack by glaciers and the ending of the large lava eruptions that periodically "healed" the erosional wounds left the mountain exceptionally high and steep by the end of the Pleistocene. The continuous fumarole and hydrothermal activity throughout the late Pleistocene also helped to create conditions that turned the mountain into a "time bomb." The escape of hot, acid gases had two important effects on the mountain: (1) Glaciers were thinner because of warm ground and therefore less erosive in the summit area. Thus reduced glacial erosion helped maintain the 16,000-foot (4880-m) elevation of the peak until about 5000 years ago. (2) The corrosive volcanic gases slowly converted the strong andesitic rock to weak, greaselike clay minerals such as *smectite* and *kaolinite*— thereby weakening the physical strength of the cone and increasing the potential of large landslides and debris flows (Frank, 1995).

Scott and others (1995) studied the *debris flow* and *mudflow* (also known as *lahars*) history at Mt. Rainier and concluded that *periodic sector collapse* (large localized landslides that remove significant localized "chunks" of mountain) is the major way that the morphology of the mountain changes. Over 55 mudflows, some of catastrophic proportions, are recorded in the sediments deposited on the mountain and in the surrounding valleys. Large segments of "missing mountain" that are apparent from the shape of the mountain flanks are the source areas of many debris flows and mudflows. If slope failure coincides with an eruption, glacier melting unquestionably occurs, and enough water is freed to turn a debris flow into a more "soupy," fluid mass—a mudflow. If large amounts (greater than 3–5 percent) of clay are included in the mudflow, the material becomes exceptionally fluid and can flow tens of miles down the surrounding valleys. Many of these clay-rich mudflows are unrelated to volcanic eruptions; thus they "can occur without warning, possibly triggered by non-magmatic earthquakes or by changes in the hydrothermal system" (Scott and others, 1995).

About 5000 years ago the combination of high elevation, steep slopes, the presence of weak clayey zones, abundant ice and water, and a volcanic eruption produced one of the earth's largest known mudflows—the Osceola Mudflow. The upper 3000 feet (915 m) of Mt. Rainier first collapsed as a distinct block or blocks, but quickly disintegrated into a giant debris avalanche. Mixing of the abundant water and clay particles transformed the debris avalanche into a wall of mud—a mudflow that raced down the White River drainage. The mud was hundreds of feet thick near the mountain but quickly thinned as it sloshed down the valley. Near the present site of Enumclaw, 35 miles (56 km) downstream, a 100-foot (30-m) wall of mud moving at about 40 miles (64 km) per hour roared over the site of an Indian village and eventually poured into Puget Sound, 75 miles (120 km) from the mountain. If a similar event occurred today, the mudflow would enter a large reservoir where it would acquire yet more water, increase its mobility, destroy the reservoir dam, and easily inundate the towns of Enumclaw, Buckley, Kent, Auburn, Sumner, and Puyallup, with a combined population over 125,000. If the mudflow should have the velocity of most mudflows, even though an immediate alert were sounded, most people would have less than 2 hours to evacuate their homes after the mudflow begins! Evacuation would be a difficult or impossible task as any law-enforcement or emergency-response official would testify.

Subsequent debris avalanches and mudflows in the five major valleys leading from the mountain were not as large as the Osceola Mudflow, but many would inflict a terrible loss of life and property if they occurred today. These moderate-size debris avalanches and mudflows occur about once every 500–1000 years on Mt. Rainier, more frequently than volcanic eruptions. Thus debris avalanches and mudflows are considered to be the most serious hazard associated with a Cascade volcano. In 1963 a series of large rockfall avalanches broke loose from Little Tahoma Peak and roared down the east flank of the mountain. It was headed for and seemed intent on wiping out the Park Service's White River Campground; fortunately it "ran out of steam" a short distance up-valley.

After the Osceola Mudflow and the associated eruption, the mountain was 3000 feet (915 m) lower, although higher parts of the crater rim, such as Liberty Cap (Fig. 4–29), were still above 14,000 feet (4270 m). The summit of the once nearly perfect cone had been transformed into a stubby, truncated cone with a broad calderalike opening at the top. The downhill process of destruction was temporarily reversed about 2000 years ago when renewed volcanic activity produced lava flows, flow breccia, and explosion debris that constructed a new cone about 1200 feet (365 m) high in the erosional caldera (Frank, 1995). A slight shift in the position of the erupting vent produced a second overlapping crater at the new summit, completing the present appearance of the summit cone (Fig. 4–29). Only brief steam and ash eruptions have occurred since. As previously mentioned, the presence of warm ground has discouraged large erosive glaciers from forming, thereby preserving much of the original volcanic topography of the summit cone.

Rainier's Historic Eruptions

Although only a few settlers, trappers, and others lived in western Washington in the 1800's, 16 apparent eruptions are recorded on Mt. Rainier from 1820 to 1894. The reports are disappointingly brief and nondetailed, but at least one of the rising steam and "smoke" events deposited a thin layer of ash on the volcano flank sometime between 1820 and 1840 (Mullineaux and others, 1969).

It is interesting to learn that the Indian guide Saluskin reported that he guided two "King George Men" (British) to the mountain in 1855. After returning from the summit they told Saluskin that there was "ice all over the top, lake in center and smoke or steam coming out all around like sweat-house." If true, it implies that sufficient heat remained after the 1820–1840 eruptions to keep the snow in the crater melted. Mountain climbers report strong sulfur odors in the crater as late as 1894—by 1936 only steam issued from the numerous summit fumaroles. In 1973 only one fumarole in the west crater emitted a small quantity (one part per million) of hydrogen sulfide gas (Kiver, unpublished measurement). The last sighting of "brown, billowy clouds" emanating from the summit was in 1882, and a possible steam eruption occurred in 1894. Beginning in the 1960s a number of well-documented small steam eruptions were recorded at various locations on the mountain. On a summit climb in August of 1970, a group of climbers, including one of the authors, witnessed a steam cloud rising from Gibraltar Rock for over 4 hours.

Snow and ice fill the two summit craters, but buried fumaroles have melted more than 7200 feet (2200 m) of ice tunnels (Fig. 4–31) along the crater floor,

FIGURE 4–31   One of a number of entrance passages leading into the extensive geothermal ice cave system inside the summit craters. (Photo by E. Kiver)

mostly along ring fractures inside the crater (Kiver and Steele, 1975). In the east crater a small crawlway leads from the large main cave to a huge grotto more than 340 feet (104 m) below the surface of the ice fill. A similar passage in the west crater leads to Lake Grotto where, beneath a thick cover of ice (160 feet; 50 m), there is a crater lake, believed to be the highest crater lake on the North American continent (elevation 14,113 feet; 4329 m). Perhaps this is the remnant of the crater lake reported by the Indian guide Saluskin in 1855.

The Future

Although Mt. Rainier has had a relatively long history (perhaps more than a million years) and might be expected to soon end its explosive behavior, the signs that future eruptions will continue are too numerous to ignore. Its location close to the Seattle–Olympia mega-metropolis and its great height and steepness have caused many to regard Mt. Rainier as our potentially most dangerous Cascade volcano. Acid gases rising from the magma chamber continue to convert the volcanic rock in the mountain core to slippery clays and further reduce the mechanical strength of the mountain. Eroding glaciers further steepen its flanks producing a "top heavy" mountain—one that is likely to collapse with a slight nudge from geologic processes. Earthquakes under the mountain (about 20 a year) indicate continuing activity in the magma chamber and on associated faults and could provide that nudge, even if no volcanic eruption occurs.

If an eruption occurs, a major debris avalanche—similar to the 1980 one on Mount St. Helens—is extremely likely. Abundant historic and geologically recent eruptions increase the likelihood of future eruptions. Numerous hot spots, that is, areas of fumaroles or unusually warm or hot ground, have persisted on the mountain since observations were first made. Some thermal manifestations are fleeting phenomena, but their number and changing locations do not bode well for the structural integrity of the mountain.

Small rockfalls and landslides are nearly continuous. Watch for rising dust from rock cliffs higher on the mountain and look for dark areas of recent slide debris on the glaciers. Historic landslides like the December 14, 1963, rock avalanche from Little Tahoma Peak and giant slides like those that produced the huge Osceola Mudflow and other dangerous mudflows will occur again, unfortunately many without warning. Small *jokulhlaups* (Icelandic term) or *outburst floods* (sudden releases of large volumes of water) from glaciers produce floods and mudflows like the Kautz Creek Mudflow in 1947 that buried a highway bridge (Fig. 4–32) under 28 feet (8.5 m) of debris. Similar small outburst floods occur in the park with a frequency of about one a year, especially from the South Tahoma, Nisqually, Kautz, and Winthrop glaciers (Scott and others, 1995). Outburst floods can be triggered by sudden release of an ice-dammed lake or by release of volcanic heat beneath a glacier. However, most of the dozens of outburst floods occurring in the twentieth century in the park were preceded by heavy rainfall, suggesting that water is somehow temporarily stored and suddenly released by the glacier. Outburst floods typ-

FIGURE 4–32   Bridge built after the 1947 mudflow buried Kautz Creek Valley. Mt. Rainier in background and standing trees killed by mudflow visible on right in middle distance. (Photo by D. Harris)

ically occur in the late afternoon on warm days or after heavy rain in the late summer or early autumn. Witnesses liken the roar of the approaching flood to that of an oncoming locomotive.

From the standpoint of active geologic processes, the mountain is "alive" and "well." Few parks provide so many exciting opportunities to experience dynamic forces at work on the landscape. Take the Wonderland Trail and others to the snout of a glacier to watch milky colored meltwater roar out as a full-fledged stream from beneath the ice; watch as rocks imprisoned in ice for decades or centuries bounce violently down the front of a glacier as tall as a 15-story building; listen and watch as rocks tumble down the over half-mile-high Willis Wall at the head of Carbon Glacier or from the vertical sides of Little Tahoma Peak; and on a cool spring or fall day watch the wispy clouds of steam rising from the crater rim—a reminder for those who live in the Pacific Northwest that there is an erupting volcano in someone's future.

REFERENCES

Burbank, D.W., 1981, A chronology of late Holocene glacier fluctuations on Mount Rainier, Washington: Arctic Alpine Research, v. 13, no. 4, p. 369–386.

Crandell, D.R., 1969, The geologic story of Mount Rainier: U.S. Geological Survey Bulletin 1292, 43 p.

Driedger, C.L., and Dennard, P.M., 1986, Ice volumes on Cascade volcanoes: Mount Rainier, Mount Hood, Three Sisters, and Mount Shasta: U.S. Geological Survey Professional Paper 1365, 28 p.

Emmons, S.F., 1879, The volcanoes of the Pacific coast of the United States: American Geographical Society Journal, v. 9, p. 45–65.

Fiske, R.S., Hopson, C.A., and Waters, A.C., 1963, Geology of Mount Rainier National Park, Washington: U.S. Geological Survey Professional Paper 444.

Frank, D., 1995, Surficial extent and conceptual model of hydrothermal system at Mount Rainier, Washington: Journal of Volcanology and Geothermal Research, v. 65, p. 51–80.

Harris, S.L., 1988, Fire Mountains of the West: The Cascade and Mono Lake volcanoes: Mountain Press, Missoula, Montana, 379 p.

Kiver, E.P., and Steele, W.K., 1975, Firn caves in the volcanic craters of Mount Rainier, Washington: National Speleological Society Bulletin, v. 37, no. 3, p. 45–55.

Molenaar, D., 1971, The challenge of Rainier: The Mountaineers, Seattle, WA, 332 p.

Mullineaux, D.R., Sigafoos, R.S., and Hendricks, E.L., 1969, A historic eruption of Mount Rainier, Washington: U.S. Geological Survey Professional Paper 650-B, p. 315–318.

Orr, E.L., and Orr, W.N., 1996, Geology of the Pacific Northwest: McGraw-Hill, New York, 409 p.

Scott, K.M., Vallance, J.W., and Pringle, P.T., 1995, Sedimentology, behavior, and hazards of debris flows at Mount Rainier, Washington: U.S. Geological Survey Professional Paper 1547, 56 p.

Veatch, F.M., 1969, Analysis of a 24-year photographic record of Nisqually Glacier, Mount Rainier National Park, Washington: U.S. Geological Survey Professional Paper 631, 52 p.

Park Address

Mount Rainier National Park
Tahoma Woods, Star Route
Ashford, WA 98304

NORTH CASCADES NATIONAL PARK (WASHINGTON)

Lying along the Canadian border in north-central Washington State is a wilderness gem that has been appropriately described as the "American Alps." Indeed, from the top of any of the hundreds of mountain peaks in the North Cascades one can view a sea of mountains, like giant frozen waves, unfolding as far as the eye can see. This broad section of the Cascades is one of the largest areas of uninterrupted spectacular scenery in the contiguous 48 states—and much of it is in the park. Any camping or hiking experience helps build character and self-sufficiency, qualities that many individuals fail to develop in our modern urban society. Here, every degree of experience and challenge awaits the park visitor. From short walks near the developed campgrounds to day-long hikes and many-day-long backpack trips into the backcountry wilderness, the choices are infinite.

The North Cascade Highway, a scenic summer road, follows along the Ross Lake National Recreation Area that bisects the park into northern and southern parts (Fig. 4–33). The Seattle Light and Power Commission constructed Gorge, Diablo, and Ross dams to harness power from the Skagit River. Ross Dam, 540 feet (165 m) high, backs water for more than 20 miles (32 km), even into the edge of Canada. The park, along with the large surrounding wilderness areas, makes up one of the largest unroaded areas in the conterminous United States. Lake Chelan National Recreation Area in the southeast part of the park includes part of the spectacular, 55-mile-long (89-km), fiordlike Lake Chelan (Fig. 4–34). Manning Provincial Park in Canada, immediately north of North Cascades National Park, preserves even more of the pristine mountain environment.

Park History

Native Americans used the area for over 11,000 years. Over 200 identified archeological sites indicate that at least seasonal use was made of all elevation zones. Winter villages were located at lower elevations along the Skagit River where climate was milder. Tribes living below the Newhalem Gorge had access to salmon runs from Puget Sound, which provided an excellent food source. The gorge, just below present-day Diablo Dam (Fig. 4–32), is a natural barrier to salmon migration. Tool-making rocks were quarried near cirques inside the park, and mountain goats were hunted for their meat and outstanding wool. Trails across Cascade Pass and elsewhere enabled the west-side tribes to trade dried salmon, stone tools, and dried clams and other products for berries, Indian hemp (used for rope making and cordage), buffalo meat, and other commodities from the east-side tribes. By the time Euro-American settlers arrived in the mid-1800s, the Native American population had already been severely decimated by smallpox epidemics unwittingly transmitted to them from previous contacts with Europeans. Many of the unfortunate Native Americans who perished never saw a person of European descent— other tribal members acted as disease carriers.

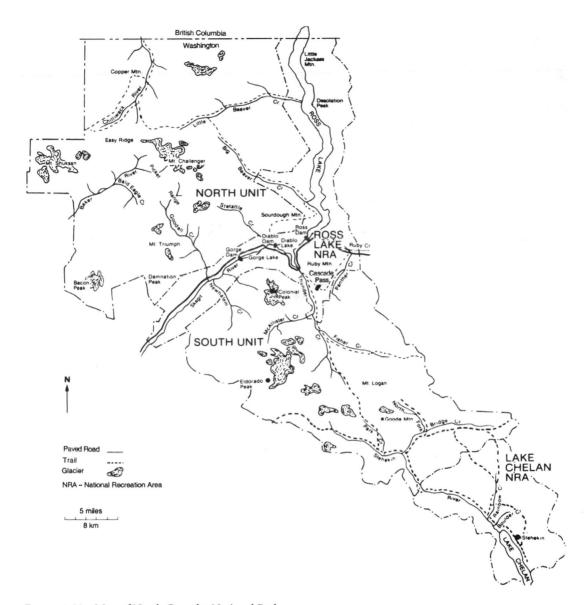

FIGURE 4–33   Map of North Cascades National Park.

Fur trader Alexander Ross in 1814 was the first outsider to cross the mountains and visit the Stehekin River and Cascade Pass areas. The treaty of 1855 with the Skagit and other tribes opened the area to settlers who initially cut timber and farmed along the lower Skagit River near Puget Sound. The short-lived gold rushes of 1858 and 1878 in the Ruby Creek area brought miners temporarily into areas

FIGURE 4–34   Lake Chelan occupies a 55-mile-long (89 km) glacial trough on the east side of the North Cascades in Washington State. (Photo by D. Harris)

now part of the park and promoted further settlement downstream along the Skagit River. The poor grade of ore, the rugged mountainous topography, the short summer season, and the lack of transportation routes ended the flurry of mining activity within a year or two. A clue to the real value of the area was given by Henry Custer in 1859 who was part of the reconnaissance survey team for the International Boundary Commission. After sputtering out long lists of adjectives Custer finally noted that "the area must be seen, it cannot be described."

The area was closed to homesteading in 1897 and made part of the Washington Forest Preserve and later, in 1924, part of Mt. Baker National Forest. A small power-generating dam constructed on the lower end of Lake Chelan in 1927 raised the lake level 21 feet (6.4 m), and dams in the upper Skagit River constructed from 1924 to 1949 created Gorge, Diablo, and Ross lakes. The cry for park status began in 1906 when the Mazama Outing Club from Portland recognized the unique character of the area. After 62 years of debate President Lyndon B. Johnson signed the North Cascades Act in 1968 that set 684,000 acres aside for the park and recreational areas. In 1989 much of the backcountry in the park was designated the

Stephen J. Mather Wilderness in honor of the tireless efforts of the first director of the National Park Service.

One ominous threat to the integrity of the greater North Cascades area is the presence of a large ore body in the Glacier Peak Wilderness area immediately south of the park. Use of the open-pit method would provide economical mining of copper, gold, and other minerals from this deposit and would furnish employment for a few hundred people for perhaps 10 years. But as mining has always treated its surroundings, it would also leave behind a huge "glory hole," piles or ponds of tailings, and pollution of water and atmosphere. How present and future generations deal with this situation will reflect how carefully voters elect and influence their government representatives.

Geographic Setting

The character of the Cascades changes dramatically north of the Interstate 90 corridor along Snoqualmie Pass in Washington State. The Middle and Southern Cascades south of the pass are devoid of rock exposures older than the Cenozoic—to the north older rocks are abundant. Paleozoic and Mesozoic sedimentary and volcanic rocks, now highly metamorphosed into gneisses and schists, and Mesozoic and Tertiary plutonic rocks form the crystalline core of North Cascades Park. The Cascades end a short distance north of the international border along the Fraser River in British Columbia where the North Cascades merge with the Coast Ranges.

The Skagit River drains the wetter western side of the park, eventually flowing into Puget Sound as the second largest river on the west coast of the contiguous United States. The temperate western rainforest gives way to alpine tundra at higher elevations. East of the Cascade crest are the drier east-side forests and the sagebrush plains on the eastern edge of the Cascades. Basalt lava flows (Miocene, about 15 million years old) of the Columbia Intermontane Province cover the older North Cascade rocks along the Columbia River valley where the two provinces meet. Here, the much smaller rivers east of the Cascade crest, like the Stehekin, Twisp, and Methow, flow into the Columbia River.

Average elevations rise dramatically from about 5000 feet (1524 m) in the Middle and Southern Cascades to 7000–9000 feet (2134–2744 m) along the Cascade crest in the North Cascades. Mt. Shuksan (9131 feet; 2783 m) is the highest peak in the park and is a favorite of many mountain climbers—even though in most years a fatal climbing accident occurs. Just west of Mt. Shuksan on national forest land is Mt. Baker (10,775 feet; 3284 m), the northernmost of the Cascade volcanoes in the United States. South of the park in a wilderness area is Glacier Peak (10,528 feet; 3209 m)—another of the magnificent Cascade volcanic sentinels. Mt. Baker erupted numerous times in the 1800s and Glacier Peak erupted during the last 300 years. In 1975 Mt. Baker abruptly began to emit large quantities of steam containing abundant hydrogen sulfide and other gases that often precede an eruption (Kiver, 1978). No magma followed these gases to the surface—this time!

Structural Framework

Making sense of North Cascade geology has taken a lifetime of work for geologist Peter Misch and others who have followed in his footsteps. The mountains did not "give up" their secrets easily—dense forests, vertical rock walls, and few roads and trails make field studies very time-consuming—progress came slowly. After reasonable geologic maps were finally compiled, it was apparent that a system of major strike-slip faults separated the region into a mosaic of northwest–southeast oriented bands of rock—each band having little or nothing in common with its neighbors (Fig. 4–35).

The plate tectonic model seemingly did not make sense here—not until it was recognized that many of these unrelated rock bands originated elsewhere and were added onto the continent as microplates or *suspect terranes*. Faults sliced up the continent edge and moved large chunks of continent northward, much like the San Andreas in California is doing today. Some of the rock slices resemble and were likely derived from the Sierra–Klamath Mountains of northern California (Burchfiel and others, 1992). During the Mesozoic (Jurassic and Early Cretaceous) these pieces of crustal rock were organized offshore into sizable blocks of continental material (*superterranes*) that were rafted and welded onto the edge of the North American continent about 100–90 Ma (Late Cretaceous). The northward movement of the oceanic plate relative to North America during the Late Cretaceous to early Tertiary (90–40 Ma) produced northwest–southeast oriented strike-slip faults that sliced the North Cascades into a series of fault-bound blocks, further complicating the already jumbled mosaic of rock terranes that make up the North Cascades. Movements along these strike-slip faults were right lateral (block on opposite side moves to the right as one faces the fault) with displacements of 50–100 miles (80–160 km) or more on major faults such as the Chewack-Pasayten, Ross Lake, Entiat, and Straight Creek–Fraser River faults (Fig. 4–35). Additional discussion of the microplate concept is presented in the Olympic Mountain section (Chapter 3) and in Chapter 1 (see Fig. 1–11). Other general summaries are available in Alt and Hyndman (1984) and Orr and Orr (1996). Articles by Burchiel and others (1992), Cowan and Bruhn (1989), Haugerud (1989), McGroder (1989), Misch (1988), and Tabor (1994) are good sources of more detail and discussions about the many as yet inadequately answered questions about North Cascade geology.

Most of the park lies between the Straight Creek and Ross Lake faults (Fig. 4–35) where intense pressure and temperature from former deep burial in the earth's crust has changed many of the original sedimentary and volcanic rocks into high-grade metamorphic rocks—gneiss and schist. Fossils and sedimentary features such as bedding and ripple marks are completely obliterated here, reducing the amount of information that can be extracted about the history of the rocks in the metamorphic core. Intense folding and large thrust faults occur in both the Methow Basin to the east of the Ross Lake Fault and in the San Juan–Northwest Cascade belt to the west of the Straight Creek Fault (Fig. 4–35). However, less severe metamorphism and local preservation of fossils and other features promote

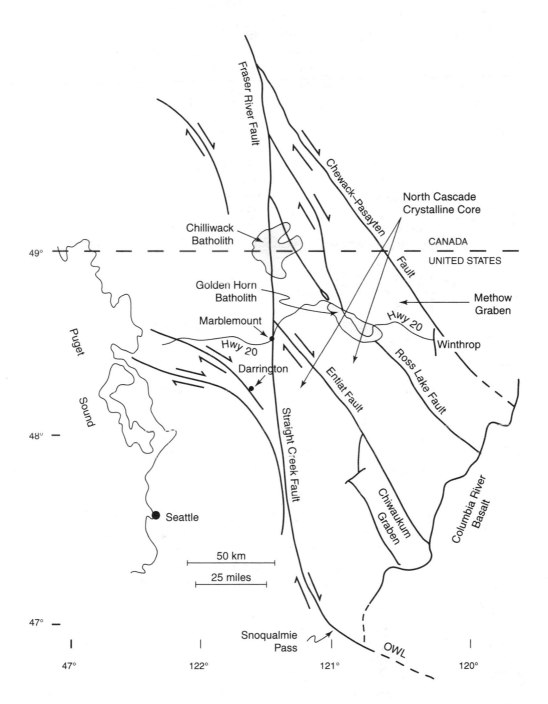

FIGURE 4–35   Major faults and selected geologic and cultural features, North Cascades. (Modified from Haugerud, 1989)

better understanding of the rock history in these outer zones. Radiometric dating of the metamorphic rocks and of the numerous batholiths throughout the North Cascades is also invaluable in helping to work out the complicated events that built one of the most rugged mountain ranges on the North American continent.

Geologic History

Paleozoic and Precambrian rocks in the San Juan–Northwest Cascade belt record an older (Paleozoic) metamorphic and mountain-building event (about 350 Ma), but most structural and metamorphic events in the North Cascades were initiated in the Mesozoic as North America began its relentless westward drift. Mid-Mesozoic (Jurassic, about 150 Ma) deformation produced a second generation of mountains that developed large overthrust faults and granitic intrusions. Erosion leveled the topography by Cretaceous time when marine sediments covered the area. However, this was merely the "lull before the storm" as the slow but unstoppable forces of subduction continued. By mid-Cretaceous time, about 100 Ma, tectonic forces increased as the offshore superterranes and microplates began to "dock"—beginning the "big squeeze" as the plates collided. This major mountain-building event (about 100–40 Ma) affected the Rocky Mountains to the east as well as coastal areas from California to Alaska. Locally it produced most of the major structural features in today's park.

Extensive thrust and overthrust faults were produced in the Northwest Cascades belt and metamorphic forces at depth were intense to the east in what was to become the metamorphic core. The Skagit Gneiss, an important and widespread rock unit in the core, contains excellent examples of *injection gneiss* or *migmatite*. Migmatites are mixed rocks, or rocks "living on the edge" between metamorphic and igneous environments. They appear to begin as a *schist* (high-grade metamorphic rock) that later has light-colored molten material injected as thin layers between bands of darker metamorphic minerals. Random veins of quartz and feldspar cut across the rock grain and intricate flowage features often occur. Road cuts at the Diablo Dam overlook provide excellent, readily accessible views of the migmatized Skagit Gneiss (Fig. 4–36).

Numerous *plutons* (massive igneous intrusions such as *batholiths* and smaller masses called *stocks*) were emplaced during the mountain-building process. Initially the Chilliwack and other batholiths were thought to be single large granitic intrusions; now, as McKee (1972) points out, it is known to consist of several intrusions, ranging from granodiorite to granite in composition, which were emplaced at different times, from Late Cretaceous to Oligocene (90–30 Ma). Many of these intrusions were emplaced along the complex system of faults in the North Cascades and were later displaced by fault movements. Of particular significance is a later intrusion, the 50-million-year-old Golden Horn Batholith. This intrusion was emplaced along the Ross Lake Fault Zone and is *not* offset by the fault (Fig. 4–35). Thus, most movement along the northwest-trending system of faults ended by 50 Ma.

FIGURE 4–36   Skagit gneiss laced by quartz veins, at Diablo Lake Overlook on Highway 20. Vertical lines are manmade drill holes. (Photo by D. Harris)

During Eocene time (50–36 Ma), the north-trending Straight Creek Fault sliced across the Ross Lake and other faults carrying the west part of the metamorphic core and other rocks into Canada. Movement along the Straight Creek Fault ended before the Chilliwack pluton (36 Ma) was emplaced along the fault (Fig. 4–35). The Straight Creek Fault continues northward from near Darrington and through the park headquarters area at Marblemount into British Columbia. The Fraser River Valley follows the fault (called the Fraser River Fault in Canada) for several hundred miles. Horizontal displacement along the fault is at least 70 miles (110 km) but could be as much as 120 miles (190 km). South of Darrington the fault bends to the southeast where it becomes part of the Olympic–Wallowa Lineament or OWL (Fig. 4–35), a major structural feature that cuts diagonally across Washington state and is discussed in more detail in Chapter 7.

The right-lateral fault movements were likely caused by *oblique subduction* as the northeast-moving oceanic plate, with the microplates embedded, collided with the west-moving North American plate. The shearing forces created by the horizontal component of plate movement caused the Methow block east of the Ross Lake Fault Zone and west of the Pasayten Fault to be pulled apart and form a large graben (Fig. 4–35). Late Mesozoic sediments and volcanic rocks filled the structural depression and were later intruded by Cenozoic plutons.

Erosion began the unroofing of the Skagit Gneiss and other deeply buried metamorphic core rocks in the Eocene. Erosion increased substantially during the Pliocene as arching of the Cascade Range and vertical movements along faults began. After at least 50 million years of burial under the pressure of perhaps as much as 21 miles (34 km) of crustal rocks (Burchfiel and others, 1992), the former sedimentary and volcanic rocks would again see the light of day—this time in a greatly changed form as the Skagit Gneiss and other related metamorphic and plutonic rocks of the North Cascade metamorphic core. Although a minor mountain-building event compared to the late Mesozoic–early Cenozoic orogeny, the substantial late Cenozoic elevation increase, erosion, and the formation of the Mt. Baker and Glacier Peak volcanoes established many of today's major topographic elements. The final touches to this incredible landscape were soon to come—the invasion of the Pleistocene glaciers.

North Cascade Glaciers

Alpine glacial landforms, mostly erosional, are king of the mountain in the North Cascades. The rugged topography for which the park is world famous is composed of an unusually large number of cirques, glacial valleys, aretes, horns, and cols (Fig. 4–37). Where the glaciers surrounded a peak they developed cirques with steep headwalls; as the cirques were enlarged, the mountain was eaten away, leaving only a sharp spine or horn, such as the Matterhorn in the Alps. The massive mountain glaciers, guided by stream-cut valleys, scoured the valleys of the Skagit and Stehekin to within 2000 feet (610 m) of sea level, even upstream close to the drainage divides. Thus, being close in distance to a high peak that is over a mile higher still means that considerable effort is necessary for those who want to "get on top of things." Glac-

FIGURE 4–37 The jagged Picket Range in northern section of the North Cascades National Park, as pictured from Skagit River below Diablo Dam. (Photo by D. Harris)

iers in the Lake Chelan area outdid themselves—they gouged the bedrock not only to sea level but 400 feet (120 m) below! The 1605-foot-deep (489 m) Lake Chelan is the third deepest in North America and the 8500-foot-deep (2600 m) valley that it sits in exceeds Hells Canyon and the Grand Canyon in depth (Fig. 4–34).

The Cordilleran ice sheet moving south from Canada occasionally invaded the mountain interior, burying northern areas with ice whose surface was as high as 6000 feet (1830 m) above sea level. Continental ice incursions into the mountains were brief and relatively ineffective compared to the longer periods of more erosive alpine glaciation (Weisberg and Riedel, 1991). However, because lower peaks and ridges were rounded by the continental ice sheets and higher peaks retained their jagged arete-horn topography, the height of the ice sheet is easily discerned throughout the range. The last ice sheet invasion maximum was about 15,000 years ago when glaciers covered about 90 percent of the park. Ice in the Puget Lowland to the west covered Seattle with nearly a mile of ice and extended about 50 miles (80 km) south. Continental ice on the east side of the Cascades flowed through the Okanogan Valley to Grand Coulee and along the Columbia River valley to the lower edge of Lake Chelan. By 12,000 years ago all vestiges of continental ice in the North Cascades were gone.

The rugged area with its present-day near-glacial climate and its moderate elevation continues to support many cirque and small valley glaciers (Fig. 4–38). Closeness to the Pacific moisture source ensures high precipitation on the west slope where 43 feet (13 m) of snow falls in an average year. Precipitation amounts drop rapidly in the east slope rain-shadow area where snowfall averages only 9 feet (2.8 m) in the Lake Chelan area.

FIGURE 4–38   Small glaciers on Magic Mountain near Cascade Pass. (Photo by E. Kiver)

The return to cooler climates during the last 4000 years (Neoglacial interval; see Fig. 1–25) caused a worldwide rebirth of many small glaciers and enlargement of existing glaciers. Presently, 318 glaciers cover about 45 square miles (117 km^2), or less than 5 percent of the park. However, the entire North Cascades has a whopping 756 glaciers that cover about 103 square miles (267 km^2)—the largest glaciated area in the United States south of Alaska (Post and others, 1971). A view from the air almost compels one to believe that the Ice Age is not over!

Beginning in the late 1800s glaciers began to shrink worldwide as climates warmed. A temporary standstill and local readvance was first noted in the North Cascades as Coleman Glacier on Mt. Baker began to readvance in the 1940s. Thickening and advance of glaciers on Mt. Rainier and the North Cascades soon followed. Unfortunately, this temporary aberration reversed and once again the glaciers of the world are losing the battle to global warming—a process being accelerated if not completely produced by human-caused changes in our atmosphere and landscape modifications. Krimmel (1994) noted that South Cascade Glacier, a small valley glacier in North Cascades National Park that has been studied since the late 1950s, has shrunk every year since 1977. Pelto (1991) predicts that 690 of the 756 glaciers in the North Cascades, or 91 percent of today's glaciers, will disappear in the next few decades due to global warming! Not only will it be sad to lose these remarkable natural features, but the costs of global warming to society and the environment will be extremely high.

Glaciers in the Skagit and other drainages in the North Cascades provide a stabilizing influence on water flow. Large volumes of meltwater are released during the summer—especially in years of drought when stream flow would be much less without a melting glacier source. About one-third of Seattle's energy needs are met by the Ross Lake hydroelectric power-generating complex. A steady flow of water, especially during the summer, is imperative. The Skagit is also an important salmon-spawning stream and supports the largest wintering bald eagle population in the continental United States. About 600 eagles arrive with the October and November Chum salmon run and remain until spring if a sufficient number of fish are present. Success of fish runs requires an adequate, year-round flow of clean water. One important factor in the prevention of sediment pollution in the Skagit River is the restriction of logging in the national park and surrounding wilderness areas. Trees hold the soil in place and prevent discharge of great quantities of sediment into the river, ensuring clear water for successful salmon spawning.

Visiting the North Cascades

The unique unspoiled area known as the North Cascades is a special place where generations of Americans can go and meet nature on her own terms—and discover, as have previous generations, the self-sufficiency that has made this a great nation. Setting off on foot with only those worldly possessions needed to survive helps develop the independence and confidence needed of a truly free person.

Some excellent opportunities for sightseeing from an automobile or from the Lake Chelan ferry are also available, but the real experience comes from meeting nature "up close" along some of the over 360 miles (580 km) of maintained trails. Access is by numerous Forest Service roads and the North Cascade Highway, a seasonal (May through October), high-maintenance-cost road constructed during the road-building frenzy of the 1960s. The popular Cascade Pass road (dead-end road) and trail system, Easy Pass, parts of the Pacific Crest Trail, and trails in the Ross Lake and Diablo Lake areas provide some outstanding opportunities for half-day and longer hikes.

Hikers reaching high vantage points are rewarded with views of a seemingly endless ocean of mountain peaks and ridges. The mountain stillness is occasionally interrupted by the unforgettable cannonlike retorts and rumbling sounds as cracks form in the glacier and ice blocks tumble from precariously perched glaciers on the steep mountain slopes. Large mammals, including the mountain lion and grizzly bear are present in small numbers but are unlikely to be seen. Some visitors are treated to a special experience—the howl of the gray wolf. This once widespread predator was exterminated from the American West by the 1920s and is now on the Endangered Species List. However, a few breeding wolves inhabit the Ross Lake and Baker Lake (just west of the park) areas.

The "unreal" green color of the waters of Diablo Lake is due to the *glacial flour*, finely ground rock produced by Boston Glacier (largest glacier in the park) and other glaciers that feed into Thunder Creek. Excellent exposures of the Skagit Gneiss and a "rock garden" of large specimens of North Cascade rocks are also present at the Diablo Lake Overlook along the North Cascades Highway—a "must stop" for visitors.

The sparsity of roads that penetrate the park make the North Cascades a special place. Knowing that unique, relatively unspoiled areas such as the North Cascades still exist in our land is justification in itself for the park. Experiencing the area on its own terms is an added benefit. The Park Service should be commended and supported in its policy to exclude the automobile. If park officials can withstand political pressure to build highways, and if they can persuade park visitors to behave like animals and not desecrate the area, the park will continue to be unspoiled.

REFERENCES

Alt, D.D., and D.W. Hyndman, 1984, Roadside geology of Washington: Mountain Press, Missoula, Montana, 282 p.

Burchfiel, B.C., Cowan, D.S., and Davis, G.A., 1992, Tectonic overview of the Cordilleran orogen in the western United States, in, Burchfiel, B.C., Lipman, P.W., and Zoback, M.L., eds., The Cordilleran orogen: conterminous U.S.: Boulder, Colorado, Geology of North America, v. G-3, pp. 407–479.

Cowan, D.S., and Bruhn, R.L., 1989, Late Jurassic to early Late Cretaceous geology of the U.S. Cordillera, in, Burchfiel, B.C., Lipman, P.W., and Zoback, M.L., eds., The Cordilleran orogen: Conterminous U.S.: Boulder, CO, Geological Society of America, Geology of North America, v. G-3, p. 169–203.

Haugerud, R.A., 1989, Geology of the metamorphic core of the North Cascades: in, Joseph, N.L. and others (ed.), Geologic guidebook for Washington and adjacent areas: Washington Division of Geology and Earth Resources Information Circular 86, p. 118–136.

Kiver, E.P., 1978, Mount Baker's changing fumaroles: The Ore Bin, v. 40, no. 8, p. 133–145.

Krimmel, R.M., 1994, Runoff, precipitation, mass balance, and ice velocity measurements at South Cascade Glacier, Washington, Water Resources Investigation Report 94-4139.

McGroder, M.F., 1989, Elements of the Cascades "collisional" orogen: Introduction to a transect from the Methow Basin to the San Juan Islands, Washington, in, Joseph, N.L., and others (eds.), Geologic guidebook for Washington and adjacent areas: Washington Division of Geology and Earth Resources Information Circular 86, p. 91–95.

Mckee, B., 1972, Cascadia; the geologic evolution of the Pacific Northwest: McGraw-Hill, New York, 394 p.

Misch, P., 1988, Tectonic and metamorphic evolution of the North Cascades—an overview, in, Ernst, W.G. (ed.), Rubey Volume, Prentice Hall, Englewood Cliffs, NJ, v. 11, p. 179–195.

Orr, E.L., and Orr, W.N., 1996, Geology of the Pacific Northwest: McGraw-Hill, New York, 409 p.

Pelto, M.S., 1991, North Cascade glaciers; their recent behavior: Dept. of Environmental Science, Nichols College, Maine, 17 p.

Post, A., Richardson, D., Tangborn, W.V., and Rosselot, F.L., 1971, Inventory of glaciers in the North Cascades, Washington: U.S. Geological Survey Professional Paper 705-A, 26 p.

Tabor, R.W., 1994, Late Mesozoic and possible early Tertiary accretion in western Washington State: The Helena-Haystack melange and the Darrington-Devils Mountain fault zone: Geological Society of America Bulletin, v. 106, p. 217–232.

Weisberg, S., and Riedel, J., 1991, From the mountains to the sea: A guide to the Skagit River watershed: North Cascades Institute, Sedro Woolley, Washington, 64 p.

Park Address

North Cascades National Park
2105 Highway 20
Sedro Woolley, WA 98284

FIVE

Sierra Nevada Province

Massive and majestic are the words that best describe the Sierra Nevada, the highest, longest, and grandest mountain range in the United States outside of Alaska (Plate 5). The Sierras form California's backbone, extending more than half the length of the state and forming one of the earth's largest fault-block mountains. In Spanish, Sierra (jagged range) Nevada (snowed upon) is an appropriate name.

To get the full impact of this massive, westward-tilted fault block, make your approach from the east across the still-active Sierra Fault—the one that raised the east edge of the fault block during the past few million years to its present dizzy heights. Stop in Owens Valley near Lone Pine (elevation 3700 feet; 1128 m) and look up, up, and up to the top of Mt. Whitney, the highest peak in the conterminous United States (elevation 14,495 feet; 4419 m). As the crow or eagle flies, Mt. Whitney is less than 10 miles (16 km) away from the west edge of Owens Valley, however, it is nearly 2 miles (3.2 km) vertically above! Then drive north to Lee Vining and up and over Tioga Pass. In the days before the road was paved, the trip up the east face was an adventure; now it is disgustingly simple! Early motorists made many stops and enjoyed the fabulous scenery, and certain ones pondered the fact that they were on the edge of the huge, still-active Sierra Fault—all while their car radiators were cooling down! Now, all too many people zip up the mountain, eager to get over it and down to the lodge in the valley in time for dinner. Take a little time. There will be a tomorrow. Stop at a few of the overlooks and let a bit of it sink in! On Tioga Pass, tarry longer than just to show your Golden Eagle Pass or pay your entrance fee into Yosemite National Park. Look to the west. You can see the Pacific, or at least you could back in the pre-pollution days. But you can still see the High Sierra with its mountain meadows and open parks and the snow-capped peaks rising high above them (Fig. 5–1). Finally, cruise down the gradual west slope of the fault block, past the many rounded granite domes and then into the main valley.

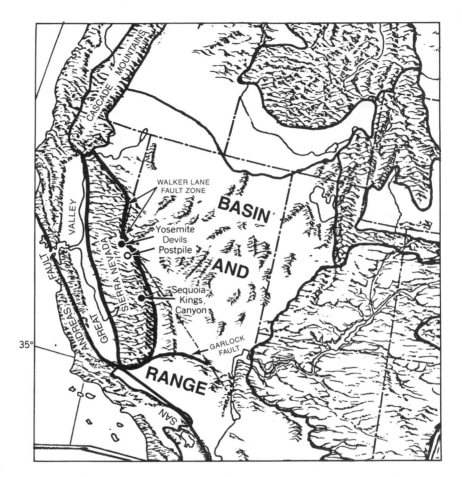

PLATE 5   Sierra Nevada Province and selected geographic and geologic features. (Base
map copyright Hammond Inc.)

Three other park areas are within the province, Sequoia–Kings Canyon National Parks, about 100 miles (162 km) to the south, and Devils Postpile National Monument, a short distance southeast of Yosemite (Plate 5). Generations of geologists have pondered the many questions about the local and regional geology raised by this "granddaddy of all fault blocks." As you gaze eastward into the Basin and Range Province, contemplate this interesting relationship—the tallest peak in the conterminous United States is located only 87 miles (140 km) from Badwater in Death Valley, the lowest point in the Western Hemisphere! As we shall see again, spectacular geology and spectacular scenery go hand in hand.

FIGURE 5–1   The High Sierra. (Photo by Peter Barth)

Geomorphic Setting

"The Sierra Nevada is a huge block of the earth's crust that has broken free on the east along the Sierra Nevada fault system and been tilted westward. It is overlapped on the west by sedimentary rocks of the Great Valley and on the north by volcanic sheets extending south from the Cascade Range"—so say Bateman and Wahrhaftig (1966). In addition, the Sierra Nevada block is chopped off at the south end by the Garlock Fault (Plate 5). The trend of the 400-mile-long (640-km), 40- to 80-mile-wide (65- to 160-km) range is north-northwest, reflecting the dominant direction of the Sierra Nevada fault system. The thick sediments in the 40-mile-wide (64-km) Great Valley conceal the west edge of the fault block. Likely it terminates along faults on the west edge of the Great Valley that place the fault block against Franciscan rocks, an important rock group in the California Coast Ranges (Norris and Webb, 1990). Rocks in the Klamath Mountains in northwestern California are of similar type, age, and structural pattern to those in the northern Sierra Nevada ("Mother Lode Country") and are likely their continuation—their intervening connection concealed by sediments in the northern Great Valley (Sacramento Valley). Even the presence of gold in the Klamath and Sierras is the same—an important geologic factor that contributed to the "gold rush days" and the rapid settlement of California in the mid-1800s.

The imposing mountain range also effectively drains eastward-moving air masses of their water content—producing the great deserts of the Southwest and resulting in the concentration of people in the cities and towns of southern California. No water from the Sierras reaches the sea. Water not extracted from the rivers flowing westward into the Great Valley either sinks into the ground or evaporates—east-side water is mostly gathered up and transported to the Los Angeles area via aquaducts.

Two major geologic stories need to be told about the Sierra Nevada: (1) the older story of the intrusion of the huge Sierra Nevada Batholith into older sedimentary and volcanic rocks and (2) the latest chapter of landform development resulting from uplift, erosion, weathering, and glaciation of the rocks during the past 2 million years.

Plate Tectonics and Geologic Sequence

The area that is now the Sierra Nevada was on the "trailing edge," or passive margin, of the North American Plate during the Paleozoic and early Mesozoic Eras. Islandlike land masses, some containing rocks as old as early Paleozoic, were added or accreted to the edge of the North American Plate as western North America became the "leading" or active edge of the plate about 200 Ma (early Mesozoic). A newly formed subduction zone began to heat and thicken the crust under western North America eventually forming significant mountain ranges during the Sevier Orogeny, particularly in what is now the Great Basin Section of the Basin and Range Province (Livacarri, 1991) immediately east of today's Sierra Nevada (Plate 5). As we shall see, the Sevier Mountains and the thick, hot crust in the Great Basin were destined to play a major role in the building of today's Sierra Nevada.

Melting of crustal rocks began on a grand scale, from California to Alaska—the most extensive episode of batholith formation recorded on the North American continent. Partial melting of mantle rocks in the subduction zone produced a basalt magma that accumulated at the base of the earth's crust (Fig. 5–2). Seawater carried down in the subduction zone greatly aided this process by lowering the melting point of already hot mantle rocks. The basalt "underplating" in turn melted crustal rocks— rocks that are much richer in silica content than are mantle rocks. Molten masses are buoyant (less dense than solid rock) and floated upward as individual *plutons* until they reached a density equilibrium with upper crustal rocks or lost sufficient heat and crystallized. Many of the plutons accumulated a mile or two (1–3 km) beneath the earth's surface in the massive complex called the Sierra Batholith (Fig. 5–2).

Some of these plutons vented—becoming the sources for the andesitic and rhyolitic volcanoes that dominated surface events during the building of the Mesozoic-age Nevadian Mountains (ancestral Sierra Nevadas). Huber (1987) and other geologists visualize a Cascade Range or Andes Mountain type of geography for the ancestral Sierra Nevada during the Mesozoic—erupting volcanoes at the surface

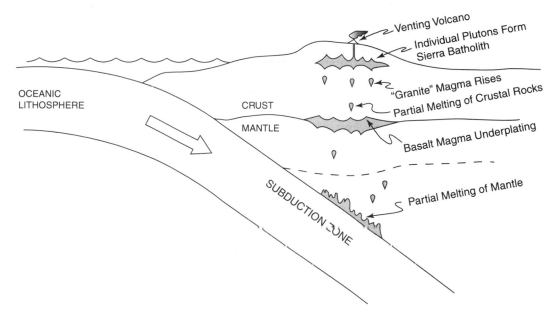

FIGURE 5–2   Formation of Sierra Batholith and ancestral Sierra Nevada mountains during Nevadan Orogeny in Late Jurassic to Late Cretaceous time. Individual plutons intrude older rocks that were previously added as accreted blocks onto the edge of the North American Plate.

and magma cooling at depth to form intrusions (Fig. 5–2). Abundant ash layers in Mesozoic rocks in interior North America were likely derived from the numerous volcanic centers associated with this great episode of batholithic and volcanic activity in western North America.

Early workers visualized one huge intrusion of molten material to form the Mesozoic-age Sierra Nevada Batholith. However, subsequent detailed work demonstrated that large batholiths are composed of numerous intrusions—hundreds of individual plutons in the Sierra Nevada. Younger magma intrusions were emplaced mostly side by side or "punched up" and cross-cut older plutons. Intrusive activity lasted for well over 100 million years and shifted eastward with time in the Sierra Nevada.

The older rocks were buried by miles of overlying strata and were mostly metamorphosed before being accreted to the North American Plate. However, rising magma during the Mesozoic further metamorphosed and folded these older layered rocks. Rocks engulfed by the rising magma were melted and assimilated into the melt—adjacent rocks were further recrystallized in this giant pressure cooker. Based on the huge volume and the coarseness of Cretaceous-age sediment shed into the Great Valley to the west, the Ancestral Sierra Nevada was a mighty range—likely topographically higher than even the present impressive range.

A long interval of erosion enabled streams to cut deeply into the Ancestral Sierra Nevada so that by the end of the Cretaceous, about 65 Ma, miles of overlying rock were removed and fragments of the once-deeply buried metamorphic and batholith rocks were washing down the river channels and into the Great Valley. At least some of the plutons were "unroofed" and the landscape was greatly subdued. Uninterrupted erosion continued well into the Cenozoic until the Miocene, about 25 Ma, when a much different set of stresses developed along the California coast.

As the westward moving North American Plate continued to override and consume the oceanic plate, it eventually reached and began to eliminate the spreading ridge off the California coast about 25 Ma (Fig. 3–2). This produced a *transform fault*, a type of *strike-slip* fault where major plates slide horizontally past one another. The newly formed transform fault, known as the San Andreas Fault, grew in length as the unrelenting plate movements continued (Fig. 3–2). The oceanic plate was moving northward now—tearing at the edge of North America and removing lateral support for the thick crust created in the Great Basin during Mesozoic mountain building. The resulting gravitational collapse and lateral spreading of the thickened crust created pull-apart (tensional or extensional) stresses and faults along the length of the Mesozoic mountains, but especially in the Basin and Range and Sierra Nevada areas (Livaccari, 1991).

Uplift along the Sierra front was slow at first but accelerated about 10 Ma and especially during the last 3 million years (Fig. 5–3). At least 5000 feet (1500 m) of displacement occurred in the northern Sierra Nevada and 11,000 feet (3350 m) in the south as this giant fault block tilted downward to the west (Norris and Webb, 1990). Perhaps the crust is stronger here—fused together by the hundreds of Mesozoic-age batholiths into one mechanically strong crustal block—helping to account for North America's longest single range of mountains. The Basin and Range Province to the east broke into much smaller, shorter blocks—mostly composed of sedimentary strata and lacking the coherence produced by a large, continuous batholith complex. Movements were episodic with long intervals of erosion and partial formation of erosional surfaces as streams developed lower gradients and migrated broadly over their floodplains. Renewed uplift rejuvenated streams producing downcutting and complete or nearly complete destruction of surfaces cut under previous less energetic stream-flow conditions.

The remarkable coincidence that Mt. Whitney, the highest point in the conterminous United States, and Badwater in Death Valley, the lowest point in the western hemisphere, are located a mere 87 miles apart requires a special tectonic explanation. Topographically high areas are usually associated with thick, low-density crust that "floats" higher in the denser mantle below—just as a thicker iceberg floats higher in the water. Not so in the Sierra Nevada. Ducea and Saleeby (1996), Park and others (1996), and Wernicke and others (1996) report that the crust under the Sierras is about 22 miles (35 km) thick—6–12 miles (10–20 km) thinner than expected and only 3 miles (5 km) thicker than adjacent low-elevation areas in the Basin and Range. Rather than floating at nearly the same elevations, the Basin and Range lies nearly 6000 feet (1800 m) below the Sierras. The "missing root problem" under the Sierra Nevada can be resolved if the light, lower crustal material was removed and replaced by hot mantle material. Such material at shallow depths would expand and produce a lower-than-normal density in the upper mantle and account for the high topographic elevation of the Sierra fault block. The location of the Sierras on the west edge of the stretching crust under the Basin and Range (a "rift shoulder" position) apparently thins the crust and permits hot mantle or *asthenosphere* (partly molten mantle material underlying the lithosphere) to rise upward beneath the Sierras, thus accounting for its higher elevation and continuing history of volcanism.

Volcanism accompanied the uplift, especially in the northern Sierra Nevada where Miocene- and Pliocene-age (20–5 Ma) volcanics buried earlier-formed Cenozoic valleys. Quaternary volcanism produced the Devils Postpile flow with its spectacular columnar jointing just southeast of Yosemite National Park, and localized volcanism occurred elsewhere, especially along the huge frontal faults on the east edge of the fault block. Particularly interesting are the eruptions that produced one of the earth's larger calderas near Mammoth Lakes along the Sierra front where monstrous explosions about 730,000 years ago produced the 10- by 20-mile (16- by 32-km) Long Valley caldera that spread ash (the Bishop Tuff) as far away as Kansas and Nebraska. The caldera's location near the Mammoth Lakes ski and recreation area is of concern, especially when swarms of earthquakes occurred in 1980 as new magma rose toward the surface. Small eruptions have occurred during the past few 100 years here, and based on the frequency of past activity, the possibility of another eruption during the next 50 years, perhaps one that dwarfs Mount St. Helens awesome show of power, cannot be ruled out.

Another part of the answer to the high elevation of the Sierra Nevada and its unusually thin crust may lie with recent and continuing plate movements. Unruh (1995) describes evidence that the Sierra Nevada block is part of a small microplate that is rotating counterclockwise (about 11 mm/yr) relative to North America. The pull-apart zone on the east side of the Sierra block is part of the Greater Walker Lane Fault Zone, a wide zone of faults about 60 miles (100 km) wide that extends northward along the east edge of the Sierras to the Lake Tahoe and Lassen Peak areas where it veers more westward toward the Pacific Ocean. A more northerly oriented branch of the Walker Lane extends along the east edge of the Cascade Range for another 300 miles (483 km) to Newberry Volcano National Monument in Oregon. Movements along the faults in the Sierra area display strike-slip (horizontal) in addition to *normal fault* (vertical) displacements, supporting the block-rotation model. Perhaps stress created by moving faults pushes upward the underlying magma or *ductile*[1] upper mantle–lower crust, which in turn pushes the mountain range upward, much like the pressure in a cylinder pushes a piston outward.

Whatever the causes and exact mechanisms are, the result is one of the most impressive mountain ranges in the conterminous United States. Fault scarps, many now much eroded, are abundant along the steep east face; there are, however, certain essentially unmodified fault scarps that cut across recent volcanics and glacial deposits, indicating that parts of the fault system are still active. More convincing

---

[1] Material that stretches or deforms without breaking—mechanically similar to Silly Putty.

to some is the faulting that produced California's largest historic earthquake in 1872 that completely destroyed the town of Lone Pine along the east edge of the Sierra block. A 13-foot-high (4-m) fault scarp remains today as a reminder.

Batholith Features

Erosion has removed the overlying metamorphosed sedimentary and volcanic rocks, exposing the Sierra Batholith throughout most of the middle and southern parts of the range. Thus an opportunity is provided to examine the exposed granitic bedrock and to use one's imagination—an experience nearly equivalent to taking a field trip into a magma chamber. Such a trip in real life would be a very uncomfortable experience—especially when one encountered 1800°F (1000°C) magma! Most of the rocks are granodiorites, diorites, and granites—all lumped together here and referred to as "granite." As would be expected of a granite pluton that cooled slowly over a few million years, most of the rocks display relatively large quartz, feldspar, and other mineral grains that are visible to the naked eye. Variations in crystal size and other features add an interesting variety to what visitors see "up close" and provide clues to what "life" for silica, potassium, oxygen, and other atoms and molecules is like in a magma chamber.

Large-scale features include the individual plutons themselves—some small, some quite large—younger ones cutting across older. *Roof pendants* are masses of older rock caught between adjacent plutons or projecting downward into a pluton (Fig. 5–3). Had the magma reached a higher level in the earth's crust or had erosion gone deeper, no evidence of the pendant would remain. Roof pendants occur in Yosemite and Sequoia–Kings Canyon National Parks; those in Sequoia contain marble (the metamorphic equivalent of limestone) with a few small but interesting solution caves.

Not all of the intrusive rocks are contained in their parent batholith. Cracks formed as plutons cool and shrink and are often filled with molten material, forming *dikes, sills,* and *veins*. Fluids released from rocks during metamorphism often contained unusual elements that produced rich ore deposits, including the famous Mother Lode deposits near Placerville in the northwestern part of the range. Here, mineral-rich *hydrothermal* (hot-groundwater) solutions formed quartz veins in faults and joints in the Mariposa Slate and other metamorphic rocks of the Foothill Metamorphic Belt. Weathering and erosion exposed many of the veins—streams then washed the heavy flakes and nuggets of gold into pockets in the stream beds and behind large rocks as *placers*. Most of these placer deposits were exhausted long ago. However, modern-day miners farther south are extracting substantial amounts of fine, disseminated gold particles in metamorphic rocks excavated from large open-pit mines.

Some other small-scale features associated with granite are visible in a single outcrop or hand specimen. The salt and pepper appearance of most granites is due to a mixture of light-colored quartz and feldspar crystals and dark-colored horn-

West   East

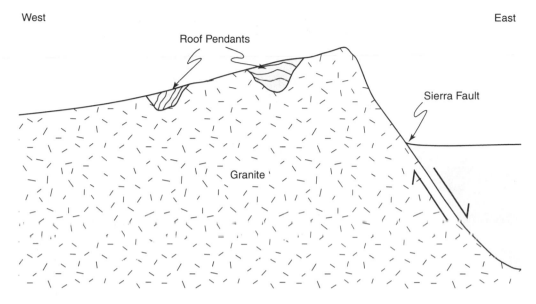

FIGURE 5–3   Long period of erosion unroofs Sierra Batholith. Uplift along east edge begins in mid-Tertiary (25–15 Ma) and accelerated greatly in late Tertiary-Quaternary time (10 Ma–present).

blende and biotite mica crystals. Other granites display large crystals surrounded by crystals of a smaller size—a *porphyry*. Two rates of cooling are often involved—large crystals characterize slow cooling rates and small crystals form with more rapid cooling. Perhaps cooling rates increased as pressure lowered when magma moved from lower to higher crustal levels or when magma was injected into cracks to form dikes and sills. Surface venting and volcanic activity also significantly lowers pressure and likely produced many of the granite porphyries found in batholiths.

Dark masses of crystalline material surrounded by lighter-colored granites are common features in many batholiths. In some cases these dark "blobs," or *inclusions,* are fragments of older rock broken from the roof of the magma chamber and engulfed and frozen in place by the solidifying magma (Fig. 5–4). Other blobs may result from the mixing of magmas of different compositions in one chamber, or from magma from two different chambers.

Granite Landforms

Of special concern are the interesting and in many cases spectacular landforms found in the Sierra Nevada. Many result from glaciation; however, bedrock features significantly influence their ultimate appearance. Faulting within the Sierra fault block produced highly fractured zones that often localize erosion and valley

FIGURE 5–4   Inclusion of dark minerals surrounded by granite. Wire cable (1 inch; 2.5 cm in diameter) is part of cable trail on Half Dome in Yosemite National Park. (Photo by E. Kiver)

formation. A spectacular example, the Kern Canyon in Sequoia National Park, is discussed in more detail later.

The number and orientation of cracks or *joints* significantly influence the weathering and erosion of the bedrock. Spacing of joints is critical—wide spacing produces a more resistant, monolithologic rock unit. High, vertical cliffs and peculiar rounded landforms known as *exfoliation domes* are associated with these more resistant rocks in Yosemite and elsewhere. Rocks with more closely spaced joints are more erodible and develop more craggy outcrops and less steep cliffs. The expression of these bedrock characteristics is greatly enhanced by the rugged topography resulting from the finishing touches—the invasion of glaciers.

Glaciers: Today and Yesterday

The rapid increase in elevation over the past 3 million years combined with worldwide climatic cooling inevitably produced the glaciers that significantly changed the appearance of this lofty mountain range. The slow-moving streams on the gentle western slope increased in volume and power during uplift, cutting V-shaped canyons. The expanding ice cap at the summit eventually sent fingers of ice down the stream-cut canyons, deepening, widening, and straightening them into today's magnificent glacial valleys. During peak times of glaciation a 250-mile-long (450-km), discontinuous ice cap blanketed the range crest above about 9000 feet (2740 m) (Norris and Webb, 1990). Locally it was about 30 miles (48 km) wide—much like some of the ice caps present today in Alaska.

Dating of glacial episodes in mountainous areas is always difficult. Here the association of radiometrically dated lava flows and volcanic ash layers with glacial deposits provides some chronologic control. The oldest recognized glacial deposit in the Sierras, the McGee till, is younger than the 2.6 Ma lava flow it rests on, and older than the overlying Sherwin till and a 730,000-year-old volcanic ash (Bishop Tuff)—present estimates are about 1.6 Ma for the McGee till. Older glaciations were more extensive than the younger Wisconsin age (10,000–75,000 years) glaciations.

Minor fluctuations of climate during the last 10,000 years (Holocene Epoch) produced three or four episodes of expanded ice, creating small cirque and valley glaciers. The most recent expansion coincided with the worldwide climatic cooling of the last 700 years called the Little Ice Age (Fig. 1–25). Small moraines a short distance downvalley from cirques and today's small glaciers were formed during the cooler intervals of the last few hundred to few thousand years.

About 60 small glaciers (*glacierets; cirque glaciers*) cling tenaciously to cirque walls, especially on rock walls facing north or northeast where sunlight is minimal on the sensitive snow and ice surfaces. Present global warming trends will take their toll on these fascinating creations of nature—this may be the last generation to enjoy glaciers in the more southern mountain ranges of the United States.

REFERENCES

Bateman, P.C., and Wahrhaftig, C., 1966, Geology of the Sierra Nevada, in, Geology of northern California: California Division of Mines and Geology Bulletin 190, p. 107–172.

Ducea, M.N., and Saleeby, J.B., 1996, Buoyancy sources for a large, unrooted mountain range, the Sierra Nevada, California: Evidence from xenolith thermobarometry: Journal of Geophysical Research, V. 101, no. B4, p. 8229–8244.

Huber, N.K., 1987, The geologic story of Yosemite National Park: U.S. Geological Survey Bulletin 1595, 64 p.

Livaccarvi, R.F., 1991, Role of crustal thickening and extensional collapse in the tectonic evolution of the Sevier-Laramide orogeny, western United States: Geology, v. 19, p. 1104–1107.

Norris, R.M., and Webb, R.W., 1990, Geology of California: Wiley, New York, 541 p.

Park, S.K., Hirasuna, B., Jiracek, G.R., and Kinn, C., 1996, Magnetotelluric evidence of lithospheric thinning beneath the southern Sierra Nevada: Journal of Geophysical Research, B, v. 101, no. 7, p. 16, 241–16255.

Unruh, J.R., 1995, Late Cenozoic tectonics of the greater Walker Lane belt and implications for active deformation in the Lake Almanor region, northeastern California, in, Page, W., Quaternary geology along the boundary between the Modoc Plateau, southern

Cascade Mountains, and northern Sierra Nevada: San Francisco, CA, Pacific Gas and Electric, Friends of the Pleistocene, field guidebook.

Wernicke, B., Clayton, R., Ducea, M., Jones, C.H., Park, S., Ruppert, S., Saleeby, J., Snow, J.K., Squires, L., Fliedner, M., Jiracek, G., Keller, R., Kiemperer, S., Luetgert, J., Malin, P., Miller, K., Mooney, W., Oliver, H., and Phinney, R., 1996, Origin of high mountains in the continents; The southern Sierra Nevada: Science, v. 271, p. 190–193.

YOSEMITE NATIONAL PARK (CALIFORNIA)

Magnificent cliffs, granite domes, waterfalls, and glacial valleys are all part of the Sierra Nevada scenery that has inspired generations of Americans. The occurrence of some of the earth's best examples of these features in one relatively small location led John Muir to describe the Yosemite Valley as the "Incomparable Valley." The valley is a small part of the park (Fig. 5–5) but a big part of the scenery that draws millions of visitors each year. The park's 1200 square miles (3109 km^2) contain not only Yosemite Valley but extensive exposures of the glaciated topography of the Sierra Nevada Batholith in the high country, large areas of alpine meadows, and forested areas at lower elevation with three groves of the world-famous giant Sequoia trees.

Creation of a Park

Joseph Walker led a group of trappers across the Sierra Nevada in 1833. Most of the copies of their journal describing the incredible waterfalls and the huge trees they encountered were destroyed in a print shop fire in Pennsylvania in 1839. The few surviving copies were not widely read, thus the Walker party discovery was known to very few. The discovery of gold in 1848 brought a swarm of miners to the Sierras who quickly disrupted the ability of the local Miwok and Paiute Indians to survive in their traditional style. Numerous conflicts and skirmishes followed. Rediscovery of the Incomparable Valley by non-Indians occurred in 1851 when a troop of soldiers pursued a group of Miwok Indians into the valley. The troop physician, Dr. Lafayette Bunnell, was awe-struck with the magnificent scenery and named it Yosemite after the local Indian tribe. James Hutchings of San Francisco guided tourists and brought artists and later photographers into the area beginning in the mid-1850s to record the magnificent scenes. Hutchings developed tourist facilities and made an excellent decision that would later help lead to the establishment of a national park—he hired a young handyman named John Muir in 1868 (National Park Service, 1989).

As word of the remarkable Yosemite country spread, public pressure mounted to prevent overdevelopment and destruction of the area—the germ of the idea that would eventually lead to the concept of a national park. Abraham Lincoln signed

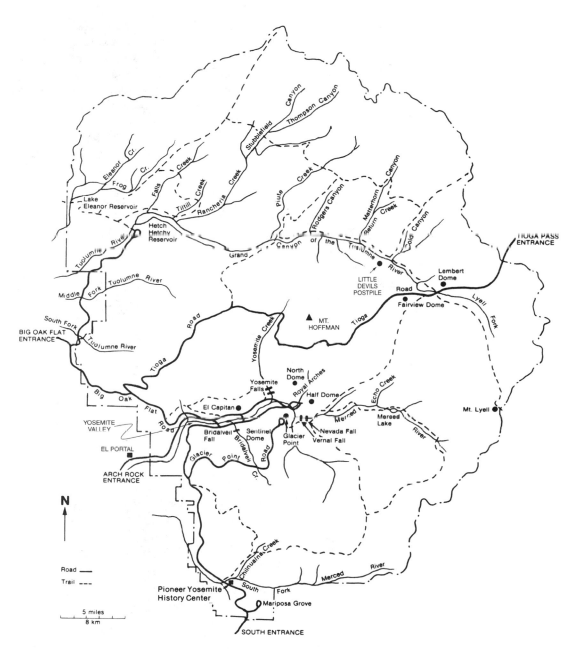

FIGURE 5–5   Map of Yosemite National Park.

a bill in 1864 that gave the valley and the Mariposa Grove of Sequoia trees to the state of California for use as the nation's first public park. "Muir of the Mountains" rambled through the area—living with nature and writing numerous essays extolling the virtues of his view of nature as opposed to the views of those who would "conquer" the wilderness for utilitarian or profit purposes. Nature was a "dynamic flow" and it was only proper that the life-giving waters born in the high country and flowing through Yosemite Valley should also be part of the Yosemite experience. Thus in 1890 our third national park, Yosemite, was established. In 1905 the state of California receded Yosemite Valley and the Mariposa Grove to the federal government—a more logical and reliable way to administer a national treasure that belongs to all citizens of the United States. Protection and administration of the park was delegated to the U.S. Cavalry in 1891 and transferred to the newly formed National Park Service in 1917. John Muir and President Theodore Roosevelt (Fig. 5–6) went on a 3-day camping trip in Yosemite in 1903. That experience helped develop strong presidential support for national parks and monuments—a commitment that Roosevelt exhibited throughout his administration. He designated 18 areas as national monuments and was instrumental in the establishment of five national parks by Congress.

A setback to Yosemite occurred in 1905 when pressure from mining and logging companies succeeded in reducing the size of the park. A boundary change eliminated more than 500 square miles (1295 km^2) from the park! A further and even more serious setback occurred in 1913 when Congress approved the con-

FIGURE 5–6   President Theodore Roosevelt (left) and John Muir at Glacier point, May 1903. (National Park Service photo)

struction of the Hetch Hetchy dam and reservoir on parkland in the Tuolumne River valley north of Yosemite Valley. The reservoir provides electric power and part of San Francisco's water supply. This "second" Yosemite Valley with its now partially flooded 2000-foot-high (610-m) canyon walls and spectacular waterfalls would have been extremely valuable today by helping to reduce the overcrowding in Yosemite. The concrete dam serves as a sad testimonial to the impact that short-sighted decisions can have on millions of people over many generations.

Geologic Development

As described in the chapter introduction in more detail, the development of a sub-duction zone along the west edge of the North American Plate during the Meso-zoic triggered a sequence of events that produced massive mountain ranges in western North America. The Nevadan, or Ancestral Sierra Nevada, mountains is a product of the longer-lived Cordilleran Orogeny that affected all of western North America. The complex of plutons that make up the Sierra Batholith was emplaced at depth (Fig. 5–2) during a period of well over 100 million years, from about 210 to 80 Ma (mostly Jurassic through Late Cretaceous). Some of these magma bodies fed a chain of erupting volcanoes that likely resembled today's Cascade Mountains. Mesozoic subduction activity also greatly thickened and heated the crustal rocks to the east where a massive range of mountains formed. The collapse of this zone of thickened crust would later play a major role in the formation of the modern Sierra Nevada.

Older sedimentary and volcanic rocks were metamorphosed and added to the edge of the North American Plate as *accreted terranes* before the Sierra Batholith was emplaced. Remnants of these older terranes are found as linear bands along the west and east edges of the park and as roof pendants in the central part of the park. Late Mesozoic (Cretaceous, about 100–65 Ma) and early Tertiary (65–25 Ma) erosion removed a few miles of overlying rock, thereby unroofing many of the plutons that make up the Sierra Batholith. By 25 Ma the Nevadan mountains had been reduced to low-elevation hills perhaps a few hundred feet high. Mid-Tertiary lava and pyroclastics buried the northern part of the Sierras, and uplift of the modern Sierra Nevada began slowly about 20 Ma along the Sierra Fault (Fig. 5–3) as stresses along the southwestern edge of North America changed. The oceanic plate (and part of southern California) began to slide northward along the newly formed San Andreas Fault (Fig. 3–2). Thus, lateral support for the thick crustal zone to the east in the Great Basin Section of the Basin and Range Province was greatly reduced enabling it to collapse and spread. The spreading produced crustal extension and the thousands of gravity (normal) faults in the Basin and Range to the east and the huge gravity fault along the east edge of the Sierra fault block.

Although volcanism accompanying uplift was significant in the northern Sierra and along the east edge close to the Sierra fault, volcanism in the Yosemite area was minor. A few volcanic deposits occur in the northern part of the park and a locally erupted lava, the Little Devils Postpile flow (Fig. 5–5), is accessible along the Glen Aulin trail.

Matthes (1930) recognized four major stages of landform development during the uplift and westward tilting of the Sierra fault block—generalizations that are still valid today. A *broad-valley* stage (about 50–10 Ma) formed by low-energy streams flowing down the gentle westward slope of the developing fault block was followed by a *mountain–valley* stage during which increased uplift further steepened slopes, increased the erosive power of streams, and caused deeper valleys to form. Greatly accelerated uplift during the past 3 million years caused even deeper erosion in the *canyon* stage and produced V-shaped valleys perhaps 3000 feet (915 m) deep. Instead of pauses in uplift between stages as suggested by Matthes, a more continuous but accelerating uplift is envisioned by Huber (1987) and other geologists. The substantial elevation increase of the westward rotating and uplifting fault block brought cooler climates to the higher mountains. When combined with the worldwide climatic cooling occurring simultaneously with uplift, the triggering of the Pleistocene-age (last 1.6 million years) *glacial* stage resulted. Glaciers greatly straightened, deepened, and widened the stream-cut valleys, producing the typical glacial valley profiles found today in the higher Sierras.

Sierra Batholith

Landforms in the park are almost entirely developed on the granitic bedrock of the Sierra Batholith complex. Over 100 individual plutons were emplaced over a 60-million-year interval. Most formed during the Late Jurassic and Early Cretaceous (about 148–132 Ma) although some, like the Half Dome Granodiorite, solidified as recently as 87 Ma (Norris and Webb, 1990; Huber, 1987). Calkins (1930) was one of the first to recognize the complex nature of the Sierra Batholith and formally named individual plutons in the Yosemite Valley. As is customary in naming rock units, they are named after prominent geographic features where one can go and examine a *type locality*. For example, the east part of the valley is composed mostly of the massive Half Dome Granodiorite; the Bridalveil Granodiorite crops out near the falls of the same name; and the El Capitan Granite is well exposed in the west part of the valley and at the formidable El Capitan cliffs (Calkins and others, 1985). Bands of northwest-trending Paleozoic and Mesozoic metamorphic rocks occur both west and east of the park as wallrock or roofrock of the batholith. Isolated outcrops occur as roof pendants (Fig. 5–2) in the Mt. Lyell, Merced Peak, and other areas in the park. Only 5 percent of the park has metamorphic rocks at the surface—granitic rocks are of overwhelming importance.

Take time to examine up close the many interesting features in the granite outcrops along trails and roadways. Granite and its relatives (granodiorite, diorite, etc.) are beautiful rocks, some of the earth's most attractive. All of the mineral grains are large and easily seen with the naked eye—the result of slow cooling over millions of years under a thick insulating blanket of rock. Light-colored quartz and feldspar mineral grains contrast pleasantly with dark-colored hornblende and biotite grains. Some of the feldspar grains are unusually large compared to surrounding mineral grains and form a *porphyritic* igneous rock (Fig. 5–7). The feldspar grains often show perfect crystal form as reflected at a right (90°) angle or nearly right angle (depending on variety of feldspar) junction of adjacent crystal faces. The flat crystal faces mimic the repetitious arrangement of atoms inside the mineral. However, minerals whose growth is restricted by interference with adjacent minerals also have an ordered internal arrangement of atoms even though the external surface is irregular. Thus while many minerals do not express *crystal form,* they still are *crystalline* materials. Porphyries develop when cooling conditions change, causing different size crystals to form. For example, when a magma chamber vents and erupts at the earth's surface, the resulting pressure reduction and

FIGURE 5–7   Porphyritic granite—large feldspar crystals embedded in coarse-grained groundmass. (Photo by D. Harris)

rapid cooling causes smaller crystals to form—accounting for many of the porphyries in Yosemite (Huber, 1987).

Dikes of fine-grained quartz and feldspar are called *aplite* and those displaying unusually large mineral grains are called *pegmatite*. A darker-colored diorite dike cuts across the face of El Capitan ("map of North America" dike, so named because of its crude resemblance to such a map) and stands out boldly against the lighter-colored El Capitan Granodiorite. *Inclusions* (many are pieces of older wall rock that are surrounded by cooled magma) are also locally abundant (Fig. 5–4). The mostly fist-size to beach-ball-size "blobs" of dark minerals surrounded by light-colored granite stand out like "sore thumbs."

Yosemite's Landscape—Joints and Fractures

Critical to understanding details of landforms is an appreciation of the characteristics of the underlying bedrock. Why do some rocks stand as tall pinnacles, such as Washington Column, and others as sheer cliffs thousands of feet high? Why do some features such as Half Dome have a rounded surface and others have straight or craggy forms? The answers lie in the arrangement and spacing of fractures in the bedrock. Nonfractured or wider-spaced fractured rocks are more resistant to weathering and erosion—closer spacing makes rock more vulnerable.

Fracturing can occur by stresses generated during intrusion of nearby plutons and by contraction produced during cooling of a pluton. Tectonic stresses generated during uplift or deformation produce intense fracturing in and near fault zones, and they can also produce geographically extensive groups of mostly vertical fractures called *master joints*[2] that often control the location of many valleys and narrow canyons. Shorter, more localized joints influence the weathering, erosion, and therefore the appearance of smaller topographic elements. *Cross joints* (intersecting sets of joints) account for the vertical pillar and columnlike landforms such as Washington Column. The chemical composition of the granitic rock, especially its silica content, also influences the amount of fracturing and therefore the rock resistance to weathering and erosion. Joints are more widely spaced in high-silica, quartz-rich granite and granodiorite rocks—therefore making these rocks more resistant (Huber, 1987). Wider joint spacing also tends to occur in coarse-grained (large mineral grains) igneous rocks.

A secondary type of fracturing in granitic rocks is produced by the process of *unloading,* or exfoliation. The weight of overlying rock exerts immense pressure and compresses buried rock. Removal of overlying rock allows the buried rock to expand slightly. This small expansion can be absorbed by highly fractured rock or structurally weak rock such as shale or slate. However, in a brittle, nonfaulted or relatively nonjointed massive rock such as many granites, the expansion is accommodated mostly at the earth's surface, producing *sheeting*. A series of fractures de-

[2]A joint in a rock is a fracture along which little or no displacement occurred.

velop parallel to the surface and slabs of rock will "peel off" as weathering proceeds. If the earth's surface were perfectly flat, expansion would occur upward and the exfoliation cracks would tend to be horizontal. If the rock is exposed along a deep canyon, as at El Capitan, the cracks form vertically and help account for the incredible 3000-foot-high (915-m) El Capitan cliff (Fig. 5–8). Near the edges of a relatively unfractured section of granitic rock, the exfoliation cracks will often curve and give a rounded form to the rock—an exfoliation dome. Numerous domes delight the visitor in the valley and elsewhere in the Sierras where rock conditions are right—silica-rich granitic rocks that are sparsely jointed. The great rounded domes of Yosemite could not have been developed in the rocks at Mt. Rainier or in Glacier National Park. Massive, brittle rocks with widely spaced joints are essential to exfoliation.

The most famous of the domes is one of Yosemite's trademarks—Half Dome (Figs. 5–8, 5–9, and 5–10) Well-developed exfoliation domes are rare in the world—a well formed half dome is rare indeed! Although initially regarded as a dome that had its northwest side removed by glaciers, it is most likely that Half Dome never had a second half! Half Dome and a number of other exfoliation domes are located along a wide ridge localized by master joints on either side. The vertical master joints that localized the formation of Tenaya Canyon on the north also control the impressive, 2000-foot-high (610-m) north wall of Half Dome (Fig. 5–10). Matthes (1930) determined that the peak was a *nunatak* during the most extensive glaciation—projecting some 900 feet (274 m) above the surrounding glaciers.

FIGURE 5–8 Yosemite Valley with El Capitan cliff (left), Bridalveil Falls and its hanging valley (right), and Half Dome in distance. (Photo by E. Kiver)

FIGURE 5–9   Half Dome as seen from Stoneman Bridge on Merced River, Yosemite National Park. (Photo by D. Harris)

There are a number of domes, however, that are distinctly elongate and asymmetrical. In the park, the steep ends are on the west, and their gradual east slopes are polished and striated. These domes—such as Lembert Dome (Fig. 5–11) and Fairview Dome—were overridden and reshaped by the ice. The up-glacier end was ground down by abrasion and the down-glacier end was quarried away, leaving a steep, stair-stepped cliff. Thus, the generally symmetrical domes are the result of exfoliation, and the distinctly asymmetrical domes—*roches mountonnees, whalebacks,* or *stoss and lee* forms—were formed by glacial abrasion and quarrying.

Arches are also among Yosemite's remarkable features. These are not the "see through" type of arches, but ones that form on the steep sides of exfoliation domes or exfoliation cliffs when large sections of an exfoliation sheet spall off

FIGURE 5–10   Wire-cable trail up Half Dome. Note the vertical exfoliation fractures maintaining cliff on right and curved exfoliation sheets on dome. (Photo by E. Kiver)

FIGURE 5–11   Lembert Dome, a roche moutonnee (stoss and lee feature); glacier rode over it from right to left. (Photo by D. Harris)

from the lower slopes. Almost invariably, the roof section forms a graceful *inset arch,* or *exfoliation arch.* The Royal Arches in upper Yosemite Valley, as classic examples, stand a thousand feet above the base (Fig. 5–12). Here, Mathews (1968) surmised that the glacier removed the lower sections of the exfoliation shells, leaving the upper sections without support.

Yosemite's Landscape—Glacial Features

In addition to the stoss and lee forms described above, a full range of alpine glacier landforms give the Sierra Nevada its rugged form. From cirques and horns in the high country where a large Pleistocene ice cap buried all but the tallest peaks to the glacial valleys and moraines in Yosemite and other high valleys, the work of ice is spectacularly displayed. However, this was not the widely accepted explanation during the early explorations of Yosemite nor, in general, during the 1800s when few studies of mountain landforms had been attempted.

The first geologic studies of the Yosemite area were by the Geological Survey of California under the direction of Josiah Whitney in 1863. Studies of glacial features were in their infancy and although Whitney recognized that the high Sierra had been glaciated during the Pleistocene, he believed that Yosemite Valley itself was too large and deep to be ice caused. He suggested that the valley was created by a "grand cataclysm." Thus faulting (the bottom dropped out!) was invoked to explain the valley.

FIGURE 5–12   Royal Arches in the Yosemite Valley have exfoliation sheets 200 feet (61 m) thick. (Photo by E. Kiver)

So began a 70-year-long controversy. John Muir disagreed and elegantly defended its glacial origin. Supporting evidence came from Muir's observations of glaciers in Alaska and during his many ramblings through his beloved mountains. Further support for the glacial hypothesis was added by U.S. Geological Survey geologist Francois Matthes who, along with Frank Calkins, studied Yosemite's geology beginning in 1913. Matthes's 1930 publication is a classic work in the geological sciences.

Older glaciers (pre-Wisconsin[3] or regionally named pre-Tahoe glaciations) flowed some 10 miles (6 km) past Yosemite Valley and deposited moraines and erratics in the El Portal area just outside of the park (Fig. 5–5). However, most of the glacial features in today's landscape formed during the Wisconsin Glaciation (regionally named Tahoe, Tenaya, and Tioga glaciations). It was during the most recent glacial period, the Tioga Glaciation (from about 30,000–10,000 years ago), that many of the finer details of today's landscapes were produced, including the abundant smooth, striated, and polished rock surfaces on many of which are found *glacial erratics* (Fig. 5–13). Tioga-age ice extended only as far as Bridalveil Meadow where it built a moraine that impounded the most recent Glacial Lake Yosemite. Like its predecessors, the lake quickly filled in with *till* and *outwash* sediment that was washed down the Merced River and its tributaries. Early explorers encountered swampy meadows that they later drained by blasting out a *recessional moraine* (moraine left by a glacier receding from its maximum or *terminal* position). The level of the valley was raised during each episode of lake filling, and the flat valley surface reflects the most recent filled-in lake (Fig. 5–14). Geophysical studies[4] and deep drill holes further supported the glacial hypothesis when it was determined that the underlying valley had a glacial profile rather than a trenchlike form as visualized by Whitney. The same studies also disclosed that the valley fill was locally

FIGURE 5–13   Erratic boulder, moved by a glacier, high up on Pothole Dome. (Photo by C.K. Harris)

[3]The Wisconsin Glaciation is the last major advance of continental glaciers in North America.

[4]A geophysical technique was used that measures the time required for shock waves to travel through the sediment, bounce off the bedrock, and return to the surface. Travel time is proportional to sediment thickness.

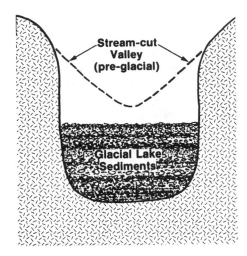

FIGURE 5–14   Generalized cross section of Yosemite Valley showing stream-cut valley (dashed line) and thick lakebed deposits. (Illustration by Gregory Nelson)

2000 feet (610 m) thick and that at least six intervals of lake deposition occurred. Thus, the total depth of the Yosemite Valley, including the 2000 feet (610 m) hidden beneath the valley floor, is nearly 6000 feet (1830 m)!

Striations high on valley walls have been used by some as reference points in measuring the thickness of the ice. This method is logical but hardly accurate because the position of the bottom of the glacier (and of the valley) at the time the striations were made remains unknown. What is known is that glaciated valleys were not cut during a single advance but during perhaps a dozen or more advances, each cutting deeper than the one before.

Yosemite's Landscape—Waterfalls

Thanks to the activity of former glaciers and variations in the spacing of bedrock fractures, an extravaganza of waterfall vistas awaits visitors to the Sierras—especially in the snowmelt season (April, May, and June). Those in the Yosemite Valley are particularly spectacular. Some are associated with hanging valleys, where smaller tributary glaciers joined the larger Merced and Tuolumne valley trunk glaciers. Others are *glacial steps,* where relatively unjointed resistant granitic rock makes up the lip of the falls and where glaciers deeply excavated the more highly fractured granite in the valley below. Vernal Fall (317 feet; 97 m) and Nevada Falls (495 feet; 181 m) are glacial steps that lead up the "Giants Stairway" to the Little Yosemite Valley. Bridalveil Falls (Fig. 5–8) is a classic example of a hanging valley waterfall—with a clear-leap drop of 620 feet (189 m) and a total drop of 850 feet (260 m).

Powerful Yosemite Falls, the highest in North America and the third tallest in the world, is composed of two vertical falls, 1435 feet (435 m) for the upper falls and 320 feet (98 m) for the lower. An intermediate cascades 675 feet high (206 m) brings the total vertical drop to 2425 feet (782 m)! The water shoots out into the valley from the upper fall and is a clear leap except for about 70 feet (21 m) where it runs over rock before reaching the intermediate cascades. Small in volume but highest in a single vertical drop is Ribbon Falls, 1612 feet (492 m) high.

Big Trees

Also among nature's "spectaculars" are the Giant Sequoia found in 75 isolated groves on the west slope of the Sierras in central California. Shorter than their coastal redwood (*Sequoia sempervirens*) cousins, the up to 300-foot-tall (91-m) Giant Sequoia (*Sequoiadendron giganteum*) is stouter and is the earth's most massive tree. They were widespread in North America some 25 Ma and were extensive in the Sierras prior to the Ice Age. Only patches survived the rigors of the Ice Age climates and many of these were logged in the late 1800s and early 1900s.

Yosemite contains three groves, the Mariposa (200 trees), Tuolumne (25 trees), and Merced (20 trees). The oldest tree (2700 years old) is the Grizzly Giant in the Mariposa Grove, the earth's fifth largest (by volume) tree. The diameter at its base is 27 feet (8.2 m) and its circumference is 89 feet (27 m)! At 96 feet (29 m) from its base it is 14 feet (4.3 m) in diameter where its first branch is located. That branch is 6 feet (2 m) in diameter, larger than most tree trunks! We can only attempt to imagine the grandeur of logged-out trees elsewhere in the Sierras that have 108 foot (33 m) circumference stumps. The famous Wawona Tunnel Tree with a wagon road (later an auto road) through its base toppled in 1969, weakened by the man-made tunnel. Now new threats have emerged as pollution from southern California cities is raising ground level ozone levels, even as far downwind from metropolitan areas as the Sierra parks. Tissue damage to the needles makes the trees more vulnerable to insects and disease. Also, global warming is increasing temperature and reducing precipitation—a condition that will further stress the Sequoia.

Today's Glaciers

The small remaining glaciers of today occupy only the highest of areas in favorable locations. North and northeast facing cirques and rock walls offer protection from the sun's rays where snowbanks and small glaciers can survive the summer melt. Muir's discovery of a living glacier on Merced Peak in the high country in 1871 further fueled the debate over the origin of the Yosemite Valley. Although Muir believed that this was a remnant of the last large Pleistocene glacier (Tioga glacial advance), Matthes later recognized that these were "new" glaciers that formed during one of the cool climatic intervals (Neoglacial events) since the Ice Age ended.

The most recent interval of cooling (last 700 years) was named the Little Ice Age by Matthes (Fig. 1–27).

Nearly 500 glaciers were present in the Sierra Nevada in 1980 (Huber, 1987) but that number is less today as global warming takes its toll. Muir's Merced Glacier no longer exists and many others have also disappeared or lost 75 percent or more of their surface area. The Lyell Glacier, the largest in the park, covers about one fourth of a square mile (160 acres).

In 1933 when park rangers hiked up to the Lyell Glacier, they saw a mountain sheep ram standing at the toe of the glacier. This discovery was truly surprising because the last mountain sheep in the Sierras was shot in about 1880. On closer inspection they learned that they had found the mummified remains of a mountain sheep that had fallen into the *bergschrund* (crack or crevasse that forms at the head of the glacier where it pulls away from the cirque headwall) about 250 years earlier. The frozen ram was carried by the slowly moving ice to the terminus where two of its feet were still frozen in the glacier at the time of discovery. One of the few positives in our thoughtless misuse of our natural environment is that a small number of Bighorn sheep were successfully reintroduced in the Yosemite area in 1986.

Yosemite Experience

Yosemite is a special place for many reasons. From the geologic standpoint many of our modern ideas about batholith emplacement, the work of glaciers, and Neoglaciation were spawned in this incredible area. Matthes's recognition of the Little Ice Age greatly influenced and encouraged studies of post-Pleistocene climatic change. Understanding these past changes will help us prepare for the serious impacts that the present and future climatic changes will have on the environment and on modern society. John Muir's new attitude toward Nature greatly influenced the way in which geologists and others see the earth and its workings. His view of Nature and the Yosemite Valley where "every rock in its walls seems to glow with life" is a dynamic view—one reflected in the process-oriented approach used today in most geologic studies. Geology is not just a study of landscapes that were created long ago, but also a study of living, changing landscapes with a future as well as a past.

Muir not only influenced the scientific community, he also understood that experiencing an unspoiled natural setting is vital to human development, and his writings encouraged others to share the glorious scenery of Yosemite, and to join him in preserving this and our nation's other remarkable areas as our national heritage. His influence spurred the environmental movement that continues to slow and may eventually reverse the deterioration of the earth's ecosystems and the lessening of the quality of life for the planet's inhabitants.

Muir's writings are not the only celebrations of our parks. The paintings of Albert Bierstadt, Thomas Moran, Wilson Hurley, and other artists, and the stunning photographs by Muybridge, Weed, and Ansel Adams reveal from various approaches the beauties of our parklands. These art treasures have emotionally impacted and given pleasure to generations of people. Even your own snapshots are precious records of your delighted reactions to your discoveries.

Each visitor to a park makes his or her own discoveries. Almost every spot in every park has been visited by someone, but to each new visitor every discovery is new. Each visitor is a modern-day explorer, discovering scenery and gathering information and getting ideas. The scenery may be overwhelming—mountains, canyon, caverns, oceans, forests, broad plains. Or it may be miniature—sea life in tidepools, exquisite wild flowers, colorful rocks, eons-old fossils of shells, leaves, insects. All of it is the visitor's to store away in memory. The value of sharing with family and friends also creates the type of experiences that last a lifetime. Changes in a national park over time are minimal—especially in comparison to the rapid changes in urban and suburban areas where technology and societal values and attitudes are rapidly evolving—often with negative effects.

Where should one start to discover Yosemite? The visitor center is a good beginning, followed by a hike on one of the many trails. Short loop trails, 2- or 3-hour trail hikes, many-day backpack trips—all are available. Come with open minds, open eyes and ears, open hearts—they will be filled.

John Muir's advice is still good today—take the 3-mile trail from Mary Lake to the top of Mt. Hoffman. Here, near the center of the park, one can see to the edges and beyond. Take Huber's (1987) U.S. Geological Survey Bulletin 1595 with you and match his excellent photos with the 360° view of the park. Perhaps you will agree with Muir's philosophy that "going to the mountains is going home."

REFERENCES

Calkins, F.C., 1930, The granitic rocks of the Yosemite region, in, Matthes, F.E., Geologic history of the Yosemite Valley: U.S. Geological Survey Professional Paper 160, p. 120–129.

Calkins, F.C., Huber, N.K., and Roller, J.A., 1985, Bedrock geologic map of Yosemite Valley, Yosemite National Park, California: U.S. Geological Survey Map I-1639.

Huber, N.K., 1987, The geologic story of Yosemite National Park: U.S. Geological Survey Bulletin 1595, 64 p.

Mathews, W.H., III, 1968, A guide to the national parks: Their landscape and geology, The Natural History Press, New York.

Matthes, F.E., 1930, Geologic history of the Yosemite Valley: U.S. Geological Survey Professional Paper 160, 137 p.

National Park Service, 1989, Yosemite: Official national park handbook: National Park Service, Division of Publications, Washington, DC, 144 p.

Norris, R.M., and Webb, R.W., 1990, Geology of California: Wiley, New York, 541 p.

Park Address

Yosemite National Park
P.O. Box 577
Yosemite National Park, CA 95389

SEQUOIA–KINGS CANYON NATIONAL PARKS
(CALIFORNIA)

In the southern Sierra Nevada about 40 miles (64 km) southeast of Yosemite are two parks that together cover an area of 864,383 acres (1350 square miles; 3500 km²). They are discussed here as a single park partly because they share a common border (Fig. 5–15) and are administered together. The huge Sierra fault block terminates about 75 miles (120 km) to the south in the Tehachapi Mountains, which in turn merge into the Transverse Ranges of southern California. Sequoia–Kings Canyon include the highest part of the Sierra with each park containing six peaks over 14,000 feet (4268 m)—including Mt. Whitney (14,494 feet; 4419 m), the highest point in the conterminous United States. Elevation changes are spectacular, especially on the east edge of the fault block where it drops precipitously from over 14,000 feet along the Sierra crest to elevations of 3500–4000 feet (1070–1220 km) in Owens Valley, less than 10 miles (16 km) to the east. The vertical change on the gently inclined west side of the fault block is even greater but occurs over a 40-or-more-mile (64 km) distance. The Great Western Divide lies west of the Kern Canyon and the Sierra Crest and is tall enough to block views of Mt. Whitney from the west side of the park. The west slope is drained by the San Joaquin, Kings, Kaweah, and Kern rivers and on the eastern, drier side of the range by small creeks that flow into the Basin and Range Province. No Sierra water reaches the Pacific Ocean. Evaporation in the dry desert climate, irrigation, or diversion to Los Angeles and other cities and towns accounts for all of the runoff from the Sierra Nevada. No road crosses the rugged mountains here. Access is from the west on roads that dead end in Kings Canyon on the north and Kaweah valley and Mineral King valley (Little Kern River) in Sequoia to the south. Those who wish to cross or to see the fabulous interior of the park will have to do it the "Indian way" using the excellent back country trail system.

Park History

Spanish explorers named the River of the Holy Kings (now Kings River) in 1805, but little was known about the upriver canyons and mountain peaks. A hunter discovered the giant sequoia trees near Yosemite in 1833 and loggers discovered some of the groves of giant trees near the present Sequoia Park in 1858. Sawmills were soon erected, and many of the giant trees were turned into stakes for California's flourishing grape industry. John Muir explored the area in 1873 and found Kings Canyon of comparable grandeur to his beloved Yosemite. More importantly from the standpoint of designating this an area worthy of national park protection, he discovered and named the Giants Grove, the earth's largest grove of sequoia trees (3500 trees—about 25 percent of all giant sequoia in the world). The grove was withdrawn from eligible homesteading land in 1880, and far-sighted individuals like John Muir and especially Colonel George Stewart, a Visalia newspaperman, led

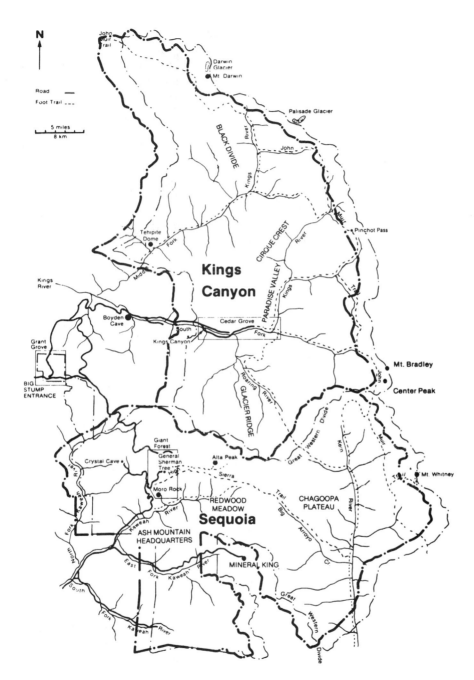

FIGURE 5–15 Map of Sequoia–Kings Canyon National Parks.

the cause to save some of these unique trees and some of the Sierra wilderness (National Park Service, 1992).

Sequoia was established as our second national park in 1890. The name honors the tree and one of the remarkable original Americans, chief Sequoyah of the Cherokee tribe. Among other accomplishments, Sequoyah invented a simple alphabet for his people in 1821 that enabled them to learn to read and write quickly. One week after the establishment of the park, it was tripled in size and nearby Grants Grove became General Grant National Park. Fortunately, Giant Forest was also added to the park—just as a timber company had nearly completed a haul road to the grove. Not so lucky were the trees in Big Stump Basin. One can wander through this area examining the huge stumps and imagine the immensity of the trees that once graced the landscape. If we are patient and take a far-sighted view (skills not yet developed by most politicians and those interested in only their "bottom lines"), others will enjoy a regenerated old-growth forest of sequoia here in about 2000 years!

General Grant National Park was later absorbed into the newly established Kings Canyon National Park in 1940. Considerable increase in area and the addition of other groves created a park closer to the early conservationist's dream of a continuous park along the Sierra Crest from Yosemite to Sequoia. Interestingly, the namesake of Kings Canyon was originally omitted along with other spectacular valleys because of their hydroelectric potential. These were later included in the parks, the last addition being Mineral King valley where a proposed ski complex would have forever changed the character of the southern end of Sequoia National Park.

Just as the writings of John Muir and the efforts of Colonel Stewart saved the area from destruction by gaining park status for it, the work of Colonel John White, the second (and fourth) civilian superintendant saved the back country wilderness and relatively uncluttered character around the sequoia groves by preventing overdevelopment. White was 50 years ahead of his time. No giant hotels, golf courses, and other human-contrived amusements obstruct the landscape inside the park. Had all of our parks followed his lead, the quality of today's park experience would be greatly improved. White's advice is still valid today—park your car and experience the park along its trails—the ultimate way to experience nature.

Geologic Story

The overall geologic picture in Sequoia–Kings Canyon is similar to that at Yosemite and elsewhere in the Sierra Nevada. Accretion of terranes containing metamorphosed Paleozoic and Mesozoic marine sedimentary and volcanic rocks was followed by intrusion of batholiths during the Late Jurassic to Late Cretaceous (120–80 Ma). The newly formed subduction zone on the west edge of the North American continent initiated a series of mountain-building events that began with orogenies in the Sierra area about 200 Ma (early Mesozoic) and advanced eastward through Nevada, Utah, and Arizona to the Rocky Mountains, finally ending about 40 Ma (middle Tertiary). "Bubbles" of magma rose from the lower lithosphere

(Fig. 5–2) and eventually experienced a "heat death" by crystallizing into the massive complex of plutons known as the great Sierra Nevada Batholith. The long interval of erosion following the building of the Mesozoic-age Nevadan Mountains removed miles of overlying rock, unroofing the batholith by the time that late Cenozoic uplift began along the huge Sierra frontal fault. More than 2 miles (3.2 km) of vertical displacement along the Sierra Fault during the past 20 million years generated North America's longest single mountain range and elevated the Sierras as the highest range in the conterminous United States.

In contrast to Yosemite where most of the overlying older metasedimentary and metavolcanic rocks were removed by erosion, roof pendants in Sequoia–Kings Canyon are more numerous and cover approximately 20 percent of the park (Ross, 1958) compared to 5 percent for Yosemite. These roof remnants are oriented northwesterly and locally display sharp boundaries with the "granite" (actually compositions range from gabbro to alaskite—mostly granodiorite); elsewhere fingers of granitic rock are injected into the older metamorphic rock. Limestones caught in the metamorphic pressure cooker during the batholith and mountain-building stage were locally mineralized and converted to marble. The slow processes of solution later attacked the marble and developed over 100 known caves in the park, including the delightful Crystal Cave, one of California's most popular commercial caves (Despain, 1996). Summer tours to the well-decorated "cave in a roof pendant" are arranged at the Lodgepole Visitor Center in the Giant Forest.

Increasing elevation of the rising fault block combined with the alternating cooling and warming climate intervals of the Pleistocene initiated a series of glacial advances and retreats in the High Sierra. Although less intense in the southern Sierra, glaciation was robust at high elevations and left its signature written in the rugged cirques, aretes, polished bedrock and other glacial features of the high country (Fig. 5–16). An ice cap covered much of the area above 9000 feet (2744 m) and valley glaciers scoured deeply through Kings Canyon, Kern Canyon, and other park valleys. The extent of ice is marked by the change from the broad glacial profile of the valley to the notchlike, V profile more typical of a stream-cut valley. Glaciers on the east side of the Sierra Crest were much shorter, less than 10 miles (16 km) in length.

Landforms

In addition to the unique marble caves discussed previously, granite landforms including exfoliation domes and *granite sheeting* along vertical cliffs are prominent features in the parks. Unloading, or removal, of the overlying load of rock permits the massive, relatively unfractured sections of the granite to expand and form cracks, or joints, that roughly parallel the earth's surface. Parallel, curved sheets of rock spall from the exfoliation domes, much like the shells of a giant onion. The quarter-mile-long, concrete-step trail to the top of Moro Rock near the Giant Forest in Sequoia provides outstanding views of the glaciated topography of the Great Western Divide and the Kings Canyon and the Kaweah River glacial valleys. Because

FIGURE 5–16   Center Peak (12,760 feet; 3890 m), with the peaks along the Kings–Kern Divide (Great Western Divide) in the distance. (Photo by Anthony Morse)

of the high elevations of the Great Western Divide, Mt. Whitney cannot be seen from Moro Rock or other areas on the west side of the park. Mt. Whitney and the Sierra Crest east of the Great Western Divide are best seen from Owens Valley along the east side of the Sierra fault block or from high trails deep in the Sierra wilderness (Fig. 5–16).

The entire east face of the Sierra block is a landform in itself—a *fault scarp* that continues to form today (Fig. 5–3). About 18 miles (29 km) due east of Mt. Whitney at the base of the fault scarp lies the town of Lone Pine at an elevation of 3700 feet (1128 m)—over 2 vertical miles below the peak. The March 26, 1872, Lone Pine earthquake is a reminder that the restless crust under the Sierra Nevada is still active. An estimated magnitude 8.0 temblor (largest historic earthquake in California) leveled the town buildings, killed 27, and left a 100-mile-long (161 km) gash in the earth's surface.

Remnants of erosion surfaces along the Kern River valley and elsewhere are also the product of continued uplift. Lateral cutting by streams produced broad surfaces that were later incised by streams that were energized by the continuous uplift and tilting of the fault block.

Kings Canyon is one of North America's deepest canyons, 8200 feet (2500 m) from the river to the top of Spanish Peak just outside of the park—deeper than Arizona's Grand Canyon or Hells Canyon along the Idaho–Oregon border! The 44-mile-long (71-km) glacier that cut the canyon was the largest and longest in the parks. The Kern River in southern Sequoia National Park flows south in a remarkable 70-mile-long (113-km), arrow-straight canyon that separates the Great Western Divide ridge from the higher Sierra Crest to the east. Running water and later glaciers reached the highly fractured rock along the Kern Canyon fault and excavated a canyon up to 6000 feet (1829 m) deep. The Kern glacier flowed 7 miles (11 km) south of the park boundary making it the southernmost Pleistocene glacier in the Sierra Nevada. The river finally leaves the fault trough, abruptly turning west down the gentle slope of the fault block. The water never reaches the Pacific Ocean but is destined to irrigate the agricultural fields in Kern County—"the salad bowl of America." Elevations to the south rapidly decrease—too low to have supported glaciers.

Each park has six peaks over 14,000 feet (4268 m) in elevation. It is here along the Sierra Crest that the cirques, horns, and other glacial landforms are at their finest. Small moraines nestled in cirques or just below cirques and today's small cirque glaciers were formed during the Neoglacial—many of the moraines were emplaced during the late 1800s near the end of the Little Ice Age (Fig. 1–25). The Palisade Glacier in Kings Canyon is the largest living glacier in the Sierras and the southernmost true glacier in the United States. The glacier is about one mile long—about one half its size in 1850. Other high areas in the parks were extensively glaciated. Many are accessible by half day and day hikes in the west side of the park. Cirque lakes and other excellent glacial landforms including abundant striated and polished rock surfaces (Fig. 5–17) and high country snowfields (Fig. 5–18) can be examined at Heather and Pear Lakes—reached by trail from the Giant Forest.

FIGURE 5–17   Glacial polish and erratics in Center Basin, looking east to Mt. Bradley (13,289 feet; 4050 m) in southeastern part of Kings Canyon National Park. (Photo by Anthony Morse)

FIGURE 5–18   Ice pinnacles (nieve penitente) characteristically form only at very high elevations. (Photo in Mt. Whitney area by Peter Barth)

Big Trees

The scrublike chaparral and oak woodlands in the western foothills give way to a mixed conifer forest near an elevation of 4000 feet (1220 m). Tree growth is greatly reduced at about 9000 feet (2744 m) in the subalpine zone and nonexistent above 11,000 feet (3354 m) because of the rigorous climate. Stripping of soil by Pleistocene glaciers also contributes to the sparsity of vegetation at higher elevations. Between elevations of 4000 and 8000 feet (1220 and 2440 m), in the mixed conifer zone in nonglaciated areas, are isolated groves of the earth's largest trees—*Sequoiadendron giganteum.* Although not as tall as their coastal redwood cousins *Sequoia sempervirens,* they have much larger diameters and are therefore more massive.

Cedar Grove in Kings Canyon is outstanding and the Giant Forest in Sequoia is particularly significant. With over 3000 mature trees (diameters greater than 10 feet; 3 m) Giant Forest is the earth's largest sequoia grove. Four of the five largest living things are found here, including the General Sherman tree, named in 1879 by a Civil War veteran for William Tecumseh Sherman. With a 103-foot (31-m) circumference and a height of 272 feet (83 m), General Sherman weighs in at 1385 tons, over eight times the mass of the blue whale (165 tons), the world's largest animal. Giant sequoia can reach an age of 3500 years or more (Cook, 1955), not as old as the bristlecone pines in Great Basin National Park and other areas of the southwest, but certainly one of the world's oldest living things. Their major enemies are toppling (especially due to ground settling or if their roots are undercut) and chainsaws. Fortunately, the latter threat is gone—but so are many of the groves that were in the Sierras in the late 1800s and early 1900s. The Converse Grove, just north of Grant Grove, formerly contained the world's largest sequoia grove until a timber company cut down nearly every mature tree.

The sequoia appears in the fossil record during the Mesozoic—contemporaneous with the dinosaurs. The tree was very widespread in North America and elsewhere with dozens of species existing the world over—up to Pleistocene (Ice Age) times—about 2 million years ago. Fossil stumps, branches, cones, and/or needles occur in a number of park areas including Petrified Forest (200 Ma), Yellowstone's fossil forests (50–40 Ma), and Florissant National Monument (36–30 Ma). The major climate disruptions of the Ice Age greatly reduced their numbers—only two species were able to reestablish following the Wisconsin Glaciation (last glacial interval in North America). A third surviving species was discovered in 1947 in central China. *Metasequoia,* or the dawn redwood, had previously been known only from the fossil record.

Park Future

As is true of other remarkable areas that remain in our parks today, as well as in many other areas in the world, Sequoia–Kings Canyon is safe—for now! The vagaries of politics and human attitudes can quickly change. The need for jobs and the perceived need for more material things can produce short-sighted actions that ignore those who come after us. Even with the best of intentions the population time bomb will negate at least some of our efforts. Our past record has bright spots, such as the establishment of our national parks, forward-looking environmental legislation, and the efforts of public-interest groups that have prevented the construction of reservoirs, a trans-Sierra highway, a tramway, and a large ski resort in the Sequoia–Kings Canyon Parks. However, losses, usually of the irreversible type, also occur.

Park Animals

One bright spot is the increase in Bighorn sheep population in the Sierra Nevada. Overhunting and diseases introduced by domestic sheep nearly exterminated the Bighorn population by the late 1970s. Two small natural herds survived in the east part of Kings Canyon and were used to restore populations in Sequoia and Yosemite. The black bear population is high, particularly after the last individual of one of their two main enemies, the grizzly bear, was shot just outside of Kings Canyon in 1922. The other enemy, man, still remains.

A small population of rare Ice Age mammals such as the wolverine, red fox, and fisher still inhabit the Sierras. These Ice Age relicts are at the extreme southern limit of their ranges. Global warming will further stress their precarious existence—as well as further restricting the distribution of the climate-sensitive giant sequoia. New threats, such as air pollution originating elsewhere, now threaten the park. Visibility is greatly reduced because of high particular content in the air. Acid rain and snow resulting from the burning of fossil fuels is stressing plants and aquatic systems in lakes, ponds, and rivers. Ground-level ozone pollutants from Los Angeles and other population centers have also dramatically risen in the park. About 40 percent of all Ponderosa and Jeffrey pines in the park display visible needle damage. Higher ozone levels are expected to reduce growth in Sequoia trees as well as damaging lung tissue in animals.

Seeing Sequoia–Kings Canyon

Sequoia–Kings Canyon is fortuitously located in southern California within a day's drive of some of the nation's most populous areas. The opportunity to leave the unnatural urban setting and to see and touch nature up close provides some stability to a way of life that can be trying and stressful. As John Muir would say, "going to the mountains is going home." Experiences in the out-of-doors, especially those shared by entire families, serve as reminders of what things are really important.

As is true for many park areas, there are two very different elements to the park experience. Road access is mainly from the west up the gentle slope of the tilted fault block to the magnificent Kings Canyon and Kaweah River valleys and the incredible sequoia giants in the Grant, Cedar, and Giant Forest groves. From there the second element of the park experience is available along short trails that take one to local viewpoints and areas of interest. Longer trails lead many miles into the backcountry across the Great Western Divide, through the deep Kern River Canyon, and onto the Sierra Crest in the fragile alpine zone where humans are but short-term visitors. The backcountry wilderness is a relatively unspoiled piece of nature that will hopefully remain in a nearly pristine state forever—if humans act responsibly.

Although this rugged alpine wilderness is accessible only to those with the energy and time to explore, it is accessible in the minds of all who know or can imagine the glistening, glacial-scoured granite surfaces, alpine lakes, and the small alpine plants clinging tenaciously to bits of soil at high elevations. Just as all of us cannot visit the Smithsonian Institution or the Louvre in Paris, we can all feel satisfaction knowing that these places exist and their special collections are protected. The road over Tioga Pass in Yosemite and many other areas in the high mountains of the west provide glimpses of the rugged environment at the roof of the mountains (Fig. 5–1) for those who are unable or unwilling to explore along trails. For those willing to be self-sufficient for a few days and leave our contrivance-dependent society behind, hiking the John Muir Trail (part of the Pacific Crest Trail from Canada to Mexico) along the backbone of the Sierra Nevada is an outstanding experience. For those seeking a further challenge, Mt. Whitney's 14,494-foot (4419-m) peak awaits those who wish to stand on the highest summit in the conterminous United States and drink in the magnificent views of the high country cirques, horns, aretes, and small glaciers as well as Nevada's Basin and Range desert stretching eastward as far as the eye can see.

REFERENCES

Cook, L.F., 1955, Giant sequoias of California: National Park Service publication, 28 p.

Despain, J., 1996, Crystal Cave: Three Rivers CA, Sequoia Natural History Association, 49 p.

National Park Service, 1992, Sequoia and Kings Canyon: Handbook 145, Washington, DC, National Park Service, 127 p.

Ross, D.C., 1958, Igneous and metamorphic rocks of parts of Sequoia and Kings Canyon National Parks: California Division of Mines, Special Report 53, 24 p.

Park Address

Sequoia–Kings Canyon National Parks
Three Rivers, CA 93271-9700

DEVILS POSTPILE NATIONAL MONUMENT (CALIFORNIA)

A narrow strip of land containing a mere 800 acres, Devils Postpile is one of our smallest national monuments (Fig. 5–19). It is located east of Yosemite National Park and a few miles southwest of the resort town of Mammoth Lakes near the east edge of the Sierra Nevada fault block. The Middle Fork of the San Joaquin River flows south through the monument before turning west to flow across the Sierras and into the Great Valley of California near Fresno. For those hiking the John Muir Trail between Yosemite and Sequoia–Kings Canyon, this tiny monument is just a short side trip, but the remarkably perfect polygonal columns ("posts") of basalt (Fig. 5–20) make a visit here a must. These great pillars mystified early observers— some of whom thought that lava "plunged over a precipice, split into prisms, and hardened in mid air!" (Huber and Eckhardt, 1995). Fortunately, we have a more logical explanation today that involves the lava coming to rest, solidifying, and contracting as the hot, solid lava loses further heat.

History

Indian campsites and trails are abundant in the area, but no record of their opinions of the postpile are recorded. Non-Indians like local hermits Red Sotcher and Tom Agnew, as well as early miners, must have known about the striking columns. However, no early written information is known other than that the area was locally called the Devils Woodpile.

In 1890 when Yosemite National Park was established, its boundaries were much more extensive than present and encompassed Devils Postpile. Pressure from the timber and mining industry was immediate to reduce the park boundaries.

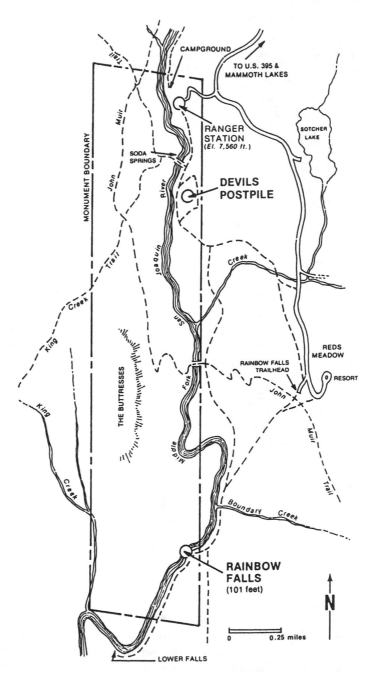

FIGURE 5–19   Map of Devils Postpile National Monument. (Huber and Eckhardt, 1995, permission by Sequoia Natural History Association)

Military officers served as the first park superintendents—many of them recommended that the southeast part of Yosemite be returned to the public domain and be made available to those who might exploit the resources. However, Captain Alex Rodgers noted in 1895 that the Devils Postpile was justification in itself to maintain park status. Subsequent army superintendents built trails and a substation in the area. In spite of growing recognition of its importance, Congress removed 500 square miles (1295 km^2) including Devils Postpile, from Yosemite in 1905. Newspaper articles about Devils Postpile soon aroused the public's interest. Intense mineral exploration began immediately in the former parklands and continued for many decades—eventually determining that the area was barren of important mineral deposits!

In 1910 an application from a mining company was received by Walter Huber, the district engineer for the U.S. Forest Service, proposing to blast the Devils Postpile and use the rock for the core of a hydroelectric dam on the Middle Fork! Huber was personally opposed to this ludicrous plan as were John Muir and other members of the newly formed Sierra Club. Letters written to officials in Washington led to President William Howard Taft designating the area as Devils Postpile National Monument on July 6, 1911.

Regional Geology

Devils Postpile National Monument, the area near Mammoth Lakes, and the nearby Ritter Range contain a spotty but adequate geologic record that outlines the broad history of the Sierra Nevada. Roof pendants containing accreted Paleozoic and Mesozoic sedimentary and volcanic rock terranes in the Ritter Range to the west and north of the monument provide evidence of long vanished seaways and island volcanoes. Significant tectonic activity began during the Mesozoic (Huber and Rinehart, 1965) as a subduction zone formed along the west edge of the North American Plate (Fig. 3–2). Rocks caught in the subduction zone were further bent, broken, and metamorphosed as they were swept against and under the North American Plate. Simultaneously, a series of rising plutons intruded the older rocks in the ancestral Sierra Nevada mountains from about 150 to 80 Ma (Fig. 5–1) forming the Sierra Batholith. Subduction activity also greatly thickened and heated the crust in the Great Basin Section of the Basin and Range Province to the east—an important factor that would later account for the pull-apart or tensional forces that produced the Sierras, the Basin and Range, and episodes of Cenozoic volcanism.

A long interval of erosion literally removed miles of overlying rock, eventually exposing the granitic core of the range. Slow uplift and tilting to the southwest began about 25 Ma (Fig. 5–3). Uplift was minor initially and rivers flowing from Nevada were able to maintain their courses across the future site of today's Sierra Nevada. Eventually the uplifting fault block diverted rivers from their paths across the range. Blockage of valleys by lava to the north also helped sever the stream connection with Nevada and helped isolate the San Joaquin drainage basin where

Devils Postpile is located. Rapid uplift along the Sierra frontal faults combined with volcanic activity and glaciation in the past few million years have produced a spectacular, fascinating landscape in the Devils Postpile area.

Recent Volcanism and Faulting

Renewed volcanism, particularly near the active faults in the eastern part of the Sierra Nevada, began most recently about 3 Ma and continues to the present. Both recent volcanism and fault movement are tied to the lateral, or strike-slip, movement of plates along the massive San Andreas Fault near California's west coast. The change from subduction (compression) to strike-slip plate movement about 20 Ma (Fig. 3–2) further removed support for the thick crust in the Great Basin, encouraging the crust to collapse and the Basin and Range Province to widen. The resulting crustal thinning produced tensional faults in the eastern Sierras and Basin and Range, which in turn caused hot mantle to rise. The resulting thermal expansion and partial melting of the mantle reduced rock density and account for the unusually high elevation of the eastern Sierra (Park and others, 1996). Localized magma bodies still rise and occasionally vent—particularly in areas where faults provide easier access to the low pressures at the earth's surface. As long as the current plate movements continue, the high elevation of the Sierras and occasional volcanism and earthquakes will persevere.

The oldest volcanic rocks in or near Devils Postpile are the Quaternary-age basalt flows that partially underlie an area of craggy peaks called the Buttresses on the west side of the Middle Fork of the San Joaquin River (Fig. 5–19). The Reds Meadow welded tuff just east of the monument is about the same age as the massive, catastrophic eruption of the nearby Long Valley caldera (760,000 years old) and is likely related to it (Huber and Eckhardt, 1995). Next in the sequence is a large andesite flow that cascaded down the eastern valley wall from the Mammoth Pass area and the formation of the impressive Mammoth Lakes plug dome at the edge of the Long Valley Caldera about 180,000 years ago. A rhyodacite lava that erupted from vents just south of the park forms the resistant rock ledge where the San Joaquin River leaps 101 feet (31 m) over Rainbow Falls in the south end of the monument. The Devils Postpile flow is less than 100,000 years old and erupted from a vent just west of the river near the Upper Soda Butte Campground. It is in this flow that the remarkable posts or columns have developed. Remnants of cinder at the vent site indicate that a cinder cone once stood here. However, the volcanic story does not end here. Even more recent eruptions have occurred on the nearby Mammoth Mountain plug dome and in the Long Valley Caldera. Just southeast of the monument the Red Cones cinder cones and basalt flow formed in postglacial time—less than 10,000 years ago! Even younger are the crunchy-underfoot, airfall pumice deposits along the trails and hillsides that fell on the area during the last few hundred years. Eruptions from one of the plug domes located along the Mono Craters fissure just east of the Sierras is its suspected source.

Devils Postpile Flow

The Devils Postpile basalt flow was at least 3 miles (4.8 km) long and ponded in the valley to a depth of at least 400 feet (122 m). Although much of the lava was removed later by glaciers and stream erosion, the thicker parts of the flow that display the peculiar fracture pattern called *columnar jointing* remain for all to ponder and admire.

The key to understanding fracture or joint patterns in lava is that most solids occupy a larger volume when hot and shrink when cooled. The general history of

FIGURE 5–20   The "posts" of Devils Postpile. (Photo by D. Harris)

cooling of a thick body of molten rock at the earth's surface is easily understood. Cooling and thus conversion to a solid occurs most rapidly at two places—one where the lava loses heat to the ground below and the other to the air above. The cooling unit at the base of the flow is called the *colonnade* and the upper cooling unit is called the *entablature* (Fig. 5–21). Eventually the liquid core in the middle part of the flow also solidifies. Once solid, further heat loss produces contraction and cracks called *cooling joints*.

Columnar Jointing Formation

Brittle, homogeneous materials in the colonnade and entablature shrink slowly and tend to crack in hexagonal patterns. Such a geometric shape "provides the greatest relief of stresses with the fewest cracks" (Huber and Eckhardt, 1995). Perfect hexagonal columns in lava are rare, indicating that ideal cooling conditions and/or homogenous material are seldom present. Cooling is slowest and least likely to be interrupted in the lower part of a lava flow. Consequently columnar jointing forms best in the colonnade. Other polygonal forms in nature occur where contraction is important—where muddy sediment contracts during the drying process and in cold areas where ice begins to contract when ground temperatures go below −20°F (−29°C).

The cracks or joints tend to form perpendicular to the cooling surface and will be vertical if the ground beneath the flow is horizontal and ideal cooling conditions prevail (Fig. 5–21). However, some of the columns display spectacular curved

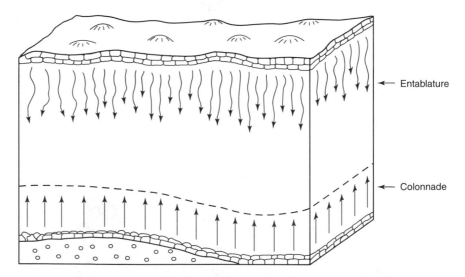

FIGURE 5–21   Propagation of vertical cooling joints in the lower (colonnade) and upper (entablature) cooling units in an idealized lava flow. (After Huber and Eckhardt, 1995)

shapes. Because cooling joints tend to form perpendicular to the cooling surface, buried hills and valleys, irregular surfaces of lava flows, filled lava channels and lava tubes, and rain or water contacting the lava may account for the nonvertical columns. Rapid cooling by water or movement during cooling may account for the random fracturing observed in some parts of lava flows (Long, 1987).

The columns at Devils Postpile are a maximum of 3.5 feet (1.1 m) in diameter and are up to 60 feet (18 m) long. According to Huber and Eckhardt (1995), 2 percent of the columns are four sided, 37 percent are five sided, 55 percent are six sided, and 5 percent are seven sided.

Pleistocene Glaciers

The finishing touches to the Devils Postpile flow came when the most recent of numerous episodes of glaciation gripped the Sierra Nevada in its icy claws. Rugged landforms were cut into the rapidly uplifting fault block as glaciers stretched from the high country to lower elevations following the stream-cut valleys. Deep canyons and valleys were cut—including the valley of the Middle Fork of the San Joaquin. The Devils Postpile flow was mostly scoured and plucked away about 12,000 years ago by the latest of the Pleistocene valley glaciers. The numerous cooling joints enabled glaciers to completely remove the entablature (upper cooling unit) and excavate the upper ends of the colonnade—especially on slopes facing downstream where the glaciers could freeze to rocks and drag and pry joint blocks away from their rocky homes. Upstream-facing slopes could be slowly ground away only by debris embedded in ice and dragged along the rock surface—equivalent to a giant file or miles of sandpaper being dragged across the rock surface. The elongate ridges with gentle upstream slopes and steep downstream slopes are called whalebacks, stoss and lee topography, or roches moutonees.

The sides of the posts are spectacularly exposed a short distance from the campground where the river has cut through the colonnade (lower cooling unit) section of the flow. Talus at the foot of the cliffs allows one to examine the fallen posts or columns up close. A hike to the top is a "must" to see the glacially polished upper ends of the columns (Fig. 5–22). Only more recently exposed surfaces show the fresh-appearing polish and striations—weathering quickly takes its toll. The importance of preserving these irreplaceable features for succeeding generations to enjoy is readily apparent.

Rainbow Falls

A short hike to the south end of the monument (Fig. 5–19) brings one to the 101-foot-high (31-m) Rainbow Falls where whitewater against the black cliffs presents a striking contrast, enhanced by a rainbow during the middle of the day. Water plummets over the upper, more resistant section of the Rainbow Falls rhyodacite

FIGURE 5–22   Glacially polished and striated tops of posts, Devils Postpile. (Photo by E. Kiver)

(a welded tuff with a silica content between that of rhyolite and dacite). The relatively unfractured rock at the lip of the falls grades downward to a more fractured, less resistant zone displaying numerous horizontal cooling cracks, an example of *platy fracture*. Water plunging over the falls erodes the fractured rock underneath causing undercutting and occasional collapse of the resistant rock. Thus, the cliff is maintained although the falls is slowly retreating headward up its valley—similar to the conditions at Niagara Falls in upper New York State.

The Middle Fork of the San Joaquin River has not always flowed in its present position. After the last glacier melted, the river flowed in a channel closer to the Buttresses on the west side of the valley and cut a cliff on its eastern bank. Later the river was diverted upstream to a new channel and plunged over the previously cut cliff. The cliff or waterfall has now retreated about 500 feet (152 m) upstream but the old channel is still recognizable.

Geologic Future

Plate motions continue to pull the crust apart in the eastern Sierra Nevada. Thinning of the crust produces additional deep melting as pressure is reduced and hotter mantle material continues to rise. The former vent system for the Devils

Postpile flow near the Upper Soda Springs Campground is inactive now. However, local springs in the monument and nearby areas continue to discharge carbon dioxide and hot water derived from magmatic sources below. The Red Cones (cinder cones) are hardly touched by erosion and frequent earthquakes occur in the Mammoth Mountain–Long Valley–Mono Craters areas. It is in these latter areas that the next of the inevitable eruptions is most likely to occur.

The causes of ice ages is still under investigation, but strong evidence currently supports a climate-forcing mechanism involving variations in the earth's orbit around the sun. Distance between the earth and the sun varies over a 100,000-year interval, and the tilt and wobble of the earth's axis of rotation vary over shorter time intervals. According to some calculations (Skinner and Porter, 1987) the Sierra Nevada and the entire world will experience another episode of glacial activity in about 23,000 years. What the world will look like and which animals will be here to greet the arrival of the next ice sheets is an interesting question to ponder.

The talus at the base of Devils Postpile bears mute testimony to the work of water entering and freezing in the cracks or rock joints. Ice expansion pries the columns apart, causing them to lean and eventually topple under the pull of gravity. Three leaning columns photographed in 1909 fell in 1980 (Huber and Eckhardt, 1995, p. 20). The increase in earthquake activity in 1980 in the Long Valley caldera (some earthquakes were magnitude 6.0) may have aided the loosening and collapsing processes. There are many uncertainties as to which processes or which combination of processes will ultimately destroy these remarkable columns. One thing is known for certain—change will occur.

REFERENCES

Huber, N.K., and Eckhardt, W.W., 1995, Devils Postpile story: Three Rivers, California, Sequoia Natural History Association, 30 p.

Huber, N.K., and Rinehart, C.D., 1965, Geologic map of the Devils Postpile Quadrangle, Sierra Nevada, California: U.S. Geological Survey Map GQ-437.

Long, P.E., 1987, Review of evidence for the quenching origin of entablatures in Columbia River Basalt flows: Geological Society of America, Program with Abstracts.

Park, S.K., Hirasuna, B., Jiracek, G.R., and Kinn, C.L., 1996, Magnetotelluric evidence of lithospheric mantle thinning beneath the southern Sierra Nevada: Journal of Geophysical Research, v. 101, p. 16,241–16,255.

Park Address

Devils Postpile National Monument
c/o Sequoia and Kings Canyon National Parks
Three Rivers, CA 93271

SIX

Basin and Range Province

The Basin and Range Province is a sprawling area that extends from southern Oregon southward into Mexico and eastward to include the Big Bend country in southwestern Texas (Plate 6). The entire state of Nevada, more than half of Arizona, about half of New Mexico and Utah, and parts of California, Idaho, Oregon, and Texas are included. As the name implies, there are mountain ranges—over 400 if the small ranges are included—with basins between them (Fig. 6–1). These distinctive north–south oriented fault-bound ranges are known as *fault-block mountains*. Many of the basins are enclosed and contain an ever-increasing accumulation of erosional debris from the surrounding mountains. Pioneer American geologist Clarence Dutton noted that a map of the province outlining the ranges resembled "an army of caterpillars crawling north from Mexico!"

How can an area this large be but one geomorphic province? In addition to the distinct fault-block topography, climate is also a significant factor that promotes similar appearance in widely separated areas; the Basin and Range Province is arid, with the exception of small "islands" that rise high enough to intercept moisture. Precipitation ranges from a few inches/year in the basins to perhaps 30 inches (76 cm) in the higher mountains. The present *rain shadow* condition is a relatively recent geologic development brought on by the rapid uplift of the Sierra Nevada and Cascade Mountains during the past few million years. Air masses from the Pacific are forced to rise, cool, and release most of their moisture as they cross the formidable Sierra Nevada–Cascade Mountains topographic barrier.

The rocks in the Basin and Range Province vary considerably, but in this dry climate the limestones are as resistant to weathering and erosion as many of the sandstones and granites. The ranges are generally jagged with bold faces and cliffs essentially devoid of vegetation. Details of the rock sequence and older geologic structures vary considerably from area to area within the Basin and Range Province. However, recent episodes of crustal stretching caused the "pull-apart" faults that help account for the similar-appearing topographic expression throughout the province. Thus, the fault-block structure and arid climate override other

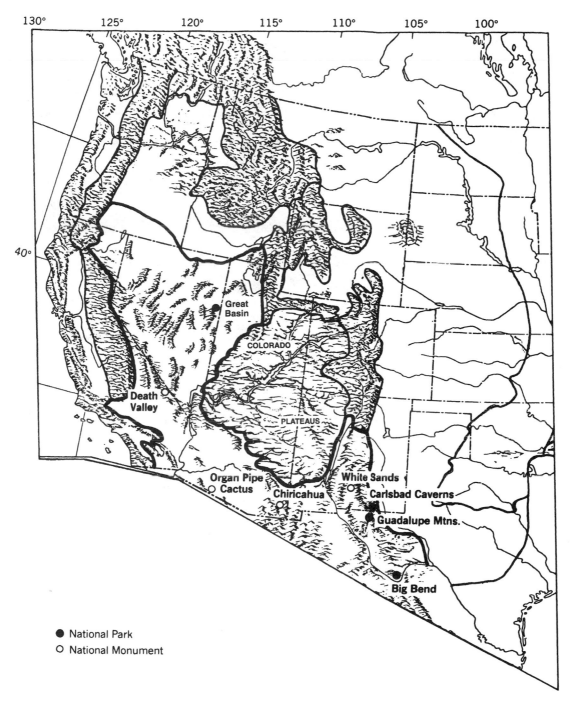

PLATE 6   Map of Basin and Range Province. (Base map copyright Hammond Inc.)

FIGURE 6–1   Basin and Range Province topography near Ajo, southernmost Arizona. As the pediments enlarge by mountainward extension, the ranges are reduced in size. (Photo by Fred Goodsell)

geologic features and enable this very large area to fit the definition of a geomorphic province reasonably well.

The basins occupy much more of the area than the ranges; in fact, in the distant future the ranges will be largely buried beneath their own waste material, especially in the northern part—the *Great Basin*—where there is only interior drainage. Even in the remaining area where the Colorado and Rio Grande used to drain to the ocean before modern settlement, there are local interior basins; the Tularosa Basin in southern New Mexico is one example.

With few exceptions the landscapes are typical of dry regions. The dominant landforms, in addition to the ranges themselves, are *alluvial fans, bajadas, pediments, bolsons,* and *playas.* Vegetation is sparse and wind erosion is active; large sand dune areas, though less common than one might expect, are found in Death Valley, in Mojave National Preserve in southernmost California, and in several other places.

The erosional process is carried on mainly by water, even in this desert region; locally the landforms are clearly the result of wind activity or at higher and especially more northerly elevations, glaciation. They are, however, local features superimposed on a fault-block topography modified by water erosion and deposition. Unlike the permanent streams in a humid area, most of these streams are dry much of the time. When intense storms occur, with little or no vegetation to retard runoff, flash floods develop that have tremendous capacity to erode and to transport materials. Therefore, during the brief periods of flood flow, the streams in the Basin and Range Province erode rapidly and quickly alter the landscape. Those fortunate (or unfortunate, depending on one's perspective) enough to safely observe one of these events can watch the landscape change before their eyes.

The Basin and Range Province is being enlarged at the expense of the Colorado Plateaus. Streams that head on the escarpment that separates the two provinces are cutting headward, enlarging their watersheds. Promontory Butte in east-central Arizona is now within this boundary zone. Actually it is a narrow projection of the Mogollon Plateau; soon, geologically, it will be separated by erosion from the Colorado Plateaus and eventually become a "true," isolated butte within the Basin and Range Province.

Rocks representing all geologic eras except the older Precambrian (Archean) are exposed within the province. Precambrian granites and metamorphic rocks are the only rocks in some of the ranges; in others, Paleozoic limestones and sandstones are dominant. Mesozoic rocks, which dominate in the Colorado Plateaus, are much less abundant here. Tertiary-Quaternary lava flows and pyroclastics cover a significant number of areas, in places capping tilted fault blocks. But the basins that occupy most of the province are floored by thick bolson deposits of Quaternary age.

Basin and Range Structures

The structures of the Basin and Range Province are difficult to decipher because they are largely buried beneath bolson, or basin, deposits that cover over one half of the province. In many individual ranges the rocks are all dipping in one direction; in some cases lava flows and ash-flow deposits that are the caprocks are now dipping steeply. Early reconnaissance geologists concluded that each of the small ranges was a fault block that had been tilted; this was the typical "Basin and Range structure." More detailed work revealed that, in addition, a few of the ranges exposed peculiar domelike forms called *metamorphic core complexes* and that there are a considerable number of more complex, even older structures such as overturned folds and large thrust faults contained within many of the individual mountain ranges (Fig. 6–2).

The Roberts Mountain Thrust in central Nevada (Stewart, 1980), the Keystone Thrust west of Las Vegas, Nevada (Hewett, 1931), and the Muddy Mountain Thrust north of Lake Mead (Longwell, 1928) are classic examples. The conclusion is that the folded and thrust-faulted structures are Paleozoic and Mesozoic in age and that most of the metamorphic core complexes similar to that in Saguaro National Monument began to form during the late Mesozoic. Mid-Tertiary [about 45–15 Ma (millions of years ago)] crustal stretching initiated a violent episode of rhyolitic volcanism that produced huge calderas and extensive ash-flow sheets—events of this frightening magnitude have not been experienced by humans in modern times. Even greater and more widespread stretching (*extension*) during the late Tertiary (beginning about

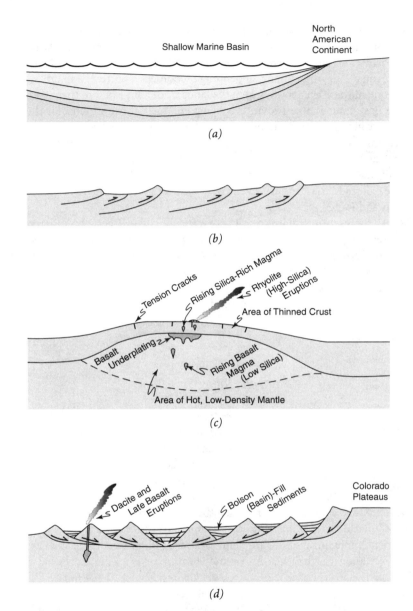

FIGURE 6–2   Generalized evolution of the Basin and Range Province. (*a*) A landmass separates from North America in late Precambrian time, thick marine sediments accumulate during Paleozoic time. (*b*) Mesozoic (about 150–80 Ma) subduction thickens and heats crust in Basin and Range area. Mountain building and overthrusting (Sevier Orogeny) migrates eastward with time where it is known as the Laramide Orogeny (about 70–40 Ma) in the Rocky Mountain area. (*c*) Stretching and doming of crust (45–15 Ma) produces magmatism. (*d*) Width of Basin and Range doubles, crust fractures, and blocks collapse inward creating fault-block mountains.

14–10 Ma) produced the north–south oriented normal faults and tilting that led to today's fault-block mountains (Wilson, 1990). In spite of the complex structure and geologic events recorded in the exposed rocks in the ranges, when "Basin and Range structure" is used to describe a certain area in the literature, it refers to the block faulting and tilting that is one of the "trademarks" of the province.

Geologic and Structural History

The bounding faults along each range front are inclined steeply (about 60° from the horizontal), but geophysical studies (seismic reflection) confirm that these faults flatten with depth as shown in Figure 6–2d. It appears that the upper, brittle layer of the earth here is detached from the hotter, more plastic material below. The lower lithosphere (lower crust and upper mantle, Fig. 1–9) is stretching and thinning much like a sheet of rubber overlain by a thick layer of peanut brittle (upper lithosphere) that cracks and collapses inward (Fig. 6–2d). The amount of stretching during the last 35 million years is enormous, perhaps 100 percent or twice its original width across the province in the last 35 million years! If all this stretching (perhaps 200 miles; 320 km) occurred in a narrow zone, western North America would look more like parts of the Mideast where plate tectonics is opening up trenches along the Red Sea and the Persian Gulf. Perhaps the Basin and Range Province is an aborted rift or perhaps, as Wilson (1990) suggests, the North American continent over the next few million years may split and separate in this area. If he is correct, then those wishing to make a "long-term" investment might want to move into the Basin and Range and eventually own a piece of seaside property!

Continuing hot-spring activity, earthquakes, Holocene volcanic eruptions, and localized surface faulting (Fig. 6–3) indicate that stretching and crustal

FIGURE 6–3   Fault scarp near Frenchman Mountains in western Nevada. New scarp was formed by the faulting that occurred in 1954 in this tectonically unstable section of the Basin and Range Province. (Photo by D. Harris)

thinning are still occurring. The Basin and Range is enlarging at the expense of its neighbor, the Colorado Plateaus. Blocks are sliding westward along the impressive Hurricane and other active faults near the eastern boundary of the Basin and Range, converting the edge of the Colorado Plateaus into the tilted fault blocks of the Basin and Range Province.

A highly generalized sequence of geologic events for this huge province is shown in Figure 6–2. Part of the North American continent edge during the early Paleozoic extended through central Nevada and the west edge of Idaho. However, because of subsequent plate rotation and migration, rather than a north–south orientation, the continent edge during very late Precambrian and Paleozoic time extended from east to west and was located close to the equator (Wicander and Monroe, 1993). The basin that was destined to become the Basin and Range Province sagged deeply during the Paleozoic, allowing many thousands of feet of sand, clay, and limey shallow marine sediments to accumulate along its landward margin and volcanic and deep-water deposits farther seaward (Fig. 6–2*a*). Paleozoic mountain building (*Antler Orogeny*) was followed by erosion and a return to marine conditions. Islandlike areas, or *exotic terranes,* were added to the continent edge during the Mesozoic as the impressive *Cordilleran Orogeny* began the massive deformation and building of mountains in western North America. The initial phase of mountain building is called the *Sevier Orogeny* and the later phase that affected areas to the east in the Rocky Mountains is known as the *Laramide Orogeny.*

The Sevier Orogeny began about 120 Ma as the rocks were uplifted and thrust eastward as if a giant accordian were being squeezed shut. The mountain building was a response to the change from a passive plate margin to one that was actively colliding with the offshore Pacific Plate system along a subduction zone. Subduction greatly heated and thickened the crust creating an *orogenic welt* beneath western North America, enabling the crust to eventually begin to partially collapse under its own weight and increased plasticity, much like silly putty flows and flattens on a table top. Erosion was considerable but widespread deposits of Mesozoic-age shallow marine, nonmarine, and abundant volcanic materials remain. Later, as the rate of subduction slowed, the compressive forces on the thick crust reduced, enabling the crust to collapse and spread laterally at an even faster rate, causing the "giant crustal accordian" to reverse and pull the crust apart (Livaccari, 1991). Lateral spreading began during the Late Mesozoic but greatly increased about 30–20 Ma as subduction along the west edge of the North American Plate was replaced by strike-slip movement (Fig. 3–2)—movement that further removed lateral support for the thickened crust allowing more stretching and thinning beneath the Basin and Range Province.

A spectacular episode of rhyolitic volcanism accompanied the stretching from about 40–20 Ma. Large calderas and huge sheets of red-hot ash clouds rolled across the future Great Basin and southern Arizona and New Mexico areas locally snuffing out life and burying the landscape with ashy sediments—many of which were hot enough to anneal into *welded tuffs.*

Geophysicists have determined that the crust here is unusually thin—as thin as 18 miles (29 km) compared to 30 miles (48 km) in the adjacent Colorado

Plateaus. A thinning crust reduces pressure on the mantle and allows more heat to reach the earth's surface. The warm lithosphere expanded and domed upward (Fig. 6–2*c*). About 17 Ma the rate and extent of lower lithosphere stretching increased (*underthrusting*) and the overlying brittle lithospheric plate began to rapidly break apart into numerous blocks that collapsed toward the center of the province (Fig. 6–2*d*).

Plate tectonics holds the key although researchers are still uncertain of the details. Some point out that significant stretching began about the same time that the North American Plate began to override the spreading ridge in the Pacific and the strike-slip motion of the San Andreas Fault began (Fig. 3–2). The change from subduction (compression) to strike-slip (lateral) motion between these huge plates further reduced the compressive forces holding together the thick crust generated during the Sevier Orogeny. Thus, the southwest edge of North America stretched further, and the thick, hot crust continued to collapse and spread. Some suggest that the Basin and Range Province stretching is due to its location above the East Pacific Rise that was overridden as North America drifted westward or that a rising plume of hot material (a hot spot) underlies the area. Tuzo Wilson (1990) suggests that the Yellowstone hot spot in the north and the New Mexico hotspot in the Rio Grande Rift are causing the Basin and Range crust to pull apart. Other papers by Dickinson and Snyder (1979), Mutschler and others (1998), and Gans and others (1989) discuss some of these ideas.

Pleistocene Glaciers and Lakes

Only a few of the highest areas in the province were glaciated; almost all are in the northern section—East Humboldt Range, the Ruby Mountains in northeastern Nevada, and the Snake Range, as examples. Sierra Blanca in southeastern New Mexico is one exception; it is 12,003 feet (3660 m) high and was extensively glaciated.

Pleistocene lakes, some very large, occupied the interior basins. The topographic and structural doming of the Basin and Range Province discussed previously (Fig. 6–2*c*) caused the largest lakes to form along the province edges in the Great Basin Section. On the eastern edge was Lake Bonneville, much larger than its puny descendant, Great Salt Lake; on the western edge was Lake Lahontan with present-day Carson Sink as its deepest point. Dozens of others, including Lake Manly in Death Valley, occupied other interior basins during the cooler climate intervals of the Pleistocene. At its maximum about 15,000 years ago, 1100-foot-deep (335-m) Lake Bonneville on the east side of the province covered over 20,000 square miles (52,400 km^2), about the same area as Lake Michigan is today. Its counterpart on the west side of the province, Lake Lahontan, covered about 8000 square miles (20,960 km^2). As the climate fluctuated repeatedly from glacial to interglacial, so did the lakes wax and wane, leaving *wave-cut notches,* or *strandlines,* marking temporary still-stands of the lakes at different elevations on the sides of the mountains. In the smaller basins the lakes have long been dry, except for short periods during wet seasons. Some of the water penetrates into the deposits, but with the extremely high evaporation rates in this climate most of the moisture returns to the atmosphere before plants can utilize it. Morrison (1991) presents a detailed summary and an excellent bibliography for the Pleistocene lake histories in the Basin and Range Province.

Caverns and Groundwater

The work of groundwater is recorded in many places in the province. Limestones are widespread and in those containing well-developed fracture systems combined with abundant water, either now or in the past, solution is most effective. Carlsbad and its recently discovered neighbor, Lechuguilla, in southeastern New Mexico are world famous. Caverns also occur in the limestones in Big Bend and Guadalupe Mountains National Parks. In the northern part of the province there are a number of caverns; Great Basin National Park in eastern Nevada contains over two dozen caves, the most significant of which is Lehman Cave.

Groundwater is limited in this arid desert but does exist in quantities that enable small towns, a few cities, and a few ranches to exist. Developers, particularly those in Las Vegas, seek to appropriate water rights in large areas to further growth in a few urban areas at the expense of the desert plants and animals and the rural life-styles of the rugged humans who make this place home. Pumping groundwater from rural areas and transporting it to cities would lower desert water tables and dry up many important springs and water sources that make life possible for many desert inhabitants.

REFERENCES

Dickinson, W.R., and Snyder, W.S., 1979, Geometry of subducted slabs related to San Andreas transform: Journal of Geology, v. 87, p. 609–627.

Gans, P.B., Mahood, G.A., and Schermer, E., 1989, Synextensional magmatism in the Basin and Range Province, a case study from the eastern Great Basin: Geological Society of America Special Paper 233.

Hewett, D.F., 1931, Geology and ore deposits of the Goodsprings quadrangle: U.S. Geological Survey Professional Paper 162.

Livaccari, R.F., 1991, Role of crustal thickening and extensional collapse in the tectonic evolution of the Sevier–Laramide orogeny, western United States: Geology, v. 19, p. 1104–1107.

Longwell, C.R., 1928, Geology of the Muddy Mountains, Nevada, with a section through the Virgin Range to the Grand Wash Cliff, Arizona: U.S. Geological Survey Bulletin 798.

Morrison, R.B., 1991, Quaternary stratigraphic, hydrologic, and climatic history of the Great Basin, with emphasis on Lakes Lahontan, Bonneville, and Tecopa: in Morrison,

R.B., ed., Quaternary nonglacial geology: Conterminous U.S.: Boulder, CO, Geological Society of America, The Geology of North America, v. K-2, p. 283–320.

Mutschler, F.E., Larson, E.E., Gaskill, D.L., 1998, The fate of the Colorado Plateau—a view from the mantle, in Friedman, J.D., and Huffman, A.C., Jr., coordinators, Laccolithic complexes of southeastern Utah: Time of emplacement and tectonic setting, workshop proceedings: U.S. Geological Survey Bulletin 2158, p. 203–222.

Stewart, J.H., 1980, Geology of Nevada: Nevada Bureau of Mines and Geology Special Publication 4, 136 p.

Wicander, R., and Monroe, J.S., 1993, Historical Geology: Minneapolis/St. Paul, West Publishing, 640 p.

Wilson, J.T., 1990, On the building and classification of mountains: Journal of Geophysical Research, v. 95, p. 6611–6628.

LAVA BEDS NATIONAL MONUMENT (CALIFORNIA)

Lava Beds may be the only place in the National Park System where geology played a key role in an Indian war. The Modoc Indians may not have known how the lava tubes and the collapse features were formed, but they knew how to make the most of the resulting terrain. In the Modoc War (1872–1873) Captain Jack's 53 braves fought—and eluded only to fight again—about 1000 U.S. soldiers. Finally, after five months of embarrassment (not to mention casualties), the soldiers had the Indians surrounded and were ready to round them up. However, as related by Waters (1981), they were a bit down-hearted the next morning when they could find not one Indian! By knowing every inch of the terrain, 160 women, children, and warriors, along with dogs and horses, had slipped through the lines and escaped, temporarily. The story has a sad ending; they were captured a short time later, and the U.S. Army passed up the opportunity to use Captain Jack to train its officers in the applications of military geology!

Interest in the area's caves grew as homesteaders moved in and roads were constructed beginning in the 1880s. Rancher E.L. Hopkins discovered and explored Skull Cave in 1892 where he found a pit containing not only antelope and bighorn sheep bones, but two human skeletons. Judson D. Howard from nearby Klamath Falls began to visit the area caves in 1916, fell in love with the area, and devoted much of his life to its study and protection. He wrote to the Modoc National Forest about the increasing vandalism and sloppy camping habits of many visitors and advocated in 1923 that the area be placed under federal protection. The area was declared a national monument in 1925, placed under National Park Service protection in 1933, and enlarged to about 72 square miles (187 km^2) in 1951. However, not all of the locals were interested in the esthetic and scientific value of the caves. Entrepreneur Jim Howard (no relation to J.D. Howard) manufactured moonshine in Merrill Cave and later in Mushpot and Crystal Cave during prohibition!

Located on the north flank of the Medicine Lake shield volcano in northernmost California (Fig. 6–4), Lava Beds National Monument consists of many lava flows, 17 cinder cones, 300 known lava tubes, and other volcanic features (Heiken,

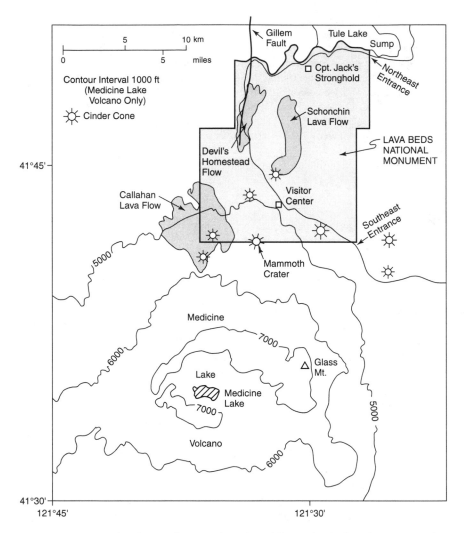

FIGURE 6–4   Generalized map of Lava Beds and Medicine Lake shield volcano area; only a few of the 17 cinder cones in the monument are shown. (Modified from Waters and others, 1990)

1981), several of which were formed during the past few hundred years. Local rancher G.W. Courtright felt earthquakes in 1910 and observed a small ash eruption and flame from vents in the top of the shield. The next day a layer of blue mud coated vegetation on the mountain flank (Finch, 1928). The probability of renewed volcanic activity here is high.

Geographic Setting

Lava Beds is part of the lava-covered Modoc Plateau desert of northeastern California and is included here in the Basin and Range Province because typical fault-block (graben) structures extend into the area from the Tule and Klamath Falls (Oregon) basins immediately north. The monument is a part of the much larger Medicine Lake shield volcano (Fig. 6–5), which dominates the local geology. Thus the story of Lava Beds must include part of the story of the Medicine Lake volcano. The area also shares geologic characteristics with the nearby Cascade Mountain volcanic province to the west and the extensive Miocene-Pliocene volcanism associated with the Columbia Intermontane Province to the northeast. Tule Basin contains a permanent lake along the monument's north border that was much larger during the ice ages and as recently as the mid-1900s before irrigators began to withdraw substantial volumes of water.

Geologic History

Separation of the Sierra Nevada and the Klamath areas about 140 Ma created a 60-mile (97-km) gap that probably allowed Mesozoic seas to cover the area with marine sediments (Orr and others, 1992). Volcanic rocks of Miocene-Pliocene age associated

FIGURE 6–5   Medicine Lake shield volcano with cinder cones on flanks; those in foreground are in Lava Beds National Monument. (Photo by E. Kiver)

with the Columbia Intermontane volcanism buried the older sediments and topography, setting the scene for the dramatic changes that the Pleistocene would bring.

A wide zone of strike-slip faulting called the Walker Lane extends northwest from the Las Vegas and Sierra Nevada area to the Lava Beds–Medicine Lake volcano area. Recent lateral fault movements are pulling the crust apart here generating earthquakes, magma, and north–south oriented fault scarps. The destructive magnitude 6.0 Klamath Falls earthquake in November of 1993 and a swarm of volcanic-style earthquakes directly beneath the Medicine Lake volcano in 1988 serve as reminders that this part of the Basin and Range Province is still very much "alive" geologically. Most of the earthquakes occur along border faults about 15 miles (25 km) or more from the summit as the massive, geologically recent volcano subsides and extension of the crust in this part of the Basin and Range continues (Dzurisin and others, 1991).

As crustal blocks are pulled apart, pressure is reduced on the hot asthenosphere below, allowing partial melting of the heavy mantle rock. The chemistry of the resulting melt is that of basalt, and, because it is lighter than the surrounding mantle rock, it rises buoyantly toward the surface. Most of the volcanic products making up the massive, 130 cubic mile (600 km^3) Medicine Lake volcano are relatively fluid basalt and andesite flows. The oldest shield-forming basalts are the 1.0–1.4 Ma lava flows exposed in Gillems's Bluff, a prominent Basin and Range fault scarp that is part of a larger system of faults that extends through the monument and the Medicine Lake volcano (Fig. 6–4). This major rift zone was utilized numerous times by rising magma, thus accounting for the alignment of many volcanic vents.

Withdrawal of magma during one or more of the numerous eruptions removed support for the summit and a 4.5- by 7.5-mile (7- by 12-km) caldera formed. Later eruptions obscured the caldera edges as Mount Hoffman, Lyons Peak, and other volcanoes formed near the caldera rim. If the rising basalt resides for a long time in the silica-rich rocks that make up the earth's crust, those rocks may be melted and assimilated into the basalt, thereby changing its chemistry. Silica-rich magma such as basaltic andesite and andesite tend to produce more explosive eruptions and cinder cones that disrupt the smooth shield profile (Fig. 6–5). Cinder cones are usually "one-shot" volcanoes in that they experience only one episode of eruption. An early gas-rich phase sends cinders and volcanic bombs into the air in a spectacular pyrotechnic display. Later, lava wells up and breaks through the rubble along one flank of the cinder cone forming either a smooth, ropy-surfaced *pahoehoe* flow or a clinkery *aa* flow. Thirteen of these "children of violence," such as Schonchin Butte, Caldwell Butte, and Cinder Butte (Fig. 6–6) are within Lava Bed's boundaries and many more occur elsewhere on Medicine Lake volcano.

As the 3000-foot-high (914-m) shield volcano was building, basaltic eruptions continued in Lava Beds and elsewhere on the flanks, and increasing amounts of higher silica lavas (andesite, dacite, rhyolite) occurred in and near the summit caldera. The extreme variation of lava types led Eichelberger (1975) to postulate the existence of both high-silica (rhyolitic) and low-silica (basaltic) magma chambers beneath the area, with different degrees of mixing to account for the extreme com-

FIGURE 6–6   Aa flow front of Calahan Flow, one of the youngest in the monument; top of Cinder Butte visible in distance; note old-growth pine in foreground where lava did not reach. (Photo by E. Kiver)

positional range of volcanic products. Donnelly-Nolan (1988) and Dzurisin and others (1991) believe that the answer involves a combination of processes including: (1) the presence of magma chambers containing zones of different chemical composition, (2) assimilation or melting of surrounding rock thereby changing the magma chemistry, and (3) mixing of two or more magma bodies beneath the volcano. A side trip to the Glass Mountain obsidian flow on National Forest land to the south is worthwhile to observe lava of very high silica content.

The nearly 8000-foot (2439-m) elevation of the shield volcano was high enough that an ice cap formed in the summit area about 20,000–15,000 years ago smoothing out some of the volcanic features and scratching or striating some of the bedrock surfaces. A distinctive andesitic ash-flow tuff found on the volcano flanks is absent from the caldera rim—a distribution that could be explained by a volcanic eruption occurring under an ice cap. Dry channels on the northwest flank were likely produced by the catastrophic flood produced when hot, searing lava contacted and melted a hole in the glacier (Donnelly-Nolan and Nolan, 1986).

During the late Pleistocene a large basalt flow erupted from Mammoth Crater and other vents and covered about two thirds of the monument as well as a large area to the east. More than one flow makes up the Mammoth Crater basalt, but the paleomagnetic directions recorded in these flows are so close that Donnelly-Nolan and Champion (1987) believe that they were all emplaced in less than 100 years. Overlying the Mammoth Crater basalt, and therefore younger in age, are the Devil's

Homestead and Callahan Flows that look as if they occurred "yesterday." The geologic recency (less than 1200 years old) of these eruptions combined with the dry desert climate accounts for the lack of weathering and soil development. Fleener Chimney, an impressive 130-foot-deep (40-m) vent was the source of the Devils Homestead Flow (Fig. 6–4) that erupted 500–1000 years ago. Another fresh-appearing flow surface occurs on the Calahan lava (about 1160 years old) in the south end of the monument. Pockets of pumice on the flow surfaces record even more recent activity that originated at the summit area of the Medicine Lake volcano.

The recency of volcanic eruptions and the behavior of seismic waves in the subsurface suggests that a small magma body exists beneath the Medicine Lake volcano. Exploratory wells drilled by an energy company indicate that unusually hot rock and fluids do indeed exist at shallow depths beneath the caldera. Future commercial development is planned. Hopefully development will not compromise the national monument or the equally spectacular features elsewhere on the volcano.

Lava Caves

Low-silica pahoehoe basalt (SiO_2 less than 54 percent) like that found in the Mammoth Crater and Valentine Cave basalts can cover large areas. An important mechanism that permits long-distance transfer of lava with very little heat loss is the roofing over of lava channels to form *lava caves* or *lava tubes*. Pieces of floating crust lodge against channel walls and eventually bridges across. If lava levels drop slowly, the crust thickens, further strengthening the cave roof and reducing heat losses. An individual tube might be tens of miles long during an eruption, but subsequent roof collapses segment the tube into numerous individual caves. Lava Beds is rich in lava caves—about 300 of them! When the supply of lava stops, the lava in the tube system may continue to flow, leaving an empty conduit behind with a fascinating array of features. According to Waters and others (1990) only 10–20 percent of the lava tubes drain and remain accessible.

Over a dozen caves are accessible near the visitors center along Cave Loop Road and many others from trails throughout the park. A few hours to a few days can easily be spent in this fascinating underground environment. Areas of roof collapse allow human entry and opportunities to trace the underground pathways of prehistoric lava (Fig. 6–7). The serious explorer should acquire U.S. Geological Survey Bulletin 1673 (Waters and others, 1990) for descriptions of the more significant caves and explanations for their unusual characteristics and features. Your powers of observation and imagination will be challenged as you trace the lowering lava river by noting *lava benches* (*bathtub rings or curbing*) on the cave walls or the remelting and sagging of the cave walls and ceilings above a lava river or pond by discovering *lavacicles* (*lava stalactites*) (Fig. 6–8) or *lava driblets* and *pull outs*.

Although most lava tubes have a simple, solitary tunnel, others such as Skull Cave have multilevels with tubes stacked one on top of the other—others such as Catacombs Cave have multiple passages at the same level. Some caves are temporarily

FIGURE 6–7   Entrance to one of the hundreds of lava tubes in Lava Beds National Monument. (Photo by E. Kiver)

FIGURE 6–8   Lavasickles and glassy lava-tube glaze on walls and ceilings. (Photo by E. Kiver)

closed by the Park Service during bat hibernations or at maternity colonies to protect these valuable insect predators. Decreasing numbers of bats are occurring on a world-wide basis as land development, pollution, and human population increase.

The abundance and accessibility of the lava tubes is a unique and major feature of Lava Beds that in itself justifies the park's existence. A walk through the maze of tunnels and passageways is a delight to visitors wishing a *spelunking* (cave exploration) experience. Chocolate lovers will find Hopkins Chocolate Cave of interest until one finds the delicate brown colors are from mud that drips through the ceiling during the wet season. Not only are these chocolate delights inedible but collecting or eating any of the natural features in a park is highly illegal!

REFERENCES

Donnelly-Nolan, J.M., 1988, A magmatic model of Medicine Lake volcano, California: Journal Geophysical Research, v. 93, p. 4412–4420.

Donnelly-Nolan, J.M., and Champion, D.E., 1987, Geologic map of Lava Beds National Monument, northern California: U.S. Geological Survey Map I-1804.

Donnelly-Nolan, J.M., and Nolan, K.M., 1986, Catastrophic flooding and eruption of ash-flow tuff at Medicine Lake volcano, California: Geology, v. 14, no. 10, p. 875–878.

Dzurisin, D., Donnelly-Nolan, J.M., Evans, J.R., and Walter, S.R., 1991, Crustal subsidence, seismicity, and structure near Medicine Lake Volcano: Journal of Geophysical Research, v. 96, no. B10, p. 16,319–16,333.

Eichelberger, J.C., 1975, Origin of andesite and dacite: Evidence of mixing at Glass Mountain in California and at other circum-Pacific volcanoes: Geological Society of America Bulletin, v. 86, p. 1381–1391.

Finch, R.H., 1928, Lassen Report no. 14: The Volcano Letter, no. 161, p. 1.

Heiken, G., 1981, Holocene plinian tephra deposits of the Medicine Lake Highland, California: in Johnston, D.A., and Donnelly-Nolan, J., eds., Guides to some volcanic terranes in Washington, Idaho, Oregon, and Northern California: U.S. Geological Survey Circular 838.

Orr, E.L., Orr, W.N., and Baldwin, E.M., 1992, Geology of Oregon: Dubuque, IA, Kendall/Hunt, 254 p.

Waters, A.C., 1981, Captain Jacks's Stronghold (the geologic events that created a natural fortress), in U.S. Geological Survey Circular 838.

Waters, A.C., Donnelly-Nolan, J.M., and Rogers, B.W., 1990, Selected caves and lava-tube systems in and near Lava Beds National Monument, California: U.S. Geological Survey Bulletin 1673, 102 p. plus maps.

Park Address

Lava Beds National Monument
Box 867
Tulelake, CA 96134

GREAT BASIN NATIONAL PARK (NEVADA)

History and Description

As noted in Chapter 1, establishment of new park areas is becoming more difficult. It may take a volcano blowing its top such as Mount St. Helens or it may take a group of concerned citizens such as those in Ely, Nevada, or it may take more quiet researchers such as Robert Waite at the University of Utah to enable public lands with outstanding characteristics to be designated as special places for the use of the public rather than a few specialized commercial interests. After an 11-year hiatus with no new national parks established, on October 27, 1986, some 77,082 acres in the Snake Range in eastern Nevada (Plate 6) was established as Great Basin National Park, our 49th park. Even then, a compromise was made in the enabling legislation that cows, cowboys, and visitors will forever share this parkland. Because of past land abuses, mostly triggered by overgrazing, rangeland is heavily damaged and the number of cows permitted now is greatly reduced.

Absalom Lehman's discovery and commercialization of a spectacular cave on his Snake Range ranch in 1885 drew considerable attention to the area. The enthusiasm of a Tonopah, Nevada, mining broker, Cada C. Boak, and support by Senator Odie Tasker helped establish Lehman Caves National Monument under U.S. Forest Service control in 1922. Their proposal to include the spectacular nearby Wheeler Peak area and other parts of the Humboldt National Forest in the Snake Range as a national park was defeated in 1924. Control of the national monument was transferred to the National Park Service in 1933, but subsequent attempts to establish a national park failed until 1986.

Interest was heightened in the 1950s when reports of crevasses in the permanent snowfield in the Wheeler Peak cirque might make this the only living glacier in Nevada. Again interest was aroused in the area in the late 1960s when a researcher, with Forest Service permission, cut down a gnarly bristlecone pine growing at timberline in the Wheeler Peak cirque to determine its age and the minimum age of the lateral moraine on which it was growing. A count of the 4950 annual growth rings revealed that the researcher had cut down the oldest known living tree in the world! According to Darwin Lambert (1992) cutting of Prometheus, the oldest living tree, enraged park supporters and provided the extra push that eventually led to the establishment of the park.

Bristlecone pines grow only in some of the higher elevations in Nevada, California, Utah, Colorado, New Mexico, and Arizona. The needles on the branches resemble the "tail of a frightened cat," and their equally gnarly Limber pine friends that share the same area have needles that resemble the tail of a dead cat! The more rigorous the climate the slower the growth and the older the tree. The oldest trees appear to be more dead than alive as living tissue dies back in years of drought and the tree supports the minimum amount of foliage. The species growing here is appropriately named *Pinus longaeva*.

One of the goals of the early explorers in the Basin and Range Province was to find the legendary San Buena Ventura River that Spanish explorers Fathers Escalante and Dominguez in 1776 predicted would exist. They reasoned that rivers flowing to the west in what is now Utah must join and eventually flow into the Pacific Ocean. After searching for this easy path to the Pacific ocean by criss-crossing the area a number of times, Captain John C. Fremont in 1844 recognized that this huge area was ringed by drainage divides. He named the northern 200,000 square miles (524,000 km^2) of the Basin and Range the Great Basin in recognition that rivers on the west flowed eastward never to escape the topographic confines of the desert. To the south the Colorado River and the Rio Grande receive a sufficient charge of runoff from the Southern Rocky Mountains that flow across the desert is maintained. However, as human population and the corresponding agricultural, industrial, and recreational needs (such as golf courses, swimming pools, and lawns) has increased, water reaching the ocean through these channels is now a rarity.

The Snake Range is ideal to demonstrate the variety of natural features present in the northern Great Basin. Leaving the sagebrush basins surrounding the Snake Range, one climbs through five life zones in 5 miles (8 km), topping out in the tundra zone above 11,000 feet (3354 m). Wheeler Peak (13,063 feet; 3983 m) commands an inspiring view through nonpolluted air across the sea of fault-block mountains typical of the province (Fig. 6–9).

FIGURE 6–9   Like a vast armada, fault-block mountains stretch across the Nevada landscape. View from 13,063-foot-high (3983-m) Wheeler Peak in Great Basin National Park. (Photo by E. Kiver)

The extreme elevation and northern location of the Snake Range promoted the development of over a dozen glaciers and produced the spectacular Wheeler Peak and other cirques with steep walls as high as 2000 feet (610 m) (Fig. 6–10). Glacial moraines, a rock glacier over a mile long, spectacular speleothem decorations in Lehman Caves, a six-story natural arch in the southeastern section of the park, and the present remoteness and relatively undeveloped nature of the park also provide much for one to enjoy in one of our newest national parks.

Geologic Evolution

The geologic development of Great Basin National Park closely parallels the sequence described for the Basin and Range in the chapter introduction and summarized in Figure 6–2. Again, major plate movements account for the "big picture"

FIGURE 6–10 Wheeler Peak cirque contains Nevada's largest icefield (small glacier?) and a large rubbly surfaced rock glacier. (Photo by E. Kiver)

of geologic development. A major crack or rift, part of which is along the Wasatch Line (north–south zone of faults along the west edge of the Wasatch Mountains in central Utah), opened up in western North America about 1000 Ma creating a new continental margin. The western plate is gone, perhaps becoming attached to Antarctica or another continent. In its place was left a giant basin—the Cordilleran Geosyncline into which thick deposits of limestone, sandstone, and mudstone accumulated during the Paleozoic Era (Fig. 6–11). Mountain building during mid-Paleozoic time affected only the west edge of the geosyncline. The Mesozoic brought mountain building and erosion in the Great Basin with deposition of marine and nonmarine sediments restricted to areas to the east. A number of granitic

Time		Material	Event
Cenozoic	Quaternary	Sand and gravel	Desert processes dominate Rock glacier forms Valley glaciers erode high country; outwash, till Speleothems form Lehman Caves begin to form (≈2 Ma)
Cenozoic	Tertiary	Volcanic rocks	Basin and Range faults begin (≈17 Ma)
Cenozoic	Tertiary	No record	Crust stretches and thins (≈30 Ma)
Mesozoic		Granite rocks emplaced	Local mineralization
Mesozoic		No record	Sevier Orogeny–large overthrust faults, erosion
Paleozoic		Limestone, shale, sandstone Pole Canyon limestone Pioche shale Prospect Mt. quatrzite	Deposited near margin of North America
Precambrian (P€)		Osceola argillite	Deposited near margin of North America
Precambrian (P€)			Plate rifting initiates Cordilleran Geosyncline

FIGURE 6–11   Simplified rock record at Great Basin National Park.

intrusions occurred during mid-Mesozoic (Jurassic, about 160 Ma) and later as western North America changed from the passive, trailing edge of a plate to the active, leading edge of a westerly moving plate. The tectonic vise closed and thrust faults cut through the thick section of Paleozoic rocks and shoved rocks eastward forming a late Mesozoic mountain system during what is known as the Sevier Orogeny. The topographic mountains of Mesozoic time have long since vanished, but remnants of the thrust faults are exposed in many of the Great Basin mountain ranges, including parts of the Snake Range (Fig. 6–12).

Geomorphic Evolution

Following a long interval of erosion, the area began to be simultaneously uplifted and stretched about 35 Ma. The rigid crust separated from the stretching plastic zone beneath (Fig. 6–2c) along a detachment surface called a *decollement*. The decollement and the upper and lower layers are well exposed in the Snake Range both to the north and south. Erosion in the park area has removed the upper plate leaving only the lower, relatively undeformed lower layer or plate. These relatively flat separation surfaces resemble overthrust faults except that a decollement leaves young rocks on top of older and overthrusts put older on top of younger.

The stretching and doming affected the entire Great Basin, but unusually large stretching might have triggered a localized domal uplift in the Snake Range producing a metamorphic core complex. The decollement, or separation, between the slightly metamorphosed core rocks and the unmetamorphosed rocks in the upper slab allowed the overlying slab to slide off the core. Bartley and Wernicki (1984), Miller and others (1987), and Whitebread (1982) are good sources for ideas about the structural and geologic history of the area.

Increased stretching about 17 Ma caused extensive cracking of the lithosphere (brittle, upper shell of the earth) creating hundreds of major faults along which

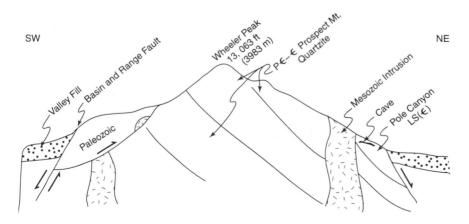

FIGURE 6–12   Highly generalized cross section in Wheeler Peak–Lehman Caves area.

blocks began to slide and rotate forming the fault-block topography of the Basin and Range (Fig. 6–2c). The Snake Range block was uplifted and rotated eastward along these bounding faults (Fig. 6–12). Streams cut vigorously into the Snake Range block and other developing ranges, removing rock layers and washing huge amounts of sediment into the intervening basins. Obviously the blocks uplifted and tilted at a much faster pace than the erosional forces tore away at the mountain blocks. The lower edges of the Snake Range block are deeply buried by sediment in the Snake Valley to the east and the Spring Valley to the west.

The erosional transfer of material from the ranges to the basins was speeded up during the cool intervals of the Pleistocene when more water, and in areas such as the Snake Range, glaciers were present to grind and gouge at its bedrock core. Large lakes occupied the Snake and Spring valleys during these cooler times, adding layers of fine sediment and evaporite materials to the thick basin fill. Someone standing on Wheeler Peak summit 15,000 years ago looking westward would have seen in the distance the eastern edge of Lake Lahontan and, looking to the east, would have seen waves on giant Lake Bonneville lapping against the east edge of the Snake Range. The abundance of moisture also speeded up the underground chemical processes that dissolve away vulnerable rocks such as limestone to leave behind the fascinating voids known as caves or caverns.

Great Basin Caves

Over 30 caves exist in the park and most have yet to be completely inventoried. Because caves are one of the most fragile environments known, an effective management plan will need to be developed as the young park matures. The potential for vandalism is high in these unprotected caves, and the archaeological record has yet to be fully evaluated. Cave entrances were used by Paleo-Indians—skeletal remains of 12 to 26 individuals at or near the entrance to Lehman Caves were discovered during the 1950s.

Mormon settlers entered the area in the 1850s. In 1869 a former miner, Absalom Lehman, settled in the area and decided that he could acquire more gold and silver by selling food to miners than by digging for ore. An 1885 newspaper article describing "Ab's" discovery of a spectacular cave on his property brought over 800 people to be guided through during the first year. Lehman installed ladders and enthusiastically developed and promoted the cave until his death in 1891. Eventually the cave became part of the Humboldt National Forest and somewhat forgotten until local mining broker Cada Boak and Nevada Senator Tasker Odie succeeded in having the cave declared a national monument under Forest Service supervision in 1922. The cave was transferred to the Park Service in 1933. Lehman Caves is only about a quarter-mile long, but its richly decorated passages and chambers, as well as a large number of rare shield formations, makes the cave special and deserving of perpetual protection.

How Caves Form

One of the more fascinating processes of nature involves the generation of sub-surface voids, or caves, usually associated with limestone bedrock. After many millenia of creating space, nature reverses the processes and begins to fill these openings with sediment and incredible displays of chemically deposited material known as speleothems. The lower slopes of the park are underlain by the Cambrian age Pole Canyon Limestone (Fig. 6–12), which was deposited in a shallow sea about 540 Ma. Microscopic organisms known as forams and other invertebrates lived in this sea and extracted calcium from seawater to build their shells. The shells rained down onto the sea floor and the limey ooze eventually lithified into the microscopic calcite crystals that form the Pole Creek Limestone. Additional pressure and heat from igneous intrusions partly recrystallized the calcite crystals into a low-grade marble. After metamorphism the calcite crystals are larger but still susceptible to the weak acids that attack and dissolve away carbonate rock. An abundant water supply, such as that found at high elevations, and the presence of abundant rock fractures further encourages the development of caves.

A common acid at or near the earth's surface is carbonic acid (H_2CO_3), which forms when water from rain or snow combines with carbon dioxide (CO_2) in the atmosphere and soil. The reaction is shown by the simple equation:

$$\underset{\text{(water)}}{H_2O} \quad + \quad \underset{\text{(carbon dioxide)}}{CO_2} \quad \rightleftarrows \quad \underset{\text{(carbonic acid)}}{H_2CO_3}$$

The arrows going both directions in the equation indicate that the process is reversible—the acid can break back apart into water and carbon dioxide. This is an important characteristic that will help explain why calcite can be redeposited as cave decorations, or *speleothems*. Carbonic acid is a weak acid but greatly increases the rate at which calcite ($CaCO_3$) dissolves. This reaction is shown by the equation:

$$\underset{\text{(calcite)}}{CaCO_3} + \underset{\text{(carbonic acid)}}{H_2CO_3} \rightleftarrows \underset{\substack{\text{(calcium} \\ \text{in solution)}}}{Ca^{2+}} + \underset{\substack{\text{(bicarbonate ions} \\ \text{in solution)}}}{2HCO_3{}^{3}}$$

The slightly acidic water descends to the zone of saturation, which is the zone where all the small openings and spaces in the rock are completely filled with water. The boundary between the zone of saturation and the overlying zone of aeration is called the water table (Fig. 6–13*a*). Some solution occurs deep in the zone of saturation but only small openings are believed to form here. The zone just below the water table is where the maximum flow of groundwater and therefore the maximum enlargement of caves occurs. Lowering the water table and stabilizing its position enables a lower cave level to form (Fig. 6–13*b*).

For many years, speleologists (scientists who study caves) working in areas with flat-lying limestones such as those at Mammoth Cave believed that bedding

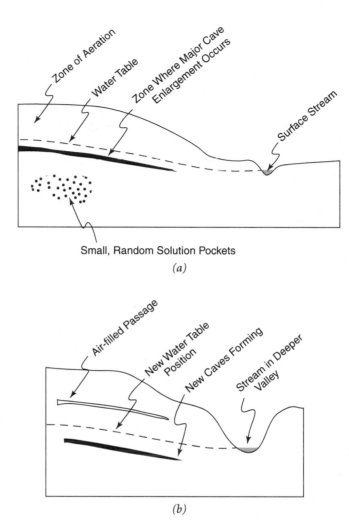

FIGURE 6–13   Formation of cave levels: (*a*) Stable position of water table enables small solution openings to integrate into larger openings near the water table. (*b*) A lowering water table and stabilization at a new level develops a younger, lower cave level.

plane weaknesses were the main cause of cave passages occurring along definite elevation levels. At Lehman, the steeply *dipping* (inclined) bedding planes obviously exert no influence on cave levels—the nearly level water table must be the main control here. The uplift and tilting history of the Snake Range during the past few million years along with the remarkable climate fluctuations characterizing the Pleistocene are intricately related to the development of this underground national treasure.

How Speleothems Form

Lowering the water table drains the flooded cave and allows air to enter (Fig. 6–13*b*). Because air is less dense than water, the pressure in the cave is reduced, and calcium-rich water entering the cave loses carbon dioxide to the atmosphere, which in turn reduces the quantity of carbonic acid and therefore the amount of calcium that the water can contain. The precipitation of calcite speleothems can begin as a small ring on the ceiling that elongates as a delicate mineral cylinder called a *soda-straw stalactite*. Additional water flows through the center of the soda straw, hangs as a water droplet on the end where CO_2 escapes, and a tiny ring of calcite is added onto the end as the cylinder slowly elongates. Blockage of the soda straw causes water to flow on the outside and a graceful, tapered shape called a *stalactite* forms.

Water droplets falling to the floor form a blunter-shaped, upward-pointing speleothem known as a *stalagmite*; and, if it joins a stalactite, a *column* is formed. Other speleothems include the carpets of *flowstone* on the walls and floors, *rimstone* formed at the edges of pools on the cave floor, gravity-defying *helictites*, needle-shaped calcium carbonate crystals (the mineral aragonite) called *frostwork*, lumpy, nodular forms called *popcorn*, which were deposited underwater when the cave was partially reflooded, and Lehman's special feature—the cave *shield* (Fig. 6–14).

Shields are rare except at Lehman where an ideal set of circumstances permitted an unusual number to form. Water under pressure flows through cracks, and two plates, separated by a narrow space, build outward at any angle from the wall. The upper plate is flat and shieldlike. Excess water emerging from the gap between the two plates flows down the lower plate depositing draperies and flowstone. The "parachute" (Fig. 6–14) is an outstanding example and has come to symbolize Lehman Caves.

FIGURE 6–14   Shields, an unusual type of speleothem, occur abundantly in Lehman Caves at Great Basin National Park. (Photo by National Park Service)

Special Features

The cave tour is available all year (bring a jacket, year-round temperature is 52°F; 11°C) but easy access to the high-country trails with their closeup views of the bristlecone forests and the spectacular array of alpine glacial landforms is limited to the summer months when the scenic automobile road is open to the 10,000-foot-elevation (3050 m) level at Wheeler Peak Campground and parking area. The small permanent icefield in the Wheeler Peak cirque has in the past displayed crevasses and a large separation crack between the rock wall and the glacier (a *bergschrund*) leading some to believe that this may be Nevada's only living glacier.

A trip over a rough road and a short hike in the wild southern part of the park will lead one to the 75-foot-high (23-m) Lexington Arch. The arch is in limestone, leading to the speculation that it is a remnant of a collapsed cave passage—a much different explanation than that proposed for similar features at Arches National Park in the Colorado Plateau to the east.

Late Mesozoic-Cenozoic igneous intrusions produced hot, chemical-rich fluids that locally deposited minerals rich in tungsten, gold, silver, lead, zinc, and copper. Typical for the west, the discovery of silver in the Snake Range in 1867 brought the usual rush of miners and fortune seekers to the area. Also typical for the west, the mining of investment money, mostly from the east, exceeded the production of metals in most of the mining camps! Numerous mining relics and structures, including the Johnson mill and mine and portions of the Osceola ditch that supplied water to the nearby Osceola mining camp in the early 1880s provide glimpses of former times. Not so romantic are the 247 mining claims that are filed on land within the park—238 of these are in the heart of the gnarled Bristlecone Forest atop Mt. Washington.

Colonel Steptoe named the highest peak Union Peak in 1855, and a small group of Lieutenant George Wheeler's explorers and surveyors renamed it Wheeler Peak in 1874. If you have good health and stamina as well as a long day to spend, follow the footsteps of John Muir and others across the Prospect Mountain Quartzite and enjoy the magnificent views from the top. Look down 2000 feet (610 m) into the cirques and glacial valleys and out over 100 miles (60 km) at one of the last areas where air pollution has not yet reached. Drink in the view of a vast sea of basins filled with an armada of giant, sinking ships listing heavily along their Basin and Range faults.

The stone hut foundations along the ridge top are remnants of one of the mountain top heliograph stations installed by the U.S. Coast and Geodetic Survey in the 1880s on a number of peaks from Colorado to California. It was hoped that mirrors would quickly send messages along a series of relay stations. The telegraph and later the radio were much more reliable—they even work in cloudy weather!

REFERENCES

Bartley, J.M., and Wernicke, B.P., 1984, The Snake Range decollement interpreted as a major extensional shear zone: Tectonics, v. 3, p. 647–657.

Halladay, O.J., and Peacock, V.L., 1972, The Lehman Caves Story: Lehman Caves Natural History Association, Baker, Nevada, 28 p.

Lambert, D., 1992, Great Basin drama: The story of a national park: National Parks and Conservation Association, v. 66, no. 11-12, p. 46–47.

Miller, E.L., Gans, P.B., and Lee, J., 1987, The Snake Range decollement, eastern Nevada: Boulder, CO, Geological Society of America Centennial Field Guide, Cordilleran Section, p. 77–82.

Whitebread, D.H., 1982, Geologic map of the Wheeler Peak and Highland Ridge further planning areas, White Pine County, Nevada: U.S. Geological Survey Miscellaneous Field Studies Map MF-1343-A.

Park Address

Great Basin National Park
Baker, NV 89311

DEATH VALLEY NATIONAL PARK
(CALIFORNIA AND NEVADA)

In 1933 almost 3000 square miles (7772 m) of southeastern California and westernmost Nevada were set aside as Death Valley National Monument (Fig. 6–15). Many attempts later, this unique area was finally enlarged and elevated to national park status in 1994. For many years special-interest groups, including those with off-road vehicles and who enjoy firearms, wanted space to pursue their activities— even if at the expense of potential parkland. With our rapidly increasing population, pressure on the remaining land for development, commercial uses, as well as many forms of recreation can only increase. Until population growth can be controlled, we should choose the least damaging land use. Unfortunately, the land and the environment always lose.

Death Valley is the hottest, driest, and lowest of our park areas. The maximum temperature ever recorded in North America was at Death Valley in 1913 when 134°F (57°C) was measured in the shade! The average annual precipitation is low, only 1.7 inches (4.3 cm), and some years have no recorded rainfall! Death Valley experiences a triple rain shadow effect in that the Sierra Nevada, Argus, and Panamint ranges all help drain most of the moisture from Pacific air masses moving inland. Local storms of high intensity cause floods that in a few hours remodel the landscape significantly. Great Basin National Park in the northern Basin and Range Province illustrates many of the features present in the high desert areas, while Death Valley illustrates the more extreme arid conditions of the southern deserts. Soils are slow in forming, and, when subjected to the warmer, more arid climatic fluctuations during the Quaternary, the mountains can become "skeletonized" as *soils* (loose sediment and organic material above the bedrock) are removed, exposing the bare rock surfaces of the mountain core. Once skeletonized,

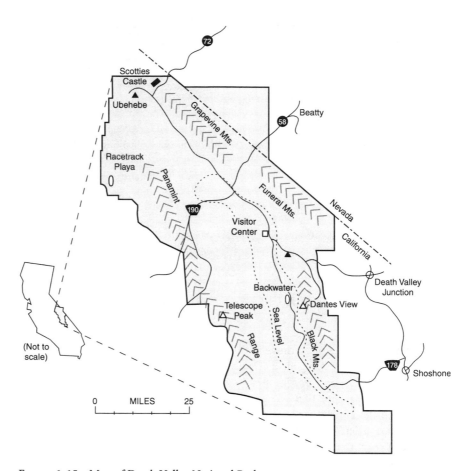

FIGURE 6–15   Map of Death Valley National Park.

only a significantly long interval of wetter climate will restore the soil, a condition that is unlikely for millions of years.

The topographic relief within the monument is notable. Telescope Peak near the western border is 11,049 feet (3369 m) high; less than 20 miles (32 km) to the east is Badwater (Fig. 6–16), the lowest point in the western hemisphere, 282 feet (86 m) below sea level! The topographic relief of 11,331 feet (3455 m) is less than it was earlier; in this enclosed basin all of the erosional debris carried down out of the mountains is deposited and trapped on the floor of the valley. The relatively flat floor in Death Valley as well as in other basins in this fault-block mountain landscape can be misleading, just as the flat surface of the ocean gives no clue as to how deep the ocean floor lies below. Only drilling or geophysical techniques such as bouncing shock (seismic) waves off the bedrock underlying the sediment fill can determine its thickness. Nearly 2 miles of sediment fill and over 2 miles of topo-

FIGURE 6–16 The lowest point in Death Valley is 282 feet (86 m) below sea level. The sign near the top of the picture is the sea level marker. (Photo by D. Harris)

graphic relief indicate that at least 4 miles of vertical change occurred along the Death Valley faults!

The early history of this unique area is colorful and sometimes painful when one considers that in 1849 the California-bound gold seekers and their 100 wagons were trapped for weeks trying to find a pass westward out of Death Valley. After consuming their oxen to stay alive and abandoning their wagons, they finally walked and staggered westward through Wingate Pass. As they escaped, one of the party reportedly remarked, "Goodbye Death Valley" and a name was born. Prospectors flooded the area later when reports of rich gold deposits circulated. The gold and silver mines were short-lived. It is ironic that it was the not-so-glamorous borax mining beginning in 1882 and the 20-mule team borax wagons that would make Death Valley famous.

General Geology

The rocks in the Panamint Range on the west and the Funeral and Black Mountains on the east record a metamorphic event during the Precambrian (1800 Ma): deposition of marine sediments along the rifted margin of North America during later Precambrian and Paleozoic time, igneous intrusions and uplift during the late Mesozoic, and volcanism and deposition of ash-rich sediments along streams and in shallow lakes during the Cenozoic Era. Footprints in the soft muds along the Pliocene lake basins and bones found in sediments speak to a time when mammoths, camels, horses, and titanotheres inhabited the area. Mineralizing fluids from the igneous intrusions produced some gold and silver ores, and large, Yellowstone-like hot-spring systems leached boron from volcanic materials and concentrated them into borax minerals such as colemanite (calcium borate) and probertite (sodium-calcium borate) in the Pliocene-age Artist Drive Formation. Ghost-town mining camps like Rhyolite, Skidoo, Ryan, and Leadville in Titus Canyon are sprinkled throughout the monument and surrounding areas. Beehive-shaped coke ovens in Wildrose Canyon in the Panamint Mountains, remnants of old mills, and a concentrator at Harmony Borax Works on the valley floor are reminders of the tumultuous late nineteenth- and early twentieth-century history of Death Valley. Stories such as stealing the gold shipment, 50 murders in four years in Panamint City, and salted outcrops that led to the building of the town of Leadville in an area with no mineral deposits are part of the total experience at Death Valley.

How was Death Valley formed? Is it actually a valley? True valleys are formed by streams, and sea level is the base level of erosion. Death Valley is a structural basin formed by down-faulting of a large block—the "valley" or graben—and up-faulting of the mountain ranges surrounding it. Death Valley is the "new kid on the block," a newcomer to the Basin and Range scene. Yesterday (3 million years ago) it didn't exist, today it is the lowest point in the Western Hemisphere.

As in other parts of the Basin and Range, the lithosphere is being pulled apart—here the plate tectonic mechanics are more discernible because of its closer position to the North America–Pacific Plate boundary and the recency of fault movements (Troxel and Wright, 1987). As shown in Figure 3–2 and explained in the Pacific Border section, when the North American Plate overrode the Pacific Plate ridge system, the subducting plate margin changed to one of lateral or strike-slip movement. As the Pacific Plate slides northwest along the main plate boundary fault (the San Andreas), that motion transfers some of the stress to a complex of strike-slip faults including the Garlock Fault located just south of Death Valley. The Panamint block is caught between the Garlock and the Furnace Creek strike-slip faults and slides westward, continuing to open the gap that we call Death Valley. The crust here is analogous to placing a square sheet of window glass on a thick carpet and walking across it (be sure to keep your shoes on!). The many cracks around the glass slivers represent the fault system. If one pushes a corner of the glass sheet toward the next corner, some of the slivers will slide past each other and others will separate and produce gaps. Death Valley is the gap where the Panamint Range block has slid along a gently inclined glide fault 50 miles (80 km) to the northwest between the lateral moving Garlock and Furnace Creek Faults. Articles by Stewart (1983) and Wernicke and others (1982) are good places to start for further information on the fault system, and the yellow pages are a good place to locate a carpet cleaner!

A number of features bear mute testimony to the recency and magnitude of the fault movement. At Zabriskie Point, 5-Ma lake sediments are no longer horizontal but are rotated to vertical and inclined positions (Fig. 6–17). The overlying Pleistocene-age Funeral Formation contains coarse gravels that record the presence of nearby mountains and the birth of Death Valley—perhaps only 2 million years ago. Both downward and lateral movement of the Death Valley block continues today as indicated by impressive fault scarps cutting across recently deposited gravels on alluvial fans (Fig. 6–18). Offset stream channels indicate that lateral motion occurs along the Furnace Creek Fault on the east side of the valley (Wills, 1989). One of the fault scarps may be less than a century old! Impressive canyons such as Titus and Golden result from running water trying to keep pace with a mountain block that episodically uplifts and a basin block that occasionally

FIGURE 6–17   Ash-rich Pliocene (5 Ma) lake sediments at Zabriskie Point rotated to near vertical inclinations along the active Furnace Creek strike-slip fault. (Photo by E. Kiver)

FIGURE 6–18   Faulted alluvial fan near Badwater in the southern part of Death Valley. Fault scarps marked by arrows. (Photo by R. Scott Creely)

lowers. Some of the canyon mouths are wineglass shaped—the broader bowl shape cut when erosive energy of the intermittent streams was less—the "stem" cut when erosion was more effective following vertical movements along the bounding faults.

Vertical faults run deep in this pull-apart graben. The crustal thinning releases pressure on the deeper areas, encouraging hot rock to expand and liquefy. Basaltic lava flows are widespread, particularly in the northern part. Volcanoes erupted in relatively recent time in the northwestern corner, near where "Death Valley" Scotty, with more than a little assistance from Chicago millionaire Albert Johnson, built "his" castle. It was probably not more than 2000 years ago that rising magma encountered groundwater in an alluvial fan and a series of steam explosions blew out Ubehebe Crater, a half-mile wide, 800-foot-deep (244-m) tuff ring. The craters here superficially resemble cinder cones because of the raised rims where laterally directed pyroclastic debris was deposited during the numerous steam explosions occurring during an eruption. Twelve of these craters are present, the youngest may be less than 200–300 years old.

Special Features

Alluvial fans are widespread and are an important landform in Death Valley. Those along the Panamint Range on the west side of Death Valley are very large; those on the east are smaller because they are dropping down more rapidly along the more active frontal faults and are being buried by playa sediments. Conditions are optimum for fan building; large amounts of coarse, weathered materials that accumulate in the mountains are flushed out by periodic torrential floods; in addition, this is a tectonically unstable area where the mountains are being uplifted, and fan building continues unabated. In contrast, in a tectonically stable area in arid or semiarid climates, *pediments* (erosional surfaces that slope basinward from a mountain front and are covered by a relatively thin veneer of gravel) are more common at the range fronts.

So extensive and so closely spaced are the alluvial fans that they overlap each other; a series of coalescing alluvial fans forms a continuous alluvial slope called a *bajada*. Thus, bajadas are the dominant landforms in Death Valley; however, where the canyons are more widely spaced, individual fans have maintained their own identities. From a distance the fans appear dark—up close, one sees that the surface rocks are coated with a mineral coating called *desert varnish* that requires a thousand or more years to form. Lighter areas mark younger sections of fans or areas where more recent floods scoured the area and deposited fresh debris.

Pleistocene climates affected the Death Valley area, albeit in a manner different from that in areas where large glaciers developed. During periods of higher precipitation, the excess water formed large lakes in the enclosed basins of the province. While at their maximum, what are today separate basins became interconnected, and fish from the Colorado River migrated into Death Valley and other basins. The lake in Death Valley, Lake Manly, expanded and was as much as 90 miles (145 km) long and 585 feet (191 m) deep. When the lake level remained constant for an extended period, wave-cut terraces or strandlines were notched into the sides of the mountains; these ancient shorelines can be clearly seen today at a number of locations including Shoreline Butte near the southern end of the park. The last major high lake levels occurred about 22,000 years ago in most lake basins in the Basin and Range Province, and plummeting lake levels occurred about 10,000 years ago (Smith, 1991) with the warming and drying that culminated in the Thermal Maximum.

As the lake disappeared, more salt deposits were added to the playa deposits, and Lake Manly's fish died except for a few minnow-size fish that found refuge in springs along Salt Creek and in the flooded cave known as Devils Hole. After 10,000–15,000 years of isolation, the fish populations evolved into closely related, but separate desert pupfish species. Each spring has perhaps a few hundred individuals—a unique species found in no other place in the world. If the proposed groundwater diversions are approved to send water to rapidly growing Las Vegas, the water table will lower and kill off these remarkable ice-age survivors.

The Devils Golf Course, between Furnace Creek and Badwater, is a remnant of Lake Manly salt deposits that continually changes as pinnacles and ridges form only to be etched by wind and rain into the bizarre forms that inspired its name (Fig. 6–19). Although the ridges and spires are usually a foot or less high, the terrain is almost impossible to traverse. For an exhaustive report on the processes involved in the formation of many fascinating features—dessication cracks, salt wedges, salt pans, and salt pools—see Hunt (1975).

Playa lakes form during storms and then vanish in a few days, mainly by evaporation. The playa deposits consist of silts and clays interbedded with salts that

FIGURE 6–19   Salt crystals and pinnacles on the Devils Golf Course. (Photo by E. Kiver)

were formed by precipitation as the lake water evaporated. Some of these salts contain the boron-rich mineral ulexite. The boron was dissolved from the Tertiary age sediments in the nearby mountains and reprecipitated in the playa. Early-day mining utilized the playa salts to extract borax for antiseptics, cleaning agents, glass making, and as a flux in jewelry making. Borate minerals are rare, and the Death Valley discovery and even larger deposits in the Kramer, California, area made the United States the world's largest producer. Mining stopped in Death Valley in 1928 with the Kramer discovery but resumed near Ryan inside the park boundaries in 1971 and continues today as an underground mining operation (Majmundar, 1985). Today boron compounds are used extensively in glass making, porcelain enamels for appliances such as refrigerators and stoves, and in the production of glass-belted tires and fiberglass.

An auto trip to Dantes View places one 5750 feet (1750 m) above Badwater almost directly below. The over 200-square-mile (500 km²), salt-floored playa surface stretches southward where the winding course of the usually dry Amargosa River is visible (Fig. 6–20). To the north, Salt Creek occasionally brings water and dissolved chemicals to the playa. The water vapor haze above is the only avenue of escape for the trapped water. Good views of alluvial fans and their surfaces darkened by desert varnish are also available at this spectacular viewpoint.

In the north end of the park, past Ubehebe Craters and over a series of dirt roads subject to frequent washouts, lies Racetrack Playa, home to some of the mys-

FIGURE 6–20   Extensive Death Valley playa surface lies over a mile below Dantes View. (Photo by E. Kiver)

terious playa scrapers (Fig. 6–21). A combination of wind, a slippery playa floor, and perhaps ice frozen to isolated rocks weighing as much as 700 pounds (25 kg) enables these rocks to skid merrily across the playa surface as much as 600 feet (202 m) in a year (Sharp and Carey, 1976)! Observations over a number of years verify that these rocks move once or twice in a decade; however, no humans have ever witnessed this unusual event (Reid and others, 1995).

Sand dunes are found in certain places in this desert area. What is puzzling is why they are not more widespread than they are—where fresh sediment is occasionally washed down the alluvial fan surface, winds are strong, and vegetation is sparse or lacking. The main dune areas are in the vicinity of Stovepipe Wells in the north-central section of the monument (Clements, 1966). Clements points out that most of the dunes are composed mainly of quartz sand, but that an unusual dune area is found about 10 miles (16 km) north of Stovepipe Wells. Here, derived from large travertine deposits, the dunes are made up of travertine sand. When wind velocity exceeds about 15 miles/hour (7–8 m/sec), the sand and the dune form move. However, the dune field remains fixed in position because seasonal winds come from different directions—north in the winter and south during the summer.

Death Valley has more to offer than its many geologic features; more than 600 species of plants and many animals have adapted to this harsh environment. Root systems over 50 feet (15 m) deep enable creosote and mesquite to survive in what seems to be an impossible environment by tapping freshwater contained in the

FIGURE 6–21   Track left by one of the enigmatic playa scrapers on the Racetrack Playa surface. (Photo by E. Kiver)

pores and spaces in the sediment below. A tour through Scotties Castle is a must. Travel back in time to another era—share the pleasant architecture, furnishings, the lives, and perhaps the thoughts of the Johnsons and the colorful western character known as Death Valley Scotty. A long day hike (14 miles; 23 km) to the summit of Telescope Peak will take one to cooler temperatures and through sparsely forested terrain that includes bristlecone pine, some of the oldest living things. The remarkable elevation changes in the southwestern United States is evident from the summit—the lowest point in North America is visible at Badwater to the east—Mt. Whitney, the highest peak in the conterminous United States is visible about 50 miles (80 km) west in the Sierra Nevada. Plan your trip carefully; the regular season is from October 15 to May 15. The summer sun is merciless and the range in temperature from day to night is great. If you are there during the right season, in a car in good condition, with more than enough drinking water, your trip into Death Valley will likely be one of the best.

REFERENCES

Clements, T., 1966, Geological story of Death Valley: San Bernadino, Death Valley 49ers, 62 p.

Hunt, C.B., 1975, Death Valley, geology, ecology, archaeology: Berkeley, University of California Press, 234 p.

Hunt, C.B., and Mabey, D.R., 1966, Stratigraphy and structure, Death Valley, California: U.S. Geological Survey Professional Paper 494-A, 162 p.

Majmundar, H.H., 1985, Borate mining history in Death Valley: California Geology, v. 38, p. 171–177.

Reid, J.B., Jr., Bucklin, E.P., Copenagle, L., Kidder, J., Pack, S.M., Posissar, P.J., and Williams, M.L., 1995, Sliding rocks at the Racetrack, Death Valley: What makes them move? Geology, v. 23, no. 9, p. 819–822.

Sharp, R.P., and Carey, D.L., 1976, Sliding stones, Racetrack Playa, California: Geological Society of America Bulletin, v. 87, p. 1704–1717.

Smith, G.I., 1991, Stratigraphy and chronology of Quaternary-age lacustrine deposits: in Morrison, R.B., ed., Quaternary nonglacial geology; Conterminous U.S.: Geological Society of America, The Geology of North America, v. K-2, p. 339–352.

Stewart, J.H., 1983, Extensional tectonics in the Death Valley area, California: Transport of the Panamint Range structural block 80 km northwestward: Geology, v. 11, p. 153–157.

Troxel, B.W., and Wright, L.A., 1987, Tertiary extensional features, Death Valley region, eastern California: Geological Society of America, Boulder, CO, Centennial Field Guide—Cordilleran Section, p. 121–132.

Wernicke, B., Spencer, J.E., Burchfiel, B.C., and Guth, P.L., 1982, Magnitude of crustal extension in the southern Great Basin: Geology, v. 10, p. 499–502.

Wills, C.J., 1989, Neotectonic tour of the Death Valley fault zone: California Geology, v. 42, p. 195–200.

Park Address

Death Valley National Monument
Death Valley, CA 92328

ORGAN PIPE CACTUS NATIONAL MONUMENT (ARIZONA)

In southern Arizona, along the Mexican border, about 516 square miles (1573 km²) of the Sonoran Desert Section of the Basin and Range Province were set aside as a national monument in 1937. Its primary function is to preserve a typical desert environment, but as an important incidental, classic examples of desert landforms are there in abundance. It is a true desert, receiving an average of about 8 inches (20 cm) of rainfall per year in the basins and higher amounts in the surrounding ranges. The adjoining Pinecate area in Mexico is now protected, establishing the first international park on our southern border.

A relatively small area of granitic rocks and associated metamorphic rocks on the southwest side of the Puerto Blanco Mountains record part of the largest episode of igneous intrusion ever to effect western North America. Melting of the upper mantle and lower crust occurred along a major subduction zone that developed from California to Alaska during the Mesozoic Era. Emplacement of huge batholiths occurred closer to the subduction zone and smaller plutons of Laramide age (late Mesozoic to early Cenozoic) are scattered inland. Conditions were just right as magma rising from the mantle in what is now southern Arizona produced an abundance of extremely rich copper ores that fueled Arizona's mining industry (Chronic, 1983, 1986).

However, the bedrock geology in the Ajo Mountains along the eastern border, the Bates Mountains in the western section, and the Puerto Blanco Mountains in between at Organ Pipe National Monument is dominated by volcanic materials erupted during two phases of crustal stretching that occurred about 22–14 Ma and 10–6.5 Ma in this part of the Basin and Range Province. The fault-block mountains are tilted to the northeast and rise abruptly above the broad, nearly flat lowlands such as the Valley of the Ajo, which occupies most of the north-northwestern section of the monument.

It might be expected that the landforms here would be similar to those in Death Valley. However, because differences of a few million years occur in fault-block mountain development from place to place in the Basin and Range Province, some areas have been stable much longer than others and more extensive erosion has taken place. Morrison (1991) notes that the Sonoran Desert Section of the Basin and Range, which includes Organ Pipe National Monument, was one of the first to become tectonically stable. Thus, erosion has not been interrupted by major tectonic events for at least 6.5 million years, and pediment-forming processes have operated here longer than in most areas of the Basin and Range. As the mountain fronts retreat by erosion, pediment surfaces (gently sloping erosion surfaces cut across bedrock, often of varying resistance; see Fig. 6–22) extend back into the mountains, and pediment passes and *inselbergs* (islands of rock projecting above a debris-capped pediment; see Fig. 6–23) form. If plate tectonics "leaves this area alone long enough," erosion will consume and obliterate the mountains. Another difference compared to Death Valley is that water and sediment occasionally leave the basins when the normally dry washes (stream channels) fill and flow through Growler Canyon to the north and into the Gila River. The normal faults along which the mountain blocks were originally uplifted have been planed off by erosion and are buried by coarse sediment washed downslope. No fault scarps or

FIGURE 6–22   Cactus-covered pediment with Ajo Mountains in the background. (Photo by D. Harris)

FIGURE 6–23　View south from Diaz Peak along eastern part of Organ Pipe Cactus National Monument, southern Arizona. Pediments have been extended into the mountains, leaving rock hills (inselbergs) rising above the pediment surface. (Photo by Fred Goodsell)

other surface expression of these former frontal faults is present—geophysical techniques such as seismic methods or measuring minute differences in the force of gravity can sometimes locate the buried faults.

The streams that flow out of the canyons are seldom loaded to capacity. When they are in flood, they have large capabilities to erode, but they have very little loose material to pick up and transport out of the mountains. Consequently, swinging from side to side, the streams plane off the bedrock at the base of the mountains. Coarse materials, the eroding tools, are spread in thin layers over the eroded bedrock, forming the pediment surface. Once this flat surface is developed, it is then perfected by sheet-flood erosion as McGee (1897) vividly described in 1897 after being caught in a desert cloudburst. The Papago country, including the Ajo Mountain area, was recognized by Bryan (1922) as being an ideal place for field research on the problem of the origin of pediments; his studies contributed much to our present knowledge of pediments.

Organ Pipe Cactus National Monument affords unusual opportunities to study plant moisture and soil–geology relationships. For example, the organ pipe cactus—one of many species in the monument—shuns the packed soils of the flatter areas but flourishes on the steep, dry, and rocky slopes of the Ajo Mountains. The giant saguaro, however, grows to great heights, as much as 50 feet (15 m), on the flats where the soils are finer and hold more of the precious moisture.

Two auto tour routes, the Ajo Mountain and Puerto Blanco drives, provide access to the mountains, basins, and trails in this fascinating desert environment. The monument is large and it has many interesting aspects, some of them colorful, especially the brilliant flame-red blossoms of the ocotillo, usually at their best in April. Wear hiking boots on the trails and carry a good supply of drinking water.

REFERENCES

Bryan, K., 1922, Erosion and sedimentation in the Papago country, Arizona: U.S. Geological Survey Bulletin 730.

Chronic, H., 1983, Roadside geology of Arizona: Mountain Press, Missoula, MT, 321 p.

Chronic, H., 1986, Pages of Stone: The Mountaineers, Seattle, Washington, v. 3, p. 128–133.

McGee, W.J., 1897, Sheetflood erosion: Geological Society of America Bulletin, v. 8, p. 87–112.

Morrison, R.B., 1991, Quaternary geology of the southern Basin and Range Province, in Morrison, R.B., ed., Quaternary nonglacial geology; Conterminous U.S.: Boulder, CO, Geological Society of America, The Geology of North America, v. K-2, p. 353–371.

Park Address

Organ Pipe Cactus National Monument
Route 1, Box 100
Ajo, Arizona 85321

SAGUARO NATIONAL PARK (ARIZONA)

Desert Environment

A delightful variety of cacti, including the park's namesake, occurs in profusion in the desert environment near the city of Tucson in southeastern Arizona. Besides the saguaro, the ocotillo, palo verde, prickly pear, and barrel are also found here. The area is also home to deer, javelina, and many small desert animals. Inhabitants that should be avoided include the rattlesnake, scorpion, and cactus needles! As reflected in the name, the giant saguaro (Fig. 6–24) was a major reason to establish the park. These vegetable giants can grow over 50 feet (15 m) tall and can weigh over 5 tons! However, the desert landforms, and particularly the presence of an unusual type of mountain structure called a metamorphic core complex make this area one of particular geologic interest (Davis, 1987).

FIGURE 6–24   Saguaro cactus, Saguaro National Monument near Tucson, Arizona. The variety of desert environments in the Basin and Range Province provides habitats for many species of cactus. (Photo by D. Harris)

Two physically separate park areas straddle the city of Tucson. The smaller and lower elevation Tucson Mountains unit on the west has Wasson Peak (4687 feet, 1429 m) at its crest. The much higher (Mica Mountain, 8666 feet, 2642 m) Rincon Mountain Unit is located about 23 miles (37 km) east on the eastern outskirts of Tucson. Together they make up the 131-square-mile (339 km^2) park. In 1933, when the area was designated a national monument, it was a comfortable 15 miles (24 km) from Tucson—now the city suburbs have grown to its doorstep bringing with it serious problems and threats to the park. This familiar type of story is being repeated many times over throughout our land.

Cactus thievery and vandalism and concern over a decrease in giant saguaro reproduction are among the special problems. All of the desert plants have evolved characteristics that enable them to survive in conditions of low and unreliable rainfall. Many have sharp spines to discourage would-be grazers, some have a thick protective skin that minimizes moisture loss, some store large quantities of water in their pulpy flesh, others have diminutive leaves, and others like the ocotillo shed their leaves and go dormant until the next rain occurs (Shelton, 1972). A comparison of old photographs with present scenes shows an alarming decrease in the number of saguaro. The reason is unclear—perhaps the climate is becoming drier or 100 years of overgrazing by cattle and the removal of mesquite and other brushlike plants that furnish shade for cactus seedlings may contribute to their reduction. Another reason, and perhaps the main one, may relate to problems with pollination. Their attractive flowers open only at night when temperatures are cooler and moisture loss is lessened. However, pollinating insects are inactive at this time—the saguaro relies exclusively on bat pollinators. Because bat populations decrease as more destructive chemicals are added to the environment and because their habitat is being destroyed by the developing Tucson metropolis, seed production and the number of new seedlings are reduced.

Metamorphic Core Complex—Rincon Unit

Like Chiricahua National Monument 60 miles (97 km) to the east, Saguaro is also in the Mexican Highland section of the Basin and Range Province and experienced the same general types of geological growing pains; intrusion of Precambrian granite, deposition of mostly marine sedimentary rocks during the Paleozoic, mostly erosion in the Mesozoic, Laramide compression, extensive volcanism, crustal stretching during the mid-Tertiary, and fault-block mountain formation during the past 10–15 million years. Unlike Chiricahua, volcanic rocks (although present) and intricate erosional forms are not important here. Rather, the Rincon and nearby Catalina and Tortolita Mountains contain an unusual type of mountain structure called a metamorphic core complex, or *gneiss dome complex.*

A number of other metamorphic core complexes occur in ranges that extend across southern Arizona in the Mexican Highland section. Each has a dome-shaped granite pluton with a shell of severely stretched and microfractured metamorphic rock such as gneiss. Above this is a separation plane called a decollement, or *detachment fault,* above which rocks moved laterally as a separate plate across the underlying gneiss (Fig. 6–25). This recently recognized structural style of mountain development is not yet completely understood. As in many geological problems, the first step is geological mapping to determine what is really in the field. What rocks are there? What are their mineralogical and structural features? Do the landforms shed light on the problem? Fortunately, this has been done at Saguaro by George Davis (1987), and the exposures here display very clearly the rock relationships described above.

The crust here was stretched under high-temperature and pressure conditions. The lower areas behaved plastically and stretched without breaking. Buoyant rock masses, usually composed of granite, rose toward the surface. The lower area continued to remain plastic much like a balloon that was being inflated. The balloon surface, and the area just below is a transition area where, in the core complexes, recrystallization under stress produces a lining up of elongate and flat minerals that define a metamorphic trait called *foliation.* Since the foliation here separates dark- and light-colored mineral bands, the rock is named *gneiss.* The transition zone is more brittle and contains countless fractures and microfractures. Above the gneiss is a curved detachment fault overlain by brittle rock layers that are highly faulted. The plastic zone punched its way upward pushing its metamorphic shell (usually gneiss) and overlying sedimentary strata upward. The heat was intense—the rock locally became doughlike, and minerals became flattened and sheared as the giant taffy pool progressed upward. Other minerals grew as the shell was being stretched and a distinctive metamorphic rock with large, egg-shaped feldspar crystals engulfed by streaky mineral layers called an *augen gneiss* formed. Watch for these at some of the pull outs along Cactus Loop Drive. High temperatures caused atoms in individual minerals to reorganize, causing those with radioactive atoms to reset their atomic clocks. Most of the reset clocks read about 25 Ma—mid-Tertiary when rock temperatures lowered and allowed the clocks to begin "ticking" again.

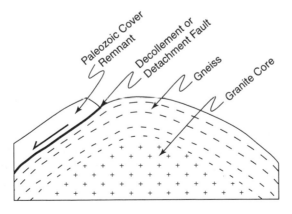

FIGURE 6–25   Generalized cross section of the metamorphic core complex in the Rincon Mountains Unit, Saguaro National Park. (After Davis, 1987)

The dome form is preserved in the Rincon Mountain Unit where three large wrinkles or antiforms, protected by their resistant gneiss shells, are reflected in the topography. The Tanque Verde ridge antiform, shaped like a mountain-size turtle shell, plunges southwesterly toward the monument headquarters. A trip along Cactus Loop Drive, preferably with Davis's (1987) field guide in hand, leads through the hardened metamorphic taffy pot. The detachment fault and the stack of Paleozoic sedimentary rocks that slid off the dome occur along the lower slopes of the ridge and can be reached by a short scenic hike.

Exactly why metamorphic core complexes form is uncertain, but all known occurrences are associated with extensional or pull-apart conditions, and many are associated with a granitic intrusion. In addition to southern Arizona, core complexes occur in southeastern California, northeastern Nevada, northwestern Utah, southern Idaho, and in the Northern Rocky Mountains in northern Idaho, northeastern Washington, and southern British Columbia. Speculation that areas of locally thickened crust due to overthrusting may trigger the core-forming process (Parrish and others, 1988) may explain why core complexes are located where they are. Oldow and others (1989) is a good place to start for additional information.

Tucson Mountain Fault Block

The western unit of Saguaro National Park centers around the Tucson Mountain fault block and at first glance appears to be completely unrelated to the metamorphic core complex in the Rincon Mountain Unit some 20 miles (13 km) to the east. The Tucson Unit contains a section of Mesozoic and Tertiary sedimentary and volcanic rocks intruded by a Tertiary intrusive (Fig. 6–26). Older rocks exposed in the Rincon Mountains and just outside the Tuscon Mountain Unit boundary extend the regional geologic story back to 1600 Ma when granite squeezed into the Precambrian crust. Gentle rise and fall of the crust along the trailing edge of the North American Plate produced deposition of mostly limestone in shallow seas punctuated by episodes of erosion through Paeozoic time. Nonmarine *redbeds* (shales and sands brightly colored by oxidized iron), arkose, and—during late Mesozoic and early Tertiary—volcanic rocks were deposited as the North American Plate began its relentless westward movement that was to turn the west into an incredible display of mountains, basins, volcanoes, and plateaus.

Compressive forces gripped the crust as microcontinents and slivers of land were crunched against North America in the newly formed subduction zone. Crustal stretching began in the Basin and Range Province about 30 Ma as the subduction zone along the west coast of North America was replaced by a fledgling fault that was destined to grow into the monster structure known as the San Andreas Fault. Large Laramide overthrust faults may have been reactivated about 30–20 Ma as detachment zones utilized by the developing metamorphic core complexes.

Removal of overlying layers caused some of the granite to bob toward the surface, forming the cores of mountains such as the Rincon and Catalina Mountains

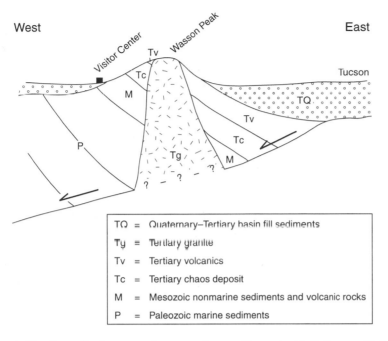

FIGURE 6–26 Generalized cross section across Tucson Mountains Unit, Saguaro National Park. The mountains are believed to be part of a detached block that slid from a nearby metamorphic core complex (gneiss dome).

located east and north of Tucson. Slices of younger rock slid from the rising dome, leaving blocks around the base of the mountain and perhaps transporting a larger block westward as the Tucson Mountains. As the block glided westward, it rotated so that the strata are like a stack of pancakes with older pancakes (strata) to the west and younger ones to the east (Fig. 6–26).

Renewed stretching about 15–10 Ma broke the crust into the jumble of broken fault blocks that make up the Basin and Range Province. Locally, the Santa Catalina Mountains are 6800 feet (2073 m) higher (Mt. Lemmon elevation is 9157 feet, 2790 m) than the surface of the Tucson Basin, and the top of the Tucson Basin block lies below sea level buried by 7000 feet (2130 m) of Miocene-Pleistocene debris. Thus, a minimum of almost 14,000 feet (4268 m) of vertical change occurred.

In the Tucson Mountains is an interesting rock layer of uncertain origin called the Tucson Mountain Chaos. Chaotic it is! House-size blocks of Precambrian schist, Paleozoic and Mesozoic sedimentary rocks, and Tertiary volcanic rocks all next to one another in a matrix of sandstone and tuff have no clear explanation. Similar features in Death Valley and the Olympic Mountains are attributed to fault activity—here they may be due to volcanism. Perhaps blocks were ripped from the walls of a volcanic conduit and incorporated into a lava flow; or perhaps they were caught up in an eruption-caused flood of water.

Landforms and Water

Our old desert friends, alluvial fans, bajadas, pediments, desert varnish, and *desert pavement* (collection of large surface rocks that armor and protect finer sediment below from wind erosion) are here in abundance. Relics of wetter times and more effective stream erosion during the Pleistocene include the upper part of the thick basin fills and remnants of Pleistocene bajada surfaces, pediments, and stream terraces (Morrison, 1991). Just south of the Rincon Unit in a detached segment of Paleozoic limestone is Colossal Cave—beautifully decorated with speleothems—but high and dry in today's desert climate. This delightful cave is a state park and well worth a side trip.

Water is still becoming more scarce here—but not because of a changing climate. Tucson was established along the banks of the Santa Cruz River—a river that drains the west side of the Tucson Basin and used to flow all year around to the Gila River to the west. Several hundred thousand people later, water only flows in the river after an unusually heavy rain or snowmelt. Water tables in the wells supplying the city are decreasing dramatically, causing the land surface to lower and increasing concern that large ground cracks will become a serious problem.

REFERENCES

Davis, G., 1987, Saguaro National Monument, Arizona: Outstanding display of the structural characteristics of metamorphic core complexes: in Centennial Field Guide, Geological Society of America, Boulder CO, v. 1, p. 35–40.

Morrison, R.B., 1991, Quaternary geology of the southern Basin and Range Province, in Morrison, R.B., ed., Quaternary nonglacial geology: Conterminous U.S.: Geological Society of America, Boulder, CO, The Geology of North America, v. K-2, p. 353–371.

Oldow, J.S., Bally, A.W., Ave Lallemant, H.G., and Leeman W.P., 1989, Phanerozoic evolution of the North American cordillera; United States and Canada: in The Geology of North America, Geological Society of America, v. A, p. 139–232.

Parrish, R.R., Carr, S.D., and Parkinson, D.L., 1988, Eocene extensional tectonics and geochronology of the southern Omineca belt, British Columbia and Washington: Tectonics, v. 7, p. 182–212.

Shelton, N., 1972, Saguaro National Monument, Arizona: U.S. Government Printing Office, Washington, D.C., 98 p.

Park Address

Saguaro National Park
36933 Old Spanish Trail
Tucson, AZ 85730

CHIRICAHUA NATIONAL MONUMENT (ARIZONA)

The Chiricahua Mountains in southeastern Arizona (Plate 6) are part of the Mexican Highland Section of the Basin and Range Province. Water is scarce, but an integrated drainage system of washes leads to the Gila River and eventually the Colorado—except for Sulfur Spring Valley west of the Chiricahua Mountains and the Hualapai Valley on the far western edge of the Mexican Highland Section where, much like Death Valley and the Great Basin, drainage is internal. Pleistocene Lake Cochise probably spilled southward to Mexico from the Sulfur Spring Valley about 13,500 years ago. The lake receded with the drier conditions at the end of the last ice age and a small remnant—Willcox Playa—remains today. Although part of a true desert landscape, the ranges and basins in this section are higher and wetter than those in Organ Pipe Cactus National Monument in the Sonoran Section to the west. With peaks as high as 9796 feet (2986 m), they receive both snow and rain sufficient for a dense growth of vegetation in the more favorable sites, particularly in the shaded canyons.

Yucca, cactus, and century plants dot the desert areas; a chaparral of scrub oak and manzanita covers the higher slopes, with the sycamore and Arizona cypress shading the deep canyons. White-tailed deer, peccaries, bobcats, and even the coatimundi are here, along with rodents and birds of many kinds. For the details on the plant and animal life of the area, see Jackson (1970).

The Chiricahuas are composed of upfaulted Precambrian granites and schists flanked by Paleozoics and younger rocks. But within the monument, volcanics carved into weird columns, spires, and balanced rocks are of greatest interest (Fig. 6–27). This maze of rocks, with its myriad columns and labyrinthine passageways, provided an essentially impenetrable stronghold for the skillful and effective Apache warriors Cochise and Geronimo, and later for "Bigfoot" Massai. The profile of Cochise can be seen from interstate 10 near San Simon or looking north from high on the monument road; Cochise lies in stony silence on the mountain north of the monument boundary (Fig. 6–28).

The geologic story might begin early in the Precambrian or during the deposition of thick Paleozoic-age marine sediments in the area; "today's" story must at least begin with the "Basin and Range polka"; squeezing of the crustal accordion during the Laramide Orogeny about 65 Ma followed by crustal stretching beginning about 30 Ma when southwestern North America began to tear apart. At that time, the ocean ridge was overrun by the westward moving North American Plate and the ocean plate began sliding northwest along the newly formed San Andreas Fault. Subduction activity terminated along the newly formed fault, but fluids contained in sediments that were previously dragged down the subduction zone continued to rise into the hot mantle and crust causing the melting point of rock to lower and basaltic magma to form.

Magma generation was further increased as the sliding plates caused the crust in the Basin and Range Province to stretch and thin, thereby reducing the confining pressure on the hot mantle below. The expansion allowed liquids to ooze from

FIGURE 6–27   Tertiary volcanics occur in a number of mountain ranges in the Basin and Range Province. The Organ Pipes, shown here, are among the many fantastic erosion forms in Chiricahua National Monument, southeastern Arizona. This welded tuff was formed when the Turkey Creek caldera erupted in the nearby Chiricahua Mountains. (Photo by D. Harris)

FIGURE 6–28   Profile of Cochise, bathed in smog from smelters in the valley. (Photo by D. Harris)

the mantle, accumulate in the lower crust as magma, and then rise upward toward the surface. The chemistry of the rhyolitic volcanic material that erupted is similar to what one would get by throwing shale, sandstone, limestone, granite, and other common continental crust rocks into a blast furnace. Thus, *igneous petrologists* (people who study internal and external features of igneous rocks and how they form) believe that melting of crustal rocks is necessary to produce the large volume of explosive rhyolitic magma necessary to fuel the massive eruptions recorded in the rocks at Chiricahua and other areas.

Relatively minor eruptions deposited the early volcanics of the Faraway Ranch Formation in a large valley that existed in the national monument area 27–35 Ma. About 27 Ma conditions changed dramatically as a second series of eruptions began. Rhyolitic magma was rising through the crust about 7 miles (11 km) south of Chiricahua under what was soon to become the Turkey Creek caldera. As magma reached shallower depths in the crust, more dissolved gases, mostly water and carbon dioxide, separated from the magmatic fluid and increased the pressure in the magma chamber. The ground above bulged, just as a car tire with a faulty sidewall bulges. The brittle crust above cracked and the pressurized magma blew out in the first of at least three giant explosions that dwarfed any of historic record.

The eruption was like the 1980 Mount St. Helens eruption in the Washington Cascades except that 1000 times more material was erupted from the Turkey Creek vent. Ash was hurled skyward and hot glowing clouds of choking gas and volcanic fragments were propelled laterally at hundreds of miles per hour with a fury equivalent to the blast of a mountain-size jet engine. The cloud engulfed everything in its path, probably killing every living thing within 20 miles (32 km) of the vent (Pallister and others, 1993). A denser boiling cloud followed that partially filled the valley at Chiricahua with ash-flow debris. The heat from the thick deposit of ash, pumice, and rock fragments was trapped [about 1100°F (590°C)] causing the deposit to become "squishy," melt, and form a dense, coherent unit called a welded tuff.

Overlying this layer, a partially welded unit is topped off by a loose, ash layer that air cooled too fast to weld. Over a period of time, even these loose layers can be hardened by mineral cements converting a sediment (ash) to a rock (*tuff*). The three (at least) violent eruptions of ash flows are recorded in the Miocene-age Rhyolite Canyon Formation—the thickest is an incredible 880 feet (268 m) and the total thickness of all the welded tuffs is about 1500 feet (457 m). The peaceful valley that existed before the Turkey Creek caldera erupted was now completely obliterated. Eruption of over 100 cubic miles (417 km³) of magma in a short time caused the surface over the magma chamber to collapse and form the 12-mile-diameter (19-km), 5000-foot-deep (1525-m) caldera. Chiricuhua would not have been a good place to bear witness to these eruptions.

The final chapter in volcanism here occurred when dacite magma welled to the surface near the north end of the caldera and flowed through a valley and capped the ash-flow deposits at Chiricuhua. This resistant lava now sits atop Sugarloaf Mountain in the monument. Erosion attacked the less resistant rocks at the flow margins leaving the former valley bottom as a ridge top—a good example of *topographic inversion.*

About 12 Ma the main episode of Basin and Range faulting began here and ended a little more than 3 Ma except for some local, infrequent faulting that continues to the present (Morrison, 1991). Earth movements added to the vertical cracks already formed during cooling of the hot tuff. The processes of weathering and erosion began their sculpture spree by attacking these cracks, or *joints,* and shaping the rhyolite deposits into the pedestals, balanced rocks, columns, and spires for which Chiricahua is famous.

The pinnacles at Crater Lake were formed by weathering and erosion of poorly welded ash from around columns of highly welded material. In Chiricahua, an entirely different origin is envisioned, one based on surface processes attacking the intersecting rock fractures and dissecting the tuff into an incredible array of tall, thin, stony giants that Chronic (1986) describes as a "pinnacle army" that marches "up Chiricahua's mountainsides."

Special mention is made of accretionary *lapilli,* sometimes called volcanic hailstones. They are generally uniform in size—about that of marbles—and most are nearly spherical. A good place to see them is along the trail from the Echo Canyon parking area to Massai Point or along the Rhyolite Canyon trail. Their origin involves the mixing of volcanic dust with water—perhaps during a rainstorm or with steam from magma mixing with groundwater.

Some of Chiricahua's wonders can be seen from a car. Be sure to see the exhibits in the visitor center and also the slide-tape presentation. Go to the Sugarloaf parking area and hike to the top of Sugarloaf Mountain, where the views are excellent in all directions. The dacite here is a lava flow—it is the caprock for the entire thickness of volcanic rock in the monument. Other trails take you to the Natural Bridge, the Big Balanced Rock in the Heart of Rocks section, and Duck-on-a-Rock. Here you can observe that not only are there vertical weaknesses, but there are also horizontal weaknesses where mud layers or less-welded layers influence how rapidly a rock will decay. Your conclusion should be that not all rocks were created equal!

REFERENCES

Chronic, H., 1986, Pages of Stone: The Mountaineers, Seattle, Washington, v. 3, p. 79–85.

Jackson, E., 1970, The Natural History Story of Chiricahua National Monument, Arizona: Globe, AZ, Southwest Parks and Monuments Assoc. (out of print).

Morrison, R.B., 1991, Quaternary geology of the southern Basin and Range province, in Morrison, R.B., ed., Quaternary nonglacial geology; Conterminous U.S.: Boulder, CO, Geological Society of America, The Geology of North America, v. K-2, p. 353–371.

Pallister, J.S., du Bray, E.A., and Hall, D., 1993, Geology of Chiricahua National Monument, a review for the non-specialist: U.S. Geological Survey Open-File Report 93-617, 17p.

Park Address

Chiricahua National Monument
Dos Cabezas Route
Box 6500
Willcox, Arizona 85643

BIG BEND NATIONAL PARK (TEXAS)

History and Description

The location of Big Bend National Park is easily found by examining Plate 6 or a map of Texas and noting the prominent "big bend" in the Rio Grande. With 1252 square miles (3205 km²) this is the fifth largest park in the conterminous United States. Authorized in 1935, it officially became a park in 1944. Because of its location and the fauna and flora migrations related to Pleistocene climate changes, Big Bend has a mixture of Rocky Mountain and Mexican Highland species and was declared an internationally recognized biosphere reserve in 1976. Vegetation changes with elevation from the lush vegetation immediately along the river to desert shrubs, desert grasses, pinyon-juniper forest to Douglas fir–ponderosa–aspen at higher elevations (Moss, 1984). Numerous cacti and yucca species exist, including the ocotillo, lechuguilla, and century plant, which lives for 25 to perhaps 100 years before it blooms and dies. Animals include peregrine falcon, mule and white tail deer, peccaries, and coyote. In addition, Mexico declared the adjacent 1.2 million acre Sierra del Carmen Mountain section a wildlife preserve in 1994 making this area one of two park areas shared by the two nations.

The area also has a fascinating human history beginning with Paleo-Indians hunting elephants, camels, and other Pleistocene animals about 10,000 years ago. More recently, Chisos, Mescalero Apache, Comanche, Spaniards, the U.S. Army, and Texas Rangers all had a hand in how the land was to be used. Mercury (quicksilver) miners discovered the area in the 1890s and settlers didn't move in until the early 1900s. Raids on settlements by outlaws and revolutionaries continued almost to World War I; the area still remains one of the most remote in the conterminous United States. Because mountain lions still survive in the park, this area fulfills noted outdoor author Edward Abbey's definition of a wilderness—"it ain't wilderness unless there's something out there that can eat you."

When Big Bend was no more than just "the Big Bend country" a young geologist became intrigued with this jumbled mass of rocks of which no one had been able to make heads or tails. This was his challenge, more attractive because of its remoteness—wildness in fact. When Ross Maxwell began the survey there were no roads; the Comanche Trail established by the U.S. Army in the 1840s was the main throughfare. A horse named Nugget was his "field assistant"; Nugget carried Ross over many miles of mountain and desert trail during the long days of the survey.

This was not the Big Bend of the 1880s as might be assumed, but the Big Bend of the 1930s and 1940s—the last frontier in the lower 48 states (Maxwell, 1968). Slowly and stubbornly, Big Bend surrendered its "past" and the puzzle was put together, minus a few details buried under Chisos volcanics. When the park was established in 1944, Maxwell became its first superintendent. He laid out the roads where they would not be hazardous; it was not by coincidence that they pass by, or are within sight of, almost every fascinating geologic feature in the vast area, which is Big Bend National Park.

Geographic Setting

Big Bend is part of the Chihuahan desert and is located in the eastern Basin and Range Province near the boundary with the Great Plains Province. The nearly 1900-mile-long (3040-km) Rio Grande begins in the Southern Rocky Mountains and flows south through a series of grabens to the El Paso area, where its direction to the Gulf of Mexico changes to southeast except for a 120-mile (192-km) segment known as the "big bend" that is apparent in Plate 6 and shown in greater detail in Figure 6–29. The river flows through three separate gorges or canyons in the park; the most spectacular is Santa Elena (Fig. 6–30), which is a "must" for visitors—either by viewpoints, a canyon trail (when weather and river conditions permit), or by float trip. Santa Elena has sheer walls 1500 feet (457 m) high and is in places only 30 feet (9 m) wide; therefore, in flood the Rio Grande is a "river on edge." Downstream, the Rio Grande has cut deep Mariscal Canyon through Mariscal Mountain. Farther downstream, along the eastern border, Boquillas Canyon is equally striking. Because the narrower canyon segments contain wall-to-wall water, a float trip is the only way to see the over 100 miles (62 km) of river along Big Bend's southern border.

The Chisos (ghost or spirit) Mountains contain Tertiary-age volcanic rocks that form the central area of the park. Elevations at river level are about 1600 feet (490 m) and rise to 7835 feet (2389 m) at Emory Peak. The Santiago and Sierra del Carmen Mountains on the east, as well as Mariscal Mountain on the south and the Mesa de Anguila on the west are all fault-block mountains with north to northwest trends (Fig. 6–29).

Long slopes covered with mesquite, creosote, cactus, and other desert plants extend out from the mountains. These are pediments and remnants of older pediments (Fig. 6–31). Badland areas are common where streams have dissected easily eroded, poorly consolidated rocks.

Bedrock Geology and Geologic History

Big Bend is composed mainly of Cretaceous limestones that in large areas are buried under Tertiary volcanics. In addition, intrusive stocks, laccoliths, dikes, and

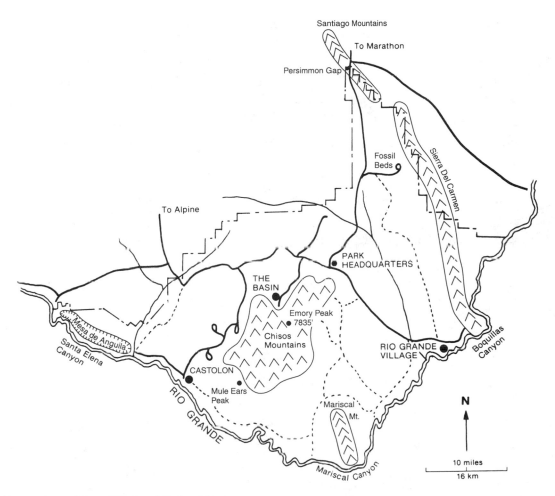

FIGURE 6–29   Map of Big Bend National Park.

sills are extensively exposed (Fig. 6–32). In limited areas, there are Paleozoic limestones and shales, and Cretaceous and Tertiary sandstones and shales. Because Big
Bend experienced many of the same plate tectonic movements as other parks and
monuments in the Basin and Range Province, similar geological features and history can be found here as well. Compression during Laramide times, crustal extension and major volcanism during mid-Tertiary time, a second interval of
extension and fault-block development during the late Tertiary, and the major
fluctuations of both warmer and cooler climate episodes during the Quaternary are
familiar themes in the Basin and Range. However, each park area has something
unique—some feature or variant of the geologic history that makes each area a special place. Big Bend is no exception.

FIGURE 6–30   Santa Elena Canyon as seen from the air. Fault block of limestones is tilted away from viewer. Mexico is to the left of the canyon: Texas is on the right. (Photo by Hunter's of Alpine, Texas)

FIGURE 6–31　One of many extensive remnants of pediments in Big Bend. Streams have cut down, leaving the pediment remnants well above present stream level. (Photo by D. Harris)

FIGURE 6–32　Volcanic spine exposed by erosion of soft volcanic tuff that encased it. (Photo by D. Harris)

The oldest rocks in the park are lower Paleozoic located in the Persimmon Gap area near the north entrance (Fig. 6–29). These marine rocks were deposited between two continental plates in the Ouachita Geosyncline—a continuation of the Appalachian Geosyncline that borders the east side of North America. As the east and south edges of this ancient plate collided with other plates in the late Paleozoic about 300 Ma, northeast–southwest oriented folds and large overthrust faults formed. As the European and South American Plates separated from the North American Plate in what is now the Atlantic Ocean and Gulf of Mexico, North America began to move west, eventually producing a long downwarp forming an elongate, early Cretaceous seaway that extended from the Arctic to the Gulf of Mexico. Thick limestones and other types of rocks, including shale and coal, were deposited in this fluctuating seaway. Big Bend received several thick limestone units including the massive and resistant Santa Elena Limestone that forms the rim rock at Santa Elena Canyon. By late Cretaceous time, about 70 Ma, the rate of plate collision and subduction along the western, leading edge of North America had increased and the Laramide Orogeny was in full sway. The late Paleozoic overthrust fault was caught in this new episode of compression and was itself folded and overthrust over pre-Laramide rocks—this time in a northwest–southeast direction, at right angles to the late Paleozoic mountain-building forces. Thus, Big Bend marks the location where the older mountain system of the southeastern and eastern United States abuts against the Laramide structures of the west. It is also one of only two sites in Texas where the buried Ouachita Mountain system is visible at the earth's surface.

The Laramide ended here about 50 Ma, and a long period of erosion was interrupted by igneous activity about 35 Ma in the Chisos Mountains area during the older phase of extension in the Basin and Range. The latest episode of extension began about 15 Ma, initiating the modern fault-block mountain development that renders the older Laramide and Ouachita structures insignificant in terms of topographic expression of mountain ranges. Otherwise, Big Bend would be an extension of the Southern Rockies. Here, as in other parts of the Basin and Range Province, fault-block mountains and desert landforms dominate.

Block faulting of large magnitude formed the mountain ranges of today. The sheer face of limestone at the mouth of Santa Elena Canyon is ample evidence; here, along the Terlingua Fault, a long block of the crust was lifted and rotated, thus forming a mountain range called Mesa de Anguila (Fig. 6–29). Sierra del Carmen to the east is one of many fault blocks in Big Bend.

The mid-Tertiary (Eocene, about 35 Ma) crustal stretching allowed hot material from the *asthenosphere* (plastic zone beneath the rigid lithosphere) to rise under the thinned crust. The thinner crust reduced the confining pressure on the asthenosphere, allowing hot rock to expand and melt, producing numerous small *plutons* (magma chambers) that rose into the Chisos Mountains as well as many other areas in the Basin and Range. The shapes of these bodies varied—domes, fingers, sills, dikes, mushroom shapes (*laccoliths*)—and help to account for some of the bizarre shapes of Chiso Peaks—Mule Ears Peak and Elephant Tusk are examples. The unusual igneous shapes help the area to maintain its Chisos, or spirit look.

Extrusive activity associated with these intrusions was dominantly explosive, indicating that the rising basalt magma likely melted large quantities of silica-rich crustal rocks to form explosive magmas of andesite and rhyolite compositions. Some volcanic centers collapsed, forming calderas and sending glowing ash clouds across the countryside and blanketing the area with thick ash deposits. Basaltic lava flows are also common and cap a number of the higher ridges and peaks. Casa Grande Mesa, which towers above the Basin, is topped by a lava flow as is Burro Mesa and the south rim of the Chisos. One unusual volcanic feature resembles a huge petrified stump and has, in fact, been erroneously identified as such. As it is composed of dense rhyolite and rhyolitic pitchstone, it is definitely of igneous origin (Fig. 6–32).

Geomorphic Features

The topographic expression of a landscape is the product of external forces such as erosion acting on geologic features exposed on the earth's crust. Many of these features such as folds, faults, and igneous bodies, as well as land elevation, are produced by internal forces. The scale of topographic features can range from small to very large and are often the product of *differential erosion* in which running water and other erosional agents tend to preferentially attack weaker or less resistant materials. An analogy might be the fate of a box of neopolitan ice cream in which, in some households, certain flavors are more likely to be attacked, leaving their chocolate and vanilla neighbors standing as vertical ice cream walls.

An example of large-scale differential erosion would be the course of the nearly 1900-mile-long (3040-km) Rio Grande, which is mostly determined by the underlying fault structure. Farther north, in southern Colorado and New Mexico, the river is trapped in a series of grabens along a major crustal break called the Rio Grande rift (Baldridge and Olsen, 1989). The rift is less than 10 Ma and has experienced recent movements.

The southeastern trend of the Rio Grande from El Paso to Big Bend is also related to structural features that may be associated with the opening of the Rio Grande rift. The fault basins were filled to the brim during mid-to-late Tertiary with debris eroded from surrounding mountains. Large amounts of basin fill were removed by larger rivers operating during the Pleistocene. The Rio Grande established its course along these structural features but occasionally was *superposed* (lowered down through sedimentary strata above an unconformity) across fault blocks or other topographic features forming what is known as *transverse drainage*. Spearing (1991) favors an alternate explanation for the transverse drainage in which the stream is in place and able to erode as fast as the fault block is uplifted.

Such a river is called an *antecedent river*. The Santa Elena, Mariscal, and Boquillas canyons are examples in which transverse drainage formed across thick, resistant limestones producing spectacular vertical canyon walls.

A "medium"-scale example of differential erosion would be the central high peaks such as Emory, Lost Mine, Casa Grande, Pummel, and Pullian peaks underlain by resistant igneous rocks in the Chisos. The rounded form of Mariscal Mountain is maintained because the resistant outer rock layers mirror the anticlinal structure of the mountain.

The origin of the Basin has been a matter of controversy. The Basin is an amphitheater-shaped topographic feature (Fig. 6–33) high in the Chisos with a narrow and spectacular 10-foot-wide (3-m) slot in the west side called the "Window." Its general cirque shape led Jenkins (1958) to the erroneous conclusion that it is a glacial cirque. Its shape led others to believe that it is a volcanic crater, but this too lacks supporting evidence. Rather, it is a basin carved out of volcanic rocks by streams. Differential erosion of the complex of rocks here left the bowl-shaped amphitheater.

Landforms resulting from stream erosion and deposition are widespread and well developed. Pediments are by far the most important of these landforms (Fig.

FIGURE 6–33   The Basin in the Chisos Mountains. Water exits through the narrow pass in the middle distance. (Photo by Ross A. Maxwell)

6–31). These long flat-surfaced slopes extend out from the mountains and are graded to the main streams of the area. The bedrock is generally covered with gravelly alluvium from 5–20 feet (1.5–6 m) in thickness; locally it is thicker. Remnants of former pediments are also present, representing earlier cycles of pedimentation. As in the Organ Pipe Cactus National Monument area, widespread pediments with few and small alluvial fans indicate that the Big Bend country is relatively stable tectonically.

Although limestones are widespread, caverns are apparently not common features in Big Bend. Maxwell (1968) mentions one—Smugglers Cave—in the wall of Santa Elena Canyon. Apparently, the fractures and abundant water that are essential in the cavern-forming process are generally lacking.

Population Explosions in Texas

During the latter part of the Mesozoic and at times during the Cenozoic, the Big Bend area of Texas was much more thickly populated than now. Most of the Cretaceous limestones are highly fossiliferous, containing both invertebrate (Fig. 6–34) and vertebrate remains. Skeletons of marine reptiles such as the mosasaur, ichthyosaur, and plesiosaur have been found. At one time late in the Cretaceous, the area was slightly above the sea, and the swamps were well populated by dinosaurs, both herbivores and carnivores. Seemingly appropriate to the "bigness" of Texas, the world's largest flying creature—a pterodactyl with an estimated wing span of 50 feet (15.5 m) was discovered in the park in the early 1970s (Lawson, 1975). Stumps composed of agatized wood also occur in the dinosaur-bone beds. Cenozoic rocks contain a large number of different species of mammals, including Hyracotherium (the dawn horse), camels, deer, and many others. In fact, from Maxwell (1968) and from Matthews' (1960) handbook on Texas fossils, it becomes evident that Big Bend is one of the more highly fossiliferous national parks.

FIGURE 6–34 Fossil clams in Cretaceous limestone, Big Bend National Park. (National Park Service photo)

Big Bend is a good park to visit at any time of the year; in summer, if it is hot down along the Rio Grande, it likely will be pleasant high in the Chisos Mountains. Hot springs along the river in the southeastern part of the park indicate that faults exist here and that perhaps hot rock from the latest episode of crustal stretching is still close to the surface. A large relief model at park headquarters will help you orient yourself. Do not miss Santa Elena Canyon; a trail leads up into the canyon where you can look up out of this narrow slit in the earth. To get the flavor of the frontier days in Big Bend, go to nearby Castolon Trading Post. Plan to be over in the eastern section just before sunset—when Sierra del Carmen is ablaze. You must take pictures in Big Bend; use a haze filter if there is dust in the air. Also, bring tweezers—in case you need to extract cactus spines from various parts of your anatomy.

REFERENCES

Baldridge, W.S., and Olsen, K.H., 1989, The Rio Grande Rift: American Scientist, v. 77, p. 240–247.
Jenkins, H.O., 1958, Glaciation in Big Bend National Park, Texas: Sacramento State College Foundation, Sacramento, California.
Lawson, D.A., 1975, Pterosaur from the latest Cretaceous of west Texas: Discovery of the largest flying creature: Science, v. 187, p. 947–948.
Matthews, W.H. III, 1960, Texas fossils: An amateur collector's handbook: Univ. Texas, Bureau Economic Geology Guidebook no. 2.
Maxwell, R.A., 1968, The Big Bend of the Rio Grande: Bureau of Economic Geology, Univ. of Texas Guidebook 7.
Moss, H., 1984, Big Bend, where mountains seam the sky: National Parks Magazine: National Parks and Conservation Association, Nov/Dec., p. 22–27.
Spearing, D., 1991, Roadside Geology of Texas: Mountain Press, Missoula, MT, 418 p.

Park Address

Big Bend National Park
Big Bend National Park, TX 79834

GUADALUPE MOUNTAINS NATIONAL PARK (TEXAS)

Guadalupe Mountains (Fig. 6–35) became Texas' second national park in 1966. The spectacular, rugged cliffs in a wilderness-like setting, the highest mountain peak in what is topographically a relatively flat state, and one of the world's best exposures of a fossil barrier reef led Wallace Pratt, an oil company geologist, to donate 5632 acres of land in the scenic McKittrick Canyon area as a nucleus for the new park. Texan Judge Hunter was way ahead of his time in 1925 when he first

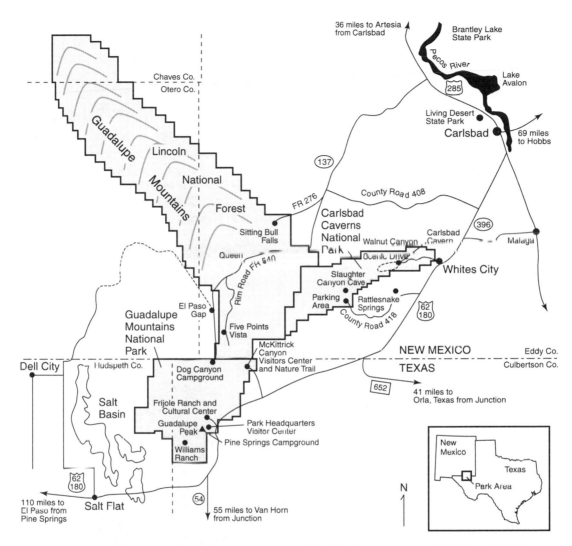

FIGURE 6–35   Generalized map of Guadelupe Mountains and Carlsbad Caverns area. (Slightly modified National Park Service map)

proposed a national park. Hunter began to buy land still clinging to his dream of a park. His son took up the cause and brought the holdings up to 67,312 acres, which he sold to the National Park Service for the bargain price of $22 an acre—thus fulfilling Judge Hunter's dream.

Artifacts, pictographs, and other archeological finds record thousands of years of use of the area by Indians. Many of these features were found in some of the over 300 caves that occur in the Guadalupe Mountain range. The last Indian group to use the area, the Mescalero Apache, prevented settlement in the area for many years and helped close the short-lived Butterfield Stage Line station located near the park entrance. A series of skirmishes and battles with the U.S. Army beginning in 1869 led to their removal to a reservation in 1879 and opened the area to settlement. However, the dry desert conditions greatly limit the number of people that the land can support.

El Capitan's bold cliff at the south end of the Guadalupes (Fig. 6–36) has long been a landmark for travelers in the El Paso extension of west Texas. With elevations rising from low desert plains to the "top of Texas" at Guadalupe Peak (8751 feet; 2668 m), several life zones are represented, as Matthews (1968) reports. A number of interesting 8- to 9-mile-long (14-km) hikes or longer backpack trips into the rugged backcountry are available. However, this is desert country, carry all the water you will need. Incredible views from Guadalupe Peak (the highest point in Texas), the sheer limestone walls of El Capitan, or a hike through McKittrick Canyon where an outstanding cross section of one of the worlds most extensive

FIGURE 6–36   El Capitan, a majestic landmark that rises abruptly above the plains in Guadelupe Mountains National Park, is composed of the Permian-age Capitan Limestone. (Photo by D. Harris)

fossil reefs is exposed are some of the outstanding hikes available to the visitor-explorer. Guadalupe's sister park, Carlsbad Caverns, lies a few miles north (Fig. 6–35) and shares the same geologic history and the same rock formations. One interesting difference is that the limestone unit seen in the eerie light of Carlsbad Cave is the same one exposed on the sun-baked vertical face of El Capitan.

Geologic Development

In the late Paleozoic Era plate tectonics activity was assembling the earth's major plates into the Pangaea supercontinent. The Appalachian area to the east bordering the proto-Atlantic Ocean was experiencing mountain building as plates squeezed together; and, along the southwest edge of what was later to become North America, a rising sea level combined with regional subsidence resulted in a shallow seaway encroaching into that area. A deep part of an arm of this Permian sea in New Mexico and west Texas occupied the Delaware Basin, which had a narrow connection to the open ocean through a relatively narrow channel to the south (King, 1948). Just as the Strait of Gibraltar restricts the connection of the Meditteranean Sea with the Atlantic Ocean, so too did what geologists call the Hovey Channel restrict water flow into the Delaware Basin. In mid-Permian time the Delaware Basin was close to the equator and was ringed by organic reefs similar to the present-day Great Barrier Reef off the Australian coast (Fig. 6–37). A *reef* is a topographic mound or ridge on the seafloor that is built up by organisms that secrete mineral material such as lime. The reef significantly influences the surrounding sedimentation. Rapid growth of the algae, bryozoans, brachiopods, and sponges, with the accompanying precipitation of calcium carbonate as they built on top of their predecessors, produced a carbonate-cemented framework that slowly grew toward the center of the Delaware Basin. As with modern reefs, growth is encouraged by warm, clear, tropical waters. According to Newell and others (1953), the youngest reef (the Capitan Reef) overlies the slightly older Goat Seep Reef (Fig. 6–37c). The reef limestones are the main rock formations in both Carlsbad and Guadalupe national parks.

A glance at the generalized cross section of the developing reef (Fig. 6–37b) should suggest that major differences existed in wave activity, current energy, and resulting sediment characteristics in the protected lagoon, the wave-swept reef core, the debris-laden front edge of the reef, and the relatively quiet waters in the deep basin to the east. Geologists refer to rocks with different characteristics deposited in different environments at the same time as representing different *facies*. Generations of students and practicing petroleum geologists have studied these remarkable reef exposures that tell a detailed story to those who learn to read the rocks. The life of the reef ended in very late Permian time when the connection to the open ocean became restricted. Evaporation exceeded the inflow of normal marine water, thereby increasing salinity and snuffing out the life forms of the reef. Evaporation enabled the thick deposits of gypsum ($CaSO_4 \cdot 2H_2O$) that make up the Castile Formation to later fill the Delaware Basin and cover the reef core (Fig. 6–37c).

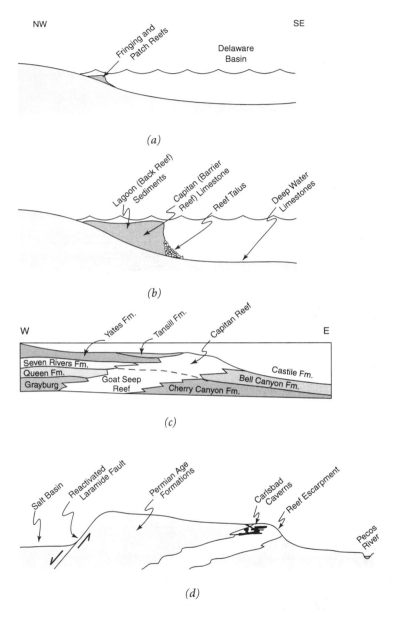

NW   SE

Fringing and Patch Reefs

Delaware Basin

(a)

Lagoon (Back Reef) Sediments

Capitan (Barrier Reef) Limestone

Reef Talus

Deep Water Limestones

(b)

W   E

Yates Fm.

Tansill Fm.

Capitan Reef

Seven Rivers Fm.
Queen Fm.
Grayburg

Goat Seep Reef

Castile Fm.

Bell Canyon Fm.

Cherry Canyon Fm.

(c)

Salt Basin

Reactivated Laramide Fault

Permian Age Formations

Carlsbad Caverns

Reef Escarpment

Pecos River

(d)

FIGURE 6–37 Generalized evolution of Guadelupe Mountains area: (*a*) Arm of the Permian sea covers part of the Southwest, deeper subsiding area forms the Delaware Basin in New Mexico and west Texas. (*b*) Capitan reef grows laterally into deeper water, distinct sedimentary facies reflects different marine environments. (*c*) Connection with ocean narrows and evaporation exceeds inflow; gypsum and other evaporites of the Castile Formation buries reef and Delaware Basin. (*d*) Laramide faulting occurs on west, several thousand feet of uplift occurs along the faults in late Tertiary (about 3 Ma) as Basin and Range Province forms. Castile Formation removed and caves form in Guadelupe Mountains.

During the Mesozoic, sediments were deposited on top of the Capitan Limestone and extensive Laramide faults developed to the west. Crustal arching and stretching during the late Cenozoic (1–3 Ma) reactivated these Laramide faults and uplifted and tilted the Guadelupe Mountain block to the east-northeast (Fig. 6–37*d*). Removal of younger rocks and part of the Castile Formation during uplift exhumed the buried reef and restored the area to a replica of its former Permian self. Take the trail to the top of Guadalupe Peak or stand on the reef crest at Carlsbad and let your imagination take hold. Fill the huge basin in front of you with seawater as far as your eye can see and behold a landscape much like it was 250 million years ago. Be careful, for if you let the waves crash onto the reef crest you will surely get your feet wet!

REFERENCES

King, P B., 1948, Geology of the southern Guadalupe Mountains, Texas: U.S. Geological Survey Professional Paper 215, 183 p.

Matthews, W.H. III, 1968, A guide to the national parks: Their landscape and geology: The Natural History Press, New York.

Newell, N.D., Rigby, J.K., Fischer, A.G., Whiteman, A.J., Hickox, J.E., and Bradley, J.S., 1953, The Permian reef complex of the Guadalupe Mountains region, Texas and New Mexico: Freeman, San Francisco, 236 p.

Park Address

Guadalupe National Park
H.C. 60, Box 400
Salt Flat, TX 79847

CARLSBAD CAVERNS NATIONAL PARK (NEW MEXICO)

Less than 5 miles (8 km) north of Guadalupe National Park two of the world's greatest caves are located in Carlsbad Caverns National Park (Fig. 6–35). Proposals to connect and unite the two parks to better administer and protect the wilderness and cave resources have so far been unsuccessful. The need to protect more of the cave resources in the Guadalupes was further emphasized as a group of weekend explorers, over a period of years, slowly removed tons of rubble from a blocked sinkhole about 3 miles (4.8 km) from the entrance to Carlsbad and just inside the park borders. A strong flow of air through the rubble indicated that somewhere below there was empty space to be discovered. Finally, in 1986 they entered the incredible depths of Lechuguilla Cave, a cave now recognized as the deepest in the conterminous United States (1593 feet; 486 m) and containing over 92 miles

(148 km) of cave passages with still more areas to explore. The new cave is over twice as long as Carlsbad's 30 plus miles (48 km) and is the fourth longest in the United States and the eighth longest in the world. Rare speleothems and cave features never before seen by humans have inspired new lines of inquiry by speleologists. About 100 known caves are found within Carlsbad's borders and more are discovered every year. How many other world-class caves lie just below the surface in the Guadalupe Mountains waiting to be discovered?

In the late 1800s, cowboys in southeastern New Mexico investigated a column of "smoke" and found millions of bats emerging from a huge hole in the ground; thus "Bat Cave"—later named Carlsbad Caverns—was discovered. Mining claims were filed and as much as 120 tons/day of bat guano, an extremely rich fertilizer, was removed from Carlsbad and other Guadalupe caves from 1901 to 1921 and shipped to the citrus groves in California. The floor of Bat Cave was lowered by as much as 50 feet (15 m) suggesting that thousands of years of accumulation are recorded here. A number of companies operated the endeavor but none successfully turned a profit.

The real profit to citizens came when one of the miners, James Larkin White, began to explore the cave and enthusiastically guided interested people through its depths. Robert Halley of the General Land Office, Department of the Interior, was so impressed with his tour and the beauty of the caverns that his report led to President Coolidge establishing a national monument in 1923. Willis T. Lee, a noted U.S. Geological Survey geologist, led a 6-month National Geographic expedition to the cave. The resulting article (Lee, 1924) in National Geographic and public interest led to elevation to national park status in 1930.

What became of Jim White? From the first explorer and unofficial guide to the cave he later became a park ranger and finally chief ranger—devoting his life to a project that is now enjoyed by nearly one million visitors from all over the world each year.

Carlsbad's Bats

The bats use a section of the cave where visitors do not go—be sure to watch the evening bat flight during the summer months—5000 bats/minute exiting the cave is impressive! The exodus of spiraling columns of bats can last for up to 2 hours. Only about 300,000 bats (mostly Mexican-freetail bats) use the cave now—down from the estimated 8 or more million present in the early 1900s. Recent population increases provide encouragement that recovery may be underway.

Past mining activities in the cave, loss of habitat to the expanding human population, and the use of chemicals, especially DDT in the 1950s and 1960s, took a large toll on these gentle, insect-eating mammals. The use of this toxic, persistent poison was banned in the United States in 1972 because its presence in the food chain was killing wildlife (including bats), and its residues were appearing in humans as well. However, neighboring Mexico, where the bats spend the winter, does not ban the use of this dangerous chemical. Thus the residues in Carlsbad bats remained high while other regions of the country experienced decreases. In spite of high residues, a slow increase of bat population from its historic low of 150,000 in the 1970s to 300,000 in the early 1990s occurred. The discovery in 1994 of a large quantity of improperly stored, and illegal DDT in a shed not far from the park may be part of the cause of higher DDT levels. Another population-reduction factor was the presence of two shafts dug by miners before 1923 that allowed warm air near the ceiling to escape, making survival of the babies in the maternity colony more difficult. The Park Service plugged the shafts in 1981 and bat populations climbed to 750,000 in the mid-1990s. Hopefully the numbers of these important mammals will continue to increase.

Bats have received "bad press"—however, they are the only flying mammal in the world and one of the only night predators of insects. Studies indicate that they can consume up to 600 mosquitoes/hour and a lactating female can eat over half her weight in insects on her nightly forage! Bats, like humans, are at the top of the food chain and are also susceptible to increased levels of chemical pollutants that were ingested and concentrated in lower plants and animals. As the total numbers and species of plants and animals continue to decline around the globe, how far behind is the human species?

Setting the Stage

The geologic story at Carlsbad involves the generation of thick sequences of late Paleozoic (Permian) limestone along a barrier reef, as described in the discussion of Guadalupe National Park and summarized in Figure 6–37. An important factor at Carlsbad is that broad folding and fracturing of the limestone accompanied the uplift (Fig. 6–37*d*) of the Guadalupe Mountains fault block as the Rio Grande Rift opened during late Tertiary time. The *joints* (fractures without significant displacement) are oriented parallel to the reef front, and a lesser set is oriented perpendicular. These zones of weakness established a blueprint for the solution processes to follow as they etched out Carlsbad, Lechuguilla, and the 80 other known caves in the park.

Formation of Guadalupe Caves

Most caves form by the interaction of the relatively soluble rock called limestone with carbonic acid (H_2CO_3), a relatively common substance in groundwater as described previously in the section on Lehman Caves in Great Basin National Park. For many years this general process was thought to be "nailed down" by speleologists as the only process that produces solutional caves in limestone. However, some of the features recognized for many years in Carlsbad were troublesome— they did not seem to fit the accepted cave formation model. The features in Lechuguilla were even more bizarre and demanding of a different explanation.

As in most geologic studies, careful observation of geologic features, understanding the sequence of development, and consideration of the limitations of natural laws and principles enables geologists to develop a reasonable model or hypothesis against which present or future evidence can be compared. If evidence is contradictory, then a hypothesis has to be modified or rejected. Any explanation of speleogenesis (cave formation) for the Guadalupe Mountain caves must explain the following:

1. The presence of a maze of small openings in the cave walls called *spongework* where irregular dissolving of the limestone below the water table produces a "Swiss cheese" effect. The Boneyard area of Carlsbad nicely displays this type of feature.

2. The preferential development of most cave passages at distinct levels (Fig. 6–38), indicating that a horizontal control (usually the water table) was important in cave formation.

3. The presence of unusually large chambers and randomly oriented passages that abruptly end.

4. The distinct orientation of cave passages along certain preferred directions. For example, cave passages at Carlsbad and Lechuguilla generally trend northeast, parallel to the late Tertiary folds and fractures in the area.

5. The remnants of gypsum (calcium sulfate; $CaSO_4 \cdot 2H_2O$) beds many feet thick and unusually abundant secondary deposition of gypsum speleothems (especially in Lechuguilla). The layers in Carlsbad were recognized by Bretz (1949) but were initially ignored or considered secondary deposits that were unimportant in speleogenesis.

6. The presence of sulfur deposits in small rock openings or associated with the gypsum deposits.

7. The presence of mineral coatings and crystals on many passage walls and ceilings that are best explained by an episode of flooding of the cave chambers.

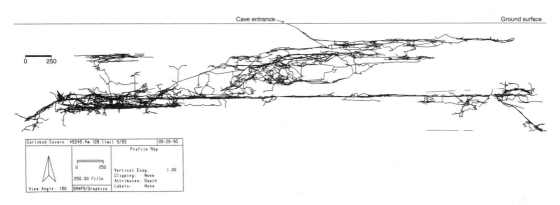

FIGURE 6–38   Cross section of Carlsbad Caverns showing dominant horizontal development of cave passages. (National Park Service illustration)

Most of these features—such as spongework, distinct cave levels related to former water table elevations, preferred orientation of cave passages, and secondary mineral coatings due to a rising water table—are common in caves formed by groundwater containing carbonic acid (H_2CO_3). However, the presence of thick gypsum beds, sulfur, and the large size and irregular shapes and connections of different chambers, as well as other evidence requires a different explanation. Jagnow (1979) studied a process where sulfur reacting with oxygen at the water table produces sulfuric acid, a much stronger acid than carbonic, and one that could dissolve limestone at a rapid rate and form very large cave chambers.

$$2S^- \text{ or } S^= + 2O_2 + 2H_2O \longrightarrow 2H_2SO_4$$
$$\text{(sulfur)} \quad \text{(oxygen)} \quad \text{(water)} \quad \text{(sulfuric acid)}$$

or

$$H_2S + O4 \longrightarrow H_2SO_4$$
$$\text{(hydrogen sulfide)} \quad \text{(oxygen)} \quad \text{(sulfuric acid)}$$

The source of the sulfur may come from minerals such as pyrite (FeS_2) in the sedimentary rocks, the gypsum beds ($CaSO_4 . 2H_2O$) belonging to the Castile Formation that buried the reef (Davis and others, 1992) in Permian time (Fig. 6–37) or from the subsurface migration of hydrogen sulfide gas associated with the abundant oil and gas accumulations (Hill, 1991) in the Delaware Basin to the east. Sulfur-rich water rose to the water table where it combined with oxygen to form sulfuric acid and dissolved huge chambers, such as the 370-foot-high (113-m), 14-acre Big Room in Carlsbad. Kane Caves in Wyoming were also formed (may still be forming!) by sulfuric acid, in this case the sulfur was derived from *hydrothermal* (hot water and gases) sources (Maslyn, 1979).

The suggested sequence of cave formation in the Guadalupe Mountains is illustrated in Figure 6–39 and explained in the caption. Lowering the water table by uplifting the Guadalupe Mountain block, changing the climate to more arid conditions, and downcutting by surface streams allowed air to enter and begin the slow process of filling the caves with incredible displays of mineral materials collectively known as speleothems. As the water table lowered, support for the cave ceilings was reduced and ceiling blocks tumbled to the floor as *breakdown*. Boulder-size blocks are common, but the massive Iceberg Rock weighing over 200,000 tons is the "big daddy" of them all.

Speleothems

Most of the types of speleothems—stalactites, stalagmites, columns, flowstone, and helictites—are also found in Lehman Caves and other cave parks. However, the abundance of speleothems, interesting variations of common speleothems,

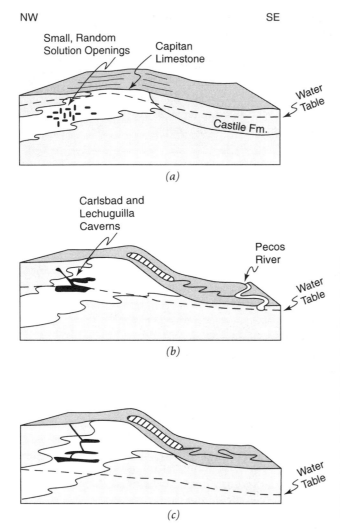

NW   SE

(a)

(b)

(c)

FIGURE 6–39   Generalized evolution of Carlsbad and Lechuguilla Caves: (*a*) Small, random solutional openings form deep in the zone of saturation. (*b*) Guadalupe block is uplifted, water table lowered as Castile Formation is removed, and surface streams cut canyons into uplifted block; sulfur is oxidized to sulfuric acid at water table, thereby causing limestone to be dissolved and cave levels to form; thick gypsum beds and sulfur form in caves; caves are reflooded, causing calcite crystals to coat walls and ceilings. (*c*) Downcutting and cave formation continues, air enters caves and speleothems form; ceiling breakdown occurs; roof collapse creates a sinkhole entrance to surface, or perhaps the natural entrance is a paleo-spring.

and the presence of rare and exotic speleothems, especially in Lechuguilla, make these caves world famous. In describing Lechuguilla Cahill and Nichols (1991, p. 37) note that:

> It's not just the immense size of the rooms that is so amazing but also their lavish decorations—glittering white gypsum chandeliers 20 feet long, walls encrusted with aragonite "bushes," rippling strands of indescribably delicate "angel hair" crystals, some 30 feet long but so fragile that a puff of air can break them. The cave's shimmering lakes, like liquid sapphires, have lain untainted for millennia.

Extremely delicate soda-straw stalactites, usually a few inches in length, grew to 6 feet (1.8 m) in length at the Kings Bell Cord in the Kings Chamber in Carlsbad and to 18 feet (5.5 m) in Lechuguilla! Stalagmites that grow to immense size are called *domes* (Fig. 6–40)—Giant Dome in the Hall of Giants is 50 feet (15 m) tall and Crystal Springs Dome, although only 25 feet (7.6 m) tall, is one of the few speleothems still growing in the present episode of dry climate. Its surface is still wet as groundwater saturated with dissolved calcium drips onto the dome, carbon dioxide (CO_2) escapes into the cave atmosphere, and slowly a thin film of new calcite forms. Touching a speleothem transfers oil from a persons hand to the calcite surface and effectively stops the growth process until the oil is slowly washed away. The estimated growth rate for Crystal Springs Dome is equal to the thickness of a coat of paint added every 90 years! The tens of thousands of years needed to form many of these speleothems is even more remarkable when one considers that the growth of many speleothems is not continuous but is rather a stop and start process.

FIGURE 6–40   One of the huge domes in Carlsbad Caverns, New Mexico. (Photo by Jack Odum)

The present dry conditions discourage speleothem growth: a return to the wetter conditions of the Pleistocene will restart growth on many of these currently dormant speleothems. Carol Hill (1991) reports radiometric dates on speleothems of over 879,000 years and also reversed magnetic fields recorded in calcite layers in domes. Because the last magnetic reversal ended about 700,000 years ago, these large speleothems must go back at least that far.

Some cave minerals and speleothems form underwater such as the elongate calcite crystals ("dogtooth spar") coating some cave walls. The rounded forms called popcorn on the Lions Tail and cave walls in the Big Room mark a former water level during an episode of reflooding. In the lower, noncommercial part of the cave are larger rounded forms called *cave clouds,* formed by deposition of thin calcite layers in an underwater environment, which also bear mute testimony to times of higher water levels.

How the gravity-defying helictite formed was a mystery to speleologists for many years. The discovery of a tiny opening in the center that allows water to move upward due to capillary action (much like kerosene moves upward through a wick in a kerosene lamp) seemed to solve the problem. However, explorers discovered helictites in Lechuguilla—some of which are growing underwater in pools where capillary forces cannot exist! Another process must form these look-alike forms. Davis and co-workers (1992) suggest that minute streamlets of calcium-rich water drain under a calcite coating at the edge and bottom of the pool and precipitate a helictite when it emerges from the bottom of the gypsum-rich pool water.

Lechuguilla and the Future

Because access to Lechuguilla is extremely difficult and dangerous and unlimited access will forever change this pristine environment, no development plans are proposed. As described in a letter from the regional director of the National Park Service: "It is highly probable that more spectacular formations and scientific wonders will be discovered as exploration of Lechuguilla Cave progresses."

As stated by Frank Deckert, Superintendent, Carlsbad Caverns, "Lechuguilla Cave is a world-class resource worthy of world-class protection." Indeed, researchers have already discovered new microbes in the depths of Lechuguilla and other caves that have never been described before. Such new compounds show promise in fighting cancer and other diseases (Bigelow, 1998). Significant changes caused by excess human intrusion could destroy this delicate underground environment.

Although most of us will not be able to directly experience the wonders of the Lechuguilla underworld, tantalizing glimpses are available in the March 1991 issue of *National Geographic* and in the excellent color photographs in *Lechuguilla, Jewel of the Underground,* edited by Urs Widmer (1998).

The 74 miles (119 km) of cave passage at Lechuguilla are not the end of the story; new leads and some small passages still need to be explored. Caves act as natural barometers (air moves into caves when high-pressure weather conditions prevail and moves out when low-pressure systems arrive). Therefore, the volume of air moving through the entrance during a measured change in atmospheric pressure is proportional to the size of the cave, and its force to the size of the opening. Wind velocities through the narrow, hand-dug entrance sometimes exceed 30 mph (50 kph). Calculations suggest that 98 percent of Lechuguilla has yet to be found!

Caves are among the most fragile environments on earth—once something is removed or damaged, it is gone forever in the context of human lifetimes. According to a shocking report in the *National Speleological Society News* (February, 1995), over *two thousand* speleothems near the trail in Carlsbad are broken annually! The Park Service now has a new tour system in which only small, guided tours will be conducted through more vulnerable areas such as the King and Queens chambers. Destruction of a speleothem is forever, using the calendars of humans. Once destroyed no human will see a new speleothem at that location for many thousands of years.

The popular self-guided Natural Entrance Route that descends 750 feet (229 m) to the Big Room is still available to those who want a more strenuous and exciting experience "discovering" Carlsbad's beauties on their own. Those wishing a primitive spelunking tour should make reservations at the visitor center to accompany a ranger on the New Cave tour, where no installed lighting or developed trails are constructed. The 89-foot-high (27-m) Monarch in New Cave is one of the tallest columns in the world.

Because of the irresponsible few, do we need to limit the number of people we allow to enter the cave? Do we need more park rangers to act as police and protect this incredible treasure? Do we need stricter laws and more severe punishment for offenders? Are the national parks becoming more of a microcosm of the American society we thought we escaped by visiting a national park?

REFERENCES

Bretz, JH., 1949, Carlsbad Caverns and other caves of the Guadalupe block: Journal of Geology, v. 57, p. 447–463.

Cahill, T., and Nichols, M., 1991, Charting the splendors of Lechuguilla Cave: National Geographic, v. 179, no. 3, p. 34–59.

Davis, D.G., Palmer, A.N., and Palmer, M.V., 1992, Extraordinary subaqueous speleothems in Lechuguilla Cave, New Mexico: National Speleological Society Bulletin, v. 52, p. 70–86.

Hill, C.A., 1991, Sulfuric acid speleogenesis of Carlsbad Cavern and its relationship to hydrocarbons, Delaware Basin, New Mexico: American Association of Petroleum Geologists Bulletin, v. 74, p. 1685–1694.

Jagnow, D.H., 1979, Cavern development in the Guadalupe Mountains: Cave Research Foundation, Columbus, Ohio.

Lee, W.T., 1924, A visit to Carlsbad Caverns: National Geographic Magazine, v. 45, no. 1, p. 1–40.

Maslyn, 1979, National Speleological Society Bulletin, v. 41, p. 31–51.

Widmer, U., 1998, Lechuguilla, jewel of the underground: Caving Publications International, Basel, Switzerland.

Park Address

Carlsbad Caverns National Park
3225 National Parks Highway
Carlsbad, NM 88220

WHITE SANDS NATIONAL MONUMENT (NEW MEXICO)

Located in the Tularosa Basin in south-central New Mexico, about 100 miles (160 km) northwest of Guadelupe and Carlsbad National Parks, White Sands National Monument was set aside in 1933 to preserve 230 square miles (596 km^2) of the largest gypsum dune field in the world. The Heart of Sands 16-mile-long (26-km) Loop Drive affords an opportunity to experience these remarkable dunes. Part of the drive is paved and part is a good road surface on hard-packed gypsum. Sections of abandoned road surfaces tell their silent story of shifting sand.

Along the east edge of the Basin and Range Province and extending northward into the southern part of the Southern Rocky Mountain Province is a 600-mile (1000-km) "arm" of fault-block structures that are part of a major structural feature called the Rio Grande Rift. This relatively wide zone of fault-block mountains and grabens extends from Mexico to West Texas, divides New Mexico in half, and extends into southern Colorado where it splits the southern end of the Southern Rocky Mountains into two prongs. It also separates the Great Plains Province to the east from the Colorado Plateaus on the west. The rift resulted from a unique combination of geologic conditions that are just starting to be recognized (Baldridge and Olsen, 1989). In addition to containing the Tularosa graben and White Sands, other important Park Service areas occurring in the Rio Grande Rift include Great Sand Dunes, Bandelier, El Malpais, Carlsbad, Guadalupe, Big Bend, and recently established Valley of Fires State Park.

The Tularosa Basin graben is about 37 miles (60 km) wide and is bounded on the west by the San Andres Mountains fault block and on the east by the Sacramento Mountains block. Its location in the continental interior and the presence of high, moisture-blocking topography to the west ensures that low precipitation (11 inches, 28 cm/year) and desert conditions will prevail. As grabens have slowly filled with debris during the past 10–20 million years in the Rio Grande Rift, through-flowing streams have slowly developed and integrated into the Rio Grande, which eventually discharges into the Gulf of Mexico. Not so in the Tularosa Basin which is still a true bolson, a desert area with internal drainage, similar to Death Valley and many other Basin and Range areas west of the Rio Grande Rift.

Within the Tularosa graben is one of nature's unique features—the glistening gypsum sands of White Sands National Monument (Fig. 6–41). The northern part of the dune field is more typical of sand dunes the world over—it is composed mostly of sand-size grains of the relatively abundant and resistant mineral called quartz (LeMone, 1987). Because of its resistance to mechanical and chemical attack, quartz is one of the last minerals left as rocks are subjected to weathering processes. On the other hand gypsum ($CaSO_4$) is very soluble in water and, on a 10-point scale (Moh's hardness scale), has a hardness of 2 compared to 7 for quartz. How can it be that over 9 billion tons of gypsum, one of the earth's softer minerals, can cover nearly 300 square miles (915 km^2) of the New Mexico desert floor?

Geologic History

To set the stage for the generation of this unique dune field, let us first back up and explore the geologic events that created the nearby rocks and topography. During the Paleozoic, shallow seaways advanced and retreated over the New Mexico area, leaving their records in the sedimentary strata and the incorporated fossils. Although the older Paleozoic record in the Tularosa Basin area is not well preserved or exposed, thick exposures of late Paleozoic sedimentary rocks in the San Andres and Sacramento Mountains indicate that an arm of the sea extended into the west Texas–New Mexico area during late Paleozoic time. Thick evaporite deposits in the Permian age Yeso Formation including over 500 feet (152 m) of gypsum were destined to play a major role in the development of the modern dune field.

Also during the late Paleozoic the ancestral Rocky Mountains were rising to the north, and by Mesozoic time floodplain and delta deposits from the eroding

FIGURE 6–41   "Ships" in the desert of white gypsum dunes. The "sails" protect visitors from the sun. (Photo by D. Harris)

mountains were burying the older strata. A brief return of the ocean occurred during the late Mesozoic and was followed by renewed mountain building in the Rocky Mountains, perhaps as a result of increased subduction rates to the west. The intense compressive forces during the late Mesozoic–early Cenozoic generated a broad anticline in the Tularosa Basin area that later collapsed as the crust began to experience tensional or pull-apart forces during the middle Cenozoic about 30 million years ago. Volcanism occurring along this pull-apart zone further weakened the crust so that when renewed tension began about 10 million years ago a long split, the Rio Grande Rift, cut northward well into the Southern Rocky Mountains. Fault scarps in the alluvial fans along the Sacramento Mountains front and lava flows a few miles north that are less than a thousand years old attest to the tectonic instability of the Rio Grande Rift.

The amazing fluctuations of climate during the last 2 million years had a unique effect in the arid regions of the Basin and Range Province. Although most of the province is too far south or not high enough to allow glaciers to form, the cooler episodes increased stream flow, erosion, and allowed playa lakes to enlarge, sometimes to a significant size. Lake Otero in the Tularosa Basin was one of these lakes. As climate warmed and more water evaporated than flowed into the basin near the end of the Pleistocene 20,000–10,000 years ago, mudflats containing gypsum precipitates formed—another key element in the story of the dunes. The remnant of Lake Otero is today concentrated in the deepest part of the basin and is now a shallow lake called Lake Lucero (Blair and others, 1990).

Source of Gypsum Sand

The ultimate source of gypsum must be in the Paleozoic (Pennsylvanian-Permian) rocks such as the Yeso (Spanish for gypsum) Formation exposed in the San Andres Mountains to the west. Eroded sediments were washed basinward, and dissolved sulfates were precipitated as gypsum crystals in the filling graben. The presence of Lake Otero allowed groundwater to permeate these sediments and redissolve some of the gypsum. As Lake Otero evaporated, a variety of gypsum that occurs in large transparent crystals called selenite formed in the muddy deposits. Some of these crystals are up to 4 feet (1.2 m) long—not likely to be whipped up and bounced along by the wind! Disturbance of these deposits near present-day Lake Lucero by wind and rain as well as temperature changes that produces cracks in the mineral enables small particles to be freed and moved by wind.

Another source of sand is the gypsum-charged groundwater beneath Lake Lucero. Evaporation of the lake draws these groundwater brines toward the surface where new gypsum grains form. When the wind exceeds 17 mph (27 kph) sand-size particles from these two sources—the selenite crystals deposited by Pleistocene Lake Otero and the grains deposited from present day Lake Lucero—begin to move.

The dune-forming processes were not possible until climates returned to a drier mode and the volume of Lake Otero was reduced. Likely the dunes are 12,000 to perhaps 24,000 years old. Prevailing winds are from the southwest and are strongest from February through March when, consequently, most of the sand movement occurs.

Dune Forms and Processes

Some of the selenite crystals are clear—others are dark brown to golden yellow—but all become progressively whiter during transport. Gray silt associated with the gypsum-rich sediments becomes airborne and is carried off in *suspension,* further concentrating the larger gypsum fragments. Sand-size particles move within a few feet of the ground by *rolling* or by a bouncing motion called *saltation.* A saltating grain impacts the desert floor and dislodges additional grains, which in turn are carried by the wind to another impact site. As these numerous collisions occur, small microfractures produce a lighter appearance until each grain is a dazzling white color.

The Lake Lucero area is the incubator where these grains begin their 20,000 or more year journey across the Tularosa Basin desert. The sand is about two thirds of the way across now, but the slow-moving edges will take many more years to reach the visitor center on the east edge.

As the sand moves away from Lake Lucero, small interruptions in air flow created by plants or topography will cause sand to accumulate as distinct mounds or hills called dunes. The classic study by McKee and Moiola (1975) describes a progression from a circular, low-lying dome form near Lake Lucero to individual dunes called *barchans* and a linear dune oriented perpendicular to the wind direction called a *transverse* dune (Fig. 6–42). Transverse dunes are produced by coalescing of numerous barchans. Near the edge of the dunes, where vegetation has a foothold (Fig. 6–43), *parabolic* dunes form that resemble barchans turned inside out. The barchan, transverse, and parabolic dunes have a characteristic gentle upwind slope and a steep downwind *slip face* where saltating sand rolls or flows down the protected lee side producing an internal layering of sand called *crossbedding.*

A trip to Zion Canyon National Park or a picture (see Chapter 8) of the rock layering will demonstrate the internal appearance of a dune. The dome dunes near the sand source move at 24–38 feet per year (7.3–11.6 m), barchans and transverse in the central part of the dune field at 4–13 feet (1.2–4 m) per year, and parabolic near the edge at 0–5 feet (0–1.5 m) per year.

Nature has put together an incredible mining and purification process at White Sands that makes human efforts seem puny—the history that each sand grain tells is remarkable. Be sure to bring your sunglasses and if you happen to see a snowplow (Fig. 6–44) pushing "white stuff" on a hot day in the middle of the summer don't panic—you're not delirious with the heat—you're at White Sands National Monument!

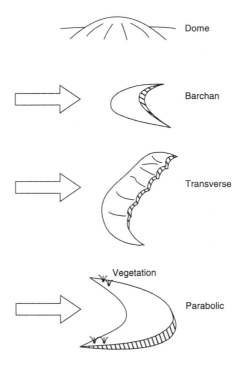

FIGURE 6–42   Common dune forms; arrows indicate wind direction.

FIGURE 6–43   Silent battle between vegetation and wind-transported sand along the east edge of the White Sands "sand sea." (Photo by E. Kiver)

FIGURE 6–44   In the middle of the summer with temperatures above 90°F (32°C), the sight of a snowplow moving "white stuff" can be disconcerting! (Photo by E. Kiver)

REFERENCES

Baldridge, W.S., and Olsen, K.H., 1989, The Rio Grande Rift: American Scientist, v.77, p. 240–247.

Blair, T.C., Clark, J.S., and Wells, S.G., 1990, Quaternary continental stratigraphy, landscape evolution, and application to archeology: Jarilla piedmont and Tularosa graben floor, White Sands Missile Range, New Mexico: Geological Society of America Bulletin, v. 102, p. 749–759.

LeMone, D.V., 1987, White Sands National Monument, New Mexico, in Beus, S.S., ed., Rocky Mountain Section of the Geological Society of America: Centennial Field Guide, Geological Society of America, Boulder, CO, v. 2, p. 451–454.

McKee, E.D., and Moiola, R.J., 1975, Geometry and growth of White Sands dune field, New Mexico: Journal of Research, U.S. Geological Survey, v. 3, p. 59–66.

Park Address

White Sands National Monument
P.O. Box 458
Alamogordo, NM 88310

SEVEN

Columbia Intermontane Province

Province Boundaries

The Columbia Intermontane Province has generally been called the Columbia Plateau. When examined, however, the province is found to include large areas of plains, hills, and mountains, in addition to plateaus. Therefore, as the province is essentially surrounded by mountains, it is logical to use the term *intermontane,* as Freeman and others (1945), Kiver and Stradling (1994), and Thornbury (1965) have done. The province extends eastward from the Cascades to the Northern and Middle Rocky Mountains (Plate 7). The southern boundary with the Basin and Range Province is less definite but is ordinarily drawn across southern Oregon and Idaho, including a narrow strip along northeastern Nevada.

The Columbia Intermontane Province has been divided into subprovinces, some of which have been divided again into sections. Here, the discussion is confined to the Snake River Plain Section, where Craters of the Moon and Hagerman Fossil Beds National Monuments are located; to the Blue Mountains Section containing John Day Fossil Beds National Monument in north-central Oregon; and to the western part of the High Lava Plains Subprovince in Oregon where Newberry Volcanic National Monument is located.

Hot Times—Volcanic Wonderland

The exposed rocks of the province are mainly basaltic lava flows and interbedded sediments ranging in age from 17.5 to 6 Ma (millions of years ago) Miocene flood basalts in the Washington and northern Oregon area (Reidel and Hooper, 1987), to Quaternary-age (last 1.6 Ma) lava in southern Idaho (Alt and Hyndman, 1989). Older rocks, some as old as Precambrian, are exposed in the adjacent Rocky Mountain provinces, in the Blue Mountains in eastern Oregon, and along the northern and eastern edges of the province as *steptoes* or *kipukas*—hills and mountain tops that were surrounded by lava. The latest volcanic activity occurred about 2000 years ago

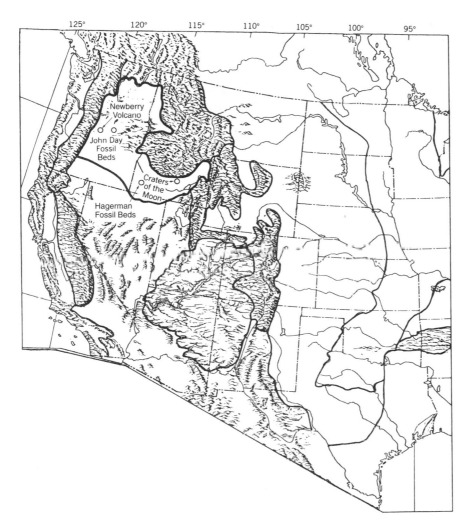

PLATE 7   Columbia Intermontane Province National Monuments. (Base map copyright Hammond Inc.)

and involved lava flows and the building of a line of cinder, shield, and spatter cones in and near Craters of the Moon in southeastern Idaho, and obsidian flows in the caldera at Newberry Volcano just east of the Cascade Mountains in central Oregon.

Volcanism began about 17 Ma in two separate areas of the Columbia Intermontane: (1) the northeastern Oregon–southeastern Washington area and (2) the northeastern Nevada–southeastern Oregon area. In the northern Nevada area a combination of tectonic factors produced a massive rise of hot mantle material—the beginning of the Yellowstone hot spot. Farther north a huge outpouring of *flood basalt* (Columbia River Basalt) sent voluminous flow after voluminous flow from one or more of the thousands of northwest-trending fissures (now *dikes*) in north-eastern Oregon and southeastern Washington (Fig. 7–1). Possible tectonic factors contributing to magma generation, according to Hooper and Conrey (1989) and Parsons and others (1994), include crustal thinning due to extension behind a subduction zone (*back-arc spreading*), the change to oblique subduction (plates collide at an angle rather than head on), and the formation of a major *transform fault* (the San Andreas Fault) in California that caused southwestern North America, and especially the Basin and Range Province, to spread apart (Fig. 3–2).

Columbia River Basalt

The fissure eruptions in northeastern Oregon and southeastern Washington produced one of the world's largest floods of lava, the Miocene age *Columbia River Basalt*. Over 63,300 square miles (164,000 km²) of eastern Oregon and Washing-

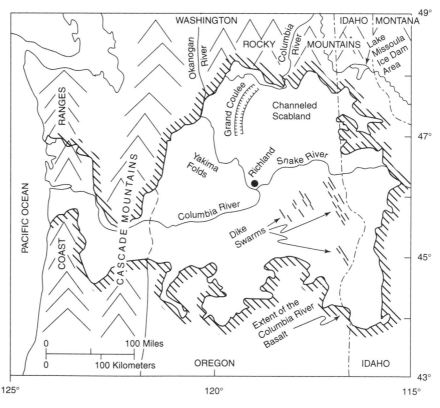

FIGURE 7–1   Map showing selected features in the Pacific Northwest including the distribution of Columbia River flood basalts and the generalized location of the thousands of known feeder dikes.

ton (Fig. 7–1) were overwhelmed by hundreds of lava flows that buried valleys and hills beneath a sea of lava between 17.5 and 6 Ma (Tolan and others, 1989). Although lava flows continued over an 11-million-year period, 90 percent of this massive outpouring occurred in a short interval of time—perhaps in less than a million years. Individual flows were huge, some over 240 cubic miles (1000 km³) in volume—some eventually pouring into the Pacific Ocean hundreds of miles away from their source vents! Volcanism of such a magnitude is fortunately rare— so rare that similar events have not occurred during historic times. The 1783 Laki eruption in Iceland is the largest (2.9 miles³; 12 km³) historic eruption—but puny compared to the much larger Columbia River Basalt flows. However, the devastation to the Icelandic people was immense, with an estimated loss of 10,000 lives, mostly due to the indirect effects of the eruption such as the loss of thousands of head of livestock, crop failures, and the resulting famine.

Rather than flowing rapidly from a vent area to its farthest point in a few days or weeks, Columbia River Basalt fissures pumped magma into the liquid core of flood-basalt flows, causing them to inflate and move at a slow pace over great distances for periods of months and perhaps years (Self and others, 1996). A protective insulating rock crust forms quickly on such a flow, thus enabling the liquid core to persist for many months or years. Thicknesses are greatest in southeastern Washington where the city of Richland sits atop an incredible 13,000-foot-thick (4000-m) stack of basalt flows!

Yellowstone Hotspot

Unlike the Hawaiian hotspot that originates deep in the earth at the core–mantle boundary, the Yellowstone Hot Spot may be a relatively shallow *passive hotspot* that has spread laterally with time away from its point of origin in northern Nevada and southern Oregon (Mutschler and others, 1998). The hotspot acted like a giant blowtorch and caused the surface in northern Nevada to swell, fracture, and eventually become the site of a large caldera (McDermitt caldera) about 16 Ma. The topographic swelling and "blowtorch" effect migrated away from its point of origin in northern Nevada along northeast- and northwest-trending fracture zones. A trail of rhyolitic volcanoes that are progressively younger to the northeast occurs along the Snake River Plain in southern Idaho. The trail leads to Yellowstone National Park where the still-active hotspot resides a few miles beneath the surface today (Fig. 7–2). A similar trail to the northwest leads toward Newberry Volcano in Oregon. The rate of migration of the passive hotspot for the past 16 million years is about an inch (2.5 cm) a year. These huge, caldera-forming rhyolitic eruptions were followed by basaltic eruptions that, in most places, buried the wreckage of the rhyolitic volcanoes beneath a relatively thin skin of lava.

Basalt and rhyolite are at the opposite ends of the chemical spectrum of volcanic materials and form under radically different conditions. How is it then that both low-silica (basalt) and high-silica (rhyolite) lavas were formed in the same

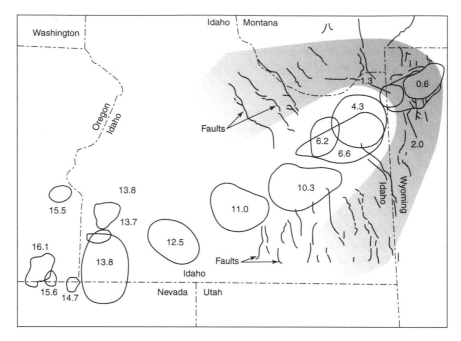

FIGURE 7–2   Track of Yellowstone hot spot through the Snake River Plain area of southern Idaho. Note the location and age (in millions of years) of rhyolitic calderas and the location of active faults associated with the present hotspot location beneath Yellowstone. (Modified from diagram by P. Kirchmeier, in Crenson, 1994)

geographic location? The process by which high-silica rhyolite magma forms is described by Leeman (1989). The pressure release associated with crustal extension or reduction of compressive forces produces a low-silica basaltic magma through partial melting of the mantle rock. The melt is less dense than overlying rocks and rises through the mantle like a drop of oil rises through a column of water. If the magma is sufficiently buoyant, it will rise relatively rapidly to the surface and produce basaltic eruptions. However, if the magma resides for a long time in the crust, the basaltic magma will melt the surrounding crustal rocks and produce lighter, silica-rich melts. These more silica-rich magmas in turn rise higher in the crust and may eventually feed surface eruptions of high-silica lava such as andesite or rhyolite.

Faulting accompanies the topographic doming above the hotspot, thus accounting for the fault-block mountains and active faults both north and south of the hot spot track in southern Idaho, northwestern Utah, and southwestern Montana. The 1959 West Yellowstone and the 1983 Mt. Borah (largest historic earthquake in Idaho) earthquakes resulted from two of the many recent fault movements to occur in the vicinity of the hotspot.

Extension behind the migrating hotspot further reduces pressure on the upper mantle, allowing more expansion and partial melting to occur. Basaltic magma, most less than 2 million years old (Quaternary age), rose through the crust and flowed out at the surface, capping the rhyolites with a cover of basalt about 1000 feet (305 m) thick. Flow volumes were small compared to their older giant flood-basalt cousins in eastern Oregon and Washington. Hundreds of low shield volcanoes and some cinder cones erupted along a series of rifts only to be buried by yet younger flows. The latest activity was in the eastern part of the Snake River Plain (Kiver and Stradling, 1994; Malde, 1991) with the most spectacular volcanic landforms located along the Idaho Rift, a 53-mile-long (85-km) vent system with the same northwest–southeast orientation as the fault-block mountains in neighboring provinces.

A small part of the northern segment of the Idaho Rift (also called the Great Rift) is located in Craters of the Moon National Monument. To the south, swarms of open fissures occur; King's Bowl Rift, first discovered in 1956, has been descended to a depth of 800 feet (244 m) where an ice floor ended the exploration of the world's deepest open rift! World-class examples of volcanic features occur here as well as in the Wapi Lava Field farther south. All three areas last erupted a mere 2000 years ago and are good candidates for more excitement. Attempts to place areas along the Idaho Rift into a national park have been thwarted so far by agricultural, gun clubs, and other special-interest groups. Although Idaho contains a number of spectacular and significant areas worthy of national park status, it remains the only state west of Kansas and Nebraska without a national park.

Northwest of McDermitt caldera where the hotspot first appeared, a similar sequence of rhyolitic volcanism followed by basalt eruptions occurred, this time migrating northwest across eastern Oregon along the Brothers Fault Zone to Newberry Volcano where recent eruptions and high heat flow indicate that the magma chamber here is also "alive" and will likely be "kicking" again in the future.

Cold Times—Gargantuan Floods

Although the results of volcanism and folding (Yakima folds in central Washington area, Fig. 7–1) are obvious influences on the present topography of the Columbia Intermontane, other great changes took place during the Pleistocene as a result of glaciation. The Okanogan lobe, a southern extension of the ice sheet in Canada, pushed southward into eastern Washington, damming and displacing the Columbia River near the present site of Grand Coulee Dam. While temporarily displaced by ice, the Columbia River—repeatedly experiencing incredibly large floods—cut the Grand Coulee (Figs. 7–1 and 7–3). The spectacular 50-mile-long (80-km) Grand Coulee gorge is locally nearly 1000 feet (305 m) deep and is itself worthy of park status. The source of the floods was Glacial Lake Missoula, an ice-dammed lake in northern Idaho and northwestern Montana (Fig. 7–1) that contained as much as 500 cubic miles (2080 km^3) of water (Weis and Newman, 1989). Rising water behind the ice dam tended to float the ice—eventually causing its catastrophic failure and releasing devastating floods of almost unimaginable

FIGURE 7–3   Dry Falls. Diversion of the Columbia River combined with catastrophic floods from glacial Lake Missoula created this now-abandoned giant cataract that dwarfs Niagra Falls in its height and extent. Located in Sun Lakes State Park near Lake Roosevelt (formerly Coulee Dam) National Recreation Area, Washington. (Photo by E. Kiver)

magnitude across eastern Washington (Bretz, 1923, 1969). At least 13 (Kiver and Stradling, 1982) and perhaps as many as 70 (Atwater, 1983) Missoula (also known as Spokane or Bretz) Floods occurred during Late Wisconsin (latest Pleistocene) time, with the latest occurring shortly after the eruption and deposition of ash from Mount St. Helens 13,000 years ago (Stradling and Kiver, 1986). After each flood, about 30–50 years were required for advancing ice to again block the Lake Missoula outlet and for the impounded water to reach the depth necessary to produce another outburst flood. J Harlan Bretz was the first to recognize that giant glacial floods produced the unusual topographic and sedimentologic features in eastern Washington's Channeled Scabland.

These catastrophic floods acted like a giant hydraulic vacuum cleaner and cut huge channels and scars in the bedrock, forming the distinctive erosional features of the famous Channeled Scabland of eastern Washington. Erosion went on at a rate difficult to comprehend; try to imagine what took place when 9.5 *cubic miles* (40 km^3) of water/per *hour* (more water than the Amazon and all other rivers of the world combined!) discharged through the broken dam (Pardee, 1942) and raced along its flow path at velocities up to 65 miles/hour (105 km/hr). Indians inhabiting the area no doubt heard the roar of water—perhaps a half hour or more before the flood wave arrived—and they also must have felt the ground shake—much like one would feel the station platform vibrate as a speeding train approached and passed. Those in the wrong place at the wrong time would unfortunately not have experienced the after-effects of one of the world's largest documented floods!

Acceptance of J Harlan Bretz's radical flood hypothesis did not come easy. After decades of lively debate the tide of opinion swung strongly in favor of the flood explanation with the discovery of sweeping ridges of gravel—giant ripple marks or *current dunes*—in some of the areas inundated by these incredible floods. No other known process can account for these unique features or can better explain the myriad of unusual landforms and other features found in the floodpath. Only one larger flood, also of glacial origin, is known to have occurred—one in south-central Siberia (Rudoy and Baker, 1993).

To have one of the earth's largest documented floods in a province is remarkable. To have *two* (the Missoula and Bonneville floods) of the largest known North American floods of water roar across one of the earth's largest floods of lava—all located in the same province—is incredible! The Bonneville Flood is again due to Pleistocene climates and a chance combination of topographic features. However, unlike the Missoula Flood, this flood was not caused by the failure of an ice dam. Bonneville flood waters raced through the Snake River Plain in southern Idaho creating catastrophic-flood features including a few large waterfalls (Fig. 7–4) along its path.

Cooler climates reduced evaporation rates and caused hundreds of closed basins southward in the Basin and Range Province to develop permanent lakes whose levels fluctuated with changes in temperature and precipitation. Higher mountain ranges like the Wasatch just east of Salt Lake City responded to lower snowlines by developing alpine glacier systems that stored and moderated water flow into the Salt Lake Basin. By 15,000 years ago waves and currents in the 20,000-square-mile (52,000-km²) Lake Bonneville (a giant Pleistocene version of the present-day Salt Lake) were building shoreline features at the 5085-foot (1550-m) level. Rising water eventually topped the lowest divide in the basin at Red Rock Pass, raced northward into the Portneuf River valley past the future site of Pocatello in

FIGURE 7–4   Shoshoni Falls along the Snake River formed during the Bonneville Flood when floodwaters cut through the relatively thin cover of basalt and encountered the more resistant hotspot rhyolite below. (Photo by E. Kiver)

southeastern Idaho, and into the Snake River valley. The water at the spillway rapidly cut through weak rock units to more resistant strata some 350 feet (108 m) lower—thereby dumping hundreds of cubic miles of water in a few weeks time. About 0.333 cubic miles (1.4 km³) of water/hour roared through Red Rock Pass moving large boulders [some up to 9 feet (3 m) in diameter!], depositing flood bars, and creating scabland topography much like that found in eastern Washington.

Case of the Missing Water

The groundwater system in the Snake River Plain is unusual. In this dry area underlain by fractured lava flows, essentially all of the water in streams flowing out of the mountains to the north disappears into the fractures in the basalt and becomes "lost rivers." Eventually this water reappears in the form of large springs along the north wall of the gorge cut by the Snake River. Of these, Thousand Springs, about 25 miles (40 km) downstream from Twin Falls, is most widely known. Here, more than 150 feet (46 m) above the river, large volumes of water gush forth in quantities sufficient to irrigate large tracts of valley land, for operating fish hatcheries, and even to collect spring water to run a small hydroelectric plant (Fig. 7–5; Kiver and Stradling, 1994)! The rate of flow of these underground waters is abnormally rapid, a puzzling problem until detailed geological investigations were made. It was

FIGURE 7–5   Thousand Springs Power Plant along the Snake River near Hagerman Fossil Beds in southern Idaho. Spring water emerging from basalt interbeds is collected in the concrete flume on the hillside and directed down the large feeder pipe (penstock) to generate electricity. (Photo by E. Kiver)

discovered that there are continuous zones of highly permeable materials (flow rubble and pillow basalt) between lava flows that allow groundwater to move rapidly through the porous materials and reappear as springs, especially the Thousand Springs.

We will begin our tour of the Columbia Intermontane Province at Craters of the Moon in the eastern Snake River Plain and will move westward to Hagerman Fossil Beds, John Day National Monument (central Oregon), and finish at Newberry Volcanic National Monument in western Oregon near the east edge of the Cascade Mountains.

REFERENCES

Alt, D.D., and Hyndman, D.W., 1989, Roadside geology of Idaho: Mountain Press, Missoula, MT, 393 p.

Atwater, B.F., 1983, Jokulhlaups into the Sanpoil arm of glacial lake Columbia, in Guidebook for 1983, Friends of the Pleistocene field trip to the Sanpoil River valley, northeastern Washington: U.S. Geological Survey Open-File Report, 22 p.

Bretz, JH., 1923, Channeled Scablands of the Columbia Plateau: Journal of Geology, v. 31, p. 617–649.

Bretz, JH., 1969, The Lake Missoula floods and the Channeled Scabland: Journal of Geology, v. 77, p. 505–543.

Freeman, O.W., Forrester, J.D., and Lupher, R.L., 1945, Physiographic divisions of the Columbia Intermontane Province, Association of American Geographers, Annals, v. 35, p. 53–75.

Hooper, P.R., and Conrey, R.M., 1989, A model for the tectonic setting of the Columbia River basalt eruptions: Geological Society of America Special Paper 239, p. 293–306.

Kiver, E.P., and Stradling, D.F., 1982, Quaternary geology of the Spokane area: Tobacco Root. Geological Society, Missoula, MT, 1980 field conference guidebook: p. 26–44.

Kiver, E.P., and Stradling, D.F., 1994, Landforms, in Ashbaugh, J.G., ed., The Pacific Northwest: Geographical perspectives: Kendall/Hunt, Dubuque, IA, Chapter 2, p. 41–75.

Leeman, W.P., 1989, Origin and development of the Snake River Plain (SRP)—an overview, in Ruebelmann, K.L., ed., Snake River Plain-Yellowstone volcanic province: Washington, DC, American Geophysical Union, Field Trip Guidebook T305, p. 4–12.

Malde, H.E., 1991, Quaternary geology and structural history of the Snake River Plain, Idaho and Oregon, in Morrison, R.B., ed., Quaternary nonglacial geology: Boulder, Geological Society of America, Geology of North America, v. K-2, p. 251–281.

Mutschler, F.E., Larson, E.E., and Gaskill, D.L., 1998, The fate of the Colorado Plateau—a view from the mantle, in Friedman, J.D., and Huffman, A.C., Jr., coordinators, Laccolithic complexes of southeastern Utah: Time of emplacement and tectonic setting, Workshop proceedings: U.S. Geological Survey Bulletin 2158, p. 203–222.

Pardee, J.T., 1942, Unusual currents in glacial Lake Missoula: Geological Society of America Bulletin, v. 53, p. 1569–1600.

Parsons, T., Thompson, G.A., and Sleep, N.H., 1994, Mantle plume influence on the Neogene uplift and extension of the U.S. western Cordillera?: Geology, v. 22, p. 83–86.

Reidel, S.P., and Hooper, P.R., 1987, Columbia River Basalt Group, in Hill, M.L., ed., Centennial field guide: Boulder, Cordilleran Section of the Geological Society of America, p. 351–356.

Rudoy, A.N., and Baker, V.R., 1993, Sedimentary effects of cataclysmic late Pleistocene glacial outburst flooding, Altay Mountains, Siberia, in Fielding, C.R., ed., Current research in fluvial sedimentology: Sedimentary Geology, Amsterdam, Elsevier, v. 85, p. 53–62.

Self, S., Keszthelyi, T.T.L., Walker, K.H., Murphy, M.T., Long, P., and Finnemore, S., 1996, A new model for the emplacement of Columbia River basalts as large, inflated pahoehoe lava flow fields: Geophysical Research Letters, v. 23, p. 2689–2692.

Stradling, D.F., and Kiver, E.P., 1986, The significance of volcanic ash as a stratigraphic marker for the late Pleistocene in northeastern Washington, in Keller, S.A.C., ed., Mount St. Helens; Five years later: Eastern Washington University Press, Cheney, WA, p. 120–126.

Thornbury, W.D., 1965, Regional geomorphology of the United States: Wiley, New York, 609 p.

Tolan, T.L., Reidel, S.P., Beeson, M.H., Anderson, J.L., Fecht, K.R., and Swanson, D.A., 1989, Revisions to the estimates of the areal extent and volume of the Columbia River Basalt Group, in Reidel, S.P., and Hooper, P.R., Volcanism and tectonism in the Columbia River flood-basalt province: Boulder, CO, Geological Society of America, p. 1–20.

Weis, P.L., and Newman, W.L., 1989, The Channeled Scablands of eastern Washington: U.S. Geological Survey and Cheney, WA, Eastern Washington University Press, 25 p.

CRATERS OF THE MOON NATIONAL MONUMENT (IDAHO)

Early History

Paleo-Indians were in the area some 12,000 years ago but left little evidence of their presence. Although Shoshoni trails lead through the area and some campsites and waterholes in lava tubes were used by Indians, the bleak landscape at Craters of the Moon held little for hunters and gatherers. Settlers and ranchers also avoided this rugged "useless" area, although in 1879 they explored, without much success, for that precious desert commodity, water. The studies of early geologists I.C. Russell (1901) and H.T. Stearns in the 1920s (summarized in Stearns, 1963) drew attention to what has proved to be the largest basaltic lava field of mostly Holocene age in the conterminous United States. A local taxidermist and adventurer, Robert Limbert, began to explore the area and lectured and wrote of his experiences. Again, one dedicated person can make a difference. Limbert almost singlehandedly created a national monument. His article in *National Geographic* and the establishment of the national monument by Calvin Coolidge both occurred in 1924. Limbert's airedale terrier dog accompanied him in 1920 on the first recorded crossing of the lava field [an 80-mile (130 km), 17-day hike!] and was apparently not as enthusiastic as his master. The sharp edges of the lava cut the dog's paws requiring Limbert to carry the animal across miles of lava. To Stearns the strange landscape suggested what a moonscape might look like—hence the name Craters of the Moon (National Park Service, 1991).

Yesterday's Volcanism

A two-mile-deep drill hole in the Snake River Plain as well as other geologic studies provide us with a clearer picture of the evolving landscape of the region. The upper 2600 feet (793 m) of the drill hole contains basalt—similar to that at Craters of the Moon. The lower 8000 feet (2440 m) is all rhyolite—similar to that recently erupted at Yellowstone. Thus, we have before-and-after examples of hotspot volcanism. As Yellowstone looks today, so once did Craters of the Moon about 10–11 Ma. As Craters of the Moon looks today, so too will Yellowstone look in the distant future!

Rather than a deep mantle plume such as that under Hawaii, a shallow plume is visualized for the Snake River Plain (Mutschler and others, 1998). The expanding plume of hot mantle material is embedded in the North American plate and has simultaneously migrated northwest toward Newberry Volcano in Oregon and northeast toward Yellowstone. As the shallow plume of heat migrates, the landsurface rises dramatically, and tension along the margins of the hotspot trail causes fault-block mountains to form (Fig. 7–2). As the hotspot migrated, the land subsided and a series of rifts or fissures opened up in the Snake River Plain. The rifts are parallel to nearby Basin and Range faults and are likely caused by the same extensional forces. The Great Rift in Craters of the Moon is the northern segment of the most recently opened rift that extends some 53 miles (85 km) to the southeast, nearly across the entire Snake River Plain.

"Today's" Volcanism

Time is of course a relative term. By most standards of geology and archaeology 2000–15,000 years—the interval of volcanism at Craters of the Moon—is not that long ago. The combination of sufficient gas pressure in the basaltic magma chamber that formed in the hotspot tail, plus pressure reduction on the magma as the lithosphere was pulled apart along the Great Rift beginning about 15,000 years ago, triggered the first of eight major eruptive episodes (Kuntz and others, 1987, 1989). Just as snapping the lid on a warm can of soda permits the dissolved gases to violently expand and escape from the fluid and drive an "eruption" of gas and liquid, so too does the escaping magmatic gas bring the molten rock to the surface. In both cases, gas and liquid escape first, followed by a frothy fluid as gas pressure lessens.

An eruption may begin with a *curtain of fire* along an open rift segment but will quickly narrow to a central vent where a shield or, more typically at Craters of the Moon, a *cinder cone* will form. Gas-charged lava droplets and clots will be thrown upward, and, when the gas expands, a clinkerlike or popcornlike particle called cinder forms. Larger airborne particles are called *volcanic bombs* and may have *spindle, ribbon,* or *breadcrust* forms. They may be only baseball size or they can be quite large (Fig. 7–6). The particles accumulate around a vent forming a cinder cone. Smaller *spatter cones* form when blobs of molten lava are ejected and then

FIGURE 7–6   Large breadcrust bomb blown from a cinder cone during a geologically recent eruption. (Photo by E. Kiver)

adhere together. Twenty-five separate cinder cones are aligned along the 1- to 5-mile-wide (1.6- to 8-km) rift zone in the park. Most are "single-shot" volcanoes that experience only one episode of eruption. However, Sheep Trail Butte (Fig. 7–7), North Crater, and Watchman experienced more than one eruptive episode. Excellent areas to view the alignment of cinder cones along the rift are from the summit of Inferno or other cones in the park. Most cinder cones are a few hundred feet high—Big Cinder Butte is over 700 feet (213 m)—very tall by cinder cone standards.

As gas pressure is reduced, magma wells up more quietly into the base of the cinder cone where it ruptures one flank and often carries away crater and cone fragments "piggyback" style. Large rafted blocks can be seen up close along the North Crater trail, which is located on some of the youngest volcanic materials (2200 years old) in the park.

Lava with low-silica content (40–50 percent SiO_2, typical basalt) dominates at Craters of the Moon and tends to form the smooth and ropy surfaces of *pahoehoe* lava. A thin glassy crust forms rapidly on a fresh flow and is bent into folds or ropes as the liquid interior moves faster than the crust (Fig. 7–8). The surface resembles a wrinkled throw rug on a linoleum floor. Late eruption movements or deflation of the crusted pahoehoe surface will often form pressure ridges. Cracks along the crest of the pressure ridge will often permit lava to flow to the surface as a *squeeze-up*—much like toothpaste squeezed from its tube (Fig. 7–9). Lava that is less fluid

FIGURE 7–7   Double crater at Sheep Trail Butte. (Photo by J. Wiebush)

FIGURE 7–8   Pahoehoe ropes on recent flow at Craters of the Moon. The flexible solid "skin" was folded into rolls as the liquid interior moved forward. Note camera case for scale. (Photo by E. Kiver)

Figure 7–9   Pressure ridge with bulbous squeeze-up at its crest, North Crater Flow Trail, Craters of the Moon. (Photo by E. Kiver)

because (1) it is cooler, (2) it has a higher SiO_2 content (greater than 50 percent), or (3) it has flowed turbulently and lost dissolved gases, forms an *aa* (blocky surface with sharp spines on blocks) or *block* flow (smooth, no spines on blocks). These flows move like a tractor tread at a speed slower than a human can walk. However, pahoehoe, especially on steeper slopes, can be more challenging to those who would closely approach an active flow.

Larger pahoehoe flows such as the Blue Dragon Flow (30 miles; 48 km long) are usually fed by a system of *lava tubes*. A constant lava supply for days or weeks is required to allow lava channels to develop thick insulating roofs. The tubes enable lava to flow tens of miles from its vent without appreciable temperature loss. Unroofed sections (skylights) or collapsed areas (sinkholes) provide access to segments of the lava tube system such as Indian Tunnel, Boyscout Cave, and Beauty Cave. However, sections of the Blue Dragon and other flows are uncollapsed and not accessible at this time. A series of lava linings, curbing, or "bathtub rings" on the cave walls record decreasing lava levels. Lava stalactites formed where the tube ceiling began to remelt and drip downward. Visitors are generally surprised after a hike across the sun-baked lava to find themselves on ice in the bottom of Boy Scout Cave, a typical *ice cave*. Caves with floors lower than their entrances can trap cold air and freeze water that enters. Ice may persist for all or part of the year in these ice caves. Some of the ice may date back to a time when a cooler climate prevailed; once formed, air circulation is often too poor to efficiently melt the ice. Also of special interest are the *lava trees,* a special type of fossil. The trees were encased in the lava, and their charred trunks often leave an unmistakable imprint in the rock (Fig. 7–10).

Tomorrow's Volcanism

A glance at most of the lava flows and cinder cones leads everyone to the same conclusion—volcanism here is quite young. Soil cover is thin or lacking, vegetation is sparse, and weathering is slight. Just how young wasn't determined with confidence until the 1980s (Kuntz and others, 1987) when charred vegetation beneath the edges of lava flows was radiometrically dated and the magnetic characteristics of the more than 60 lava flows in the park were measured. Because the earth's magnetic pole location changes slightly over time, each time interval has a different pole location associated with it. When lava cools magnetite crystals form in the rock and act like tiny bar magnets. As the rest of the rock solidifies, the magnetite is locked in place, faithfully recording the earth's magnetic field at the time that temperature lowers below the *Curie point* (temperature below which the magnetic field is "locked in"—about 930°F; 500°C). Thus, knowing the magnetic field changes over the past 15,000 years and determining the paleopole location of each flow pinpoints the age of the flow.

The Great Rift first opened up about 15,000 years ago, about the same time that the last glaciers in the nearby mountains reached their maximum extent. The last major eruption occurred about 2000 years ago, about the same time as the heyday of the Roman Empire. Each of the eight eruptive intervals recorded lasted for 1000 years or less. Dormant intervals, or the time needed to recharge the magma chamber, lasted from 500 to as long as 3000 years. By calculating the volume of the

FIGURE 7–10   Tree mold, the mold of a charred log. (Photo by D. Harris)

60 lava flows and knowing their ages, Kuntz and his colleagues (1987) determined that 0.35–0.67 cubic miles (1.5–2.8 km³) of lava is expected every 1000 years. Since the last eruptions occurred about 2100 years ago, the area is at most 900 and more likely less than 100 years away from its next eruptive interval.

Craters of the Moon fascinates many people; geologists and volcanologists are enthralled there. For many years individuals and groups have supported expansion and redesignation of the area as a national park. More of the Great Rift would be included in an expanded park, including the deepest known open volcanic rift on earth (800 feet; 244 m) and the once commercialized Crystal Ice Cave. Opposition comes from grazing interests, hunters, and those who see no beauty in the rough and twisted surfaces of some of the youngest lava flows on the North American continent. Perhaps Washington Irving was restraining his enthusiasm for lava when he wrote in 1868 that it was an area "where nothing meets the eye but a desolate and an awful waste; where no grass grows nor water runs, and where nothing is to be seen but lava." This is a beautiful area, like no other on earth; it is somewhat bleak and lonely, but so are most areas on the moon!

REFERENCES

Kuntz, M.A., Champion, D.E., and Lefebvre, R.H., 1987, Geology of the Craters of the Moon lava field, Idaho, in Beus, S.S., ed., Rocky Mountain Section of the Geological Society of America: Boulder, CO, Geological Society of America Centennial Field Guide, p. 123–126.

Kuntz, M.A., Champion, D.E., and Lefebvre, R.H., 1989, Geologic map of the Inferno Cone quadrangle, Butte County, Idaho: U.S. Geological Survey, Map GQ-1632.

Limbert, R.W., 1924, Among the "Craters of the Moon": National Geographic, v. 45, no. 1, p. 303–328.

Mutschler, F.E., Larson, E.E., and Gaskill, D.L., 1998, The fate of the Colorado Plateau—a view from the mantle, in Friedman, J.D., and Huffman, A.C., Jr., coordinators, Laccolithic complexes of southeastern Utah: Time of emplacement and tectonic setting—Workshop proceedings: U.S. Geological Survey Bulletin 1258, p. 203–222.

National Park Service, 1991, Craters of the Moon: U.S. Dept. of Interior, Handbook 139, 64 p.

Russell, I.C., 1901, Geology and water resources of Nez Perce County, Idaho: U.S. Geological Survey Water-Supply Paper 53, part 2, p. 87–141.

Stearns, H.T., 1963, Geology of the Craters of the Moon National Monument, Idaho: Arco, Craters of the Moon Natural History Association.

Park Address

Craters of the Moon National Monument
P.O. Box 29
Arco, ID 83213

JOHN DAY FOSSIL BEDS NATIONAL MONUMENT (OREGON)

John Day was a member of the party sent into the Pacific Northwest in 1812 to help develop the fur-trading industry for John Jacob Astor. The party became widely separated and John Day and another fur trader were "relieved" of their equipment, including their clothes, by a group of hostile Indians along the Mah-hah River. The river, one of the longest in Oregon, was later renamed the John Day. Rescue came later for the two traders when they encountered a group of trappers in canoes along the Columbia River.

John Day country lies in northeast Oregon north of the Brothers Fault Zone (discussed in the section on Newberry Volcanic National Monument) along the upper reaches of the John Day River (Plate 7). Settlement began in 1862 with the discovery of gold in Canyon Creek. Pioneer minister Thomas Condon, who later became Oregon's first state geologist in 1872, soon discovered some of the rich fossil beds in the area and alerted prominent geologists and paleontologists to the incredible fossil record preserved here. Far-sighted state leaders directed Oregon to purchase some of these outstanding areas in the 1930s to establish state parks. Congress realized the national significance of the area and authorized the national monument in 1974. The 14,000-acre area consists of three widely separated units—Sheep Rock, Painted Hills, and Clarno—with monument headquarters in the town of Kimberly and a visitor center and fossil museum located in the historic Cant Ranch House in the Sheep Rock unit. The area is world famous for the extraordinarily complete and nearly continuous record of over 40 million years of the terrestrial plant and animal life of the 65-million-year-long Cenozoic area. If Dinosaur National Monument in the Rocky Mountains can informally be called "Jurassic Park," then the John Day and Badlands of South Dakota areas are some of the world's best examples of a "Cenozoic Park."

Older Geologic History

The nearby Blue Mountains, including the Ochoco, Aldrich, and Strawberry Mountains just south of the John Day valley, are a series of fault blocks and fold structures containing older rocks. In the Ochoco Mountains are Oregon's oldest rocks—380-million-year-old limestones of Devonian age. However, these rocks, as well as most of the late Paleozoic and early Mesozoic rocks in the Blue Mountains are not "homegrown" or found where they originally formed (Orr and others, 1992). Rather, they are part of *accreted terranes* transported by plate tectonic processes and welded onto the edge of the North American continent during late Mesozoic time. At least three slices were added, shifting the edge of the continental plate and its associated subduction zone from near the Oregon–Idaho border to the edge of the present-day Cascade Range. Early Cretaceous magmas probably gave rise to the gold veins in the Canyon Mountains area—the gold that caused the rush of 1862 and hastened the settlement of the area. During the latter part of the

Mesozoic, there was marine and nonmarine deposition; marine deposition ended with the retreat of the sea late in the Cretaceous and the uplifting of the Rocky Mountains to the east (Thayer, 1976). The scene was set for the rapidly evolving mammal populations and Cenozoic plants to populate the area and, fortunately for science, some of them became entombed in the river, lake, swamp, and volcanic deposits in the John Day Basin.

The Cenozoic Story

With a new subduction zone and the edge of the sea much farther to the west, streams, rivers, and intermittently erupting volcanoes took charge of landscape development in the John Day area. Three themes are discernible from the rocks and fossils found in the area: (1) a long history of volcanic episodes that began about 44 Ma (Eocene time), (2) an excellent record of the plants and animals inhabiting the area, and (3) the use of these fossils and other geologic features to document the profound change from the near tropical climates of Eocene time to today's colder desert environment.

The Cenozoic layers shown in Figure 7–11 are relatively widespread in eastern Oregon and are exposed in one or more of John Day's three units. The Eocene age Clarno Formation in the Palisades Cliffs area spectacularly displays mudflows generated by nearby andesite volcanoes of the Mount St. Helens type. Finer layers entombed bones and plant stems that were slowly impregnated with mineral material, thereby fossilizing and preserving a rich sample of former life. Leaves, including palm leaves over 2 feet (60 cm) long (Orr and others, 1992), are preserved as impressions in the hardened mud. The world famous Clarno Nut Beds are unique—they contain at least 173 species of hardwood nuts, fruits, and seeds

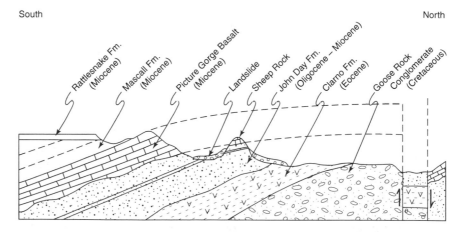

FIGURE 7–11   Generalized cross section through the Sheep Rock unit. (Modified from Thayer, 1976)

(Retallack and others, 1996). Other layers contain the bones of extinct tapirs, a cow-size rhinoceros, titanotheres (no modern descendants), oreodonts (sheep size, no modern descendants), crocodiles, a condorlike bird much larger in size than the giant California condor, and *Orohippus,* a dog-size, four-toed ancestor of the modern horse. Likely, *Hyracotherium* (*Eohippus* or dawn horse), the first horse, also roamed through the area, but fossil remains have yet to be found here. The first recognition of the profound evolutionary changes of the horse and the best known fossil record of the rhinoceros is associated with the John Day area. Figure 7–12 is a sampling of some of the fossil mammals unearthed from the Clarno Formation in John Day National Monument.

The palm, avocado, fig, banana, and cinnamon trees, as well as the deeply weathered *paleosols* (former soils buried by younger material), indicate that a subtropical rain forest, much like that in Central America, existed here during deposition of the Clarno Formation. Nearby Clarno beds in eastern Oregon even contain fossil banana trees—more evidence of a climate much different from the present. Clarno deposition ended about 40 Ma, and erosion occurred until deposition of the overlying John Day Formation began about 39 Ma (Retallack and others, 1996).

In the Sheep Rock and Painted Hills units the base of the John Day Formation is marked by iron-rich red soils produced by the intensified chemical weathering of the subtropical climate. Episodic explosive volcanic activity from the fledgling Cascade Range to the west, as well as more local volcanoes, spread air-fall ash, ash-flow layers, and lava flows over the landscape from about 39–18 Ma (later Eocene through lower Miocene). Streams washed the ashy sediment into valleys where locally over 1000 feet (305 m) of debris accumulated over a 21-million-year interval. Mixed in are ash-flow deposits that contained enough heat to melt particles together to form *welded tuffs*. Older soils were buried and new ones formed on top of succeeding layers. The spectacular color bands in the formation record these ancient soils (Getahun and Retallack, 1991). Delicate shades of red, green, buff, and cream are best displayed in the badlands of the Painted Hills unit.

Later on during deposition of the John Day Formation the change to a warm temperate climate signalled an increasing rain shadow and cooling effect as the Cascade Range to the west began to grow taller, and the gradual worldwide cooling trend began that would ultimately lead to the Pleistocene ice age. Forests with species related to modern birch, oak, beech, and chestnut replaced the earlier tropical species. Conditions were still humid, but not tropical. Similar plant genera no longer occur in the Pacific Northwest but are found in east Asia and eastern North America (Manchester, 1987). The nearly 21 million years of almost continuous sedimentation of the John Day Formation also provides an unusually long record of mammal evolution.

Saber-toothed cats, doglike carnivores, giant pigs, camels, two-horned rhinoceros, and other mammals roamed through the forests and the increasingly more open woodland environments. *Mesohippus,* the three-toed horse, was a forest dweller about six hands high, the size of a modern sheep. An episode of erosion followed before the next volcanic event affected the area, the eruption of the Columbia River Basalt.

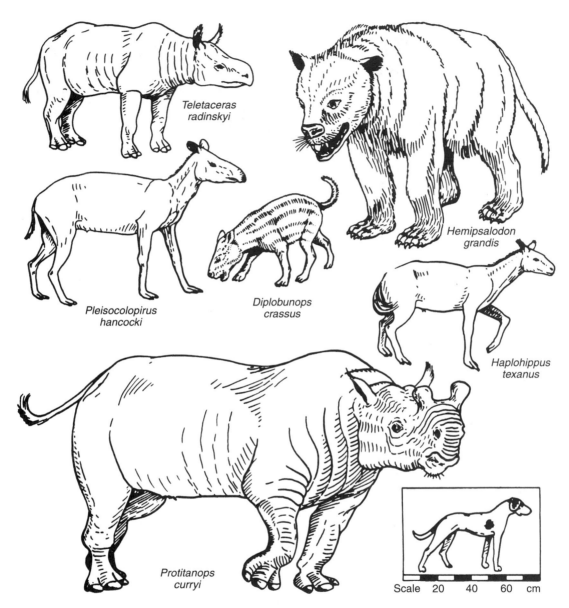

FIGURE 7–12   Reconstructions of fossil mammals from the upper Clarno Formation (late Eocene) at John Day Fossil Beds National Monument. A sketch of a modern-day dog provides a scale for comparison. (From Retallack and others, 1996; Oregon Department of Geology and Mineral Industries)

Starting about 17 Ma (Miocene), the first of the very fluid Columbia River flood basalts (Imnaha and Picture Gorge basalts) poured from long linear vents a few tens of miles to the north and northeast. The lava filled deeper valleys and basins, and by 12 Ma nearly two dozen flows of the Picture Gorge Basalt had covered the John Day area with as much as 1500 feet (457 m) of basalt. Life must have been completely eliminated; animals that could migrate did so—other living things perished as the slow but persistent lakes of lava burned their way across the landscape. Temporary refuge might be found on hills higher than the lava surface, but most of these too were eventually overwhelmed.

Ash from the erupting Cascade volcanoes during the Miocene mixed with local volcanic sources from the nearby Blue Mountains to form the Mascall Formation that overlies the Picture Gorge Basalt (Fig. 7–11). Rivers and streams reworked the ash and entombed a variety of organisms. The dominance of cool temperate plants like oak, hickory, sycamore, maple, redwood, gingko, box elder, and elm reflect the even greater rain-shadow effect of the growing Cascade Range and the seasonal frost of the colder climate. Grasslands were increasing and a rich mammal assemblage, including antelope, camel, deer, horse, and oreodonts, along with predators like dogs, bears, and weasels, is contained in the Mascall sediments.

Horses were taller and a change in tooth structure enabled the group that would later evolve to the modern horse to graze on the grasslands that were beginning to stretch across the eastern Oregon landscape. Raising up on their toes lengthened their legs and turned the horse into an efficient running machine. Speed in the open grassland was an important defense, and horses with longer legs and fewer working toes were less likely to become a meal for a hungry predator. Much of the folding and faulting evident in Figure 7–11 came after Mascall Formation deposition and before the overlying Rattlesnake Formation was deposited.

Uplift during the late Miocene of the nearby Blue Mountains along faults and folding of rocks in the basins revitalized streams that now carried gravel, sand, and silt that contributed to the Rattlesnake Formation. A welded-ash cap records a giant ash flow that emanated 6.6 Ma from a large caldera about 90 miles (145 km) to the south. This monster eruption dwarfs those recorded in the John Day Formation—nothing remotely equal in magnitude to this eruption has occurred in historic times. Again life was snuffed out as this huge glowing cloud raced across eastern Oregon. The thick, resistant welded tuff layer caps the skyline in the Picture Gorge area.

Again, bones preserved in the sandy sediments of the Rattlesnake Formation provide a glimpse of yet younger life forms. Newer species of horse, camel, rhinoceros, and antelope are joined by mastodon, sloths, and other more modern forms. Some of the last of the North American rhinoceroses occur in the Rattlesnake beds, and the horses, soon to become extinct in North America, are closely related to the modern one-toed horse. By this time the horse had reached the size of a pony and had feet and teeth like those of the modern horse. Plant fossils indicate that the climate had changed to one similar to that of today's. For those interested in the Pliocene, a visit to Hagerman Fossil Beds in southern Idaho will continue the evolutionary drama of changing life forms.

Most of the final touches to the landscape were finished in the Pleistocene when glaciers scoured the higher Blue Mountain uplifts and streams continued their slow removal of millions of years of accumulation of rock and sediment. A stop at the spectacular Picture Gorge where the John Day River and its tributaries have removed the Rattlesnake and Mascall Formations is a good place to allow your imagination to restore the missing Cenozoic layers and visualize how much material was removed to form the canyon and the rest of the present landscape. How much has been removed is much more impressive than the height of a hill or the depth of a canyon. As material was removed hillsides lost support and landslides, such as those at Cathedral and Sheep rocks (Fig. 7–11), occurred. The now lofty Cascade peaks to the west dried out the land even more. During the Holocene (last 10,000 years of earth history) the climate became semiarid and lower elevations could support only sagebrush, juniper, and grass.

Be sure to visit the fossil display at the historic Cant Ranch buildings (early 1900s) near Sheep Rock. Photographers should try to visit the Clarno unit in the morning, the Sheep Rock unit in early afternoon, and the Painted Hills in the late afternoon. If your timing is fortunate, a visit to the Painted Hills after a rain, followed by sunshine, will challenge even the best artist who dabbles in oil. Remember that the fossils here, as well as the vertebrate fossils on surrounding government land, are legally protected so that future generations can study and enjoy Oregon's Cenozoic Park.

REFERENCES

Getahun, A., and Retallack, G.J., 1991, Early Oligocene paleoenvironment of a paleosol from the lower part of the John Day Formation near Clarno, Oregon: Oregon Geology, v. 53, no. 6, p. 131–136.

Manchester, S.R., 1987, Oligocene fossil plants of the John Day Formation: Oregon Geology, v. 49, no. 10, p. 115–127.

Orr, E.L., Orr, W.N., and Baldwin, E., 1992, Geology of Oregon: Kendall/Hunt, Dubuque, IA, 254 p.

Retallack, G.J.; Bestland, E.A.; and Fremd, T.J., 1996, Reconstructions of Eocene and Oligocene plants and animals of central Oregon: Oregon Geology, v. 58, no. 3, p. 51–69.

Thayer, T.P., 1976, The geologic setting of the John Day Country, Grant County, Oregon: U.S. Geological Survey Information Circular 69-10, Washington, DC, U.S. Government Printing Office, 23 p.

Park Address

John Day Fossil Beds National Monument
Kimberly, Oregon 97848-9701

HAGERMAN FOSSIL BEDS NATIONAL MONUMENT (IDAHO)

Hagerman is one of our newest national monuments, joining the Park Service ranks as an important paleontological area in 1988. The rocks here contain the world's best record of terrestrial lifeforms that existed in the late Pliocene, about 3.5 Ma. In combination with the fossil record at nearby John Day National Monument that ends in late Miocene, Hagerman provides a glimpse into the next chapter of the earth's changing lifeforms. Hagerman is also the first Idaho area to be added to the park system since 1924.

Geographic and Geomorphic Setting

The national monument is located near the town of the same name in south-central Idaho (Plate 7) near the geomorphic and structural boundary between the northwest-trending western Snake River Plain and the northeast-trending eastern Snake River Plain. Topographically the two areas appear similar and continuous, but their geologic origins differ. Hagerman and the western Snake River Plain are part of a long Basin and Range fault trough (*graben*) that formed from 16 to 3 Ma, eventually filling with volcanic rocks and thick lake and stream sediments, including the Pliocene age Glenns Ferry Formation (Fig. 7–13) that contains the world-famous Hagerman fossil beds (Malde, 1991). In contrast, the southwest corner of Idaho and the eastern Snake River Plain mark the migration track of the Yellowstone hotspot as explained in the introduction to this chapter.

Geologic Story

Lava dams along the ancestral Snake River near Hells Canyon during Miocene and Pliocene time (16–3 Ma) created an immense lake, Lake Idaho. Hagerman is near the southeastern lake margin where both stream and lake sediments accumulated as the position of the lake shifted. Thus, different sedimentary *facies* are recorded in the Glenns Ferry Formation (Malde, 1991).

Following deposition of the Glenns Ferry Formation, Quaternary-age sediments and basalt flows continued to cover parts of the Snake River Plain faster than the downcutting Snake River could remove them. The Bonneville Flood, caused by spillover of the giant Pleistocene lake in the Salt Lake Basin, swept through the area about 14,500 years ago. The up to 300-foot-deep (100-m) flood dwarfed all floods of historic record in the world. The Melon Gravel flood deposits are found the length of the Snake River Plain and contain many "watermelon-size" basalt boulders as well as some up to 9 feet (3 m) in diameter! Boulders near Hagerman are up to 3 feet (1 m) in diameter and were deposited only on the east side of the valley. The Bonneville Flood eroded and steepened the bluffs on the west side, thereby

FIGURE 7–13   The Glens Ferry Formation at Hagerman Fossil Beds forms the west wall of the Snake River valley. Rocks at water level are about 3.7 Ma and those at the top of the bluff are 3.15 Ma. Note the collapsing fossil-bearing beds in the large landslide triggered by irrigation on the plain behind the bluff. (Photo by E. Kiver)

better exposing the upper layers of the Glenns Ferry Formation and some of its paleontologic treasures (Fig. 7–13). Radiometric dates of volcanic ash layers near the top (3.15 Ma) and bottom (3.7 Ma) of the bluff indicate that 550,000 years of earth history are exposed here. The bones and other fossils that had not been exposed to the sun's rays for at least 3.15 million years began to appear in the ravines and slopes along the Snake River valley as modern slope processes slowly removed rock material.

History

Humans have used the area for at least 10,000 years beginning with the Folsom paleo-Indian culture and later the Great Basin Archaic hunters and gatherers. The Fremont (2000–700 years ago) and later the Shoshoni peoples also made this area home. Wagons on the Oregon Trail rolled through the southern end of the monument from 1840 to 1870 leaving deep ruts that are still visible today. Settlers moved in shortly after, but surprisingly it was not until 1928 that local rancher Elmer Cook led geologist H.T. Stearns to the 550-foot-high (168-m) bluffs above the Snake River containing the bone beds. Stearns returned the next year with Smithsonian paleontologist J.W. Gidley whose team excavated three tons of bones that first summer—mostly of one species of an extinct zebralike horse. Over 120 horse skulls and 20 complete skeletons were excavated through the 1930s. A large quarry was opened up called the Horse Quarry, or Gidley Quarry, where major digs continued through 1934 and again in the 1950s, 1960s, and most recently in 1997. Qualified researchers still continue various research projects using the excavated fossils and those that continue to be exposed by wind and rain. Over 200 scientific papers have been written about the Hagerman fossils.

Paleontology and Paleoenvironments

Over 100 mostly extinct species of both large and small vertebrates are described from the fossil beds. The best examples in the world of 39 of these species come from Hagerman. The abundance and variety of animal and plant fossils as well as clues from the sediments themselves enable the *paleoecology* of the area 3.5 Ma to be reconstructed with a high degree of confidence. The area had a climate twice as wet as the present semiarid conditions and a lush savanna grassland with patches of trees surrounding a Great Lakes–size lake called Lake Idaho that occupied the western Snake River Plain structural basin. Streams and rivers, especially in the Hagerman area, flowed into Lake Idaho and fish, ducks, geese, pelicans, cormorants, beaver, and an extinct variety of muskrat were abundant in the relatively lush, watery environment. Horses, camels, antelope, mastodon, peccary, sabre-tooth cats, and many other animals roamed across the grasslands and through the patchy stands of trees (McDonald, 1993).

The horse remains found in the main fossil layer are of one species, *Equus simplicidens,* otherwise known as the Hagerman horse, Idaho's official state fossil (Fig. 7–14). This species is the earliest representative of the modern genus, *Equus,* and is more closely related to the African zebra (typically larger hoofed and distinct tooth characteristics) than the modern horse. An unusual aspect is that the horse fossils range from yearlings to adults to older adults and many are almost complete skeletons with no evidence of predators disturbing the carcasses. Many of these outstanding specimens are on display in major museums around the country.

Whether the horse fossils from this layer represent an accumulation over many seasons, a single herd that clustered around a drying water hole during a drought, or perhaps an entire herd that drowned crossing a flooded river is unknown. The alignment of the bones in the same general direction suggests deposition by river currents and supports but does not prove the flooded-river hypothesis. The hypothesis presently favored by paleontologist Greg McDonald is that the animals died during a drought, and their bones were washed to their present location by a seasonal flood, probably during the following fall. Future excavations, application of the newest techniques, and use of future techniques yet to be discovered will add even more to our understanding of the past.

FIGURE 7–14   The Hagerman horse, about the size of a pony, is the oldest representative of the modern genus, *Equus*. (Photo by E. Kiver)

Threats to the Fossil Beds

The steep bluffs along the Snake River valley had stabilized after the giant Bonneville Flood swept the valley about 14,500 years ago and subsequent river downcutting slowed. However, pumping water into leaky irrigation canals and irrigating fields just west of the bluffs behind the monument have increased groundwater flow into the sediments. Groundwater increases the mass or weight of the slope, lubricates sediment layers, and thereby causes the fossil beds along the bluff to slide (Fig. 7–13). The first large landslide occurred in 1983. Pumping was discontinued at that site in 1987 after another slide occurred. Water pumping was increased to another leaky canal to the north located behind the historic quarries, triggering a new slide in 1989. Over 75 acres of fossil beds are lost so far—whether the quarries where beds are particularly rich with fossils will be endangered remains to be seen. Irrigation water also contributes to seasonal seeps and springs that discharge small streams that flow down the slopes and further erode and steepen the bluffs. Amateur collectors also pose a potential threat to the fossil treasures preserved at Hagerman.

   Increasing development in southern Idaho is withdrawing water from the Snake River system at an alarming rate, causing those with a farsighted view to recognize that the Snake River, like the Columbia, Colorado, and other rivers around the world, will soon become an endangered river. Water use and development are done at the expense of the environment—the cost is borne by the land, plants, wildlife, and eventually humans. Sharing of resources by all living things is necessary if all are to survive. However, when wasteful uses of resources and ever-growing demands (with no end in sight!) are made by the species that should know better—the future seems uncertain.

REFERENCES

Malde, H.E., 1991, Quaternary geology and structural history of the Snake River Plain, Idaho and Oregon, in Morrison, R.B., ed., Quaternary nonglacial geology; Conterminous U.S.: Boulder, CO, Geological Society of America, The Geology of North America, v. K-2, p. 251–281.

McDonald, H.G., 1993, Hagerman Fossil Beds, Hagerman, Idaho: Rocks and Minerals, v. 68, p. 322–326.

Park Address

Hagerman Fossil Beds National Monument
P.O. Box 570
221 N. State Street
Hagerman, Idaho 83332

NEWBERRY VOLCANIC NATIONAL MONUMENT (OREGON)

History and Description

Newberry Volcano, the largest Quaternary-age volcano (covers over 500 miles^2; 1300 km^2) in the conterminous United States, is located 25 miles (40 km) east of the Oregon Cascades and a few miles south of the town of Bend (Fig. 7–15). The gentle slopes of this giant shieldlike volcano rise from about 3900 feet (1190 m) elevation near its base to 7984 feet (2434 m) atop Paulina Peak—about 4000 feet (1220 m) of relief. The 4- by 5-mile-diameter (6.4 by 8 km) caldera at its top is nearly as large as the one at Crater Lake. Inside the caldera are two caldera lakes (Paulina and East Lake) and a volcanologist's paradise of volcanic features, including Oregon's youngest lava flow.

Like Medicine Lake Volcano at Lava Beds National Monument in the Basin and Range Province to the south, Newberry lies just east of the Cascade Range and cannot be pigeon-holed into neat volcanic classifications. Rather than one chemical type of lava, both volcanoes produce mostly low-silica lava (basalt and basaltic-andesite) on their flanks and high-silica (rhyolite) lava near their summits. Both volcanoes have the gentle slopes characteristic of a *shield volcano* produced by

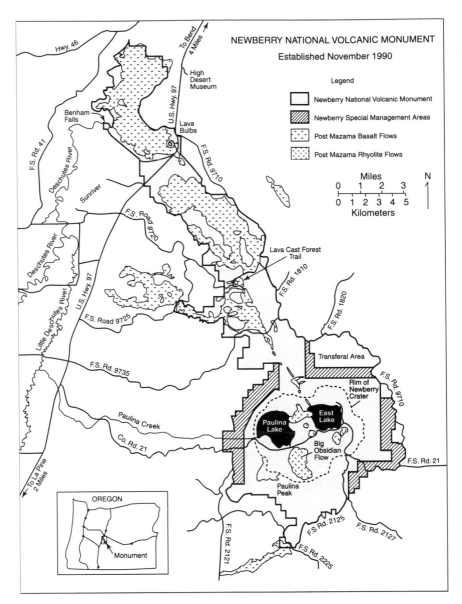

FIGURE 7–15   Location map and selected features in Newberry Volcanic National Monument. (Slightly modified from R.A. Jensen, 1995)

outflows of fluid lavas—however, both also have pyroclastic layers on their flanks that is more typical of *strato* or *composite volcanoes*. Because Newberry's slopes lack the steepness of a Mount St. Helens or a Mt. Rainier, it is commonly referred to as a shield-shaped composite volcano.

Central Oregon has a rich archeological history and evidence of occupation may go back over 12,000 years ago, perhaps a few thousand years after humans entered the New World over the Bering Land Bridge. Native Americans occupied sites on the mountain at least 11,000 years ago and no doubt witnessed (hopefully from a distance) Newberry's more recent eruptions.

As summarized in Garrett (1991), the first Euro-American visitor (November, 1826) was Peter Skene Ogden from the Hudson Bay Company. His half-starved men and thirsty horses rested along the caldera lakes before continuing their exploration to determine whether eastern Oregon could support a fur industry. The U.S. Army (Williamson-Abbot) expedition of 1855 had John Strong Newberry as scientist. Newberry became an important figure in understanding Oregon's geology and went on to become Professor of Geology and Paleontology at Columbia University. Settlers began to move into the area and some of the Indians, including Chief Paulina, began raiding the settlements. Paulina was shot in 1867 following a series of raids; however, his name lives on in the peaks and water bodies of the area.

One of John Newberry's students, Israel C. Russell, studied the area and provided the first general description of the geology. Russell named the mountain after his old professor in 1903.

Newberry Volcano and Crater Lake were considered for park status in 1903—less was known about Newberry and it was passed over. Professor W.O. Crosby investigated the area in 1919 and stated, "We feel that Newberry Crater is comparable in scenic and geologic interest with Crater Lake and recommend its designation as a national park or monument." Without special protection developments in the area of course proceeded—the spectacular old-growth ponderosa pine forests were decimated at a rapid rate, roads were established, and a health resort was constructed at East Lake hot springs. A mining claim was filed in 1931 on Lava Butte, a magnificent cinder cone that today is an integral part of the national monument. Fortunately the railroad used a different source for rock ballast to build their road bed—otherwise we would now have perhaps half a cinder cone or less.

Again in the 1940s and 1970s local citizens unsuccessfully pushed for park status. Pressure intensified as energy companies became interested in exploring and using the geothermal energy in the dormant volcano to generate electricity. Pressure from local citizens again increased, and after a number of concessions to developers, industry representatives, and various recreation groups, a small national monument, under U.S. Forest Service direction, was signed into law by President Bush in 1990. The lesson is clear—the longer it takes to establish a park area the smaller and less pristine the area will be. That very philosophy is held by small but influential "wise-use" groups that believe that scenery and environment should take a back seat when it comes to development and other "quick-pay" economic activities. Restrictive rules and regulations, and even some or all of the national parks, should be removed according to these groups.

Regional Geology—The Big Picture

That part of the Columbia Intermontane Province from southeastern Oregon northwest to the Bend area is part of the High Lava Plains Section and is characterized by a wide belt of west-by-northwest-trending normal faults and associated volcanic centers that make up the Brothers Fault Zone. The Blue Mountains Section lies to the north, the Owyhee Upland volcanic area to the southeast, the Cascade Mountain Province to the west, and the Basin and Range Province lies to the south of the High Lava Plains (Kiver and Stradling, 1994). The Basin and Range fault blocks trend north to northeast but rotate or bend westward as they merge with the Brothers Fault Zone.

As described in the introduction to this chapter, the generation of the Yellowstone hotspot some 17 Ma at McDermitt caldera in southeastern Oregon initiated the Brothers Fault Zone as well as the Snake River Plain rift. These giant rifts opened progressively away from their point of origin in southeastern Oregon—much like a crack in a log lengthens as a wedge is driven. Here the splitting force is tension created by plate movements. Lateral slippage of plates along the San Andreas Fault and oblique subduction along the West Coast continue to stretch and pull apart sections of the western United States. The shallow-mantle heat plume (hotspot) and its associated volcanism thus migrate outward along the opening rifts.

Newberry's location and huge volume are not accidents. Here three large fault zones intersect—the Brothers Fault Zone from the east, the Walker Rim Fault on the south, and the Tumalo-Green Ridge Faults from the northwest. This prime location continues to release pressure on the hot mantle rock below, allowing partial melting and generation of basaltic magma that rises buoyantly toward the surface. The result over the past 1.2 million years is the production of the 20- by 30-mile-diameter (32- by 48-km) cone that covers over 500 square miles (1295 km^2)—the largest Quaternary volcano in the conterminous United States.

Newberry's "Older" Geologic Story

Activity at the fault junction beneath Newberry may have started 1.2 Ma (Jensen, 1995) when basalt, chemically similar to Newberry's, flowed across a relatively flat older lava landscape. By 600,000–700,000 years ago summit and flank eruptions of basalt and basaltic andesite (relatively low-silica content) were building a shield cone such as those found today on the Big Island of Hawaii.

The mood abruptly changed about 500,000 years ago as a rhyolitic (high-silica content) magma chamber explosively discharged large clouds of red-hot ash down Newberry's flanks. Likely, the partial emptying of the magma chamber triggered the first of a series of collapses that eventually produced the present 4- by 5-mile-diameter (6- by 8-km) caldera. At least three other eruptions produced additional subsidence as the mountain top dropped further below the caldera rim. In contrast to the spectacular catastrophic collapse that occurred during a single eruption at Mount Mazama (Crater Lake), Newberry's collapse occurred in smaller increments over a few 100,000 years.

Thanks to a U.S. Geological Survey research well located a short distance northeast of the Big Obsidian Flow (Fig. 7–16), we now know that by 200,000 years ago the depth of the caldera was enormous—similar to that of Crater Lake. The drill bit pierced 1640 feet (500 m) of caldera-filling rhyolite, ash deposits, and lake sediments before encountering basalt—presumably the former top of the mountain. The distance from the bottom of the drill hole to Paulina Peak is 3253 feet (992 m)—a minimum estimate for displacement along the system of ring fractures outlining and paralleling the caldera rim.

Recent Volcanic Activity

The caldera floor and the northwest rift zone are covered with recent "textbook examples" of volcanic features (Fig. 7–16). Most areas in this part of Oregon are covered by an ash layer from the huge Mazama (Crater Lake) eruption that occurred

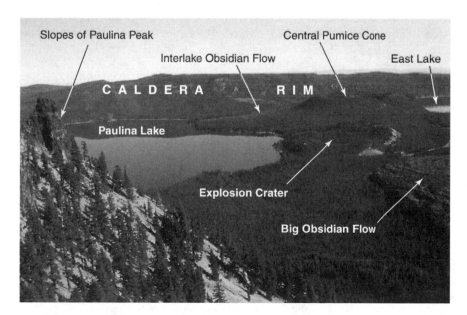

FIGURE 7–16   Central part of Newberry caldera looking north from Paulina Peak. (Photo by E. Kiver)

6800 radiocarbon years (7600 calendar years) ago.[1] The absence of Mazama ash in Newberry's caldera and on the numerous fresh-appearing lava surfaces on the volcano's flank indicate that considerable volcanic activity occurred *after* the Mazama eruption. Radiocarbon dates on plant materials scorched by lava and ash eruptions also indicate that numerous eruptions occurred during the past 6000 years.

Caldera Events

According to Bob Jensen (1995), a large lake occupied the caldera floor before being separated into two lakes during a volcanic eruption sometime before 6400 years (7270 calendar years) ago. Eruption of the Interlake Obsidian Flow (6400 years old) and the giant 700-foot-high (213-m) Central Pumice Cone further increased the topographic barrier between the two caldera lakes (Fig. 7–16). In addition, *explosion craters* (depressions formed where rising magma contacts groundwater and produces a violent steam explosion) and five other rhyolite flows—including the 1300-year-"young" (1260 years ago) Big Obsidian Flow (Oregon's youngest lava flow, Fig. 7–17)—form spectacular features inside the caldera.

The Big Obsidian eruption, and other high-silica magma eruptions, begin with large volumes of gases escaping from the magma as it encounters lower pressure near the earth's surface. The high-silica content allows long chains and networks of silica to form that intertwine and produce a viscous, pasty magma that retards gas escape. Thus, high-silica magmas are more prone to violent explosions than low-silica magmas such as basalt. Such explosions typically shower the area with pumice and ash. A destructive cloud of hot gas and pumice fragments, called an *ash flow,* may roar away from the vent. The Big Obsidian eruption did just that 1500 years ago, producing a vertical eruption cloud followed by an ash flow and the Big Obsidian Flow 1300 years (1260 calendar years) ago. Over 12 feet (3.7 m) of ash covered the upper east flank of Newberry. The pyroclastic phase was followed by the emergence of a thick, pasty rhyolite flow. Rhyolite flows are zoned, with a bottom *brecciated* (highly fractured) zone overlain by obsidian, crystalline rhyolite, and additional obsidian and *vesicular* (full of gas-bubble holes) pumice layers (Jensen, 1993).

Northwest Rift Events

As mostly high-silica (rhyolitic) volcanic products were erupting in the caldera during the last 7000 years, lower silica (basalt) eruptions burst out along the northwest rift zone. Eleven basalt flows and a number of cinder cones formed—including Lava Butte (Fig. 7–18), one of the most accessible volcanoes in the United

---

[1]Because atmospheric radiocarbon varies, calendar and radiocarbon years are not the same and may be a few hundred years different. Dates given here are radiocarbon years with calendar years in parentheses.

FIGURE 7–17 South wall of Newberry caldera with the Big Obsidian plug dome and flow—Oregon's youngest lava flow. (Photo by E. Kiver)

FIGURE 7–18 Lava Butte cinder cone with steep edge of an aa lava flow in foreground. (Photo by E. Kiver)

States. Basalt eruptions are similar to those of rhyolitic composition in that rising magma degasses and initiates a gas and pyroclastic eruption followed later by liquid rock. However, the violence of the basalt eruption is less devastating and, as was the case at Lava Butte 6160 years (7120 calendar years) ago, it often begins with a curtain of fire along a segment of the rift. The eruption quickly narrows to a *lava fountain* that produces rock fragments, volcanic bombs (molten masses larger than 2 inches in diameter), and a spray of smaller foamy liquid blobs that form cinder. The pyroclastic material accumulates around the vent to form a cinder cone. The process is spectacular and resembles a giant popcorn machine except that the red-hot particles are not tasty—even after they cool down!

As the magma degasses further, the pyroclastic phase subsides and lava may well up into the base of the rubble pile (cinder cone) and push out *through* the volcano flank (*not* over the crater rim!). If one follows the extensive lava flow back to its source at Lava Butte (best reached by a trail from the Lava Lands Visitor Center near the highway), he or she will see *lava levees* that merge headward into the volcano's south flank. The other end of the flow extends westward where the lava blocked the Deschutes River channel with as much as 100 feet (30 m) of lava and created a series of delightful waterfalls, including Benham Falls that is accessible by trail from the Benham Falls picnic area.

Special Features

In addition to the highlights mentioned above, many other features await the modern explorer, both in the park and in surrounding areas. Begin your tour at Lava Lands Visitor Center where maps and other publications can be obtained. Jensen's (1995) *Roadside Guide* is particularly useful and will provide many days of discovery for those wishing to follow the trails and roads in the Newberry area. The geologic map of the volcano by MacLeod and others (1995) is also very useful. The monument includes only a small part of the giant shield-shaped volcano—many more interesting features like Hole-in-the-Ground (an explosion crater), numerous lava caves, and Fort Rock (a state park) occur in or near the surrounding Deschutes National Forest.

A rough but passable military road (no trailers!) leads to Paulina Peak (7984 feet; 2434 m) where a former radar station was built in 1958 (abandoned because of difficult winter conditions at this elevation). The view from Paulina Peak of "Lava Lands" on a clear day is spectacular. Everything one can see is of volcanic origin—the Cascade Range to the west from Mt. Adams in Washington State to Mt. Shasta in northern California; Walker Mountain in the northwestern part of the Basin and Range; the High Lava Plains; Newberry's cinder cone-studded flanks; and the lakes, tuff cones, and rhyolite flows in the huge caldera below the peak.

East Lake (Fig. 7–16) is 40 feet (12 m) higher than Paulina Lake and has no surface-stream connection. Paulina Lake overflows to the west and plunges over 80-foot-high (24-m) Paulina Falls. A mass of rock fell from the face of the falls in

1983 reminding us that waterfalls, and all topographic features including entire mountain ranges, are temporary aberrations of the landscape. The falls is now a little closer to Paulina Lake. In a few thousand years it will reach the lake and the water level will lower a few feet. Eventually the lake will completely drain—assuming that volcanism does not drastically change the face of the land before then. However, from what we know of Newberry's recent volcanic past, this is not a good assumption to make.

Along the northwest rift is an older pahoehoe flow (smooth, ropy surface) containing Oregon's longest lava tube, Lava River Cave (Fig. 7–15). As discussed in more detail in the section on Lava Beds National Monument in Chapter 6, lava tubes are the major mechanism by which lava can flow many miles from a vent without losing significant heat and gases. Their development requires a fluid pahoehoe lava and a uniform flow rate to enable a channel to "roof over." Lava River Cave is 6180 feet (1884 m) long and locally has walls as wide as 50 feet (15 m) and a roof 60 feet (18 m) high. It crosses under U.S. Highway 97 where the roof rock is 50 feet (15 m) thick.

As mentioned in the introduction, interest in developing Newberry's geothermal energy to generate electricity provided the final incentive to protect as many of Newberry's spectacular features as possible. Indications were abundant that significant geothermal energy, at relatively shallow crustal depths, existed at Newberry. The fresh-appearing volcanic features and drowned hot springs in the caldera lakes where hydrogen sulfide gas bubbles to the surface are unmistakable. The East Lake Health Resort piped the 175°F (79°C) spring water into tubs for tourist use until 1923 when the resort was destroyed by fire. The resort was rebuilt and burned again in 1941. A shallow well in the Little Crater Campground pumped 86–97°F water for many years before a pressurized water system was built.

Exploratory wells revealed that they must penetrate the *rain curtain*—the zone where cold circulating groundwater cools rock. The true rate of temperature increase with depth (*geothermal gradient*) can only be determined by drilling below the rain curtain. A well drilled in 1981 northeast of the Big Obsidian Flow penetrated the rain curtain 2000 feet (610 m) beneath the surface. Drilling stopped at 3058 feet (932 m) where temperatures reached 509°F (265°C), one of the highest drill hole temperatures recorded in North America and well above the 350°F (177°C) needed for commercial development.

No drilling will be permitted in the monument, only slant drilling from sites on the mountain flank. However, each 30-megawatt power plant built on Newberry would occupy an 18.5-acre site with 3.4 acres for each of 10 well pads (34 acres). When the 52.5 acres per site is multiplied by a possible 30 power plants, and considering the noise generated, the possibility of destroying the hot springs in the caldera, visual impacts, and other problems, one wonders what the now serene area will be like in the future. Society might be willing to sacrifice some outstanding areas (we have already done that to most of the landscape) to solve the important material and energy needs of our citizens. However, each new source of energy or material provides only a *temporary* solution for our insatiable appetite for more things for more and more people. The direction in which we are going is very clear. Will we do something about it?

REFERENCES

Crosby, W.O., 1919, Report on the Benham Falls project on the Deschutes River drainage, Oregon: U.S. Reclamation Service, 99 p.

Garrett, S.G., 1991, Newberry National Volcanic Monument: Pacific Northwest Books, Medford, Oregon, 123 p.

Jensen, R.A., 1993, Explosion craters and giant gas bubbles on Holocene rhyolite flows at Newberry Crater, Oregon: Oregon Geology, v. 55, no. 1, p. 13–19.

Jensen, R.A., 1995, Roadside guide to the geology of Newberry volcano: CenOreGeoPub, Bend, OR, 155 p.

Kiver, E.P., and Stradling, D.F., 1994, Landforms, in Ashbaugh, J.G., ed., The Pacific Northwest: Geographical perspectives: Kendall/Hunt, Dubuque, IA, p. 41–75.

MacLeod, N.S., Sherrod, D.R., Chitwood, L.A., and Jensen, R.A., 1995, Geologic map of Newberry volcano, Deschutes, Klamath, and lake counties, Oregon: U.S. Geological Survey Map I-2455.

Park Address

Newberry Volcanic National Monument
1230 N.E. Third Street, Suite A-262
Bend, Oregon 97701

EIGHT

Colorado Plateaus Province

The Four Corners, where Arizona, Utah, Colorado, and New Mexico meet, is near the center of the vast Colorado Plateaus Province, one of our most scenic provinces. The province was so named because most of the region (about 90 percent) is drained by the Colorado River and its main tributaries—the Green, Little Colorado, San Juan, and Virgin rivers. A few rivers in the High Plateaus Section on the west edge drain northward and then westward into the huge water trap known as the Great Basin Section of the Basin and Range Province. A small part on the east drains into the Rio Grande (Plate 8; Fig. 8–1).

With nearly 30 separate parks, the Colorado Plateaus contain the highest concentration of parklands in North America. Most of these special areas are noted for their spectacular scenery and geology—however, many also celebrate the fascinating culture and civilization developed by the Native Americans who settled and flourished in the area for thousands of years.

The Plateaus region is particularly significant because many important geologic concepts were developed here. Studying the excellent exposures of rock and contemplating how rivers evolve led to the recognition of new geologic ideas. The region is mostly arid or semiarid and is largely devoid of the vegetation and thick soil that obscure the rock record in more humid areas. Deep canyons and folds and faults have helped expose an unusually clear record of the earth's history developed over vast eons of time. An expedition led by Major John Wesley Powell, a geologist and one-armed Civil War veteran, used fragile wooden boats to explore the unknown depths of the Colorado River canyons in 1869 and again in 1872. Carefully observing the sequence and characteristics of the rock layers, Powell remarked in his journal that "the book is open and I can read as I run." In this chapter we too will read parts of the stories locked in the rocks and landforms of the Colorado Plateaus.

The Colorado Plateaus are rimmed on the north by the Uinta and Wasatch mountains (Middle Rocky Mountains), on the east by the Southern Rocky Mountains, and on the southeast, south, and west by the Basin and Range Province.

PLATE 8   Map of Colorado Plateaus showing park areas discussed in this chapter.

Lower plateaus have elevations near 5000 feet (1500 m) and the western part, the High Plateaus Section, contains eight lofty plateaus with elevations as high as 11,530 feet (3575 m). The High Plateaus are blocked out by north–south oriented normal (gravity) faults such as the Grand Wash, Hurricane, Sevier, and Paunsaugunt, which were activated during Tertiary time by the same tensional forces that produced the pull-apart faults in the nearby Basin and Range Province. The High Plateaus Section is in fact transitional, having the plateau topography of the Colorado Plateaus and the tensional faults similar to those in the adjacent Basin and Range. Significant volcanic fields are concentrated near its edges, and the interior contains such isolated igneous mountains as the La Sal, Abajo, Henry, and Navajo mountains (Fig. 8–1). These mountains are visible for over 100 miles (60 km) and make excellent landmarks for travelers. Broad domal uplifts like the San Rafael Swell and Monument Upwarp and basins such as the Uinta, San Juan, Paradox, and Kaiparowits also occur in the province (Fig. 8–2). Unusual folds called *monoclines* (a single flexure connecting relatively flat-lying strata) often form boundaries between adjacent plateaus. Salt-cored anticlines are prominent in the Canyonlands area. However, in most areas on the Colorado Plateaus the rocks are horizontal or very gently inclined.

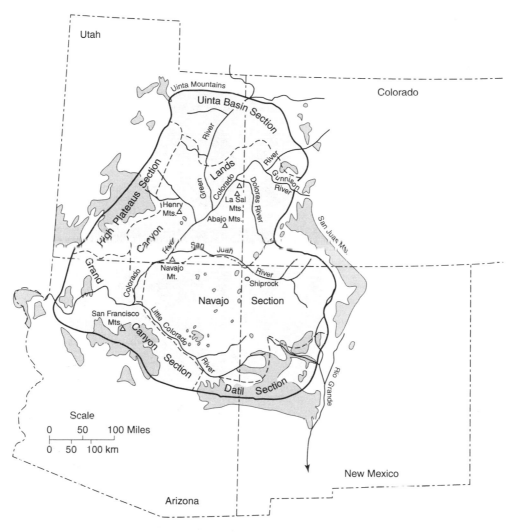

FIGURE 8–1   Generalized map showing selected features in and near the Colorado Plateaus. Darker shaded areas contain Cenozoic-age igneous rocks. (Modified from Hunt, 1974)

Mineral resources include locally abundant coal beds, oil and gas, and uranium. The uranium boom of the 1950s brought thousands of fortune seekers temporarily into the Colorado Plateaus, as described in Ringholz's (1989) book, *Uranium Frenzy*. As a result of the burgeoning world population and as more people acquire resource-intensive life-styles, the earth's oil and gas reserves will be rapidly depleted, and coal mining activity and the number of nuclear power plants will increase dramatically in the next few decades. Increased population and resource development will negatively impact not only the Colorado Plateaus, but earth environments everywhere.

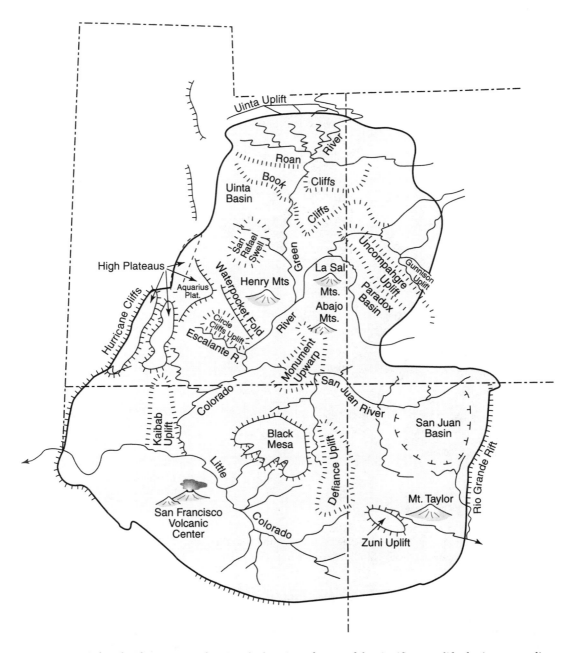

FIGURE 8–2　Colorado Plateaus map showing the location of some of the significant uplifts, basins, monoclines, and faults. (Modified from Hunt, 1974)

Fenneman (1931) divided the province into six sections (Fig. 8–1), including the High Plateaus Section previously discussed. The Datil section in the southeast has a thick lava cover of mid-Tertiary [about 40–26 Ma (millions of years ago)] and late-Cenozoic (10–0 Ma) age (Mutschler and others, 1998; Luedke and Smith, 1991). High structural uplift of the Grand Canyon section tilted the rocks gently to the northeast and enabled erosion to strip off most of the Mesozoic rocks in the Grand Canyon area. Thick resistant Mesozoic sandstones were eroded into cliffs that retreated northward into the Navajo and Canyonlands sections (Fig. 8–1). The resulting series of east–west trending escarpments extending northward from the Grand Canyon is called the "Grand Staircase." Each escarpment is named for the color of its dominant rock strata and is younger than rocks located immediately to the south. The Chocolate Cliffs (Triassic-age Moenkopi Formation and Shinarump Conglomerate) are succeeded by the younger rocks in the Vermillion Cliffs (Chinle Formation and Wingate Sandstone), the White Cliffs (Navajo Sandstone of Jurassic age in Zion National Park and elsewhere), the Gray Cliffs (Cretaceous-age Mesa Verde Group and equivalent strata, mostly sandstone in cliffs), and finally the Pink Cliffs (early Cenozoic Claron Formation, part of the Wasatch Group in Bryce Canyon and Cedar Breaks). The cross sections in Plates 8A, B, C, and D illustrate the relatively simple structure and stratigraphy in the Colorado Plateaus and suggest why this area has been called an area of "layer-cake geology."

It is a land of lonesome beauty where, from many vantage points, the mesas, plateaus, and canyons seem to extend to infinity. It is a magic land that inspires a feeling of vastness and wilderness that refreshes all, but especially those of us caught in more crowded living conditions and complicated life-styles. For those wishing a taste of a time in the not-too-distant past when the Colorado Plateaus were even less cluttered with the "improvements" of modern society, Edward Abbey's (1990) *Desert Solitaire* and David Lavender's (1956) *One Man's West* are highly recommended. It is our most colorful province; rocks of all colors—brilliant reds, salmon pinks, yellows, browns, grays, and white—are exposed, layer-cake fashion, in the high cliffs. Even the name Colorado (Spanish for red colored), although originally applied to the river because of the color imparted to the water by the large volume of sediment carried by its turbulent waters, suggests the brilliance of color associated with the entire region. Just as the Nile River in Egypt receives sufficient nourishment from its headwaters in central Africa to enable it to flow across the Egyptian desert, the Colorado River receives most of the water needed to flow across the Colorado Plateaus and Basin and Range deserts from its Rocky Mountain headwaters. Before Euro-Americans developed the West, the river even flowed into the Pacific Ocean!

This is desert country except for higher elevations and a narrow strip along the wetter canyon and valley bottoms. The region is in the Sierra Nevada rain shadow and receives only 6–16 inches (15–40 cm) of precipitation each year. The high plateaus, such as the Kaibab north of Grand Canyon and the High Plateaus Section of Utah, have the highest precipitation and are covered with pine, spruce, and fir forests. Lower elevations have pinyon pine, sage, and grass. With low amounts

GEOLOGIC CROSS SECTION OF THE GRAND CANYON –

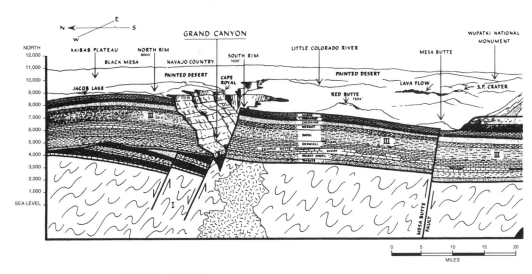

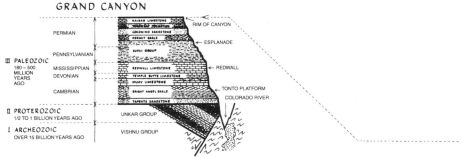

In the region between the Grand Canyon and the Verde Valley, earth history is revealed on a gigantic scale. As revealed in the bottom of Grand Canyon, two episodes of mountain building and erosion occurred during the Archeozoic and Proterozoic Eras from two billion to one-half billion years ago. From the beginnings of the Paleozoic Era to the end of the Mesozoic Era — five hundred million to sixty million years ago — this area was essentially a low-lying plain, sometimes submerged under the sea, at other times a flood plain crossed by sluggish rivers and on occasion a desert with blowing sand dunes. During this time period over 10,000 feet of sediment accumulated — rock present today at Grand Canyon and in the Zion Canyon region to the north and Black Mesa

(*a*)

PLATES 8*a,b*   Geologic cross section of the Grand Canyon–San Francisco peaks–Verde Valley region. (Zion Natural History Association)

SAN FRANCISCO PEAKS ~ VERDE VALLEY REGION

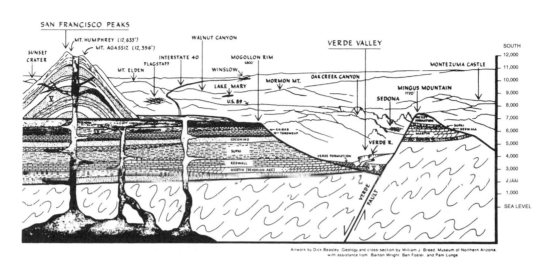

Artwork by Dick Beasley. Geology and cross-section by William J. Breed, Museum of Northern Arizona, with assistance from Barton Wright, Ben Foster, and Pam Lunge.

and the Navajo Reservation to the east.   In Cenozoic times from sixty million years ago to the present this region was uplifted, the Mesozoic rocks were removed by erosion and eventually canyons such as Grand Canyon and Oak Creek Canyon were formed. During the latter part of the Cenozoic, outpourings of lava built the volcanic field near Flagstaff, including the San Francisco Peaks. These peaks were later modified by glaciation. This area of the Colorado Plateau is only slightly disturbed by faulting and folding and is mainly underlain by horizontal beds — in great contrast to the Basin and Range regions south of the Mogollon Rim.

(b)

GEOLOGIC CROSS SECTION OF THE CEDAR

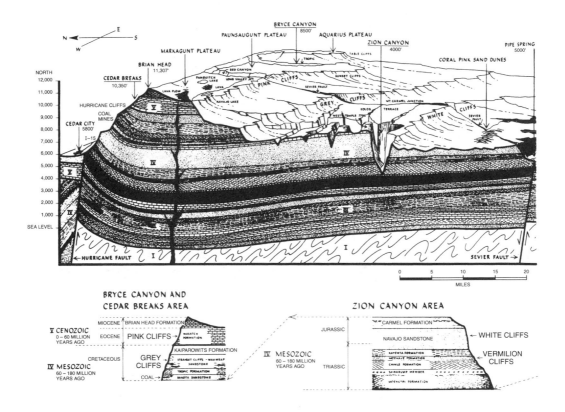

In this region, the forces of erosion have laid bare 1 billion 500 million years of earth history.   The oldest rocks, those of the Archeozoic, Proterozoic and Paleozoic, are found in the walls of the grand Canyon. The Mesozoic forms the temples and towers of Zion. The most recent, the Cenozoic, is exposed at Cedar Breaks and Bryce. Presumably all the layers of the Cenozoic and Mesozoic at Cedar Breaks and Zion once extended over the region of the Grand Canyon. The relentless wearing of the waters has stripped the layers back to the north forming the celebrated Great Rock Stairway of the Vermilion Cliffs, the White Cliffs, the Grey Cliffs and the Pink Cliffs.

(c)

PLATES 8c,d   Geologic cross section of the Cedar Breaks–Zion–Grand Canyon region. (Zion Natural History Association)

BREAKS — ZION — GRAND CANYON REGION

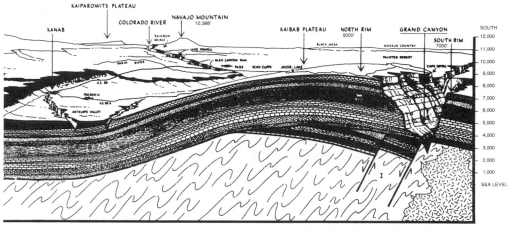

Artwork by Dick Beasley. From an original by Peter Coney, Middlebury College.
Revised by William J. Breed, Museum of Northern Arizona.

GRAND CANYON

Published by the Zion Natural History Association, Zion National Park, Springdale, Utah 84767,
in cooperation with the National Park Service, U.S. Department of the Interior.
ONE DOLLAR. © 1975 Zion Natural History Association.
Lithography by Northland Press, Flagstaff, Arizona.
ISBN 0-915630-02-8

(d)

of precipitation come not only sparse vegetation, but also bare rocks, and best of all, a small population of humans.

Trouble in Paradise

But the scene is changing. Wherever humans go, they always have an impact. Minimal change occurred when the Anasazi or Hisatsinom (Hopi word for "people who lived here long ago") inhabited the landscape from about 2000 to 700 years ago or when their descendants, the Pueblo people, grew their crops and hunted deer and other game. More intense use of the land followed as early Euro-American settlers brought livestock into the area. However, much more serious environmental change would soon occur. Now we see strip mines and coal-fired power plants springing up in various places; the Four Corners Power Plant near Shiprock, New Mexico, is fired with coal from the world's largest strip mine.

The air that was once clear is now polluted in large areas, including that in our national parklands. Depending on which way the wind blows, visibility at Grand Canyon, Zion, or even Canyonlands or Mesa Verde, as well as in other parks can be drastically reduced. To visit Grand Canyon and not be able to see the canyon bottom or barely see the opposite rim because of human-generated pollutants is more than a little disappointing. Those who came great distances to see one of the world's greatest spectacles could as well have visited a large city to "enjoy" the polluted air component of their park visit. Rangers are now using instrumentation to measure visual range and haze at many parks. The apparent trend toward decreasing visibility as measured since 1987 when measurements began is not encouraging. With over 300 sunny days each year, the Colorado Plateaus would be a logical place to locate solar-collector farms as we rapidly use up our petroleum fuels.[1] However, one of the trade-offs for renewable, nonpolluting solar power will be the visual pollution produced by row after row of solar collectors occupying many acres of land! Haze and polluted air greatly reduce the energy reaching a solar collector and increase certain health hazards.

How can this be possible? The Clean Air Act, passed in 1977 and reauthorized in 1990, specifies that the air over the parklands will be kept clean, restoring where necessary the visibility that has been impaired. Because essentially all of the pollution originates outside park boundaries, this law requires a general cleanup of the atmosphere. Although most Americans are strongly in favor of clean air, certain industrial interests are much less so. Amendments have been proposed that would render ineffective this law and other environmental laws that protect public health and public lands. Although the assaults have mostly failed so far, they did so by only small congressional margins or presidential vetoes. Pressure to reduce or eliminate regulations of all kinds and allow decisions to be made at more local levels is increasing and is hastening the attack on federal lands, including our parks, particu-

---

[1]Exhaustion of petroleum as a widespread energy source will occur in about the year 2050.

larly those in the West. These actions serve as warnings that the protection of public health, of our parks, national forests, and other public lands is not assured.

Construction of Glen Canyon and other dams has drastically changed downstream environments along the Colorado River and its tributaries. The 86 million tons of sediment washing through the Grand Canyon each year was suddenly reduced to 20 million tons after Glen Canyon Dam was finished[2] in 1963. The river that was "too thick to drink and too thin to plow" now runs like a clear mountain brook—its immense load of sediment dumped into the huge reservoir behind the dam. Baars (1972, p. 43) points out that "the river is now thin enough to drink in Grand Canyon and it will soon be thick enough to plow in Lake Powell above Glen Canyon Dam!" What an interesting "gift" we are leaving to our children and grandchildren.

Without the large sediment volume and the seasonal floods, sandbars and beaches disappear as the river washes sediment from the river edges into the channel. Spawning areas for endemic fish (only five of the eight original species remain) are greatly reduced, and the invasion of the remaining sandbar surfaces by the exotic saltcedar (tamarisk) tree has further reduced the native vegetation and wildlife. Without floods to rebuild the sandbars and remove the stubborn tamarisk, the character of the canyon floor is rapidly changing.

In 1996 scientists succeeded in convincing decision makers of the beneficial effects of flooding. The flood gates at Glen Canyon were opened for 12 days that spring. Rather than the 12,000–15,000 cubic feet per second (cfs) (340–425 m^3) flows that the dam operators would normally release, a larger flood discharge of 45,000 cfs (1275 m^3) was released during the experimental flood.[3] Sediment in the channel was redistributed to its edges, beaches grew larger, and 55 new sandbars were created—mostly in the first 20–48 hours! Similar restoration floods are recommended every 5–10 years.

Structural Development and Plate Tectonics

The enigma of a relatively large geomorphic province composed of mostly horizontal or nearly horizontal strata surrounded by mountains on all sides has long mystified geologists. How can it be that this "raft" of relatively stable crust is surrounded by a rolling sea of mountains? Intense tectonic activity occurred everywhere surrounding the Colorado Plateaus—but only relatively minor (but interesting) folding and faulting affected the Plateaus (Plates 8A, B, C, and D). Thus the Colorado Plateaus are a unique feature—an "island of tranquility" in a sea of tectonic upheaval. Also remarkable is the fact that the area was at sea level some 65 Ma and that today it locally stands over 11,000 feet (3350 m) high. A significant portion of the over 2 miles (3.2 km) of uplift occurred in geologically recent time—the past 5–10 million years.

[2]Most of the 20 million tons of sediment each year comes from the Little Colorado River, a major tributary downstream from Glen Canyon (Fig. 8–1).

[3]Floods of over 300,000 cfs (8500 m^3) occurred in predam days!

Structural features (mostly ancient faults) in the Precambrian basement rocks underlying the Colorado Plateaus played a major role in its later geologic history. For example, the west edge of the Colorado Plateaus is also part of a geologic zone called the Wasatch Line, which has significantly influenced geologic events in this part of the West for the past 2 billion years (Stokes, 1986). Part of the line to the north in the Middle Rocky Mountains currently forms the 225-mile-long (362-km) active frontal fault along the Wasatch Range. A series of parallel faults extends south from the Wasatch Fault, forming the boundaries between individual plateaus in the High Plateaus Section. The Uncompahgre structure in southeastern Utah and southwestern Colorado and the Uinta Basin area on the north edge of the Plateaus (Fig. 8–2) both follow structural weaknesses formed in Precambrian time. All of these features have affected subsequent mountain building and sedimentation. Other uplifts on the Colorado Plateaus are also likely reactivated geologic structures formed long ago in the Precambrian basement rocks (Baars, 1983, 1995).

The present topographic expression of the Colorado Plateaus began to take shape during the Laramide Orogeny (75–42 Ma). The eastward-migrating Cordilleran Orogeny that began in California during Jurassic time (about 200 Ma) was fueled by the collision of the North American and Pacific plates and perhaps later by the collapse of a thick *orogenic welt*[4] in Nevada. The compressional forces generated were destined eventually to form the Rocky Mountains. Apparently the lithospheric block beneath the Colorado Plateaus was too thick to deform into mountains but merely acted as a "middle man" and transmitted these stresses eastward to the Rocky Mountains where major thrust faults and anticlines formed. Reactivation of Precambrian basement structures in the Plateaus produced much smaller flexures (anticlines, synclines, and monoclines) and fault displacement during the Laramide. These structures played a major role in forming today's scenery.

Shifting of basement blocks beneath the Colorado Plateaus permitted hot mantle material to rise and partially melt, initiating igneous activity about 70 Ma that has continued intermittently up to the present. An episode of igneous activity during the middle Tertiary (about 40–26 Ma) produced the topographically high intrusive igneous mountains such as the La Sal, Henry, Navajo, and Abajo (Fig. 8–2) as well as numerous lava fields. Late Cenozoic volcanism continued intermittently, particularly along the western and southern margins of the Plateaus where *passive hotspots*[5] are still active (Mutschler and others, 1998). Holocene

---

[4]An orogenic welt is an area of thickened crustal rocks produced by mountain-building processes.

[5]A passive hotspot is a heated mass that moves with the lithospheric plate in which it is embedded. Extension or stretching of the lithosphere around the Plateaus' edges, mostly in Cenozoic time, caused hot mantle material to partially melt and rise toward the surface. In contrast, thermal plumes, such as the Hawaiian hotspot (Chapter 2), originate deep in the mantle (well below the lithosphere) and remain fixed in position as the lithospheric plate moves over its top.

FIGURE 8–3   Earthquake-generated landslide destroyed a number of houses near the entrance to Zion National Park in September, 1992. (Photo by E. Kiver)

eruptions, some of which are less than 1000 years old, occurred at Sunset Crater and El Malpais National Monuments and in the St. George area of Utah (Luedke and Smith, 1991).

Continuing tectonic adjustments, particularly along the western (High Plateaus Section) and southeastern (Rio Grande Rift) margins makes future earthquakes and volcanism here a certainty. The Hurricane, Sevier, and other faults along the High Plateaus margins[6] experience numerous minor to moderate earthquakes, many of magnitude 5 and 6. Stokes (1986, p. 13) notes that "in terms of destructive potential this belt is rated second only to the San Andreas Fault zone of California." Much larger earthquakes are likely in the distant future. The September 1992 magnitude 5.9 earthquake along the Hurricane Fault rattled residents from St. George to Zion National Park. The quake triggered a landslide that destroyed a number of houses (Fig. 8–3) in the Balanced Rock Subdivision just west of Springdale near the entrance to Zion National Park. Uplift and further tectonic adjustments continue today and will continue well into the future as the restless earth tends to its unfinished business.

---

[6]The margin of the High Plateaus section is part of the Wasatch Line, which in turn is part of the much longer Intermountain Seismic Belt—a major zone of faults and earthquake activity in the West.

Geomorphology

In addition to the extensive plateau or table-top-like surfaces, other major land-forms in the Colorado Plateaus include the canyons produced by the Colorado River and its tributaries, plateau edges and basins localized by fault scarps and folds, igneous mountains produced by both intrusive and extrusive processes, and cinder-cone-studded lava fields. The colorful sedimentary rocks brilliantly exposed along the cliffs and canyons in this desert climate contribute to the scenic panorama of one of our most dramatic geomorphic provinces. For those who have experienced first hand the breath-taking scenes and quiet beauty of the Colorado Plateaus, it is not surprising to learn that the province is home to nearly 30 National Park Service areas—the greatest concentration of parks in the United States.

In addition to the larger topographic elements, smaller landforms including rock arches, bridges, and *badlands* are present. The intricately eroded rocks (badlands) in Bryce Canyon and the Painted Desert badlands around Petrified Forest National Park are outstanding. In the Arches and Canyonlands area, parallel fractures (Fig. 8–4) have resulted in the formation of parallel drainage patterns which, in turn, have led to the creation of high rock walls, or *fins*. Later, as the fins are destroyed, features such as arches, standing rocks, and balanced rocks come into being.

FIGURE 8–4   Parallel fractures, common in parts of the Colorado Plateaus, are well developed in the Arches National Park area; view westward over fins and arches to Salt Valley, a graben. (Photo by National Park Service)

A large part of the province is underlain by sedimentary rocks that are mostly flat-lying; consequently, where one nonfaulted plateau rises to a higher one, the latter is capped by younger strata. Spectacular examples of this stair-step layering is encountered as one progresses northward from the Grand Canyon across the Grand Staircase. In a number of places vertical faults separate plateaus of different elevations, particularly in the western part of the province. Where basement faults have moved slowly and have moderate displacement, the overlying sedimentary rocks drape over the fault and produce a single flexure type of fold known as a monocline. This interesting type of fold is well displayed in Colorado National Monument, Capitol Reef National Park, Grand Staircase–Escalante National Monument, Grand Canyon, and elsewhere on the Colorado Plateaus.

Abrupt topography is the rule, with escarpments, or *scarps,* of resistant sandstone or limestone rising almost vertically from one plateau level to a higher one. In contrast, easily eroded shale generally forms gradual slopes that extend out from the base of the scarps or cliffs. Once a scarp forms, its fate is to erode and retreat headward. Thus, although the rates of erosion are slow by human standards, each plateau is getting smaller when many millenia are considered.

Individual plateaus reduce in size as blocks of rock fall from the cliffs, either pried loose by frost-wedging or by the removal of the underlying shale support by erosion. It would seem that there should be an abundance of fallen blocks—*taluses* perhaps—along the base of the cliffs. Talus deposits, however, are essentially nonexistent and even scattered boulders are rare. The scarcity of boulders prompted the assumption that active scarp retreat had ceased, perhaps when there was a change in climate. This assumption appears to be invalid. Schumm and Chorley (1964) found that (1) many sandstone boulders essentially disintegrate by impact and (2) others weather rapidly once they are partially buried in material that tends to hold moisture. Therefore, scarp retreat continues today although the boulders disappear about as rapidly as they are replaced by others falling from the cliff above.

As the extensive surfaces of plateaus reduce in size by scarp retreat or by widening of river and tributary valleys due to stream erosion, the original plateau is dissected into a series of smaller tablelands called *mesas* (Fig. 8–5). Continued erosion forms jagged-topped *buttes* (Fig. 8–6) or needlelike rocks called *pillars,* or *spires* (Fig. 8–7). Numerous Colorado Plateaus areas display these natural sculptures including Arches, Canyonlands, Canyon de Chelly, and Monument Valley. Broad *pediments* (gently sloping surfaces cut across bedrock in an arid climate) in Monument Valley and elsewhere extend out from the mesas and buttes, forming a new surface that may in turn be dissected in the future to form yet another generation of flat-topped landforms (Fig. 8–8). The Navajo Nation has done its part with the parks concept in setting aside the Monument Valley area as a Tribal Park. It is, in essence, a Native American national park.

The geomorphic history of the Colorado River presents some fascinating problems that have been debated for well over a century. The river and its tributaries flow across every major uplift on the plateau rather than flowing around

FIGURE 8–5   Monument Valley, southern Utah, with mesas and buttes standing above the pediment surface. (Photo by D. Harris)

FIGURE 8–6   The "Priest and Nuns," a butte located about 17 miles (27 km) northeast of Moab in the Colorado River valley. (Photo by E. Kiver)

FIGURE 8–7   Canyon de Chelly with Spider Rock, a pinnacle or spire (a type of butte) that rises about 800 feet (244 m) above the canyon floor. (Photo by D. Harris)

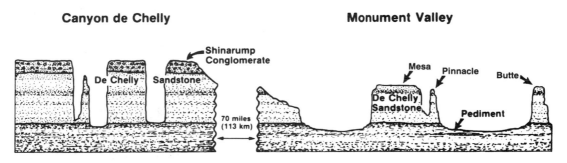

FIGURE 8–8   Generalized cross sections, showing early stage (left) of the arid cycle in Canyon de Chelly area, Arizona, and an advanced stage (right) in Monument Valley, Utah. Note how headward retreat of a plateau will eventually form a mesa that can further erode into a butte. (Illustration by Gregory Nelson)

these topographic and structurally high areas. How is it that the rivers on the Plateaus came to flow across uplifted areas rather than around their lower edges? Land did not simply "pop up" out of the ocean 65 Ma (end of Mesozoic) complete with a well-adjusted system of rivers and tributaries—particularly when land, even today, adjusts elevations as differential movements along faults and folds continue. The rivers no doubt changed courses and integrated their channels over time—exactly how and when are the areas of disagreement. The record is one of erosion—usually difficult to interpret. The sparsity of extensive, definitive, datable sediments makes parts of the proposed development models subject to debate. Early greats such as J.W. Powell (1875), G.K. Gilbert (1877), C.E. Dutton (1880), and W.M. Davis (1901) delved into the problem but did not solve it completely. Ideas involve an *antecedent* river (river predates the structure it crosses), *superposed* river (river cuts through sediment or rock that buries an earlier-formed structure), or *stream piracy* (headwater channel cuts through a drainage divide and "captures" the waters from another stream).

The evidence is clear that interior drainage existed in the Colorado Plateaus through at least early and middle Tertiary time (perhaps up to 10 Ma). According to Stokes (1986), headward erosion of streams produced stream piracy that forced the continental divide to gradually migrate to the northeast and the upper Colorado–Green River system to eventually integrate into nearly its present form. However, water continued to pond in Utah or perhaps Nevada or Arizona until a path to the Pacific Ocean was established. The connection between the lower Colorado River and the Pacific Ocean occurred about 5.5 Ma as indicated by the presence of Colorado River sediments in deposits of that age in the Gulf of California. By headward erosion the lower Colorado River later cut through the Kaibab Uplift (stream piracy again!), thereby connecting the upper and lower Colorado into one river system (Lucchitta, 1990). The increased water flow of the combined rivers enabled the channel to deepen rapidly.

By 1.18–1.16 Ma (Hamblin, 1990) the channel near Toroweap Point in the Grand Canyon was about 50 feet (15 m) above the present channel. Even to the casual visitor the evidence is clear and unmistakable. Lava from nearby vents cascaded down the north rim of the Grand Canyon 1.1 Ma and plunged into the river. What a turmoil the steam and red hot lava must have made! For more detailed discussions of these ideas and others see Baars (1983), Stokes (1986), and Lucchitta (1990).

The high elevations of some of the igneous peaks such as those in the La Sal Mountains and of the High Plateaus like the Aquarius (over 11,000 feet; 3350 m) in combination with the cooler and wetter climates of the Pleistocene promoted the local development of ice cap and valley glaciers, especially during the last (Wisconsin) glaciation. Hunt (1974) attributes the sparsity of older (pre-Wisconsin) glaciation to lower elevations of the Plateaus at the beginning of the Pleistocene (1.6 Ma). Uplift in the High Plateaus Section during the past few million years has been unusually rapid—as much as 1280 feet per million years (390 m per million years) according to Mayer and Condit (1991). Lucchitta (1990) believes that the entire region was similarly uplifted (nearly 0.6 mile, 0.9 km) during the past few million years. Such uplifts would not only place areas into climatic zones where glacier buildup is more likely, but would also further energize the streams and rivers to cut their canyons deeper.

Geologic Sequence

This highly generalized summary of the 2 billion years of earth history recorded in the Colorado Plateaus provides an overview of some of the highlights of the area's geologic evolution. Most of the individual pieces of the time puzzle are found in one or more of the parks covered later in this chapter. Summaries of parts of the province are available in the *Roadside Geology Series* (Chronic, 1980, 1987, 1990, 1993), in Baars (1995), Beus and Morales (1990), Hintze (1988), and in Stokes (1986). More complete summaries for the entire province are found in Baars (1983), Hunt (1974), Patton and others (1991), and Thornbury (1965).

Precambrian and Paleozoic

The Colorado Plateaus are more akin to the tectonically stable *craton* that forms the interior of the North American and other continents than to the highly deformed mountain areas that completely surround the province. Gently warped Precambrian basement rock overlain by a few thousand feet of mostly Paleozoic rock characterizes the North American craton in the conterminous United States from east of the Rockies to the Appalachians. Mesozoic and Cenozoic strata in turn overlie these older rocks just east of the Rockies in the Great Plains and along the southern craton margin in the Coastal Plain—much like conditions in the Colorado Plateaus.

Mountain ranges that formed along the margin of the Precambrian craton about 2 billion years ago were eroded to a flat plain (beginning about 1700 Ma) and covered by sediments during later Precambrian time (about 1250–1070 Ma). A range of fault-block mountains then formed that was in turn leveled by erosion before an early Paleozoic seaway encroached upon the passive western edge of the continent. In late Paleozoic time a branch of the Ancestral Rockies in Colorado and New Mexico curved northwestward into southwestern Colorado and southeastern Utah (the Uncompahgre Mountains). The estimated 12,000- to 15,000-foot (3660- to 4575-m) high Uncompahgre Uplift was accompanied by a correspondingly deep basin (the Paradox Basin) that filled with nearly 4 miles (6.4 km) of debris from the eroding mountains. The sediment fill contains a thick (6000- to 7000-foot; 1830- to 2130-m) evaporate deposit (the Paradox Formation containing salt beds) that plays an important part in the geology of Arches and Canyonlands national parks. Iron-rich minerals deposited in the late Paleozoic and Mesozoic stream, beach, and dune sediments were strongly oxidized—accounting for the spectacular reddish colors in the "red rock country" of the Colorado Plateaus.

Mesozoic

Mountain building to the west in eastern California, Nevada, and the western Utah region (Cordilleran Orogeny, which includes both the Nevadan and Sevier Orogenies) during the Mesozoic produced a rain shadow and a great inland desert—much like present conditions. Rivers wound their way into this vast Sahara-like landscape and dinosaurs roamed through the land. The last of the great shallow seas crept across the landscape, advancing and retreating, finally retreating for the last time from the province during the Late Cretaceous. The Laramide Orogeny was now in control (about 75 Ma) and began to significantly rearrange the crust to the east in what was soon to become the Rocky Mountains.

Cenozoic

Compared to the severe Laramide deformation in the Rockies, relatively mild structural changes occurred in the Colorado Plateaus. The thicker, cooler crust here was apparently rigid and was able to transmit stresses eastward into the Rocky Mountains without suffering severe folding and faulting itself. Except for the deep Uinta, San Juan, Piceance, and Kaiparowits basins, only broad anticlines, monoclines, and shallow basins formed. The province was encircled by mountains that shed water and sediment that became trapped in extensive landlocked lakes in the Colorado Plateaus Province during the early Tertiary (65–40 Ma). Collapsing and stretching of the thick crust to the west of the province during the middle Cenozoic began to create the tensional (normal) faults that characterize the Basin and Range Province and the west margin of the Colorado Plateaus. Intrusive igneous activity began as *stocks* (massive intrusions smaller than batholiths) and associated *laccoliths* (flat-floored, lens-shaped intrusions) were emplaced in the Henry Mountains and elsewhere. A number of cylindrical intrusions, such as that at Shiprock (Fig. 8–9) and elsewhere, were previously interpreted as the necks of eroded volcanoes. Now most are known to be *diatremes,* gas-charged releases of magma from the mantle that rapidly migrate to upper crustal levels. Those that originate in deeper levels of the mantle, such as those in South Africa, often bring high-pressure minerals, including diamonds, closer to the earth's surface. The apparent lack of diamonds in the Colorado Plateaus indicates a shallow mantle source for these diatremes.

With increasing elevation of the Colorado Plateaus and topographic collapse of the crust in the adjacent Basin and Range, rivers began to carry water away from the Plateaus. Streams began to integrate into drainage systems that remained landlocked until a path to the ocean was established about 5.5 Ma (very late Miocene). Headward erosion of streams produced further integration of stream systems and deepened the remarkable canyons for which the Plateaus are famous.

Intermittent volcanic activity along the western and southeast margins of the Plateaus during the past 10 million years produced significant lava fields (Luedke and Smith, 1991). Eruptions have persisted up to a few hundred years ago and can

FIGURE 8–9   Air view of Shiprock, a rapidly emplaced igneous intrusion of middle Tertiary age (about 30 Ma) near the Four Corners area. (Photo by Felix Mutschler)

be expected to continue into the future. Recent fault scarps and frequent earthquakes, especially along the High Plateaus Section and the Rio Grande margins, indicate that at least some of the dynamic processes that formed the Colorado Plateaus are still at work.

This chapter is loosely organized around the stratigraphic units found in each park. Recall that the province is an area of "layer-cake geology." With that analogy in mind we will begin to climb the stratigraphic staircase beginning with the lower layer or step (the Precambrian rock) and proceed upward through younger and younger layers until finally reaching the youngest—the frosting (Cenozoic equivalent). The most recent events are recorded in the topmost layers—the finger marks (if your house is like ours!) or gouges on the top of the cake. Our tour will begin with rocks and events about 1.8 billion years old in the Black Canyon of the Gunnison and in the depths of the Grand Canyon, proceed up the stratigraphic staircase through parks with mostly Paleozoic and Mesozoic layers exposed, and end with the Cenozoic record in Bryce, Cedar Breaks, and the 700- to 800-year-old lava flows at El Malpais and Sunset Crater. At Sunset Crater Sinagua Indians had to flee their homes as mighty eruptions in 1066 built one of North America's newest volcanoes and covered their homes with volcanic ash and dust. The enlarged maps (Plate 8 and Figs. 8–1 and 8–2) will help locate geographic features mentioned in the text as we tour through the fascinating region known as the Colorado Plateaus.

REFERENCES

Abbey, E., 1990, Desert solitaire: Simon and Schuster, New York, 269 p.

Baars, D.L., 1972, Red Rock Country: Doubleday, Garden City, NY.

Baars, D.L., 1983, The Colorado Plateau—a geologic history: Univ. of New Mexico Press, Albuquerque, 279 p.

Baars, D.L., 1995, Navajo Country: A geology and natural history of the Four Corners region: Univ. of New Mexico Press, Albuquerque, 255 p.

Beus, S.S., and Morales, M., 1990, Grand Canyon geology: Oxford University Press, New York, 518 p.

Chronic, H., 1980, Roadside geology of Colorado: Mountain Press, Missoula, MT, 322 p.

Chronic, H., 1987, Roadside geology of New Mexico: Mountain Press, Missoula, MT, 255 p.

Chronic, H., 1990, Roadside geology of Utah: Mountain Press, Missoula, MT, 325 p.

Chronic, H., 1993, Roadside geology of Arizona: Mountain Press, Missoula, MT, 321 p.

Davis, W.M., 1901, An excursion to the Grand Canyon of the Colorado: Harvard Museum Comparative Zoology Bulletin, v. 38, p. 106–201.

Dutton, C.E., 1880, Report on the geology of the High Plateau of Utah: U.S. Geographical and Geological Survey of the Rocky Mountain Region (Powell) 32.

Fenneman, N.M., 1931, Physiography of western United States: McGraw-Hill, New York.

Gilbert, G.K., 1877, Report on the geology of the Henry Mountains: U.S. Geographic and Geologic Survey of Rocky Mountain Region.

Hamblin, W.K., 1990, Late Cenozoic lava dams in the western Grand Canyon, in Beus, S.S., and Morales M., eds., Grand Canyon geology: Oxford University Press, New York, p. 385–433.

Hintze, L.F., 1988, Geologic history of Utah: Provo, Utah, Brigham Young University Geology Series, Special Publication 7, 202 p.

Hunt, C.B., 1974, Natural regions of the United States and Canada: Freeman, San Francisco, 725 p.

Lavender, D., 1956, One man's west: Doubleday, Garden City, NY, 316 p.

Lucchitta, I., 1990, History of the Grand Canyon and of the Colorado River in Arizona, in Beus, S.S., and Morales, M., eds., Grand Canyon Geology: Oxford University Press, New York, Ch. 15, p. 311–332.

Luedke, R.G., and Smith, R.L., 1991, Quaternary volcanism in the western conterminous United States, in Morrison, R.B., ed., Quaternary nonglacial geology; Conterminous U.S.: Geological Society of America, Boulder, CO, The Geology of North America, v. K-2, p. 75–92.

Mayer, L., and Condit, C.D., 1991, Quaternary tectonics and erosion along the western margin of the Colorado Plateau, in Morrison, R.B., ed., Quaternary nonglacial geology: Conterminous U.S.: Geological Society of America, Boulder, CO, The Geology of North America, v. K-2, p. 379–381.

Mutschler, F.E., Larson, E.E., and Gaskill, D.L., 1998, The fate of the Colorado Plateau—A view from the mantle, in Friedman, J.D., and Huffman, A.C., Jr., coordinators, Laccolithic complexes of southeastern Utah: Time of emplacement and tectonic setting—Workshop proceedings: U.S. Geological Survey Bulletin 2158, p. 203–222.

Patton, P.C., Biggar, N., Condit, C.D., Gillam, M.L., Love, D.W., Machette, M.N., Mayer, L., Morrison, R.B., and Rosholt, J.N., 1991, Quaternary geology of the Colorado Plateau, in Morrison, R.B., ed., Quaternary nonglacial geology; Conterminous U.S.:

Geological Society of America, Boulder, CO, The Geology of North America, v. K-2, p. 373–406.

Powell, J.W., 1875, Exploration of the Colorado River of the west and its tributaries: Smithsonian Institute Annual Report 291.

Ringholz, R.C., 1989, Uranium Frenzy: Norton, New York, 310.

Schumm, S.A., and Chorley, R.J., 1964, The fall of threatening rock: American Journal of Science, v. 262, p. 1041–1054.

Stokes, W.L., 1986, Geology of Utah: Utah Museum of Natural History, University of Utah, Occasional Paper No. 6, 280 p.

Thornbury, W.D., 1965, Regional geomorphology of the United States: Wiley, New York, 609 p.

BLACK CANYON OF THE GUNNISON NATIONAL MONUMENT (COLORADO)

The Black Canyon of the Gunnison River is located in western Colorado in the transitional zone between the mostly flat-lying topography of the Colorado Plateaus and the rugged topography of the Southern Rocky Mountains (Plate 8, Fig. 8–1). Numerous canyons bless parts of our American landscape, but no other canyon combines the narrow opening, sheer walls, and depths of the Black Canyon. Here the Gunnison River has cut such a deep, narrow canyon through dark, foreboding crystalline (mostly metamorphic) rocks that it appears rather gloomy, hence the name Black Canyon (Figs. 8–10 and 8–11).

Prehistoric Indians and later the Utah Indians (Utes) visited and hunted along the canyon rim but apparently did not venture into its depths. Early explorers missed seeing the canyon, including John W. Gunnison! In 1853 Gunnison was scouting for possible routes for a transcontinental railroad. He detoured around the canyon while looking for a river crossing—reasoning no doubt that the canyon was not a promising route for a railroad! It was not until the 1873–1874 Hayden expedition that the first of the new Americans experienced this startling canyon. Later surveying parties declared that the canyon was inaccessible. Local citizens, including the Reverend Mark T. Warner, strongly promoted protection for the area and were instrumental in molding its future. Clearly, Black Canyon belongs in the National Park System and it was so designated by President Herbert Hoover on March 2, 1933.

Establishing the River Course

The Gunnison River heads in the volcanic peaks of the West Elk Mountains and flows westward into the Colorado Plateaus directly along the axis of a northeastward-tilted fault block (the Gunnison Uplift), which was uplifted during the Laramide Orogeny and during late Tertiary and Quaternary time (Hansen, 1987b). As is common in the Colorado Plateaus, the river flows across rather than around a

FIGURE 8–10   View of the Black Canyon and the Gunnison River. (Photo by D. Harris)

significant topographic and structural feature in a *transverse valley*. As discussed in the chapter introduction, such a valley might be explained by the processes of antecedence, *superposition,* or stream piracy. Here the explanation involves superposition with a unique twist related to volcanic activity. The Gunnison River isn't where it used to be; until middle Tertiary time it flowed westward several miles to the north of Black Canyon—where the West Elk Mountains are now. Nothing is more diverting to a stream than to have a volcano burst out in the middle of its valley! That is precisely what happened to the Gunnison River when the first of the West Elk volcanoes began to erupt.

Thick sheets of volcanic debris shed from the growing West Elk volcanoes buried the lower elevations, including the beveled remains of the Gunnison Uplift. Similar eruptions to the south in the San Juan Mountains trapped the river between the two piles of volcanic materials from these volcanic centers. The ances-

FIGURE 8–11   Aerial view of the foreboding Black Canyon. (National Park Service photo)

tral Gunnison River cut downward only to be overwhelmed and pushed around by layer after layer of volcanic sediment, lava, and later on up to 600 feet (183 m) of ash-flow material (welded tuff). The last volcanic activity to affect the river was the outpouring of a series of basalt flows during the Miocene about 18.5 Ma (Hansen, 1987).

A shallow *syncline* (downfold) that formed on top of the buried Gunnison Uplift produced a trough-like structure that provided a natural pathway for the ancestral Gunnison River to follow (Hansen, 1987b). The story from here goes "downhill." The Gunnison River had a lot of "catching up" to do. The Colorado River canyons and valleys had deepened during the time when the West Elk and San Juan volcanoes were building a new landscape. Also, late Tertiary regional uplift of the Rocky Mountains and Colorado Plateaus combined with the establishment of a through-flowing Colorado River and the cutting of the Grand Canyon further encouraged tributary rivers, such as the Gunnison, to deepen their valleys. Now the Gunnison was over 8000 feet (2440 m) above sea level and a few thousand feet above its *local base level,* the Colorado River. Thus the energy of the flowing waters was greatly increased and the turbulent flow of water and the violent bashing and grinding of loose rock against the bedrock floor of the channel deepened the canyon. Vertical cutting greatly exceeded widening thereby producing one of North America's most spectacular canyons.

Black Canyon Rocks

The gorge is about 2700 feet (823 m) deep (twice as high as the Empire State Buildings) in one place; at the Narrows, it is 1750 feet (534 m) deep, but at this point it is only 1100 feet (335 m) wide at the rim and a mere 40 feet (12 m) wide at the bottom! Clearly, in order to maintain nearly vertical walls of this height, the rocks must be both mechanically strong and resistant to weathering. All of the rocks in the canyon walls are Precambrian, and they were formed deep in the earth's crust as the roots of mountains. These are *high-grade* metamorphic rocks—rocks that only form at high temperature and pressures equivalent to the weight of 10–20 miles (16–32 km) of rock. They are mainly gneisses and quartz-mica schists, with lesser amounts of black hornblende schist and amphibolite. Granitic magmas later intruded to form irregular masses as well as dikes and sills (Fig. 8–12), many of which interlace to buttress the metamorphics, further strengthening them.

FIGURE 8–12   Granite dikes and sills (light colored) that cut through the metamorphic rocks in the Painted Wall. The 2250-foot-high (686-m) wall is the highest cliff in Colorado. (National Park Service photo)

Similar-appearing Precambrian rocks occur deep in the Grand Canyon, in Colorado National Monument, in the cores of many of the ranges in the Rocky Mountains, and underlying the stable interior (craton) of the North American and other continents. They record part of the earth's early history when geosynclines and exotic terranes accreted onto the edges of the growing continents. All that remain of the numerous majestic mountain ranges of Precambrian time are their roots—the once deeply buried gneisses, schists, and other metamorphic rocks. The oldest rocks at Black Canyon, the gneisses and schists, produce radiometric ages of 1.7 billion years, the time of the last metamorphic event. The sandstones and shales from which the gneisses and schists were derived are yet older, perhaps as much as 2 billion years old. The granites intruded the metamorphic rocks and must, of course, be younger. Radiometric ages of the granites are 1.4–1.5 billion years (Hansen, 1987a and b).

A short distance back from the rim, the Jurassic-age Entrada Formation rests directly on the Precambrian, with a major unconformity between them. Younger Mesozoic and Tertiary sedimentary rocks exposed near the margins of the Gunnison Uplift are in turn locally covered by the West Elk volcanics.

Black Canyon History

By using the rock record at Black Canyon and surrounding areas, geologists have pieced together the last 2 billion years of earth history in this part of the Colorado Plateaus. The following brief summary is further abbreviated in Table 8–1. More detail is available in the excellent publications by Hansen (1965, 1987a and b) and Baars (1983).

The record begins with deposition of sediments and volcanic rocks about 2 billion years ago. The formation of primeval mountains produced high-grade metamorphism at depth and magmatism, forever changing the appearance of the original rocks. Major faults sliced through the ancient terrain—faults that would be reactivated and play an important role in the area's later geologic history. A very long period of erosion leveled the mountainous topography before early Paleozoic seas inundated the area.

The sandstones, shales, and limestones of early Paleozoic age were removed later by erosion when crustal stresses reactivated the ancient Precambrian faults and formed the Ancestral Rockies in the Colorado and New Mexico area. A northwest branch (late Paleozoic Uncompahgre Uplift) of this massive range extended through the Black Canyon area following the trend of the ancient faults in the Precambrian basement rocks below. A long time was required to reduce the uplifted area to sea level in the Black Canyon area. The sea finally encroached onto the area again in Jurassic time depositing the beach and intertidal sediments of the Entrada Formation that are exposed in the park near the northwest rim of the Black Canyon. Marine sediments continued to bury the Black Canyon area as well as much of interior North America until the crust again began to stir during Late Cretaceous time.

TABLE 8–1   Geologic History—Black Canyon of the Gunnison

GEOLOGIC ERA	SIGNIFICANT EVENT
Cenozoic	Canyon deepened as Colorado Plateaus uplift
	River superimposed onto buried Gunnison Uplift
	Canyon area buried by sheets of volcanic debris
	Middle to late Tertiary volcanism pushes river to south
	Erosion levels Gunnison Uplift
	Uncompahgre and Gunnison structural uplift, monoclines form
	Early Tertiary lake sediments accumulate
Mesozoic	Last of vast shallow seas retreat
	Laramide Orogeny begins about 75 Ma
	Jurassic seas deposit Entrada and other formations in canyon area
	Mountains eroded to sea level
	Long interval of erosion
Paleozoic	Late Paleozoic Uncompahgre Uplift (ancestral Rockies)—early Paleozoic rocks eroded away
	Early Paleozoic sediments (mostly marine) deposited
Precambrian	Erosion levels mountains
	Granitic rocks intrude (about 1400 Ma)
	Mountain building, metamorphism (about 1700 Ma)
	Sandstones, shales, volcanic rocks deposited

The huge inland sea began to withdraw from the Black Canyon area during latest Cretaceous time (about 75 Ma) as the last phase of the eastward-migrating Cordilleran Orogeny began to affect the Rocky Mountains (the Laramide Orogeny). Water ponded in huge lakes in the Colorado Plateaus during the early Tertiary. Again the ancient Precambrian faults moved and a newer Uncompahgre Uplift formed in the southeastern Utah–southwestern Colorado area as a result of Laramide crustal stresses. The Gunnison Uplift (a part of the Uncompahgre Uplift) bowed the overlying strata upward and erosion began its relentless but unfulfilled task of removing all land above sea level.

The early Tertiary lakes drained as rivers began to flow into the Basin and Range Province to the west during the middle Tertiary (about 20 Ma). The ancestral Gunnison River began to establish its course only to have it interrupted time and time again as the erupting West Elk volcanoes nearby and the San Juan volcanoes to the south forced the river to change its course. Thick sheets of volcanic debris buried the lower elevations and transformed the area into a sloping plain where the river began to establish its present course. At first the river cut slowly through the Tertiary volcanic materials and the Mesozoic rock before encountering the hard Precambrian rock in the core of the Gunnison Uplift—an example of the process of superposition as discussed in the chapter introduction. The river was now imprisoned by its resistant valley walls and continued its downward incision. If the river had been shifted a few miles farther to the south, beyond a major fault, it would have cut its valley into weak sedimentary rocks—forming a valley like many others, one entirely unlike the Black Canyon.

The regional uplift of the Colorado Plateaus in the past few million years energized the flow of rivers and maintained or increased their rates of downcutting. The formation of glaciers in the West Elk Mountains assisted the canyon-cutting process by supplying swollen meltwater rivers with abundant gravel and sand to help grind, batter, and lower the bedrock channel.

Because of the depth of the canyon and the superior resistance of the rocks, it was once assumed that canyon cutting required many millions of years. Now that the sequence of events is known, it is obvious that canyon cutting did not begin until well after most of the paroxysms of the West Elk volcanoes had died down and sometime before local glacier meltwater cut through late Tertiary gravels, perhaps 2 Ma. Indeed, Hansen (1987b) determined that the canyon was about 1200 feet (430 m) deep about 1.2 Ma and that another 1200 feet of canyon cutting has occurred since—about one foot (30 cm) every thousand years. Thus, much if not all of the Black Canyon was probably cut within the last 2 million years.

Humans have, once again, temporarily changed the natural processes, including the rate of downcutting of the Gunnison River. The 53-mile-long (85-km) canyon along the Gunnison River now has three dams upstream from Black Canyon in the Curecanti National Recreation Area. The river still flows free and is still deepening its canyon through the 12-mile (19-km) stretch of river in the national monument, but not with the energy or downcutting rate of a few decades ago. Flows during the spring are still impressive, but not as violent as the 12,000 cubic feet (340 m^3) per second floods in the predam days. However, Hansen (1987a) cites tongue-in-cheek evidence that the canyon is still deepening. He notes that through the years, each time he descends to the canyon floor he finds that the climb out is invariably "steeper and more arduous than the time before!"

Black Canyon is not a large monument—only about 22 square miles (57 km^2)—but not all of the choice park areas are large. The canyon overlooks are accessible by automobile and trails in the warmer months and by snowshoes or cross-country skis in the winter. Those in excellent physical condition and with the necessary backcountry skill wishing to explore the canyon depths should discuss their plans with a park ranger.

REFERENCES

Baars, D.L., 1983, The Colorado Plateau—a geologic history: University of New Mexico Press, Albuquerque, 279 p.

Hansen, W.E., 1965, The Black Canyon of the Gunnison: Today and yesterday: U.S. Geological Survey Bulletin 1191, 76 p.

Hansen, W.E., 1987a, The Black Canyon of the Gunnison, Colorado: Geological Society of America Centennial Field Guide, Rocky Mountain Section, Geological Society of America, Boulder, CO, p. 321–324.

Hansen, W.E., 1987b, The Black Canyon of the Gunnison; in depth: Tucson, Arizona, Southwest Parks and Monuments Association, 58 p.

Park Address

Black Canyon of the Gunnison National Monument
c/o Curecanti National Recreation Area
102 Elk Creek
Gunnison, CO 81230

GRAND CANYON NATIONAL PARK (ARIZONA)

"Do nothing to mar its grandeur—keep it for your children, your children's children, and all who come after you" President Theodore Roosevelt said, adding that Grand Canyon was "the one great sight which every American should see" (Fig. 8–13). In 1908 he set it aside as a national monument to protect it from exploita-

FIGURE 8–13   View across Grand Canyon, John Wesley Powell's "Book of Geology." Kaibab Limestone caps the Kaibab Plateau on the north rim. (Photo by D. Harris)

tion until the time Congress could act. Grand Canyon, an area of about 1100 square miles (2850 km²), was upgraded to national park status in 1919, primarily because both the scenery and the geologic history exposed in the canyon are unparalleled. Certain geologic events are recorded only in the area downstream from the park, and in 1932 this adjacent area of almost 310 square miles (106 km²) was established as Grand Canyon National Monument. Later, in 1969, about 41 square miles (106 km²) upstream from the park became Marble Canyon National Monument. In 1975, these two monument areas were added to the park; thus, Grand Canyon National Park consists of the canyon country along a 278-mile (448-km) stretch of the Colorado River, from the southern terminus of Glen Canyon National Recreation Area (Lake Powell) down to the eastern boundary of Lake Mead National Recreation Area (Fig. 8–14).

From Lake Powell to Lake Mead the river flows past seven individual plateaus—each separated from its neighbor by either a fault or a monocline. Access to the park is on paved roads that funnel visitors to the North and South Rims in the east part of the park and bumpy, unimproved roads that provide access to the Toroweap-Tuweap area in the west side of the park. Lack of roads elsewhere maintains the wilderness character that helps make this area special. The ultimate way to see the canyon, however, is by boat. Those having the skills and equipment or those willing to spend a few days with a professional river guide are in for an outstanding adventure.

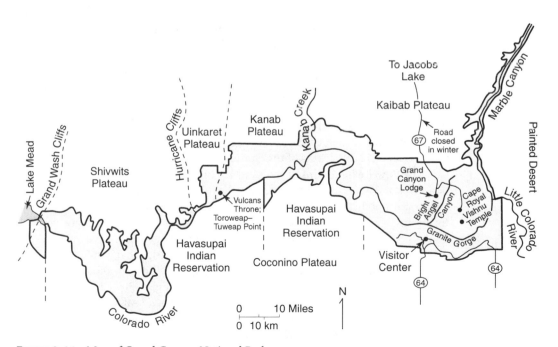

FIGURE 8–14   Map of Grand Canyon National Park.

The maximum depth from rim to river is 6000 feet (1829 m), and from Cape Royal on the North Rim to Zuni Point on the South Rim it is only 7.5 miles (12 km) across. The elevation of the heavily visited South Rim is 6000–7500 feet (1829–2286 m). The less visited North Rim is 1000–1200 feet (305–365 m) higher. Forests of juniper and pinyon pine cover the South Rim and a ponderosa, fir, and aspen forest covers the cooler, wetter North Rim. The South Rim is accessible year round but the North Rim closes from late October to mid-May.

History

Native Americans have trod through the depths of the canyon for at least the last 10,500 years. Over 3500 known archeological sites exist in the park including Stanton's Cave where intriguing split-willow animal figures were fashioned by a long-vanished, pre-Anasazi culture. The mysterious Anasazi or Hisatsinom ("people who lived here long ago") followed only to abandon the canyon and many other traditional homes on the Colorado Plateaus about 700 years ago. More recently, Paiute, Hualapai, Navajo, and Hopi peoples inhabited the surrounding plateaus. Indians still have a hold on the canyon. The Havasu Village, in the western part of the enlarged park, is probably one of the oldest settlements of continuous occupation in the conterminous United States. Its inhabitants live in much the same way that their ancestors did for centuries.

Europeans first viewed the canyon in 1540 when a group of 13 Spaniards under Captain Don Lopez de Cardenas (one of Coronado's officers) was directed there by a group of Hopi Indians. The soldiers were unable to descend the formidable limestone and sandstone cliffs in their path.

Geologic observations were first made by Jules Marcou in 1856 as part of the Pacific Railroad Survey. A study of the Paleozoic stratigraphy was made by John Strong Newberry in 1857–1858. Newberry was a pioneer geologist who accompanied the Ives military expedition searching for railroad routes to the Pacific Coast. Newberry's report helped encourage fellow geologist John Wesley Powell to attempt to explore the depths of the Grand Canyon.

The Powell expeditions of 1869 and 1871 involved 1000-mile-long (1600-km) boat trips and were undoubtedly among the most significant in advancing geologic exploration of the West. Why anyone would even consider plunging into such a forbidding canyon, without any maps or other information—in fragile wooden boats—is almost beyond comprehension. Major Powell was a Civil War veteran who lost his right arm in the battle of Shiloh (Fig. 8–15). What would have meant the end of an active career for many apparently spurred him on, and he became the country's most daring and tireless exploration geologist. Powell saw the Grand Canyon as the "book of geology." In his diary he wrote: "All about me are interesting geologic records, the book is open and I read as I run."

Powell left Green River, Wyoming, on May 24, 1869 with four boats and nine men. He quickly lost one boat and a significant part of his supplies, making the last

FIGURE 8–15 Major John Wesley Powell, first geologist to explore and interpret the canyons of the Colorado. (Photo courtesy of National Park Service)

part of the trip particularly arduous. After successfully negotiating most of the difficult rapids, three men left the expedition at what is now called Separation Canyon. They walked out only to be killed by Shivwits Paiute Indians who mistook them for miners who had recently molested a squaw.

Powell became a national hero and in 1881 the second director of the prestigious government scientific organization known as the U.S. Geological Survey. Later work by G.K. Gilbert, C.E. Dutton, E. McKee, and many others pieced together nearly 2 billion years of earth history—almost one half of the earth's existence—as revealed in the rock walls and landforms of the Grand Canyon.

Senator Benjamin Harrison unsuccessfully introduced a bill in 1887 to establish a national park at the Grand Canyon. However, later as president, he established the Grand Canyon Forest Preserve—which still left the area open to exploitation by mining and lumber companies. President Theodore Roosevelt visited the area in 1903 and was so impressed with its superb scenery and scientific potential that he established monument status in 1908—beginning the creation of one of the world's most popular parks.

The unfolding of the geologic history of the Grand Canyon region is presented in intriguing and understandable fashion by Shelton (1966); his book *Geology Illustrated,* replete with pictures and reconstructions, is a classic and well worth reading. Beus and Morales's (1990) *Grand Canyon Geology* is an excellent collection of articles on most phases of canyon geology, and Baars's (1995) *Navajo*

Country presents a good summary of Grand Canyon geology. For color, and for a collection from Powell's diaries, Porter's (1969) *Down the Colorado* is suggested.

With apologies to this remarkable region, only a brief geological summary can be included here. Recorded in the "book of rocks" in the Grand Canyon is a physical and paleontological record of our changing earth covering nearly 2 billion years of time. Erosion has removed some of the pages here, but nearby these pages remain intact—thus enabling geologists to piece together, step by step, the empirical record of our changing earth and its inhabitants.

Mountains of Grand Canyon

The mountains here are of two general types—the Vanished Mountains of Precambrian time and the Mountains of the Future. Two episodes of late Precambrian (Proterozoic Eon, 2500–550 Ma) mountain building, each followed by long periods of erosional leveling, are recorded here. The record of these Vanished Mountains is exposed only in the canyon depths where erosion has cut completely through the sequence of overlying Paleozoic strata. The Mountains of the Future are beginning to be formed now—down in Grand Canyon.

The older Proterozoic mountains were high and extensive, but their precise dimensions may never be known. Radiometric dating indicates that volcanic and marine sedimentary rocks were swept against and accreted onto the edge of the enlarging North American continent from at least 1800–1600 Ma (Ilg and others, 1996). Intense metamorphism recrystallized the rocks in the core of the forming mountain range into the Vishnu schist and other metamorphic rock units. The metamorphosed rocks were in turn intruded by granitic rocks up to as recently as 1660 Ma. Lenses of marble in some of the metamorphic rocks were probably formed by colonies of algae—some of the only known life forms that lived during ancient Precambrian times.

An exceptionally long episode of erosion, perhaps 500 million to 1 billion years, leveled the once-majestic range to a low, flat surface with only small hills a few tens to a few hundreds of feet high (Hendricks and Stevenson, 1990). The deep *crystalline* core rocks (igneous and metamorphic rocks such as the Vishnu Schist) were exposed at the earth's surface by erosion when deposition of the Grand Canyon Supergroup rocks began about 1250 Ma. Marine basins extended along the edge of the ancient North American continent from at least Lake Superior to Glacier Park in Montana to the Uinta Mountains and Grand Canyon. These large rift basins formed as plate tectonics pulled a large plate or plates away from North America during late Precambrian time (Ford, 1990). Powell called the buried erosion surface the "Great Unconformity," one of the best exposed *nonconformities*[7] in the world.

---

[7]A nonconformity is a type of unconformity where bedded rocks overlie crystalline rocks such as metamorphic and plutonic rocks.

A thickness of well over 2 miles (3.7 km) of sediments and lavas was deposited as part of the Grand Canyon Supergroup in a shallow seaway from about 1250 to 1070 Ma. Shales, some brilliant red, are the dominant rocks, but more significant perhaps are the limestones because they contain "heads" of colonial algae—similar in age to those in the Siyeh Limestone in Glacier National Park. Although these Precambrian rocks have been combed carefully by Walcott (1895) and many others, no unquestioned multicellular fossils have been found (Ford, 1990). Interesting markings and impressions in the sediment surfaces are of uncertain origin and may have been produced by ancient, soft-bodied invertebrates. These rocks are best seen along the river or from the North Rim; they are well exposed in the upstream end of the park in the Inner Gorge.

The Grand Canyon Orogeny occurred about 800 Ma and reactivated older Proterozoic faults to form a new range of mountains—this time fault-block mountains. Erosion attacked the younger Proterozoic mountains, eventually forming another *peneplain* or erosion surface. The erosion surface cut across the upturned edges of the sedimentary and lava layers of the Grand Canyon Supergroup as well as the crystalline Vishnu Schist and other Precambrian rocks. Here and there *monadnocks* of hard rocks such as quartzite extend into the overlying Paleozoics (Fig. 8–16); otherwise the peneplain was a strikingly smooth surface, as we saw in the Black Canyon of the Gunnison and will encounter again in Colorado National Monument. Where the flat-lying Paleozoic rocks bury the truncated edges of the Grand Canyon Supergroup, a type of unconformity called an *angular unconformity* (Fig. 8–16) is present. By placing one's hand along the sharp line that is the

FIGURE 8–16   Sedimentary strata exposed in the Grand Canyon of the Colorado River record a span of a billion years of earth history. Note that the wavy contacts in the stratigraphic column to the left of the photograph represent unconformities. (Photo by U.S. Geological Survey)

unconformity, one can simultaneously touch rocks that are 1 billion years old and overlying rocks that are 540 millions of years old! About 460 million years of rock pages are missing from the geologic history book along this unconformity.

The Precambrian metamorphic rocks and granite of the Inner Gorge are much more resistant to weathering and erosion than the overlying mile-thick Paleozoic sedimentary rock section. As the river cut deeper and deeper, the walls became progressively higher, but the rocks of the Inner Gorge, being much stronger, were better able to maintain steep, in places near-vertical, cliffs.

Sequence of Events—Paleozoic Rocks

Many visitors are intrigued by the stair-step-like character of the Grand Canyon above the Inner Gorge (Fig. 8–13). Why aren't the sideslopes relatively uniform from the rim down to the river? The steepness of the slopes is a direct reflection of rock resistance. In this dry climate, chemical weathering is slow, and limestones such as the Kaibab and the Redwall maintain near-vertical cliffs; in later chapters we will see that in humid regions, limestones weather rapidly and tend to form valleys and other lowlands.

Shales weather readily in any climate, forming the slopes that extend out from the base of limestone and sandstone cliffs. The thick Bright Angel Shale atop the hard Tapeats Sandstone (Cambrian) forms the broad, flattish slopes known as the Tonto Platform above the Inner Gorge (Plate 8A and Fig. 8–13) in the eastern Grand Canyon. The canyon appears quite different to the discriminating eye in the less-visited western Grand Canyon. Here the dominant topographic platform is developed on the stratigraphically higher (Permian) Hermit Shale atop the hard sandstone of the Supai Formation. The resulting erosional surface here is called the Esplanade (Plate 8A; Fig. 8–16). The excellent exposures of rocks in the canyon walls enable geologists to follow individual rock layers without interruption for nearly 200 miles. In this distance rock layers often experience lateral or *facies* changes. In this case the Supai Formation contains more resistant sandstones in the western Grand Canyon, thus accounting for the Esplanade. Also, the river is west of the Kaibab Uplift and has to cut still deeper to reach the Precambrian-Cambrian rocks and widen its valley to develop a "Tonto surface" in the western Grand Canyon.

To detail the geological history of the Grand Canyon would fill a book such as Beus and Morales's (1990) *Grand Canyon Geology*. Here only time-lapse glimpses of the history are given. During the Paleozoic Era western North America was on the *trailing edge* or *passive margin* of the moving plate. It was also located on or near the equator and experienced warm, mild climates. The geologic history involved mostly slow encroachment (*transgression*) of the seas over the continent, oscillation of shoreline positions, and retreat (*regression*). Transgressions are marked by a distinct sequence of layers reflecting increasing water depth. Sandstones often mark beach and near-beach environments, shales (mud-size sediment) indicate slightly deeper water offshore, and limestones form in relatively sediment-free, warm-water environments. Stacked sediments such as the Tapeats Sandstone overlain by the Bright Angel Shale and the Muav Limestone (Fig. 8–16) record such a transgression. Regressions expose rocks to weathering and erosion. When deposition later resumes, the buried erosion surface is called an unconformity. Paleozoic seas differ from those of Precambrian time in that they contained abundant lifeforms (invertebrates and later fishes) that had hard shells and bone that lent themselves to preservation in the rock record.

As the inland sea encroached over the late Precambrian erosion surface, sands and gravels were laid down that, when cemented, became the middle Cambrian (about 530 Ma) Tapeats Sandstone. With the deepening of the water, the muds that became the Bright Angel Shale were deposited. The sea was teeming with invertebrate animals, such as the trilobite (Fig. 8–17). After making remarkable strides up the evolutionary ladder during the Paleozoic, the trilobites suddenly disappeared extinct at the end of the Paleozoic, for reasons as yet unknown.

Also during the Cambrian period, the Muav Limestone was deposited, resting on the Bright Angel Shale. Next, there should be Ordovician rocks, but they are not there; nor are there any Silurian rocks. Obviously, Powell's "book of geology" had several pages torn out. Just what happened in the Grand Canyon area during these

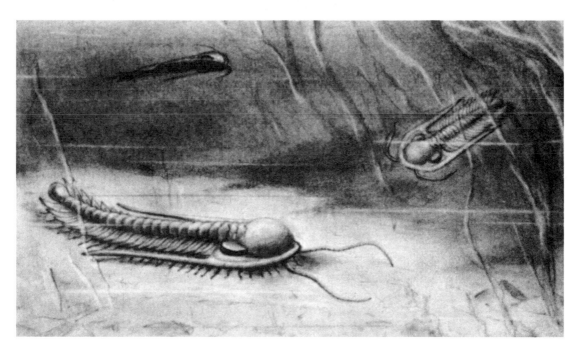

FIGURE 8–17   Restoration of the Cambrian trilobite, *Paradoxides harlani,* one of the giants of its time, nearly 12 inches (30 cm) long! (Reprinted from Dunbar, 1966, by permission of John Wiley & Sons, Inc.)

two geologic periods will never be known; it is likely that some deposition took place, but whatever was deposited was eroded away, prior to the next encroachment or transgression of the seas. The Devonian Temple Butte Limestone rests on the eroded surface of the Muav Limestone along a type of unconformity between parallel layers, often showing erosional relief (hills and valleys), called a *disconformity* (see stratigraphic column in Fig. 8–16).

Erosion again removed rock material before the next transgression, producing another disconformity between the Temple Butte and the overlying Mississippian age Redwall Limestone. In fact, in places all of the Temple Butte was eroded away in the Grand Canyon area as in the area shown in Fig. 8–16. Because of parallel bedding above and below the unconformities, recognition of the disconformities by merely viewing the canyon walls from the rim is difficult. Up close one finds the buried hills and valleys and, especially on the upper surface of the Redwall, filled sinkholes and caves that formed when the limestone was exposed at the earth's surface. During Redwall time, the seas were warm and shallow, and marine life was abundant, with brachiopods, clams, and corals of many kinds. By this time, about 330 Ma, the trilobites had increased in size and complexity, at that time unlikely candidates for the mass extinction that would affect most of the world's invertebrates at the end of the Paleozoic. The Redwall is thick and generally massive, and it forms bold cliffs, mainly red colored. Actually, the limestone is gray; only the surface is red, colored with iron oxide (hematite, the pigment in red barn paint) that streamed down the cliff from the red Supai Formation above.

The Supai is a nonmarine formation that was laid down during the Pennsylvanian and Permian periods, on the eroded Redwall Limestone. Mountain uplifts, part of the Ancestral Rockies in the Utah, Colorado, and New Mexico area were the major source of clastic sediment for the Supai and overlying Hermit Shale. Although the Supai is largely nonmarine shale, there are sandstone layers that contain tracks of amphibians, or perhaps primitive reptiles. In the western Grand Canyon the Supai contains marine limestone indicating the presence of a shallow sea.

The Permian-age Hermit Shale (about 280–250 Ma) overlies the Supai and is also red-colored and nonmarine. In addition to more vertebrate tracks, some "new" fossils appear in the Hermit—worm trails, insects, ferns of many kinds, and primitive coniferous plants.

The white cliffs that shine out from a distance are made of Coconino Sandstone. Eolian (wind) cross-bedding and well-rounded, well-sorted, frosted sand grains are distinctive features exhibited in sand dunes. During this part of the Permian, the Plateau country was a desert and the wind piled up dunes over a wide area. Then once again the sea prevailed, and the marine Toroweap limestones and sandstones were laid down.

Above the Toroweap, the bold, generally unscalable cliffs of Kaibab Limestone rise vertically to the top of the canyon (Fig. 8–18). Because of its massive character

FIGURE 8–18 Colorado River, looking down from Toroweap Point, western Grand Canyon. Note the Esplanade surface on top of the Supai Sandstone in the middle distance. (Photo taken by J.K. Hiller in about 1872. U.S. Geological Survey photograph)

and high resistance to erosion, the Kaibab is the surface rock over a broad area, the Kaibab Plateau to the north and the Coconino Plateau on the south side of the canyon. Thus, when one stands on either canyon rim in the eastern Grand Canyon, one can amaze your companions by telling them that they are standing on the 250-million-year-old Kaibab Limestone! Marine invertebrate fossils are abundant in certain layers of the Kaibab; also, a few shark's teeth have been found.

Thus, we come to the end of the Paleozoic Era during which the Grand Canyon region was both above and below the sea, at times being eroded for long periods. Where only primitive algae and perhaps soft-bodied worms and other invertebrates characterized the late Precambrian seas, by Paleozoic time marine invertebrates had developed the ability to build hard, protective shells that are more readily fossilized. The much-improved fossil record tells of thousands of new species appearing to fill environmental niches and many individual species becoming extinct. Land plants appeared sometime during the middle of the era, a prerequisite for the land vertebrates, some large and well advanced, that were present by the end of the era.

Early fish appear in rocks of middle Paleozoic age, including the tidal channel facies of the Temple Butte Limestone. Footprints of lizardlike reptiles are found on the bedding layers of late Paleozoic dunes. The transition from water to land by vertebrates was completed during the middle Paleozoic, establishing conditions that would permit a fascinating variety of amphibians, reptiles, mammals, and birds to evolve over the next 400 million years. Some of the Mesozoic-age dinosaurs undoubtedly roamed in the Grand Canyon area, but Mesozoic rocks have long since been eroded away. We will see their footprints in parks such as Arches and Zion and will examine their incredible story more closely in our visit to Dinosaur National Monument just north of the Colorado Plateaus.

The mystery of what caused most of the invertebrate life on the planet to go extinct at the end of the Paleozoic (about 245 Ma) remains unsolved. Perhaps over 75 percent of land species and 85 percent of all marine invertebrates, including many important groups such as the trilobites, disappeared from the roster of earth's inhabitants. It is known that the six major crustal plates of the earth were assembling through the Paleozoic Era into one supercontinent called Pangaea (Greek for "all earth"). Formation of one single, giant continent would severely change world climates and would restrict living space for invertebrates dependent on shallow marine shelf environments. Less living space creates greater competition and stress in order to survive. Some scientists suggest that changes in ocean salinity would further stress marine invertebrates, most of which have low-tolerance limits to salinity change. Huge eruptions of flood basalts in what is now Siberia occurred at the end of the Paleozoic and may have changed the composition of the atmosphere and contributed to or perhaps caused the mass extinction. However, the cause of the earth's greatest known mass extinction remains uncertain and may be due to a combination of factors. Whatever the cause or causes, the early Mesozoic seas lacked the diversity of life that characterized the Paleozoic.

Mesozoic-Cenozoic Events

Interpreting the Mesozoic history would at first seem bleak because of the sparsity of rocks of that age in the Grand Canyon area. However, two small erosional remnants, or *outliers,* of Mesozoic rocks bear mute evidence that Mesozoic strata once covered the area. Likewise, Mesozoic rocks elsewhere on the Plateaus must have formed extensive sheetlike deposits that had to extend over the area now occupied by the Grand Canyon. Thus, the over 5000 feet (1525 m) of Mesozoic strata preserved elsewhere must have been stripped by erosion from the Grand Canyon area. A glance at the cross section in Plates 8C and 8D further indicates that the structural elevation of the Kaibab Upwarp at Grand Canyon is much greater than in areas to the north, accounting for the deeper level of erosion. Stand on the canyon rim and look down into the deep abyss of the Grand Canyon and consider the amount of erosion here. Now swing your head skyward and look up that same distance. The mile-deep (1.6-km) Grand Canyon pales in comparison when one contemplates that the nearly same thickness of rock above the Kaibab Limestone has been removed by erosion. One must further realize that this missing thickness of rock extended as far as the eye can see in all directions—an almost incomprehensible volume of material.

Mesozoic deposition began with the laying down of reddish sands and silts by the sluggish streams that wandered across the Grand Canyon area and into the sea to the west. In places the Triassic Moenkopi was as much as 1000 feet (305 m) thick. The Shinarump Conglomerate at the base of the Chinle Formation and other younger beds were also laid down; however, all but a few small remnants of Mesozoic rocks have been stripped off. Cedar Mountain, a mesa near the southeastern boundary of the park, provides direct proof that at least the Moenkopi covered the Kaibab Limestone. Red Butte south of Grand Canyon Village is composed of Moenkopi overlain by Shinarump and a much younger lava flow. We will learn considerably more about Triassic times when we visit Canyon de Chelly and Petrified Forest National Park later in this chapter.

The uplift that rejuvenated the streams that stripped off the Mesozoic rocks began during the Laramide Orogeny (about 75 Ma) (Morales, 1990). The development of a so-called *stripped plain* is probably uncommon, at least on a widespread basis. Here, it was the superior resistance of the Kaibab Limestone that made it possible. The streams shifted back and forth on top of the Kaibab, removing the poorly resistant overlying materials, much like hosing mud from your driveway. Remember that the Grand Canyon wasn't there at that time.

With no outlet to the sea until about 5.5 Ma, *temporary base levels* caused sediment and water in the ancestral Colorado River to be trapped in large lakes in the Colorado Plateaus during early Tertiary time and perhaps in the Basin and Range Province during middle Tertiary time. The rivers adjusted their courses by shifting laterally and by *headward erosion,* especially in the upper Colorado River drainage where most of the modern river courses were likely established by

Miocene time (Stokes, 1986). When plate movements opened a large rift (Gulf of California) about 5.5 Ma, *base level* for the lower Colorado River became sea level, and the river greatly deepened its channel and cut headward. The river eventually cut through the Kaibab Upwarp and pirated the waters of the upper Colorado. The increased volume of water and the steep drop (*gradient*) of the stream resulted in the vigorous downcutting that formed the Grand Canyon (Lucchita, 1990). The river now had a complete, integrated drainage system similar to today's. Rapid uplift of the Colorado Plateaus and Rocky Mountains during the last 10 million years further energized the river to cut the deep, foreboding canyons that help make the Plateaus a special place. Thus, although the rocks in the canyon are nearly 2 billion years old, the Grand Canyon itself is young.

As downcutting progressed through the sedimentary rocks for a few million years, concurrent mass wasting and cutting along side valleys widened the canyon. Many tributaries developed their valleys along fault zones and major fractures. Bright Angel Creek developed its valley in the Bright Angel Fault Zone, which angles across Grand Canyon; on the south side, Garden Creek follows the same fault and flows into the Colorado River against the current. These are both *subsequent streams*[8]; Garden Creek is a *barbed tributary* because downstream from the junction, the angle formed by the two streams is distinctly acute. A glance at a drainage map or the space image view in Figure 8–19 shows clearly the location of the Bright Angel and other faults. Also apparent in Figure 8–19 is that the tributaries north of the river are longer than those on the south. Recall that the rocks here are *dipping* (sloping) southwest—away from the crest of the Kaibab Upwarp (Plates 8A–D). Thus the northern tributaries flow along the slope of the beds and the land surface and are longer than their counterparts on the south side of the canyon. Also recall that the North Rim is more than 1000 feet (305 m) higher than the South Rim—also accounted for by the south-dipping strata.

Those fortunate enough to experience the canyon from a boat will readily identify the many fracture and fault zones that slice into the Colorado River valley by noting where rapids are located. In the 280-mile (450-km) run from Lee's Ferry to Lake Mead the river drops over 2300 feet (700 m) and roars over 161 rapids (Dolan and Howard, 1978). Most tributaries form along fracture zones and flush large quantities of sediment into the Colorado River as a result of short, intense intervals of rain that produce *flash floods*. The resulting debris fans produce significant areas of turbulent water called *rapids*. Rapids produced by the outcrop edges of exceptionally resistant rocks are uncommon in the Grand Canyon.

When the canyon was cut almost to its present depth, volcanoes began to erupt from vents in the Toroweap Valley area in the western Grand Canyon. For the adventuresome, and those with rugged vehicles who enjoy looking into a nearly

---

[8]A subsequent stream is one that adjusts its course to coincide with structural weaknesses in the rocks.

vertical inner gorge almost 3000 feet deep (900 m) and less than 1 mile (1.2 km) wide, the western Grand Canyon is ideal (Fig. 8–18). Those visitors with children who listen carefully and obey their parents need not fear. Others beware! The geologic story here is unusual and well-worth a visit.

Thanks to the detailed studies of Kenneth Hamblin (1994), who worked under difficult conditions in the rugged canyon, we have a better understanding of the geologic history of the Toroweap area. The river is a one-way street. Once through a rapids it is difficult or impossible to quickly return to compare lava outcrops in one area with those downstream. Numerous float trips, mountain climbers, and fly-overs were needed. Hundreds of lava flows, all less than 2 Ma, cover the fault-bound Uinkaret Plateau to the north. Volcanic vents also popped up in the Grand Canyon (Fig. 8–20) and along faults that extend south of the canyon. Lava spilled from the Uinkaret Plateau into the Toroweap Valley on the east and Whitmore Wash to the west. Some flows reached the Grand Canyon—cascading down the 3000-foot-high (900-m) walls and into the roaring river below (Fig. 8–21). As Powell noted in his journal entry of August 25, 1869 (Powell and Porter, 1969, p. 131):

> What a conflict of water and fire there must have been here! Just imagine a river of molten rock running down a river of melted snow. What a seething and boiling of waters; what clouds of steam rolled into the heavens.

The lava dominated, but only temporarily, in its battle with the raging river. At least 13 times lava plugged the valley, forming 200- to 2000-foot-deep (60- to 600-m) lakes as water ponded behind. The oldest dam is about 1.8 million years old and the most recent formed about 0.5 Ma. Shorelines extended far upstream into Utah well beyond the present shoreline of Lake Powell. As water deepened, waves occasionally lapped at the base of the Redwall Limestone near where park headquarters is located today in the eastern Grand Canyon. The river persisted and each lava dam was overtopped creating waterfalls and rapids that retreated head-ward—much like Yellowstone Falls and Niagara Falls are doing today. Just as eventually Yellowstone Lake and Lake Ontario will be drained by retreating waterfalls, so too were each of the 13 dams on the Colorado River overtopped and de-stroyed—some releasing their waters rapidly as huge floods. Only patches of black lava plastered on the canyon walls today stand in mute testimony of the turbulent times of the past 2 million years.

Of particular significance in dating the cutting of the Grand Canyon is the Toroweap lava dam that formed about 1.2 Ma. Lava reached a canyon bottom about 50 feet (15 m) above the present canyon floor, thus the canyon was cut nearly to its present depth 1.2 Ma. Therefore, most of the deepening of the Grand Canyon was accomplished between the time when the Colorado River became a through-flowing river 5.5 Ma and when the Toroweap lava dam formed about 1.2 Ma.

Mountains of the Future

As the Colorado River cut down through the rocks, its main tributaries cut many side canyons. Tributaries of the large tributaries also cut canyons. The sedimentary rocks above the Tonto Platform were therefore intricately dissected, and tall "temples" were left standing high above the platform—Shiva Temple, Osiris Temple, Isis, Vishnu, Brahma, and Zoraster temples are a few of the higher ones. These temples are erosional mountains—the mountains of the future. If they rose above a large, flat surface instead of being lost in the vastness below the canyon rim, they would be immediately recognized as mountains. Several million years from now, when the entire Colorado Plateau area is completely dissected—like the area immediately adjacent to the canyon—erosional mountains by the thousands will be there, with peaks rising up to present or near present plateau heights. Eventually these mountains in turn will be eroded down, as countless mountains have been in the past.

Grand Canyon—A Natural Barrier

The cutting of the canyon effectively isolated the Kaibab from the Coconino Plateau, insofar as certain lifeforms are concerned. Birds such as the eagle fly nonstop from rim to rim, but small birds apparently develop acrophobia looking down

FIGURE 8–19 (opposite)   A view from space.

For centuries geologist mapped geologic formations and structures on foot; later they observed them from the air and used aerial photographs. Now, space and computer technologies provide long-range views that often reveal structures too large to see from the ground—structures essential in the new plate tectonics interpretations.

On the facing page we observe a section of about 30 by 40 miles (50 by 67 km) of the Grand Canyon, as it appeared from the Landsat satellite 570 miles (920 km) out in space. We see the Colorado River flowing southward in Marble Canyon (top) located parallel to the East Kaibab Monocline at 3 (see inset), to be joined by the Little Colorado (6). As it bends to the west across the Kaibab Upwarp (5), the river plunges into the narrow and steep-walled Inner Gorge. Here, as we look deeper and deeper into the earth, we are traveling farther and farther back into the ancient history of our planet.

Grand Canyon Village on the South Rim is only barely visible, but Bright Angel and Vishnu faults are seen distinctly at 1 and 2, cutting across the canyon and adjacent plateaus. The West Kaibab Monocline and Fault are located at 4.

This view is a small part of a full Landsat frame taken June 19, 1976, and designated Scene number 2514-17200. It was processed at the Environmental Research Institute of Michigan from U.S. Geological Survey data tapes. Made available by Anthony Morse of Synoptic Views, Box 193, Fort Collins, Colorado 80522.

FIGURE 8–20   Vulcans Forge, an eroded volcanic neck in the middle of the Colorado River! (Photo by L. Hymans)

through a mile of air. They fly down and across the river and up again—or elect to stay home. For many animals the canyon bottom is uninhabitable because of the summer heat—up to 120°F (49°C)—down along the river. Also, food sources available up on the plateaus are lacking in the desert in the canyon bottom. Squirrels, for example, depend upon the ponderosa pines for their mainstay diet. Thus,

FIGURE 8–21   View down canyon from Toroweap Point in western Grand Canyon. Note lava flows (arrows) that formed dams across the lower section of the canyon. (Photo by E. Kiver)

the Kaibab squirrel is confined to the Kaibab Plateau, and another subspecies, the Abert squirrel, is found only on the south side of the canyon. Although only 7.5 miles (12 km) apart as the crow flies, the story for them is that "you can't get there from here." Both subspecies evolved from a common ancestor a few million years ago and became separated as the canyon was cut below the ponderosa pine life zone. The canyon itself acts as a migration corridor. Desert plants and animals from the Mojave Desert have extended their range northward, establishing footholds in the canyon and, to a lesser extent, in hotter parts of the Painted Desert.

Seeing Grand Canyon

Some people must see the canyon without getting very far from their cars, looking down into the beautiful but awesome abyss from the rim; others look up from their boats (when they aren't fighting the rapids); still others go down the trail, on foot or mule back.

For those who can't do better, the views from the various overlooks are superb. The intriguing glimpses of the canyon below also provide an opportunity for those who want to attempt "long distance geology" and try to identify some of the rock layers shown in Figures 8–13 and 8–16. For those who want less crowded conditions a trip to the North Rim, 7.5 "crow miles" (12 km) and about 135 road miles (215 km) away, is an escape to the more wilderness-like atmosphere that our national parks are supposed to have.

Others take the trail down the canyon on foot—theirs or a mule's. For those in good physical condition, the way to get the full impact of the Grand Canyon is on foot. Each downward step takes you farther back into the history of the earth as you descend through a vertical mile of rock layers. At the river you walk on rocks nearly 2 billion years older than the ones you left behind on the rim. To sleep in the starlight under the tall ponderosa pines on the North Rim, shivering and waiting for first light, then to hike down the trail through one life zone after another before sweltering in the desert of the canyon floor—this is an experience never to be forgotten. If, however, you become completely exhausted and have to be taken out on muleback—a "dragout," as they call it—you won't forget that either, nor the expense. In other words, it is well to seek the advice of a park ranger before making the decision. If you hike down (of course carrying a large volume of water), with luck you may glimpse the fox-faced cacomistle, distant cousin of the raccoon and coatimundi; with luck you won't make contact with any rattlesnakes, scorpions, or catclaws.

With over 5 million visitors each year, Grand Canyon as well as many other parklands are being overloaded. There are those who would cram more people in and build roads that destroy our last remaining open spaces—much like those who would cut more trees to make parking spaces for those who want to see the forest! Viable solutions to the problem might include mass transit, more national parks, and ultimately a permit system at overloaded parks. Construction of Glen Canyon

Dam has been detrimental to the Colorado River ecology, and now the park is being attacked from the sky. Air pollution from urban and industrial areas in California and southern Arizona, as well as from the large coal-fired power plant at Page, Arizona, has significantly impacted visibility on dozens of days each year. The impact of the coal-fired plant will be greatly lessened when better air pollution equipment is installed. Reducing air pollution from other sources is not as easy. Now noise pollution is a significant problem. To enjoy a magnificent view accompanied by the drone of hundreds of sight-seeing aircraft on a daily basis further degrades the visitor's experience.

Commercial interests push strongly for unlimited flights over the canyon so that tourists can have a second-rate experience of viewing the canyon from hundreds or thousands of feet up from a noisy airplane or helicopter rather than a one-on-one experience with the canyon and what it has to offer. Increasingly, silence is not one of the virtues of the Grand Canyon. As Miller (1997) observes, the silence of the canyon is being drained—"the way the Great Plains were emptied of bison or the rivers of the Northwest were emptied of salmon."

A new experiment, if successful, will add to the park experience. Watch for birds of prey soaring about the limestone and sandstone cliffs. The giant California condor (8 foot wingspan!) has been missing from the Grand Canyon for over 70 years. Whether the reintroduction of these unique, endangered birds is successful remains to be seen. Certainly the attempt to "fix" the errors of the past is commendable and provides hope that perhaps humans can restore some of the special things that we have not yet completely destroyed.

Grand Canyon has many moods—as the light changes and as the seasons change. Goldstein (1977) has captured most of them in *The Magnificent West: Grand Canyon*. But you will certainly want a few pictures of your own. Go out on one of the overlooks, perhaps Yaki Point, when the sun is low; record the scene as the sun slowly surrenders the canyon to the moon.

REFERENCES

Baars, D.L., 1995, Navajo Country: A geology and natural history of the Four Corners region: Univ. of New Mexico Press, Albuquerque, 255 p.

Beus, S.S., and Morales, M., 1990, Grand Canyon geology: Oxford University Press, New York, 518 p.

Dolan, R.A., Howard, A., and Trimble, D., 1978, Structural control of the rapids and pools of the Colorado River in the Grand Canyon: Science, v. 202, p. 629–631.

Ford, T., 1990, Grand Canyon Supergroup: Nankoweap Formation, Chuar Group, and Sixtymile Formation: in Beus, S.S., and Morales, M., eds., Grand Canyon geology: Oxford University Press, New York, p. 49–70.

Goldstein, M., 1977, The Magnificent West: Grand Canyon: Doubleday, Garden City, NY.

Hamblin, W.K., 1994, Late Cenozoic lava dams in the western Grand Canyon: Geological Society of America Memoir 183, 139 p.

Hendricks, J.D., and Stevenson, G.M., 1990, Grand Canyon Supergroup: Unkar Series: in Beus, S.S., and Morales, M., eds., Grand Canyon geology: Oxford University Press, New York, p. 29–47.

Ilg, B.R., Karlstrom, K.E., Hawkins, D.P., and Williams, M.L., 1996, Tectonic evolution of Paleoproterozoic rocks in the Grand Canyon: Insights into middle-crustal processes: Geological Society of America Bulletin, v. 108, p. 1149–1166.

Lucchita, I., 1990, History of the Grand Canyon and of the Colorado River in Arizona, in Beus, S.S., and Morales, M., eds., Grand Canyon geology: Oxford University Press, New York, Ch. 15, p. 311–332.

Miller, T., 1997, Tourist flights spoil canyon's spell: USA Today, Letters to the Editor, Jan. 10.

Morales, M., 1990, Mesozoic and Cenozoic strata of the Colorado Plateau near the Grand Canyon: in Beus, S.S., and Morales, M., eds., Grand Canyon geology: Oxford University Press, New York, p. 247–309.

Powell, J.W., and Porter, E., 1969, Down the Colorado: E.P. Dutton, New York, 168 p.

Shelton, J., 1966, Geology Illustrated: Freeman, San Francisco and London, 434 p.

Stokes, W.L., 1986, Geology of Utah: Utah Museum of Natural History, University of Utah, Occasional Paper no. 6.

Walcott, C.D., 1895, Algonkian rocks of the Grand Canyon of the Colorado: Journal of Geology, v. 3, p. 312–330.

Park Address

Grand Canyon National Park
P.O. Box 129
Grand Canyon, AZ 86023

CANYON DE CHELLY NATIONAL MONUMENT (ARIZONA)

Like Grand Canyon, Canyon de Chelly's (pronounced "duh shay") youngest sedimentary rocks are the Triassic-age Shinarump Conglomerate, located at the base of the Chinle Formation. The Shinarump caps and helps protect the thick, pink-colored De Chelly Sandstone of Permian age—the main scenery maker here. The spectacular canyons (Fig. 8–22) have provided a home for Native Americans for at least the last 1700 years. Geologically similar to Grand Canyon in that late Paleozoic (Permian, about 250 Ma) rocks are exposed by canyon cutting and that miles of overlying Mesozoic rock have been removed by erosion, Canyon de Chelly is dissimilar in that one merely looks westward across the Chinle Valley ("type locality" for the Chinle Formation) to Black Mesa where thick Jurassic and Cretaceous strata stand out in bold relief. At Grand Canyon one must drive many miles and hours to find extensive remnants of younger Mesozoic strata.

Canyon de Chelly, just east of the town of Chinle in northeastern Arizona, is near the center of the Navajo Reservation—an area about the size of West Virginia.

FIGURE 8–22   Rio de Chelly, flowing a sinuous course through Canyon de Chelly. Massive De Chelly Sandstone here is underlain by the weak Supai redbeds and capped by the resistant Shinarump Conglomerate. (Photo by D. Harris)

From a handful of tourists in the early 1900s, visitation grew rapidly from 1917 on as more people learned of the cliff dwellings and the spectacular scenery. In 1931 some 83,840 acres of tribal land were set aside as a national monument. No federal land is involved—visitors are guests of the Navajo Nation. Many Navajo families still make the monument their home, some as employees of the Park Service. As at Mesa Verde, the park was established primarily to preserve archeological and historical records. Also similar to Mesa Verde, the ideal arrangement of rocks and landforms made Canyon de Chelly a desirable place of habitation for ancient peoples. The niches beneath huge overhangs provided protected places for the Anasazi (archeological name for the "ancient ones" or "ancient enemies") or Hisatsinom (Hopi for "those who lived long ago") to build their cliff houses—apartment houses belonging to a well-developed and complex civilization.

Geographic Setting

Canyon de Chelly is located on the northwest flank of the Defiance Uplift, a northerly elongated anticline that extends southward from the Four Corners area straddling the Arizona–New Mexico border (Plate 8). The uplift is 100 miles (160 km) long and 30 miles (50 km) wide and separates the very deep San Juan Basin to the northeast from the shallower Black Mesa Basin to the west (Fig. 8–2). Rio del Muerto (Canyon of the Dead) and Rio de Chelly (Spanish for Indian word *Tsegi,* which means "rock canyon") flow westward from the crest of the Defiance Uplift—their flat-bottomed canyons joining in the monument near Spider Rock (Fig. 8–7). When water flows in the channels during spring runoff and after heavy rainstorms, it enters Chinle Wash near the town of Chinle and flows north toward Monument Valley and the San Juan River (a tributary to the Colorado). The occasional floods are hazardous, but they add much needed moisture to the floodplain soils on which the Navajos still farm today. Although these are intermittent streams, their flows are perennial where impervious shale is exposed in their channels. This year-around flow makes Canyon de Chelly highly desirable.

History

The first known inhabitants lived in circular pit houses on the valley floor; later they built masonry apartments (cliff dwellings) up on the ledges, well out of reach of floods. But as in a number of other places, the Big Drought late in the thirteenth century and perhaps other factors drove them out of Canyon de Chelly, forcing them to find homes elsewhere.

The northern Athabascan Indians migrated into Canyon de Chelly and the Four Corners region in the 1500s where the Dine ("The People") became known as Navajos. Fortunately, they have great respect for dead spirits and left the abandoned Anasazi dwelling and artifacts intact. The Navajo were inherently aggressive and their raids on the Pueblos and the Spanish villages along the Rio Grande brought reprisals; hence Canyon de Chelly was the scene of many fierce battles. In 1805, Spanish conquistadores under Lieutenant Antonio Narbona killed 115 Navajos, mostly old people, women, and children, near a rock-shelter in Canyon del Muerto; it was named Massacre Cave (Baars, 1995). Canyon del Muerto (Canyon of the Dead) was named later, when Smithsonian scientists discovered the mummified remains of two unrelated burials.

In 1864 Kit Carson was sent in with a detachment of U.S. Cavalry to subdue the Indians and to relocate them on a reservation in New Mexico. They cut down their well-kept peach orchards to help starve out and demoralize the Navajo people. The move was ill-conceived, however, and after 4 years of suffering and starvation the Navajo were allowed to return to Canyon de Chelly where their descendants now live, some in traditional hogans on the canyon floor (Fig. 8–23). Most of the hogans

FIGURE 8–23   Navajo Hogan on flat-floored valley in Canyon de Chelly. (Photo by D. Harris)

are six-sided and are used as residences and for ceremonial purposes. The door always faces east so the inhabitants can more easily welcome each new day.

Geologic Sequence

The Defiance Uplift is underlain by faults first formed during the Precambrian and intermittently active since. The present-day, elliptical-shaped, north-trending anticline is one of the many uplifts that characterize the Colorado Plateaus. Monoclines, faults, and basins are also major structures present on the Colorado Plateaus (Fig. 8–2). The Defiance Uplift is asymmetrical with a steep inclination (dip) of the beds on the east along the Defiance Monocline and a gentle slope westward in the monument area toward the Black Mesa Basin.

During the Paleozoic the persistent Defiance Uplift was present as a low-lying island in the midst of shallow seas. Consequently either no early Paleozoic rocks covered the ancient Defiance structure or they were subsequently eroded off before Middle Permian redbeds and the De Chelly Sandstone were deposited directly on top of the Precambrian basement rocks (Baars, 1983; 1995). The De Chelly sands were part of an extensive dunefield that covered parts of Arizona, Utah, and New Mexico. The geologic evidence for dunes is unmistakable—large (some 50-foot-high; 15 m) cross beds where sand cascaded down the lee sides of dunes and sand grains that show the well-sorted, rounded, and frosted characteristics typical of modern dunes. Occasional tracks of amphibians in the sand or in the mud deposits in former interdune areas indicate that the desert was not sterile of lifeforms. Rather, conditions for fossilization in this environment were poor.

The De Chelly Sandstone is over 1000 feet (305 m) thick in the Black Mesa Basin to the west, 800 feet (245 m) thick in the monument, and only 200 feet (60

m) thick at the crest of the Defiance Uplift. Apparently the dunes filled lower areas and barely extended over the top of the uplift.

During Late Triassic time rivers washing down from recently uplifted mountains in central and southern Arizona carried sand and rounded pebbles that buried the De Chelly sands and would ultimately form the Shinarump Conglomerate. This thin conglomerate layer is extremely well-cemented and forms a resistant cap on the De Chelly Sandstone, helping to slow erosion and maintain the steepness of the canyons. The Shinarump Conglomerate Member forms the base of the Chinle Formation. The upper shaley part of the Chinle was either eroded or not deposited on this part of the Defiance Uplift. We will see a lot more of the Chinle Formation when we visit Petrified Forest National Park some 75 miles to the south.

Cliffs, Caves, and People

The conditions that would make Canyon de Chelly a desirable home for the Native Americans began with the uplift of the plateau in late Tertiary time (about 10–5 Ma) and the establishment of a through-flowing Colorado River. Tributaries deepened their channels in response to the uplift and downcutting of the Colorado River—eventually headward cutting caused the tributaries of the tributaries to respond in like manner. Rio de Chelly cut down through the Shinarump Conglomerate and through the De Chelly Sandstone, in places into the redbeds below. Locally the canyon is 1000 feet (305 m) deep, exposing the full 825-foot (252-m) thickness of De Chelly Sandstone. When the weaker redbeds exposed below are undercut the canyon walls retreat, producing the wider, flat-floored, sediment-covered valley bottom where crops could be grown (Fig. 8–22).

The steep to vertical canyon walls are maintained by the resistant sandstone and its conglomerate caprock (Figs. 8–7 and 8–22). Outward expansion of the brittle canyon walls (*exfoliation,* or *unloading*) produces vertical fractures along which slabs occasionally peel off, further maintaining the sheer cliffs. When groundwater moving downward through the sandstone encounters an impermeable shale layer (one of the interdune deposits), water becomes *perched* and moves laterally to the canyon wall and removes some of the calcium cement binding sand grains to one another in a process called *spring sapping.* Frost action, wind, and gravity removes the loosened materials from the canyon wall forming an overhang, or *shelter cave.* In these niches the ancient people built many of their cliff houses, safe from floods and sheltered from wind and rain.

Unlike many of modern man's intrusions, these cliff dwellings blend in with their surroundings. The building blocks are De Chelly Sandstone and the mortar is mud from along Rio de Chelly; both are natural ingredients.

Park visitors can see some of the ruins from the North and South Rim drives, and they are free to hike down the trail to White House Ruin. However, in order to protect the fragile ruins and also the privacy of the Navajos, all other travel is under the watchful eye of an authorized guide or park ranger. Do not forget to look down on Spider Rock, a sandstone spire which rises about 800 feet (245 m) above the floor of Canyon de Chelly near its junction with Canyon del Muerto (Fig. 8–7).

Colorful pictographs and petroglyphs are found at several points; those at Standing Cow Ruins are particularly outstanding. Rock art consists of *petroglyphs* (mechanical chipping of rock to create a figure) and *pictographs* (actual rock paintings). Pictographs are rare—here they are unusually common. Perhaps the abundance of mineral pigments made this the preferred medium.

The stains of red and black on the canyon walls are called *desert varnish* and form where bacteria have chemically locked iron, manganese, and wind-blown dust onto the canyon walls as delightful streaks of color and "painted" walls (Dorn, 1991). These surfaces were nature's blackboards for the Native Americans. However, these blackboards can only be used once. When the dark crust is skillfully pecked away, the pink sandstone is exposed in the form of a petroglyph—the meaning(s) of which is unknown, but may be a record of a shaman's trancelike journey into the spirit world. The healing process of desert varnish is so slow that it has not obliterated petroglyphs that are well over a thousand years old. However, thoughtless humans could quickly destroy these national treasures if they are not adequately protected.

Canyon de Chelly represents an early stage in the evolution of arid landscapes, as illustrated in Figure 8–8. In time—a few million years—we will see here another monument valley (Fig. 8–5); "shortly" thereafter, we will have to change the name Colorado Plateaus to Colorado Plains.

Canyon de Chelly also provides continuity between the historic past and the present. The ancient ruins record the early history; the Navajos are there to tell us of the more recent past, during evening programs near the visitor center; and a modern community of Navajo families continue to live in their traditional homeland.

REFERENCES

Baars, D.L., 1983, The Colorado Plateau—a geologic history: Univ. of New Mexico Press, Albuquerque, 279 p.

Baars, D.L., 1995, Navajo Country: A geology and natural history of the Four Corners region: Univ. of New Mexico Press, Albuquerque, 255 p.

Dorn, R.I., 1991, Desert varnish: American Scientist, v. 79, p. 542–553.

Park Address

Canyon de Chelly National Monument
P.O. Box 588
Chinle, AZ 86503

COLORADO NATIONAL MONUMENT (COLORADO)

In the northeastern Colorado Plateaus in western Colorado, about 28 square miles (73 km²) of the Uncompahgre Plateau was set aside as Colorado National Monument by President William Howard Taft in 1911. With subsequent boundary changes the monument now contains 32 square miles (83 km²). The deep canyons draining the northeast edge of the Uncompahgre Plateau (Fig. 8–2) expose an impressive monocline and a relatively complete section of Mesozoic rocks. The canyons also provide one of the few opportunities in the Colorado Plateaus to glimpse the underlying Precambrian metamorphic and igneous basement rock. Spectacular views of the canyons, the Grand Valley (Colorado River valley), and the Book Cliffs and Roan Cliffs beyond to the northeast are available from trails and the turnouts along Rim Rock Drive through the park. Access to the park is by the west entrance a short distance from Fruita or through the east entrance near Grand Junction (Fig. 8–24).

The Roan Cliffs are composed of middle Tertiary rocks and are the northernmost and the stratigraphically highest of the Grand Staircase—a series of cliffs that contain progressively younger strata that are laid out, stepping-stone-style, as one proceeds north from the Grand Canyon.

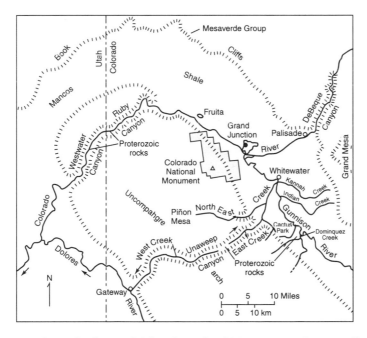

FIGURE 8–24 Generalized map of Colorado National Monument and surrounding area. (Modified slightly from Lohman, 1981)

The Colorado River flows northwest along the north flank of the Uncompah-gre monocline in a soft shale valley. The river turns southwest and flows across hard Precambrian rocks in Ruby and Westwater canyons (Fig. 8–24) before continuing its journey to Canyonlands National Park and beyond. This is an excellent place to get acquainted with Colorado Plateaus geology, especially the types of folds, faults, and the important mappable and formally named rock units (*formations*) of Mesozoic age that are encountered again and again in and near other parks in the province. Stan Lohman's well-illustrated Bulletin, *The Geologic Story of Colorado National Monument,* tells the story of the monument area, fitting it neatly into the broader area, the Colorado Plateaus.

History

Petrogyphs (rock drawings) scratched into the desert varnish and rock walls of the park are silent reminders of groups of Indians that had occupied this area for thousands of years. Over 70 archeological sites are known in the monument. Artifacts excavated from rock shelters and former campsites in and near the park indicate that the first inhabitants were basket makers—an older hunter-gatherer culture that dates back a few thousand years before the birth of Christ. The basket makers were later replaced by pottery makers. Most recently this was home to the Ute Indians.

John Otto, an early settler, became so entranced by the fantastic erosion forms in the canyons that he single-handedly carved out trails up the canyon walls, so that he could share this wild country with others. He enlisted the support of the chamber of commerce in nearby Grand Junction, and they successfully petitioned the government to set this area aside as a national monument. President William Howard Taft signed the declaration on May 24, 1911, establishing Colorado National Monument. To celebrate, Otto cut steps and handholds up the side of a 450-foot-high (137-m) sandstone monolith and on Independence Day planted the American flag on top of what was soon to be named Independence Monument (Fig. 8–25). The climbing steps and holes are still visible and occasionally used by climbers.

Geologic Story—Precambrian

The Precambrian metamorphic–igneous rocks exposed deep in the canyons are similar in appearance and age (about 1.7–1.3 billion years) to that at the Black Canyon of the Gunnison, the Inner Gorge of the Grand Canyon, and in the cores of the Rocky Mountain ranges to the east. Major late Precambrian rifting tore apart what was then the edge of the continental plate and produced a number of fault-block mountains. The northwest–southeast oriented basement faults generated during that event played an intermittent, but important role, in the subse-

FIGURE 8–25   Independence Monument rising 450 feet (137 m) above canyon floor, Colorado National Monument. (Photo by E. Kiver)

quent geologic development of the Uncompahgre area. Faults are mechanically weaker than nonfaulted crustal areas. Thus, when the crust is later subjected to stresses of the proper orientation and magnitude, faults will reactivate and can have a profound affect on the geologic record.

A long episode of erosion (perhaps a billion years) produced a major gap in the rock record and a nearly flat surface that is part of the Great Unconformity exposed here and at Grand Canyon, Black Canyon of the Gunnison, and elsewhere. As the early Paleozoic seas slowly encroached onto the continent, highly erosive waves and currents no doubt contributed to the remarkable flatness of the erosion surface.

Geologic Story—Paleozoic

Because no Paleozoic rocks are present on the Uncompahgre Plateau, the Paleozoic history is determined by studying Paleozoic rocks in surrounding areas. During the late Paleozoic (Middle Pennsylvanian, about 300 Ma) as continents were beginning to collide and assemble into one large supercontinent called Pangaea, an impressive range of mountains, the Ancestral Rockies, formed in roughly the same location as the present Rockies. The westernmost range, called the Uncompahgre Highland by Lohman (1981), developed along the Precambrian basement faults beneath the Uncompahgre area. Paleozoic rocks, if present, were eroded down to the ancient Precambrian surface shortly before Mesozoic sediments began to accumulate.

Geologic Story—Mesozoic

The stratigraphy of the "Age of Reptiles," the Mesozoic, is well represented at Colorado National Monument. This is an excellent place to become acquainted with the important Mesozoic-age rock units on the Colorado Plateaus. So far in our tour of the Colorado Plateaus we have already met the Late Triassic Chinle Formation at Grand Canyon and at Canyon de Chelly. Here again, and not for the last time in the Colorado Plateaus, we encounter these striking redbeds.

During the Late Triassic, streams draining from the eroded remnants of the Ancestral Rockies and from highland areas in southern Arizona deposited mud and sand in channels, floodplains, and swamps. These bright red Chinle Formation rocks rest unconformably on the dark-colored Precambrian rocks (Fig. 8–26).

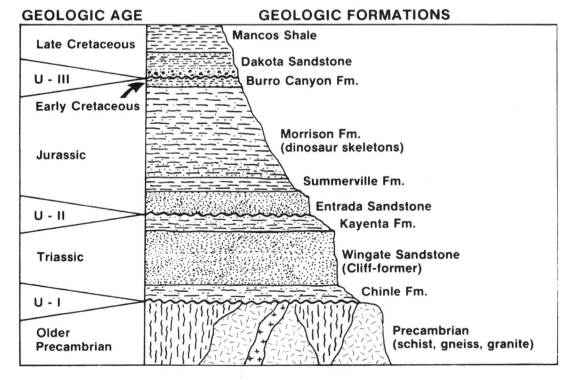

FIGURE 8–26   Columnar section of rock units in and near Colorado National Monument. U-I; major unconformity; U-II and U-III; disconformities. (Adopted from Lohman, 1981)

The Chinle Formation here is relatively soft and does not hold up well under erosion. However, the 350-foot-thick (107-m) resistant dune sands of the overlying Wingate Sandstone form bold cliffs that protect the Chinle and account for much of the scenic splendor of the park (Fig. 8–27). Critical to the maintenance of these cliffs is the relatively thin Kayenta Formation overlying the Wingate. These stream-deposited sands and gravels are cemented by silica in the form of quartz and are extremely resistant to weathering and erosion. Where removed by erosion, the underlying Wingate erodes to more rounded forms such as those at the Coke Ovens and Balanced Rock.

A significant omission of important Colorado Plateaus strata occurs above the Kayenta at Colorado National Monument. Where the famous Navajo Sandstone dune deposits should occur, there is an erosional surface, suggesting another uplift and episode of erosion. The Navajo and the overlying marine deposits of the Carmel Formation can be seen at Zion National Park and elsewhere on the Colorado Plateaus to the west. Apparently a regional uplift and westward tilting of the Plateaus stripped these formations as well as the upper part of the Kayenta from the monument area (Lohman, 1981). The Entrada Sandstone above this second unconformity (Fig. 8–26) represents coastal dunes formed along the eastern edge of a shallow sea during Middle Jurassic time (about 170 Ma). It is about 150 feet (46 m) thick and forms a secondary cliff upslope from the Wingate cliffs along Rim Rock Drive.

Climbing the stratigraphic ladder farther, we find the overlying Summerville and Morrison formations. These varicolored mudstones and siltstones are relatively soft and are therefore slope formers. Where unprotected by resistant overlying layers, they become intricately eroded and form a *badland topography* like that in the Redlands area just east of the monument. During Morrison time, dinosaurs slogged about in the shallow lakes, swamps, and along the sluggish rivers draining the plateau and surrounding areas. A number of famous dinosaur quarries, such as that at Dinosaur National Monument, are located in this important rock unit.

FIGURE 8–27 View east down U-shaped, unglaciated canyon floored by Precambrian rocks; Wingate Cliffs rise above Chinle Shale slopes. (Photo by D. Harris)

Isolated finds of bones and skeletons are common. Although none have been dis-covered so far in Colorado National Monument, a number of discoveries have oc-curred nearby. In 1900 at Riggs Hill just east of the monument, the first skeleton of *Brachiosaurus,* the giant plant-eating behemoth was discovered. This specimen was medium-sized, about 76 feet (23 m) long. Larger specimens, some about 100 feet (30 m) long and weighing in at about 136 metric tons, have been discovered elsewhere. No doubt more bones are resting peacefully, deep in one of the hills in-side the monument boundary—to be resurrected some day by erosion.

Two Cretaceous formations, the Burro Canyon Formation and the Dakota Sandstone, are exposed in the higher hills of the monument. Erosion followed the deposition of the Burro Canyon Formation and produced yet a third unconfor-mity before the advancing edge of the sea deposited the beach sands of the Upper Cretaceous Dakota Sandstone. Deepening water allowed the finer offshore sedi-ments of the Mancos Shale to cover the Dakota Sandstone and to accumulate to nearly 4000 feet (1220 m). Shallower water followed and deposited the beach and coal-bearing swamp deposits of the Mesaverde Group, subsequently eroded, along with the Mancos Shale, from the Monument area. The Mancos Shale underlies the Grand Valley, and sandstone of the Lower Mesaverde Group caps the Book Cliffs to the northeast. In very late Cretaceous time the Colorado Plateaus began to warp into uplifts (arches and anticlines) and downfolds (synclines and basins). More in-tense bending of the Uncompahgre and other structures on the Colorado Plateaus was soon to occur—during the Cenozoic.

Geologic Story—Cenozoic

With the rapid, and perhaps sudden, extinction of the dinosaurs at the end of the Mesozoic the earth entered a new age—the Age of Mammals. Their story is well recorded in parks where Tertiary-age rocks are abundant, particularly in certain parks in the Great Plains and at John Day National Monument in the Columbia Intermontane Province. Primitive mammal bones associated with reptile fossils near Fruita, just northeast of the monument, indicate that early mammals coex-isted with the dinosaurs during the Mesozoic Era.

Here at Colorado National Monument, early Tertiary (Paleocene and Eocene, 66 to about 50 Ma) stream and lake sediments that may have covered the monu-ment are gone from the Uncompahgre Plateau. However, a thick section of Eocene lake sediments are present nearby in the Roan Cliffs just north of the Book Cliffs. The gentle bending and faulting during Late Cretaceous time increased in severity sometime after these lake sediments were deposited. Much of the folding, faulting, and uplift probably occurred during the Pliocene and Pleistocene (Lohman, 1981). The restless Precambrian basement faults were on the move again—this time forming today's 125-mile-long (200-km), 30-mile-wide (48-km) Uncompahgre Plateau. Thousands of feet of sedimentary rock were stripped away as rocks were up-arched to form the Uncompahgre Plateau and its associated monocline, which was locally broken by faults. Where fault movement was slow, the overlying sedimentary beds slowly bent to conform to the new shape of the basement rocks (Fig. 8–28)—much like a blanket on a bed bends around the shape of its occupant. The rocks are essentially flat-lying except along the monocline along the northeastern border of the monument where they dip steeply toward the northeast (Fig. 8–29). The rocks across the Colorado River Valley in the Roan Cliffs are also flat-lying, or nearly so. But those rocks are Tertiary in age, much younger than those in the monument. With horizontal rocks on both sides of a dipping or inclined section of rock layers, the structure is a monocline (Fig. 8–28).

How much vertical uplift occurred on the Uncompahgre Uplift since Late Cretaceous time when the area was at sea level? A minimum amount is the present elevation of the top of the uplift (about 7100 feet; 2165 m). However, the 6700 feet (2040 m) of marine strata that formerly covered its top need to be added. Thus the area has risen over 13,800 feet (4200 m) above sea level in the past 65 million years. Chronic (1988) notes that the present surface of the Uncompahgre Uplift and the top of the Book Cliffs across the Grand Valley are at about the same elevation. Thus the 6700 feet (2040 m) of strata between the top of the Uncompahgre Uplift and the top of the Book Cliffs are also a measure of local uplift along the Uncompahgre Monocline. Lohman (1981) believes that the amount of Pleistocene uplift (last 1.6 million years) was 1700–1900 feet (518–579 m). If Lohman's time and vertical uplift estimates are correct, then streams must have made major recent adjustments of their courses to the developing Uncompahgre Uplift. Indeed, certain topographic features in the area appear to be recent relics of former river courses and help in reconstructing the drainage history.

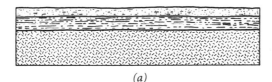

(*a*)

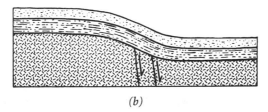

(*b*)

FIGURE 8–28 Diagrammatic explanation of the formation of a monocline. (*a*) Cross section showing flat-lying sedimentary rocks resting on granitic basement rocks. (*b*) Cross section of rocks shown in (*a*) after faulting in basement and folding of younger rocks—a monocline. (Line drawing by Gregory Nelson)

(a)

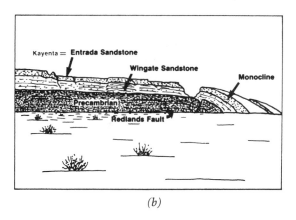

(b)

FIGURE 8–29   Photo and sketch of the monocline in Colorado National Monument; Redlands Fault is located along the base of the cliff. (Photo by D. Harris, line drawing by Gregory Nelson)

Monument Canyons

Spectacular evidence of the former position of the Colorado River is found at Unaweep Canyon a few miles southeast of the monument. Unaweep Canyon cuts completely through the Uncompahgre Arch and lacks a through-flowing stream (Fig. 8–24). The river cutting through the uplifting arch was initially an *antecedent river,* one that was present before the uplift that it crosses. However, downcutting was greatly slowed when it reached the hard Precambrian basement rock in the core of the uplift. The small upstart tributary marked "A" in Figure 8–30 was not so encumbered. It continued to erode downward and headward in the soft shale zone between the harder rocks of the Uncompahgre Monocline and the Book Cliffs. Eventually the tributary extended its valley around the north flank of the Uncompahgre Arch, past the future site of the town of Grand Junction, and to the

Colorado River and later the Gunnison River. Because the tributary was at a lower elevation, it diverted the waters of the Colorado and Gunnison rivers into its own channel (Fig. 8–24). Such a process is appropriately called stream piracy.

A comparison of the ancestral drainage in Figure 8–30 with the present one shown in Figure 8–24 shows the magnitude of the drainage rearrangement. Further examination of Figure 8–24, especially in the Ruby–Westwater canyons area where the Colorado has again cut down into Precambrian rock, indicates that another stream piracy event will likely occur in the distant future.

After the new segment of the Colorado River was established about 2 million years ago, the river rapidly deepened its channel, thereby lowering the base level of streams draining the Uncompahgre Plateau. With greater vertical drop and increased *stream gradient,* streams possess more energy and can cut spectacular canyons such as those in the monument. Lohman (1981) believes that most of the canyon cutting in the monument occurred during the past few hundred thousand years.

Some of the canyons are U-shaped in cross section (Fig. 8–27), superficially resembling canyons carved out by valley glaciers. In the Colorado Plateaus, however, only limited areas—areas much higher than any in the monument—were glaciated. No glacial deposits occur here, further casting doubt on the glacial hypothesis. Rather, running water and the physical characteristics of the rock account

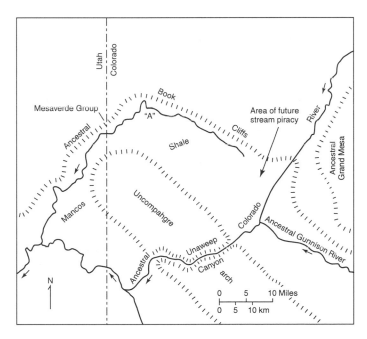

FIGURE 8–30 Map of ancestral drainage in Uncompahgre area just prior to stream piracy. Stream diversion or piracy likely occurred near the beginning of the Quaternary, about 1.8 Ma. Compare with present stream locations shown in Figure 8–24. (Modified from Lohman, 1981)

for the U-shaped valley. Initially, the valleys were likely V-shaped until the resistant Precambrian rock was encountered. Downcutting slowed and undercutting of the soft Chinle Formation above the Precambrian caused the Wingate cliffs to recede and the valley to widen. Vertical fractures in the massive Wingate play a part in maintaining these near-vertical cliffs. The broken slabs of sandstone soon disintegrate, and this sand forms the rounded slopes at the base of the cliff, resulting in the U-shaped valleys or canyons.

The canyon-widening process is assisted by the process of exfoliation, or unloading. Massive, brittle rocks such as the Wingate Sandstone are under great compression due to its weight and the weight of overlying rocks. When confining pressure is reduced or removed, as along a canyon wall, the brittle rock expands toward the canyon and cracks, forming a vertical joint. Gravity pulls at these walls and encourages vertical slabs to peel off the canyon wall, thereby helping to maintain its near-vertical inclination.

As the canyons are widened, the mesas between them are narrowed, until only high narrow rock walls remain. Then, when sections of the wall tumble down, only monoliths or monuments are left. Independence Monument (Fig. 8–25), the 450-foot-high monolith in Monument Canyon, is the sole survivor of such a wall, as Lohman (1981) points out.

Undercutting of the Wingate or exfoliation of lower parts of the Wingate often produce archlike features called *inset*, or *exfoliation*, arches in the massive sandstone. The graceful archways north of Artists Point are most impressive. Large overhangs called *rock shelters*, or shelter caves, often form along the Chinle–Wingate contact.

It may seem strange that this stream-eroded landscape lacks one important component—running water! A visitor is likely to drive from entrance to entrance without seeing a drop of running water, only dry streambeds. But if one is there at the right time, during a high-intensity summer storm, erosion will take place, and at a high rate. Dry streambeds turn into raceways as walls of sediment-choked water form. Streams will likely resemble mudflows, as the loose, weathered material that has accumulated since the last flood is flushed out of the watershed.

One can see many intriguing features from the rim; however, trails that lead down into the canyons afford opportunities to see at close range not only the geologic features but the plants and wildlife—including, with luck, the bison. Take a full canteen of water and save some of it for the long pull up out of the canyon, particularly in the heat of the summer. Colorado National Monument is open throughout the year.

REFERENCES

Chronic, H., 1988, Pages of stone: Geology of western national parks and monuments: Seattle, The Mountaineers, v. 4, p. 78–83.
Lohman, S.W., 1981, The geologic story of Colorado National Monument: U.S. Geological Survey Bulletin 1508, 142 p.

Park Address

Colorado National Monument
Fruita, CO 81521

PETRIFIED FOREST NATIONAL PARK (ARIZONA)

In our time-travel tour through the Colorado Plateaus we next proceed up one step from the late Paleozoic rocks forming the rimrock at the Grand Canyon and the sandstone cliffs at Canyon de Chelly to the lower Mesozoic rocks, our next stop being the Upper Triassic rocks (about 220 Ma) at Petrified Forest National Park. Recall that we saw small remnants of the basal conglomerate of the Chinle Formation (the Shinarump Conglomerate above the Moenkopi Formation) in an isolated outlier (erosional remnant) located on top of the Kaibab Limestone at Grand Canyon, about 120 miles (195 km) to the northwest and as the caprock at Canyon de Chelly about 75 miles (120 km) to the south.

The Chinle Formation extends across much of the Colorado Plateaus and surrounding areas and is composed of river, swamp, lake, and volcanic ash sediments. Entombed by its sedimentary particles are the stars of the show—one of the largest and most famous concentrations of colorful petrified wood in the world. Many of the petrified logs are exceptionally large; some are beautifully colored—jasper and carnelian—an outdoor mineral museum. The petrified wood, along with fossils ranging from microscopic plant remains up to large amphibians and some of the earliest known dinosaurs, provides an incredible 220 million-year-old snapshot in time—one of the earth's only nearly complete reconstruction of a Late Triassic (early Mesozoic) ecosystem.

Petrified Forest National Park is located in east central Arizona about 37 miles (60 km) west of New Mexico (Plate 8; Fig. 8–31). The Painted Desert area here is drained by the westerly flowing Puerco River and its intermittently flowing tributaries. Drainage is into the Little Colorado River near the town of Holbrook, 19 miles (31 km) west of the park. The topography is flat to hilly and sparsely covered by grass, sagebrush, saltbush, greasewood, cactus, yucca, and other plants possessing biological adaptations that minimize water loss in this hostile, arid environment. With only 9–11 inches (23–28 cm) of precipitation each year, occasional intervals of several months with no rainfall, and summer temperatures ranging into the low 100s, survival is difficult. You may not see even a drop of running water; yet, streams and sheet wash were responsible for the dissection of the badlands. Should you be there during a torrential rainstorm, however, you will see the erosional process in action, with streams muddy and fast-flowing.

Downcutting by the Colorado River has in turn caused tributaries to cut deeper. Where streams encountered resistant limestones and sandstones in the Little Colorado and Grand Canyon areas, steep cliffs and canyons result. Where weak shale rocks, such as those in the Triassic-age (225 Ma) Chinle Formation at Petrified Forest, are exposed and are being vigorously eroded, a badland topography

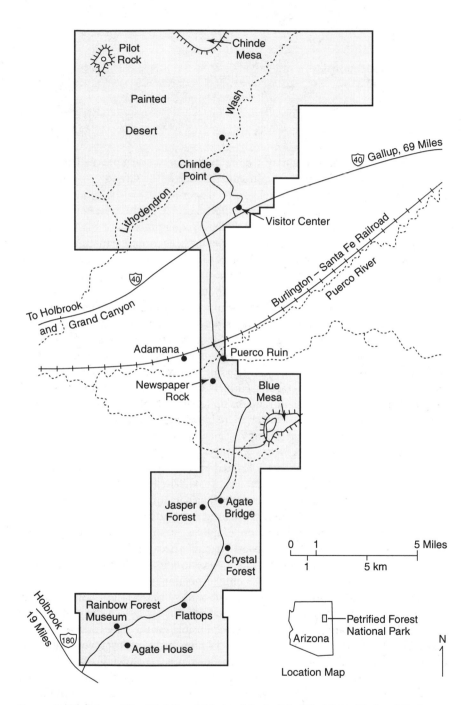

FIGURE 8–31   Map of Petrified Forest National Park. (Modified from National Park Service brochure)

forms. The Chinle's red, gray, green, and brown, with their various tints and hues, change with sun angle and weather, creating moods that can be observed only over a period of weeks or months. Beds of resistant sandstone, conglomerate, and limestone are sandwiched between the shale layers; being more resistant to weathering and erosion than the shale, they form mesas, cliffs, ledges, and overhangs.

Volcanic ash is abundant—an important constituent that provided a ready source of silica to impregnate and petrify the buried wood. As the glassy volcanic particles weathered and released silica into the groundwater, a clay mineral called *montmorillonite* was formed. This clay has the interesting characteristic that when wet it expands and when dry it contracts and occupies less volume—excellent material to form a watertight seal at the bottom of a pond or swimming pool, but poor for plants trying to establish roots on this alternately smooth and popcorn-like surface. Thus, the combination of low rainfall, rapidly eroding slopes, and an "unfriendly" surface all help to maintain the relatively barren desert environment.

History

Native Americans have long used the valley of the Puerco River as a trade route between east and west. Later explorers and pioneers followed the same route because it lacked deep canyons, high mountains, or other difficult topographic obstacles. This route was a "natural" and later became part of the route of the transcontinental railroad during the 1880s. More recently the famous Route 66 highway and now Interstate 40 follow the same route across the central part of the park (Fig. 8–31).

At least 300 archeological sites, including a number of Indian ruins, have been found in the park (Ash, 1990; Lubick, 1996). The Anasazi ("ancient ones") or Hisatsinom (Hopi for "those who lived long ago"), the presumed ancestors of the modern Pueblo Indians, built pit houses over 1500 years ago on Flattop Mesa (Fig. 8–31) and farmed nearby areas during an interval when climates were wetter than the present. Agate House in the southern part of the park is a small, eight-room pueblo constructed by the Anasazi in an interesting manner—beautiful blocks of petrified wood set in an adobe mortar. Drier climates (the Big Drought) forced abandonment of these higher sites about 700 years ago. Later Anasazi settlements were restricted to areas along the Puerco River and other more reliable waterways. One of these is Puerco Ruin near the main park road a short distance south of the Burlington Northern–Santa Fe Railroad tracks. Here, about 125 masonry rooms enclosed a large plaza; as many as 60–75 people may have made the village their home (Lubick, 1996). There appears to be a question as to when the site was first occupied, but it was abandoned about 600 years ago, about 100 years after the higher sites were abandoned and the Mesa Verde people farther to the north had to leave their homes.

Although the Big Drought drove many Indians from their villages in the Southwest, the inhabitants at Puerco Village apparently had sufficient water to continue raising corn, squash, and beans. However, during the 1300s Puerco River began to actively erode its banks and in time destroyed most of the farmland in the valley. Thus, after surviving the drought, they had to surrender to the malevolent river. By the time that the Spanish first explored in 1540, only the colorful hills of Chinle Formation and desert plants and animals greeted them in the Petrified Forest area. Newspaper Rock is only a short distance from Puerco Ruin. Here the Anasazi chipped and scratched away the iron oxide–manganese coating (desert varnish) and exposed the light-colored sandstone beneath, thus creating their enigmatic designs and drawings of humans, animals, and geometric figures.

The first discovery of petrified wood was made during a U.S. Army reconnaissance of Arizona and New Mexico by Captain Lorenzo Sitgreaves in 1851. Geologist Jules Marcou examined the area in 1853 as part of the U.S. Army exploring expedition under Amiel Whipple. Lithodendron Wash ("Stone-Tree Wash"), in the northern part of the park, was named by the Whipple expedition for the abundance of petrified wood. As a result of General W.T. Sherman's western visit in 1878 and his enthusiasm over the petrified wood, two logs were loaded on wagons the next year—one of which was sent back to the Smithsonian Institution in Washington where it is still on display.

Completion of the Atlantic and Pacific Railroad in 1883 (predecessor of the Santa Fe) brought many people to what was then called "Chalcedony Park." Unfortunately many were not content to just look and admire but took some of the forest with them. Collecting became more serious later as commercial collectors began to haul wagonloads of petrified wood away. Not content with surface samples, some brought dynamite to blast logs apart in the Crystal Forest where beautiful crystals of clear quartz had grown in cracks and vugs (openings or spaces in rock). The crowning insult, however, came in the form of a huge rock crusher constructed nearby to crush the logs and manufacture sandpaper and other abrasives!

At the request of the Arizona Territorial Legislature, Lester Ward, a paleobotanist with the U.S. Geological Survey, was sent to examine the fossil forests in 1899. After his visit Ward recommended withdrawing the area from homesteading and making it a national park. At the urging of conscientious Arizona territorial residents, President Theodore Roosevelt used the recently passed Antiquities Act legislation to establish in 1906 our second national monument—Petrified Forest National Monument—before the rock crusher went into action. The boundaries were enlarged and in 1962 Petrified Forest became our thirtieth national park. The wisdom of the move is indisputable. A walk around the protected borders reveals an area "picked clean" of petrified wood. So too would be Petrified Forest today if forward-looking people had not acted in the interest of all who would follow.

Triassic Landscapes

During the Triassic, Petrified Forest was in a quiet tectonic setting located on the southwest coast of the supercontinent, Pangaea. World geography would be a delight for students who like simplicity—one giant continent with all of today's individual land masses stuck together! The region was near the equator (about the latitude of Panama today) and enjoyed a tropical climate—much different from today's location some 1700 miles (2740 km) farther north in a dry, desert regime.

Sloping from highlands to the south and the remnants of the Ancestral Rockies to the east, was an extensive coastal plain with streams washing sediment lazily westward to an ocean located in western Nevada and southern California. Swamps and lakes were common in this tropical environment (Fig. 8–32). Forests covered higher elevations upstream and low hills in or near the swamps and floodplains. Volcanoes along the southwestern edge of the Pangaea supercontinent and more local volcanic vents occasionally showered the area with ash, adding to the thickening sheet of sediment. The resulting deposit of mudstone, ash, freshwater limestone, sandstone, and conglomerate is called the Chinle Formation. Although it is as much as 1970 feet (600 m) thick in some areas, at Petrified Forest it is only 932 feet (284 m) thick (Ash, 1987; Ash and Creber, 1992). The rocks are gently tilted to the north and planed off by erosion to a rolling surface, much like the upper edges of leaning books on a shelf would appear. Thus, the rocks and their fossils in the north end of the park rest on those in the middle and southern part and are therefore younger.

All was not always calm as the occasional sandstones, conglomerates, and huge accumulations of logs testify. Floods, perhaps associated with typhoons or thunderstorms, sent brief but intense flows of water down the stream channels. Large numbers of uprooted or broken logs crashed against one another breaking limbs and stripping bark from every tree[9] during their tumultuous journey. Wood and other fossils are randomly distributed throughout the Chinle Formation wherever it is found. However, at Petrified Forest they are much more abundant and many of the logs occur in geographically separate, massive accumulations referred to as "fossil forests." Going north from the Rainbow Forest (multicolored wood) in the southern part of the park are the Crystal Forest (relatively abundant quartz crystals formed in the hollow interiors of buried logs), Jasper Forest (the red mineral jasper is common), and the Black Forest (black, microcrystalline quartz) (Fig. 8–31).

Some floating logs were grounded against trees or on floodplains along the rivers; others were perhaps part of logjams or floating mats of logs in swamps or lakes that later became waterlogged, sank, and were buried by sediments. The "in place" (in situ) stumps are vertical and have root systems fanning from their bases. These rooted stumps have spent the past 220 million years here—never leaving the spot where they first grew and later died. Isolated stumps and stump fields are present in the park but are not common. Of the fossil forests, only the Black Forest in the northern part of the park contains in situ stumps (Ash, 1992). The transported logs are horizontal and many have their larger butt ends facing southerly, in the direction from which the rivers flowed.

---

[9]Of the hundreds of logs examined by paleobotanist Sidney Ash (1987), only *one* still retained its bark!

(a)

(b)

FIGURE 8–32   Artist's conception of the Late Triassic landscape at Petrified Forest National Park. (Painting by William Chapman, permission to use courtesy of National Park Service)

Log Preservation

The usual fate of organisms after death is decay, returning their chemical compounds back to nature where they are reused in some way. However, conditions here were particularly favorable for fossilization of wood. Favorable factors include: (1) the presence of a hard material (wood), (2) rapid burial by sediments or immersion in oxygen-poor water, and (3) an abundant source of silica in the overlying sediments and groundwater.

Groundwater chemically attacks the unstable glass shards in the buried volcanic ash, dissolving large amounts of silica. The silica-rich groundwater soaks the entombed wood and precipitates silica–oxygen compounds in the cells and pore spaces of the wood, thereby preserving, often in exquisite detail, the original cellular structure of the ancient wood. This process, called *permineralization,* makes the petrified wood extremely useful for tree identification and other detailed microscopic studies. For example, careful examination of permineralized logs that have features originally described as annual growth rings (Fig. 8–33) were found to be merely growth interruptions, features that form today in the humid tropics unrelated to seasonal temperature changes (Ash and Creber, 1992). Wood that experiences complete *replacement* by silica, retains the original external form, but loses the cellular detail of the original wood, making it less valuable to

FIGURE 8–33   Bands in some log cross sections are growth interruptions (*not* annual growth rings) similar to those forming today in humid-tropical areas. (Photo by E. Kiver)

paleobotanists—paleontologists who specialize in the study of fossil plants. Most of the wood at Petrified Forest experienced replacement fossilization.

The chemical compounds in petrified wood are silica-rich substances of three types. Each type is determined by the nature of their crystalline, or lack of crystalline, structure. Those with large crystals easily seen under a microscope and composed of silica and oxygen are called *quartz,* those with tiny submicroscopic crystals are *microcrystalline quartz,* and those silica-rich substances lacking an internal crystalline arrangement of atoms are called *opal.* Small amounts of impurities like iron (red, yellow, brown, and even some blue colors), manganese (black, gray), copper (green, blue), or other elements account for the artist's palette of colors in the petrified wood. Some of the varieties of quartz include rock crystal (clear, no impurities), milky (white or cream), amethyst (violet), smoky (gray to black), and citrine (yellow). Microcrystalline quartz is commonly called chalcedony and has many color varieties including jasper (red) and flint (gray). Color-banded forms are called agate.

Other types of fossilization occur when materials such as leaves, cones, seeds, pollen, spores, and fish scales are buried and squeezed into *compression fossils.* The resulting thin carbonaceous film often preserves minute details and further assists paleobotanists in their quest to completely reconstruct an ancient environment.

Fossil Record

Thanks to the excellent and abundant preservation of former lifeforms and the careful studies by researchers over many years, a nearly complete picture of what must have been a majestic Late Triassic forest-swamp ecosystem has emerged—one of the best known in the world. Life was plentiful, but very little would seem familiar to us since all of the Triassic species here are now extinct, although most have living relatives. Tree fossils are of great interest. However, other plant and animal fossils are of even greater scientific value. Over 200 species of plants and animals have been identified—only seven are from trees represented in the fossilized wood. Giant conifers up to 200 feet (60 m) tall and 10 feet (3 m) in diameter (*Araucarioxylon* in Fig. 8–32) grew on localized high ground in the swamps (*hammocks,* such as those in the Everglades) and other better-drained areas.

The understory contained other smaller species of conifers as well as ferns a few inches tall to tree size. A type of tree fern (*Itopsidema* in Fig. 8–32), resembled a modern palm tree although it is not a close relative. Primitive plants such as horsetails (*Neocalamites* in Fig. 8–32) were abundant and some were up to 30 feet (9 m) tall. The cycadophytes were abundant. They had short squat columnar trunks with pinnate leaves at the top. Also present in the rocks are amphibian and reptile bones and teeth, freshwater fish and clams together with insect and fungus traces. The only known specimens of Triassic-age *amber* (hardened tree sap) were discovered in the park (Litwins and Ash, 1991). Of the few specimens discovered so far, none had entrapped and preserved any of the abundant insects that undoubtedly populated the area. Missing from the scene are the hardwood forests, flowering plants, and chirping birds—organisms that had yet to appear on the earth. Primitive mammals are known from Triassic rocks elsewhere in Arizona, but none has been found in the Chinle Formation in the park.

The crocodile-like *Phytosaur,* commonly about 17 feet (5 m) long but some reaching 30 feet (9 m) in length and weighing over one ton, were at the top of the food chain (Fig. 8–32). Dragonflies about the size of a small bird soared about, and snails and freshwater clams lived in or near the abundant wetland environments. Some of the last of the 10-foot-long (3-m) giant amphibians such as *Metaposaurus* that had dominated land areas during the late Paleozoic were present—rapidly being replaced by the up-and-coming, better-adapted reptiles. The discovery at Petrified Forest of bones of some of the earliest known dinosaurs, including the 8-foot-long (2.5 m) *Coelophysis* (Fig. 8–32) with a row of sharp pointy teeth, and "Gerti," a new species of dinosaur the size of a large dog discovered in the park in 1984, are miniature versions of what was yet to come later in the Mesozoic Era. Many of the primitive dinosaurs, amphibians, phytosaurs, and other animals are *identical* to fossils found in eastern North America, Europe, and Asia—further support for land connections between continents and the existence of the Pangaea supercontinent.

Most of the petrified logs examined were from healthy trees that were likely victims of floods. However, some beds contain a large number of logs that show extensive damage similar to that produced by modern pocket-rot fungus. Creber and Ash (1990) suggest that the Triassic episodes of infestation may have their counterpart in the Dutch elm disease that destroyed many elm trees in recent years in Europe and North America. Other trees show channels and tunnels likely produced by insects such as bark beetles.

The quartz and microcrystalline quartz logs are solid rock and therefore brittle. Many of the logs are broken into sections, appearing as if cut by some mysterious chain saw (Fig. 8–34). Perhaps earthquake shocks, sediment compaction, or stresses generated during uplift of the Colorado Plateaus may be the cause.

Later Geologic History

Several thousand feet of later Mesozoic sediments buried the Chinle Formation—including Cretaceous marine sediments deposited when the seas last covered the area. Laramide mountain-building activity beginning about 75 Ma initiated the uplifts that would eventually lift the land from sea level to its present elevation of about 6000 feet (1830 m). Layer after layer of rock was stripped from the area. By Pliocene time (about 5 Ma), after over 200 million years, the Chinle Formation burial grounds were once again bathed in sunlight. A temporary pause in erosion occurred when the east-flowing river system was blocked 8–4 Ma creating a lake—Lake Bidahochi (Ash, 1990). Several hundred feet of lake and stream sediments as well as local lava flows covered the partially eroded Chinle Formation.

FIGURE 8–34   Petrified logs commonly show regularly spaced cracks or joints. (Photo by E. Kiver)

Headward erosion by the lower Colorado River to the west began to cut the Grand Canyon, eventually draining Lake Bidahochi and initiating another episode of downcutting. Stream erosion and canyon cutting likely increased during the wetter intervals of the Pleistocene—intervals corresponding to the growth of mountain glaciers and ice sheets. Today, only small remnants of the Bidahochi Formation remain as the caprock on mesas and divides in the northeastern part of the park. Erosion continues a process that creates beauty from destruction. Slowly the badlands change and more logs and other fossils are uncovered. By human calendars this process is incredibly slow—by geologic standards the removal of the entire Chinle and its petrified forests here is inevitable in the next few million years—one of a number of reasons to visit this delightful area sooner rather than later!

Future of the Past

As Lubick (1996) skillfully points out, this area is not known for its spectacular, eye-popping scenery. Rather, this is a scientific park where the educational values are there for those who take time to learn and appreciate our continuity with the past. The Chinle record permits a relatively complete picture of an entire ecosystem to be reconstructed. The Triassic was a transitional world where the flora was changing from domination by ferns and horsetails to conifers and other advanced plants, such as the cycadophytes. Also the amphibians of the late Paleozoic world were losing their dominance to the reptiles who would soon inherit the world, at least during the remaining 160 million years of Mesozoic time! Even more than at most other parks, Petrified Forest cannot be truly appreciated through an automobile windshield.

Begin your exploration at the visitor center in the north part of the park or at the Rainbow Forest Museum in the south. Be sure to take as many trails through the various fossil forests as you can. Agate Bridge in the south is a unique type of *natural bridge* (a rock span across an erosional channel or valley). The rock bridge here is an unfractured petrified log over 100 feet (30 m) long that spans a 40-foot-wide (12-m) *arroyo*, or gully. Reportedly, a local cowboy named Tom Paine rode his horse across the petrified log in 1886 to win a $10 bet. Jeffers (1967) suspects that the money was later spent at the Bucket of Blood Saloon in nearby Holbrook! As old bridges eventually collapse, new ones will form where streams are diverted to flow under resistant, nonfractured logs (Fig. 8–35) or under fractured logs whose sections precariously support one another. Since we define unfractured rocks as monoliths—perhaps unfractured logs should be called monologs! In the Blue Mesa as well as other areas, sections of large logs form the protective caps of *pedestal rocks*. Weathering will eventually remove the shale beneath, and the log will fall down the slope, as others here have done for the past few million years.

Walking among the huge, silent logs is a moving experience for those sensitive to our earth's amazing history. Picking up and examining bits of petrified wood and placing them back on the ground is a delight and an experience that most people never have, except for those who visit Petrified Forest. However, some of the selfish and unthinking individuals among us find the lure of a souvenir too great to ignore. After all, what significance is the loss of a few ounces or a pound of fossil wood? Aren't there hundreds of thousands or more pieces lying about near the trails—how could one piece be missed? By observing visitors who were unaware of being watched, it was determined that a small percentage of the visitors placed one or more samples in their pockets or in their automobiles. However, a small percentage of the approximately one million visitors each year is a large number—especially when their actions add up to an estimated 12–16 tons of stolen fossil wood each year! Unfortunately, stealing from the experience of every future visitor does not seem to be of concern to thoughtless individuals. As you approach the exit, you will see a large sign warning you that your car will be searched. This area between the sign and the exit might be called the "repentance area" because several hundred pounds of souvenirs have been retrieved from along this short section of road. Removing materials from a national park is a federal offense punishable by fines, imprisonment, or both. Legally collected samples from outside the park are available in local curio shops for those who must have a souvenir.

FIGURE 8–35　Onyx Bridge, a natural bridge before and after its partial collapse. (Photos by Sidney Ash)

REFERENCES

Ash, S.R., 1987, Petrified Forest National Park, Arizona, in Beus, S.S., ed., Rocky Mountain Section of the Geological Society of America: Geological Society of America, Boulder, CO, Centennial Field Guide, p. 405–410.

Ash, S.R., 1990, Petrified Forest: The story behind the scenery: Petrified Forest Museum Association, 48 p.

Ash, S.R., 1992, The Black Forest Bed; a distinctive unit in the Upper Triassic Chinle Formation, northeastern Arizona: Journal of the Arizona–Nevada Academy of Science, nos. 24–25, p. 59–73.

Ash, S.R., and Creber, G.T., 1992, Palaeoclimatic interpretation of the wood structures of the trees in the Chinle Formation (Upper Triassic), Petrified Forest National Park, Arizona, USA: Palaeogeography, Palaeoclimatology, Palaeoecology, v. 96, p. 299–317.

Creber, G.T., and Ash, S.R., 1990, Evidence of widespread fungal attack on Upper Triassic trees in the southwestern U.S.A.: Review of Palaeobotany and Palynology, v. 63, p. 189–195.

Jeffers, J., 1967, Petrified Forest: Arizona Highways, v. 43, no. 6, p. 2–39.

Litwins, R.J., and Ash, S.R., 1991, First early Mesozoic amber in the Western Hemisphere: Geology, v. 19, p. 273–276.

Lubbick, G.M., 1996, Petrified Forest National Park: A wilderness bound in time: Tucson, University of Arizona Press, 212 p.

Park Address

Petrified Forest National Park
Petrified Forest National Park, AZ 86028

CANYONLANDS NATIONAL PARK (UTAH)

A proposal was made in the 1930s to establish an Escalante National Park that would extend from Moab, in southeastern Utah, to the Grand Canyon in northwestern Arizona. What an incredible park that would have been! Rather than one "superpark," a number of smaller parks separated by nonpark areas were established, including Utah's largest national park, Canyonlands. Bates Wilson, Canyonlands first superintendent, worked tirelessly to develop the proposal that led to the establishment of the park in 1964 and guided its initial development. The area was enlarged to 525 square miles (1360 km^2) in 1971 and includes the canyons of both the Green and Colorado rivers (Fig. 8–36) and an amazing variety of mesas, buttes, pinnacles, grabens, and arches that surround the canyons. In addition, there is a small remote area, Horseshoe Canyon Detached Unit, about 10 miles (16 km) west of Green River.

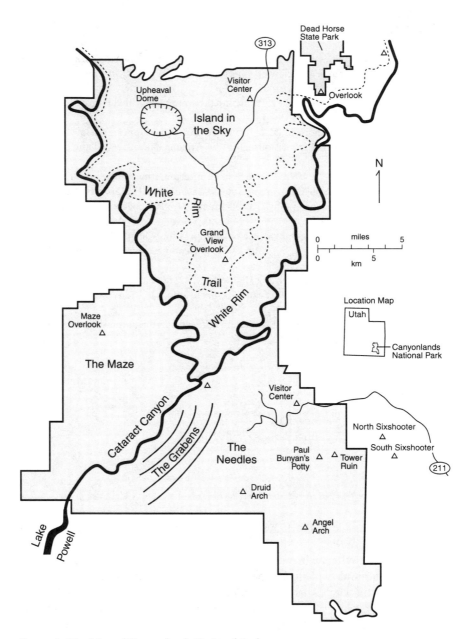

FIGURE 8–36   Map of Canyonlands National Park.

History

Native Americans have lived here for thousands of years. The Fremont people hunted and farmed in the area some 1100 years ago. Later the Anasazi people cultivated crops and constructed pueblo-style homes—mysteriously abandoned about 800 years ago, apparently as the Little Ice Age (Fig. 1–25) began. In the high mountains climates cooled and glaciers grew. However, in the Colorado Plateaus water became less abundant, particularly during the critical growing season. Increasing crop failures drove the Anasazi away from their well-organized homes and farms (Petersen, 1994). The Paiute (and Navajo farther south) were relative newcomers who entered the deserted land only a few hundred years ago. Granaries, cliff dwellings, spectacular pictographs such as the All American Man and the Great Gallery, and the petroglyphs at Newspaper Rock just outside of the eastern park boundary were left by the first Americans. These cultural features are by themselves national treasures worthy of protection. When combined with the outstanding scenery of the "Red Rock Country," the area provides an opportunity for a never-to-be-forgotten experience.

Major John Wesley Powell's 1869 and 1871 pioneering boat expeditions through the canyons of the Colorado took his group through one of the last uncharted wilderness areas in the United States. Powell named Cataract Canyon and many other geographic features along the river. His picturesque description of the area is still appropriate (Powell and Porter, 1969, p. 76).

> The landscape everywhere, away from the river, is of rock—cliffs of rock; tables of rock; terraces of rock; crags of rock—ten thousand strangely carved forms. Rocks everywhere, and no vegetation; no soil; no sand. In long, gentle curves, the river winds about these rocks.

Powell was followed into the Canyonlands area by interesting groups of people including cattlemen, bank robbers like Butch Cassidy and the Sundance Kid, and later by oil men and uranium miners. The current invasion, which shows signs of getting out of hand, is by tourists and recreationists. Their numbers have increased by 14 percent every year since 1982!

The uranium boom from 1946 to 1954 brought thousands of "strike-it-rich" prospectors into the Colorado Plateaus. Moab, a few miles east of Canyonlands, was one of the hubs of activity as described in Ringholz's (1989) book *Uranium Frenzy*. Sedimentary rocks with a high content of organic material such as the Chinle and Morrison formations were particularly sought after for their local concentrations of uraniferous ores like "yellow cake" (carnotite, a uranium oxide). As we rapidly use up (in far too many cases, waste) the world's supply of oil and gas, the current generation now being born will, out of necessity, become increasingly dependent on nuclear energy. Perhaps another episode of uranium prospecting is in the wings.

Geographic Setting

The Colorado River heads in the spectacular high country of Rocky Mountain National Park in Colorado. It flows past the scenic Colorado National Monument to Canyonlands where it joins the Green River. Below Canyonlands, the Glen Canyon National Recreation Area extends some 115 miles (185 km) downstream through Lake Powell to Glen Canyon Dam in Arizona. From there it flows through the mile-deep canyon at Grand Canyon National Park and into Lake Mead (National Recreation Area)—an impressive string of outstanding areas—a source of national pride.

Canyonlands is a time-consuming park. It is really four parks in one, each one requiring considerable driving time and effort to see adequately. The two major rivers of the Colorado system (the Colorado and Green rivers) join here at Canyonlands and divide the park into three sections. The pie-shaped wedge of plateau land between the rivers is called the Island in the Sky (Fig. 8–37) and is the most easily accessible section of the park. It is reached by a paved access road that begins 8 miles (12 km) west of Moab along U.S. Route 191. The Needles Section is east of the Colorado River in the southeast corner of the park and is accessible from U.S. 191 south of Moab. The Maze Section is in the southwest corner of the park and can only be reached by 4-wheel drive roads. The Maze has been described as a "30-square-mile puzzle in sandstone." The exceptionally rugged topography here makes parts of the Maze practically inaccessible except for those willing to climb and scramble.

The fourth "park" is the canyons themselves, 2000-foot-deep (610-m) gorges that can be fully appreciated only by those who float the mighty rivers through the park. The float to the confluence is delightful but relatively unchallenging for those wishing to test their boating skills. From there on all bets are off. A short distance from the confluence, the river enters Cataract Canyon, the wildest 14-mile (23-km) section on the Colorado—more treacherous than the impressive rapids in the Grand Canyon! With names like Capsize Rapids and Big Drop and over 50 nearly continuous rapids, this is indeed a major challenge for river runners. Seven rapids

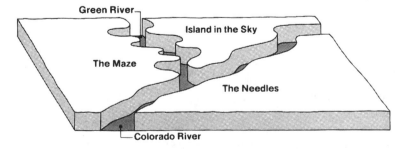

FIGURE 8–37　Diagrammatic sketch showing division of Canyonlands National Park into three major sections. (From National Park Service brochure)

are rated from 8 to 10 for a 16-foot (4.9-m) boat (Baars and Molenaar, 1971). A rating of 10 means that there is a 50–50 chance that an experienced boatman will capsize! Unfortunately, only about one half of the rapids in the lower Cataract Canyon that Major Powell experienced remain today. Construction of Glen Canyon Dam eliminated forever (by human time scales) the beautiful Glen Canyon and drowned half of the rapids in Cataract Canyon as well.

Canyonlands is not for the casual tourist with a couple of hours to spare. It is largely undeveloped, and advance preparations are necessary if the visitor is to do more than catch glimpses of what Canyonlands has to offer. The main roads are passable by highway cars, but 4-wheel-drive vehicles or, preferably, a good pair of hiking shoes and good physical condition are needed to get into some of the more remote and spectacular areas. Adventures in the back country are not without hazard or difficulty. A sudden summer downpour can quickly make 4-wheel-drive roads impassable and self sufficiency is a must, particularly on hot summer days with no water available in the back country except that in your canteens. Check with a ranger before attempting the longer hikes in the park.

Bedrock Geology and Geologic History

Mesozoic rocks are the most commonly exposed surface rocks on the Colorado Plateaus except in areas where uplifts have brought older rocks closer to the surface. At Canyonlands, a giant, gentle anticline called the Monument Upwarp extends northward through the park (Fig. 8–2) where, along the axis of the fold, late Paleozoic rocks are within reach of the awesome power of the Colorado and Green rivers (Lohman, 1974). The nose of the anticline slopes gently (technically called *plunge*) under the Island in the Sky and Arches National Park immediately to the north where only younger, Mesozoic rocks are exposed. Thus we continue our "layer cake" tour of the Colorado Plateaus national parks by generally progressing to younger and younger strata. Yet older rocks beneath the late Paleozoic strata are known from oil wells in the vicinity and from areas such as the Grand Canyon, Black Canyon of the Gunnison, the Uinta Mountains near Dinosaur National Monument, and other areas where erosion has exposed the older chapters of the book of earth history. Here we will pick up the story in late Paleozoic (Pennsylvanian) time. The generalized cross section in Figure 8–38 will be helpful in following the geologic history discussion and Figure 8–36 locates geographic features mentioned in the text.

The geologic history chapters exposed in the Canyonlands area tell of the formation of a basin, the Paradox Basin, when the Ancestral Rockies (Uncompahgre Highlands) were formed during Pennsylvanian (late Paleozoic, about 300 Ma) time. Evaporation of seawater produced thick deposits of salt (mostly the mineral halite) and other evaporite minerals in the Paradox Basin. Later, debris shed from the nearby Uncompahgre Highlands interfingered with coastal deposits and marine sediments as the basin and surrounding areas were buried under thousands

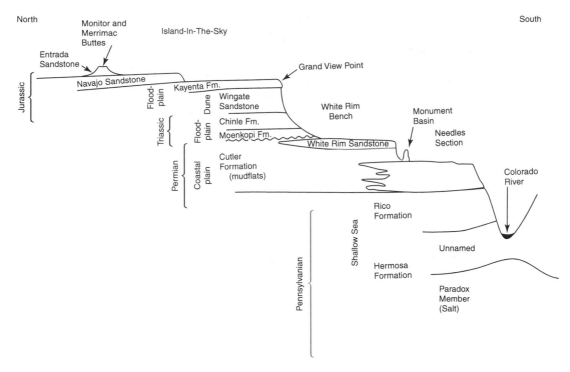

FIGURE 8–38   Highly generalized cross section in Canyonlands National Park (not to scale).

of feet of sediment. Extensive deserts formed in Jurassic time and thick deposits of dune sands accumulated. Marine conditions briefly returned during the late Mesozoic (Cretaceous) only to be terminated for the last time as the Laramide Orogeny uplifted and deformed the lithosphere in the Rocky Mountain and Colorado Plateaus areas. Movements along ancient faults produced downwarps and local uplifts—one of which, the Uncompahgre Uplift, occurred at the same site as the Pennsylvanian-age ancestral Rocky Mountain uplift. Apparently deep-seated faults of Precambrian age underlie this part of the Colorado Plateaus and are reactivated when crustal stresses of sufficient magnitude and orientation occur.

The uplifted edges of individual plateaus on the Colorado Plateaus are often broken by faults or by a type of single-flexure fold called a monocline. More sediment buried the relatively low-elevation Colorado Plateaus in early Tertiary time only to be removed later as vertical uplift in Miocene and Pliocene time energized the then sluggish rivers crossing the Plateaus. Thousands of feet of younger sediment were removed and rivers cut deeply into the remaining strata.

The oldest exposed rock unit, the Pennsylvanian Hermosa Formation (Fig. 8–38), is exposed at the confluence of the Green and Colorado rivers and locally downstream. The overlying Rico Formation is also exposed near the confluence (Fig. 8–39), below Dead Horse Point in the Island in the Sky section (Fig. 8–40), and elsewhere in the canyon bottoms. The overlying Cedar Mesa Sandstone Member of the Cutler Formation (Permian age) consists of interfingering white sand layers deposited as offshore bars and coastal dunes along a seaway to the west and red *arkosic* (rich in feldspar grains) stream-deposited sediments derived from the Uncompahgre Highlands to the north. This colorful, candy-striped unit makes up much of the delightful scenery in the Needles and Maze sections.

The brilliant white color of the White Rim Sandstone Member at the top of the Cutler Formation makes it easy to identify. It is a marine sandbar-coastal dune unit that forms an intermediate plateau surface best seen about 1200 feet (366 m) below Grand View Point (Fig. 8–41). The 100-mile-long (60-km) White Rim 4-wheel-drive trail was established in the 1950s to provide access to the hordes of prospectors and miners who combed the area searching for uranium deposits. The road follows the persistent White Rim bench from the salt-mining facility at Potash along the Colorado River to the northwest edge of the park.

FIGURE 8–39 The confluence of the Green (left) and Colorado rivers tells a changing story about storms in the two drainage basins. The murky, turbulent flow in the Green River shown here records more recent or severe storms days or weeks before this photo was taken. (Photo by E. Kiver)

FIGURE 8–40   Deep, entrenched meanders of the Colorado River are clearly visible from Dead Horse Point, just outside of Canyonlands. Rocks from Pennsylvanian (Rico Formation) through Jurassic (Kayenta Formation) are exposed in the canyon walls. (Photo by E. Kiver)

Following an episode of erosion, the first of the nonmarine Mesozoic strata, the Triassic age, brownish-red stream and deltaic sediments of the Moenkopi and Chinle formations were deposited. These mostly silt and sand layers are slope formers and are best seen from the White Rim road, just east of the park at the base of Six Shooter Peaks and along parts of the Needles section access road that leads west from U.S. 191. The Chinle Formation locally contains petrified wood, but not as abundantly as in its southern extension at Petrified Forest National Park. Many of the uranium deposits in the Colorado Plateaus occur in the Late Triassic Chinle Formation. A conglomerate layer at its base is particularly rich in radioactive wood fragments.

The next layer in our Canyonlands part of the Colorado Plateau layer cake is the Jurassic-age dune deposits that make up the spectacular Wingate Cliffs (Figs. 8–39 and 8–41) along the edges of the Island in the Sky. The slabby layers of the stream-deposited Kayenta Formation form most of the flat, tablelike surface of the Island in the Sky.

Desert conditions persisted in the Lower Jurassic as hundreds of feet of dune sands accumulated to form the Navajo Sandstone. Spectacular exposures of the Navajo occur along the access road to Island in the Sky and at Upheaval Dome where the edges of the rock layers are bent upward in a striking circular pat-

FIGURE 8–41   Kayenta–Wingate cliff with White Rim Sandstone in distance above Green River Canyon. Photographed from near Grandview Point at south end of Island in the Sky, Canyonlands National Park. (Photo by D. Harris)

tern. The overlying Jurassic-age Entrada Sandstone is mostly eroded away from Canyonlands but does form the impressive battleship-like buttes called Monitor and Merrimac buttes along the north entrance road.

During the Mesozoic, plate collisions along western North America were compressing, thickening, and deforming the Cordilleran Geosyncline sediments into the mountain ranges of the interior west. Mountain building migrated eastward, finally bending, breaking, and uplifting the Rocky Mountains during latest Cretaceous

(75 Ma) and early Tertiary time (ending about 40 Ma). This latter episode of mountain building in the Rocky Mountains is called the Laramide Orogeny. Canyonlands and the Colorado Plateaus were caught in the middle but were spared the violent upheavals of surrounding areas, perhaps because of the unusually thick, cool *lithosphere* plate beneath. The bending and faulting that occurred in the Colorado Plateaus during the Laramide Orogeny produced some of the present network of fractures and joints—important weaknesses that would strongly influence the development of the remarkable landforms of the future land of canyons.

Geologic Structures

At first view, Canyonland rocks appear to be horizontal. With low dips in most places, it is generally necessary to view broad areas in order to visualize these structures. The Monument Upwarp, a broad anticline (Fig. 8–2), extends northward through the park and is best seen from Deadhorse Overlook (located in Deadhorse State Park) or Grandview Point. A number of smaller northwest-trending, salt-cored folds occur nearby.

An unusual structure found in Canyonlands is the Meander Anticline whose crest follows the twists and bends of the Colorado River for over 25 miles (40 km) (Huntoon, 1982; Potter and McGill, 1978). How is it that a river should run exactly at the top of uplifted strata? The unique answer here may lie in the explanation that the anticline is younger than the river canyon! Another part of the explanation lies in the thick Pennsylvanian salt beds below the canyon floor that are part of the Hermosa Formation. When under pressure from the many tons of overlying rock, salt will tend to flow like Silly Putty to areas of lower pressure. Thus, the Meander Anticline is considered an unloading structure resulting from salt flowage toward an area where pressure is less—in this case where thousands of tons of rock have been removed in the canyon.

The Grabens area in the Needles section contains a spectacular series of arcuate fault blocks and valleys that parallel Cataract Canyon (Fig. 8–36). Again, the underlying salt beds are part of the answer. The rocks here are on the northwest flank of the Monument Upwarp and are gently inclined about 4° toward the Colorado River. When the river cut through the sandstone layers into the mechanically weak mudstone and evaporite layers below, the lateral support for the brittle sandstone layers was removed. The brittle sandstone slowly glided toward Cataract Canyon on top of the evaporite layers, breaking into a series of parallel fault blocks. Groundwater slowly dissolves the evaporite minerals further encouraging collapse, lateral sliding, and faulting of the sandstones. The process is very recent and began perhaps 300,000–500,000 years ago when Cataract Canyon was deep enough to remove lateral support. Thus, solution collapse and lateral movements of the brittle block formed the unique structures and topography in the Grabens area.

Fault scarps along the sides of the grabens are untouched by weathering and erosion; moreover, streams have not had sufficient time to completely adjust to the newly formed fault-block topography and often flow into depressions lacking outlets. The process continues today as the newest tensional fault is producing "baby graben" near Devil's Kitchen. As the rock continues to split and crack, the loose soil and desert plants fall into the opening abyss—about 40 feet (12 m) deep in 1993. Another indicator that salt is still being dissolved is that river water in and downstream from Cataract Canyon is undrinkable because of the high salt content.

Another outstanding structure that is easy to describe at the surface is the famous Upheaval Dome in the Island in the Sky section. It is a nearly perfect structural dome about 3 miles (4.8 km) in diameter that has been eroded into a topographic basin, an inversion of topography (Fig. 8–42). The Navajo Sandstone and older rock layers form a bullseye pattern around Upheaval Dome except that the center, some 1500 feet (457 m) below the rim, contains a confusing array of highly fractured and faulted Permian- and Triassic-age sandstone and mudstone blocks. Not so easy to describe and interpret is what lies below this intriguing structure. Without direct access to the rocks below, geologists use geophysical techniques such as measuring the force of gravity and magnetic fields (these vary with different types of rocks beneath the surface) as well as interpreting the behavior of shock or seismic waves as they travel through the deeper rocks and strata. Both gravity and magnetic high values occur directly over the structure. Multiple ideas are being considered, including igneous processes (deep gas explosion or igneous intrusion), salt flowage, meteor impact, or a combination of processes. The idea of a 1700-foot-diameter (518-m) asteroid or comet slamming into the earth at about 20,000 miles per hour (32,200 km/hr) certainly does not lack of the sensational. However, such an explanation currently lacks strong supporting evidence. As investigations continue, new information will eventually be forthcoming and will either confirm, cast doubt, or reject some of the current ideas. Such are the ways of science.

FIGURE 8–42   Air view into Upheaval Dome, with loops of the Green River in the distance. (Photo by J.A. Campbell)

Erosional Landforms

Under this general heading, the processes of weathering and mass-wasting are implied and in some cases are fully as important as erosion by running water. Within the confines of the park, fins, columns, needles, balanced rocks, potholes, and weathering pits have formed in untold numbers by a combination of processes.

Special mention must be made of the fascinating landforms known as natural arches—rock spans formed mainly by weathering rather than running water processes. Such features usually require the presence of narrow, linear rock ridges called fins. The abundance of cross joints at Canyonlands favors, instead, the formation of pinnacles such as those in the Needles and Land of Standing Rocks. However, about 25 arches do occur in the park with the huge Angel Arch and Druid Arch (Fig. 8–43) being outstanding examples. A 4-wheel-drive trail leads to the Angel Arch trailhead and a trail to Druid Arch involves an 8.5-mile (14-km)

FIGURE 8–43 Druid Arch stands 310 feet (95 m) tall—one of the most massive in North America. (Photo by E. Kiver)

round-trip hike that is well worth doing for those in good physical shape. A side trip to Paul Bunyan's Potty (Fig. 8–44), a so-called horizontal arch, is also worthwhile.

Although the canyons are for those who "go down to the river in boats," spectacular views by "land lubbers" are available near auto roads at Dead Horse State Park (Fig. 8–40) and Grand View Point Overlook at the south end of the Island in the Sky section. Four-wheelers, bicyclists, and hikers have excellent views along the White Rim trail and at the Colorado and Green River Confluence Overlook (Fig. 8–39).

A peculiar feature along the rivers that is best seen on a park map (Fig. 8–36) and at Dead Horse Point and other viewpoints is the anomalous condition of tight meander bends enclosed by deep canyons. Meanders are features usually found in more gently flowing rivers on broad floodplains such as the lower Mississippi River. Such features in an area where river flows descend steeply downstream in canyons seems contradictory. The likely explanation is that the meanders "are at least partly inherited from ancestral streams" (Harden, 1990) that flowed across the area before the most recent uplift. Rivers perhaps established their meanders when thick, relatively weak Tertiary-age strata still covered the area. When erosion lowered the stream beds to the base of the Tertiary sediments, some of the stream segments continued to maintain their meander pattern and grind their way into the resistant Mesozoic and Paleozoic strata below forming *entrenched meanders*. Thus, the streams created resistant rock walls that trapped the streams between. Just as river meanders experience *cutoffs* when water flows across narrow necks of land in the Mississippi River basin and elsewhere, so to do entrenched meanders experience

FIGURE 8–44   Paul Bunyan's Potty, a pothole arch in the southern section of Canyonlands National Park. (Photo by J. Gregory)

cutoffs as rock walls separating nearby channel segments are eventually tumbled by erosion—enabling the stream to take a short cut on its way to the sea. The abandoned river channel left high and dry is called a *rincon* in the southwestern United States and is analogous to an abandoned *oxbow* in a meandering stream.

Unquestionably Canyonlands is one of our finest parks. At first glance one might think that a layer cake can't have too many interesting embellishments. At second glance the variety of landscapes, their fascinating history, and the rugged beauty of the area continues to lure many of us back for yet another visit. Its "wildness" is one of its major attractions. Yet, how can we maintain the feeling of remoteness with a burgeoning population and rapidly increasing numbers of visitors? Perhaps there are no answers, or more likely, we will do as society has so often done in the past when long-range, difficult decisions should be made—let the next generation try to solve the problem. Unfortunately, problems like these intensify and options become fewer with passing time.

REFERENCES

Baars, D.L., and Molenaar, C.M., 1971, Geology of Canyonlands and Cataract Canyon: Four Corners Geological Society, Sixth Field Conference, Cataract Canyon River Expedition, 99 p.

Harden, D.R., 1990, Controlling factors in the distribution and development of incised meanders in the central Colorado Plateau: Geological Society of America Bulletin, v. 102, p. 233–242.

Huntoon, P.W., 1982, The Meander anticline, Canyonlands, Utah: An unloading structure resulting from horizontal gliding on salt: Geological Society of America Bulletin, v. 93, p. 941–950.

Lohman, S.W., 1974, The geologic story of Canyonlands National Park: U.S. Geological Survey Bulletin 1393, 126 p.

Petersen, K.L., 1994, A warm and wet Little Climatic Optimum and a cold and dry Little Ice Age in the Southern Rocky Mountains, U.S.A.: Climatic Change, v. 26, p. 243–269.

Potter, Jr., D.B., and McGill, G.E., 1978, Valley anticlines of the Needles District, Canyonlands National Park, Utah: Geological Society of America Bulletin, v. 89, p. 952–960.

Powell, J.W., and Porter, E., 1969, Down the Colorado: E.P. Dutton & Co., New York, 168 p.

Ringholz, R.C., 1989, Uranium frenzy: Boom and bust on the Colorado Plateau: W.W. Norton, New York, 310 p.

Park Address

Canyonlands National Park
125 West 200 South
Moab, UT 84532
801-259-7164

ZION NATIONAL PARK (UTAH)

The Grand Staircase section of the Colorado Plateaus leads stratigraphically upward, stair-step fashion, from the Precambrian and Paleozoic rocks of the Grand
Canyon, across the Triassic-age rocks of the Chocolate Cliffs (Moenkopi Formation and Shinarump Conglomerate), Vermillion Cliffs (Chinle Formation and
Wingate Sandstone or locally the lower Navajo Sandstone), and up to the Jurassic-
age White Cliffs (upper Navajo Sandstone) at Zion National Park. Beyond Zion lie
the Pink Cliffs of Bryce—the uppermost and youngest layer in this giant geologic
staircase.

Zion is an all-year park, each season having its own attractions. Utah Highway
15 through the southern section of the park and a branch road up Zion Canyon
(Fig. 8–45) enable the motorist to view many of the wonders of Zion. Those with
the spirit of exploration who stand at the rim of Grand Canyon and look into its
depths long to reach the Colorado River at its bottom. Here one stands along the
Virgin River (a tributary to the Colorado River), engulfed by sheer sandstone walls
over 2000 feet (610 m) high (Fig. 8–46) and a canyon rim 3000 feet (915 m) above,
and looks longingly at the canyon rim—imagining what the views from the rim are
like! Fortunately, there are trails here for all, ranging from a half-mile (0.8-km)
loop trail, to 3.4 miles (5.6 km) to the East Rim, 6.2 miles (10 km) to the West Rim,
and up to 18 miles (29 km) to Lava Point and Firepit Knoll from the canyon bottom. Unimproved roads also provide access to Lava Point and Firepit Knoll as well
as to trailheads in the wilder Kolob section to the north.

Zion is located on the Markagunt Plateau on the west edge of the Colorado
Plateaus (Fig. 8–45, Zion region inset map). Here, the separation from the fault-
block mountains of the Basin and Range Province to the west and the High
Plateaus section of the Colorado Plateaus is abrupt—marked by the huge normal
fault along Hurricane Cliffs. The Hurricane Fault has moved vertically as much as
6100 feet (1860 m) in the past 5–6 million years, tilting the plateau slightly to the
east and initiating the deep erosion that produced today's landscape (Grant, 1987).
Quaternary lava that flowed across the position of the fault perhaps 1000 years ago
has been displaced along the fault by as much as 2000 feet (610 m)! The Sevier Fault
to the east separates the Markagunt from its neighbor, the Paunsagunt Plateau
where Bryce Canyon National Park is located (Fig. 8–45, inset map). It too is a
young, still-active fault.

History

Gregory's (1950) classic study not only provides an excellent summary of the geology of the Zion region but also provides an excellent synopsis of the human history of the area beginning with the early Spanish explorations. Eardley and Schaack
(1994) review both the recorded history and prehistory of the area. Individual
family units of the Basketmaker culture sparsely inhabited the area in about A.D.

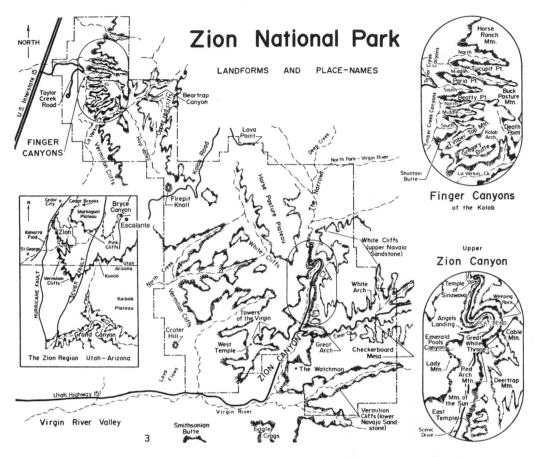

FIGURE 8–45   Map of Zion National Park and nearby region. Note the regional inset map and the enlargement maps of Zion Canyon and the Finger Canyons. (Illustration slightly modified from Hagood, 1985)

500. These hunter-gatherers also raised small plots of corn to help survive. In about A.D. 750 small Anasazi pueblos were constructed in the southern part of the present park and pit houses of the Fremont peoples were inhabited simultaneously in the north (Eardley and Schaack, 1994). Apparently, these were fringe settlements for each of these cultures with very little evidence of interactions between the two. By 1200 the climate changes associated with the Little Ice Age (Fig. 1–25) produced numerous crop failures—forcing this area and other drier sites on the Colorado Plateaus to be abandoned. The Paiutes settled the unoccupied lands during the nineteenth century. They compared the canyon with a "loogoon," or quiver of arrows—one comes out the way it goes in—an appropriate description of the almost unscalable cliffs that wall in the canyon of the Virgin River in the Zion area. The

peaceful Paiutes were there when the first Spanish explorers traveled through the area.

While the American Revolution was occurring (1776), Padres Dominguez and Escalante crossed the Virgin River near the park but avoided the formidable canyons. They were returning from their unsuccessful search for an easy route to connect Spanish settlements in Santa Fe with Monterey in California.

Mountain man Jedediah S. Smith with the American Fur Company explored some of the downstream areas in 1825. He originally named the river the Adams River (after President John Q. Adams) but reportedly changed it to the Virgin River after one of his men, Thomas Virgin, was wounded by Indians. However, some historians believe that the name is a modification of the Spanish Rio Virgen (Eardley and Schaack, 1994).

The Mormon settlement of the Great Salt Lake area in 1847 soon led to the church sending scouts out to seek additional areas where water and arable land were available. Nephi Johnson explored the canyon up to the Great White Throne and possibly to the narrows in 1858 (Fig. 8–45). A few Mormon settlers followed, establishing small farms on the flat valley floor by 1861. Isaac Behunin built his cabin on the fine-grained lake sediments close to where Zion Lodge is located today. One of his polygamist neighbors built a number of homes, each to house one of his separate families.

John Wesley Powell visited the southern Utah region in 1872 on returning from one of his Colorado River expeditions and used the Indian name Mukuntuweap ("straight canyon") for the spectacular canyon along the upper Virgin River. Pictures by Powell's expedition photographer Jack Hiller and paintings by artist Frederick Dellenbaugh would soon help ignite the interest of others in these magnificent canyons and pave the way for its eventual protection under the watchful eye of the National Park Service. To the few Mormon settlers, such as Isaac Behunin who farmed in the canyon, the grandeur and beauty of the spectacular white and pink monoliths and canyon walls reminded them of a great cathedral or heavenly place—Behunin called it Little Zion.

The area became Mukuntuweap National Monument in 1909. Subsequently the locally unpopular name was changed to Zion in 1918 and enlarged to its present 230 square miles (596 km^2) in 1956.

Bedrock Geology

As is true of any area with canyons—the canyons are young and the rocks that they cut are old. Zion is essentially a Mesozoic park; all of the sedimentary rocks are Mesozoic, but Quaternary lava flows and pyroclastics cover several areas. The Permian Kaibab Limestone, the same Paleozoic rock unit that forms the spectacular rim of the Grand Canyon, is exposed in the bottom of Timpoweap Canyon and in the Hurricane Fault cliffs a short distance outside the park boundary (Gregory, 1950). Thus, Zion Canyon is cut into rocks that rest on Paleozoic rocks. Exposed to view at Zion are layers belonging to the Mesozoic chapter in the Colorado Plateaus' book of rocks. A short distance north at Bryce Canyon and Cedar Breaks are Cenozoic rocks—a further continuation of the story. The stratigraphic position of Zion relative to other parks in the Colorado Plateaus is shown in the generalized north–south cross section in Plates 8C and 8D. The general geographic location of Zion relative to surrounding parks and features in the region is shown in the inset map (lower left in Fig. 8–45).

The oldest Mesozoic rock formation exposed is the Moenkopi near the Springdale entrance of the park, the same rock unit that forms the Chocolate Cliffs and Belted Cliffs farther south. This thick formation at Zion is a slope former that consists of mostly river-deposited sandstones and shales that formed on a coastal plain as streams washed lazily westward to the Triassic sea. A few gypsum and marine limestone beds are also present. The Moenkopi is separated from the overlying Shinarump Conglomerate (lower member of the Chinle Formation) by an unconformity formed as streams once again flowed across the area.

Next in the stratigraphic stack is the Chinle, mainly shale about 1000 feet (303 m) thick that is bright red and easily identified by its color. As in Petrified Forest National Park, the Chinle in Zion contains silicified logs and the remains of amphibians and reptiles. Much of the petrified wood and bone material is radioactive and was prospected intensely outside of the park during the uranium boom of the 1950s. The apparent tropical conditions recorded in the fossils in these Triassic rocks initially seemed mysterious to early geologists. Plate tectonics now provides reasonable answers. Recall that the crustal plates had assembled during the late Paleozoic into the supercontinent that geologists have named Pangaea. The supercontinent began to break apart during the early Mesozoic and the fragment containing North America "inched" its way northward from near the equator through warm tropical latitudes during the Triassic Period. Later, during the Jurassic, this slower-than-a-speeding-snail movement would carry the plate into and through the dry latitudes north of the equator, creating desert conditions much like those found in today's low-latitude deserts.

On top of the Chinle is a formation, the Moenave, which we haven't seen before. It is a reddish stream-deposited sandstone, about 350 feet (107 m) thick, which in its lower part contains lake sediments and fossil fish of Jurassic age, including an 18-inch (46-cm) fossilized fish skeleton—definitely a keeper! Resting on the Moenave (Springdale Sandstone Member, well exposed near the Springdale entrance) is about 200 feet (60 m) of silts and sands of the slope-forming Kayenta Sandstone. Dinosaurs were in Zion during this part of the Jurassic, as their tracks in these stream-deposited sediments testify. Soaring high above the Kayenta are the bold cliffs of Zion, cliffs of Jurassic-age (about 170 Ma) Navajo Sandstone (Fig. 8–46).

During most of Jurassic time, the Colorado Plateaus area was distinctly arid, and wind erosion and deposition were widespread. This area was located about 30° north of the equator in the dry horse latitudes and in addition was cut off from a moisture source to the west by the newly formed mountain highlands produced during the early Cordilleran (Sevier) Orogeny (Stokes, 1987). In some places sed-

FIGURE 8–46   Sheer walls of Navajo Sandstone rise majestically some 2000 feet (610 m) from the floor of the Virgin River canyon. (Photo by E. Kiver)

iments were deposited by both wind and water; in the Zion area, the Navajo Sandstone was, with minor exceptions, wind deposited. Stokes (1987) appropriately describes the Navajo Sandstone as the "greatest scene-maker of the western United States." The area covered was large, a virtual sand sea covering parts of six states from Nevada to Wyoming in a desert at least the size of today's Sahara. The sand accumulated about 175 Ma, burying the sandy desert floor with sand layer upon sand layer—as much as 2200 feet (670 m) at Zion. Curved and thick wedges of cross-bedding (Fig. 8–47) and the frosted grains of quartz sand testify to an eolian rather than fluvial origin. *Cross-bedding* is the fossilized remains of sand layers parallel to the former graceful lee side of a dune.

Sand blown over a dune crest rolls down the lee side forming layers inclined at the *angle of repose*—the maximum angle that particles will hold without slipping or rolling. The sand rolls or cascades down the dune face, producing concave-upward layers whose maximum slope is 34°. As Hagood (1985) appropriately describes these pleasing geometric lines, they represent "motion frozen in time." The direction of inclination of these layers further tells the paleo-wind direction—in this case from the north and west.[10] A side trip to nearby Coral Pink Sand Dunes

[10]The wind direction is given relative to today's land orientation. In addition to the north-ward movement of the North American Plate during the past 200 million years, considerable rotation of the plate also occurred.

FIGURE 8–47   Eolian (wind) cross-bedding in Navajo Sandstone, the main rock formation in Zion National Park. (Photo by D. Harris)

State Park provides a glimpse of a small area of active dunes and eolian processes. The processes and the laws of physics that are producing them are the same ones that controlled the Navajo dunes of Jurassic time some 170 Ma. One difference is that many of the tracks in the sand at Coral Pink Sand Dunes are not of dinosaurs, but of modern-day machines called off-road vehicles (ORVs)!

Later in the Jurassic, the wind gave way to running water as streams deposited the red muds of the Temple Cap Formation followed by marine limestones of the Carmel Formation about 145 Ma. Remnants of the still younger Cretaceous shales and sandstones that formerly covered the park are present in the Gray Cliffs (part of the Grand Staircase) to the north. The early Tertiary lake and stream sediments of the Claron Formation, which dominates the scene in Bryce Canyon, probably covered the Zion area also. However, much of the Carmel and all of the younger formations were stripped from the area as the spectacular canyons of Zion were being carved out.

Late Mesozoic–early Tertiary times brought crustal compression (a continuation of the Sevier Orogeny to the west) that buckled some of the strata into a sharp fold and thrust rocks eastward forming the Taylor Creek Thrust Fault in the Kolob section (Grant, 1987). Later Tertiary uplift and reactivation of ancient Precambrian faults blocked out the 100-mile-long (160-km) by 30-mile-wide (48-km) Markagunt Plateau (Fig. 8–45, Zion region inset map).

Uplift during the Tertiary, and especially in early Pliocene time (about 5 Ma), initiated a new cycle of erosion, one of such widespread consequences that it is known as the Great Denudation. Also contributing to this profound episode of erosion was lowering base level as the Colorado River established a path to the Pacific Ocean and began to cut the Grand Canyon. Thus, the combination of vertical uplift and lower base level increased the vertical drop (potential energy) of the Colorado and tributary streams. The response was to erode vigorously downward—a powerful process that continues unabated today in the Colorado Plateaus (except where dams have been constructed). Deepening of Zion Canyon by slightly less than an inch (2 cm) every 100 years may seem insignificant by human time standards. By the geologic clock this is rapid—Zion Canyon will be another 600 feet (180 m) deeper in another 900,000 years! However, waiting for the scenery to become even more spectacular with time is not a good reason to delay your visit.

Volcanoes erupted at various times during the uplift and spread lava flows and pyroclastics over parts of the Plateaus, including Zion National Park. The tensional stresses that produced the Basin and Range Province to the west and the High Plateaus boundary faults also caused local faults to move in the Coalpits Wash area of Zion. Melting of mantle material occurred at depth and basalt erupted along the faults. The Firepit and Spendlove Knolls cinder cones along the West Cougar Fault are accessible from the Kolob Reservoir road (Fig. 8–45). The Crater Hill cinder cone (accessible by trail) still retains its crater and is relatively unaltered, indicating that it is of recent origin, perhaps only a few hundred years old (Eardley and Schaack, 1994). Lava filling in washes and valleys in the Coalpits Wash and North Creek areas displaced creeks to valley edges, creating conditions where *inversions of topography* will occur in the future. In such situations previous valley floors become topographic ridges, and former valley walls and stream divides become valleys!

Geomorphic Development—Fractures

The Plateaus were uplifted differentially, causing folding, fracturing (*jointing*), and faulting. The Markagunt Plateau, of which the Zion area is a small part, was uplifted to about 10,000 feet (3050 m) above sea level. The uplift caused the Virgin River to begin cutting its canyon—Zion Canyon. Tributaries to the Virgin in turn started to downcut, and new tributaries also began to develop—mostly along the north- to northwest-oriented vertical fractures. That these major joints are similar in orientation to the nearby Hurricane Fault that bounds the west edge of the Colorado Plateau is probably not coincidental. Since the main fractures are parallel, many of the tributary streams are parallel. Where the heavily fractured rock is removed by erosion from the main joints and cross joints, the unfractured core rock eventually forms monoliths such as the 2450-foot-high (747-m) Great White Throne and the various mesas, buttes, and rock monuments ("temples") that help make Zion so scenically attractive.

Fractures played another role in the formation of Zion's topography. The fractures are near vertical and so are the cliffs. The streams cut down through the sandstone and formed narrow, vertical-walled, slotlike gorges. Once the highly fractured rock along the joint is removed, the process of exfoliation or unloading becomes an important process acting on the 2000-foot-thick (610-m) massive sandstone body. As the many thousands of tons of pressure on buried rock are removed, the brittle rock will expand slightly toward an area of low pressure—in this case toward the canyon wall. Vertical fractures form parallel to the free face of the cliff and help maintain the vertical cliff.

Many of these slotlike canyons remain today in all stages of development. Some near the stream headwaters are only a few tens of feet deep, Hidden Canyon is hundreds of feet deep, and the Deep Creek area in the Narrows has 1000-foot-high (305-m) nearly vertical walls that are less than 20 feet wide (6 m) at the bottom! Streams are usually wider than they are deep. Here, however, during flood flows the Virgin River is on edge, so deep is the water. It is at these times that many tons of sand and gravel up to boulder-size rocks are flushed down the channel, turning the streams into "ribbons of sandpaper" (Eardley and Schaack, 1985). For those who make the 16-mile (26-km) trek (hike and wade) through the Narrows, the sight of logs wedged into rock crags 40 feet (12 m) above the stream and the knowledge that summer flash floods can swell the seemingly peaceful river to flood proportions in 15 minutes or less can be unnerving. Compensation for the physical effort is the rugged beauty of the canyon, parts of which are bathed in sunlight for less than an hour a day.

Geomorphic Development—Valley Widening

Immediately downstream from the Narrows, the character of the valley changes. The area between the parking lot by the Temple of Sinawava and the entrance to the Narrows is swampy with numerous springs emerging from the lower canyon wall. The canyon is much wider here and continues to widen downstream toward the Springdale entrance—yet the vertical walls of the Navajo Sandstone are maintained through the canyon length. Why should the canyon width change so suddenly and dramatically here? The answer lies in the layers of strata and their physical characteristics.

Where the Virgin River cuts downward only in the Navajo Sandstone, the canyon remains narrow. Where the stream cuts below the Navajo into the underlying Kayenta mudstones and sandstones, the canyon widens. The mechanically weak Kayenta is removed by the lateral wanderings of the Virgin River and by the many springs that flow out at the Navajo–Kayenta contact—a process called spring sapping. Groundwater descends vertically through the porous Navajo Sandstone and then moves laterally on top of the impermeable Kayenta sandstones and mudstones and emerges as a springline along the canyon walls. Spring outflow further contributes to undercutting by removing the calcium carbonate cement in the sandstones and by washing sediment away. The undercutting removes support for the massive sandstone walls, eventually triggering the catastrophic collapse of large slabs along the vertical joint and exfoliation fractures, thus maintaining the vertical valley walls.

Smaller Landform Elements

Springs emerging from the Navajo–Kayenta contact and along localized impermeable interdune deposits in the Navajo cause rates of weathering to increase and indentations or overhangs to develop in the canyon wall. From this destruction comes beauty at areas such as Weeping Rock and Hanging Gardens where springs nestled in overhanging alcoves support a verdant growth of water-loving plants.

Where large slabs fall from the canyon walls, great arches are often formed. The Great Arch of Zion (Fig. 8–48) is similar to the Great Arches in Yosemite. These are mostly inset arches with rock headwalls, not the see-through type of arch such as many of those at Arches and Canyonlands National Parks. However, enlargement of a joint behind an inset arch can produce a free-standing type of arch—such as the Kolob Arch in the northwest end of the park. This structure is 310 feet (95 m) across—the largest free-standing natural arch in the world.

Rockfalls occur from time to time. Numerous scars are apparent on the rock walls of the canyon. Depending on their age, some are very fresh appearing—many are highly weathered. A cohesive mass of rocky debris on the east flank of The Sentinel (Fig. 8–45) slid into the valley about 4000 years ago (Eardley and Schaack, 1994). Tens of feet of sediment accumulated locally in a lake behind the landslide dam, producing the flat terrace surface at Zion Lodge and as far upstream as Angels Landing. This huge mass of loose debris is still very unstable and is described by Eardley and Schaack (1985) as a "mountain on the move." Loose rock, on a daily basis, rolls and tumbles downslope. Small slides from the Sentinel Mountain debris occurred in 1923, 1941, 1966, and 1996. The 1996 slide dammed the lower part of the canyon, and water backing up forced the temporary evacuation of Zion Lodge. Other significant historic landslides include the Red Arch landslide that buried a cornfield in 1880, a 5000 ton block that fell near the end of the Narrows Trail in

FIGURE 8–48   The Great Arch, an inset arch in cliff of Navajo Sandstone at Zion National Park. The arch is 600 feet (183 m) across at its base and 400 feet (122 m) tall. (Photo by D. Harris)

1968, and a larger mass crashing down near the Mount Carmel Tunnel on Utah Highway 15 in 1958. The St. George earthquake in 1992 caused a large landslide just outside of the park entrance that destroyed a number of homes (Fig. 8–3).

The sandstone in the Temple Cap Formation and the limestone in the Carmel Formation are resistant to erosion and help maintain the flat-topped mesa and butte landforms. When the Temple Cap is removed by erosion, the Navajo tends to erode to more rounded forms such as the Beehives located about 1.2 miles (2 km) northwest of the visitor center.

In areas where slopes are less steep and the cross-bedding in the Navajo is well developed, the many cross-bedded layers are planes of varying hardness that experience *differential weathering*. Differential weathering along bedding planes and on gentler slopes produces a stair-step surface—a surface more to the liking of the amateur climber than the near-vertical walls of the Great White Throne. The famed Checkerboard Mesa in the eastern part of the park south of the main highway (Fig. 8–49) is a classic example. Here, closely spaced vertical fractures and more horizontal bedding planes and cross beds give the rounded landform a checkerboard appearance.

During dry periods there are a few waterfalls, as at Emerald Pool; there are many waterfalls immediately after a thunderstorm. Tributaries, dry most of the time, have been unable to downcut as rapidly as the Virgin River; consequently, they flow out of *hanging valleys* along the main canyon. Excellent examples occur in the Angels Landing area and elsewhere.

The variety of colors is delightful. Small amounts of oxidized iron give the Navajo Sandstone a red color, especially in the lower part of the formation. The upper thickness is a tan color that appears as a brilliant white in the sun—separated from the lower red section by a sharp, undulating boundary that may represent an

FIGURE 8–49   Checkerboard Mesa, Zion National Park. Weathering along the vertical fractures and the more horizontal cross-bedding produced the checkerboard pattern on the cliff. (Photo by E. Kiver)

ancient water table—a groundwater level that existed when the canyon was much shallower. If this hypothesis is correct, then chemically active groundwater may have leached the iron oxide coloring from the sand above the water table.

Other causes for the panorama of colors in Zion include the red streaks produced as the red muds of the Temple Cap Formation wash down the canyon walls, the white and gray streaks on cliff faces where dissolved calcium cement is reprecipitated on the rock face, and the black deposits of desert varnish, a biochemical precipitate.

Visiting Zion

Most tourists visit one or more parks in the Grand Staircase. Those who begin their tour at the visitor center and examine the exhibits will be better able to "plug" Zion into its proper position in the Colorado Plateaus layer cake. The views from the roads are outstanding and unforgettable for those who stop and drink in the magnificent scenery. All of the canyons are box canyons—dead ends for motorists. One exception is the Mt. Carmel Road up Pine Creek where an engineering marvel, especially for its time, is a 5600-foot-long (1707-m) tunnel completed in 1930 that provides a road connection between the Virgin River valley and the east entrance. For those who can, the ultimate Zion experience is found along the trails to viewpoints and more remote areas.

Angels Landing Trail is excellent for those who are in good physical condition and who enjoy high places and looking down 1500-foot (457-m) vertical drops. The Cable Mountain Trail takes one to the rim where an old platform marks the site where early settlers constructed a tram to lower lumber on a 2136-foot (650-m) cable to the valley floor. Observation Point, reached by the East Rim Trail, provides one of the best views of the entire canyon. Easier trails to the Gateway to the Narrows, Weeping Rock, Canyon Overlook near the Mt. Carmel tunnel, and the Emerald Pools are reached by trails a mile (1.6 km) or less in length.

With its multicolored cliffs Zion is something to see in the daylight; it is a different world in the moonlight. But try it in the blue-blackness just before first light.

REFERENCES

Eardley, A.J., and Schaack, J.W., 1985, Zion: The story behind the scenery: KC Publications, Las Vegas, Nevada.

Eardley, A.J., and Schaack, J.W., 1994, Zion: the story behind the scenery: KC Publications, Las Vegas, 48 p.

Grant, S.K., 1987, Kolob canyons, Utah: Structure and stratigraphy, in Beus, S.S., ed., Rocky Mountain Section of the Geological Society of America: Geological Society of America, Boulder, Colorado, Centennial Field Guide, v. 2, p. 287–290.

Gregory, H.E., 1950, Geology and geography of the Zion Park region, Utah and Arizona: U.S. Geological Survey Professional Paper 220, 200 p.

Hagood, A., 1985, This is Zion: Springdale, UT, Zion Natural History Association, 73 p.

Stokes, W.L., 1987, Geology of Utah: Salt Lake City, Utah Museum of Natural History, Occasional Paper Number 6, 280 p.

Park Address

Zion National Park
Springdale, Utah 84767-1099

CAPITOL REEF NATIONAL PARK (UTAH)

The central feature that makes possible the spectacular scenery in Capitol Reef is the presence of the Waterpocket Fold—a monocline that exposes the Wingate (Fig. 8–50) and Navajo sandstones—the two most important scenery makers in the Colorado Plateaus. Recall that a monocline is a single-flexure fold that connects flat-lying beds on either side—much like a rug covering adjacent steps on a staircase (Fig. 8–28). The fold brings rocks as old as Permian within reach of the deeply entrenched, easterly flowing Fremont River and Sulphur Creek (Fig. 8–51). A relatively complete section of Mesozoic strata (Triassic, Jurassic, and Cretaceous) is lined out in bands exposed along the monocline in and near the park (Smith and others, 1963). Capitol

FIGURE 8–50   Capitol Reef. Bold Wingate Cliffs at Capitol Reef are underlain by the Chinle and Moenkopi formations in the foreground. The term "reef," as used here, describes a high cliff or escarpment. (Photo by E. Kiver)

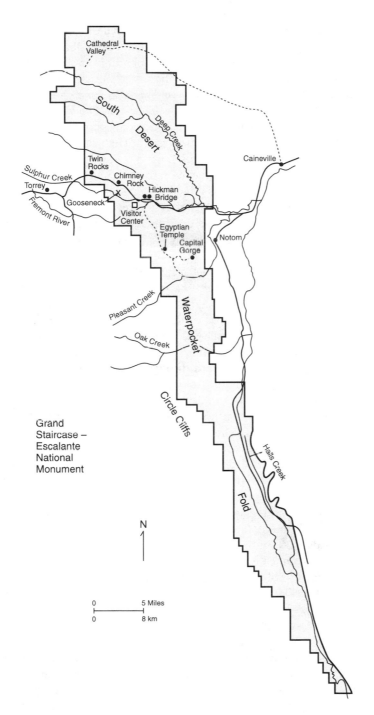

Figure 8–51   Map of Capitol Reef National Park. (Modified from National Park Service brochure)

Reef is outstanding for its combination of geology, archeology, and ecology. It is a highly photogenic park with rock formations of many colors—brilliant reds, yellows, purples, greens, and grays. Capitol Reef is also colorful in another way—its recent history. The exploration and settlement of the area, along with its questionable but immortal characters such as Butch Cassidy and his Wild Bunch, are included in Virgil and Helen Olson's (1990) comprehensive publication *Capitol Reef: The Story Behind the Scenery,* brilliantly illustrated with outstanding photographs.

Capitol Reef is near the boundary between the High Plateaus and the Canyonlands sections of the Colorado Plateaus (Figs. 8–1 and 8–2) and adjoins Grand Staircase–Escalante National Monument on the west and Glen Canyon National Recreation Area to the south. This grouping of national lands with outstanding scenery provides space for future generations to experience important parts of the American heritage. Elevations of the Aquarius and Fish Lake Plateaus to the west are over 11,000 feet (3354 m), the highest in the Colorado Plateaus. A lower plateau surface underlain by horizontal rocks near the town of Torrey and near the crest of the Waterpocket Fold are near 7000 feet (2134 m). Elevations decrease rapidly to the east, where the strata are bent sharply downward along the Waterpocket Monocline some 2000 feet (610 m) to near 5000 feet (1524 m) elevation where the strata again become horizontal. The Henry Mountains, one of the prominent intrusive-igneous mountains on the Colorado Plateaus, rise majestically from the desert floor to the east. The Fremont River heads on the Fish Lake Plateau (High Plateaus Section) and flows eastward through the park and eventually into the Dirty Devil River, a tributary to the Colorado.

Waterpockets, or *potholes,* are depressions formed in bedrock by turbulent streams. Other depressions that hold water for a short period of time form by weathering of slickrock surfaces and are called *weathering pits.* Water persists in the larger waterpockets, or *tanks,* for many weeks following a storm providing an important, even vital water source for wildlife and early settlers and explorers. The abundance of waterpockets combined with the folded strata accounts for the naming of the Waterpocket Fold. The monocline creates a rocky barrier to travel—whether by feet or wheels (Fig. 8–52). The miner's term for a linear topographic obstruction is *reef*—a term first used by sailors who left the sea to participate in an Australian gold rush (Smith and others, 1963). The other half of the park name comes from the rounded beehive-like forms of the Navajo Sandstone at the crest of the Waterpocket Fold between the Fremont River and Pleasant Creek (Fig. 8–51). These domelike forms reminded early explorers of the Capitol Building in Washington D.C.—hence the term Capitol Reef is applied to this part of the fold.

Capitol Reef is one of a string of parks from Grand Canyon to Arches that preserve some of the most remarkable scenery on the continent. These parklands as well as all others of federal and more local government administration reflect favorably on the human desire to do what is good. A respect for nature, pride in our heritage, and the unselfish desire to maintain parts of our earth in as pristine a condition as possible for those who come after us are noble human attributes—attributes not shared by all.

FIGURE 8–52   Looking southwest along Waterpocket Monocline. Note flat-lying beds on horizon and dipping beds in foreground. (Photo by D. Harris)

History

The hunter-gatherer-farming Fremont people occupied the floodplains along the Fremont River and some of its tributaries as early as 700. They constructed small granaries to store their seeds, corn, beans, and squash. They lived in pit houses— a less sophisticated life-style than their Anasazi neighbors to the south and east. Centuries later Paiute Indians who found the granaries with their small access openings attributed them to a race of tiny people they called the Moki or Moqui! Even today the granaries are called "Moki huts."

During the thirteenth century as the Little Ice Age gripped the higher elevations worldwide, colder and drier conditions at lower elevations in the Southwest brought disaster to the Anasazi and Fremont peoples farming in the Colorado Plateaus area (Petersen, 1994). Their traditional homes and farms were abandoned as people migrated to areas of more reliable sources of water or became nomadic. Today's 7 inches (18 cm) of annual rainfall continues to make conditions difficult for those wishing to derive a living from the land. Managing the area as parkland is unquestionably the best use of this desert land. The delicate environment is protected and the local economy is fueled in a sustainable manner.

Western explorer Colonel John C. Fremont discovered the river that now bears his name in 1854 and may have crossed the Waterpocket Fold along the Fremont Gorge. After perhaps three centuries of abandonment, Utes and Southern Paiute nomads began using the area again during the 1600s and were there when the first Mormon explorers arrived in 1866. The first settlers moved into the valleys in the 1870s, wresting a living from the incredibly rugged and forbidding landscape. Others followed. Some of the new settlers in the 1880s were polygamists who sought the solitude of this remote section of the Utah Territory. Occasionally the "feds" would try to raid the "polygs" as they were called in those days. Boys on fast horses would warn of their approach, and the polygs would hide out in Cohab Canyon and elsewhere inside the present park boundaries (Olson and Olson, 1990).

Many small farms and towns were unable to survive under the near-desert conditions. The small town of Fruita (now part of the park) survived, although the floodplain of the Fremont River there could not support more than 10 families at a time. Overgrazing made soils vulnerable to the flash floods and many small communities were washed away. The Fruita orchards and lawns survive today in the picnic, campground, and park headquarters area.

The Capitol Reef and surrounding areas were and are some of the most remote in the United States. The Fremont and nearby Escalante rivers were some of the last to be discovered and the Henry Mountains to the east were the last range of mountains to be named in the conterminous 48 states. No wonder Butch Cassidy and other outlaws found this sparsely populated and unmapped rugged area to their liking. Modern people also escape here—for very different reasons.

Local storekeeper Ephraim Pectol recognized the rugged beauty of the Capitol Reef area and established a "boosters club" in nearby Torrey to help spread the word to the "outside world." His brother-in-law and local Wayne County High School principal, Joseph S. Hickman, assisted Pectol in his efforts. Hickman was elected to the Utah State Legislature in 1925 and was instrumental in establishing a small state park at Capitol Reef in 1926. Pectol was later elected to the legislature and through his efforts a request was made to establish a national monument. On August 2, 1937, President Roosevelt signed the proclamation establishing Capitol Reef National Monument. The area was a national monument in name only, no funding was allocated for a number of years. President Nixon signed the bill establishing Capitol Reef National Park in 1971.

A special individual figuring prominently in the early success of Capitol Reef was "Charlie" Kelley, a writer who loved the desert, the archeology, and the more recent history of the area. Because of the lack of funds, the National Park Service was unable to assign any personnel to the new monument for many years. Kelley's dedication to the area led to his appointment as "custodian-without-pay" from 1943 to 1950. At the age of 62 when many people might think about retiring, he received his first civil service job and became Capitol Reef's energetic superintendent in 1950. As Charlie retired in 1958, *one* ranger was assigned to manage what would ultimately become a 241,904-acre (378 square mile; 979 km^2) park!

All citizens of the world should be appreciative of individuals such as Pectol, Hickman, and Charlie Kelley for their dedication in preserving for "all who come later" the special attributes of a special landscape. Hopefully, posterity will look back and say that twentieth- and twenty-first-century humans were motivated by doing what's right. Population pressures and the pressure to sacrifice areas to maintain the "economy" or to make them look like everywhere else by building golf courses, restaurants, and other amenities must be avoided in park and other open space areas in order to keep the world livable and worth living in.

Stratigraphy

For those who have followed this chapter in sequence, the stratigraphic units at Capitol Reef are "old friends"—rock units that we have met in other parks. As we saw in parks such as Grand Canyon and Canyonlands, exposures of a thick section of rock can occur where rivers have cut thousands of feet downward. At Grand Canyon over 5000 feet (1524 m) of rock is exposed to view—at Capitol Reef 10,000 feet (3048 m) can be examined without the benefit of an uncommonly deep canyon. Here erosion has cut deeply into the steeply tilted rocks along the Waterpocket Fold (monoclinal fold). Like the edges of books leaning on a bookshelf, many edges can be examined in a short horizontal distance as seen in the cross section in Figure 8–53. Strata range from small exposures of late Paleozoic (Permian) exposed in canyon bottoms through a nearly complete section of Mesozoic-age rocks. The stratigraphic column in Figure 8–54 and the cross section in Figure 8–53 provide an overall view of the structure and stratigraphy of the park.

Deep in two of the three perennial stream canyons that slice through the 100-mile-long (160-km) monocline are the oldest rocks exposed to view—the Permian–age Cutler Formation and Kaibab Limestone. The Gooseneck Viewpoint along Sulphur Creek is an excellent area to see these older rocks. Recall that the Cutler Formation is exposed in the Needles district at Canyonlands National Park and the Kaibab Limestone forms the rimrock at Grand Canyon.

The overlying river and tidal-flat sediments of the Triassic-age Moenkopi and Chinle formations were followed by deserts in Jurassic time. The star scenery makers in the Colorado Plateaus were deposited in these ancient deserts—the Wingate

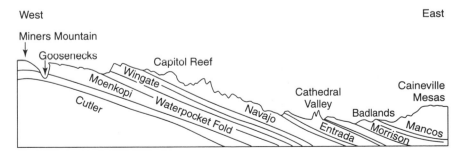

FIGURE 8–53 Typical cross section across the Waterpocket Fold at Capitol Reef National Park. (From National Park Service brochure)

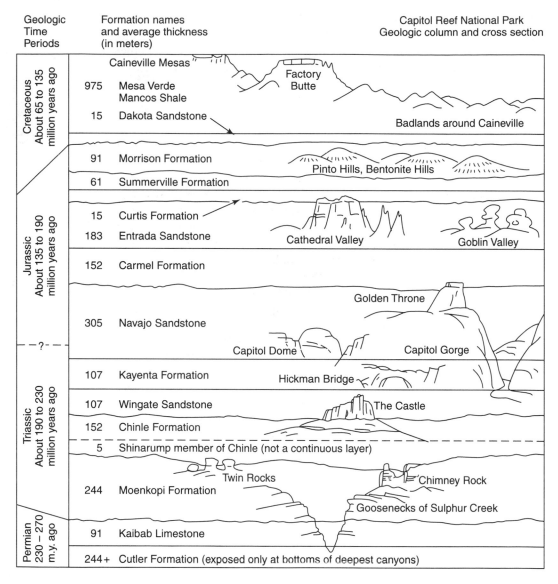

| Geologic Time Periods | Formation names and average thickness (in meters) | Capitol Reef National Park Geologic column and cross section |

FIGURE 8–54   Stratigraphic column at Capitol Reef National Park. Irregular lines between formations indicate unconformities. (From National Park Service brochure)

and Navajo sandstones. The reddish-brown Wingate dune sands are about 350 feet (107 m) thick here and the buff-colored Navajo is about 1000 feet (305 m). The tell-tale signs of wind deposition are abundant. Well-sorted, frosted sand grains deposited in thin layers that form the graceful flowing curves of tall crossbeds are distinctive dune sand features. The Wingate forms the steep cliffs on the west near the visitor center (Fig. 8–50), and the Navajo forms the domes and white cliffs at the crest of the Waterpocket Fold on the east. The arid Jurassic climates were a product of plate activity—the slow northwestward drift of the North American plate placed the area in the dry latitudes near 30° north latitude—in a position comparable with the Sahara today. Additionally, mountain building along the west edge of the continent produced a topographic barrier and a rain shadow for inland areas.

Rivers, dunes, and a return of the sea characterized the slow changing environments of late Mesozoic time. The last encroachment of the sea is marked by the nearshore deposits of the Dakota Sandstone and marine shales (Mancos Shale) as one of the most extensive and the last of the vast shallow seas covered much of the North American continent. During latest Cretaceous time, the sea edge oscillated and left deposits equivalent in age to those in the Mesa Verde Group—a group of formally named formations or recognizable sedimentary units that will be examined in more detail at Mesa Verde National Park. The path to the sea was now "forever" blocked to the west—the seas retreated eastward toward the center of the continent where they would soon leave—perhaps to return in A.D. 100,000,000.

Early Tertiary time likely brought deposits of lake sediments such as those found at nearby Bryce Canyon and elsewhere and probably middle and late Tertiary volcanic rocks. Erosion removed much of the evidence of Tertiary events in the park except for isolated dikes and sills in the Cathedral Valley and South Desert areas in the northern section of the park. Large basalt boulders along the Fremont River are part of outwash sediments derived from Quaternary glaciers that mantled the nearby lava-capped Aquarius Plateau upstream from Capitol Reef.

Structure: The Waterpocket Fold

The breakup of the Pangaea supercontinent in early Mesozoic time sent plate fragments merrily on their way across the globe. The west edge of North America was now the leading plate edge and began to experience the effects of collision with the oceanic plate to the west. Mountain building began to the west in California and Nevada and migrated eastward through the Mesozoic. By latest Cretaceous (70 Ma) and early Tertiary (65–50 Ma) time, the Laramide Orogeny was forming the Rocky Mountains to the east and causing crinkling of the crust in the Colorado Plateaus. Ancient north- and northwest-trending Precambrian faults were reactivated, locally bending the overlying sedimentary rocks into folds—including the Waterpocket Fold (Fig. 8–2). The 100-mile-long (160-km), northwest-trending monocline extends from just north of the park near Thousand Lake Mountain to the Colorado River near Rainbow Bridge National Monument at Lake Powell National Recreation Area to the south—one of the longest and best displayed folds in the Colorado Plateaus. Over 60 miles (97 km) of one of our best "textbook examples" of a monocline are included in the park.

The rocks are tilted as much as 60° to the east over a distance of about 3 miles (4.8 km). The west side of the fold near the visitor center is marked by the bold Wingate cliffs rising above the green and purple shales and sands of the Chinle and Moenkopi formations (Fig. 8–50). The Navajo Sandstone forms the crest of the ridge and the steeply tilted east flank. The eroded upturned edges of the remaining Jurassic and Cretaceous units stretch eastward where once again the rocks become horizontal (Figs. 8–52 and 8–53). Where the rocks are tilted at 45° or more the resistant rock layers form distinctive topographic ridges known as *hogbacks.*

The late Mesozoic compression was succeeded by Tertiary-age tensional forces—both of these stress conditions were accompanied by vertical uplift. The area began its vertical journey at sea level about 65 Ma and reached its present elevation of about 7000 feet (1829 m) in spite of the erosional removal during uplift of 6000–9000 feet (1830–2744 m) of late Mesozoic and Cenozoic strata. The present elevation of the crest of the Waterpocket Fold (7000 feet; 1829 m), plus the 5000–8000 feet (1524–2440 m) of strata removed by erosion (up to the last marine sediments—latest Cretaceous in age) is a measure of the total vertical uplift. Thus, vertical uplift here was at least 12,000 feet (3660 m)—a significant part of which occurred during the last 5 million years—a major factor in the present phase of erosion.

Erosional Landforms

Streams rearranged their courses during the recent regional uplift. Larger, perennial streams such as Fremont River, Pleasant Creek, and Oak Creek that flow eastward down the regional slope of the land continued to maintain their paths across the Waterpocket Fold during uplift and cut deep canyons through the upturned edges of the sedimentary rock layers. Other smaller streams could not keep pace with uplift and many were deflected by resistant rock ridges. Today they flow along the soft rock layers between more resistant layers in *strike valleys* (parallel to the edges of the inclined layers) like that of Halls Creek (Fig. 8–51) in the south end of the park and Deep Creek in the north. As layers of strata were stripped from the area, upstream erosion slowed greatly when streams encountered the resistant Navajo and Wingate sandstones along the Waterpocket Fold. These rock units armor the fold and the plateau surface to the west, maintaining their topographic position some 2000 feet (610 m) above the flat-lying strata immediately east of the Waterpocket Fold.

The broad valley along Halls Creek with the high ridge of the Waterpocket Fold to the west is one of the truly spectacular features in the park. Where streams were too small or were intermittent such as that at Capitol Gorge, their canyons are now abandoned except during times of flash flood. Some streams, such as Sulphur

Creek, had a meandering pattern that was maintained during downcutting, creating the *incised meanders* so spectacularly displayed at the Goosenecks Viewpoint (Fig. 8–51).

Until 1962 when state road 24 was completed along the Fremont River canyon, the narrow slot called Capitol Gorge was the major road connecting towns on the west side of the Waterpocket Fold with those on the east. The old road was barely wide enough in places for wagons and later automobiles to pass through. Each heavy downpour would send a flash flood through Capitol Gorge, requiring frequent rebuilding of the road. Today a hiking trail follows the route of the old road. Very little trace of the road is apparent as nature has violently rearranged the boulders and other flood debris along the canyon bottom—not a good place to be during a thunderstorm when a 9-foot-high (3-m) wall of water, mud, and boulders could roar through the canyon bottom.

Erosion, over many millenia, has sculptured the landscape into new shapes. As the weaker mudstones of the Chinle and Kayenta are removed, support for the massive sandstones above are reduced causing occasional slabs to fall and cliffs to retreat. Plateaus are slowly cut into mesas, mesas are reduced to buttes, and buttes eventually become small hills and then more flat surfaces on the desert floor. Continued downcutting will then form another generation of similar landforms. Rock units such as the Moenkopi at Chimney Rock, Twin Rocks, and Egyptian Temple are not cliff formers except where capped by a resistant rock such as the Shinarump Sandstone (base of the Chinle Formation). Where the Shinarump is absent, the soft Moenkopi becomes a slope former. Likewise, the Navajo Sandstone at Golden Throne displays steep walls where capped by the resistant Carmel Formation. When the resistant cap is eroded away, rounded landforms such as Capitol and Navajo domes typically form.

The Entrada Sandstone here is stratified rather than massive as it is in Arches National Park; consequently it does not form arches. In Cathedral Valley in the northern area, high steep-sided sharp peaks or "cathedrals" typically form as erosion takes its toll (Fig. 8–55). Where a hard cap of Curtis Sandstone protects the Entrada, impressive monoliths develop. (Those who wish to see the Entrada Goblins on parade should visit Goblin Valley State Reserve located between Capitol Reef and Green River, Utah).

The Morrison Formation, of many colors including deep purple, forms low, beautifully rounded hills and ridges, particularly along the road east of the monocline. The drab Mancos Shale is in many places completely dissected by streams and rivulets, thus forming classic badlands.

In this arid climate the vegetative cover is generally sparse except locally along the streams. The weathered materials from the Morrison and the Mancos shales are inherently sterile, and the outcrops of these two formations are almost completely barren. Therefore, on the Morrison and the Mancos areas the rate of erosion is particularly high.

Smaller-scale erosional features such as Hickman Natural Bridge (Fig. 8–56) form when streams cut through thin ridges of rock. Hickman Bridge is in sandstone

FIGURE 8–55   Temple of the Moon (right) and Temple of the Sun (left), formed in Entrada Sandstone by weathering and stream erosion in Cathedral Valley. (Photo by Fred Goodsell)

of the Kayenta Formation and is 133 feet (40 m) wide and 125 feet (38 m) high. Other smaller *natural bridges* sometimes form when streams find their way under a resistant rock layer in a stream channel or when water in deepening potholes encounters a permeable bedding plane or joint connection to the channel downstream. *Natural arches* such as Cassidy Arch superficially resemble natural bridges except that weathering is the main arch-forming process whereas running water accounts for a natural bridge.

FIGURE 8–56   Capitol Dome, as viewed through Hickman natural bridge. (Photo by D. Harris)

Turbulent streams cutting into bedrock often have local swirling currents that move cobbles and other rock material in a circular fashion and excavate depressions called potholes. These stream-cut depressions (a type of waterpocket) are abundant along the channels of many of the intermittent streams and help give the Waterpocket Fold its name. Other waterpockets are weathering pits, or *weathering depressions,* that form on slick rock surfaces. Water collecting in shallow depressions remains in contact with the rock surface for longer periods of time than on better-drained surfaces. Chemical weathering removes the cement holding the sand grains together. When the water in the weathering pit evaporates, the wind blows away the loosened sand grains, slowly deepening and widening the depression. In addition to providing a temporary water supply for desert animals, microscopic plants and animals as well as tiny fairy shrimp often inhabit the pools. Their acid secretions further speed up the removal of the calcium cement that holds the quartz sand grains together.

Another example of differential weathering occurs where cementation of rock is variable—areas with less cement weather faster, producing *honeycomb weathering,* or when the indentations are more than a few feet wide or high, *cavernous weathering* (Fig. 8–57). Many of these holes or indentations in the rock face are aligned along bedding planes where groundwater seasonally flows to the surface. Here the calcium cement is removed more rapidly. Loosened sand grains are washed, fall, or are blown by wind from the developing honeycombs. Our old friend desert varnish is here too—coating the walls of sandstone with black and red

FIGURE 8–57 Honeycomb and cavernous weathering of sandstone in Capitol Reef National Park. (Photo by E. Kiver)

deposits of manganese and iron. The Indians pecked through this "blackboard" to the light-colored sandstone beneath and left pictures and symbols called petroglyphs. Less commonly they applied pigments to the rock surface to draw pictures or pictographs. Their significance is unknown. Ideas range from "doodling" by some bored individual to deep religious significance. Bower (1996) summarizes the results of a number of researchers who suspect that petroglyphs are the work of shamans who recorded spiritual sightings and experiences when in a trance condition. A short hike along the Petroglyph Trail about a mile (1.6 km) east of the visitor center will take you to the base of the Wingate Sandstone and some of the more interesting petroglyphs in the park.

When cattle and sheep no longer tramped through the park area, after a few decades a complex intergrowth of organisms called *cryptobiotic crust* has reestablished itself. This algae–fungi–lichen growth forms a surface coating or crust that reduces soil moisture loss and soil erosion and promotes the growth of plant communities. Survival of all lifeforms is marginal in this harsh desert environment. Thus it is essential for park visitors to stay on established trails in order to preserve the "crypto" and preserve desert health.

A variety of features awaits the visitor—especially those who are willing to explore the trails and back roads of one of our lesser visited park areas. Begin your tour at the visitor center and take one or more of the trails and the scenic drive to various viewpoints. Inquire about the condition of the dirt roads and longer trails if you want to really get to know this special area. The Cathedral section in the north features 500-foot-high (152-m) erosional remnants of the Entrada Sandstone, the Escarpment section near the visitor center is dominated by the bold Wingate Cliffs, and the monocline section some 30 miles (48 km) to the south provides the best display of the tilted strata of the Waterpocket Fold. Weather, especially thunder storms, can change the roads rapidly making travel impossible. Some roads are for high clearance or 4-wheel-drive vehicles only.

REFERENCES

Bower, B., 1996, Visions on the rocks: Science News, v. 150, p. 216–217.

Olson, V.J., and Olson, H., 1990, Capitol Reef: The story behind the scenery: Las Vegas, KC Publications, 48 p.

Petersen, K.L., 1994, A warm and wet Little Climatic Optimum and a cold and dry Little Ice Age in the Southern Rocky Mountains, U.S.A.: Climatic Change, v. 26, p. 243–269.

Smith, J.F., Jr., Huff, L.C., Hinrichs, E.N., and Luedke, R.G., 1963, Geology of the Capitol Reef area, Wayne and Garfield Counties, Utah: U.S. Geological Survey Professional Paper 363, 102 p.

Park Address

Capitol Reef National Park
Torrey, UT 84775

GRAND STAIRCASE–ESCALANTE NATIONAL MONUMENT (UTAH)

"A maze of cliffs and terraces lined with stratifications, of crumbling buttes, red and white domes, rock platforms gashed with profound canyons, burning plains, barren even of sage—all glowing with bright color and flooded with blazing sunlight." Thus did pioneer geologist Clarence Dutton (1880) describe the stunning grandeur of this still remote and desolate landscape of our most recently established national monument. Senate Bill 884 of the 104th Congress (1994–1996) would have opened this area and others forever to development. Environmental groups were alarmed, and President Clinton responded by using the Antiquities Act to establish the Grand Staircase–Escalante National Monument on September 18, 1996. It is heartening to note that concerned citizens can still make a difference in ensuring that our world remains livable and worth living in for ourselves and for those who come later.

The huge 1.7 million-acre (2656 square mile; 6881 km²) monument fills the gap between Bryce Canyon National Park to the west, Capitol Reef National Park to the northeast, and Glen Canyon National Recreation area to the southeast, making this a continuous parkland. The monument is large—as large as the area of Delaware and Rhode Island combined. The new monument moves us a big step closer to the early dreams of park enthusiasts of the 1930s who hoped to establish Escalante National Park, a continuous parkland extending from Grand Canyon in northern Arizona to Canyonlands National Park in southeast Utah. Administration of the new monument is the responsibility of the Bureau of Land Management (BLM) rather than the National Park Service.

Prominent sections of the park from the southwest to northeast include the Grand Staircase (the Vermillion, White, and Gray cliffs), the upper Paria River canyons, the Cockscomb (the part of the East Kaibab Monocline that extends northward from Grand Canyon), the Kaiparowits Plateau, the Straight Cliffs, canyons of the Escalante River, and the Circle Cliffs that end in the sharp monoclinal bend at Capitol Reef National Park called the Waterpocket Fold (Fig. 8–58).

Geographic Setting

This rugged collection of plateaus, mesas, and canyons is located near the edge of the Canyonlands and High Plateaus sections of the Colorado Plateaus and is surrounded by federal lands. Fortunately these are all public lands owned by every U.S. citizen—lands that should be used for their best long-term benefit. Significant state parks adjacent to the monument and well worth visiting are Petrified Forest, Kodachrome Basin, and Anasazi Village.

Elevations range from near 5000 feet (1524 m) in the Escalante and Paria canyon bottoms near Glen Canyon National Recreation Area to over 7000 feet (2134 m) on the Kaiparowits Plateau. Northwest of the Grand Staircase–Escalante, elevations in the High Plateaus Section at Bryce Canyon and elsewhere rise to over 10,000 feet (3050 m). Dominating the skyline northeast of the monument is the

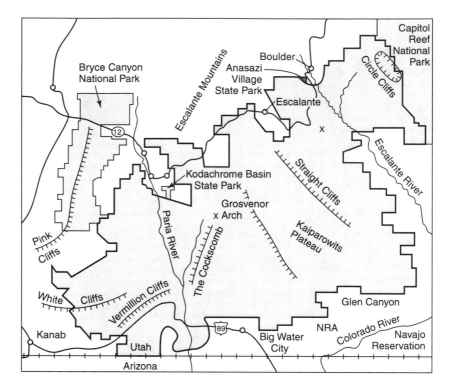

FIGURE 8–58   Generalized map of Grand Staircase–Escalante National Monument.

rugged, intrusive-igneous complex of the Henry Mountains with Mt. Ellen (11,615 feet; 3541 m) towering 6000 feet (1830 m) above the surrounding plateau surface.

Large-scale topographic features in the monument reflect the influence of the broad "wrinkles," or *folds.* (anticlines and synclines), in the underlying Mesozoic and older rocks (Fig. 8–59). The Escalante and Paria rivers follow the structural "grain" and topographic slope of the land and flow southeast from the High Plateaus to the Colorado River, paralleling the *strike* of the gently inclined strata. Sharper bends or monoclines at the Cockscomb, the Straight Cliffs, and the Waterpocket Fold produce spectacular cliffs and ridges—hogbacks where resistant rock layers are inclined steeply—*cuestas* (asymmetrical ridges) where the inclination or dip is more gentle.

Exposed along these cliffs and canyons are rocks as old as Permian. However, this is primarily a Mesozoic park with most of the rock strata exposed at Grand Staircase–Escalante the same as that at adjacent Capitol Reef National Park (see geologic column and cross section in Figs. 8–53 and 8–54). The most extensive geologic studies of parts of the area are by Davidson (1967), Doelling (1975), Doelling and Davis (1989), and Smith and others (1963). The area's remoteness has left it relatively unspoiled. Opportunities for future geological, biological, and archeological studies are substantial as are the potential adventures of visitors who wish to explore—and especially those who want to immerse themselves in the magnif-

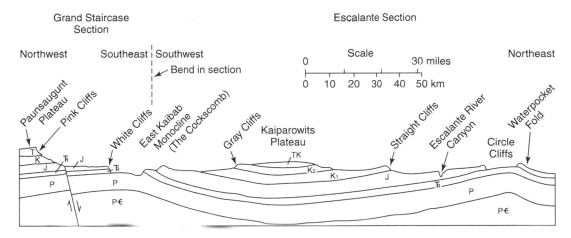

FIGURE 8–59 Generalized cross section across part of Grand Staircase–Escalante National Monument. (Modified from Hintze, 1975)

icent scenery in the primitive back country. The opportunity for such experiences is rapidly disappearing in the United States and around the world as population steadily increases. Doubling the U.S. population every few decades as we are now will seriously impact open spaces and the general health of the environment and the physical and psychological health of our people.

History

The Escalante was an area where the pueblo-building Anasazi and the more nomadic Fremont cultures intermingled and is therefore of considerable archeological interest. Anasazi or Hisatsinom cliff dwellings were constructed perhaps 1500 years ago. The entire area was mysteriously abandoned about 700 years ago, probably because of a slight but significant change in climate. Gathering of pinon nuts, cactus fruits, and sunflower seeds as well as raising squash and corn provided the staples of the Indian diet. Hunting of deer, Bighorn sheep, and other animals also occurred, especially later when the abandoned area was reoccupied by the Navajo and nomadic Paiute during the last few centuries. Kaiparowits (also spelled Kaiparowitz) is the Paiute word for "mountain home"—appropriate for the summer but at elevations near 7500 feet (2300 m), plateau habitation is not very pleasant during the winter months. The Indians wintered in the more moderate conditions found along the Colorado River and the lower Escalante and Paria valleys. Hundreds of significant sites are known in the monument and include rock-art panels, granaries, and occupation sites. However, the true extent of the archeological resources is not adequately known (Madsen, 1997).

Fathers Francisco Dominguez and Sivestre Escalante passed south of the area through what is now Glen Canyon National Recreation Area in 1776. Although never passing through the monument area, Escalante's name is honored by numerous landmarks in the area, including the Escalante River that he discovered.

The first scientific exploration was by the Powell survey from 1872–1875. From the orderly sequence of easily recognized strata and "giant stepping stone" arrangement of cliffs and rock layers the term Grand Staircase was born. Powell's brother-in-law and survey topographer, Professor Almon H. Thompson, was mistakenly led into the canyons under the belief that this was the drainage of the Dirty Devil River, which is located some 50 miles (80 km) farther up the Colorado River. Thompson soon realized the error and named the river the Escalante.

Mormon settlers moved into the area in the 1880s, mostly concentrating in nearby communities such as Hanksville, Henrieville, Escalante, and Kanab. Farming was greatly restricted by the terrain and the climate; ranching was possible if huge areas were used. The experiences, the ruggedness of the people, and the colorful history of the area are reflected in some of the interesting place names such as Tarantula Mesa, Scorpion Gulch, Death Hollow, No Mans Mesa, Last Chance Creek, and many more. Historic places include the ghost town of Paria, cowboy line camps, about 60 miles (100 km) of the Hole-in-the-Rock Trail, and the Burr cattle trail.

Intrigued with the desolate country, a young artist and poet named Everett Ruess left the town of Escalante in November 1934 with two months of food packed on two burros. As reported in Rusho's (1983) book, *Everett Ruess: A Vagabond for Beauty*, Ruess was never found, although his two half-starved burros were found a few months later and a cabin full of his food was found 20 years later. Perhaps the mystery of Ruess's disappearance will be solved as more people explore this vast wilderness.

The uranium boom of the 1950s brought hundreds of prospectors to the Colorado Plateaus, including the Escalante and Capitol Reef areas. The rocks were the right ones—the Triassic-age Shinarump Conglomerate and the Jurassic-age Morrison Formation. The uranium was there but not in sufficient concentrations to justify the high cost of mining and transportation. The Cretaceous sediments in the remote Kaiparowits Plateau contain vast amounts of coal, an estimated 62 billion tons, and perhaps some oil and gas. However, mining is a "one-shot," short-term event that lasts for a few decades and can severely change the landscape—a landscape that will be with us forever. As we rapidly increase our population and deplete the earth's resources, will we push ourselves into a corner and tear up every reachable part of our planet to maintain and increase our material wealth? Can we extract resources for a short-term "fix" and still preserve unspoiled vistas and the rich archeological remains in our parks and monuments?

Geologic Story

The relatively gentle Laramide-age (early Tertiary, about 60–50 Ma) wrinkling of crustal rocks here has enabled erosion to develop a system of ridges and cliffs where resistant strata are encountered by erosion (Fig. 8–59). The cliffs face south where the strata are inclined (dip) to the north in the Grand Staircase section in the southwest part of the monument. In the Escalante–Kaiparowits Plateau section cliffs face southwest, where the rock layers slope to the northeast. Cliffs face northeast along the Straight Cliffs where rocks slope to the southwest. Thus structural basins such as that underlying the Kaiparowits Plateau have outward-facing cliffs. The Circle Cliffs are at the crest of an anticline and form an oval ring of inward facing cliffs or scarps (short for escarpment), hence the name Circle Cliffs. The northeast edge of the Circle Cliffs Upwarp is part of the Waterpocket Fold—the same monocline found in adjacent Capitol Reef National Park. The two parks touch here, completing the total protection of one of the finest exposed monoclines in North America.

The sedimentary rock units at Grand Staircase–Escalante National Monument are the same as those in its sister park, Capitol Reef. The cross section in Figure 8–53 is the same as that along the northeast edge of the Circle Cliffs in structure and stratigraphic units. Permian-age Cutler Formation and a local, thin layer of Kaibab Limestone are the oldest rocks exposed in the Circle Cliffs area (Smith and others, 1963).

During Triassic time streams flowed lazily from the late Paleozoic Ancestral Rocky Mountain uplifts in New Mexico and western Colorado and from uplifts in southern Arizona. The gravels, sand, and mud that were deposited were later lithified into the multicolored beds of conglomerate, sandstone, and mudstones of the world-famous Chinle Formation. Just as in Petrified Forest National Park, local beds here contain brightly colored silicified logs of impressive dimensions—some up to 3 feet (1 m) in diameter and 40 feet (12 m) long (Davidson, 1967).

The stars of the scenery are once again the dune deposits of the Jurassic-age Wingate and Navajo sandstones that make up the spectacular Vermillion, White, Straight, and Circle Cliffs (Fig. 8–58). Where the rocks tilt gently a few degrees, linear cliff lines develop such as those in the Grand Staircase Section where one climbs the Vermillion (Wingate Sandstone), White (Navajo Sandstone), Gray (Cretaceous sandstones), and finally the Pink (early Tertiary age) cliffs in nearby Bryce Canyon National Park. Where the strata are steeply tilted at 45° or more from the horizontal as at the Cockscomb and the Waterpocket Fold, a distinctive, nearly symmetrical landform ridge known as a hogback forms (Thornbury, 1969).

Other Jurassic and Cretaceous rocks discussed in Capitol Reef National Park (Fig. 8–53) and many other Colorado Plateaus park areas are the same as those at Grand Staircase–Escalante. The rock strata above the Cretaceous-age Mancos Shale contain coal-bearing units that underlie the huge Kaiparowits Plateau in the central part of the monument. As the Late Cretaceous seaway fluctuated in extent, alternating marine and nonmarine conditions prevailed. The beach and nearshore sandstones are resistant to erosion and form the Gray Cliffs on the west edge and the Straight Cliffs on the east edge of the Kaiparowits Plateau. Although topographically a plateau, the Kaiparowits is structurally a basin, thus accounting for the outward facing escarpments.

The Cretaceous coal beds are locally thick and have long aroused the interest of the coal-mining industry. The coal reserves are large, but the high mining and transportation costs in this remote area have prevented their development. In addition, increased coal burning would further contribute to air quality degradation in the area. Establishment of the monument makes mining unlikely. Rather, it is hoped that the frontierlike, unspoiled atmosphere of the Grand Staircase–Escalante National Monument will be maintained.

Fossils of considerable importance are found in the Late Cretaceous strata in the Kaiparowits Plateau area, including numerous terrestrial vertebrates that include crocodilians, dinosaurs, fish, and mammals. This sequence of strata contains one of the best and most continuous records of Late Cretaceous terrestrial life in the world (Clinton, 1996).

With the close of the Mesozoic Era the last of the huge inland seas retreated toward the continental interior and eventually vanished. High mountains loomed above the Colorado Platcaus on the west and the equally high mountains of the Rocky Mountains to the east were just beginning to rise—eventually to become the backbone of North America. Rocks in the Colorado Plateaus during the Laramide Orogeny were also bent and uplifted—but in a relatively subdued manner compared to the mountain-building turmoil of surrounding areas. Broad basins such as that beneath the Kaiparowits Plateau and the Henry Basin east of the park, as well as the sharper bends of strata along the Waterpocket Fold and Cockscomb monoclines in and near the monument, were formed during the early Tertiary Laramide Orogeny. The vertical uplifts of late Tertiary and Quaternary times further energized the forces of erosion that stripped many thousands of feet of strata from the region—processes that continue today.

Erosional Features

Major topographic features here strongly reflect the underlying structure. The cliff lines follow the edges of resistant strata inclined from the horizontal (Fig. 8–59). Where the dip or inclination is steep, spectacular linear cliffs and hogbacks such as those at the Cockscomb and the Waterpocket Fold are formed. No river cuts west to east across these spectacular cliffs. Rather, rivers such as the Paria and Escalante flow to the southeast, parallel to the strike or eroded edges of the inclined layers. As Thornbury (1965) notes, these rugged canyons, as well as others in the Canyonlands section, are "for the most part nontraversable canyons."

The Escalante River is a meandering river set down in a deep canyon—an example of an *incised meander* stream pattern. The prominent meanders extend through nearly the entire river length, causing some to describe it as "the crookedest river on earth" (Stokes, 1986). Likely, the stream pattern was established as the ancient Escalante River wandered lazily across gentle slopes. Pronounced vertical uplift and northward tilting of the Colorado Plateaus triggered renewed downcutting and canyon formation. Some streams such as the Escalante have retained or inherited their previous patterns as they have cut downward and trapped themselves between resistant walls of sandstone.

Erosion of narrow sandstone ridges along stream channels occasionally perforates the ridge and forms features such as the 130-foot-high (40-m) Escalante Natural Bridge. Elsewhere weathering processes rather than stream erosion attack narrow ridges or fins and produce natural arches such as Stevens Arch (Fig. 8–60),

FIGURE 8–60   Stevens Arch is one of the numerous arches in the monument. (Photo by D. Harris)

Straight Arch in the Grand Staircase Section, the Grosvenor Arch near the Cockscomb, and the Woolsey Arch on the Kaiparowits Plateau. Grosvenor Arch is unusual in that it is a double arch. Inventory of the numerous arches and bridges and other features within the monument is not yet complete.

Unusual-appearing outcrops are found in the Burning Hills and elsewhere on the Kaiparowits Plateau. Here lightning strikes have ignited coal seams. Such fires can burn for many decades and leave behind a brick-red, clinkery rock.

Conclusion

Because of its remoteness, rugged topography, and harsh climate this was one of the last areas in the conterminous United States to be explored and mapped.

Except for a few small towns on its edges, the area is uninhabited by humans. However, mountain lion, bear, desert bighorn sheep, as well as eagles and peregrine falcons still make the Escalante region their home.

Access to the area is from Escalante in the north or from Kanab and U.S. Route 89 in the south. Roads are primitive and most require high clearance or even 4-wheel drive vehicles. Inquire at the BLM field office in Escalante for more information on roads and other activities in this gem of a park area.

It is indeed a rare experience in modern times to look out from a high ridge or peak and know that as far as the eye can see humans are only temporary visitors. On an increasingly crowded planet, open space, especially large chunks of open space, is quickly becoming a rare commodity. To provide the opportunity to all who come after us of experiencing large, relatively undisturbed landscapes is a thoughtful and unique gift to mankind.

REFERENCES

Clinton, W., 1996, Proclamation establishing the Grand Staircase–Escalante National Monument: Washington, DC, Sept. 18.

Davidson, E.S., 1967, Geology of the Circle Cliffs area, Garfield and Kane counties, Utah: U.S. Geological Survey Bulletin 1229, 140 p.

Doelling, H.H., 1975, Geology and mineral resources of Garfield County, Utah: Utah Geological and Mineral Survey Bulletin 107, 175 p.

Doelling, H.H., and Davis, F.D., 1989, The geology of Kane County, Utah—Geology, mineral resources, geologic hazards: Utah Geological and Mineral Survey Bulletin 124 and Map 121, 192 p.

Dutton, C.E., 1880, Report on the geology of the high plateau of Utah: U.S. Geographical and Geological Survey of the Rocky Mountain Region, Washington, DC.

Hintze, L.F., 1975, Geological Highway Map of Utah: Brigham Young University, Provo.

Madsen, D.B., 1997, A preliminary assessment of archaeological resources within the Grand Staircase–Escalante National Monument, Utah: Utah Geological Survey, Circular 95, 60 p.

Rusho, W.L., 1983, Everett Ruess: A vagabond for beauty: Gibbs-Smith, Salt Lake City.

Smith Jr., J.F., Huff, L.C., Hinrichs, E.N., and Luedke, R.G., 1963, Geology of the Capitol Reef area, Wayne and Garfield Counties, Utah: U.S. Geological Survey Professional Paper 363, 102 p.

Stokes, W.L., 1986, Geology of Utah: Salt Lake City, Utah Museum of Natural History, University of Utah, Occasional Paper No. 6, 280 p.

Thornbury, W.D., 1965, Regional geomorphology of the United States: Wiley, New York, 609 p.

Thornbury, W.D., 1969, Principles of geomorphology: Wiley, New York, 594 p.

Park Address

Grand Staircase–Escalante National Monument
Bureau of Land Management

Escalante Field Office
P.O. Box 225
Escalante, Utah 84726

EL MORRO NATIONAL MONUMENT (NEW MEXICO)

Carved in Jurassic Zuni Sandstone are the messages of ancient and not-so-ancient travelers, on the face of a high cliff in western New Mexico, several miles east of the town of Zuni (Fig. 8–61). The name El Morro, which means "the bluff" or "the headland" in Spanish, is appropriate because a 200-foot-high (61-m) bluff here rises to the top of a mesa. On top there are ruins, largely unexcavated, which record Indian occupation of the area long before it was invaded by the Spaniards. To the Zuni this was A'ts'ina—"place of writings on the rock." Hundreds of petroglyphs carved in the rocks record the passing of history for the past 1500 years.

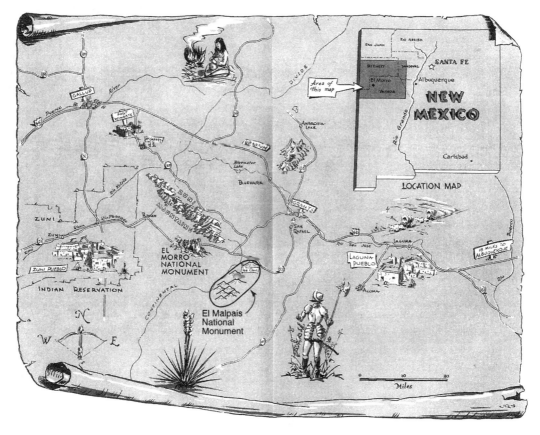

FIGURE 8–61   Location map of El Morro and El Malpais monuments. (Slightly modified from West and Baldwin, 1965)

Inscription Rock just north of the visitor center was used not only by the long-vanished Anasazi but also by Spanish Conquistadors who added their names and later by Anglo-Americans who followed their example. What attracted these people was one of the fundamental necessities of life—water. The El Morro oasis had nurtured an Indian village, and later its life-sustaining waters were used by Spanish and Anglo-Americans in their explorations. At El Morro we will primarily explore the geologic evolution of the landscape. However, also apparent are the limitations of the environment to support humans—a connection often overlooked by modern humans.

History

People of the Pueblo culture were the first to make extensive use of the isolated but dependable source of water at El Morro. The foundations of their ancient village high atop Inscription Rock are accessible by trail. Wetter climates throughout the Southwest at that time were favorable for settlement and marginal agricultural activities. About 700 years ago a slight change in climate produced numerous crop failures causing abandonment of the settlements at El Morro, Mesa Verde, and elsewhere in the Colorado Plateaus. During their long stay in the El Morro area, the Pueblo people discovered that the poorly cemented Zuni Sandstone here was relatively easy to carve. Petroglyphs of animal and human life, as well as other shapes and forms whose meanings are obscure to modern people, decorate the sandstone walls along the base of the cliff.

The Spanish search for gold led to colonization and exploration and eventually to the discovery of the isolated desert paradise at El Morro. Exploration and travel in the desert is limited by the distance between sources of water for both human and beast. The Spanish used the El Morro area for 200 years, first discovering the permanent waterhole in 1583. The oldest Spanish carving on Inscription Rock dates from 1605—15 years before the Pilgrims landed at Plymouth. The last recorded Spanish visit occurred in 1774.

The first Anglo-Americans to document the existence of El Morro were three soldiers from the U.S. Army Topographical Engineers led by Lieutenant James H. Simpson. They took a side trip with an Indian trader to see "half an acre of inscriptions" (West and Baldwin, 1965) in 1849. They documented and carefully recorded the inscriptions and followed the tradition by leaving the first inscription in English on the soft rock walls of Zuni Sandstone. In 1858 Edward Beale made a stop at El Morro while surveying a wagon road from Fort Defiance to the Colorado River. Of special interest was that Beale used a caravan of camels—an early experiment in transportation in the American desert.

The historical significance of the area was apparent to all, and the area was quickly set aside by President Theodore Roosevelt in 1906 when the Antiquities Act was passed by Congress. Construction of the transcontinental railroad and U.S. Highway 66 about 25 miles (40 km) to the north temporarily reduced interest in the watering place. Road improvements, a rapidly increasing national population, and a more mobile society has greatly increased visitation.

Geographic Setting

El Morro is a tiny monument of only 1278 acres (2 square miles; 5.2 km^2). Its geology can only be understood by placing it into a broader context with the nearby San Juan Basin and Zuni and Defiance uplifts (Baars, 1995). The monument is located on the southwest flank of the Zuni Anticlinal Uplift in western New Mexico, about 40 miles (64 km) east of the Arizona–New Mexico state line (Fig. 8–61). Laramide (70–50 Ma) "wrinkling" of crustal rocks in the Colorado Plateaus (Fig. 8–2) produced the Zuni Mountains—a northwest-southeast elongated domal uplift. The gently sloping Paleozoic and mesozoic rocks on the northeast side of the uplift are inclined into the San Juan Basin. Strata on the southwest flank are bent sharply into a monoclinal flexure that flattens out under the desert floor near El Morro. To the northwest in nearby eastern Arizona is the Defiance Uplift, a similar Laramide structure with Canyon de Chelly National Monument, another important historical site, on its west flank.

The Continental Divide extends through the Zuni Mountains and is located a few miles east of El Morro in El Malpais National Monument, also on the flank of the Zuni Uplift (Fig. 8–61). Drainage is mostly underground in this dry climate with surface streams more common farther from the Divide. The Zuni River to the west flows into the Little Colorado River south of Petrified Forest National Park, and drainage east of the Divide eventually flows into the Rio Grande along the eastern edge of the Colorado Plateaus. The crest of the Zuni Mountains is about 9000 feet (2745 m), about 2000 feet (610 m) above the visitor center (elevation 7218 feet; 2200 m).

Geologic Story

The oldest exposed rocks in the area are in the center of the Zuni Mountains. Here erosion has stripped away younger rocks and exposed the Precambrian rocks in the core of the domal uplift. The rocks are similar to the Precambrian rocks in the depths of the Grand Canyon and tell a similar story of episodes of primeval mountain building and ancient erosion surfaces. The major faults that formed the late Precambrian fault-block mountains were mildly active during Paleozoic time, keeping the block above the level of the encroaching seas. The Zuni block was finally inundated during late Paleozoic time when sandstones and limestones equivalent in age to those at the rim of the Grand Canyon were deposited directly on the Precambrian basement rocks. The late Paleozoic rocks are exposed around the mountain core flanked by concentric bands of Mesozoic rocks, including the Zuni Sandstone.

The Upper Triassic Chinle Formation (about 210 Ma) mudstones of Petrified Forest fame form a "racetrack" valley around the uplift. The overlying Jurassic-age (about 180 Ma) Wingate here consists of stream-deposited sand, clay, and gravel rather than the massive cliff-forming dune sands so prevalent elsewhere in the Colorado Plateaus. The overlying wind-deposited Zuni Sandstone accounts for the spectacular mesas and buttes that adorn the desert landscape—including the special one named El Morro by the early Spanish explorers (Fig. 8–62).

Younger Mesozoic rocks deposited in the area have been eroded from the El Morro area but are still intact farther from the Zuni Mountains in the San Juan Basin and elsewhere. The Dakota Sandstone (Cretaceous, about 90 Ma) overlies the Zuni Sandstone forming a resistant caprock that protects the softer sandstone below. Reactivation of the buried northwest-trending Precambrian faults during Laramide (about 70–50 Ma) time produced the bending and folding in the overlying strata. Vertical uplift of many thousands of feet, much of it during the past few million years, coupled with a lowering base level as the Colorado River sliced downward, contributed to the removal of an almost unimaginable quantity of weathered rock in the Colorado Plateaus. The transition from *plateau* to mesa to butte to desert floor is evident in the local landforms. Pediments enlarge and scarps or cliffs continue their extremely slow retreat as illustrated in Figure 8–8. Perceptible changes occur only over many tens of millennia.

Erosional Forms

The 200-foot-tall (60-m) cliffs of Zuni Sandstone here are a product of the resistant sandstone and the cliff-maintaining processes. Occasional thunder storms

FIGURE 8–62   El Morro, the "bluff." Zuni Sandstone here is more resistant than the underlying Wingate Formation. (Photo by D. Harris)

send sheets of water down rock faces, causing erosion next to bluffs and areas of deposition farther downslope (Fig. 8–8). Runoff locally concentrates into channels and forms valleys and canyons.

Where the weaker Wingate and Chinle rocks beneath the Zuni Sandstone are partially removed by erosion, undermining contributes to the collapse of large slabs (Fig. 8–63). Vertical fractures, or *joints,* are planes of mechanical weakness that control where the slabs separate from the rock wall. Many of these joints formed during the Laramide folding event, which occurred about 70–50 Ma. Other joints are exfoliation fractures produced by the brittle sandstone expanding toward the free face of the cliff. Rock failure often forms arcuate inset arch or exfoliation arch forms similar to those found at Zion, Canyonlands, and elsewhere where massive brittle rocks are exposed in vertical faces.

FIGURE 8–63   El Morro; note slab of Zuni Sandstone that will one day crash down. (Photo by D. Harris)

Water on top of Inscription Rock collects into rock-walled channels and pours down the cliffs during heavy rainfall or melting snow conditions. A waterfall cascading down Inscription Rock has excavated a 12-foot-deep (4-m) *plunge pool* at its base. This natural depression holds about 200,000 gallons of water and was the source of the precious fluid sought by early inhabitants and travelers. The pool is protected from the direct rays of the sun and persists the year round. Ironically, humans came to the rock for its life-giving water—now water must be brought to the rock to supply the needs of its many visitors!

A rockfall in 1942 dropped many tons of broken slabs of sandstone into the beautiful pool at the base of Inscription Rock. Removal of the rock restored the oasis to its former appearance.

Water trickling down the cliff faces has locally produced a growth of black lichens and a biochemical deposit called desert varnish. Some of the mineral deposits are associated with horizontal zones where groundwater seeps out of the cliff face. Finer layers of interdune deposits cause descending groundwater to move laterally along these impermeable layers, producing seeps and small seasonal springs.

Conclusion

The historic value of this tiny area clearly justifies its status as a national monument. It also clearly illustrates the close connection between humans, the underlying geology, and the delicate balance of nature that makes life possible. The lessons of environmental restrictions and human dependence on nature were clear to the Indians who lived and used the land for many centuries. Similar restrictions apply on a global scale, a fact that is ignored by those who believe that we can perform unlimited manipulation and "improvements" to nature.

REFERENCES

Baars, D.L., 1995, Navajo Country: Univ. of New Mexico Press, Albuquerque, 255 p.

West, S.W., and Baldwin, H.L., 1965, The water supply of El Morro National Monument: U.S. Geological Survey Water-Supply Paper 1766, 32 pages.

Park Address

El Morro National Monument
Route 2, Box 43
Ramah, NM 87321-9603

NATURAL BRIDGES NATIONAL MONUMENT (UTAH)

Within the National Park System each park area has its own particular role in portraying the variety of features that comprise our total natural environment. Although natural bridges are found in other park areas—Capitol Reef and Rainbow Bridge, for example—nowhere else are they as completely displayed as here. In this small area of 7636 acres (12 square miles; 31 km^2) are young, mature, and old bridges; the second and third largest bridges in the world; and, in addition, there is at least one natural bridge of the future—one that may be born about the time the oldest one collapses (Fig. 8–64).

Natural bridges can be formed by more than one geologic process. The famous Natural Bridge in Virginia is the only section of a cavern roof that did not collapse. Elsewhere running water along steep channels may be diverted along fractures beneath a resistant ledge of rock and leave the ledge as a rock span over the stream. In the Colorado Plateaus the more prominent bridges such as those at Natural Bridges National Monument form by stream erosion through thin walls of rock.

This lonely part of Utah is in the Cedar Mesa area. Cedar Mesa is part of the Monument Upwarp—a Laramide (70–50 Ma) structure that extends north into Canyonlands National Park (Fig. 8–2). With uplift comes increased erosion of the elevated region and removal of younger strata—in this case any Cenozoic rocks that may have been present and practically the entire Mesozoic section. Exposed in the core of the uplift and along the White and Armstrong canyons in the monument are the clean, white quartz sands of the Permian-age Cedar Mesa Sandstone—one of the integral components that enable natural bridge formation to occur here. Natural bridges are rare features. To have three of the larger bridges in North America form within walking distance of each other is unique and special.

FIGURE 8–64   A natural bridge of the future in Natural Bridges National Monument. When the stream punches a hole through the rock wall between X and Y the looping channel will be abandoned. (Photo by D. Harris)

Obviously, conditions for bridge formation are ideal at Natural Bridges. Here we will explore the coincidence of factors that encourage bridge formation on such a grand scale.

History

Paleo-Indians likely wandered through the area some 10,000 years ago, but episodes of permanent habitation began in about 620 when the first of three occupations by the Anasazi ("ancient ones") or Hisatsinom ("those who lived long ago") began (Petersen, 1990). The area was abandoned for about 200–300 years between each episode of colonization, perhaps because of small changes in climate. Other factors may have contributed to the mass exodus, including overpopulation, soil depletion, overhunting, and deforestation. Climate changes, especially those that increase aridity such as the 28-year drought during the late thirteenth century, would be particularly devastating to the agricultural-dependent Anasazi life-style. The last abandonment occurred in about 1270—roughly coinciding with similar mass migrations from Mesa Verde and other Anasazi sites in the Southwest. The Anasazi did not simply vanish but are believed to be the ancestors of the modern Hopi and Pueblo peoples.

Remnants of more primitive pit houses and later pueblo and cliff dwellings are present in and near the monument. The Horsecollar Ruin is located in a particularly scenic shelter cave along a ledge in White Canyon. Here one can "feel" the spirit of the ancient ones. The ruins, as well as the pictographs and petroglyphs in the monument, are delicate and irreplaceable and belong to all those who follow in our footsteps. They should not be touched or disturbed in any way. The nomadic Paiute and Ute Indians reoccupied the land during the past 500 years and were here to guide some of the newly arrived prospectors and explorers in the late 1800s.

Prospector Cass White was guided into the area by Indian Joe, a local Paiute Indian in September of 1883 (Peterson, 1990). White was shown a feature that the Indians called "under a horse's belly"—an imaginative description of a natural bridge. The present bridge names, Kachina, Sipapu, and Owachomo, were given to the bridges in 1908 by William Douglas, a government surveyor. During the same year, requests by public-minded citizens in nearby Blanding and Monticello led President Theodore Roosevelt to establish Utah's first national monument at Natural Bridges.

Setting the Geologic Scene

The edge of the Permian (about 260 Ma) sea varied widely in southeastern Utah as streams brought sediment southward and westward from the Ancestral Rocky

Mountains. Sands deposited on nearshore sandbars and coastal dunes are now part of the Cedar Mesa Sandstone, a part of the more extensive Cutler Group (Stokes, 1986). Both depositional environments produced abundant cross-bedding, which is inclined wedges of sand layers that represent the front edges of sand dunes or sandbars covered by shallow water. The Cedar Mesa Sandstone is up to 1200 feet (366 m) thick and is the same rock that forms the candy-striped Needles in Canyonlands National Park 35 miles (56 km) to the north. While the forces of the Laramide Orogeny (about 70–50 Ma) were building the Rocky Mountains to the east, relatively minor folding and faulting occurred in the Colorado Plateaus, including the broad uplift called the Monument Upwarp (Fig. 8–2) where Natural Bridges is located. By late Tertiary time the area was still only a few thousand feet above sea level, and streams meandered lazily across the area in tight looping bends that doubled back on each other, much like parts of today's Mississippi River.

About 10 Ma vertical uplift of the region and downcutting by the ancestral Colorado River initiated the canyon cutting that continues today. Some of the streams such as those in White and Armstrong Canyons were able to maintain their wiggly, meandering patterns as the streams began to cut deeply through the overlying weak mudstones of the Chinle and Moenkopi formations and into the hard Cedar Mesa Sandstone below. Once the streams began to cut into the sandstone, they became trapped by the walls of the resistant rock and formed entrenched meanders—deep canyons with a meandering pattern reminiscent of a strip of ribbon candy. By Quaternary time narrow walls of rock separated one meander bend from another, establishing conditions that would lead to the development, on a grand scale, of the features known as natural bridges.

Bridge-Forming Processes

Natural bridges are easily confused with their look-alike cousin, the natural arch. Both are formed by "holing-through"—but the processes forming each are quite different. In natural arches weathering processes slowly loosen individual sand grains, and gravity, wind, and rain remove the loosened particles. An arch can form in practically any location in a landscape, whereas a natural bridge forms only in valley bottoms and spans the stream that produced it. The narrow rock walls of *meander necks* in an entrenched meander valley are particularly vulnerable to erosion because of their position on the inside bend of a channel. The intermittent streams here carry large volumes of sediment during flood stage—some particles up to bowling-ball size! Because of centrifugal force, streams tend to continue flowing straight. Thus when the channel bends, the energy of the debris-laden water is directed against the rock walls. Pounding and grinding by this hydraulic drill against a thin meander neck may eventually undercut the rock wall and perforate it, thus creating a natural bridge as illustrated in Figure 8–65. The stream follows its newfound shortcut—deepening its new channel and abandoning its former path.

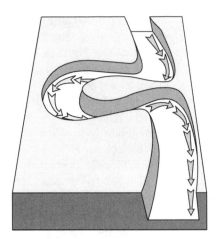
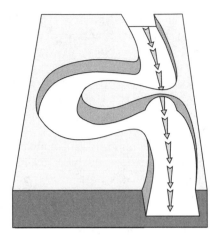

FIGURE 8–65   Simplified sketch illustrating how a stream can "punch" its way through a narrow meander neck to form a natural bridge. (From National Park Service brochure for Rainbow Bridge National Monument)

The evidence is all here for those who let their eyes read the story in the landforms. The occasional flash floods still operate the hydraulic drill at a number of "bridge-in-the-making" sites in and near the monument (Fig. 8–64). Kachina Bridge is a "young" bridge where the stream is actively eroding the relatively small bridge opening and the rock span is still quite thick and bulky. Changes are more frequent here than at some of the "older" bridges. Over 4000 tons of rock collapsed catastrophically from the north abutment in June of 1992. Huge boulders now litter the north valley wall below Kachina.

Sipapu Bridge (Fig. 8–66) is older and has been greatly enlarged; the rock span is about 50 feet (15 m) thick, roughly half as thick as that of Kachina. Streams no longer erode its abutments as the base of the rock span is 167 feet (51 m) above the stream bed. The 268-foot (82-m) rock span is second only in width to the world-champion Rainbow Bridge about 60 miles (100 km) to the south. Enlargement continues now by weathering processes that slowly loosen sand grains and rock fragments. The Owachomo Bridge is at the junction of White and Tuwa canyons and is in a late or "old-age" stage of development (Fig. 8–67). White Canyon is well below the floor of the bridge and the rock span is quite thin, a mere 9 feet (3 m) thick at its thinnest point. Gravity will eventually win and the frail bridge will cease to exist, perhaps within a few centuries, conceivably much sooner. At least two sites in the monument are likely areas where bridges were formerly located.

The age in years of the bridges is unknown but could be determined by finding the rate of stream downcutting and the vertical distance of the original bridge base above the present channel. Kachina is perhaps less than 10,000 years and Owachomo is more than 100,000 years old.

FIGURE 8–66 Sipapu Bridge has a 268-foot (82-m) span, second in the world only to Rainbow Bridge. (Photo by E. Kiver)

FIGURE 8–67 Owachomo Bridge—an "old-age" bridge whose continued enlargement will lead to its ultimate destruction. (Photo by E. Kiver)

Visiting Natural Bridges

Access to this remote area is from scenic roads leading from Blanding or Mexican Hat. Be sure to stop at Goosenecks State Park near Mexican Hat for a view of a spectacular, textbook example of entrenched meanders (Fig. 8–68). Notice that the late Paleozoic strata at the Goosenecks lack a thick, resistant sandstone such as that of the Cedar Mesa Sandstone. Thus, one of the prerequisites for bridge formation, that of a mechanically strong rock layer to hold the weight of a bridge, is lacking.

At the monument is the earth's finest outdoor museum of natural bridges. Visitors who take time to visit and contemplate each bridge and its surroundings will clearly recognize the story recorded in the landforms. An initial stop at the visitor center to look at the exhibits and see the slide show will make your exploration of one of nature's gems even more rewarding. Drive around the 8-mile-loop (12.9-km) and view the bridges from afar. But take time to hike down the trails for a closer look. If you hike the trail between the bridges you will see some of the Indian cliff dwellings.

The lonely vastness of the country here is inspiring and helps to remind us of what things are really important. The incredible landscapes, the ghosts of the Anasazi, and the spirit of those who came later have helped mold our people into

FIGURE 8–68   Goosenecks of the San Juan River, near Mexican Hat, southeastern Utah. (Photo by J.A. Campbell)

a self-sufficient lot who are proud of our heritage. Here at Natural Bridges we can be proud of the special gifts of nature that have been trusted to our safekeeping.

REFERENCES

Petersen, D., 1990, Of wind, water, and sand: The Natural Bridges story: Moab, UT, Canyonlands Natural History Association, 21 p.

Stokes, W.L., 1986, Geology of Utah: Salt Lake City, Utah Geological and Mineral Survey, 280 p.

Park Address

Natural Bridges National Monument
Box 1
Lake Powell, UT 84533

RAINBOW BRIDGE NATIONAL MONUMENT (UTAH)

The world's largest natural bridge spans Bridge Creek, 8 miles (12.9 km) upstream from its confluence with the Colorado River in southeastern Utah. Architecturally, Rainbow Bridge is inspiring; it is huge—278 feet long (85 m) and 309 feet high (94 m)—and it is almost symmetrical (Fig. 8–69). Its location in some of the most remote country in the conterminous United States limited its visitation, especially before the 1970s. A total of about 4000 people visited the bridge from 1909, when the first Anglo-Americans set eyes on the bridge, to the early 1970s when filling of Lake Powell behind Glen Canyon Dam permitted tour boats from Page, Arizona, to bring hundreds of people each day to the tiny, 640-acre (one square mile; 2.6 km^2) national monument. Thus, about 60 people each year, with only 5 or 10 visitors in some years, visited this remarkable area before 1970.

Today with Lake Powell flooding 186 miles (300 km) of the Colorado River canyons, as well as the lower ends of its many tributaries, tour boats dock within easy walking distance of the majestic rock bridge, bringing over 300,000 visitors each year. The original long and difficult trails from the Navajo Mountain Trading Post (24 miles; 39 km) and from Rainbow Lodge (14 miles; 23 km) are still there for those who wish to obtain a permit and experience some of the spectacular, unspoiled backcountry of the Navajo Reservation. Just as climbing a mountain by trail is much more satisfying than driving to the top by automobile, so too the long hike to Rainbow Bridge provides an even more rewarding experience than the long boat ride.

The monument is on the flanks of 10,388-foot-high (3167-m) Navajo Mountain, one of the sacred mountains of the Navajo people. The middle Tertiary (about

FIGURE 8–69   Rainbow Bridge is the largest natural bridge in the world. The U.S. Capitol Building would fit beneath this massive rock span! (Photo by E. Kiver)

40–30 Ma) intrusion beneath has domed the overlying strata and is not yet exposed at the surface by erosion (Stokes, 1986). Likely it consists of a group of flat-floored, upward-bulging intrusions called laccoliths—similar to those in the Henry, La Sal, and Abajo mountains farther north.

The processes that formed Rainbow Bridge are the same as those that formed the bridges in its sister park to the north at Natural Bridges National Monument. The rock span at Rainbow Bridge is the Jurassic age (about 160 Ma) Navajo Sandstone rather than the Permian-age (about 250 Ma) Cedar Mesa Sandstone found at Natural Bridges.

Historical Background

For how many centuries or millennia the Native Americans knew and used Rainbow Bridge, or *Nonnezoshe* ("rainbow turned to stone"), as a religious site may never be known. Holes in the ground are special places—gateways through which souls enter or leave the earth. When President Theodore Roosevelt visited Rainbow Bridge in 1913, he noted that his Indian guides would say a particular prayer before going under the bridge. One of the guides rode around the end of the bridge rather than crossing under it.

John Wetherill, an amateur archeologist who ran a trading post with his wife Louisa at Kayenta, was extremely interested in his Indian customers and their culture. He was instrumental in discovering the pueblo village at Keet Seel—now Navajo National Monument. Louisa Wetherill heard stories of the sacred bridge from a Navajo elder as well as from a Paiute chief. An exploratory trip in 1909 included John Wetherill, Professor Byron Cummings, government surveyor William Douglass, and Indian guide Nasja Begay. Silence gripped the expedition members when Nonnezoshe finally came into view. The remains of an ancient Indian altar or shrine were present at Rainbow Bridge when first discovered.

The trail to Rainbow Bridge descends a maze of canyons that were described by noted western author Zane Grey on his trip in 1913 as a "chaos of a million canyons," where "a man became nothing." This was the same area where hundreds of Navajos eluded Kit Carson's infamous roundup of Indians during the winter of 1863–1864. Those who were captured were forced to make the Long Walk to Fort Sumner Indian Reservation. There they endured 4 years of starvation and hardship before being allowed to return to their homeland.

President Theodore Roosevelt established the monument in 1910 and the area remained extremely remote until the filling of the Lake Powell reservoir and the establishment of a tour-boat service. Formerly people would descend the long trail from Navajo Mountain or float the Colorado River through beautiful Glen Canyon, beach their rubber boat at Bridge Creek, and walk 8 miles (13 km) to what has been described by some as one of the seven natural wonders of the earth.

Geologic Setting and Bridge Formation

The Jurassic-age Navajo Sandstone is the star scenery maker at Rainbow Bridge and many other areas in the Colorado Plateaus. Its most spectacular display is at Zion National Park where the frozen dune sands are as much as 2200 feet (670 m) thick. Other parks including Canyonlands, Capitol Reef, Grand Staircase–Escalante, and nearby Glen Canyon National Recreation Area also display the shining cliffs of cross-bedded Navajo dune sand. Rainbow Bridge formed in this thick, resistant sandstone. Also important are the underlying sands and silts of the Kayenta Formation. These river- and delta-deposited sediments are extremely well-cemented—forming excellent abutments for Rainbow Bridge.

A series of events resulted in the formation of the bridge. Eons before the creation of Lake Powell, when the Colorado River's channel was much higher up—even above the present surface of Lake Powell—meandering tributaries, including Bridge Creek, emptied into the river. When the Colorado Plateaus were uplifted late in the Tertiary (beginning about 10 Ma) and the Colorado River developed an outlet through Grand Canyon and on to the Pacific Ocean, the Colorado River entrenched its channel, causing its tributaries in turn to entrench theirs. Bridge Creek was flowing then at perhaps the level of the present top of the bridge, and one of its horseshoe-shaped meanders encircled the end of a high wall of Navajo Sandstone.

The stream, undercutting at the ends of the horseshoe, finally broke through the high rock wall; abandoning the horseshoe for a shorter course—the same process that formed bridges at Natural Bridges National Monument (Fig. 8–65). From then on Bridge Creek flowed through the hole in the wall. As Bridge Creek deepened its channel beneath the new bridge, the walls adjacent to the stream became unstable and fell in block by block, widening the opening and lengthening the bridge. Blocks also fell from above until finally a stable symmetrical form was attained. Fortunately, in this dry region chemical weathering is very slow, and Rainbow Bridge is changing at an extremely slow rate.

The Navajo Sandstone is composed mostly of quartz sand with a small amount of iron-bearing minerals. Rain washing down the sides of the bridge in combination with a specialized bacterium and wind-blown dust particles produces streaks of red and brown—a form of desert varnish. The brilliant coloring, especially in the afternoon sun, may be the basis for the Indian description of a rainbow turned to stone.

Controversy at Rainbow Bridge

As is increasingly true of all public lands, different groups want to use our scarce land in different ways. Problems often arise when extractive uses such as mining and logging are proposed—activities that could damage the land for decades or centuries. However, even longer-lasting environmental impacts occur when rivers such as the Colorado are dammed. Construction of Glen Canyon Dam initiated the first of many drastic changes to the ecosystem of the river and surrounding areas. Waters rose behind the new dam beginning in 1963. Nearly 200 miles (320 km) of delightful and spectacular canyons along the Colorado River were eliminated forever (by human standards) as Lake Powell deepened and crept upstream as far as the lower end of Canyonlands National Park.

As Eliot Porter (1963) points out in his classic book, *The Place No One Knew*, the elimination of the irreplaceable scenery along Glen Canyon was unnecessary. The purpose of the dam was to act as a focal point for dividing water between the Upper and Lower Colorado River basin states, a bookkeeping task that could have as easily been done by using the already existing Lake Mead just downstream from the mouth of the Grand Canyon. The sad conclusion is that Glen Canyon Dam benefits very few and is unnecessary. In addition, because of the 60 million tons of sediment washed annually into the reservoir, its useful lifespan is about 200 years, of which nearly 50 years, or 25 percent, of its useful life is gone. No one has even a foggy notion of what our descendants will do with a 200-mile-long canyon filled to the brim with sediment. A glimmer of hope has surfaced, more people are seriously proposing that Glen Canyon Dam be removed!

The initial engineering plan in the 1960s was to raise reservoir levels to 3700 feet (1128 m). That level would back water about one-half mile past Rainbow Bridge and would lap against its footings. This in turn would increase weathering and produce wave erosion on the bridge abutments—hastening the time when gravity would exceed the mechanical strength of the rock span and a final roar and cloud of dust would mark Rainbow Bridge's final demise. The courts ruled in favor of limiting water depths and prolonging the life of the bridge—perhaps by an extra thousand years.

Now a new problem has arisen as some six regional Indian tribes have expressed concern for the heavy visitation and nonchalant disregard of some visitors for their sacred bridge. Some would argue that a trip to Rainbow Bridge is similar to visiting a cathedral. Viewing a cathedral is an inspiring experience. Walking its aisles is different from walking around its altar—the equivalent of walking beneath the bridge as many Indians would argue.

Others see only the spectacular scenery and believe that this national treasure should not be managed as a religious shrine. For many non-Indians who are sensitive to nature and have strong feelings, areas such as this can also fulfill a religious or deep philosophical need when one explores or views such magnificent natural beauty. The National Park Service is caught in the middle and has attempted to appease all views by placing a sign asking visitors to voluntarily refrain from approaching too close to the bridge. No matter how you view the bridge, it will be an inspiring experience.

REFERENCES

Porter, E., 1963, The place no one knew: Glen Canyon on the Colorado: San Francisco, Sierra Club, 72 p.

Stokes, W.L., 1986, Geology of Utah: Salt Lake City, Utah Museum of Natural History, Occasional Paper No. 6, 280 p.

Park Address

Rainbow Bridge National Monument
c/o Glen Canyon National Recreation Area
P.O. Box 1507
Page, AZ 86040

ARCHES NATIONAL PARK (UTAH)

Any other name would not be appropriate for Arches National Park. There are other interesting features, but arches clearly dominate the landscape. The graceful curving lines of arches towering high above are here in superabundance and provide time-lapse views of stages of development from "birth" to their eventual demise. What quirks of nature have permitted the most dense concentration of arches in the world to form here in southeastern Utah? While natural arches are rare, they do occasionally form elsewhere, but not in these numbers. From the unique geologic conditions that obviously prevail at Arches a landscape like no other has evolved—the "stuff" that national parks are made of.

Historical Background

The area was used by many generations of Native Americans beginning with hunters and gatherers about 12,000 years ago. Extinction of the large prey animals such as mammoth and ground sloth likely reduced human populations to the small nomadic groups characteristic of the archaic culture. About 2000 years ago Fremont and Anasazi agricultural peoples flourished in the Colorado Plateaus, although populations near Arches were low because of marginal growing conditions. A shift to even drier climates in the late thirteenth century likely contributed to the mysterious disappearance of these heretofore successful people. The Utes inhabited the area during the past few hundred years and were the group that the first Euro-Americans encountered. These early inhabitants left excellent examples of their artwork, the enigmatic pictographs (rock paintings) and petroglyphs (pictures chipped from desert varnish on rock walls), and they also quarried nodules of chalcedony from sites in today's park for tool and weapon making. Petroglyphs are observable near the old Wolfe Ranch and some of the abandoned chalcedony quarries are along the arduous but rewarding trail that climbs to Delicate Arch, the hallmark of Arches National Park.

The first settler was John Wesley Wolfe, a civil war veteran who built his cabin near a dependable year-round spring that made life possible in the harsh Utah desert. A few cows, an irrigated garden, and a government pension enabled him to eke out an existence here for 20 years before leaving the ranch. Dr. Lawrence Gould (University of Michigan) was the first to recognize the importance of the area's fascinating scenery and geology. After touring and enthusiastically proclaiming its beauty to Marv Turnbow, the then owner of the Wolfe Ranch, Turnbow remarked that he "didn't realize that there was anything unusual about it"! Gould urged seeking national park status and a local physician, J.W. "Doc" Williams and Moab newspaperman and Lion's Club member Bish Taylor strongly supported the cause. However, Hungarian born Alexander Ringhoffer is given the title "Father of the Arches" (Hoffman, 1985). Ringhoffer was a prospector who discovered the Klondike Bluffs area in 1922 and fell in love with the scenery. He introduced the area to railroad officials who in turn contacted Stephen T. Mather, the first director of the National Park Service. Herbert Hoover signed the bill that set 7 square miles (18 km^2) aside as Arches National Monument in 1929. The boundaries were enlarged and later reduced to 114 square miles (73,379 acres) by the time national park status was approved by Congress and signed by President Nixon in 1971. Located 5 miles (8 km) north of Moab and with the Colorado River as its southern boundary, Arches is in the heart of Utah's redrock and slickrock country (Fig. 8–70).

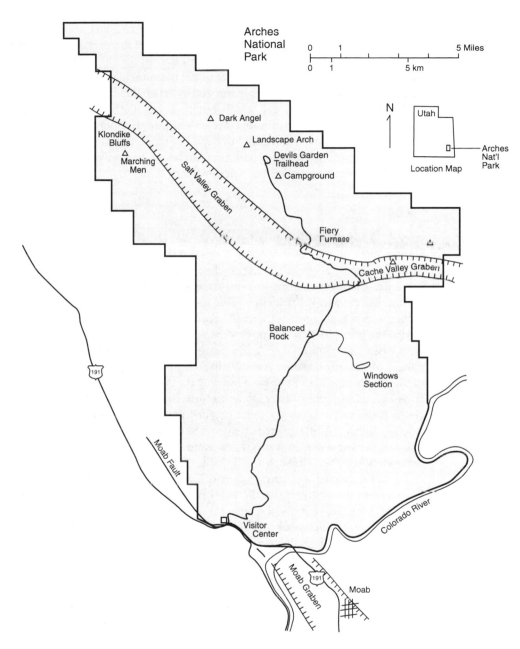

FIGURE 8–70   Map of Arches National Park.

An attractive visitor center is located at the park entrance on U.S. Highway 191, and paved roads now lead into several sections of the park. Those of us who were there in the early days are immediately aware of the change from the primitive to the modern. At that time a visit to the monument was an adventure, particularly for the visitors who were marooned by a flash flood. For a fascinating story of the Arches of yesterday, read Stan Lohman's account (1975). Among those who think back wistfully to the primitive Arches was Edward Abbey (1968) who, in his *Desert Solitaire,* speaks eloquently on behalf of wilderness. Fortunately there remain today several areas of the park that can be reached only on foot and still retain a flavor of the unspoiled and uncrowded past.

Geologic History

Hidden beneath the late Paleozoic (Pennsylvanian, about 300 Ma) to middle Mesozoic (Jurassic, about 150 Ma) rocks exposed at the surface in the Moab area are older Paleozoic sediments and some buried Precambrian-age fault-block mountains that play key roles in understanding why the world's greatest density of arches occurs in this relatively small area (Doehling, 1985). Paleozoic sediments buried the fault-block mountains—especially in the Pennsylvanian Period when a deep basin, the Paradox Basin, and a nearby range to the north called the Uncompahgre Mountains began to form. Plate collisions along the south and east edges of the North American Plate were forming the Appalachian and Ouachita mountains and transferring stresses inland to form the Paradox Basin and uplifts associated with late Paleozoic mountains called the Ancestral Rockies. Lesser uplifts (the Monument Uplift) formed a platform to the southwest that helped restrict the circulation of seawater in the Paradox Basin, enabling evaporation and chemical precipitation of minerals such as salt (halite), gypsum, and in places potash (sylvite and other minerals) to occur. Water with a fresh supply of dissolved chemicals flowed from the sea to replace the water evaporated in the basin for at least hundreds of thousands of years, thus enabling evaporite minerals in the Paradox Basin to accumulate to thicknesses of over a mile. As we shall see in the next section, deposition of these thick salt layers provides another geologic coincidence that helps account for the unusual abundance of arches.

Debris shed from the nearby Uncompahgre Uplift during Permian time (about 250 Ma) buried the salt under debris fans that graded into beach and marine deposits to the west. Later, some episodes of marine deposition occurred but more and more stream, lake, and desert deposits such as the eolian sands of the Navajo Formation dominated the scene. These "frozen dunes" formed in a vast desert that covered a large area from Wyoming to Nevada during early Jurassic times about 210 Ma. The Navajo Sandstone at Arches is particularly well exposed in the Windows area in the southeastern end of the park and is part of the same unit that dominates the scenery at Zion National Park.

The overlying Entrada Sandstone, deposited in streams and as desert dunes about 140 Ma, is another important component of the Arches story. Almost all of the arches in the park are in the Entrada Sandstone. The lowermost member of the Entrada is called the Dewey Bridge and is easily recognized in the Windows area and elsewhere by its wavy, contorted bedding. The overlying Slickrock and Moab Sandstone members are thick cliff-forming sandstones that form the body and top of the arches. In the Devils Garden area the arches are entirely in the Slickrock and Moab sandstones. Iron oxides coat the sand grains giving the rock a red or reddish-brown color. That it is merely a coating is verified by observing the color of wind-blown sand. Where it is near its source, a red sandstone cliff, the sand is distinctly red. If it has been carried for several miles, it may be coral pink. Far from the source, the sand will be essentially white, the red coating having been worn off as the quartz grains bounced along over the desert floor.

Over 5000 feet (1524 m) of younger sediments were later eroded off, especially during the last few million years, leaving the Entrada and the Navajo as the domi-nant formations currently exposed in the park. Localized outcrops of rocks as young as the Upper Cretaceous Mancos Shale occur as down-dropped masses associated with the Salt Valley and Cache Valley anticlines discussed in the next section.

Laramide uplift, folding, and faulting during early Tertiary time initiated the erosion that eventually exposed the Navajo and Entrada. However, vertical uplift of thousands of feet during the past 5–10 million years greatly speeded up the ero-sion process in the Colorado Plateaus. Arches can and do form in many sandstone units in the Colorado Plateaus other than the Entrada, but not as abundantly. Ad-ditionally, why do only a few arches form in the Entrada outside the park—but not anywhere near the concentration found in the park? Another missing piece to the puzzle is found in the unusual structural features in the area—the salt anticlines.

Salt Structures

Salt behaves plastically (flows without rupturing) when sufficient pressure is ap-plied. It is lighter (density of 2.2) than sedimentary rocks such as sandstone and shale (density of about 2.55) and tends to push and ooze upward to areas of less pressure when buried by heavier material—just as a drop of oil rises through a col-umn of water. Where salt beds are deeply buried irregularities of thicknesses or weaknesses in the overlying rock usually initiate a centralized upward movement that produces a *salt dome*. Burying the northwest–southeast trending fault-block topography in the Paradox Basin resulted in thicker salt deposits over the buried valleys and thinner deposits over the buried ridges. The thicker salt over the buried valleys moves upward near the buried northwest–southeast trending ridges, buck-ling the overlying beds into northwest–southeast trending *salt anticlines* (Fig. 8–71a). The salt thickens under the anticlines and thins in adjacent areas, eventu-ally breaking and penetrating the overlying strata. Salt domes such as those found in the Gulf Coast area of the southeastern United States are the features most commonly formed by salt flowage—the Arches area and the Paradox Basin are one of the few areas in the world where salt-cored anticlines are the rule rather than the exception. Upward movement of the Paradox Salt began in Pennsylvanian time and ended by the close of Jurassic time.

As the brittle sandstone overlying the salt was bent upward, a series of parallel fractures or joints formed parallel to the axis of the anticline (Fig. 8–71*b*). Later, after erosion when the salt is near the earth's surface, groundwater dissolves and removes the salt, particularly along the crest of the anticline, leaving the rocks without support. They collapse in a manner not unlike the collapse of rocks in a caldera. Here, however, the process is much slower, and the collapsed blocks are distinctly elongate, forming a structure called a *graben* (German for grave). This unusual sequence of events holds the key to understanding many of the landforms, including the arches, in Arches National Park.

A number of these collapsed salt structures occur in southeastern Utah and adjacent Colorado, including the Spanish Valley–Moab graben just south of the park. The Moab Fault continues northwest from the graben along the canyon by the visitor center where vertical displacement of 2600 feet (793 m) has dropped the younger Jurassic rocks at Arches downward against Permian and Triassic rocks on the south side of the canyon. To the northeast of the visitor center and the Moab Fault are the next salt structures—the Salt Valley and Cache Valley anticlines where the park's arches are concentrated.

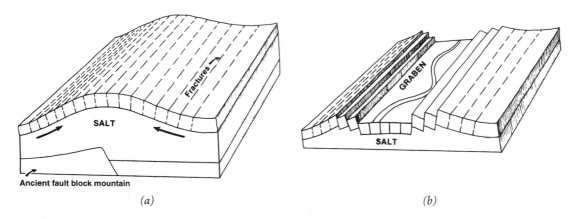

(a)	*(b)*

FIGURE 8–71   (*a*) Schematic block diagram showing the development of a salt anticline. Salt, plastic due to the weight of overlying rocks, is forced laterally to arch up above the buried fault-block structure below. (*b*) Later stage than that shown in (*a*). After salt has been dissolved and carried away by groundwater, collapse of roof rocks forms a graben. (Diagrams by Gregory Nelson)

Fins, Arches, and Other Landforms

When the highly fractured sandstone of the anticline is exposed at or near the earth's surface, weathering is more intense along areas of weakness, such as along the joints. Water persists longer in the joints and removes the calcite cement that locks the sand grains together. Plants may grow in the loose soil along these wetter zones producing organic acids that further attack the rock. Removal of loose sand by wind and occasional rain or melted snow develops long narrow ridges or fins (Fig. 8–72). As rock is removed and the confining pressure on the thick sandstone is reduced, vertical slabs and rock chunks fall away from the cliff face—a process called unloading, or exfoliation. Overhangs produced by this process as well as other rock characteristics including the narrowness of fins and the presence of localized rock weaknesses such as cross joints, bedding planes, poor cementation of sand grains, and fracturing further influence the developing landforms. A variety of possible landforms produced by weathering of the vertically fractured rock near the salt grabens is shown in Figure 8–73.

If weathering is most rapid along vertical cross joints, fins can be reduced to rock pinnacles such as those found along the Park Avenue Trail, the Dark Angel at the end of the Devils Garden Trail, or the Marching Men at Klondike Bluffs. Weathering along both vertical joints and horizontal weaknesses in the rock could

FIGURE 8–72   Devils Garden area with fins in different stages of development. (Photo by E. Kiver)

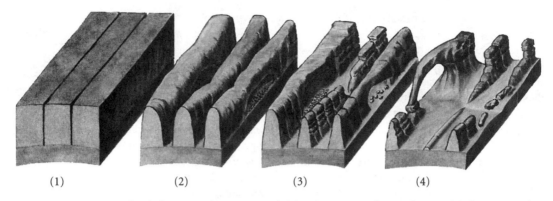

FIGURE 8–73   Sequence of arch formation beginning with (1) enlargement of vertical joints, (2) formation of fins, (3) weathering along weak zones to produce an arch and other landforms, and (4) enlargement of arch eventually leading to collapse. (Diagram from Arches National Park brochure)

produce a balanced rock such as the 3577-ton giant at the Windows turnoff (Fig. 8–74). An infinite variety of forms and shapes challenges the brain to store the details of this visual extravaganza.

How is it that arches can form in this complex of landforms? The Native Americans believed that the arches were built by the Great Sky Father, and early settlers thought they were built by Native Americans! Geologists quickly recognized that fins, especially those that were narrow, were likely to be perforated by geologic processes to form arches. Localized wind and water erosion were initially considered important, but more recent ideas favor differential weathering—selective weathering of weaker or more vulnerable areas of a rock. Herein lies the major difference between two look-alike forms—natural arches and natural bridges. Geologists insist on two designations because they are formed by different processes. Water does most of the work in creating both landforms. However, a natural bridge spans a valley of erosion and is therefore produced by running water. A natural arch is formed by "holing through" of thin walls or fins mostly by weathering processes—processes that operate more rapidly in the presence of water.

Of particular importance to arch formation is the boundary between the relatively weak, crinkly appearing Dewey Bridge Member and the overlying resistant Slickrock Sandstone Member (Fig. 8–74). Many of the arches in the Windows section and other areas form along this discontinuity. The Dewey Bridge (rhymes with "gooey bridge"!) is easy to recognize because of its characteristic contorted beds that formed shortly after deposition of river sediments during the Jurassic Period and prior to their cementation into hard stratum. Indentations of the fin begin on both sides—eventually uniting to form a *window*. Enlargement to a natural arch (a natural opening greater than 3 feet (1 m) in any one direction) continues until the arch thins (Fig. 8–75) and eventually collapses (Fig. 8–76).

The calcium carbonate cementing material in the Entrada is not uniformly distributed, and chemical weathering is most rapid where the sandstone is least ce-

FIGURE 8–74   Balanced rock (center) near the Windows turnoff is 55 feet (17 m) high and weighs 3500 tons! The massive balanced sandstone block belongs to the Slickrock Sandstone Member and the crinkly layer below is part of the Dewey Bridge Member. (Photo by E. Kiver)

FIGURE 8–75   Landscape Arch, old and frail. (Photo by D. Harris)

FIGURE 8–76   Arches that have fallen; note window or Baby Arch in the fin to the left. Rock span likely connected the fin on the left with the pinnacle on the right. (Photo by D. Harris)

mented or highly fractured. The arches in the Devils Garden area formed along such weak zones. The weak zone at Delicate Arch is along the contact of the Slick-rock and overlying Moab member (Fig. 8–77) and is the "Achilles heel" that will lead to its eventual downfall. A careful study of the fins and arches discloses secondary fractures that formed as the salt anticline developed and joint blocks collapsed (Cruikshank and Aydin, 1994). Fractures are most common where rocks contain impurities or weaknesses such as shale lenses, bedding planes, or pre-existing joints. Rock that is excessively fractured may not even be strong enough to form a fin—rocks that are highly fractured only in local zones are good candidates for arch development.

The complete sequence of development of arches begins with a narrow opening (less than 3 feet in diameter; 1 m) called a window; its enlargement to a small arch (Baby Arch in Fig. 8–76); further enlargement by the downfall of sandstone blocks such as those that fell at Skyline Arch in 1940; and eventual thinning to the next-to-last stage such as that at Landscape (Fig. 8–75) or Delicate Arch (Fig. 8–78). Visitors are no longer allowed on top of any arch in the park, in order to postpone the downfall of the arch, as well as the visitor. In many places, only the abutment rocks still stand, representing the completion of the cycle.

The Future

As we have just seen, an amazing number of geologic coincidences occurred in the Arches area to account for the world's highest density of arches. An ancient fault-block topography must be buried by thick salt beds that flow upward and fold and fracture an overlying brittle sandstone rock such as the Entrada. Erosion must lower

FIGURE 8–77   Abutment of Delicate Arch showing its "Achilles heel" where its ultimate downfall will be initiated. View is to west. (Photo by E. Kiver)

the earth's surface close to the salt beds so that groundwater can remove some of the salt and permit the long joint blocks to collapse into the graben. Weathering and erosion then form fins in the Entrada, making arch formation possible. Many other collapsed salt anticlines occur in southeast Utah, but none has the density of arches shown here. However, none of these collapsed anticlines now has Entrada Sandstone at the edge of the graben! How many "arches national parks" have existed in the geologic past in the Paradox Basin area will never be known. The present Arches National Park will exist until it runs out of Entrada Sandstone!

The Arches story is an ongoing one that is occurring before our eyes but is difficult to see because of the slow (by human standards) rates of change. All the signs are there for us to read. A balanced rock nicknamed "Chip-Off-the-Old-Block" formerly stood next to Balanced Rock (Fig. 8–74) by the Windows turnoff. Sometime

FIGURE 8–78   Delicate Arch, an arch in an advanced stage of development. View east toward La Sal (igneous) Mountains. (Photo by E. Kiver)

during the winter of 1975–1976 gravity and erosion won the snail-pace battle that had persisted for tens of thousands of years. In 1940 a major rockfall literally doubled the size of Skyline Arch in the Devils Garden Campground in a few seconds, and in 1991 a 60-foot (18-m) section of rock falling from Landscape Arch was recorded on film by a visitor. Landscape Arch is one of the world's longest rock spans (434 feet, 132 m; light opening is 306 feet, 93m) and is the most precarious arch in the park (Fig. 8–75). It is a mere 16 feet (5m) thick at its thinnest point. Its fate was of great concern in the 1960s and early 1970s as the U.S. Air Force flew supersonic aircraft at low altitudes over Arches and other parks. Were some of the rock masses loosened by sound vibrations, thereby encouraging the rockfall in 1991? Fortunately the Air Force was persuaded to avoid future supersonic overflights over Arches and other national parks.

In 1940 the National Park Service ordered 100 metal caps so that all arches in the park could be "tagged." By 1973 there were 124 known arches in the park, by 1982 there were 200, in 1989 it increased to 1000, and by 1994 nearly 2000 arches were cataloged! How can this be? Are arches forming at this amazing rate? The answer lies in the complicated topographic maze of fins and canyons and dedicated individuals spending many hours exploring areas perhaps never visited before by humans. Of the over 2000 known arches, collapse has eliminated 42. Likely, the rate of arch development will equal the rate of destruction for many hundreds of thousands of years until narrow fins in the Entrada Sandstone cease to form.

More frightening than these natural processes are what humans, mostly well-meaning humans, in ever-increasing numbers are doing. Those who are not well meaning also visit the area. This small number of destructive individuals care not for future generations or this special heritage belonging to all the citizens of the earth. Those who would carve their initials into the sandstone base of Delicate Arch or those, who in 1980, threw battery acid on the spectacular pictographs of the Moab panel that had survived the elements for over 1000 years should forever lose the privilege of entering a national park, concert hall, or a museum. Art, historical treasures, and relatively unspoiled natural environments enrich the lives of those who take the opportunity to enjoy them and help separate civilized beings from others.

Park visitation here, as well as in nearby Canyonlands National Park, has increased at the alarming rate of 14 percent per year since the early 1980s. The environment of a park, country, or the world for that matter cannot survive such unchecked population growth. The Park Service is trying an interesting experiment at some of the heavier used areas such as Delicate Arch, the Windows, and Devils Garden. Visitors at Delicate Arch prefer sharing the view with fewer than 12 people, although most people find as many as 30 acceptable. The quality of the experience decreases rapidly as "crowding" increases. That feeling of vast landscapes and closeness to nature is significantly impaired on a crowded trail or when one has to wait for 20 or more minutes to take a picture of an arch without a "mob" in the way. The Park Service is limiting parking to 64 spaces at Delicate Arch as an experiment to reduce crowding. This approach may provide a quality experience at specific sites but cannot solve the problem of too many automobiles on roads and too many visitors at more accessible sites. Ultimately more drastic measures will need to be taken.

Even well-meaning people are often unaware of the damage created by walking off trails and crushing the protective crust on the desert soil called *cryptobiotic soil*. Such crusts are complex combinations of several species of bacteria, lichens, mosses, and algae that provide nutrients for plants, restrict water loss from the desert soil, reduce erosion of the fragile desert soil, and discourage nonnative plant growth. Stepping off the trail breaks the crust and may require decades for nature to heal the scar. Even the old wilderness adage "take nothing but pictures, leave nothing but footprints" has to be restricted here to photographs with footsteps only on established trails.

REFERENCES

Abbey, E.F., 1968, Desert Solitaire: Simon and Schuster, New York.

Cruikshank, K.M., and Aydin, A., 1994, Role of fracture localization in arch formation, Arches National Park, Utah: Geological Society of America Bulletin, v. 106, p. 879–891.

Doehling, H.H., 1985, Geology of Arches National Park: Utah Geological and Mineral Survey, Department of Natural Resources, 15 p.

Hoffman, J.F., 1985, Arches National Park, an illustrated guide: Western Recreational Publications, San Diego, 128 p.

Lohman, S.W., 1975, The geologic story of Arches National Park: U.S. Geological Survey Bulletin 1393.

Park Address

Arches National Park
P.O. Box 907
Moab, UT 84532

MESA VERDE NATIONAL PARK (COLORADO)

Some of the most striking and best preserved pre-Columbian Indian ruins are found in the high plateau country in the southwestern corner of Colorado at Mesa Verde National Park. At its peak, thousands of Anasazi or Hisatsinom (Hopi word for "those who lived here long ago") farmed much of the surface of Mesa Verde and lived in relative comfort as Cliff Dwellers in well-constructed rock and adobe-mud apartment houses high on the cliffs, beneath overhangs (Fig. 8–79). By 600 years ago (about 1400) the flourishing civilization had mysteriously abandoned this and numerous other sites in the Southwest, probably migrating to the New Mexico area where their modern-day descendants, the Hopi, Zuni, and other Pueblo peoples live today.

FIGURE 8–79   Spruce Tree House below park headquarters. Cliff House Sandstone (Cretaceous) forms the rock roof above the Hisatsinom (Anasazi) ruins. (Photo by E. Kiver)

Here we will investigate the special combination of geology and geography that enabled the Hisatsinom civilization to flourish for nearly a thousand years. Why it was unable to survive at Mesa Verde is, unfortunately, uncertain. No doubt the reasons are lessons from which we could learn—so that we do not also follow a similar path to cultural oblivion.

Physiography and Geologic Development

We begin the story behind the scenery at Mesa Verde with the Late Cretaceous rocks exposed in and near the park. A long interval of dry desert conditions during the Jurassic Period came to an end as shallow inland seas once again crept across the North American continent. The expanding sea encroached from the midcontinent area across the western states, depositing near-shore sands followed by muddy sediment as the water deepened. The Rocky Mountains were a few million years away from beginning their rise to spectacular heights and presented no topographic obstruction at this time to the spread of the last of these great inland seas.

The Dakota Sandstone exposed just north of the park near the town of Cortez (Fig. 8–80) and elsewhere in the Rocky Mountains and Great Plains area are the beach and near-shore sands, the leading edge of what was to be one of the most extensive seaways to cover North America during the Paleozoic and Mesozoic eras. The thick gray muds of the Mancos Shale overlying the Dakota Sandstone is exposed along the entrance road that ascends the steep slopes to Morefield Campground along the north end of the park (Fig. 8–80). This mechanically weak material is subject to frequent slope failure, requiring considerable maintenance of

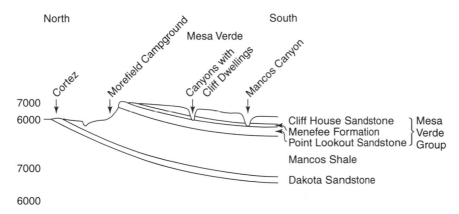

FIGURE 8–80   Generalized cross section of part of Mesa Verde.

the roadway. The lowest unit of the Mesa Verde Group, the Point Lookout Sandstone, caps the long Mancos slopes and forms the spectacular northern cliff face. These near-shore sands as well as overlying sediments indicate that numerous shifts of the edge of the sea occurred during the Late Cretaceous (Baars, 1995).

The middle formation of the Mesa Verde Group, the coastal swamp sediments of the Menefee Formation, contains sand and mostly carbon-rich shales, silts, and thin coal beds. The Menefee is poorly exposed on the surface of Mesa Verde but is exposed in some of the canyons that cut deeply into the south end of the park. Here they are overlain by the Cliff House Sandstone, the youngest formation in the Mesa Verde Group, the one in which the spectacular cliff dwellings are located. These near-shore sands contain occasional thin shale and silt zones. These finer layers and zones would seem to be unimportant, but in fact play a major role in the development of the shelter caves and thus where and how the Indians would use the land. Younger marine and nonmarine sediments were deposited but have since been eroded from the Mesa Verde area.

In latest Mesozoic and continuing into early Cenozoic times, the deep stirrings of plate movements that had initiated mountain building farther west in early Mesozoic time had migrated eastward and began to affect the Colorado Plateaus and especially the area to the east that was soon to become the Rocky Mountains. Uplift and doming of the San Juan and La Plata mountains to the north of Mesa Verde gently tilted the rock layers in the Mesa Verde area to the south. By late Tertiary time, perhaps 10 Ma, a gently sloping erosion surface was cut from the San Juan Mountains across the Mesa Verde area. Such an eroded bedrock surface that forms in dry climates, often covered by a thin layer of gravel, is called a pediment. The edges of the gently inclined strata are truncated by the pediment, establishing conditions that would eventually lead to the formation of the modern landforms.

Later in the Tertiary and Quaternary periods, perhaps 10–5 Ma, regional uplift of the Colorado Plateaus and Rocky Mountains, including the nearby San Juan Mountains, occurred. The establishment of a through-flowing Colorado River to the Pacific Ocean also helped create a greater vertical drop and more erosive power for the Colorado River and its tributaries such as the nearby Mancos and San Juan rivers. Thus an episode of canyon cutting was initiated that continues today. The Tertiary pediment surface was deeply dissected by streams forming isolated, topographically detached areas such as Mesa Verde. South-flowing streams on the surface of Mesa Verde cut a series of some 15 parallel valleys. Deep canyons formed, especially on its south end.

Mesa Verde is relatively flat, hence the name mesa (table in Spanish) is appropriate. However, Mesa Verde is also a cuesta, a wide asymmetric ridge with its steep end facing north and a gentle, gradual slope oriented south. Park Point on the north end is at an elevation of 8500 feet (2590 m), and the cuesta surface above the cliff dwellings on the south end are at an elevation of about 7000 feet (2134 m).

The nearly vertical canyon walls in the south end of Mesa Verde are composed of the Cliff House Sandstone. It is here where the large shelter caves, or "overhangs," formed and the Hisatsinom (Anasazi) built their magnificent apartment house complexes. On the cuesta surface between the canyons were sites of extensive garden complexes that supplied most of their food needs. The soil here is a highly productive silt—likely blown in as dust from the nearby mountains during times of glaciation and from the deserts and near-desert areas on the plateau during times of increased aridity.

Formation of Shelter Caves

Shelter caves form abundantly in the Cliff House Sandstone and are the result of differential weathering. During deposition of the shore and near-shore sands of the Cliff House Sandstone, the lateral shifting of the shoreline caused occasional deposition of mud as bay and marsh sedimentation occurred between periods of sand deposition. These mud layers later became thin shale or mudstone interbeds. The tiny mud-particle size creates even tinier, but abundant, spaces between particles, giving the rock a high *porosity*. Thus there is a high proportion of space compared to rock volume. However, such fine-grained sediments have a low *permeability* that prevents water from flowing easily through them. The overlying sandstone has medium porosity and high permeability. Thus, water percolating slowly downward through the overlying sandstone reaches these impermeable shale zones and moves laterally, often toward the canyon wall where a *spring* or *seep* forms.

The prolonged flow of water along these spring and seep zones effectively removes the calcium carbonate (calcite) cement causing individual sand particles and blocks of sandstone to tumble from the weakened zone, thus forming an indentation that enlarges with time. These horizontally elongated openings enlarge over many millenia with some of the larger ones, such as that at Spruce Tree House by the museum and visitor center, being 216 feet (66 m) wide and 89 feet (27 m) deep (Fig. 8–79). Some of the shelter caves even today have small, active springs along the backwall, the same springs that provided a convenient source of water for the Hisatsinom. Important too, at least to the archeologists of today, is that they had giant "dispose-alls," the canyons, right at their front doors.

Enter Humans

Early nomadic humans hunted and gathered food in the area as long ago as 10,000 years ago (Martin, 1993). Not until about 1450 years ago (about A.D. 550) did the first groups of Hisatsinom (Anasazi) begin to use the area extensively. These early settlers are called the Modified Basketmakers by archeologists. The Basketmakers lived in pithouses, some of which were constructed beneath the protection of the sandstone roofs in the shelter caves. They raised corn, squash, and beans and had domesticated turkeys and dogs. They used skillfully woven baskets but were also beginning to make and use pottery for cooking and food storage. About 1250 years ago (750) the Hisatsinom constructed groups of connected rooms (pueblos) on the cuesta surface at Mesa Verde. The walls were straight and true, and sandstone blocks were often fashioned into rectangular blocks of uniform size—skillful construction techniques showing a high degree of engineering competence and cultural development. Underground rooms evolved into *kivas*—roofed, circular chambers presumably for communal gatherings and rituals. Their dead were often buried with tools and jewelry, objects that might be of use in another life.

The "golden age" of Hisatsinom or Anasazi activity occurred between 900 and 700 years ago (1100 and 1300). A population of several thousand inhabited Mesa Verde. Their skills in construction, art, and community living were passed on and improved upon from generation to generation.

From about 1190 to the early 1200s the surface pueblos were abandoned and people built and moved into the shelter caves in the canyons in the south end of Mesa Verde. These skilled artisans built three-story structures and occasional towers four stories high. Cliff Palace, the largest of the approximately 600 cliff dwellings known in the park, had 217 rooms, 23 kivas, and probably housed about 250 people. Why they moved from the surface to the shelter caves is unknown—perhaps protection from weather or enemies? More mysterious is that these structures were lived in for less than 100 years and then were slowly abandoned over a 25–50 year period (Martin, 1993). By 1290 the area was uninhabited. The Hisatsinom probably migrated to the Rio Grande area where their descendants can be found today among the Hopi, Zuni, and other Pueblo people. Farming Indians would never return to Mesa Verde. After a number of silent, empty centuries, groups of nomadic people from the north migrated into the Southwest where they are known today as the Ute, Navajo, and Apache.

The Spanish in the 1760s and 1770s were the first Euro-Americans to pass through the area. They viewed and named Mesa Verde from afar while camping in the nearby Mancos Valley. The cover of the pinyon-juniper forest reminded them of a "green table," hence the name Mesa Verde. Nearby cliff dwellings were first photographed in 1874 by William Henry Jackson, the famous photographer who accompanied the Hayden Survey of the Yellowstone area in 1871.

The Utes told stories of "lost cities," but it wasn't until 1888 when local cowboys Richard Wetherill and his brother-in-law Charlie Mason were searching for stray cows that they discovered Spruce Tree House and Square Tower House. They descended a large Douglas fir log (called spruce in those days) and entered a prehistoric world that had not been explored up close for about 600 years. The Wetherill family was excited about their discovery and for the next few years collected artifacts and guided tourists into the area. National recognition of its significance was slow in coming. Virginia Donaghe McClurg, a journalist from New York, became enthusiastic about the significance of the area and organized the Colorado Cliff Dwellers Association to try to stop the removal and sale of artifacts and to lobby for a national park. As Martin (1993, p. 28) describes her, "she became a one-woman bandwagon to preserve the cliff dwellings." Twenty years of effort finally paid off and Theodore Roosevelt signed the bill in 1906 that established Mesa

Verde National Park. Once again, the efforts of one individual or a group of people working for the common good can make a difference.

A number of archeologists including Jesse Walter Fewkes and Jesse Nusbaum began studying the area as early as 1907. Fewkes excavated and stabilized many of the ruins and gave the first campfire programs to tourists in the national park—a tradition that continues in all of our parks today. Nusbaum surveyed and photographed the ruins and later became an extremely effective park superintendent from 1921 to 1931. The most ambitious study to date was conducted from 1958 to 1963 under the direction of the Park Service and *National Geographic*.

Visiting Mesa Verde

Start your visit at either the Far View Visitor Center (open summer only) or the Chapin Mesa Museum (open year round). The sequence of Hisatsinom architectural development can be examined at sites along the Ruins Road on Chapin Mesa. The cliff dwellings on Chapin and Wetherill mesas are open to the public but can only be visited with a ranger guide. Tours are limited during the winter months. Protection is paramount. These priceless, historical treasures should be treated with great respect, just as the Utes and later Indians did. They believed that "when the spirits of the dead are disturbed, then you die too" (Martin, 1993, p. 26). Thus, until Euro-Americans entered the picture, the sites remained as the Hisatsinom had left them, except for the crumbling of the walls. Views from Park Point Fire Lookout and elsewhere in the park enable one to see the nearby Rocky Mountains and the vast Colorado Plateaus to the south and west. When winds carry air pollution from the coal-burning Four Corners power plant away from the Mesa Verde area, the brecciated igneous spine called Shiprock is visible to the south.

Mesa Verde was our first park to be set aside to protect the works of people. This important concept has bloomed and expanded to include many sites of historical and cultural interest including battlegrounds, cemeteries, forts, and birthsites of people important in the history of our country. The greatness of a country and its people can be assessed on the basis of what things they deem important and what things they preserve for those who come later.

One of the haunting questions that most visitors come away with is why did the Hisatsinom or Anasazi abandon their centuries-old homes? They apparently left over an extended period of time, one or two generations; thus they were likely not defeated by an outside enemy. Without a written record we will never know much more than we presently do about the Hisatsinom motives. However, our controlled imaginations can consider the current hypotheses and perhaps discover a new one or add a new twist to one of the more popular ideas. Without more supporting evidence the hypotheses cannot be elevated to the status of theories—such is the way of science.

One appealing hypothesis is that an extended drought recorded in the tree ring record for the thirteenth century in the Southwest would cause crop failures and perhaps force people to migrate to areas along perennial streams. A worldwide cooling trend (the Little Ice Age, Fig. 1–25) began about the same time as the exodus and may have shortened the growing season, again contributing to crop failures. Other ideas are that the centuries of farming had depleted soil productivity and that overpopulation, overhunting, and overcutting of forests destroyed the balance between people and their environment. Could modern society be following in the footsteps of the Hisatsinom?

REFERENCES

Baars, D.L., 1995, Navajo Country: Univ. of New Mexico Press, Albuquerque, 255 p.
Martin, L., 1993, Mesa Verde, the story behind the scenery: KC Publications, Las Vegas, 48 p.

Park Address

Mesa Verde National Park
Mesa Verde National Park, CO 81321

BRYCE CANYON NATIONAL PARK AND CEDAR BREAKS NATIONAL MONUMENT (UTAH)

Bryce Canyon and Cedar Breaks are covered together because their geologic stories are similar. The spectacular erosional forms at Cedar Breaks resemble those at Bryce except they are not as extensively developed. The colors are almost unbelievable. The contrast of the deep green of the plateau forest with the reds, oranges, yellows, and white of the many whimsical erosional features—all against Utah's deep blue sky—nearly overwhelms many park visitors. Cedar Breaks is smaller but relatively uncrowded—a definite positive compared to the 1.5 million plus visitors to Bryce each year. Emphasis in this section is on Bryce Canyon National Park.

Regional geologic studies in the Colorado Plateaus began with John Wesley Powell and his intrepid group of explorers, topographers, photographers, and scientists in the 1870s. Clarence Dutton, a geologist with the Powell Survey, recognized the Grand Staircase, a stair-step arrangement of cliff-forming sedimentary rock layers that were oldest near the Grand Canyon and became progressively younger to the north (Plates 8C and D). In this chapter we have generally followed the rock record, starting with parks where Precambrian rocks are particularly significant and moved forward through time into the Paleozoic and Mesozoic. We now enter the Cenozoic world—the last 65 million years of earth history. The top of the Grand Staircase and the geologic layer cake, the "frosting," is composed of the Pink Cliffs that cap the High Plateaus Section of the Colorado Plateaus. The

Pink Cliffs were formed from the Claron Formation (part of the Wasatch Group) of early Tertiary age. Where the formation is exposed in cliffs along the edges of the High Plateaus Section, erosion has produced the spectacular, colorful badland topography that makes Bryce and Cedar Breaks special (Fig. 8–81).

Mormon pioneers displaced the Paiute people and ventured into this relatively inhospitable but beautiful country attempting to wrest a living from the land. Ebenezer and Mary Bryce and others homesteaded in the Paria Valley in 1874 (Barnett and Follows, 1971). Many, including the Bryces, moved on after a few years. The valley behind their cabin and the pink and white cliffs and *hoodoos* at the edge of the Paunsaugunt Plateau were called Bryce's Canyon, a name that later would be given to the park. The hoodoos are pinnacles that have delightful, whimsical shapes. To the Paiute they were "legend people," people turned to stone by Coyote, an important spirit to many western tribes. Among other things, the area to Ebenezer Bryce was also a "hell of a place to lose a cow!"

Visitors began to discover the area in the early 1900s, and articles written for Union Pacific and Santa Fe railroad magazines in 1916 were the first mention in the public press of this scenic area. Accommodations were built on the rim for visitors, and in 1923 President Warren G. Harding established Bryce Canyon

FIGURE 8–81   Paunsaugunt Plateau edge at Bryce Canyon showing Bryce amphitheater and badland topography. (Photo by E. Kiver)

National Monument. In 1928 the area received national park status and now includes 35,835 acres (55 square miles; 142 km^2) in an 18-mile-long (29 km) north–south strip on the plateau edge. The park contains perhaps the most colorful and unusual erosional landforms in the world.

Regional Setting

The High Plateaus Section is a transition zone between the Basin and Range Province and Colorado Plateaus Province (Fig. 8–1). It is broken into nine separate plateaus that are blocked out by a series of major north–south trending faults, some of which extend to the Colorado River in Grand Canyon National Park. The westernmost fault, the Hurricane Fault, is at the west edge of the Markagunt Plateau and forms the topographic boundary between the Basin and Range and Colorado Plateaus (Fig. 8–2). Cedar Breaks is perched on the west edge with its rim at 10,400 feet (3170 m) elevation. A short distance north of the monument, basalt-capped Brianhead Peak is 11,315 feet (3450 m) high—at or slightly above timberline in this area. Zion National Park is also on the Markagunt Plateau, 18 miles (29 km) south of Cedar Breaks and much lower—about 4000–8000 feet (1220–2440 m) above sea level.

As shown in the regional inset map in Figure 8–45, the Sevier River valley follows the Sevier Fault and separates the Markagunt from the Paunsaugunt Plateau, the next plateau to the east. The Paunsaugunt Fault and Paria River valley in turn mark the east edge of the Paunsaugunt Plateau where Bryce Canyon is located. Elevations at Bryce range from 6620 to 9105 feet (2018 to 2776 m). The pinyon-juniper forest at lower elevations grades into a ponderosa and finally a spruce–fir forest at the plateau top. The tops of these and nearby plateaus are capped by the once continuous Claron Formation, offset as much as 2000 feet (610 m) along the Sevier and Paunsaugunt faults.

Geologic Setting

The exposed geologic record at Bryce begins with rocks of Late Cretaceous age that are exposed in the southeastern corner of the park below the pink cliffs of the Claron Formation. These drab shales and sandstones record the advance and retreat of the last shallow inland sea to cover North America and were followed by river-deposited sands and gravels (Bezy, 1995). The crunching and grinding of plates along the west edge of North America finally caught up with more inland areas as tall mountains formed just west of the Colorado Plateaus in what is now the Basin and Range Province. The Cordilleran Orogeny began in California and Nevada during earlier Mesozoic time (about 200 Ma) and migrated eastward—by latest Mesozoic and early Tertiary time (70–50 Ma) the Rocky Mountains were building to the east of the Colorado Plateaus during the Laramide Orogeny phase.

Deformation in the Colorado Plateaus was considerably less intense than in the mountains to the west produced during the Sevier Orogeny phase, and the Rockies to the east deformed during the Laramide Orogeny phase of the Cordilleran Orogeny. The Plateaus remained close to sea level in early Tertiary time and became a trap for large volumes of sediment and water washing down from the surrounding mountain ranges.

A system of lakes formed from southwest Wyoming and northwest Colorado to southwest Utah during Paleocene and Eocene time. Lake Claron covered several thousand square miles and persisted for several million years, probably between 60 and 40 Ma (Ludin, 1989). Claron Formation (part of the Wasatch Group) sediments are thickest to the west near Cedar Breaks and thin to 500–800 feet (150–245 m) at Bryce. The mostly limy mud sediment contains layers of white limestone deposited in quieter areas of the lake and silty and sandy sediments deposited in streams and deltas. Lake levels fluctuated rapidly, producing a stacking of sediment layers of different composition. Small amounts of iron in the mud were later oxidized to limonite and hematite producing the reds, browns, and pink colors. Significant differences in resistance to weathering and erosion of adjacent layers would later play an important role in producing the hoodoos and other erosional forms that are Bryce's trademark.

Fossils are not abundant in the Claron, making age determinations difficult. The Claron is older than the 35-million-year-old basalt flows that cover it on the nearby Aquarius Plateau and elsewhere. The middle member of the Claron contains middle Eocene (about 45 Ma) fossils of subtropical aquatic snails, turtles, and reptiles. The Pink Limestone Member that forms much of the scenery at Bryce is younger than the underlying Cretaceous rock and could be either Paleocene or Eocene. Thus, the best current estimate places the Claron as 40–60 million years old.

About 15 Ma the Colorado Plateaus began the spectacular mile-high vertical uplift that continues today. At the same time, the Basin and Range Province experienced tensional forces as the crust stretched and fault-block mountains formed, dropping the high ranges produced during the Sevier Orogeny to elevations mostly below those of the Colorado Plateaus. Where water from the high mountains to the west once drained eastward into the area of the Colorado Plateaus, now water from the Colorado Plateaus drains westward into the area occupied by the fault-block mountains of the Basin and Range. The High Plateaus is a transitional zone where extensive north–south faults broke the crust into separate blocks or plateaus at about the same time. Cedar Breaks is located just east of Hurricane Fault—the boundary fault that separates the Basin and Range from the Colorado Plateaus. The Paunsaugunt Plateau is outlined by the Sevier and Paunsaugunt faults along which the Sevier River valley on the west and the Paria River valley on the east form topographic boundaries.

The once-continuous Claron Formation is offset as much as 2000 feet (610 m) from plateau to plateau (Lindquist, 1977). An impressive cross section of part of the Sevier Fault is exposed along the Red Canyon entrance to Bryce where black-colored Quaternary basalt flows, about 500,000 years old (Chronic, 1988), are in fault contact with the pink-colored Claron Formation. The displacement here is about 900 feet (275 m). Fault displacement during an earthquake is in inches or perhaps a few feet. Thus, many earthquakes over millions of years must have occurred as these giant blocks were shifted to their present positions. Earthquakes continue today in the area, indicating that more geologic excitement is on its way.

Rapid vertical uplift of the Colorado Plateaus during the past 10 million years (late Miocene, Pliocene, and Quaternary) was accompanied by increased downcutting and headward erosion of the Colorado River and its tributaries. The Paria River is extending itself northward, and its western tributaries are rapidly cutting into the edge of the Paunsaugunt Plateau forming long scalloplike indentations or amphitheaters containing the world-famous Bryce badlands.

Erosion: The Ultimate Leveler

Nature abhors inequality and seeks to equalize the distribution of energy. Water falling on high elevations has more potential energy to dissipate on its trip back to the sea than water falling on lower elevations. All other factors being equal, more erosion will occur on the higher landscape because the stream channels must be steeper and the water must therefore flow more rapidly.

Such is the case at the cliff edges of the Paunsaugunt Plateau and elsewhere that similar conditions exist. The intermittent streams draining the east side of the Paunsaugunt Plateau drop 1400 feet (427 m) in about 2 miles (3.2 km). Once the soil and vegetation are removed from steep slopes, it is difficult to reestablish. Thus erosion is king here and will continue unabated until it runs out of Claron Formation. Rock materials are loosened by *physical weathering* processes such as *frost action* (water freezing and expanding in cracks and spaces in the rock) and by *chemical weathering* processes such as dissolving the calcite cement that "glues" many sedimentary rock particles together. The loosened particles and rock fragments are removed either by gravity pulling them downslope or by runoff from rain or melting snow washing the particles to lower elevations. The intermittent streams act as conveyor belts, transferring upslope materials to lower elevations.

Headward erosion of the Paria River over time has enabled its western tributaries to deepen their channels and in turn cut headward into the edge of the Paunsaugunt Plateau. Elongate, shallow indentations or amphitheaters are carved into the plateau edge. Technically Bryce Canyon is not a canyon at all. A canyon is a deep valley with steep parallel or nearly parallel walls—not flaring scalloplike indentations such as those at Bryce and Cedar Breaks. Three major amphitheaters occur at Bryce, the Tropic in the north, Bryce in the middle, and the Sheep Creek–Ponderosa in the south.

The rim at Bryce is formed on a particularly resistant bed of dolomite, a limestone-like rock containing magnesium in addition to calcium. Rocks containing smaller amounts of calcium are correspondingly less resistant and weather and erode at a faster rate, undercutting resistant layers and forming cliffs and other erosional forms. Linear ridges projecting from the amphitheater wall are called fins. Fins are temporarily capped by resistant layers, eventually being dissected by erosion into pinnacle forms that are called hoodoos (Fig. 8–82). Joints or fractures in rock form during uplift and faulting when the earth's crust is distorted and stressed. Joints are more susceptible to weathering and erosion and play a major role in determining the location of fins. *Cross joints* formed at nearly right angles to the trend of the fins determine where each pinnacle or hoodoo will form along the fin.

The sides of pinnacles are further marked by horizontal grooves and protrusions—again proving that not all rocks were created equal! Recall that the Claron is composed of numerous layers with different abilities to resist the processes of weathering and erosion. The protrusions mark the more lime-rich layers, and indentations follow lime-poor layers. Resistant caprock over nonresistant rock layers sometimes forms balanced rocks, such as those at the Hat Shop along the Under-the-Rim Trail. The result is a wonderfully complex badland topography with delicate and bizarre hoodoo landforms—the product of differential weathering and *differential erosion* (Fig. 8–81). If the Claron Formation were composed of thick, resistant homogenous material, it would likely form a massive cliff akin to those at Zion, Canyon de Chelly, Natural Bridges, and elsewhere. As the Claron Formation mud is freed by weathering, it is washed down the sides of cliffs and

FIGURE 8–82   Bizarre hoodoo landforms known together as the Three Gossips. The single hoodoo in the foreground is also called Thor's Hammer. (Photo by E. Kiver)

hoodoos forming a pink-colored stucco coating of clay. Thus, many of the white limestone layers appear pink when viewed from a distance.

Characteristically, erosion is rapid in badland areas, faster than soil can form or the roots of plants can grow and stabilize slopes by binding loose soil together. Trees that germinated and grew on the plateau rim are now perched precariously on the edge with 3 or 4 feet of their roots exposed. Soil and rock were subsequently washed away from the plateau edge exposing the roots of these *tiptoe trees*. By taking a core of the tree trunk (nondestructive to the tree if done carefully) and counting the growth rings, the age of the tree can be determined. Thus, the minimum rates of cliff retreat can be calculated. The rim retreats from 9 to 48 inches each century (23 to 122 cm/100 yr) with the average rate of retreat about 1.5 feet per century (46 cm/100 yr). If this rate continues, areas where the Paunsaugunt Plateau is only 6 miles (10 km) wide will be breached in a little more than 2 million years. Yet another reason why one should visit areas like Bryce Canyon as soon as possible!

Features like natural arches form where weathering penetrates through the lower part of a fin without destroying the caprock above. Such features are even shorter-lived than hoodoos. Ostler's Castle, a favorite natural arch for visitors since it was first discovered, collapsed catastrophically under its own weight in 1964, leaving behind pinnacles, much like the thousands of others found in the park. As natural arches and hoodoos are destroyed, new ones are forming to take their places. Here one can see them in all stages of development.

Visiting Bryce Canyon and Cedar Breaks

The fairyland of landforms at both Bryce and Cedar Breaks is spread before the visitor from numerous viewpoints along the plateau rim. Distant views of the foreboding Pink Cliffs on the Aquarius Plateau, some 2000 feet (610 m) higher and about 12 miles (19 km) northeast of Bryce, and of Navajo Mountain 80 miles (129 km) to the southeast near Rainbow Bridge National Monument on clear days are also impressive.

Cedar Breaks is located 35 miles (56 km) due west of Bryce at the west edge of the Markagunt Plateau near Hurricane Fault (Plate 8C). The Claron Formation here is thicker and sandier because of its location closer to the now-vanished mountains produced during the Sevier Orogeny. These mountains still stood high in early Tertiary time and furnished sediment to the rivers flowing into Lake Claron. Elevations at Cedar Breaks are now near 11,000 feet (335 m) compared to near sea level when the Claron Formation was deposited. The Claron Formation on Brian Head Peak (11,307 feet, 3447 m) to the southwest is capped by middle Tertiary lava flows (about 40 million years old). Elsewhere on the Markagunt Plateau Quaternary lava flows are barely touched by erosion and look as if they should still be warm! Earthquakes still rattle along Hurricane Fault. Thus, the dynamic forces that built and elevated this fascinating area are not yet through.

To the early explorers and settlers, a "break" was a clifflike edge of a plateau. The juniper was called "cedar," hence the name Cedar Breaks. This cooler, snowier area closes during the winter months. Geologic conditions are similar to those at Bryce—alternating resistant and nonresistant Claron beds at the edge of a plateau generated by vertical fault movements. Dissection by intermittent streams cutting headward has developed fins, hoodoos, and other erosional forms similar to those at Bryce.

In addition to the spruce and fir trees on the rim, a scattering of bristlecone pine occurs in both parks. In California and Nevada, some of the bristlecones are over 4000 years old—the oldest living trees in the world. Here they are mere "teenagers," perhaps no more than 1600 years old!

For those who can, views of the escarpments and hoodoos are more impressive from the trails below the rim at Bryce. The trails provide an opportunity to have a "close encounter with a hoodoo" and other features in the park. Explore the Silent City of rock walls, the balanced rocks at the Hat Shop, the hoodoo statue of Queen Victoria (Fig. 8–83), and many other features. Watch for Sinking Ship where the Paunsaugunt Fault cut across and tilted a mesa that now resembles the stern of a ship ready to plunge beneath the waves. Along the trails one is more likely to become immersed in the scenery and contemplate the really important things in life.

FIGURE 8–83   Eroded hoodoo with a striking resemblance to Queen Victoria along the Queens Trail. (Photo by E. Kiver)

REFERENCES

Barnett, J., and Follows, D., 1971, Bryce Canyon: Bryce Canyon Natural History Association, 32 p.

Bezy, J., 1995, Bryce Canyon: The story behind the scenery: KC Publications, Las Vegas, 48 p.

Chronic, H., 1988, Pages of stone: Geology of western national parks and monuments: The Mountaineers, Seattle, v. 4, 158 p.

Lindquist, R.C., 1977, The Geology of Bryce Canyon National Park: Bryce Canyon Natural History Association, 52 p.

Ludin, E.R., 1989, Thrusting of the Claron Formation, the Bryce Canyon region, Utah: Geological Society of America Bulletin, v. 101, p. 1038–1050.

Park Addresses

Bryce Canyon National Park
P.O. Box 17001
Bryce Canyon, UT 84717-0001

Cedar Breaks National Monument
P.O. Box 749
Cedar City, UT 84720

EL MALPAIS NATIONAL MONUMENT (NEW MEXICO)

The name El Malpais (ell-mal-pie-ees) is Spanish for "the badlands," areas where travel is difficult, especially in areas of volcanic terrain (Fig. 8–84). These are not the stream-eroded type of badlands such as those found at Petrified Forest, Bryce, or the Badlands of South Dakota, but bad for travel because of the moonscapelike volcanic landscape. Here in west-central New Mexico the smooth sandstones of the Colorado Plateaus meet the jagged Hawaii-like volcanic landscape of El Malpais.

Native Americans used the lava area as a source of water, for hunting, and traveling along its ancient trails for over 10,000 years. As in other areas of the Southwest, populations peaked from about 950 to 1350. Hisatsinom (Anasazi) people built apartment house complexes and associated smaller detached dwellings. The Ventanas site on the east edge of the monument stands impressively on top of the sandstone bluffs that overlook the El Malpais volcanoes and lava flows. The ruins here contain a tower kiva and a great kiva—structures of significance in religious and other activities. The people were part of the Chaco civilization that flourished in the area from about 900 to 1150. Chaco Canyon was the hub of cultural, economic, and social activity in what must have been a complex society. Traces of a road that connected the Ventana site with Chaco Canyon and other villages are still visible. The area was abandoned during the mid-1300s as were

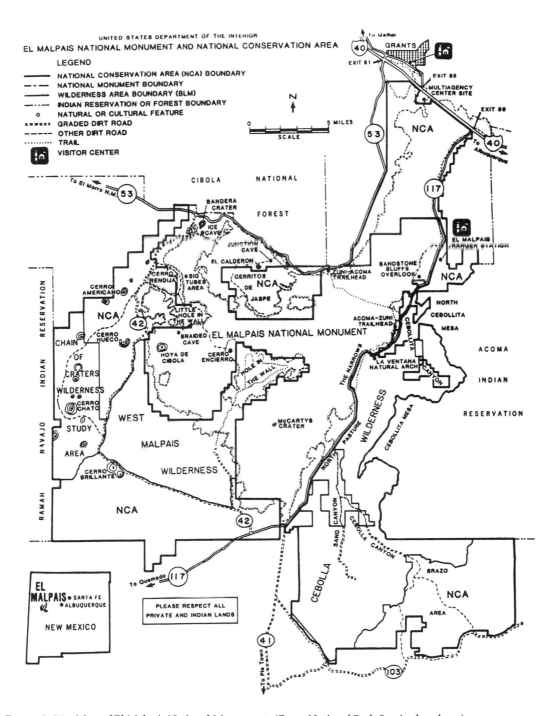

FIGURE 8–84   Map of El Malpais National Monument. (From National Park Service brochure)

many similar settlements on the Colorado Plateaus. When Coronado explored the area in 1540, he encountered only two major pueblos—Acoma and Zuni—towns still in existence today.

Anglo-Americans didn't settle in significant numbers in the area until the 1920s. Remnants of abandoned homesteads and sheepherder cabins still remain in some of the surrounding wilderness areas. President Reagan signed the bill in 1987 that preserves the geological and archeological resources for visitors and for further scientific study. There are two units here, each administered by a separate agency. El Malpais National Monument is under the Park Service and includes 114,272 acres (179 square miles; 463 km^2). An additional 263,000 acres (410 square miles; 1065 km^2) surrounding the monument is part of the Natural Conservation Area administered by the Bureau of Land Management as wilderness and multiple-use areas (Fig. 8–84).

The Native American history continues here today. The physical presence on nearby reservations of contemporary Indian groups including the Acoma, Laguna, Zuni, and Ramah Navajo ensures that their cultural and religious activities in the monument will continue. All of these groups have cultural and spiritual ties to El Malpais. In planning for the development of the national monument one Acoma elder remarked, "yes, see the land and appreciate it, but know there is something else, that we share the responsibility to love it, to protect it, to respect it."

Regional Setting

El Malpais is in the extreme southeast part (Datil Section) of the Colorado Plateaus (Fig. 8–1) about 40 miles (64 km) east of the Arizona–New Mexico border and about 60 miles (100 km) west of the Rio Grande Rift, the eastern boundary of the Colorado Plateaus (see Fig. 8–61 for the regional setting). The Rio Grande Rift is a major structural feature or crustal "crack" that forms a topographic and tectonic boundary between the Colorado Plateaus and the adjacent Basin and Range Province in the New Mexico area. The crustal splitting and associated decompression melting of the mantle and lower crust beneath the Rio Grande and associated rifts produced intense igneous activity during the past 35 million years, and especially during the past 5 million years. The Rio Grande follows this tectonic feature along its path to the Gulf of Mexico.

The Zuni Mountain Anticlinal Uplift (Fig. 8–2; 8–61) lies immediately northwest of El Malpais. High elevations in the Zuni Mountains milk additional moisture from the atmosphere and support a ponderosa–Douglas fir forest in contrast to the pinyon, juniper, and desert plants at lower elevations (6200–8400 feet; 1890–2560 m) in the national monument. Tiny El Morro National Monument is on the southwest flank of the Zuni uplift about 13 road miles (21 km) west of El Malpais. The El Malpais visitor center is located along historic Route 66 in the town of Grants, just north of the Conservation Area boundary (Fig. 8–84). The Continental Divide extends southeast along the Zuni Mountain crest and turns southwest following near the west edge of the monument. Intermittent streams and groundwater east of the divide drain to the Rio Grande and eventually into the Atlantic Ocean. Drainage west of the divide is into the Little Colorado–Colorado River system and into the Pacific Ocean.

Cutting across the south end of the Zuni Mountains and the southeast corner of the Colorado Plateaus is another major fracture zone—the Jemez Lineament.[11] This northeast–southwest trending fracture zone is huge, extending over 500 miles (800 km) from east central Arizona to northeastern New Mexico! The Jemez structure formed in Proterozoic (late Precambrian) time and is being reactivated as tensional forces continue to pull apart the southwestern United States and parts of Mexico (Mutschler and others, 1998). Volcanism along this structure has been prolific during the past 5 million years—with much of the activity at areas such as El Malpais occurring mostly during the Quaternary (last 1.6 million years) (Luedke and Smith, 1991).

About 15 miles (24 km) northeast of El Malpais, 11,301-foot-high (3445-m) Mount Taylor, a Pliocene-age (about 2–4 million years old) stratovolcano complex, towers over the cinder cones at El Malpais. This Mount St. Helens type of volcano is itself worthy of national monument status, especially as we feel the pressure of increasing travel demands and the need for park areas as population rapidly grows to unprecedented numbers. Parklands and other open space will be among our most valuable assets as this frightening population trend continues.

Geologic Background

Older rocks in the Zuni and surrounding areas record older geologic events—clear back to Proterozoic time. Granitic rocks formed during Precambrian mountain building are exposed in the core of the Zuni Mountains. As we have seen elsewhere on the Colorado Plateaus, Precambrian faults play major roles in influencing younger geologic events and today's major topographic elements. The Zuni Mountains trend was topographically high during most or all of Paleozoic time, preventing or restricting sediment deposition at its crest. Finally, during Early Permian time (about 290–270 Ma), debris shed from the Ancestral Rockies to the north covered the exposed Precambrian rocks with red bed sediments creating a major *unconformity*, a stratigraphic break—in this case with hundreds of millions of years of earth history missing. Oil wells drilled on the flanks and surrounding basins encounter the missing Paleozoic strata, enabling geologists to fill in the gaps in the record. Younger Mesozoic rocks ring the Zuni Uplift on its flanks. The spectacular tan-colored Zuni Sandstone (Jurassic age) rimrock rises abruptly above the lava-floored valley at El Malpais.

During the Laramide Orogeny (70–50 Ma) the ancient Precambrian faults again moved, forming the northwest-trending Zuni anticline and other folds and

---

[11]A lineament is an alignment of structural and/or topographic features.

basins on the Colorado Plateaus. Streams began to dissect the topography, cutting a shallow valley in the Malpais area along the fractured rocks in the Jemez Lineament. Volcanism affected the northeastern end of the lineament about 35 Ma (Mutschler and others, 1998), the Mt. Taylor area a few miles to the northeast of El Malpais about 4 Ma, and practically the entire length of the lineament during the past 5 million years (Luedke and Smith, 1991). Pliocene (about 4 Ma) basalts erupted on top of the vast Cebollita Mesa to the east of the Monument (Maxwell, 1986). These lava flows poured out long before those down in the broad valley between the mountains and the mesa—the Quaternary-age (1.6 Ma to present) El Malpais lavas.

Volcanic Features at El Malpais

El Malpais is a veritable museum of who's who, or more appropriately, what's what in the world of basaltic volcanic features. As at other basalt fields such as those near Sunset Crater, Craters of the Moon (Columbia Intermontane Province), and Lava Beds National Monument (Basin and Range Province), eruptions began with a gas-rich phase that sprayed lava fragments about the vent area, building a cinder cone. Reduction of gas pressure allowed upwelling magma to remain cohesive and burst from the edge of the cinder cone as a lava flow. More fluid lava (pahoehoe) often flows for tens of miles in tubes below the surface crust without losing appreciable heat or dissolved gases. The El Calderon lava flowed over 22 miles (35 km) through a 17-mile-long (27-km) system of tubes to the present location of the town of Grants (Fig. 8–84). Over 100 cinder and shield volcanoes occur in the El Malpais field.

Maxwell (1986) identified five major lava flows that erupted up through the fractures and rifts beneath El Malpais during the Quaternary. Because each flow overlaps at least one of the others, their *relative ages* are easily recognized. He determined their order of stacking—the oldest flow is at the bottom, the youngest is on top, those in between are intermediate ages. Radiometric ages determined by using the decay of radioactive potassium to argon for some of the flows further allows *absolute ages* to be established. The oldest major flow is from El Calderon (Fig. 8–84) cinder cone (115,000 years old) followed by La Tetra, Hoyade Cibola, Bandera (about 10,500 years old), and most recently McCartys flow (3200 years old) (Marinakis, 1997). Radiometric dates place the youngest McCartys flow as a mere 700 years old. The flow locally buries sediments containing pottery fragments and molds of corncobs, lending credence to Acoma Indian legends that their ancestors were there to witness the fiery exposé.

More fluid lava cooled as *pahoehoe* basalt, characterized by a smooth or ropy surface. As pahoehoe loses dissolved gases, it becomes more viscous and develops a *blocky* or a jagged *aa* surface. Pahoehoe is more common closer to the vent, and blocky or aa surfaces are often near the flow edges where lava tends to be more viscous. Students not careful about spelling sometimes write "vicious lava"—not a bad description if one stumbles and falls. The spiny, sharp edges of the aa make travel across the lava surface difficult and potentially painful! Falling into *skylights* (holes in lava tube ceilings) or *collapse pits* or *collapse trenches* that dot the flow surface would be more than a little painful.

Older hills that are surrounded by lava create a feature known in Hawaii and elsewhere as a *kipuka* or *steptoe*. Many are present along the Zuni–Acoma Trail and a large one, Hole-In-The-Wall, is also accessible by trail. Just as in Hawaii, the vegetation on these kipukas is isolated by the hostile surrounding lava surfaces, and the kipukas contain relatively undisturbed ecosystems untrammelled by sheep, cows, and humans.

Visiting El Malpais

El Malpais is very easy to get to but difficult to get very far into! Begin your visit at the visitor center in Grants, located along a section of historic Route 66. Paved roads skirt the north and east edges of the monument—trails and a few high-clearance roads penetrate into some of the interior areas. Often in rugged terrain, a horse will take you where a car won't—but not here! This landscape is off-limits for a horse or even a mule. Best to stay on the trails—even there tough hiking boots and a gallon of drinking water per day during the summer are recommended. Don't venture out into El Malpais unless you are certain that you're surefooted. A tumble here will bring you into intimate contact with the sharp spikelets and glassy edges of lava blocks or worse—a header down one of the 20-foot-deep (6-m) collapse sinks or pits. The Sandstone Bluffs Overlook and La Ventana Natural Arch and archeological area are accessible by short trails and provide excellent views into the volcanically jumbled interior of El Malpais. Here also in the sandstone bluffs golden eagles, prairie falcons, and owls have their nests, preying on rodents and other small game that venture into their domain.

The 7-mile-long (11-km) Zuni–Acoma Trail is a part of the old Indian trade route that connects the Zuni and Acoma pueblos. Rock bridges and *cairns* (piles of rock that mark trail routes) were constructed many centuries ago by Indian travelers. Here you will literally follow in their footsteps and experience the same landscape that the first Americans did. Beginning from state road 53, the trail begins on the El Calderon flow (a few hundred thousand years old) and crosses three younger flows including McCartys flow that overran Indian farms and artifacts about 700 years ago. Here one can practice his or her skills in recognizing *superposition* (the older rocks are buried by younger) and thereby determine the relative ages of the lava flows. Even without the obvious burial of older flows by younger, the thinner soil and sparser vegetation cover on younger flows imply less time and therefore also provide a measure of time. Remember to leave all artifacts and rocks where you find them so that those who come after can also discover our past. In addition, it's against federal law to collect anything in an area under National Park Service protection!

Regional Setting and Geology

The San Francisco Mountain area is located in the Grand Canyon Section (Fig. 8–1) in the southern Colorado Plateaus near its boundary with the Basin and Range Province. Late Tertiary and Quaternary (15 Ma to present) volcanism is prevalent along the edges of the Colorado Plateaus. Here the crustal stretching that formed the adjacent Basin and Range produced passive hotspots and partial melting of upper mantle rocks. Pockets of basalt magma then accumulated near the base of the crust. The buoyant fluid utilized ready-made fault-zone conduits produced during Precambrian mountain building and Basin and Range faulting to rise toward the surface. Of particular importance at Sunset Crater is its location along a wide and lengthy northeast-trending Precambrian fault system known as the Colorado Lineament. This ancient fracture system extends northeast for over 400 miles (670 km) from the San Francisco Mountain area through the Four Corners and the Rocky Mountains in central Colorado. Localized igneous activity and mineralization occurred along the Colorado Lineament during the past 75 million years (Mutschler and others, 1998). A parallel northeast-trending lineament to the south, the Jemez Lineament, played a similar role in localizing volcanism in areas such as that at El Malpais National Monument in New Mexico.

San Francisco Mountain is an eroded stratovolcano complex of late Tertiary age that dominates the skyline in the Flagstaff area. Humphrey Peak, at 12,680 feet (3866 m) elevation, is Arizona's tallest peak, tall enough to be extensively glaciated during the Pleistocene. Surrounding San Francisco Mountain and about a mile below on the Coconino Plateau are more than 600 cinder cones and numerous basalt flows ranging from Pliocene to Holocene in age (5 Ma to about 900 years ago), many of which show minimal weathering and erosion. Those in the east half of the volcanic field are the most recently active—most are less than 2 million years old (mostly Quaternary age) (Luedke and Smith, 1991). Buried faults localized igneous activity in the San Francisco Mountain area. Based on the recency of eruptions at areas such as Sunset Crater as well as occasional earthquakes that still occur locally, we can expect more excitement in the not too distant future in the San Francisco volcanic field. Luedke and Smith (1991) note that eruptions in the north part of the volcanic field over the last million years have migrated to the northeast at about 1.2 cm/year (0.5 inches), controlled in part by the underlying fracture zones of the Colorado Lineament and perhaps the spreading of a shallow or passive hotspot (Mutschler and others, 1998). Cinder cones tend to be "one-shot" volcanoes—subsequent eruptions will likely form over another fissure or another part of the same fissure—forming an entirely new volcanic cone.

The Coconino Plateau is the same surface that extends southward from the south rim of the Grand Canyon, some 50 miles to the north. The rock record buried beneath Sunset Crater is similar to that at Grand Canyon: a Precambrian-age geosyncline, two ages of now-vanished mountain ranges, and deposition of nearly a mile of marine and nonmarine sediments during the Paleozoic (see cross section in Plate 8A and B). Mesozoic rocks are mostly eroded away except for isolated outliers (erosional remnants) of Triassic-age rocks (Moenkopi Formation). Sandstones of the Moenkopi Formation are exposed in shallow valleys 15 miles (24 km) to the north in Wupatki National Monument where the Hisatsinom or Anasazi people used the sandstone blocks to construct apartment house structures.

The Coconino surface near Sunset Crater slopes gently to the northeast where intermittent streams and the subsurface flow of groundwater are toward the Little Colorado River. To the east of the Little Colorado lies the Painted Desert and the Triassic-age Chinle Formation so vividly exposed in Petrified Forest National Park and other areas.

Sunset Crater Eruption

Thanks to careful studies of tree rings in timbers used in the construction of Indian pit houses that were destroyed during the eruption, as well as determination of when damage occurred to growing trees, archeologists established that the eruption occurred between the growing seasons of 1064 and 1065! Over 800 square miles (2070 km^2) of countryside downwind from the Sunset volcano were buried beneath a blanket of sand and smaller-size pyroclastic particles. The deposit was over 3 feet (1 m) thick nearer to the volcano. Intermittent eruptions for about 150 years contributed to the "instant" soil and built a cone of pyroclastic debris around the vent (Holm and Moore, 1987). The Hisatsinom (Anasazi) fled their houses and farms as the earth trembled and pyroclastic debris fell from the sky. The weight of ash on the rooftops of their pit-house homes, especially those closer to "ground zero," was too much and they collapsed.

Typically, the first magma to reach the surface during an eruption is highly gas charged, much like a can of warm soda that is shaken and quickly opened. A curtain of fire perhaps 6 miles (10 km) long initiated the "ground-breaking ceremonies" at Sunset Crater. Activity rapidly concentrated to a narrow vent that initially discharged a spray of gas and liquid blobs. Larger pyroclastic particles falling near the vent began to build a *cinder cone* and finer particles were carried downwind where they fell like black snow across the landscape as an *ashfall*.

As gas pressures reduced, rising gas no longer blew the rising magma apart; instead the fluid welled up into the rubbly base of the cinder cone and broke through its flanks as a lava flow. The Kana-a flow broke through the cone on the east side and flowed 6 miles (10 km) down a small valley. Most of this flow is located outside of the monument in the Coconino National Forest. The Bonito flow occurred later, in 1180 from the western base of the cone. It is more massive and is entirely within the monument boundaries. The Bonito flow displays numerous flow features including ropy pahoehoe surfaces, aa surfaces at its perimeters, mounds of fragments torn from the cone and rafted away piggyback style on the surface of the moving flow, long fissures up to 2900 feet (884 m) long and 24 feet (7 m) deep that formed as the flow surface subsided, and lava tubes—some of which trap cold air and retain ice through much of the year.

Renewed pyroclastic activity finished building today's cone—about 1000 feet (305 m) tall and 1 mile (1.6 km) in diameter at its base. Sunset Crater is one of the larger cinder cones in North America. The cone is slightly lopsided with its crater offset to the west—a result of the prevailing westerly winds blowing more debris to the east. The crater rim sags a bit above the points where lava broke through the base of the cone. Volcanic gases vented from fumaroles near the summit for many years, oxidizing the iron-rich cinders to yellows and reds and depositing white and gray gypsum—giving the cone its light-colored, perpetual "sunset" top.

The Hisatsinom discovered that the new volcanic soil was a natural mulch, able to store precious moisture that produced outstanding crops of corn, beans, and squash. A prehistoric land rush brought perhaps 8000 people into the area. The increased farming activity increased the vulnerability of the soil to erosion by wind and water, eventually reducing the 800-square mile (2070-km^2) pyroclastic blanket to today's 122 square miles (316 km^2) (Holm and Moore, 1987). By 1225 no one lived here—the Hisatsinom had retreated to areas with more dependable water supplies. Many of their descendants live today in the Hopi communities in Arizona and among the Pueblo people in New Mexico.

Visiting Sunset Crater

The monument was established in 1930 and with 3040 acres (4.8 square miles; 12 km^2) and is one of the smaller natural areas in the park system. Surrounding areas are mostly federally owned and provide excellent opportunities for those eager to explore one of the larger cinder cone fields in North America.

Excellent views and photo stops abound along the roadways in and near Sunset Crater. A stop at the visitor center and short self-guided nature trails will introduce you to the local natural history. Climbing the cinder cone is no longer permitted. Construction of a stable trail on the loose cinder is impossible, and damage occurred when well-meaning hikers were allowed to scramble to the top. The scar of the old trail will persist for many years as the fragile and spotty vegetation slowly re-establishes itself.

For those who must "get on top of things," a visit to nearby O'Leary Peak (a silica-rich plug dome related to San Francisco Mountain) in the Coconino National Forest has outstanding views of the San Francisco Mountain stratocone and the army of cinder cones around its flanks on the Coconino Plateau. Count the almost overwhelming number of volcanoes one can see from the top and imagine the sequence of events one would have witnessed from here during the past 5 million years. Other interesting cones worthy of exploration and summit hikes include SP Mountain (Ulrich, 1987), Merriam Cone, and others. Inquire at the visitor center for directions and road conditions.

Closely associated with the volcanic events was the fate of the local people. Their story is intricately tied to Sunset Crater and can best be appreciated by visiting Wupatki National Monument about 15 miles (24 km) to the north. During their heyday following the eruption, the native people built three-story apartment buildings using the Moenkopi Sandstone exposed along the shallow washes or valleys. Other structures include a ballcourt similar to those found in Mexico and Central America and a large circular amphitheater—an intriguing and unique feature whose purpose has so far mystified archeologists.

REFERENCES

Holm, R.F., and Moore, R.B., 1987, Holocene scoria cone and lava flows at Sunset Crater, northern Arizona, in Beus, S.S., Rocky Mountain Section of the Geological Society of America: Geological Society of America, Boulder, CO, Centennial Field Guide, v. 2, p. 393–397.

Luedke, R.G., and Smith, R.L., 1991, Quaternary volcanism in the western conterminous United States, in Morrison, R.B., ed., Quaternary nonglacial geology; Conterminous U.S.: Geological Society of America, Boulder, CO, The Geology of North America, v. K-2, p. 75–92.

Mutschler, F.E., Larson, E.E., and Gaskill, D.L., 1998, The fate of the Colorado Plateau—A view from the mantle, in Friedman, J.D., and Huffman, A.C., Jr., coordinators, Laccolithic complexes of southeastern Utah: Time of emplacement and tectonic setting—Workshop proceedings: U.S. Geological Survey Bulletin 2158, p. 203–222.

Ulrich, G.E., 1987, SP Mountain cinder cone and lava flow, northern Arizona, in Beus, S.S., ed., Rocky Mountain Section of the Geological Society of America: Geological Society of America, Boulder, CO, Centennial Field Guide, p. 385–388.

Park Address

Sunset Crater Volcano National Monument
Route 3, Box 149
Flagstaff, AZ 85730

NINE

Rocky Mountain Cordillera: Northern Rocky Mountain Province

The Rocky Mountains, as outlined by Thornbury (1965), consist of the mountain ranges that extend northward from New Mexico through Canada and westward across northern Alaska (Plate 9). The general sequence of geologic events is much the same throughout, but the "style" of mountain building and the more recent geologic events are sufficiently different to justify the recognition of at least four geomorphic provinces. Because of space limitations, the Arctic Rockies of Alaska are not covered in this edition. The Rocky Mountains in the conterminous United States are discussed in this and the next two chapters.

Physiography

The east edge, or "Front Range," of the Rocky Mountains forms a sharp boundary where it meets the rolling surface of the Great Plains. The west edge is similarly clearcut where it abuts the basalt flows of the Columbia Intermontane Province, the mildly deformed strata of the Colorado Plateaus, and the fault-block mountains of the Basin and Range Province (Plate 1). Maximum elevations rise from near 10,000 feet (3050 m) near the Canadian border to over 14,000 feet (4270 m) in Colorado. Life zones range from alpine tundra through a forest zone to grasslands in Montana and Wyoming to upper Sonoran desert vegetation in New Mexico.

The Rockies in the conterminous United States are divided into three provinces based on topographic boundaries (Plate 1). However, Miller and others (1992) point out that Rocky-Mountain-style mountains locally occur beyond the accepted topographic boundaries of the provinces. The Southern Rockies as defined here extend from north-central New Mexico northward through Colorado into southern Wyoming. The Middle Rockies include the Yellowstone Plateau, the nearby Beartooth and Absaroka mountains, and the Bighorn, Teton, Wind River,

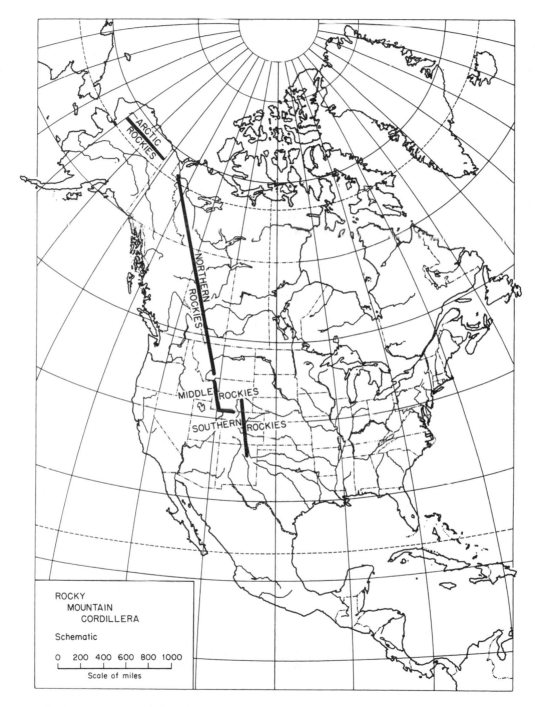

PLATE 9   Rocky Mountain Cordillera. (Base map copyright Hammond Inc.)

Wasatch, and Uinta ranges in Wyoming, Utah, Idaho, and Montana. As here defined, the Northern Rockies are a huge elongated province that extends northward from Yellowstone Park through western Canada and then westward across northern Alaska. Most of western Montana, Idaho, and northeastern Washington are part of the Northern Rocky Mountains.

Building of the Rockies: General Sequence

Although each of the three Rocky Mountain provinces in the conterminous United States has distinct geologic differences, they also have many similarities, especially in the sequence and timing of events. All geologic events beginning with the Precambrian have contributed in one way or another to the present conformation of the Rocky Mountains. The geologic evolution of all continents involves the process of *accretion* where materials are added to central nuclei, enabling a continent to enlarge in areal extent with time. Continents in turn can merge and later be fragmented, providing geologists with challenging histories to unravel.

Glimpses of one of the ancient Precambrian (Archean, 2.5 billion years and older) continental nuclei (the Wyoming Craton) occur in the cores of mountain ranges in the Middle Rockies of Wyoming and the Beartooth Mountains of Montana. These ancient roots of mountains are highly metamorphosed—mostly gneisses and schists. A later geosyncline and another Precambrian episode of mountain building from 1.8 to 1.6 billion years ago (Early Proterozoic) added yet more continental crust to the growing continent. These younger metamorphosed "basement" rocks are in turn exposed in the cores of ranges in Colorado and New Mexico and in the Madison and Gallatin ranges northwest of Yellowstone.

Western North America became a *passive plate margin* as the continent moved eastward during Middle and Late Proterozoic time. Major splits, or rifts, developed a series of basins about 1600–800 Ma (millions of years ago) that were filled to a depth of as much as 12 miles (20 km) of sand, mud, and carbonate sediment. The northern basin (Belt Basin) in Montana, northern Idaho, northeastern Washington, and southern Alberta and British Columbia will be revisited when discussing Glacier National Park. The middle or Uinta Basin is important at Dinosaur National Monument where the Green River cuts through the Precambrian rocks that form the core of the Uinta Mountains. A third basin containing Middle Proterozoic sediments is exposed in the depths of the Grand Canyon in the Colorado Plateaus Province.

Following a long interval of erosion, Paleozoic seas invaded the stable interior, or *craton,*[1] of North America including the Rocky Mountain area. Layer after layer of sand, mud, and limey sediments were deposited in relatively tranquil marine and near-marine environments. Major changes occurred during the late Paleozoic

---

[1] A craton is a relatively stable section of crust that has not been significantly deformed for a very long period of time.

(Pennsylvanian, about 300 Ma) as the earth's continents were assembling into one huge supercontinent called Pangaea (Greek for "all earth") by geologists. Structural weaknesses that were likely developed during Precambrian time were reactivated and a major range of mountains, the Ancestral Rockies, formed in New Mexico and Colorado in the approximate position of today's Rocky Mountains. The Appalachian Mountains in the east, the Ouachita Mountains in the south, and the Ancestral Rockies on the west formed a partial ring of mountains around the plate fragment that was soon to become North America. Erosion took its toll, and by early Mesozoic time the topographic expression of the Ancestral Rocky Mountains was eliminated. Streams washed lazily across the area and later desert sands further buried the eroded remnants of the Ancestral Rockies.

The giant supercontinent acted as a massive insulating blanket that trapped heat that continually rises from the mantle below. Pangaea became unstable and began to break apart, producing a new generation of spreading ridges and subduction zones that would drastically change the face of the continents. Western North America became the leading edge of the moving plate, initiating the compressive forces of the Cordilleran Orogeny—the major mountain-building events to affect western North America during Mesozoic and Cenozoic time. The resulting deformation marched slowly eastward across the West leaving in its path large numbers of thrust-fault mountains and granitic batholiths to the west and an unusual style of anticline thrust-fault mountain uplift in the Middle and Southern Rockies. This continuous episode of mountain building is often divided into three phases—each of which partially overlapped its successor in time. The Nevadan Orogeny began in Late Jurassic (about 180 Ma) in California, the Sevier Orogeny began in Nevada and Utah during the Cretaceous (about 140 Ma), and the Laramide Orogeny in the Rocky Mountains began in Late Cretaceous (about 80 Ma) and ended in middle Tertiary (about 50 Ma) (Wicander and Monroe, 1993).

The relatively quiet craton existence of the Rocky Mountain area changed drastically again in mid-Mesozoic time. The Rocky Mountain Geosyncline, which extended from the Gulf of Mexico to Alaska, became the site of thick sedimentary accumulations, especially during the Cretaceous when a continuous seaway covered the area. The area of thick deposition included not only the area that is now the Rocky Mountains but also the adjacent areas on both the east and the west. Sediments—clays, silts, sand, and gravel—were carried in by streams from land masses on both sides of the geosyncline. Limestones were laid down at certain times, by chemical and biochemical precipitation. With continued deposition the geosyncline continued to sink, thus providing space for additional sediment. Deposition was more rapid in some parts of the geosyncline; there, thicker sequences of sedimentary rocks were laid down. Deposition continued through much of the Cretaceous period, and in parts of the geosyncline as much as 30,000 feet (9150 m) of sedimentary rocks accumulated.

Squeezing during the Laramide Orogeny produced folds and thrust faults—forcing the deformed sections of crust upward as topographic mountains and intervening areas downward as basins. Magmas formed near the base of the earth's crust and intruded toward the surface in various places. Large batholiths formed in parts of the Northern Rockies and numerous igneous centers developed throughout the Rocky Mountains, especially along major linear weak zones (ancient faults) in the crust known as *lineaments*. Important northeast–southwest trending lineaments localized igneous activity along the Great Falls Lineament in Montana and Idaho and the Colorado Mineral Belt in Colorado. Much of the mining activity that led to the rapid settlement of the western United States occurred as a consequence of Laramide and post-Laramide mineralization associated with igneous activity.

As land appeared above the sea, streams began to erode the newly formed mountains. Folding and faulting ceased sometime during the Eocene (about 50 Ma), and by the end of the Eocene (36 Ma) the mountains were essentially leveled and a surface of low relief developed, a surface referred to by some as a peneplain. Elevations were only a few thousand feet above sea level—considerable uplift was yet to come as the backbone of the North American continent continued to develop.

The compressive forces of the Laramide Orogeny were followed by a mid-Tertiary episode in which hot mantle rock moved upward—arching the Southern and part of the Middle Rocky Mountains into a giant uplift reminiscent of an oceanic ridge (Eaton, 1987). The center of the uplift was stretched and overprinted with normal faults, grabens, and locally volcanic centers such as those along the Rio Grande Rift in New Mexico. Vertical adjustments, or *isostatic adjustments*, occurred as the density of mantle rock beneath the Rocky Mountain crust lessened due to heating. The low-density crust and upper mantle beneath the Rockies continued to float higher than the denser mantle rock below—much like a larger marshmallow sticks up higher in a cup of hot chocolate than a smaller marshmallow. The uplifts steepened stream gradients and initiated a new erosion cycle, and by Pliocene time (about 5 Ma) a second erosion surface, a pediplain,[2] was formed, leaving remnants of the older surface standing, benchlike, high above the pediplain. The erosion record described here is most clearly exposed in the Middle and Southern Provinces of the Rocky Mountains.

A late Tertiary (7–4 Ma) uplift pushed the mountains to near their present vertical elevation (Eaton, 1987); again a new erosion cycle was initiated—the cycle that is in progress today. Stream erosion in the higher mountains was greatly influenced when, several times during the Pleistocene, glaciers became widespread. Early Pleistocene glaciers were generally broad lobate tongues or lobes, according to Richmond (1965). Later, after distinct canyons were cut, valley glaciers became the dominant type. In the Northern Province, valley glaciers near the Canadian border flowed into and became part of the continental ice sheets that extended south from Canada.

---

[2]A pediplain is generally similar to a peneplain, but pediplains are developed under dry rather than humid climate conditions.

The general sequence given above is greatly simplified and condensed; not all of the igneous activity was confined to the times specified, and the uplifts in various areas were probably not perfectly synchronized. Although there is not complete agreement as to the number and age of the erosion surfaces, the sequence as outlined will serve as a general framework for our discussion of each of the three geomorphic provinces. Other general summaries are available in Alt and Hyndman (1986, 1989), Chronic (1980, 1987), Lageson and Spearing (1988), and most historical geology textbooks. More detailed information concerning the structural evolution of the Rocky Mountains is available in Christiansen and Yeats (1992), Cowan and Bruhn (1992), and Miller and others (1992).

Plate Tectonics and the Northern Rocky Mountains

Continental growth, or accretion, is relatively straightforward where a continental plate overrides an oceanic plate. Reworked debris from the continental plate is deposited on the sea floor and carried into the subduction trough. There it is deformed and stuffed against and beneath the overriding plate, eventually forming the folds, faults, volcanoes, and elevations typical of mountains—much like those forming today in western Washington and Oregon. However, the mechanical stresses necessary for mountain building become more difficult to unravel in an area such as the Rocky Mountains where a subduction zone was, in some areas, 800 or more miles (1300 km) away. Ideas include a nearly flat angle of subduction that carries the deforming stresses much farther inland (Miller and others, 1992) or *gravitational spreading* or collapse of thickened sections of crustal material created by earlier subduction activity (Jones and others, 1996). Heat trapped beneath the crust changes rock density and mechanical strength, further explaining certain vertical uplifts and making gravitational spreading possible (Eaton, 1987; van der Pluium and Marshak, 1997).

Other factors also played a role in the development of the Northern Rocky Mountains. Here, large segments of crustal material, or *microplates,* have slid into place along large transverse faults (faults similar to the San Andreas in southern California) or were islandlike masses that have been rafted against the continent edge. The "docking" of some of these microplates corresponds in time with the Laramide Orogeny.

Cutting across the Northern Rockies are some locally prominent alignments of tectonic and topographic features called lineaments that reflect major crustal flaws—likely ancient faults that were reactivated during the Laramide and some during post-Laramide time. Such flaws helped localize erosion and some of the igneous activity. Thus, extensive topographic features such as prominent valleys, or "trenches," are the basis of subdividing the province into five sections (Fig. 9–1). The western sections have *metamorphic core complexes* similar to those discussed in the section on Saguaro National Monument (Basin and Range Province, Chapter 6) and were produced by hot mantle rock rising upward. Numerous batholiths,

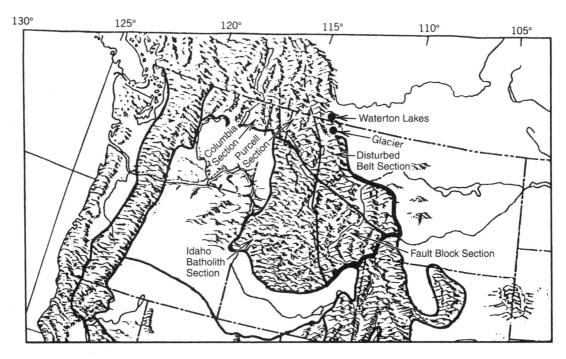

FIGURE 9–1   Northern Rocky Mountain Province in the conterminous United States showing geomorphic sections. Divisions are based on the location of major trenches and lineaments. (After Kiver and Stradling, 1994).

including the Idaho Batholith that occupies most of central Idaho, formed along a former plate boundary during late Cretaceous to early Tertiary time, about 100–60 Ma.

The north–south trending Rocky Mountain Trench extends some 4350 miles (7000 km) from Alaska through Kalispell just west of Glacier National Park and through Missoula along the front of the Bitterroot Mountains (Fig. 9–1). The trench separates the western and eastern sections. The Lewis and Clark Lineament follows the southern edge of the Rockies in Washington State and extends southeastward through Missoula (where it crosses the Rocky Mountain Trench) to Helena. These two lineaments outline the western and southern boundaries of the Disturbed Belt or northeast section of the Northern Rockies in the contiguous United States. In this mass of thrust faults lies Glacier National Park, the only national park in the vast region of the Northern Rockies. Other areas in the Northern Rocky Mountains, including the Sawtooth Mountains in Idaho and Hells Canyon along the Idaho–Oregon border, are also more than well qualified to be national parks. Congress has so far been unfriendly to such proposals.

REFERENCES

Alt, D.D., and Hyndman, D.W., 1986, Roadside geology of Montana: Mountain Press, Missoula, MT, 427 p.

Alt, D.D., and Hyndman, D.W., 1989, Roadside geology of Idaho: Mountain Press, Missoula, MT, 393 p.

Christiansen, R.L., and Yeats, R.S., 1992, Post-Laramide geology of the U.S. Cordilleran region, in Burchfiel, B.C., Lipman, P.W., and Zoback, M.L., eds., The Cordilleran Orogen: Conterminous U.S.: Geological Society of America, Boulder, CO, The Geology of North America, v. G-3, p. 261–406.

Chronic, H., 1980, Roadside geology of Colorado: Mountain Press, Missoula, MT, 322 p.

Chronic, H., 1987, Roadside geology of New Mexico: Mountain Press, Missoula, MT, 255 p.

Cowan, D.S., and Bruhn, R.L., 1992, Late Jurassic to early Late Cretaceous geology of the U.S. Cordillera: in Burchfiel, B.C., Lipman, P.W., and Zoback, M.L., eds., The Cordilleran Orogen: Conterminous U.S.: Geological Society of America, Boulder, CO, The Geology of North America, v. G-3, p. 169–203.

Eaton, G.P., 1987, Topography and origin of the southern Rocky Mountains and Alvarado Ridge, in Coward, M.P., Dewey, J.F., and Hancock, P.L., eds., Continental extensional tectonics, Geological Society of London Special Publication No. 28, p. 355–369.

Jones, C.H., Unruh, J.R., and Sonder, L., 1996, The role of gravitational potential energy in active deformation in the southwestern United States: Nature, v. 381, p. 37–41.

Kiver, E.P., and Stradling, D.F., 1994, Landforms: in, Ashbaugh, J.G. ed., The Pacific Northwest, geographical perspectives, Kendall/Hunt, Dubuque, IA, p. 41–75.

Lageson, D.R., and Spearing, D.R., 1988, Roadside geology of Wyoming: Mountain Press, Missoula, MT, 271 p.

Miller, D.M., Nilsen, T.H, and Bilodeau, W.L., 1992, Late Cretaceous to early Eocene geologic evolution of the U.S. Cordillera, in Burchfiel, B.C., Lipman, P.W., and Zoback, M.L., eds., The Cordilleran Orogen: Conterminous U.S.: Geological Society of America, Boulder, CO, The Geology of North America, v. G-3, p. 205–260.

Richmond, G.M., 1965, Glaciation of the Rocky Mountains, in Wright, Jr., H.E., ed., Quaternary of United States, Princeton University Press, Princeton, NJ, p. 217–230.

Thornbury, W.D., 1965, Regional geomorphology of the United States: Wiley, New York.

van der Pluim, B., and Marshak, S., 1997, Earth structure: An introduction to structural geology and tectonics: WCB/McGraw-Hill, New York, 495 p.

Wicander, R., and Monroe, J.S., 1993, Historical geology: West Publishing, St. Paul, MN, 640 p.

GLACIER NATIONAL PARK (MONTANA)

Glacier National Park and Canada's Waterton Lakes National Park are adjacent (Fig. 9–2), and together they form an International Peace Park, a natural monument to friendship between neighbors. About 200 miles (320 km) farther north are Banff and Jasper national parks where jagged mountains, glaciers, and glacial features are also on display. Those planning a trip to Glacier should include Canada's parks and sufficient time to see them the right way; they are truly outstanding.

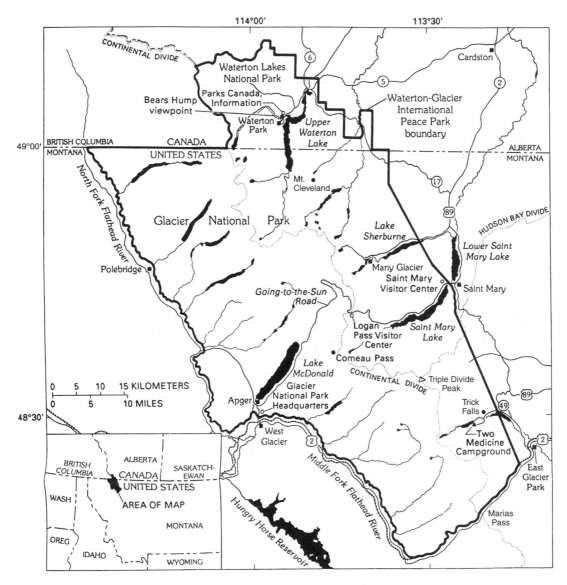

FIGURE 9–2   Map of Waterton–Glacier International Peace Park. (Modified slightly from Alpha and Nelson, 1990)

The Front Range here was dubbed the "Land of Shining Mountains" by the first French explorers who saw the snow-covered peaks from the rolling prairies of the Great Plains to the east. Glacier National Park occupies over 1600 square miles (4145 km^2) of the Disturbed Belt Section of Montana, which lies north of the Lewis and Clark Lineament and east of the Rocky Mountain Trench (Fig. 9–1). Glacier contains some of Montana's best mountain and glacial scenery. Gigantic cirques and deep glacial troughs, as well as the name Glacier National Park, suggest that large glaciers are there. However, this is not the case, for although there are about 50 glaciers, they are all relatively small. Rather, Glacier is an outdoor museum containing spectacular glacial landforms practically unchanged since the Ice Age glaciers that formed them vacated the area about 11,000–12,000 years ago.

Astride the Continental Divide, the park sheds its waters in three directions: streams on the west side flow into the Pacific; those on the east flow into either the Gulf of Mexico or Hudson Bay. Triple Divide Peak in the southern section of the park has the distinction of providing water to all three. The Continental Divide enters from Canada along the Livingston Range in the west part of the park and swings east to the Lewis Range where it follows the crest southward through Logan Pass to Marias Pass (Fig. 9–2). Elevations peak at Mt. Cleveland (10,468 feet; 3191 m) and decrease to about 3200 feet (975 m) in the Rocky Mountain Trench (Flathead Valley) to the west and 4500 feet (1370 m) at St. Mary on the east side of the park.

Park History

Indians traveled through the area for thousands of years but left a sparse record of their activities for archeologists to decipher. The Kalispell, Flathead, Kootenai, and Blackfeet (Piegan) were relative newcomers but were the Indians living in the valleys and plains area when David Thompson (1780) and other fur traders such as Finian McDonald (1810) visited the area. Lewis and Clark passed about 25 miles (40 km) east of the area in 1806 on their return trip east but very little was known about its geography. The Blackfeet acquired horses and firearms and dominated the upper Great Plains through much of the 1800s, discouraging settlement and exploration of the area. The annihilation of the bison, their main food source, and smallpox epidemics greatly weakened the Indian population.

Transcontinental railroad surveys rediscovered the legendary Marias Pass used by Indians for thousands of years, and by 1891 the Northern Pacific Railroad was completed, bringing settlers and fortune seekers westward. Unable to keep squatters and miners from their land, the Blackfeet gave up a substantial part of their reservation to the government in 1889. Settlers soon found that the rigorous climate made agriculture and grazing unprofitable (Ahlenslager, 1988). Discovery of copper sent hundreds of miners into the area—they soon realized that the low-grade ore associated with igneous dikes was unprofitable to mine and they moved on.

The real value of the area was soon to be recognized as the railroad and visitors with vision publicized the area. Professor Lyman B. Sperry of Carleton College recorded the scenic features, and the efforts of George Bird Grinnell, an influential and energetic conservationist and editor of *Forest and Stream* (later *Field and Stream*), were instrumental in establishing the park (Diettert, 1992).

Grinnell was an accomplished biologist and helped found both the Audubon Society and, along with such notables as Theodore Roosevelt, the Boone and

Crockett Club. Grinnell first visited the area in 1885. His love of the area and concern with the plight of the Blackfeet tribe (many had starved during the winter of 1883–1984) would grow as he returned often over the years. Grinnell wrote in 1886 (in Diettert, 1992, p. xii) that he was saddened that "the mountain life of today is not the life of twenty, nor even of ten years ago. . . . I regret the changes that have come and others that I see near at hand"—a thought that is even more appropriate today. Glacier's neighbor to the north, Waterton Lakes, had been designated as a "Forest Park" in 1895, and Grinnell saw that the only way to save the area for the public and prevent uncontrolled development was to establish a national park. He wrote a series of articles and essays promoting park status. His efforts came to fruition in 1910 when President William Howard Taft signed legislation creating Glacier National Park. Through the efforts of Rotary International, the area was further authorized in 1932 as the Waterton–Glacier International Peace Park, the first park to commemorate goodwill between neighboring countries.

Glacier's Rocks

The dominant rocks at Glacier are Precambrian (Middle Proterozoic) sedimentary rocks of the Belt Supergroup, a thick series of rocks first described in the Belt Mountains near Helena. The continent began to split apart about 1600 Ma, indicating that plate tectonics was alive and well during the Precambrian. Sediments poured into the resulting marine or shallow lake basin (Winston and others, 1984) until about 800 Ma from land areas located to both the west and east. The whereabouts of the western part of the Belt Basin and its associated fragment of the continent are uncertain—perhaps today it is part of Antarctica or western Australia where similar rocks of Middle Proterozoic age are found. About 21,000 feet (6400 m) of Belt rocks are exposed in the park—an incredible 50,000 feet (15,240 m) are present farther west in the deeper part of the Belt Basin.

Traversing the park along the engineering marvel known as The Going-to-the-Sun Road takes one from the oldest exposed Precambrian rocks at the base of the sedimentary stack to progressively higher and younger rocks (Fig. 9–3). Green (Appekuny Formation) and red (Grinnell) *argillites* are slightly metamorphosed mudstones that give way to the 4000-foot-thick (1220-m) Helena (Siyeh) Limestone at Logan Pass and along the famous cliffs of the Garden Wall that rise above Grinnell Glacier. Younger Snowslip and Shepard Formations form the higher peaks beyond the reach of roads (Fig. 9–3).

Clues to conditions in the Precambrian streams, deltas, and seaways in the Belt Basin occur in the rocks and are easily seen along trails and roads. Raup and others' (1983) *Geology along Going-to-the Sun Road* will guide the inquisitive visitor to stops where different stratigraphic layers and rock features can be observed in road cuts.

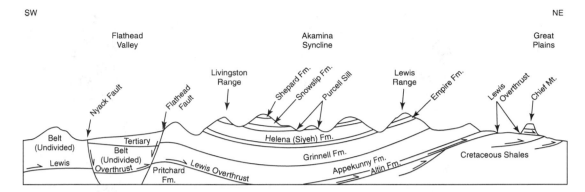

FIGURE 9–3 Highly generalized cross section showing structure and stratigraphy in the Glacier National Park area. (Based on mapping by Whipple, 1992)

The red color of the Grinnell and of parts of the Snowslip Formation is due to the presence of *hematite,* a form of oxidized iron, present in small but sufficient amounts to vividly color the rocks. Iron in the form of hematite is most likely to occur in sediments deposited in shallow water and exposed to oxygen or oxygenated groundwater after deposition. The abundance of *mud cracks* and *rain-drop imprints* in the Grinnell and other red argillites supports this interpretation. Exposure to air and contraction and cracking of the drying mud occurred frequently along the streams and tidal flats at the edges of the shallow lakes and seaways of the Belt Basin. The green-colored argillites contain iron in its reduced form—an indication that these sediments were isolated from an oxygen source, particularly after deposition.

Stromatolites are thin layers in carbonate rocks (Fig. 9–4) that formed by the activity of *cyanobacteria* or blue-green algae. Sticky algal mats trap sediment grains, and photosynthesizing cyanobacteria use carbon dioxide from the immediately surrounding water and cause calcium carbonate (the mineral calcite) to precipitate. Stromatolites occur in rocks as old as 3.5 billion years; they are the most abundant—and usually the only—evidence of life found in Precambrian rocks. These primitive lifeforms occupied these stable environments for billions of years, apparently with little or no genetic changes. The billion-year-old forms at Glacier are indistinguishable from those occurring in rocks 3.5 billion years old or those living today. Stromatolites are abundant in the Helena Limestone as mats and rounded column ("cabbage heads") forms (Fig. 9–4).

An eye-catching feature on the massive cliffs of the Garden Wall is a conspicuous dark band extending from one side to the other; below and above the dark band are narrower bands of white (Fig. 9–5). The dark-colored band is a sill formed by the cooling of magma that intruded along bedding planes in the Helena Limestone, perhaps about 1000–800 Ma. Clearly it is a sill and not a lava flow that

FIGURE 9–4   Proterozoic colonial cyanobacteria (blue-green algae) in Helena (Siyeh) Limestone along Going-to-the-Sun Road, Glacier National Park. (Photo by E. Kiver)

was later buried. Evidence comes not only from radiometric dates of the sill (younger than the rocks above and below), but also from the observation that the limestone *both* above and below it was metamorphosed by heat to marble, the white bands shown in Figure 9–6. Dikes that cut across the bedding planes are also present but are less conspicuous than the sill. The reason is that the dike rock

FIGURE 9–5   Garden Wall, showing thick diabase sill (arrows). (Photo by D. Harris)

FIGURE 9–6   Purcell Sill on the headwall of Grinnell cirque is about 200 feet (60 m) thick. (Photo by E. Kiver)

weathers more rapidly than the adjacent rocks, thus forming recesses and little valleys that are snow and debris-filled much of the time.

In several places, basalt lava flows are exposed, some of which contain pillow structures—indicating that some of the lavas poured out into the sea. Lavas are found in the Boulder Pass area, near Swiftcurrent Pass, and in Granite Park. (If a geologist had done the naming, it would be Basalt Park rather than Granite Park!)

Other rocks in the park include Late Cretaceous marine shales and thick Tertiary-age stream and lake deposits in the North Fork valley of the Flathead River. These rock units at first glance raise difficult problems: How is it that 100-million-year-old Cretaceous rocks lie *beneath* billion-year-old Belt rocks? Why did a basin form in the North Fork area of the park that permitted nearly 2 miles (3.2 km) of Tertiary-age sands and gravels to accumulate? The answers lie with how the mountains of Glacier formed. The explanation again supports the observation that behind spectacular scenery lies spectacular geology.

Glacier's Mountains

Paleozoic marine rocks similar to those both north and south of the park were likely deposited and at least partially removed by erosion before Late Cretaceous (about 100 Ma) marine muds covered the area. The earth's crust began to buckle about 70 Ma in the Northern Rocky Mountain area as the Laramide Orogeny compressed and raised the area above sea level. Numerous *thrust* faults as well as nearly flat, low-angle *overthrust* faults of large extent, such as the 280-mile-long (452-km) Lewis Overthrust, were thus formed (Figs. 9–3 and 9–7). How a relatively thin section of crustal rock can be displaced horizontally 40-or-more miles (65 km) has long puzzled *structural geologists*—those who study the bending and breaking of rocks.

One part of the answer lies in the reduced friction along such faults due to high fluid pressures produced when fluids such as water are trapped and squeezed in a rock. Such trapped water helps "float" the thrust-fault slab and reduces the amount of horizontal force required to move the overlying rock slab. Horizontal pressure produced by *gravity sliding* or *gravity spreading* accounts for the lateral force that drives thrust-fault slabs such incredible distances (van der Pluijm and Marshak, 1997). Overthrust movement began well after the Laramide Orogeny was underway—beginning in the late Paleocene (60 Ma) and continuing through early Eocene (50 Ma). Movement was "jerky" and likely produced tens of thousands of small slips and minor earthquakes. Rather than the entire block moving simultaneously, the movement was similar to that of a caterpillar where only part of the body moves at a time.

FIGURE 9–7   Lewis Overthrust Fault at the base of a steep cliff of Precambrian rocks in the Lake Sherburne area; Cretaceous shales lie beneath the fault. (Photo by E. Kiver)

Where the lower end of the fault broke the surface, the rigid Belt rock slab rode up and over the Cretaceous shales. The overthrust plate was gently folded into a broad syncline during movement—peaks along the eroded west edge form the Livingston Range, and the eastern range is the Lewis Range (Fig. 9–3). Later uplift may have tilted the block to the west so that the Lewis Overthrust now slopes (dips) about 7° to the west. Like snow sometimes does when sliding off a barn roof, as the huge plate moved eastward it fractured and a 13-mile-wide (21-km) crack or pull-apart valley—the North Fork valley—formed. Thus, overthrusting satisfactorily accounts both for the billion-year-old Belt rock resting in fault contact on top of 100-million-year-old Cretaceous rock, and for the over-2-mile-thick (3.5-km) section of Tertiary (Eocene to Oligocene, about 45–35 Ma) sediment in the North Fork valley (Whipple, 1992).

Erosion removed thousands of feet of Paleozoic and Mesozoic sediments. The east edge of the overthrust plate was removed and deeply dissected, leaving remnants of the overthrust plate as islands of older Belt rock surrounded by younger Cretaceous rock. Such features are called *klippe*—Chief Mountain is a spectacular example (Fig. 9–8).

Renewed uplift in the past 5–10 million years along the entire Rocky Mountain chain triggered a new stage of erosion. The increasing elevation also produced a colder mountain climate that, when combined with the general worldwide cooling of climates that was underway, would soon subject the area to the climatic ex-

FIGURE 9–8   Chief Mountain, about 5 miles (8 km) east of the Rocky Mountain front in Glacier National Park. (Photo by E. Kiver)

tremes of the Pleistocene Ice Age. Climates cooled and warmed numerous times—alternately producing intervals of glaciation interspersed with warm intervals (*interglacials*) such as that of the present. Evidence for each episode of glaciation was eliminated or nearly destroyed by younger glaciers. As expected, the last glacial advance, the late Wisconsin, is easiest to reconstruct.

Glacier's Glaciers

Glaciers initially developed along the Continental Divide as *ice caps* whose lower edges formed lobes that extended down the broad mountain slopes. Valleys cut during the interglacials caused later ice to concentrate into *valley* or *alpine* glaciers. It was at this time that the deep troughs formed (Fig. 9–9) as ice locally reached thicknesses of over 3000 feet (1000 m), scouring out bedrock depressions that hold many of today's glacial lakes.

Moraines and *outwash* from the melting glaciers partially blocked some valley segments forming other lakes such as Lake McDonald and St. Mary (Fig. 9–10). Older ice streams leaving their mountain valleys merged with the huge Laurentide

FIGURE 9–9   View of deep McDonald Valley glacial trough from Going-to-the-Sun Road on a frosty September day. The line of snow coincidentally marks the approximate location of the upper elevation of the last glacier to occupy the valley. Early September snow closed the road the next day. (Photo by E. Kiver)

FIGURE 9–10   St. Mary Lake on the east side of the mountains is impounded behind thick outwash sediments. (Photo by D. Harris)

Ice Sheet to the east and the Cordilleran Ice Sheet to the west (Karlstrom, 1987). About 20,000 years ago during the maximum of the last glaciation, larger valley glaciers such as the Two Medicine Glacier on the east side of the park extended beyond the canyon mouth and spread out in the form of a *piedmont glacier* but did not merge with the Laurentide Ice Sheet farther north and east. The presence of an ice-free corridor from Alaska to Montana during late Wisconsin time is believed to be critical in allowing human migrations and the peopling of southern North America.

Hanging valleys with waterfalls tumbling from them can be seen along several of the main valleys. The trunk glaciers deepened their valleys more rapidly than the smaller tributary glaciers; when the glaciers melted away, the tributary valleys were left hanging high above the main valley floor. Grinnell, Virginia, and Birdwoman falls are three of the many that pour out of hanging valleys.

Large cirques developed at the head of the valleys, and with prolonged cirque enlargement from two or more directions, *matterhorns* or *glacial horns* and sharpened ridges called *aretes* were formed. From one vantage point, Kinnerly Peak in the northwestern corner of the park closely resembles the famous Matterhorn in the Swiss Alps. Also formed by glacial erosion—abrasion and quarrying—elongate roches mountonnees, large and small, are abundant. On these and other abraded surfaces, *glacial grooves* and *striae* (*striations*) can be observed in many places.

Glacial deposits, abundant in and at the ends of the valleys, clearly represent a number of advances. Several techniques that have been used to determine the relative ages of the deposits are topographic location, degree and depth of weathering, and carbon-14 dating. The general lack of organic fragments in glacial *till* (direct deposit by ice) and *outwash* (glacially derived sediment deposited by meltwater) in alpine areas has limited the usefulness of radiocarbon dating. However, *tephrochronology,* the study of datable volcanic ash beds and their relation to geologic and geomorphic features such as moraines and outwash, has proved valuable at Glacier (Carrara, 1989). Ash erupted from Glacier Peak volcano in the North Cascades of Washington 11,200 years ago and ash from a Mount St. Helens eruption 11,400 years ago occurs in the Marias Pass area and indicates that the large valley and piedmont glaciers were 90 percent or perhaps entirely gone by that time. Thus, probably by 12,000 years ago the Glacier area appeared much as it does today except for the lack of a thick cover of trees, grasses, and brush. Revegetation advanced rapidly after ice retreat by utilizing the seed crop from isolated clumps of vegetation that had survived the rigors of the Ice Age—much like the rapid return of vegetation presently underway following the devastating eruption of Mount St. Helens in 1980.

Two sets of moraines occur high up in many valleys in cirques and a short distance from today's small glaciers (Carrara, 1989). The outer moraines have a thicker soil and are covered by ash from the massive eruption of Mount Mazama (Crater Lake) 6845 years ago and are, therefore, early Holocene or late Pleistocene (12,000–7000 years) in age. The inner moraine lacks an ash cover, is barely weathered, and therefore likely dates from the last few hundred years. Most of the small remnants of Pleistocene glaciers did not survive the heat of the Hypsithermal Interval or Thermal Maximum (Fig. 1–25) that peaked about 6000 years ago. However, cooler conditions during the past 4000 years, and particularly during the past 700 years (the Little Ice Age), promoted the rebirth of cirque and small alpine glaciers around the world, including those in Glacier National Park.

By mapping these Little Ice Age moraines and by studying the growth rings and ages of trees just outside of the moraines, Carrara (1989) determined that the peak of Little Ice Age glaciation occurred in the mid-19th century. Over 150 glaciers existed in the Park in the mid-1800s, about 70 were present in 1950, and by 1958 only 53 remained. The rapid retreat has now slowed and ice fronts are presently stabilized although the current move toward global warming does not bode well for the glaciers of Glacier and elsewhere in the world. The largest glacier in the park, the Blackfoot Glacier, covered about 3 square miles (7.6 km^2) in the mid-19th century—by 1979 it had an area of only 0.67 square miles (1.74 km^2), about a 75 percent reduction (Fig. 9–11)! The Grinnell Glacier, the most frequently visited glacier in the park, has receded about 3100 feet (945 m) since the 1850s. As Vice President Al Gore noted in a speech given in the park in September, 1997: "Global warming is no longer a theory—it's a reality," and "if this trend contin-

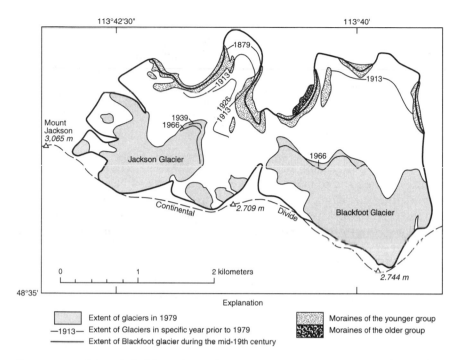

FIGURE 9–11   Map of Blackfoot Glacier showing its progressive retreat and thinning that caused its separation into the present-day Blackfoot and Jackson glaciers. (From Carrara, 1989)

ues, in about thirty years (about the year 2030), there won't be any glaciers left at all"—views shared by most scientists.

Visiting Glacier

Here at Glacier National Park are some of the last vestiges of the vast wilderness that covered western North America a little more than 100 years ago. All of its original mammal inhabitants, except the bison and mountain caribou, are still here in this vast outdoor museum. A 52-mile (99 km) trip across the Going-to-the-Sun Road and making the stops outlined in Raup and others' (1983) excellent road guide enables the inquisitive visitor to explore and discover the remarkable stories locked in the stack of rocks at Glacier. The rocks become younger in age (but still Precambrian) as one proceeds upward in elevation along the road (Fig. 9–3). The incredible story of the Lewis Overthrust is best experienced on the east side of the park where one can stand on 100-million-year-old (Cretaceous) rocks along Lake

Sherburne or St. Mary Lake and look up to where the billion-year-old Belt rocks stand in spectacular cliffs above the plane of the fault (Fig. 9–7). The trip from the sage-grassland near the 3000-foot (1000-m) elevation at the base of the mountain to the alpine tundra at 9607 feet (2929 m) at Logan Pass is equivalent to an 1800-mile (2900-km) journey northward to the Arctic tundra of northern Canada and Alaska. A visit during the short growing season when tundra flowers carpet the high country is an experience never to be forgotten by those who sense the plants' urgency to bloom and produce seeds before the winter, which at most is only a few weeks away.

The "real" park is along the over 800 miles (1290 km) of trails that cover parts of the wilderness and connect with Waterton Lakes Park to the north. Many short and intermediate-length hikes are available from Logan Pass, Swiftcurrent Lake, and elsewhere in the park and are described in publications available at the visitor centers at Logan Pass (summer only) and St. Mary Lake. Sunrift Gorge, a narrow pull-apart gorge formed when part of the bedrock slope slid downslope, is reached by a short trail. Trick Falls in the Two Medicine Lake area appears normal during heavy runoff (Fig. 9–12) but is not nearly as high in the late summer when water pours from a cavern in the lower limestone wall but not over the higher lip (Fig. 9–13). During high-water flows the cavern behind the falls is obscured. The Avalanche Lake trail passes by deep *potholes* drilled into the valley floor by debris swirled about in the channel of Avalanche Creek. Spectacular Avalanche Lake lies in a glacier-carved rock basin below a rock step near Sperry Glacier. The turquoise color of Avalanche and other lakes fed by active glaciers is due to the presence of *rock flour* (fine, glacially ground rock). Hidden Lake overlook is reached by trail from Logan Pass and provides a panoramic view of the Avalanche Basin, Sperry Glacier, and the large *col* that is Logan Pass. Abundant glacial striae, stromatolites in the Helena Limestone, and ripple marks and mudcracks in the overlying Shepard Formation are visible along the trail. Here especially one needs to stay on the trails because fragile tundra requires one or more human lifetimes to recover from trampling by unthinking visitors. Both the Granite Park hike from Logan Pass and the Grinnell Glacier hike from Swiftcurrent Lake are "all-dayers" that are geologically and scenically rewarding. An extra roll or two of film is strongly recommended.

Although today's glaciers here are a bit less impressive than some elsewhere, the opportunities for climbing around on a live glacier are good. The chances of returning from such an adventure are even better if a glacier hike is under the guidance of an experienced person such as a park naturalist. Crevasses are common on most active glaciers, certainly on Grinnell and Sperry, two of the larger and more frequently visited glaciers in the park. The glaciers advance about 6–50 feet (2–15 m) each year—mostly balanced by an equal amount of meltback. Different parts of the same glacier move at different rates and produce stresses that in turn produce fractures or *crevasses* in the brittle upper zone of the glacier. At times, crevasses are bridged over by snow too thin to support a person's weight. The depth of crevasses varies but is never more than 200 feet (61 m) and usually much

FIGURE 9–12   Running Eagle Falls (also called Trick Falls), near Two Medicine Lakes with heavy, early summer runoff. (Photo by D. Harris)

FIGURE 9–13   Trick Falls during low runoff. Note water emerging from the cavern on the rock wall. (Photo by E. Kiver)

less; at about that depth the ice is rendered plastic by the weight of the overlying ice and flows inward to fill any gaps that might form.

Glacier National Park, like other parks as well as towns and cities throughout the United States, is experiencing pressures that seriously threaten the quality-of-life experiences available to humans. Visitors to Glacier in 1996 numbered over 2 million—3 million are expected by 2020. Impacts come not only from the increasing number of visitors in the park but also the impacts created by increasing commercial and residential developments on private lands surrounding the park. More roads and campgrounds are not the answer. The answer will come when and if society decides what it wants its long-term future to be and takes conscious steps in that direction.

REFERENCES

Alpha, T.R., and Nelson, W.H., 1990, Geologic sketches of Many Glacier, Hidden Lake Pass, Comeau Pass, and Bears Hump Viewpoint, Waterton-Glacier International Peace Park, Alberta, Canada, and Montana, U.S.A.: U.S Geological Survey Miscellaneous Investigations, Map I-1508-E.

Ahlenslager, K.E., 1988, Glacier: The story behind the scenery: KC Publications, Las Vegas, NV, 48 p.

Carrara, P.E., 1989, Late Quaternary glacial and vegetative history of the Glacier National Park region, Montana: U.S. Geological Survey Bulletin 1902, 64 p.

Diettert, G.A., 1992, Grinnell's glacier: George Bird Grinnell and Glacier National Park: Mountain Press, Missoula, MT, 128 p.

Karlstrom, E.T., 1987, Zone of interaction between Laurentide and Rocky Mountain glaciers east of Waterton-Glacier Park, northwestern Montana and southwestern Alberta, *in* Beus, S.S., ed., Rocky Mountain section of the Geological Society of America: Boulder, Centennial Field Guide, v. 2, p. 19–24.

Raup, O.B., Earhart, J.R.L., Whipple, J.W., and Carrara, P.E., 1983, Geology along Going-to-the Sun Road, Glacier National Park, Montana: Glacier Natural History Association.

van der Pluim, B., and Marshak, S., 1997, Earth structures: An introduction to structural geology and tectonics: WCB/McGraw-Hill, New York, 495 p.

Whipple, J.W., 1992, Geologic map of Glacier National Park, Montana: U.S. Geological Survey Map I-1508-F.

Winston, D., Woods, M., and Byer, G.B., 1984, The case for an intracratonic Belt-Purcell basin: Tectonic, stratigraphic, and stable isotopic considerations: Montana Geological Society Field Conference, p. 103–118.

Park Address

Glacier National Park
West Glacier, Montana 59936

TEN

Middle Rocky Mountain Province

T he Middle Rocky Mountain Province is made up of the mountains, plateaus, and basins of western Wyoming, northeastern Utah, and a small section of Montana and northwest Colorado (Plate 10). In addition to the various mountain ranges, there are two large plateaus—one is a volcanic area in Yellowstone National Park, the other is the Beartooth Plateau immediately northeast of the park.

Geologic and Geographic Features

In the Bighorn Mountains of northern Wyoming, in the Beartooth Plateau, in the Tetons south of Yellowstone, and in the Wind River Mountains to the southeast, the Precambrian Archean (older Precambrian) basement complex is extensively exposed. This is part of the Wyoming Craton—one of the earlier nuclei around which the North American continent would eventually form. The overlying Paleozoic and thick Mesozoic (over 20,000 feet; 6000 m) sedimentary rocks (Miller and others, 1992) are exposed along the flanks of the anticlinal fold–thrust fault ranges of the Uintas, Beartooths, and central Wyoming and in the extensive stack of overthrust sheets of westernmost Wyoming and easternmost Idaho. The overthrust sheets are similar to those in the Northern Rockies and the fold–thrust mountains are similar to those in the Southern Rockies. The Wasatch Range is a large fault block, generally similar to the Sierra Nevada and the much smaller ranges immediately west of the Wasatch in the adjacent Basin and Range Province. Thus, the Middle Rocky Mountains are a mixture of structural styles of mountains found in adjacent geomorphic provinces.

 The general pre-Tertiary sequence outlined for the Rocky Mountains in Chapter 9 is also recorded in the Middle Rockies. The compressive forces of the Sevier and especially the Laramide Orogenies are reflected in the abundant anticlines and

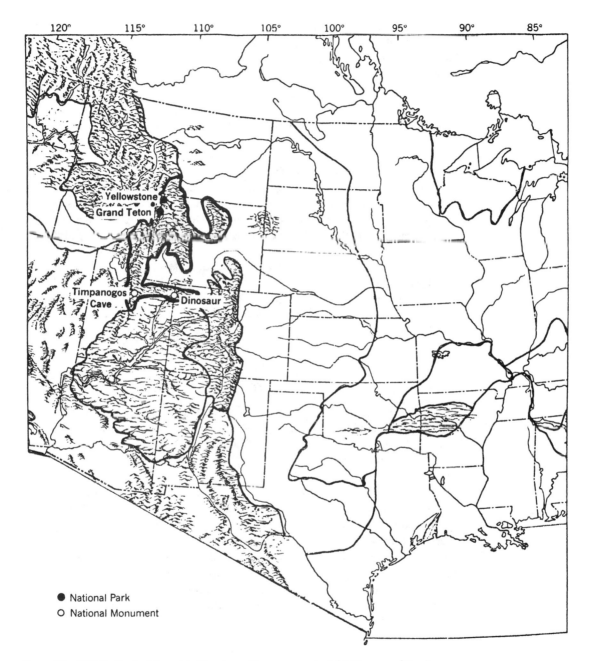

Legend:
● National Park
○ National Monument

PLATE 10   Middle Rocky Mountain Province. (Base map copyright Hammond Inc.)

thrust faults of the province. Certain of the later events—such as the extensional forces (crustal tension) that produced the Wasatch and Teton fault-block ranges were confined to the western part of the province—major volcanism was confined to Yellowstone.

The Yellowstone Hotspot

The tie-in between Yellowstone and plate tectonics is still uncertain. One hypothesis suggests that plate tectonic events produced local thinning of the crust in the Cordillera and formation of *passive hotspots* (Mutschler and others, 1998). Hot mantle material rose upward, doming the land above and causing decompression melting of the mantle and magmatism along pre-existing or newly formed fracture zones near the hotspot. Rhyolitic (high-silica) eruptions above the Yellowstone hotspot began about 16 Ma in northern Nevada. Rhyolitic volcanism and doming of the land surface migrated northeastward along the Snake River Plain in southern Idaho, leaving a trail of calderas, faults, and other volcanic features in its wake. The extraordinarily high heat flow, large active faults, recent volcanism, and a plateau over 8000 feet (2440 m) high at Yellowstone mark the present location of the hotspot.

Erosion and Deposition—Running Water

The post-Laramide erosion surfaces—peneplains and/or pediplains—are recoded by extensive remnants, particularly in the Bighorn Mountains, Beartooth Plateau, and the Uintas. As Thornbury (1965), Bradley (1987), and Mears (1993) point out, their precise origin and age have been variously interpreted; regardless, the general consensus is that these flattish upland areas are the result of long periods of erosion. Debris eroded from the Laramide-age mountains filled in the intervening basins and was carried eastward into the Great Plains. The flat, plainslike topography in southwestern Wyoming along Interstate 80 is part of the Wyoming Basin Section that separates the Middle from the Southern Rocky Mountains. The Wyoming Basin is considered an extension of the Great Plains because of its similar prairie topography—even though Cenozoic sediments here bury uplifts and mountain ranges, such as the Rock Springs and Rawlins Uplift and the Ferris, Seminole, and Green Mountains, some of which are barely exposed.

Erosion and Deposition—Glaciers

Pleistocene glaciers—valley glaciers, piedmont glaciers, and the massive Yellowstone ice cap—relandscaped extensive areas within the province, as will be shown in the discussions of the four parks and monuments within the borders of the Middle Rocky Mountains.

REFERENCES

Bradley, W.C., 1987, Erosion surfaces of the Colorado Front Range: A review, in Graf, W.L., ed., Geomorphic systems of North America: Boulder, Geological Society of America, Centennial Special Volume 2, p. 215–220.

Mears, B., Jr., 1993, Geomorphic history of Wyoming and high-level erosion surfaces, in Snoke, A.W., Steidtmann, J.R., and Roberts, S.M., eds., Geology of Wyoming: Geological Survey of Wyoming Memoir No. 55, p. 608–626.

Miller, D.M., Nilsen, T.H., and Bilodeau, C.L., 1992, Late Cretaceous to early Eocene geologic evolution of the U.S. Cordillera, in Burchfiel, B.C., Lipman, P.W., and Zoback, M.L., eds., The Cordilleran Orogen: Conterminous U.S., Geological Society of America, Boulder, CO, The Geology of North America, v. G-3, p. 205–260.

Mutschler, F.E., Larson, E.E., and Gaskill, D.L., 1998, The fate of the Colorado Plateau—A view from the mantle, in Friedman, J.D., and Huffman, AC., Jr., coordinators, Laccolithic complexes of southeastern Utah: Time of emplacement and tectonic setting—Workshop proceedings: U.S. Geological Survey Bulletin 2158, p. 203–222.

Thornbury, W.D., 1965, Regional geomorphology of the United States: Wiley, New York, 464 p.

YELLOWSTONE NATIONAL PARK (WYOMING)

Enthralled by the performances of Old Faithful and other geysers and by the beauty of Yellowstone Canyon and Yellowstone Falls, park visitors may leave without becoming aware of other attractions, less spectacular but actually more important in the geologic story of Yellowstone. The variety and uniqueness of geologic features is perhaps greater here than in any other park.

The area is wonderfully unstable from a geologic perspective. Hot, seething magma lies but a few miles below the surface, and the resulting ground adjustments stress the hundreds of active faults in and near the Yellowstone area. Thousands of tremors shake the area on an annual basis as the magma chamber rumbles about—akin to the gut sounds of a hungry animal. Occasionally a big one hits, like the 1959 West Yellowstone earthquake (magnitude 7.5), which sent a mountainside down on top of an occupied Forest Service campground, and the 1984 Mt. Borah earthquake, the largest historic earthquake (magnitude 7.3) in Idaho.

As outlined in Chapter 1, Yellowstone was where it all began—where, around a campfire, the National Park concept took root in the minds and hearts of people—people who placed the good of all citizens above personal gain. In 1872, about 3472 square miles (8995 km^2) were officially set aside as the world's first national park.

History

Indians had used the area on a seasonal basis for over 11,000 years, but few took up permanent residence because of the severe winter conditions. A group of Shoshoni and Bannocks, the Sheepeater Indians, had a village and were leading a rigorous life-style in the north end of Yellowstone when the first Euro-American explorers arrived in the 1800s. This resourceful, timid group lacked horses but used dogs for hunting and packing. They were removed to the Wind River Reservation in 1879 by Philetus Norris, the park's second superintendent.

The first white explorer to see the area (about 1807) was mountain man John Colter, who reported "fire and brimstone." The area was quickly dubbed "Colter's Hell" by those who heard his tale. Other mountain men visited the area and also emerged with unbelievable descriptions of the area. Fact and tall tale were sometimes hard to separate—especially when told by mountain men like Jim Bridger. Catching a trout, dunking it into a pool of boiling water behind him, and eating it without taking it off the line was hard for some folks to believe!

As described in Chapter 1, the visit and reports by General H.D. Washburn and other Montanans in 1870 and geologist F.V. Hayden in 1871 led Congress to conclude that the area did have unique characteristics and besides, as one legislator put it, "there is nothing valuable here so why not make it a national park?" Its early history was turbulent. The park was established in 1872 with little funding to develop and protect the area. Visitors damaged geysers and hot pools—some used axes to break off sections of hot spring terrace formations as souvenirs—poachers hunted at will—some visitors stuffed rocks and debris into geysers such as Old Faithful to see what might happen. Many thermal features were damaged and some destroyed by thoughtless acts.

Finally, in 1886, the U.S. Army was given responsibility for the administration of the park, and it was saved from complete destruction until the fledgling National Park Service could take over in 1917. Now, again, the park is seriously threatened, this time by large numbers of tourists and their vehicles, by developments along the borders of the park, by insensitive politicians, and by those who would exploit public lands, subscribing to the manifest destiny conviction prevalent in the early history of the country—we must use the land and resources, even those in the national parks, without regard to the future. The negative impacts of increasing population and economic gain for a select few at the expense of federal land owned by all Americans are obvious. The need to elect leaders who place principle and quality of life for this and future generations above nonsustainable, short-term economic benefits is paramount if we hope to preserve a semblance of the open spaces and values that have made this nation great.

Geomorphic Setting

Most of the park is a broad, undulating volcanic plateau (Fig. 10–1) about 8000 feet (2440 m) above sea level. Near the eastern and northern boundaries the plateau

FIGURE 10–1   Landform map of Yellowstone National Park showing location of major geographic features. (From Keefer, 1971)

rises abruptly to the Absarokas, a high range of volcanic mountains. A north-central range, the Washburn Range, is geologically related to the Absaroka Mountains. The Beartooth Mountains to the north played an important role in Yellowstone's development, particularly its glacial events. The Gallatin Range, which extends northwestward out of the park, is actually part of the Northern Rocky Mountain Province; thus Yellowstone occupies parts of two geomorphic provinces—the Northern and Middle Rocky Mountains. The Teton Range extends southward from the southwestern corner of the park, into Grand Teton National Park. There are several large lakes on the plateau; by far the largest is Yellowstone Lake in the southeastern part of the park. The Yellowstone River flows northward into the lake and then through the Grand Canyon of the Yellowstone and eventually into the Missouri River. South of the Continental Divide, which crosses the southwestern part of the plateau, the Snake River flows southward into and out of Jackson Lake in Grand Teton National Park and eventually into the Columbia and the Pacific Ocean.

Older Geologic Story

Up through Laramide mountain building the geologic development in the Yellowstone area was much like that in other Rocky Mountain areas. A notable difference is shown by the over 2.8-billion-year-old Precambrian gneiss and schist exposed in the Gallatin Range along the north-central edge of the park (Fig. 10–2) and in surrounding mountains such as the Beartooth and Madison ranges.

Relatively quiet, mostly marine conditions prevailed through the Paleozoic and early Mesozoic as "trailing edge" conditions existed on the west edge of the eastward-moving North American Plate. Westward plate movement in early Mesozoic time produced mountain building in the West, which culminated in the Laramide Orogeny and the broad anticlinal and overthrust structures of the Middle Rockies. Remnants of Paleozoic-Mesozoic strata occur in the Gallatin Range, along the north edge of the park, and in an area south of Heart Lake (U.S. Geological Survey, 1972b), where we will pick up their story again where the strata extend southward into nearby Grand Teton National Park. Elsewhere the Paleozoic-Mesozoic strata are covered or engulfed by Cenozoic-age volcanic rocks—rocks that record the turbulent events leading to the birth of this unique volcanic area.

Tertiary Volcanism

Near the end of the Laramide Orogeny an episode of volcanism affected many areas in the Rocky Mountains, including the Yellowstone area. The compressional forces of Laramide time were replaced by tension—enabling melting and volcanism to occur in Yellowstone and other Rocky Mountain areas about 50–40 Ma dur-

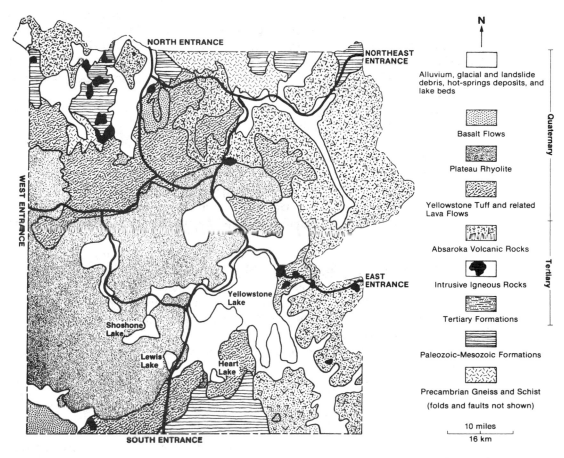

N

Alluvium, glacial and landslide
debris, hot-springs deposits, and
lake beds

Basalt Flows

Plateau Rhyolite

Yellowstone Tuff and related
Lava Flows

Absaroka Volcanic Rocks

Intrusive Igneous Rocks

Tertiary Formations

Paleozoic-Mesozoic Formations

Precambrian Gneiss and Schist

(folds and faults not shown)

10 miles
16 km

Quaternary

Tertiary

FIGURE 10–2   Simplified geologic map of Yellowstone National Park. (Adapted from Keefer, 1971)

ing the Eocene Epoch. The andesitic breccias, basalt lava flows, and other volcanic
deposits making up the Washburn and Absaroka ranges are part of the Absaroka
Volcanics (Fig. 10–2). Magmas that intruded the pre-Tertiary sedimentary rocks
formed stocks, sills, dikes, and laccoliths. Bunsen Peak near Mammoth is a small
intrusion. A hike to the summit of Mount Washburn, the eroded stump of an Absaroka volcano, takes one past both volcanic breccias and also dikes and sills.

The intervening area between the Washburn and Absaroka ranges was a large
valley that received considerable debris from the nearby erupting and eroding
Eocene volcanoes. These deposits contain an incredible variety of fossilized Eocene
plant life. Petrified wood and fossil leaves are well exposed in the north end of the
park in the Lamar Valley and Specimen Ridge areas. The initial hypothesis to
explain layer upon layer of petrified wood was that at least 27 individual forests

grew and that each was killed and buried in place by volcanic debris (Dorf, 1981). Paleobotanists were uncertain as to why some of the semitropical species such as dogwood, fig, and magnolia were mixed with cooler climate species such as walnut, oak, maple, and redwood in the same deposit. That a warm climate existed here 50 million years ago (Ma) is understandable. Continents had not yet reached their present more northerly locations and elevations were relatively low, perhaps 2000 feet (610 m) compared to 7000 feet (2134 m) or more today. But why should warm climate species be mixed with cool species?

After studying the events and examining the deposits left by the 1980 eruption of a similar type of volcano, Mount St. Helens in the Washington Cascades, Fritz (1985) recognized that mudflows descended the steep slopes of the Absaroka volcanoes carrying trees from the cooler, upper slopes and mixing them with warmer varieties at lower elevations. Groundwater containing silica derived from the ashy debris then flowed through the buried tree fragments slowly petrifying the woody material.

Examining the Washburn Volcano and the system of radiating dikes indicates that the center of one of the Washburn Range volcanoes was located south of Mount Washburn, in the area now occupied by much younger, flat-lying rhyolite flows of the Yellowstone Plateau. Why should a mountain range abruptly end along such a distinct topographic boundary? The answer lies with a much younger episode of volcanism that occurred during the last 2 million years—an event coincident with the positioning of the Yellowstone hotspot directly beneath the park.

Yellowstone Calderas

Even at the beginning of the Pleistocene, perhaps 2 million years ago, the area bore little resemblance to the Yellowstone of today. Laramide forces in the early Tertiary, the building of the Absaroka volcanoes during the Eocene, and later tensional forces that produced fault-block mountains together created a rugged mountainous topography. Lacking were the rhyolites and volcanic tuffs that cover today's vast Yellowstone Plateau. Lacking also were the geysers, hot springs, and mudpots. And of course none of the glacial features were there. Clearly, colossal changes took place during the past few winks of geologic time.

As described in the introduction to this chapter, a progression of rhyolite volcanoes which are younger to the northeast, have been forming across southern Idaho during the past 17 million years (Smith and others, 1980). A shallow, or passive, hotspot, the Yellowstone hotspot, formed in southwestern Idaho and began its slow migration to the northeast. This blowtorch-like area raised the ground surface thousands of feet, produced numerous faults and earthquakes, and generated partial melting of the mantle and crust and accompanying volcanism (Parsons and others, 1994). About 2 million years ago the first of three of the world's most violent known eruptions occurred in the Yellowstone area. Other major eruptions would follow, one about 1.3 Ma and the most recent one about 600,000 years ago.

Each eruption began with magma forcing its way up toward the surface, arching up and fracturing the overlying rocks. This magma was high in silica—indicating that the low-silica basaltic magma derived from the mantle had moved upward and resided in the earth's crust for a long time—long enough to melt the high-silica metamorphosed crustal rocks to form a gas-charged rhyolitic magma. As gas pressure increased, the confining pressure of thousands of tons of rock overlying the magma chamber was exceeded—producing an eruption whose violence was far in excess of anything ever experienced by humans. When a warm can of carbonated soda (a mixture of liquid and dissolved gas) whose contents have been shaken is opened, gases escape violently bringing liquid with them, often giving someone an unwanted bath! Magma is also a mixture of liquid and gas—it too explodes violently if the confining pressure is suddenly reduced, allowing gases to come rapidly out of solution. Huge quantities of ash, pumice, and blocks of rock are hurled into the air. Propelled by hot, expanding gases, ash spread at hurricane velocities across the Yellowstone area, devastating forests and animal populations for 100 miles (160 km) around the resulting caldera rim. Thick *ash-flow* deposits— some over 1000 feet (305 m) in depth—must have looked like a "valley of ten *million* smokes" as hot gases trapped in the deposit emerged at the surface for decades afterwards. The trapped heat also remelted or fused the ashy material together, forming a *welded tuff*. With the almost instantaneous ejection of many cubic miles of material from the magma chamber, the roof "unzipped" and collapsed inward forming a huge circular- to elliptical-shaped volcanic depression called a *caldera*. When the infall of wall rocks ceased, the caldera had an area of hundreds of square miles. Subsequently, however, large volumes of rhyolite were periodically extruded into the caldera, filling it and spilling out over the rim in places.

The exact dimensions of the first (Huckleberry Ridge) caldera are not known because the most recent eruption 600,000 years ago produced an overlapping caldera with an area of about 1000 square miles (2600 km^2)! Collapse of this latest caldera also cut through the Washburn Range, swallowing its southern extension. A similar, but smaller caldera formed in the Island Park area just west of the park boundary 1.3 Ma and produced the Mesa Falls Tuff. The Lava Creek Tuff was produced by the most recent caldera eruption 600,000 years ago and whose associated features dominate today's Yellowstone area (Fig. 10–3). One measure of eruption magnitude is the volume of ejected material. A measure of estimated volume of ejecta of some prominent explosive volcanic events around the world (Fig. 10–4) shows that the Yellowstone eruptions were equivalent to 280 to as many as 2500 Mount St. Helen's eruptions!

Following the Lava Creek eruption, a large caldera lake formed, only to be squeezed and pushed around by large, viscous, postcaldera rhyolite flows that form the surfaces of the Madison, Central, Pitchstone, and Mirror plateaus (Fig. 10–1). The flows filled and obscured the caldera rim in places making recognition of the

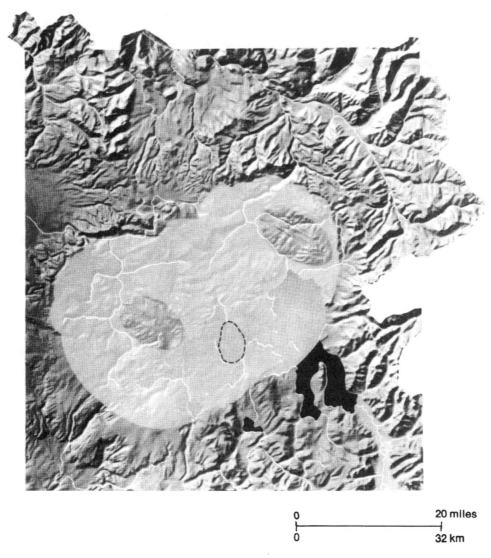

0   20 miles
0   32 km

FIGURE 10–3   Outline of Yellowstone caldera (Lava Creek Caldera) produced by an enormous volcanic eruption 600,000 years ago. The caldera partially overlaps the 2.0-million-year-old Huckleberry Ridge Caldera. The 1.2-million-year-old Mesa Falls Caldera (also known as Island Park Caldera) is located in Idaho, just west of the map. The two oval-shaped areas are resurgent domes that formed during the last 150,000 years as pressure in the underlying magma chamber increased. The domes have in recent time both enlarged and subsided as much as 1 inch (2.5 cm)/year. The area outlined by the dotted line shows the "Baby Caldera" formed during the 155,000-year-ago eruption in the West Thumb basin of Yellowstone Lake. (Illustration from Keefer, 1971)

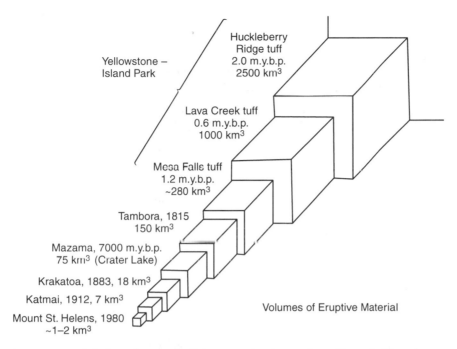

FIGURE 10–4   Relative volumes of well-known volcanic eruptions. (From Smith and Braile, 1984)

world's largest known caldera difficult. Only after an intense study of the area in the 1960s and 1970s was the remarkable 36- × 45-mile-diameter (58- × 72-km) Lava Creek Caldera recognized (Fig. 10–3). Not only do the edges of flows control the shape and dimensions of Yellowstone Lake, but the location of smaller lakes and streams, such as the Firehole River, is a direct consequence of lava flows. The space between the 117,000-year-old West Yellowstone flow and the 150,000-year-old Central Plateau flow is occupied by the valley of the Firehole River. The last rhyolite flow to "ooze" into the caldera occurred 70,000 years ago on the Pitch-stone Plateau in the southwestern part of the park (Fig. 10–1). Although rhyolitic lavas are dominant, a few basalt flows are known. Along the canyon walls near Tower Falls two buried basalt flows, about 1.5 million years old, display spectacular examples of columnar jointing.

Some of the later flows isolated portions of Yellowstone Lake and formed Shoshoni, Lewis, and Heart lakes. A smaller caldera, a "baby caldera," blasted out the West Thumb basin of Yellowstone Lake about 155,000 years ago. This "small" caldera is small only by Yellowstone standards. This 4- × 6-mile-wide (6.4- × 9.7-km) depression equals in size the impressive caldera that contains Crater Lake in the Oregon Cascades!

Yellowstone's Glaciers

With the cooling of the climate at the beginning of the Pleistocene, glaciers developed in the high mountains surrounding the Yellowstone Plateau, particularly in the Absaroka and Beartooth mountains (Fournier and others, 1994). Each episode of glaciation followed a definite sequence of glacier development. This sequence will again occur here and in many other areas of the world during the next period of world cooling, perhaps 23,000 years away according to climate models based on known variations of the earth's orbit (summarized in Skinner and Porter, 1987). The return to glacier climates will see small localized ice masses called *cirque glaciers* form in the surrounding high mountains and eventually expand to *valley or mountain* glaciers. The larger glaciers may eventually form *piedmont* glaciers if they flow out of the mountains onto the flatter areas along mountain fronts—areas such as the Yellowstone Plateau. Merging of piedmont glaciers form *ice caps,* like the giant ones that covered all of the Yellowstone Plateau with as much as 2000 feet (610 m) of ice.

The dating of glacial events is usually difficult because of the lack of datable materials associated with glacial and interglacial sediments. Yellowstone is again unique—here volcanic and glacial events overlapped. The relationship of glacial and interglacial sediments to radiometrically dated ash and lava flows provides the best dated glacial sequence, particularly of older glacial and interglacial events, in North America. Richmond (1986) describes a pre-Pleistocene (Pliocene) till and numerous layers of glacial materials sandwiched between younger rhyolite ashes and flows and locally associated with organic materials that are radiometrically dated.

Later glaciations are divided here and in other Rocky Mountain areas, according to the sequence recognized in the Wind River Mountains of Wyoming. The Bull Lake glaciations occurred about 200,000–130,000 years ago, and the Pinedale glaciation (late Wisconsin age) occurred about 30,000–13,000 years ago. In addition, an early Wisconsin glaciation occurred about 70,000 years ago (Richmond, 1986).

An ice cap glacier covered the Yellowstone Plateau during at least Bull Lake and early Pinedale time and sent tongues of ice (*outlet glaciers*) spilling northward down the Yellowstone River valley, westward through West Yellowstone in the Madison Valley, and south across the Continental Divide into the Snake River valley. We will pick up the trail of this southern glacier when Grand Teton National Park is discussed.

Along the way, the glacier deposited its materials (U.S. Geological Survey, 1972a), including erratics such as the giant one near the rim of Yellowstone Canyon (Fig. 10–5). The waning stages of the Pinedale advance are discussed by Richmond (1986) and Pierce (1979) and are too complex to discuss here. Thick layers of glacial outwash and other glacial-derived sediment cover much of the Yellowstone Plateau. Such material is relatively loose and *permeable* (fluids and gases flow readily through them) and, in this condition, do not produce the concen-

FIGURE 10–5   Glacial erratic boulder, about 24 feet (8 m) long and weighing more than 500 tons, near Inspiration Point on the north rim of Yellowstone's Grand Canyon. (Photo by D. Harris)

trated flow needed to produce the geysers and other thermal features for which Yellowstone is world-famous. Some transition is needed to form these bizarre, delightful geothermal features.

Geothermal Features

Areas of low elevation between flow fronts and where underlying fractures in the caldera rock permit heat and gases to rise to the surface are those magical areas where most of Yellowstone's geysers and other thermal features are located. About 70 percent of the world's active geysers and many of its hot springs are located along the Firehole River valley along the west side of the park. Others occur at Norris, Heart Lake, Shoshoni Lake, and at isolated areas throughout the park, mostly within the confines of the Lava Creek caldera. About 10 natural geyser areas occur in the world—three have been destroyed and four have been severely affected by geothermal development. New Zealand was once the geyser capital of the world with over 300 active geysers. Geothermal development there and elsewhere have proved how delicate these thermal systems are—fewer than 10 active geysers now remain in New Zealand! Iceland and other areas have suffered a similar fate—ominous warnings of what happens when humans tamper with geothermal systems.

Surface manifestations of thermal energy can take many forms, including geysers, spouters, hot springs, mudpots, and fumaroles. Understanding why they are different requires an understanding of the three components of a thermal feature—heat, water, and the nature of the "plumbing" system.

Rain and snow entering the groundwater system furnish the supply of water. Groundwater soaks downward where it encounters heat and acid gases rising through the fractured caldera blocks below. The groundwater is heated, becomes more buoyant, and rises to the surface, usually in topographically low areas. As the acidic fluids rise through the rhyolite and other rocks in the caldera, the minerals are decomposed and changed by chemical reactions. Silica is carried upward in solution to reservoirs as hot as 390°F. (200°C) that occur less than 1500 feet (460 m) from the surface. As the chemically charged fluids rise and cool, silica is precipitated and cements the relatively loose glacial gravel covering the Yellowstone Plateau into a concrete-like rock that restricts water movement to the surface. The groundwater below this self-sealing zone (about 80–160 feet; 25–50 m) becomes overpressurized, and superheated water escapes only through those areas with cracks or where the gravels are as yet uncemented.

Where water is less abundant only steam and other gases escape—such a feature is known as a *fumarole*. Where water is more abundant but insufficient to fill and overflow the surface vent, a thick pasty mixture of water, fine clay, and silica bubbles delightfully, like a pot of oatmeal on a stove, in a feature called a *mudpot*. Where hot water is more abundant and rises through relatively open conduits, overflowing at the surface, a feature called a *hot spring* forms. Flowing water flushes out the fine clay and silica producing crystal-clear pools. A *spouter* continuously ejects water in a column like a broken fireplug, and a *geyser* is a special type of hot spring that erupts but has an interval of repose.

The plumbing system of a geyser with its restrictions and complex side passages, along with some simple principles of physics, explain why geysers erupt. One important principle is that given a column of water, the boiling temperature of the water at the bottom, where the pressure is higher, is greater than at the top. Surface water at Yellowstone's high elevation (about 7700 feet; 2350 m) boils at 198°F (93°C). For every 3 feet (1 meter) of water column, the boiling point is about 3.5°F (2°C) higher (Rinehart, 1976). When the geyser tube is filled to the top and at least the lower parts of the water column are at the boiling point, steam bubbles are generated and build at a geyser tube constriction, sending surges of steam toward the surface. Preliminary splashing reduces the pressure on deeper parts of the water column allowing water there to flash into steam. Like an explosion in a gun barrel, the expanding gas drives the missile or water column out the "barrel." In the case of a geyser, one of the most amazing displays of nature is created. Verification that narrow restrictions in geyser plumbing do indeed act as triggering mechanisms for eruptions was determined by Hutchinson and others (1997) when they lowered a miniature video camera into Old Faithful and discovered a major narrowing of the geyser vent about 23 feet (7 m) from the surface.

Old Faithful (Fig. 10–6) is but one of over 200 geysers in the park, but it is, as the name suggests, one of the most reliable. Whereas most other geysers erupt infrequently and at their own convenience, Old Faithful can be counted on to go off again within an hour and a half, and generally within about an hour. Historically, the interval has varied from about 45 to 105 minutes. Hutchinson and others (1997) discovered that Old Faithful is developed over an east–west oriented crack that narrows to a mere 5.9 inches (15 cm) wide about 22.3 feet (6.8 m) from the surface.

Geysers are many and varied in Yellowstone, each with its own habits that are related to peculiarities in its plumbing system. The geyser basins are located in the western part of the park, from south of Old Faithful north to Norris Junction. Grand, Grotto, Riverside, Castle, and Lone Star (Fig. 10–7) are among those that are likely to perform for the visitor who spends several hours in the geyser area.

FIGURE 10–6   Old Faithful geyser, hallmark of the National Park System, inspired those who gave birth to the national park concept more than a century ago. Note the apparently disinterested, cud-chewing local inhabitants. (Photo by E. Kiver)

FIGURE 10–7   Lone Star Geyser in eruption. Cone is one of the largest in the park (over 9 feet tall; 3 m). The cone builds when silica-rich water splashes during the quiet phase. The Lone Star Cone is built on top of a mound form indicating a change from mound to cone activity. Compare with mound geyser in Figure 10–6. (Photo by E. Kiver)

As indicated above, most geysers are unpredictable. After a long period of inactivity, a geyser may perform in a spectacular fashion. The world's largest geyser, Steamboat Geyser, was first noted to erupt in 1878 and occasionally erupted up until 1911. After 50 years of quiet, it burst into activity in 1961. Since then, intervals between eruptions can be many years or perhaps a few weeks apart. When they do occur, they are awesome events—water columns as high as 380 feet (116 m), sometimes dropping small rock fragments into the parking lot some 800 feet (244 m) away (Rick Hutchinson, personal communication). The water phase is followed by a roaring vapor or steam phase that lasts for hours. As Bryan (1995, p. 273) notes: "At times it is literally impossible for people to yell at one another and be heard—as if anybody had anything sensible to say."

Siliceous sinter, or *geyserite* (a form of opal), is deposited around geysers and hot springs (Fig. 10–8). Silica is dissolved from the rhyolite down below and gradually deposited in the form of rings and mounds (Figs. 10–6 and 10–7) around the vent. By this process the "pipes" may be so enlarged as to reduce the violence of the eruptions; eventually, the water merely overflows as a hot spring. Morning Glory Pool, shaped like a giant morning glory and filled with blue water, is a thing of great beauty (Fig. 10–8). Conversely, plugging or restricting the conduits by deposition of sinter or debris falling, or being thrown into hot pools by the thought-

FIGURE 10–8   Morning Glory Pool, a particularly attractive hot spring in the Upper Geyser Basin (Old Faithful basin), has temperatures close to the boiling point. Temperatures have reduced since at least 1950 as thoughtless visitors have thrown coins and other debris into the pool, apparently interfering with water circulation. (Photo by E. Kiver)

less few who live in our society, reduces water circulation and temperature, ultimately causing its extinction. Coins and other debris thrown into Morning Glory Pool over the years have reduced its temperature and brilliant blue color—a change recognizable in the past few decades. Geysers and hot springs experiencing violent eruptions often self-destruct. Thus any change in the delicate plumbing system often spells extinction for a geothermal feature.

In addition to the gray colors of siliceous sinter, algae and cyanobacteria colonies inhabit the warm thermal waters and provide elegant splashes of color to the thermal features. The bright yellow colors are bacteria that can survive in water up to 195°F (76°C), and the darker colors are algal mats that often exist at the edges of pools where water temperature is cooler. Thermus aquaticus, a microbe that lives in 176°F (66°C) water has been extracted from Yellowstone's pools and is used to duplicate DNA molecules in both research and commercial laboratories. Without it the recent advances in DNA research and technology would be impossible.

The Hot Springs Terrace area at Mammoth Hot Springs near the north entrance is a "must" at Yellowstone. These huge and beautiful terraces (Fig. 10–9) rise steplike high above the valley. The springs have been building Terrace Mountain along the Gardner River valley for at least 400,000 years and have deposited hundreds of feet of *travertine* or *calcareous tufa,* a type of limestone, rather than the

FIGURE 10–9   Travertine terraces building in Mammoth Hot Springs area of Yellowstone National Park. (Photo by E. Kiver)

siliceous sinter found at most of Yellowstone's thermal areas. Thermal waters rising through Paleozoic limestone below transfer over 2 tons of calcium to the surface every day—making this the world's largest carbonate-depositing spring system! Water emerging at the surface loses carbon dioxide (CO_2) to the atmosphere, thus reducing acidity and the water's ability to contain calcium in solution. By this process, calcium carbonate is added to the stony mass of terrace deposits as ridges, cones, and graceful terracettes (Bargar, 1978).

Many of the terraces are beautifully colored by different colored algae; when the hot waters shift to another place the abandoned section soon turns white. Liberty Cap is a striking feature in this travertine area. It stands high, like a 37-foot (11.3-m) drinking fountain. The hot spring that formed it is now dead, but evidently it maintained a small flow that deposited all of its dissolved material around the pipe rather than out beyond the base. Across the road, near park headquarters, a new spring appeared in 1926 and is now threatening a historic home built in 1908. Sand bags have temporarily diverted the spring away from the building. Where limestone exists, acidic groundwater can form caves and sinkholes. Some of these trap heavier-than-air CO_2 and H_2S gases and become death traps for unwary animals.

Although most thermal area features are within the confines of the 600,000-year-old Lava Creek caldera, the hottest and most unstable of the basins (closest to the magma chamber below?), Norris Basin and Mammoth Hotsprings are located along a zone of faults that extend northward out of the park along the Yellowstone River valley. Norris experiences an as yet unexplained phenomenon, the fall hydrothermal disturbance. During this annual event, usually clear thermal waters become turbid, and geyser activity, temperature, and water chemistry change dramatically.

Walk the trails and boardwalks in the thermal areas with Bryan's 1995 publication in hand that describes each geyser and major hot spring and its history. The documented changes in behavior of many thermal features since records were first kept is astounding and leave no doubt as to the fragility of these surface features and why this world-significant area needs continued protection. Those who are proposing to pump thermal waters just north of the park from the fault system that is connected to Mammoth and Norris should not be permitted to jeopardize such priceless features. The American public had an opportunity in 1981 to purchase the ranch where the hot spring, as well as a critical grizzly habitat, is located, but the federal government failed to allocate funds. The money "saved" in 1981 by "thrifty" politicians and bureaucrats has been spent many times over trying to rectify a potentially serious situation that could disrupt this unique, world-renowned area.

Yellowstone's Grand Canyon

The Grand Canyon of the Yellowstone River (Fig. 10–10) is a youthful canyon that formed after the last glacier retreated from the Lamar Valley in the north end

FIGURE 10–10   Grand Canyon of the Yellowstone and the 308-foot-high (94-m) Lower Falls. (Photo by E. Kiver)

of the park. Water flowing northward from Yellowstone Lake flows lazily across the high, central plateau only to pick up speed as it descends to the lower elevations in the Lamar Valley. The canyon first formed in the north and extended headward some 23 miles (37 km) where it now terminates at the 109-foot-high (33-m) Upper Falls. The spectacular 308-foot-high (94-m) Lower Falls (another of Yellowstone's "must see" features) is located about 2500 feet (760 m) downstream from the Upper Falls. Whether the lake drained northward by this route immediately after the Lamar Glacier melted is questionable. Richmond (1987) discusses the complicated rock, sediment, and erosion record in the area.

In any event, the Yellowstone River cut rapidly through the thick, highly altered (by rising acidic hydrothermal solutions), 150,000-year-old Canyon Rhyolite flow, until it encountered the unaltered rocks that form the lip of the waterfalls. The altered rhyolite or "yellow stone" is well exposed in the walls of the 1200-foot-deep (366-m) canyon (Fig. 10–10). Similar brightly colored rock undoubtedly underlies today's geyser basins and other thermal areas where acidic solutions continue their chemical attack on the fractured rock below.

How rapidly the canyon was cut is unknown. However, the presence of the native cutthroat trout in Yellowstone Lake above the falls requires that these fish arrive there by swimming upstream. Thus, they arrived in the caldera lake *after* the Lamar glacier had disappeared and *before* they had to make a 308-foot (94-m) leap to get over the Lower Falls! The falls continue their slow, headward retreat. Given time they will ultimately reach and drain Yellowstone Lake. Another plausible scenario is that the magma chamber repeats what it has done every 600,000–700,000 years in the recent past—then all bets are off!

Obsidian Cliff

Along the north-trending fault zone known as the Norris–Mammoth Corridor is a 180,000-year-old lava flow that makes up Obsidian Cliff. This glassy lava flow likely lacked sufficient water content at the time of eruption that would enable chemical elements to migrate and form the tiny mineral grains typical of a rhyolite. Rather, the chemical mixture is frozen as a noncrystalline volcanic glass. Indians used the obsidian to make points and tools—some of which found their way a thousand miles away to the Mississippi River valley. The abundance of loose fragments of the sparkling glass now present is greatly diminished along the flow edge compared to personal visits made in the 1950s. Thoughtless visitors have carried away a great many of the loose pieces (and others not so loose!) and changed Bridger's "mountain of glass" to a barren graveyard of what is left after the more attractive pieces have disappeared. Such souvenirs are likely now discarded or "living" in someone's attic or garage.

West Yellowstone Earthquake

The first geologist to visit the Yellowstone area, F.V. Hayden, reports feeling earthquakes in 1871. These have continued, sometimes in great numbers, as in the 1500 tremors recorded just west of Madison Junction during the week of July 4, 1995. The August 17, 1959 magnitude 7.5 earthquake was a major event that rocked the region. The epicenter was near the park boundary, about 12 miles (19 km) north of the town of West Yellowstone. In the Madison River Canyon just outside the park boundary, the earthquake triggered a huge landslide (Fig. 10–11) that buried an occupied U.S. Forest Service campground and dammed the valley, forming Quake Lake. A crack appeared in the concrete core of Hebgen Lake Dam, a dam that impounds a large reservoir a few miles upstream of the landslide. *Seiche* waves ("sloshing" back and forth of the reservoir water) raced from end to end in the reservoir, first draining away from the dam and, 17 minutes later, returning to crest over the top of the dam. Four spillovers occurred, but fortunately for those downstream the dam held.

FIGURE 10–11 Madison Canyon landslide resulting from the 1959 West Yellowstone earthquake. Quake Lake, just upstream (to the left), drains through the canyonlike spillway in middle distance. (Photo by E. Kiver)

All of the signs for an exciting future are present—the Yellowstone hotspot is alive and well! The broad uplift and cracking of the crust continue as heat continues to rise from the mantle. Is magma continuing to accumulate and build pressure? Two localized domal areas, the Sour Creek and Mallard Lake domes (Fig. 10–3) experienced rapid inflation of as much as an inch/year up to 1984 when rapid deflation began. Geophysical anomalies such as the low density of crustal rocks, high heat flow, low seismic velocities, and a lack of seismic activity below about 3 miles (5 km) under the caldera indicate that magma or partially molten rock is within a few miles of the earth's surface. Just as walking on top of a waterbed is an interesting experience, walking on the roof of a magma chamber is even more interesting and will someday provide more excitement than we might like!

REFERENCES

Bargar, K.E., 1978, Geology and thermal history of Mammoth Hot Springs, Yellowstone National Park, Wyoming: U.S. Geological Survey Bulletin 1444, 55 p.

Bryan, T.S., 1995, The Geysers of Yellowstone: Associated University Press, Boulder, CO.

Dorf, E., 1981, Petrified forest of Yellowstone: National Park Service pamphlet.

Fournier, R.O., Christiansen, R.L., Hutchinson, R.A., and Pierce, K.L, 1994, A field-trip guide to Yellowstone National Park, Wyoming, Montana, and Idaho—volcanic, hydrothermal, and glacial activity in the region: U.S. Geological Survey Bulletin 2099, 46 p.

Fritz, W., 1985, Roadside geology of the Yellowstone country: Mountain Press, Missoula, MT, 144 p.

Hutchinson, R.A., Westphal, J.A., and Kieffer, S.W. 1997, In situ observations of Old Faithful Geyser: Geology, v. 25, p. 875–878.

Keefer, W.R., 1971, The geologic story of Yellowstone National Park: U.S. Geological Survey Bulletin 1347, 92 p.

Parsons, T., Thompson, G.A., and Sleep, N.H., 1994, Mantle plume influence on the Neogene uplift and extension of the U.S. western Cordillera: Geology, v. 22, p. 83–86.

Pierce, K.L., 1979, History and dynamics of glaciation in the northern Yellowstone National Park area: U.S. Geological Survey Professional Paper 729-F.

Richmond, G.M., 1986, Stratigraphy and chronology of glaciations in Yellowstone National Park: in, Quaternary glaciations in the northern hemisphere, Pergamon, New York, p. 83–98.

Richmond, G.M., 1987, Geology and evolution of the Grand Canyon of the Yellowstone, Yellowstone National Park, Wyoming: in Geological Society of America Field Guide, Rocky Mountain section, Geological Society of America, Boulder, CO, p. 155–160.

Rinehart, J.S., 1976, A guide to geyser gazing: Hyperdynamics, Santa Fe, New Mexico, 64 p.

Skinner, B.J., and Porter, S.C., 1987, Physical Geology: Wiley, New York, 750 p.

Smith, R.B., and Christiansen, R.L., 1980, Yellowstone Park as a window on the Earth's interior: Scientific American, v. 242, p.104–117.

Smith, R.B., and Braile, L.W., 1984, Crustal structure and evolution of an explosive silicic volcanic system at Yellowstone National Park: National Research Council, Washington, DC, National Academy Press, p. 96–109.

U.S. Geological Survey, 1972a, Surficial geologic map of Yellowstone National Park: Map I-710.

U.S. Geological Survey, 1972b, Geologic map of Yellowstone National Park: Map I-711

Park Address

Yellowstone National Park
P.O. Box 178
Yellowstone National Park, WY 82190

GRAND TETON NATIONAL PARK (WYOMING)

The Teton Range, located just south of Yellowstone National Park in northwestern Wyoming (Plate 10) is unquestionably one of the most spectacular mountain ranges in North America. The range is small by Rocky Mountain standards, only 40×15 miles (64×24 km), but huge in its visual impact on visitors (Fig. 10–12). This fairytale-book range rises over 7000 feet (2134 m), nearly straight up from the flat-floored 6300-foot-high (1920 m) Jackson Hole to the top of Grand Teton peak at 13,700 feet (4177 m). The three central high peaks are visible from miles away and served as landmarks for the Indians and early explorers and mountain men. The Snake River begins its 1000-mile (1600-km) journey in southern Yellowstone Park where it flows south to Jackson Hole—the first leg of its tortuous path to the Columbia River in southeastern Washington State.

Ironically, the youngest mountain range in the Rockies is composed of some of the oldest rocks in North America! Obviously, the rocks and landforms have much to tell to those who take time to read their stories. The spectacular Teton peaks began their upward journey a mere 9 Ma—a journey that is continuing today at a rate much faster than the forces of erosion can lower the peaks. The finishing touches were etched into the landscape during the Ice Age (Pleistocene), which ended here about 10,000 years ago. Love and Reed's *Creation of the Teton Landscape* (1971) and Good and Pierce's *Geology of Grand Teton and Yellowstone* (1996) are recommended to those interested in Grand Teton National Park. With excellent photographs, including aerial views and block diagrams, they are good books for both the layman and the geologist. For a general look at the natural and human history of the park the National Park Service Handbook 122 (1984), available at the visitor center, is also recommended.

Humans and the Park

Clovis-style points on the shore of Jackson Lake (Fig. 10–13) indicate that humans camped on its shores about 10,000 years ago. Historically, a number of tribes used the Jackson Hole area for hunting and gathering activities during the summer season but retreated to milder environments in the winter. The Sheepeater Shoshoni wintered in northern Yellowstone and were the last tribe to leave the area when in 1879 they were moved to the nearby Wind River Reservation in Wyoming.

The Indians left the land unmarred—a tradition that was not followed by the Euro-Americans who came later. Mountain man John Colter, a member of the

FIGURE 10–12   Grand Teton (13,770 feet; 4198 m) and the Teton fault block rise majestically above the Snake River and the outwash plains of Jackson Hole. Note the distinct stream terraces cut during nonglacial intervals. As a result of recent tilting of the Jackson Hole fault block, the high outwash surface across the river slopes downward to lower elevations along the Teton Fault at the front of the range. (Photo by Antoinette Lueck)

Lewis and Clark expedition (1804–1806), was the first of the newcomers to see the area. Indians knew the mountains as "tee-win-ot" or "the three pinnacles." The French voyageurs named the three prominent peaks Les Trois Tetons, now called the Grand Teton, Middle Teton, and South Teton. The area became one of David Jackson's favorite places and was named Jackson's Hole by his fur-trader partner William Sublette in 1820. To mountain men a "hole" was a deep valley surrounded by mountains. By the 1840s the beaver were nearly exterminated in the West when, fortunately for the beaver and modern man alike, men's hat styles changed from the stovepipe beaver hat to one of silk.

A contingent of the Hayden survey under geologist James Stevenson explored the Teton country in 1872 while Hayden was studying Yellowstone. Stevenson and Nathaniel Langford (later to become Yellowstone's first superintendent) climbed the highest mountain peak during their stay. This first reported climb of Grand Teton was later strongly contested by William Owen who climbed the peak in 1898. Owen apparently "won" the battle when the Wyoming Legislature declared in 1929 that Owen was first!

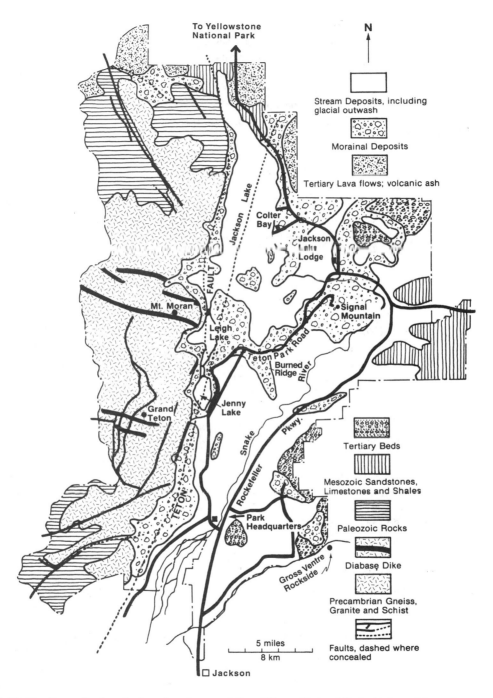

N

To Yellowstone
National Park

Jackson Lake

Colter
Bay

Jackson
Lake
Lodge

Signal
Mountain

FAULT

Mt. Moran

Leigh
Lake

Teton Park Road

Burned
Ridge

Snake River

Grand
Teton

Jenny
Lake

TETON

Rockefeller Pkwy.

Park
Headquarters

Gross Ventre
Rockside

Stream Deposits, including
glacial outwash

Morainal Deposits

Tertiary Lava flows; volcanic ash

Tertiary Beds

Mesozoic Sandstones,
Limestones and Shales

Paleozoic Rocks

Diabase Dike

Precambrian Gneiss,
Granite and Schist

Faults, dashed where
concealed

5 miles

8 km

Jackson

FIGURE 10–13   Generalized geologic and park map of Grand Teton National Park. (Adapted from Love and Reed, 1971)

Settlers trickled into the area beginning in 1884 where they took up the difficult existence of ranching under severe winter climate conditions. However, the area was too beautiful and spectacular to be ignored by the public for long. The lure of one of the most spectacular mountain ranges in North America would prove irresistible—increasing numbers of people came to visit. Some of the marginal ranching operations began to operate dude ranches and hunting lodges as early as 1903—the beginnings of the tourist industry. A group of local citizens in 1923 recognized that the quality of a visit to the area and the desirability of living in the area would depend on how the area was managed and how much commercial exploitation occurred. This far-sighted group of local citizens met with Horace Albright, then superintendent of Yellowstone National Park, to propose a plan for setting aside a portion of Jackson Hole to be protected by some government agency for the enjoyment of the public. A small national park was established in 1929 to protect the high peaks, but very little of Jackson Hole was included.

Unbeknown to Congress and the rest of the nation, John D. Rockefeller, Jr., had visited the area in 1926 and had been inspired by Albright's enthusiasm for a Teton park that included much of Jackson Hole. Rockefeller apparently said nothing of his intentions to Albright or others but established a land company in 1927 that quietly began to purchase private ranches and other lands in northern Jackson Hole with the intent to donate them to the public. The over 32,000 acres (50 square miles; 130 km^2) purchased by the Rockefeller family became part of a greatly enlarged park (310,000 acres; 485 square miles; 1256 km^2) in 1950. What a remarkable gift to society!

Precambrian Rocks

Special mention should be made of the unusually old rocks in Grand Teton National park and some of the nearby mountain ranges in Wyoming, Montana, and Idaho. Ancient sedimentary and volcanic rocks were metamorphosed about 2800 Ma (Reed and Zartman, 1973), thus deposition of the rocks must be older than 2800 Ma, perhaps 3000 Ma. Such rocks are nearly 70 percent as old as the earth and are some of the oldest rocks on the North American continent! The Teton and nearby areas, as well as another block of land in the upper Great Lakes area, were apparently part of the nuclei around which land was added by episodes of geosyncline development and mountain building to eventually forge today's continent.

These ancient metamorphic rocks (Fig. 10–13) consisting mostly of banded gneiss and schist (Fig. 10–14) were formed during one of these ancient mountain-building episodes under pressure and temperature conditions found at depths of 5–10 miles (8–16 km) beneath the earth's surface. About 2500 Ma a large mass of granitic magma intruded the older rocks, now exposed extensively in the center of the present range. Later, still during the Precambrian Era (about 1400 Ma), dark-colored basaltic magma was injected into cracks and fissures that cut through both the metamorphic and granitic rocks (Fig. 10–15). The magma cooled into

FIGURE 10–14   Contorted gneiss, or *migmatite* (mixed rock containing both metamorphic and igneous components), Grand Teton National Park. (Photo by D. Harris)

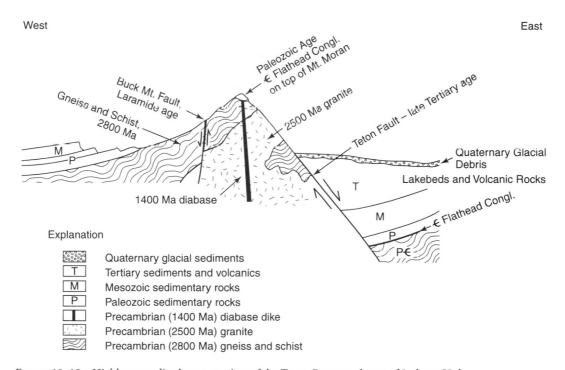

FIGURE 10–15   Highly generalized cross section of the Teton Range and part of Jackson Hole.

numerous *diabase* and basaltic dikes. The prominent 150-foot-wide (46-m) dike on the face of Mt. Moran is an outstanding example (Fig. 10–16).

Geologic History

Following a long period of erosion that leveled the Precambrian landscape to a flat or perhaps rolling surface, seas again invaded, and marine deposits of Paleozoic age buried the Teton area with about 4000 feet (1220 m) of sediment. The Flathead Sandstone, the oldest Paleozoic (Cambrian) rock layer, will be important in a later discussion regarding the estimation of how much movement has occurred along the Teton Fault.

Mesozoic-age sediments covered the area with another 10,000–15,000 feet (3050–4570 m) of debris. Mesozoic mountain building that began far to the west gradually encroached on the Rocky Mountain region. Marine sedimentation gave way to coarse terrestrial sediments as the transition occurred. By early Tertiary time (about 60 Ma) the compressive forces of the Laramide Orogeny were deforming the Teton area.

A broad, northwest–southeast trending anticlinal uplift, similar to those found in many areas in the Middle and Southern Rocky Mountains, formed in the Teton region during Laramide time. Sediments derived from the eroding uplift, as well as volcanic rocks and sediments from both local sources and from vents in and near Yellowstone, accumulated in the area during the Tertiary. The Laramide Mountain structure was mostly leveled by erosion when sediment accumulations began to thicken appreciably in the Jackson Hole area beginning about 9 million years ago. Movement along the north–south trending Teton Fault had begun to form the newest mountain range in the Rocky Mountains. As the giant 40-mile-long (64-km) Teton fault block began its episodic uplift, the Jackson Hole block dropped downward relative to the uplifting block, creating the space for many thousands of feet of late Cenozoic sediment to accumulate. Tectonic conditions had changed significantly—compression had ended and local stretching or extension began to affect the Rocky Mountains. The northwest-oriented Laramide-age anticline is now sliced and exposed along the north–south trending face of the latest mountain range, the Tetons.

The amount of vertical displacement along the Teton Fault is mind-boggling—from 3.7 to 5.6 miles (6 to 9 km) (Smith and others, 1993)! Such estimates are based on past drill holes made by oil companies in Jackson Hole, geologic studies, and geophysical measurements. As shown in Figure 10–15, the contact between the Flathead Sandstone and Precambrian rocks on top of Mt. Moran is the same once-connected stratigraphic level that is now deeply buried beneath Jackson Hole. The vertical offset here is between 5 and 6 miles (8 and 10 km)! Because each movement probably involved a few inches or at most a few feet of displacement, many tens of thousands of earthquakes must have occurred during the past few million years. If these movements were distributed equally over the 9-million-year interval of fault movement, the fault would move on the average at about 3.3 feet (1 m) every thousand years. A few feet every thousand years may seem trivial, but the results in the scenery are more than a little spectacular!

What processes of plate action account for this remarkable mountain uplift? Recall that fault-block mountains are the dominant structure in the nearby Basin and Range Province. Undoubtedly the same extensional forces that produced the Basin and Range (oblique subduction and crustal collapse?) are at work here—as well as the unsettled conditions associated with the hotspot centered beneath nearby Yellowstone National Park.

Teton's Glaciers: Past and Present

As in Glacier National Park, the finishing touches of scenery came from the effects of Pleistocene glaciers—not today's small diminished glaciers. The combination of a mountain rising to higher, cooler elevations and the general worldwide cooling of climates during the Pleistocene (Ice Age) produced conditions in which winter snows persisted through the following summer melt season. A number of such years in a row would initially produce small glaciers high in such sun-protected places as cirques (cirque glaciers). Additional accumulations of snow and ice would enable a cirque glacier to thicken and expand beyond the confines of the cirque and form a valley glacier. Continued growth might form a piedmont glacier if the ice flowed out of its valley onto the piedmont surface (Jackson Hole) at the mouth of the canyon. Coalescing of piedmont or adjacent valley glaciers would form *ice sheets* at lower elevations or ice caps higher in the mountains. All of the above have affected the Teton area during the past 2 million years.

Although the record of glaciation in the Yellowstone–Teton area goes back over 2 million years (Richmond, 1986), evidence of older glaciers has been mostly removed where overridden by younger glaciers. At least some of the pre-Wisconsin glaciers were huge. About 200,000 years ago ice flowed south from the Yellowstone ice cap, through Jackson Hole where ice was about 2000 feet (610 m) thick, and west into Idaho along the Snake River canyon (Lageson and Spearing, 1988). Valley glaciers flowed eastward from the Tetons and westward from the Gros Ventre Mountains on the east side of Jackson Hole into the large ice mass occupying Jackson Hole. The later Bull Lake Glaciation about 140,000 years ago was less extensive but covered the present townsite of Jackson with 1400 feet (427 m) of ice and extended well beyond (Good and Pierce, 1996). Late in Wisconsin during the Pinedale Glaciation (about 30,000–10,000 years ago) the last major episode of glaciation occurred, the one that would leave its mark most prominently on the present landscape. Ice covered the north end of Jackson Hole in the Jackson Lake area but reached only a short distance to the south to the vicinity of Burned Ridge (Fig. 10–13), which is part of the Pinedale *terminal moraine*. The glacier scoured the valley floor leaving a 386-foot-deep (118-m) hole that later filled with water to become Jackson Lake. The lake was enlarged in 1906 when a small irrigation–flood control dam was built, raising the lake level by 39 feet (12 m) and deepening the lake to 425 feet (130 m).

Farther south, valley glaciers from the Tetons barely extended into Jackson Hole as small piedmont glaciers. Ice gouged the floor of Jackson Hole along its west edge and deposited distinct terminal moraines that today enclose Leigh, Jenny, Bradley, Taggart, and Phelps lakes. These moraines contain *rock flour* produced by rubbing and scraping of rock dragged and carried in the base of the moving glacier. The fine sediment retains water and permits a dense growth of lodgepole pines to occur. In contrast, Jackson Hole is floored by coarse gravel that drains rapidly, causing drier conditions and permitting growth only of sage and some grass.

Meltwater washed huge volumes of glacially fragmented rocks from the glacial front and formed *outwash* plains and *valley trains* (outwash deposits confined by valley walls). During ice retreat and nonglacial times streams cut through the outwash leaving an impressive sequence of *stream terraces* along the Snake River (Fig. 10–12). The higher terraces formed first—lower terraces correlate with younger glacial events.

Worldwide warming beginning about 12,000 years ago spelled the beginning of the end for this most recent glacial episode. Ice thinned and glaciers "retreated." Recall that retreat simply means that ice melted back up glacial valleys faster than the ice moved downvalley. An analogous phenomenon occurs to one's checkbook when money is withdrawn from the account at a faster rate than it is deposited!

The large piedmont glacier in northern Jackson Hole retreated from the Burned Ridge moraine to the position of the recessional moraine around Jackson Lake. Ice masses that detached during retreat were buried by outwash sediments. As these buried ice blocks melted, depressions, or *kettle holes,* formed in the sediment surface, many of which now contain water. Locally this area is known as the Potholes, a descriptive but geologically inaccurate name. Technically, potholes are depressions in bedrock formed by the swirling action of gravel-laden waters. A few kettle holes can be seen from the Teton Park Road south of Jackson Lake; all are visible from the top of Signal Mountain, an excellent viewpoint in the park.

The warm times of the Holocene began to reverse about 5000 years ago (Fig. 1–25) and ice and snow became more persistent in higher mountainous areas of the world. Small cirque glaciers reformed in the abandoned cirques, and some enlarged and developed into valley or mountain glaciers. About 12 of these small glaciers still remain in spite of considerable melting since global warming began in the late 1800s. Falling Ice Glacier on the front face of Mt. Moran is in a re-entrant where the black dike has been eroded back of the mountain face (Fig. 10–16). Teton Glacier, about 3500 feet (1067 m) long in 1970, sits in the shade of Grand Teton (Fig. 10–17). Although small, they are true glaciers and not merely icefields.

FIGURE 10–16   Telephoto view of Mt. Moran from Jackson Hole on an early autumn morning. Diabase dike (1400 Ma) is overlain by light-colored Cambrian-age (about 550 Ma) Flathead Sandstone along a major unconformity. Falling Ice Glacier lies in a cirque cut along the dike. (Photo by E. Kiver)

Tetons: Mountains on the Move

Much of the movement along the Teton Fault has occurred within the past few million years—there is no reason to believe that the process has stopped. On some days sensitive seismographs detect hundreds of microearthquakes as small adjustments between adjacent fault blocks occur. Evidence of recent movement is abundant. Fault scarps locally stand in bold relief (Fig. 10–18)—some of which cut through recently formed alluvial fans! Others displace Pinedale moraine (about 14,000 years old) surfaces as much as 115 feet (35 m)! Smith and others (1993)

FIGURE 10–17   Telephoto of Teton Glacier in its sun-protected location between Grand Teton and Mount Owen. Note that the glacier has retreated from its Little Ice Age moraine (arrow) that formed about 1870. (Photo by E. Kiver)

describe radiocarbon-dated sediments offset by fault movements and estimate that large, scarp-forming earthquakes of magnitudes up to 7.2 occur every 1600–6000 years.

As a result of the continued westward tilting of the Jackson Hole block, flood control engineers have a real battle keeping the Snake River from shifting westward to a lower route at the base of the mountains (Love and Reed, 1971). Rivers usually occupy the lowest elevations in a valley—here the marshy ground just north of the town of Jackson and some distance from the Snake River is along the Teton Fault and is the lowest elevation in the valley. The ground slopes *up* from the marsh to the Snake River!

FIGURE 10–18   Recent fault scarp (arrow) along Teton front cuts across an alluvial fan that is at most a few hundred years old. (Photo by E. Kiver)

The Teton Fault is considered by geologists to be overdue for reactivation. A displacement of 2 or more feet (0.6 m) along the fault could occur at any time. Unfortunately, scientists so far have not discovered a reliable way to predict exactly when a catastrophic movement will occur.

Gros Ventre Rockslide

On the east side of Jackson Hole are the Laramide-age Gros Ventre (pronounced "grow vaunt") mountains. A combination of factors produced a huge landslide here in 1925. Paleozoic rocks on Sheep Mountain along the Gros Ventre River are tilted toward the river. The river had undercut the slope, reducing support for the valley wall. Snowmelt and rain had saturated the sedimentary rocks, including a shale layer beneath the Tensleep Sandstone. At 4:20 p.m. on June 23, 1925, a small earthquake caused the sandstone on the valley wall to detach and begin to toboggan down on the slippery shale layer into the valley below. Local rancher Guil Huff observed the less-than-3-minute-long spectacle—from atop a galloping horse that barely got out of the way of 50 million cubic yards (46 million m³) of crashing slabs and boulders (Lageson and Spearing, 1988)!

Geologists and engineers were to learn yet another lesson from the landslide. A large, 5-mile-long (8-km) lake formed behind the debris dam. Water topping the dam 2 years later in 1927 eroded rapidly and sent a wall of water downvalley, devastating the small town of Kelly. Six people and hundreds of livestock lost their lives. A side road a few miles south of park headquarters leads to the impressive rockpile that was once the valley wall.

Visiting Grand Teton

Most visitors are content to view the story-book mountain range from the comfort of their automobiles. Views from the roads, Signal Mountain, and other viewpoints are magnificent. For some, the challenge of the cliffs and peaks is overwhelming, and they must conquer one or more of the peaks. This is a mountain climber's Mecca where the peaks are easily accessible and tough Precambrian rocks will forgive minor mistakes.

There are many hiking trails that give the park visitor the opportunity to examine at close range the handiwork of the glaciers. Cascade Canyon Trail takes you up a classic glaciated canyon, past Hidden Falls where the water tumbles out of a hanging valley, into the land of spires, horns, and aretes. Along the way, look back from a vantage point and admire little Jenny Lake and the artistically arcuate moraine that holds back the water. The scar of the massive Gros Ventre rockslide is visible across Jackson Hole from many of the high trails in the park. Then, be up on top of Signal Mountain a little before sunrise; the sun on the very tip of Grand Teton and then suddenly on the whole face of the Tetons makes this first-light safari worthwhile. And, if you are there in the fall, you will hear the "rusty-gate" bugling of the bull elk in the trees below you!

Grand Teton National Park is small. It has no spouting geysers and it has no Grand Canyon. But as Love and Reed (1971) have said, few places "can rival the breath-taking alpine grandeur of the eastern front of the Tetons."

REFERENCES

Good, J.D., and Pierce, K.L., 1996, Interpreting the landscapes of Grand Teton and Yellowstone National Parks: Grand Teton Natural History Association, 58 p.

Lageson, D., and Spearing, D., 1988, Roadside geology of Wyoming: Mountain Press, Missoula, MT, 271 p.

Love, J.D., and Reed, J.C., 1971, Creation of the Teton landscape: Moose, Wyoming, Grand Teton Natural History Association.

National Park Service, 1984, Grand Teton, National Park Service Handbook 122, 95 p.

Reed, J.C., and Zartman, R.E., 1973, Geochronology of Precambrian rocks of the Teton Range, Wyoming: Geological Society of America Bulletin, v. 84, p. 561–582.

Richmond, G.M., 1986, Stratigraphy and chronology of glaciations in Yellowstone National Park, in Sibrava, V., Bowen, D.Q., and Richmond, G.M., eds., Quaternary glaciations in the northern hemisphere: Pergamon, New York, p. 83–98.

Smith, R.B., Byrd, J.O.D., and Susong, D.D., 1993, The Teton fault, Wyoming: seismotectonics, Quaternary history, and earthquake hazards, in Snoke, A.W., Steidtmann, J.R., and Roberts, S.M., eds., Geology of Wyoming: Geological Survey of Wyoming Memoir No. 5, p. 628–667.

Park Address

Grand Teton National Park
P.O. Drawer 170
Moose, WY 83012

DINOSAUR NATIONAL MONUMENT
(COLORADO AND UTAH)

Dinosaur National Monument straddles the Utah–Colorado border in the rugged Uinta Mountains—an area where the geology and its resultant rugged topographic expression remain a barrier to roads and development. Dinosaur is really two parks in one: the main park that most people see with the world-famous quarry (Fig. 10–19) where thousands of dinosaur bones were entombed 150 Ma during the Late Jurassic Period, and the major part of the park where the spectacular canyons of the Green and Yampa rivers expose folded and faulted rocks formed during the past 2 billion years. The strata here are the same as the flat-lying formations exposed in the Colorado Plateaus immediately to the south. Here in the Uinta Mountains the Laramide Orogeny has deformed the layers into a broad, east–west trending anticline (Plate 10) through which the Green and Yampa rivers have sliced. The rivers seem to defy logic by uniting in the *middle* of a mountain range and by flowing *across* the mountain structure rather than taking the seemingly easy path around the topographically lower east end of the Uinta Mountains—an anomalous condition that requires a special explanation.

FIGURE 10–19   Dinosaur bones in the Morrison Formation. (Photo by National Park Service)

Although most of the 325-square-mile (842 km²) monument is unrelated to the dinosaur story, the quarry is the main attraction that draws over a thousand people each day during the summer season. It is one of the eight national park units established primarily for the protection of significant fossil resources—an important aspect of earth history that belongs to all citizens of the world—and a logical inclusion in a system that preserves important parts of our heritage.

Park History

Paleo-Indians (Archaic culture) hunted big game in the Dinosaur area at least 7000 and perhaps 10,000 years ago. The later Fremont culture appears about 1800 years ago. In addition to hunting and gathering activities, they grew corn, beans, and squash—until the droughts and crop failures about 700 years ago drove them from the area. They left evidence of their presence in rock shelters and numerous other sites in today's monument. They also left spectacular pictographs (rock paintings) and petroglyphs (forms scratched and chipped into the desert-varnish coatings on rock surfaces) on rock walls throughout the park. Unfortunately, the true significance of this fascinating art work will never be known with certainty. Were they some form of writing, religious art, territorial boundary markers, or merely interesting doodles?

The Shoshoni reoccupied the area a few hundred years ago following a hunting-and-gathering life-style. Spanish padres Silvestre Escalante and Francisco Dominquez camped along the Green River near the present monument in 1776 on their historic quest to find an overland route connecting the missions of Santa Fe with those in California. Mountain men in search of beaver soon arrived and in 1825 William Ashley and a small group of trappers became the first known individuals to boat through the dangerous canyons of the Green River—a feat later duplicated by John Wesley Powell in 1869. Powell and his small group continued down the Green River to the Colorado River. They became the first to conquer both the canyons in the Green and Colorado rivers. The loss of one boat with valuable supplies to a large boulder in Disaster Rapids in the Uintas made the remainder of their journey extremely difficult. Months later they emerged, half-starved, from the Grand Canyon.

Settlement in the rough Uinta country came slowly, as a few pioneer families found isolated patches of bottom land to farm. Pat Lynch, a Civil War veteran, became the hermit of Echo Park where he lived alone for nearly 40 years talking to himself, "spirits," and a mountain lion who shared his domain (Hagood and West, 1994). Butch Cassidy (and the Sundance Kid) and other outlaws occasionally used the area on their way south to the Robbers Roost country in southern Utah.

The search for the remains of the monstrous beasts called dinosaurs first captured the interest of American paleontologists and the general public in the late 1800s. The dinosaur hunting grounds were in the West where rocks of the proper age (Mesozoic) and paleoenvironments (ancient swamps and streams) were abundant.

Earl Douglass of the Carnegie Museum in Pittsburgh was one of these avid bone hunters. In 1909 he discovered "Dinosaur Ledge," a bone bonanza and one of the richest dinosaur deposits ever discovered. Douglass lived among the dinosaurs for a number of years while the Carnegie Museum, and later other museums and the University of Utah, further excavated the fabulous bone record. The isolated sand layer containing the bone bonanza is part of the famous Morrison Formation of the Rocky Mountain region—a rock unit that has produced other significant dinosaur quarries. Twenty four nearly complete skeletons were excavated as well as partial skeletons of 300 other individuals.

Still the bone layer continued, prompting Douglass to recommend that the area be set aside as a park or monument—a suggestion that was followed when Woodrow Wilson established an 80-acre (0.13-square miles; 0.2-km^2) national monument in 1915. The monument was greatly expanded in 1938 to include the rugged canyons of the Green and Yampa rivers. Douglass had also visualized that the bone could be exposed on the quarry face and become part of the wall of a large building to protect the fossils and for the public to view "one of the most astounding and instructional sights imaginable." His dream came true in the 1950s when the present-day visitor center was completed (Fig. 10–20). Today, over 2000 bones stand in relief on the quarry face—a sight never to be forgotten by those who see it.

In the 1950s water developers set their sights on the Green River—exerting strong political pressure to flood the canyons in the park. Public outcries finally caused Congress to eliminate possible dams within the park. A tragic compromise was reached when opposition to a dam at Glen Canyon on the Colorado River was dropped by environmental groups in return for protecting the Green River through the Uinta Mountains. Thus the well-known canyons of the Green were saved and Glen Canyon, a little-known gem, was lost forever to "progress," unless society decides to right some of the environmental wrongs of the past.

FIGURE 10–20   Visitor Center at Dinosaur National Monument is built against the rock wall of the historic bone quarry. (Photo by D. Harris)

Geologic History

The Laramide Orogeny here produced a 160-mile-long (258-km) range of folded and faulted mountains—the Uintas. Its east–west orientation among the north–south trending western mountains is unusual. The range occupies an ancient crustal weakness established during Precambrian times (Hansen, 1969; Stokes, 1987). Northward movement or rotation of the crustal block containing the Colorado Plateaus during early Tertiary times (Miller and others, 1992) compressed and uplifted the current range. The major structure here is a large asymmetrical anticline broken by several major thrust faults (Fig. 10–21*a*) that parallel the trend of the range—a structural style prominent in many of the ranges in the Middle and Southern Rocky Mountains. Several smaller anticlines and synclines, including Split Mountain near the dinosaur quarry, formed on the south flank. The range is highest in its western part, and the anticline decreases in amplitude to the east (Hansen, 1969).

Erosion breached the anticline, exposing the billion-year-old Uinta Mountain Group *metasediments* (sedimentary rocks that are slightly metamorphosed) in the core of the range. These mud, sand, and gravel sediments are similar in age and origin to the late Precambrian (Proterozoic) deposits found in Glacier National Park and in Grand Canyon. Much to the disappointment of the mining industry, no major igneous activity with its associated mineralization occurred in the Uinta area.

Paleozoic and Mesozoic strata are laid out like pages of a book on the flanks of the eroded anticline and are especially visible in the walls of the Green and Yampa river canyons. A special rock layer in this sequence is the Jurassic-age (about 150 Ma) Morrison Formation. Rivers flowing eastward from rising mountains to the west deposited sands and muds on the Morrison coastal plain. Dinosaurs lived and died here, their bones occasionally concentrating in areas such as that at "Dinosaur Ledge." The seas returned during the Cretaceous, later retreating eastward as the crust in the Rocky Mountain area began to experience the forces of the Laramide Orogeny.

Major John Wesley Powell was the first geologist to see this outdoor geological museum as he was fighting the rapids in Lodore and other canyons of the Green River in 1869. He and his party survived the rapids and reached Steamboat Rock, a huge rib of rock that rears up high at the junction of the Yampa and the Green (Fig. 10–22). Finally they emerged from Split Rock Canyon where the Green leaves the mountains. Powell and his group (1875) had seen the inside of an anticline; the same strata that they had left behind as they entered the canyon reappeared as they came out at the lower end, but here the beds were dipping in the opposite direction.

Riddle of the Canyons

Powell recognized that the course of the Green River across a mountain range was unusual and required a special explanation. He reasoned that the river had been there first and that the anticline had risen up from below athwart the course of the

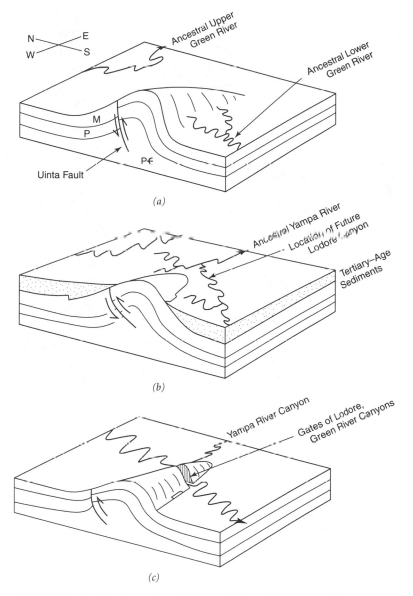

FIGURE 10–21   Generalized sequence of diagrams illustrating superposition as a mechanism to account for the transverse valleys in the Uinta Mountain (Dinosaur National Monument) area. (*a*) Uinta anticline forms during Laramide Orogeny (about 60 Ma). (*b*) Sediments shed from higher elevations bury lower parts of Uinta Mountains. Streams acquire new paths over buried mountains during middle to late Tertiary time. (*c*) Sediment fill removed by erosion, streams lowered onto buried mountain structures during Pliocene-Pleistocene time, and rising Continental Divide to the northeast diverts upper Green River to a southerly course.

FIGURE 10–22   Steamboat Rock, a monolith at the confluence of the Yampa and Green rivers, Dinosaur National Monument. (National Park Service photo)

stream. The uplift was gradual, he said, and the river had cut down through the uprising rocks and maintained its course. He coined the term *antecedent* for this type of stream, a term that was immediately accepted. At that time no detailed geologic work had been done in the Rocky Mountain region, and therefore he did not know when the Uinta Mountain anticline had developed. Much later it was determined that it is a Laramide structure and that it had been there before the Green and Yampa rivers began to cut their present canyons. Therefore this segment of the Green River is *not* an antecedent stream but more likely a *superposed* stream (Hansen, 1986) that had cut down through flat-lying Tertiary strata that had buried the old structure, the anticline (Fig. 10–21). Even though Powell was wrong about the Green River, he contributed a truly important concept that has been applicable in other areas.

Drainage changes also occurred by the process of *stream piracy* where a headward-eroding stream captures or diverts the drainage of an adjacent stream. The ancestral upper Green River likely flowed eastward into the Platte River and ultimately the Mississippi River in mid-Tertiary time. Uplift along the Continental Divide combined with superposition, and stream piracy helped divert drainage from the upper Green to the lower Green River (Fig. 10–21). The Green River now had a shorter course to a lower *base level* in the Colorado River drainage and more water coursing through its channel. Erosion increased and the canyons began to deepen.

The Yampa River became trapped in an east–west trending, grabenlike valley (Browns Park) that formed after the compressional forces of the Laramide Orogeny ceased. The Yampa initially flowed eastward (Fig. 10–21*b*) but, like the Green River, the rising Continental Divide to the east restricted its flow, and the Yampa was diverted westward (Fig. 10–21*c*). The graben was filled to overflowing with sediments of the Browns Park Formation, and much of the lower elevations of the Uintas was buried by debris shed from the highlands. The Yampa acquired a sinuous, meandering course (superposition) on this sediment fill, a course that was later maintained after the Green River found its way across the Uinta crest and began to cut Lodore Canyon. The Yampa then flowed westward and was able to maintain its meandering pattern as it too cut a deep canyon— thereby creating today's unusual *entrenched meander* stream pattern.

Dinosaurs—Who Are They?

Primitive reptiles first appear in the rock record during Late Pennsylvanian time, about 300 Ma (Fig. 10–23). Although primitive compared to later reptiles, the ability to lay eggs in a protective shell was a major biologic revolution. Reptiles, unlike their amphibian ancestors, could roam widely over the terrestrial landscape and need not return to a swamp or water environment to lay eggs. This innovation permitted reptiles to expand rapidly as new variations led to new reptilian groups. The dinosaur line first appeared in Late Triassic, about 210 Ma, and a few million years before the mammal-like reptiles gave rise to the first mammals (Fig. 10–24).

The dinosaur line had the advantage and quickly dominated the terrestrial world for an incredible 150 million years! Based on the structure of the pelvic bones, two lines of dinosaurs are recognized (Fig. 10–23). The sauropods (birdlike pelvis) are represented in the quarry at Dinosaur by the giant plant eaters like *Apatosaurus* (*Brontosaurus*), *Diplodocus,* and *Camarasaurus*. Ornithiscian (lizardlike pelvis) dinosaurs like *Stegosaurus* with its two rows of bony back plates and bony spikes at the end of its tail (Fig. 10–25) are also found in the scrambled bone deposit in the quarry. Although most of the remains are of herbivores, carnivores such as *Allosaurus* and *Ceratosaurus* (cousins of *Tyrannosaurus*) are also present. Flat-crested teeth ideal for grinding vegetation distinguishes herbivores; a mouthful of recurved, serrated teeth resembling rows of steak knives and up to 1-foot long (30 cm) characterizes the carnivores!

How is it that so many bones are concentrated at Dinosaur Ledge? In this one layer of sandstone in the Morrison Formation, skeletons of dinosaurs, both large and small, occur in numbers yet unknown, because the supply is still good. This

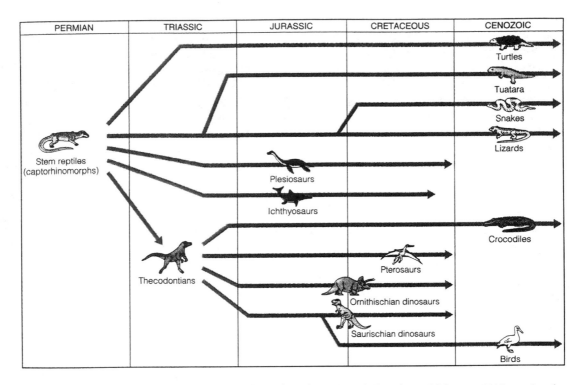

FIGURE 10–23   Major reptile groups and birds through geologic time. (Wicander and Monroe, 1993; reprinted by permission of Wadsworth Publishing Company)

baffled the early workers; it appeared that in this one area there had been an unbelievable concentration of dinosaurs during the Late Jurassic. No area could have had the carrying capacity to support such a large number of voracious eaters! In time, it became evident that this was a large sand bar, or perhaps a delta, and many dinosaurs, caught in floods, had been rafted or washed in from other areas. This then was the burial ground for the entire drainage area—a virtual sand-bar cemetery. Carrion eaters feasted on the dinosaur delectables delivered to their doorstep. Currents also helped disperse the skeletal debris into the jumbled form apparent on the quarry face today (Fig. 10–19).

Intriguing discoveries and new ideas about dinosaurs have surfaced in recent years. Unusual to most scientific studies, the general public has followed many of these discoveries and the scientific debates as these new ideas are tested. Acceptance of new ideas is usually slow in science as evidence is brought forth for and against. Often, the evidence is not overwhelming and different degrees of acceptance result. The status of new ideas (postulates) can be raised to hypotheses and perhaps theories as supporting evidence increases.

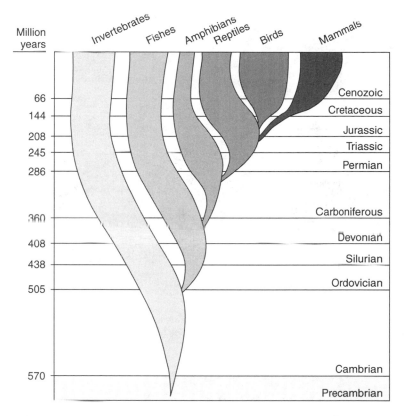

FIGURE 10–24   First appearances of major animal groups through geologic time. (Wicander and Monroe, 1993; reprinted by permission of Wadsworth Publishing Company)

Dinosaur size has long been recognized as ranging from the size of an overfed chicken to the 80-foot-long (27-m) *Brachiosaurus*. However, recent discoveries of partial skeletons in Colorado, New Mexico, and elsewhere indicate that even larger forms, nicknamed "Supersaurus" and "Ultrasaurus," existed. Similar but larger bones than those of the *Brachiosaurus* require a "scaling up" of animal size. Such animals may have weighed in at 150 tons, by far the largest land animals to ever have walked the face of the earth. By comparison, the average African elephant weighs a mere 6.6 tons! Thus, some dinosaurs were the equivalent of nearly 23 elephant units!

Because all living reptiles are cold blooded, it was long assumed that dinosaurs were the same. Evidence from bone structures in *Stegosaurus* and other forms, as well as the small percentage of carnivores to herbivores (2–3 percent) in the fossil record, suggest that at least some of the dinosaurs were warm blooded. If the

FIGURE 10–25   Life-size reconstruction of *Stegosaurus* located in front of the visitor center at Dinosaur National Monument. As is true in all national parks, no feeding of wildlife is permitted! (Photo by E. Kiver)

carnivores were cold blooded (therefore not required to eat as often), their numbers could reach 40 percent of the total population. Thus many scientists argue that at least some of the dinosaurs were warm blooded.

Social behavior of the dinosaurs is also of interest. Where Mesozoic-age muds were of the right consistency, dinosaur footprints or trackways are often preserved. The tracks are often going in the same direction, suggesting that at least some of the dinosaurs traveled in herds, probably for protection.

The discovery in Montana of nests of at least three different dinosaur species by paleontologist John Horner indicates that some dinosaurs nested in colonies, much like penguins and certain other bird species. Some, like the duck-billed dinosaur (genus *Maiasaura*, meaning "good mother lizard"), kept their young in the nest where they received parental care and protection for some time after hatching (Wicander and Monroe, 1993).

Dinosaurs—Where Did They Go?

Other than the amazing fact that such a unique, interesting group called dinosaurs once existed, perhaps the most intriguing aspect of their history is their sudden disappearance from the face of the earth 65 Ma, at the Cretaceous-Tertiary (K–T) time boundary. Dinosaur extinction is of interest not only to biologists and paleontologists, but also to those who thoughtfully contemplate today's rapid extinctions—about 100 species a day disappearing from the roster of earth's inhabitants. The rock record is clear—*all* species of life eventually become extinct. Some extinctions involve slow decreases in numbers over tens of millions of years—others, such as that of the dinosaurs, appear to be extremely rapid. Why is it that after 150 million years of dominance the most successful group of animals to ever rule the planet's land areas should suddenly disappear?

Science deals with such problems by using a method that considers all possible explanations that have been proposed—the method of multiple-working hypotheses. Evidence accumulates, and one or more ideas eventually become more acceptable to the scientific community. Should new, contradictory evidence surface, old ideas will either be modified or thrown out. Such is the way of science.

Extinction of species and larger groups is continuous through geologic time. However, some time intervals have rates of extinction much greater than the average. These are episodes of *mass extinctions* where in a relatively short time an unusually large number of species are lost forever to the biological kingdom. The most serious mass extinction to affect the earth occurred at the end of the Paleozoic Era (about 240 Ma) when paleontologists estimate that 96 percent of all living species disappeared! This was the closest that life has come to complete extermination since its origin some 3.5 billion years ago.

The mass extinction at the end of the Mesozoic was also very severe, involving perhaps 75 percent of all living species, including the dinosaurs, marine reptiles, flying reptiles, and various groups of marine invertebrates. More chilling is the present human-caused episode of mass extinction—promising to be the most severe in the geologic record! Although many of the 100 species/day currently going extinct are in the tropical rain forests that are being cut at the rate of 32 acres every minute, extinctions are occurring elsewhere in the world—also mostly due to a of loss of environment or living space as human population and development grow like an out-of-control cancer. The possibility of eliminating key components of the ecosystem and causing major food-chain collapses is a frightening scenario—one that should not be taken lightly.

What caused the demise of the dinosaurs? We may never know the cause or causes for certain, but we do know that great environmental upheavals were occurring during the Mesozoic and that large numbers of animal and plant groups did not survive the environmental stresses. The appearance and growing importance of a new plant group—the flowering plants (angiosperms)—required considerable adaptation to the new food source. Seaways were reducing in extent, and climates were becoming more varied and rigorous as mountain ranges grew in extent and elevation in many areas of the world. Huge eruptions of flood basalt occurred in India, sending large volumes of climate-altering gases into the atmosphere. Small, scrawny mammals had persisted throughout the Mesozoic as a relatively unimportant group. Perhaps they decided that dinosaur eggs were their favorite food; they were abundant by that time, and were probably clever and conniving!

However, one of the more recent theories is far more spectacular. It involves the splash-down of a giant asteroid or meteor near the end of the Cretaceous period (Alvarez and others, 1980). The dust, pulverized rock debris, and molten rock fragments were thrown high into the air. Airborne debris may have been thick enough to prevent sunlight from reaching the earth. A giant firestorm would almost instantaneously have ignited entire forests—further sending many tons of soot into the atmosphere. With the sudden blackness, photosynthesis was quenched and the food supply drastically diminished for months and perhaps years.

That a large meteor impacted the earth 65 Ma has been established; a thin layer of sediment at the Cretaceous-Tertiary contact contains an abundance of iridium, an element generally rare except in meteorites. Numerous scientific papers have appeared on the subject; the collection of articles in Ryder and others (1996) is a good place to start for those wishing to delve further. Small quartz grains with microfractures (shocked quartz) in associated deposits and small glass spheres splattered around the globe (*tektites*) due to melting of rock at the impact site further support the asteroid hypothesis. Ground zero has now been identified, a 110-mile-diameter (180-km), 65-million-year-old crustal scar—the Chicxulub Crater off the Mexican coast by the Yucatan peninsula. A six-mile-diameter (10-km) rock going at 20,000–45,000 miles per hour would do the trick—releasing unimaginable amounts of energy on impact.

Similar size objects still race through space—some occasionally entering our solar system where they move harmlessly through—most of the time. Smaller objects such as the meteor that struck Arizona about 35,000 years ago and formed a mile-wide crater (Meteor or Barringer Crater near Flagstaff) about 800 feet (244 m) deep. The large comet that exploded over Siberia in 1908 incinerated thousands of reindeer and flattened and scorched hundreds of square miles of forest. The world watched in awe as the Shoemaker-Levy comet impacted the planet Jupiter in 1994. Similar and even larger events will occur in the future. The "when" is unknown!

Whether the impact was the main cause of the K–T extinctions or one of many factors is still being debated. However, the coincidence in timing and the magnitude of the event has moved the impact hypothesis to the top of the plausible explanation lists of many scientists.

Are the Dinosaurs Really Gone?

Paleontologists identify fossil species by similarities in bone structure—the only animal parts that are usually preserved. A particular type of fossil originally found in Germany has a definite reptilian tailbone, a mouthful of teeth, sharp claws on the forelimbs, and the long legs of a running animal—characteristics that place the animal into the dinosaur family. When the impressions of feathers were found in association with the bones, *Archaeopteryx* was immediately recognized as a link between reptiles and birds! A number of specimens, including specimens from China with primitive protofeathers have been discovered (Ackerman, 1998), leaving little doubt that the birds in your bird feeder—or the chicken in the basket from your favorite fast-food restaurant—are really feathered dinosaurs!

Should we be sad to lose such an incredible group of animals? Perhaps not. Elimination of the ruling land vertebrates opened the door for another group to take over—the mammals—some of which are now reading these pages!

REFERENCES

Ackerman, J., 1998, Dinosaurs take wing: National Geographic, v. 194, July, p. 74–99.

Alvarez, L.W., Alvarez, W., Asaro, F., and Michel, H.V., 1980, Extraterrestrial cause for the Cretaceous/Tertiary extinctions: Science, v. 208, p. 1095–1108.

Hagood, A., and West, L., 1994, Diosaur: the story behind the scenery: KC Publications, Las Vegas, NV, 48 p.

Hansen, W.R., 1969, The geologic story of the Uinta Mountains: U.S. Geological Survey Bulletin 1291, 144 p.

Hansen, W.R., 1986, Neogene tectonics and geomorphology of the eastern Uinta Mountains in Utah, Colorado, and Wyoming: U.S. Geological Survey Professional Paper 1356.

Miller, S.M., Nilsen, T.H., and Bilodeau, W.L., 1992, Late Cretaceous to early Eocene geologic evolution of the U.S. Cordillera, in Burchfiel, B.C., Lipman, P.W., and Zoback, M.L., eds., The Cordilleran orogen: conterminous U.S., Boulder, CO, Geological Society of America, The Geology of North America, v. G-3, p. 205–260.

Ryder, G., Fastovsky, D., and Gartner, S., eds., 1996, The Cretaceous-Tertiary event and other catastrophes in earth history: Geological Society of America, Special Paper 307, 569 p.

Stokes, W.L., 1987, Geology of Utah: Salt Lake City, Utah Museum of Natural History, Univ. of Utah, Occasional Paper No. 6, 280 p.

Wicander, R., and Monroe, J.S., 1993, Historical geology: West Publishing, St. Paul, MN, 640 p.

Park Address

Dinosaur National Monument
P.O. Box 210
Dinosaur, CO 81610

TIMPANOGOS CAVE NATIONAL MONUMENT (UTAH)

Nestled against the steep south wall of the American Fork Canyon in the Wasatch Range of north-central Utah lies a small (5600 feet long; 1707 m), well-decorated limestone cave complex—Timpanogos Cave. The cave is actually three caves connected into one by man-made tunnels that were excavated in the 1930s. The name is an anglicized version of the Ute Indian word that describes themselves (Horrocks,

1994). The Utes hunted in the area for centuries but apparently did not enter the cave.

Woodcutter Martin Hansen tracked a cougar to the westernmost cave in 1887. Noticing a number of bones in the entrance, he decided wisely that entering the cave with only his axe for a weapon and no light could have unpleasant consequences! Hansen later explored the cave and began giving tours for a small fee until 1891. A group of miners removed at least two freight-car loads of flowstone in 1892–1893 while Hansen was absent. The plan was to use the flowstone as decorative onyx. The quality was poor and the operation was unprofitable. However, major damage had been done to Hansen Cave. Other damage was done by early visitors who felt compelled to break off mineral souvenirs.

A second cave was discovered nearby in 1914 and rediscovered in 1921 by a local hiking club. This cave would later be named Timpanogos by local Forest Service officials. A third cave, Middle Cave located between Hansen and Timpanogos, was discovered in 1922. Quick action by Forest Service officials resulted in cave protection and national monument status in 1922. Transfer to the National Park Service occurred in 1933.

Geologic History

The Wasatch Mountains are located along a persistent north–south trending structural feature called the Wasatch Line. This ancient (Precambrian) feature has significantly influenced the geology in the area for over 800 million years (Stokes, 1987). Splitting of the ancient continent about 1 billion years ago along the Wasatch Line and areas to the north and south caused sand and mud to accumulate from Arizona (Grand Canyon area) to Montana (Glacier Park area) and Canada. The Timpanogos visitor center at the bottom of American Fork Canyon is located on the Mutual Formation, equivalent in age to the Belt Rocks of Montana and the Grand Canyon Series in Arizona.

The hike from the visitor center (elevation 5665 feet; 1727 m) to the cave entrance not only takes the visitor up 1065 feet (325 m) to the cave entrance, it is also a journey through time as progressively younger strata are encountered (Fig. 10–26). Thick accumulations of lower Paleozoic sediments occurred in the deep geosyncline to the west of the Wasatch Line and thinner sediments accumulated to the east. The Cambrian-age Tintic Quartzite, Ophir Formation, and Maxfield Formation are laid out like pages in a book that can be examined as one hikes up the canyon wall to the cave. About 125 million years of history are missing between the Maxfield Formation and the overlying Missippian-age limestones. The trail ends at the Deseret Limestone where the cave and its entrance are located. The stack of sediments continues up the canyon wall to the top of Mount Timpanogos (11,750 feet; 3582 m), a glacial horn located about 3 miles (1.9 km) southeast of the park.

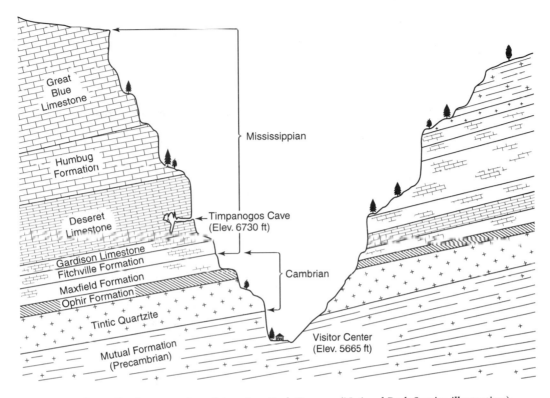

FIGURE 10–26 Generalized cross section of American Fork Canyon. (National Park Service illustration)

Older Mesozoic (200–80 Ma) mountain building involved extensive thrust faulting west of the Wasatch Line in western Utah and Nevada. Younger Laramide-age mountains to the east are mostly anticlinal uplifts with thrust-faulted flanks. The western extension of the east–west trending Uinta anticline intersects the Wasatch Line in the Timpinogos area. The tilted layers so evident in the American Fork Canyon near Timpanogos Cave (Fig. 10–26) are on the south flank of this anticline. A northeast-trending fault, the Timpinogos Fault, also formed during the Laramide Orogeny and would eventually play a major role in cave formation.

About 40 Ma the vast area between the Wasatch Line and the east edge of the Sierra Nevada began to be pulled apart, forming the hundreds of ranges in the Basin and Range Province. Thousands of feet of vertical movement along the Wasatch Fault produced the spectacular mountain backdrop visible today behind Salt Lake City, Provo, and other Wasatch front cities. The Wasatch towers over 7100 feet (2165 m) above the Great Salt Lake Basin—one of the steepest mountain fronts on the earth (Horrocks and Tranel, 1994). The large fault movements of the recent geologic past are expected to continue. With spectacular scenery comes spectacular geology—in this case with potentially serious consequences to the 70 percent of Utah's population that lives within a few miles of the Wasatch Fault.

Cave Formation

Uplift of the Wasatch Mountains speeded up the slow but inevitable process of erosion, which eventually destroys all land above sea level. Westerly flowing streams dissected the Wasatch into a series of isolated peaks separated by deep canyons. Sometime before the American Fork Canyon was cut, slowly circulating groundwater charged with carbon dioxide from the air and soil attacked the highly fractured rock along the Timpanogos Fault and dissolved a series of caves in the Deseret Limestone. As the canyon deepened, undoubtedly helped by the continuing uplift of the Wasatch Mountains and the increased runoff from snowmelt during episodes of glaciation, the canyon acquired a V-shape. Those who hike the steep, 1.5-mile-long (1-km) trail from the visitor center will have magnificent views of a textbook example of a V-shaped, stream-eroded canyon.

Downcutting lowered the regional water table permitting air to enter the cavern. The chemical environment changed from one of solution to one in which deposition of calcite, aragonite, and other secondary minerals formed *speleothems*. The slow process of cave beautification was underway.

Mineral-charged water flowing down the cave walls and on the cave floor formed *flowstone*. Water dripping into the cave formed an abundance of *stalactites* that cling to the ceiling and *stalagmites* that nestle on the floor, often beneath a corresponding stalactite on the cave roof. The Heart of Timpanogos is a particularly spectacular heart-shaped stalactite complex about 6 feet (2 m) long (Fig. 10–27). *Helictites* are remarkable curved speleothems that defy gravity in their seemingly random patterns of growth. They and needlelike growths of aragonite called *frostwork* or radiating crystal sprays called *anthodites* (Hill and Forti, 1997) are unusually abundant in Timpanogos. Helictites often start as stalactites but later grow horizontally or even back toward the cave ceiling (Fig. 10–28). Apparently a small central canal enables water to move by capillary force in any direction. Hydrostatic pressure and the formation of wedge-shaped crystals may also influence growth patterns. Speleothem colors vary widely from the usual white and gray of relatively pure $CaCO_3$ to pink, yellow, and green as small impurities of iron, nickel, and other elements are included in the crystal structure.

Widening of the American Fork Valley eventually caused the canyon wall to intersect the cave and provide an entrance for Martin Hansen and other discoverers of the Timpanogos Cave system. Notice the remnants of the fault plane preserved on cave walls and remember to bring a light jacket or sweater to wear in the cool, 45–48°F year-round-temperature (7.2 to 8.8°C) cave atmosphere. Because the cave is small, the number of tours and the number of visitors each day is limited. Tickets are usually sold out each day during the tourist season (May 1 to October 31). Call ahead to purchase tickets.

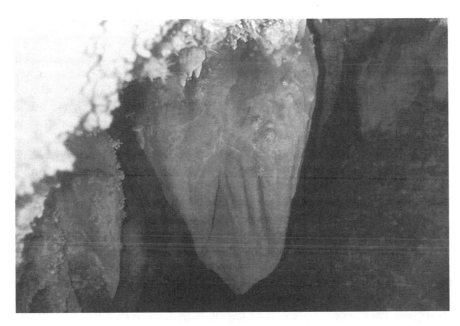

FIGURE 10–27   "Heart of Timpinogos," a major attraction and the largest stalactite in the cave—measuring 7-feet long (2-m) and 5 feet wide (1.5 m). (Photo by E. Kiver)

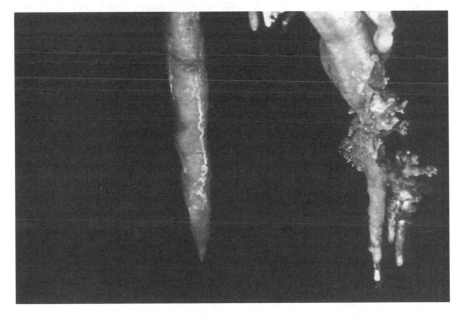

FIGURE 10–28   A stalactite-helictite speleothem in Timpanogos Cave. Field of view is about 18 inches (46 cm). (Photo by E. Kiver).

Cave "Improvements"

Early commercialization of Timpanogos Cave followed the pattern practiced in most tourist caves—change the cave to make it more convenient for visitors. A new entrance and exit were constructed and debris was piled in low areas to make the floor level for visitors to walk on. As a result, the cave meteorology changed drastically. Cave winds dried the cave out and large amounts of dust blew in and covered and dulled the once pristine speleothems. Speleothem development was greatly reduced or stopped as a result of these and other changes. Black powder was used during the 1920s to enlarge an entrance in order to install a protective door to prevent further vandalism. The powder instantly turned nearby walls and speleothems an unsightly black color.

Some of the reversible damage problems have a happy ending. The National Park Service and hundreds of volunteers began a program in 1991 to restore the cave as much as possible to its original condition. Double doors (airlocks) were installed and a bat-friendly gate now covers the original entrance. Humidity has again increased and speleothem growth has returned. Careful, tedious cleanup of many of the delicate speleothems required thousands of hours of labor by volunteers. For many people, volunteering to help short-handed Park Service staff take care of a park is impossible. However, we can take care of the parks when we visit them, and more importantly, we must elect officials who will adequately fund, staff, and legislate for the benefit of our heritage and environment.

Removal of 250 tons of debris dumped in one room of the cave during the early mining days and from commercialization work has also been undertaken by volunteers (Horrocks, 1995). Through the efforts of the volunteers and the Park Service this gem of a cave is slowly regaining much of its former glory. Unfortunately, the early damage by vandals and mining has robbed forever some of the cave's beauty. Millions of people, those living today and those yet unborn, will never see what was lost because of the ignorance or thoughtlessness of a few individuals. Fortunately, much still remains at Timpanogos to be seen and enjoyed.

REFERENCES

Hill, C., and Forti, P., 1997, Cave minerals of the world: Huntsville AL, National Speleological Society, 463 p.

Horrocks, R.D., 1994, The story of Timpanogos Cave: National Speleological Society News, January, p. 6–14.

Horrocks, R.D., 1995, Artificial fill removal project: Timpanogos Cave system, Timpanogos National Monument: National Speleological Society News, April, p. 102–107.

Horrocks, R.D., and Tranel, M.J., 1994, Timpanogos Cave resurvey project, 1991–1992: National Speleological Society News, January, p. 15–21.

Stokes, W.L., 1987, Geology of Utah: Salt Lake City, Utah Geological and Mineral Survey, 280 p.

Park Address

Timpanogos Cave National Monument
RR 3, Box 200
American Fork, UT 84003
Phone: 801-756-5238

ELEVEN

Southern Rocky Mountains

The Southern Rockies consist of a series of ranges, generally parallel, that extend southward from southeastern Wyoming across Colorado into northern New Mexico (Plate 11). The early explorers came across the flat rolling Great Plains to the east; hence the eastern side of the Rockies, known as the Front Range, was the first major range they encountered. It extends from near Canon City, in southern Colorado, northward into southeastern Wyoming, where the name Laramie Range is also used. Its southern extension is called the Sangre de Cristo Range in southern Colorado and northern New Mexico. Had the explorers come from the west, the Sawatch-Park ranges or San Juan Mountains might have been called the Front Range.

The northern province boundary is well-defined by an extension of the Great Plains called the Wyoming Basin, which divides the Southern Rockies from the ranges of the Middle Rockies. The fault-block ranges of the Basin and Range Province just south of Santa Fe mark its southern boundary.

Prominent geomorphic elements of the Southern Rockies are the "parks"— broad open areas that separate the Front Range or eastern ranges from the western ranges. The Laramie Basin in Wyoming, the North, Middle, and South parks in Colorado, and the San Luis Valley in southernmost Colorado are the principal ones. These are mostly aligned structural basins, downfolded and/or downfaulted blocks of the earth's crust that follow the trend of major faults. Northward movement of the Colorado Plateaus of as much as 37–74 miles (60–120 km) along this zone of weak crust may have occurred during the Laramide Orogeny (Miller and others, 1992). Post-Laramide [Oligocene, about 30 million years ago (Ma)] extension along this same trend formed the Rio Grande Rift from at least southern Colorado to central New Mexico—a major pull-apart zone characterized by numerous faults and volcanism (Christiansen and Yeats, 1992; Chronic, 1987). Even today, a shallow, or passive, hotspot underlies New Mexico where active faulting, recent volcanism, and magmatism continue along the rift (Luedke and Smith, 1991; Mutschler and others, 1998).

PLATE 11   The Southern Rockies. (Base map copyright Hammond Inc.)

West of the parks the main ranges are the Medicine Bow, Park, Mosquito, and Sawatch ranges in Wyoming and Colorado and the Jemez-Nacimiento Mountains in New Mexico. In southwestern Colorado, west of the San Luis Valley, the San Juan Mountains, composed mainly of Tertiary volcanics, occupy an irregularly shaped area.

The Southern Rockies are higher than the Middle or Northern Rocky Mountain provinces and form the roof of the continent. Going west of Denver (the "mile-high city") and other Front Range cities, everything goes up—to 2 or more miles (3.2 km) above sea level! Fifty-four peaks are more than 14,000 feet (4270 m) high, Mt. Elbert (14,431 feet; 4400 m) in the Sawatch Range being the highest. There are several prominent peaks in the Front Range; among them, from north to south, are Longs Peak (14,255 feet; 4346 m) in Rocky Mountain National Park, Mt. Evans (14,264 feet; 4348 m) west of Denver, and Pikes Peak (14,110 feet; 4300 m) west of Colorado Springs.

There are numerous vantage points from which to view Front Range scenery and geologic features—one at the top of the well-worn Longs Peak Trail. The trail was blazed by Major John Wesley Powell and his party in 1868. Major Long, early explorer for whom the peak was named, merely saw the peak and was satisfied, without climbing it. Air views can be particularly instructive—especially along the east edge of the Front Range (Fig. 11–1).

Plate Tectonics and Mountain Building

The Southern Rocky Mountains contain broad anticlinal uplifts with thrust faults on one or both flanks that formed during the Late Mesozoic–early Tertiary Laramide Orogeny. Downfolded and faulted blocks divided the province into a series of eastern and western mountains during mountain building. The Rio Grande Rift later followed this same trend, especially in the New Mexico area. Volcanism was widespread and particularly intense in the San Juan and Jemez-Nacimiento mountains in southern Colorado and northern New Mexico. Volcanism also occurred in the Never Summer Range in Rocky Mountain National Park and in the South Park area where there are extensive lava flows and welded tuffs. The prevailing westerlies carried fine particles of volcanic ash eastward over the plains; in the southern Colorado Front Range, ash buried many plants and animals in Florissant Fossil Beds National Monument.

The structural development of the Southern Rocky Mountains has long intrigued geologists. One particularly interesting element of its mysterious past is that the present range is the most recent of perhaps four ranges that once stood majestically at nearly the same location. Apparently crustal flaws or faults established during Precambrian mountain building and plate tectonic activity were reactivated both in late Paleozoic time (Pennsylvanian, about 300 Ma), forming the Ancestral Rocky Mountains, and again during the Laramide Orogeny about 75–40 Ma when the present range rose in nearly the same geographic location. A similar

FIGURE 11–1 Looking south along the Front Range to Pikes Peak, with Red Rocks State Park west of Denver in foreground. The lower area to the left, including the hogbacks, is the Colorado Piedmont, which is composed of sedimentary rocks, less resistant to erosion than the Precambrian metamorphic and igneous rocks of the core of the Front Range. Dissection of the Subsummit erosion surface has removed most of the flat and rolling surface but left ridges between valleys at similar elevations. (Photo by John Shelton)

structural history occurred in the adjacent Colorado Plateaus where Laramide folds and faults developed along similar reactivated Precambrian structures.

Also of interest is the location of the Southern Rocky Mountain ranges nearly 800 miles (1300 km) from the area of active plate collision (subduction zone) near the coast of North America. How is it that compressional forces were transferred this great distance to the continent interior during Laramide time?

Some suggest that rapid plate movements during Laramide time flattened or lessened the angle of subduction and permitted compressive stresses to be transmitted far inland from the plate edge (Miller and others, 1992). Another possibility involves a great thickening of crustal material in a subduction zone to the west. Subsequent collapse of this thick pile of crustal rocks produced eastward migrating compression that initially formed the Mesozoic mountains of California,

Nevada, western Utah, and finally the Rocky Mountains of Wyoming, Colorado, and New Mexico (Livacarri, 1991). Some ideas consider the Southern Rocky Mountains as a giant, elongate bulge, perhaps produced by the rise of hot, less dense mantle material (asthenosphere). Tension along the axis of the bulge produced extensional strain and the formation of the Rio Grande Rift and associated pull-apart valleys to the north. Eaton (1986) considers that the Southern Rocky Mountains are analagous to a midoceanic ridge and are produced by similar *thermotectonic* processes.

Whatever its origin, compression squeezed the crustal rocks and produced a peculiar type of broad, flat-topped anticlinal uplift with thrust faults along at least one of its flanks. In the process, one of the highest and most rugged ranges of North America was created.

Geologic Evolution

North of a major northeast-trending fault that cuts across the Laramie Range and Medicine Bow Mountains, and extends across the Great Plains to the south end of the Black Hills in South Dakota, Archeozoic-age Precambrian metamorphic rocks form the cores of the mountains. These ancient rocks are almost 3 billion years old—over half the age of the earth and the oldest rocks in western North America (Lageson and Spearing, 1988). These rocks are part of the Wyoming Craton, one of the nuclei of the growing Precambrian continent that ultimately became North America. Immediately south of the fault zone and extending through Colorado and New Mexico are Proterozoic metamorphic and igneous rocks that are about 1.6–1.8 billion years old—the result of later Precambrian geosyncline and mountain-building activity that added more crustal area to the growing continent. Some of the Precambrian rocks are still recognizable as sedimentary (Fig. 11–2), but most were recrystallized as schist and gneiss. Large batholiths such as the Sherman Batholith in the north and the Pikes Peak Batholith in the south and other smaller intrusions squeezed into and melted their way into the crystalline complex during later Precambrian time (about 1000–1400 Ma). Major faults produced during these Precambrian events would later be reactivated and play important roles in localizing both late Paleozoic and Laramide-age mountain building.

A long interval of erosion followed Precambrian mountain building. Many thousands of feet of rock were removed—thereby exposing the deeper crystalline mountain roots. Paleozoic seas crept across the area burying the ancient erosion surface with sediment and producing a major *unconformity*. Late Paleozoic mountain building (Pennsylvanian, about 300 Ma) formed the Ancestral Rockies as continents were assembling into the late Paleozoic Pangaea supercontinent. The Ancestral Rockies paralleled and coincided with some of the major faults in the crystalline basement rock. Thus, the location of mountain building is not necessarily determined by random processes, especially when pre-existing weaknesses

FIGURE 11–2 Proterozoic-age stromatolites produced by algae now lying in the upturned layers of metamorphosed dolomite in the Medicine Bow Mountains, southeastern Wyoming. Algae are similar to those found in Precambrian rocks at Glacier National Park and elsewhere. (Photo by E. Kiver)

occur in the basement rocks. Erosion accompanied uplift and removed the earlier Paleozoic strata from the mountain uplifts by early Mesozoic time—again exposing the Precambrian basement complex and erasing any topographic traces of the Ancestral Rockies.

The area remained above sea level well into the Jurassic (about 150 Ma), receiving abundant river, swamp, and dune deposits. Dinosaurs scampered about and retreated westward as the Rocky Mountain Geosyncline formed and seas in the interior of North America flooded westward. Thick marine sediments accumulated, especially during the Cretaceous. Also during the Mesozoic Era, subduction and other plate tectonic processes were squeezing western North America in a giant tectonic vise. Mountain ranges formed first in California and western Nevada. By Late Cretaceous time the Rocky Mountain area began to respond to the relentless forces of mountain building—producing the latest of several mountains to occupy the site. The episode of mountain building in the Rockies from Late Cretaceous (about 75 Ma) to about middle Eocene (about 40 Ma) is called the Laramide Orogeny—named from geologic evidence first recognized in the Laramie Basin in southeastern Wyoming.

Geomorphic Sequence

Laramide-age compression in the Southern and Middle Rockies caused many areas to bend upward into giant, flat-topped anticlinal ranges with large thrust faults on one or both of their flanks. The upturned edges of the late Paleozoic and Mesozoic strata outline many of the edges of the mountains (Fig. 11–1) as *hogbacks* (steeply inclined strata) and *cuestas* (gently inclined)—elsewhere thrust faults have slid Precambrian rocks over the top of the steeply inclined sedimentary layers thereby obscuring them from view. These early Tertiary mountains were only a few thousand feet above sea level and by Eocene time (about 35 Ma) were flattened to an erosion surface of low relief—a *peneplain* (Epis and Chapin, 1975). Richmond (1974) estimates that the surface stood perhaps 3500 feet (1100 m) above sea level with peaks and ridges standing prominently above the peneplain surface. Such peaks are called *monadnocks,* after Mt. Monadnock in New England. Much was yet to happen as remnants of this erosion surface now occur as much as 12,000 feet (3660 m) above sea level, with monadnocks such as Longs Peak at Rocky Mountain National Park rising still higher!

During the mid-Tertiary (Oligocene and Miocene, about 38–15 Ma), the Rockies were being arched upward, and tensional forces produced localized collapse, numerous normal faults, and localized decompression melting at depth. Volcanism was particularly severe in the San Juan Mountains of southwestern Colorado and spotty elsewhere in the Southern Rockies. Oligocene-age (about 28 Ma) ash flows from volcanoes in the Never Summer Range along the western edge of Rocky Mountain National Park rest on the Eocene erosion surface (Summit Surface) providing some chronological control of geologic and geomorphic events.

Volcanism continued intermittently through the remainder of the Cenozoic Era, especially during the past 5 million years in northern New Mexico along the Rio Grande Rift. The northeast–southwest trending Jemez Lineament cuts across northern New Mexico from Mt. Capulin volcano in the Great Plains to Sunset Crater in the Colorado Plateaus in east–central Arizona (Luedke and Smith, 1991). The Valles Caldera is located at the intersection of the Jemez Lineament and the Rio Grande Rift. Catastrophic eruptions of thick rhyolitic ash flows about 1.1 and 1.4 Ma from the Valles Caldera played a major role in the geology of Bandelier National Monument.

According to some interpretations, uplift during the late Miocene epoch initiated a second "cut-and-fill" fluvial cycle that continued well into the Pliocene (Mears, 1993). A second, lower erosion surface, the Subsummit Surface (Fig. 11–3), left remnants of the first peneplain standing above, as high benches. The interpretation of Rich (1938) and others that this younger surface was a pediplain—a dry-climate peneplain—is accepted here. In granite areas, notably in southern Wyoming and northernmost Colorado and in the Pikes Peak area, a thick *regolith* (a blanket of physically weathered, rocky granite debris) is indicative of a dry climate. Others accept only the formation of a single early Tertiary erosion surface and consider that subsequent offset along faults accounts for surfaces at different levels (Bradley, 1987: Epis and Chapin, 1975; Richmond, 1974).

Uplift near the end of the Pliocene brought the mountains up to their present height and initiated a third fluvial cycle, the one that is in progress today. The streams cut canyons—as much as half a mile deep—below the pediplain surface, which in many places was narrowed to mere ridges of accordant heights (Fig. 11–1).

It was not until this latest uplift occurred that the boundary between the Southern Rockies and the Great Plains was delineated. When uplifting began, the relatively smooth pediplain surface extended eastward from well within the mountains across part of the area that would become known as the Great Plains Province. The evidence is clear; in southernmost Wyoming there is a narrow remnant of the pediplain surface—the Gangplank—which slopes gradually up from the plains to the crest of the Laramie (Front) Range (Fig. 11–3). The higher part of the surface cuts across the resistant Precambrian Sherman Granite that forms the core of the Laramie Range and the upturned edges of the Paleozoic and Mesozoic sedimentary rock layers (Mears, 1993). The lower part of the Gangplank is a gentle ramp of eroded debris (Tertiary-age sedimentary rock) extending well out into the Great Plains. To the south in Colorado, the resistant-rock area became mountains, and the area of weak sedimentary "gangplank" rocks was removed by erosion to form the Colorado Piedmont Section of the Great Plains Province. Thus, it was primarily a lithologic break that determined the boundary between the two

FIGURE 11–3 The Gangplank, an extensive remnant of the Pliocene pediplain along Interstate 80 in the Laramie Range. (Photo by D. Harris)

geomorphic provinces. Unlike the abrupt eastern boundaries of the Tetons and the Sierra Nevada, where geologically recent faulting was responsible, the eastern boundary of the Front Range is primarily the result of differential erosion.

In the Southern Rocky Mountains there are, at canyon-top-level, ridges and flattish areas at 9500–10,000 feet (2900–3050 m)—remnants of the late Pliocene Subsummit Pediplain (Fig. 11–1). High above in the alpine zone are benches from 11,500 to 12,000 feet (3500–3660 m) that are older—remnants of the Eocene Summit Peneplain, above which rise the many monadnocks. For those who find hiking in the alpine tundra exhilarating, a stroll along trails on these flat to gently rolling surfaces well above timberline is an experience never to be forgotten. The steep-walled youthful canyons were cut after the latest uplift, near the end of the Pliocene. The uplift elevated the mountains to their present heights—high enough for glaciers to form.

Glaciation

The first glaciers were ice caps that formed along the high divides and extended in lobes down the then-undissected pediplain surface. During the long interglacials, canyons were cut by the streams, and valley glaciers became the dominant type during later episodes of glaciation. The evidence of the early glaciers is fragmentary, essentially erased by later advances; almost all of the prominent glacial features of today were formed during the several Wisconsin advances within the last 70,000 years. About 11,000–12,000 years ago these great glaciers began their last retreat with the warming of the climate, which culminated in the Thermal Maximum about 5000 years ago. The cirques that were enlarged during Wisconsin time were occupied only by cirque lakes; a few of the cirques were reoccupied by glaciers during the Neoglacial (Fig. 1–25). Ten small glaciers or glacierets are known to occupy the more sheltered cirques; icefields—potential glaciers—occupy others.

Rock glaciers are glacier-shaped masses of bouldery debris (Fig. 11–4) that are found in many mountainous areas, but concentrations are confined to only those localities where optimum conditions exist. A near-glacial climate is the primary requirement. Rocks that weather readily into large and durable fragments also contribute to the process. Glaciers in retreat often develop a thick debris cover and form one type of rock glacier—advancing downslope at perhaps a few inches each year. Another type forms when water freezes in spaces between rock accumulations, or *talus,* at the base of cliffs. Over a period of time, the ice-cemented rocky debris also flows slowly, forming a second type of rock glacier.

The rock glacier shown in Figure 11–4 and many others in the Rocky Mountains are active today, as determined by examining their frontal sections or measuring their movements over periods of a few years (White, 1976). Forward movement of a few inches or less (1–6 cm) each year renders the frontal section exceptionally steep and highly unstable (White, 1976). (Try if you must to climb up the front of an active rock glacier!) Perhaps more convincing, lichens have not had

FIGURE 11–4   Rock glacier in southern Medicine Bow Mountains. (Photo by E. Kiver)

time to grow on newly exposed surfaces of the boulders. Those rock glaciers that are now inactive are the ones in which the ice core has been so reduced by melting as to make it ineffective in moving its load of rocks.

REFERENCES

Bradley, W.C., 1987, Erosion surfaces of the Colorado Front Range: A review, in Graf, W.L., ed., Geomorphic systems of North America: Boulder, CO, Geological Society of America centennial special volume 2, p. 215–220.

Christiansen, R.L., and Yeats, R.S., 1992, Post-Laramide geology of the U.S. Cordilleran region, in Burchfiel, B.C., Lipman, P.W., and Zoback, M.L., eds., The Cordilleran Orogen: Conterminous U.S.: Geological Society of America, Boulder, CO, The Geology of North America, v. G-3, p. 261–406.

Chronic, H., 1987, Roadside geology of New Mexico: Mountain Press, Missoula, MT, 255 p.

Eaton, G., 1986, A tectonic redefinition of the Southern Rocky Mountains: Tectonophysics, v. 132, p. 163–193.

Epis, R.C., and Chapin, C.E., 1975, Geomorphic and tectonic implications of the post-Laramide, late Eocene erosion surface in the Southern Rocky Mountains: Geological Society of America Memoir 144, p. 45–74.

Lageson, D.R., and Spearing, D.R., 1988, Roadside geology of Wyoming: Mountain Press, Missoula, MT, 271 p.

Livacarri, R.F., 1991, Role of crustal thickening and extensional collapse in the tectonic evolution of the Sevier-Laramide orogeny, western United States: Geology, v. 19, p. 1104–1107.

Luedke, R.G., and Smith, R.L., 1991, Quaternary volcanism in the western conterminous United States, in Morrison, R.B., ed., Quaternary nonglacial geology: Conterminous U.S.: Geological Society of America, Boulder, CO, The Geology of North America, v. K-2, p. 75–92.

Mears, B.B., Jr., 1993, Geomorphic history of Wyoming and high-level erosion surfaces, in Snoke, A.W., Steidtmann, J.R., and Roberts, S.M., eds., Geology of Wyoming: Geological Survey of Wyoming Memoir No. 5, p. 608–626.

Miller, D.M., Nilsen, T.H., and Bilodeau, W.L., 1992, Late Cretaceous to early Eocene geologic evolution of the U.S. Cordillera, in Burchfiel, B.C., Lipman, P.W., and Zoback, M.L., eds., The Cordilleran Orogen: Conterminous U.S.: Geological Society of America, Boulder, CO, The Geology of North America, v. G-3, p. 205–260.

Mutschler, F.M., Larson, E.E., and Gaskill, D.L., 1998, The fate of the Colorado Plateau—a view from the mantle, in Friedman, J.D., and Huffman, A.C., Jr., coordinators, Laccolithic complexes of southeastern Utah: Time of emplacement and tectonic setting—Workshop proceedings: U.S. Geological Survey Bulletin 2158, p. 203–222.

Rich, J.L., 1938, Recognition and significance of multiple erosion surfaces: Geological Society of America Bulletin, v. 49, p. 1695–1722.

Richmond, G.M., 1974, Raising the roof of the Rockies: Rocky Mountain Nature Association, Estes Park, CO, 81 p.

White, S.E., 1976, Rock glaciers and block fields, review and new data: Quaternary Research, v. 6, p. 77–97.

ROCKY MOUNTAIN NATIONAL PARK (COLORADO)

High mountain scenery, mainly the work of alpine glaciers, is preserved in Rocky Mountain National Park in the northern Front Range of Colorado (Fig. 11–5). The park sets astride the Continental Divide in an exceptionally high section of the Southern Rocky Mountains where peaks over 14,000 feet (4270 m) are commonplace. This region, as Richmond (1974) notes, is the "roof of the Rockies"—it is also the "roof of the continent." The northernmost of the "fourteeners," Longs Peak (14,256 feet; 4346 m), is located in the southeastern part of the park. Longs Peak rises majestically above lower but nearly equally rugged surrounding peaks (Fig. 11–6) and the broad meadow, or "park" (grassy area rimmed by forest), called Estes Park (elevation about 7000 feet, 2134 m) just east of the national park. Remnants of a gently rolling upland surface, the Summit Erosion Surface (or Summit Peneplain), are major elements that give a special appearance to the alpine topography (Fig. 11–7). These erosional remnants provide clues to the geologic evolution of the Southern Rocky Mountains and demand a special explanation. Here the Colorado River begins its 1400-mile-long (2250-km) tortuous journey that takes it through the Grand Canyon and on to the Pacific Ocean. Melting snow and summer rains change from mere trickles on the high western slopes to a significant river where it leaves the park near Grand Lake.

FIGURE 11–5 View westward toward the high glaciated peaks along the Continental Divide in Rocky Mountain National Park. (Photo by E. Kiver)

FIGURE 11–6 Longs Peak from Trail Ridge Road. Clark's Nutcracker in foreground is a common bird in the park. (Photo by E. Kiver)

FIGURE 11–7   Summit (Flattop) Peneplain surface along Tundra Curves section of Trail Ridge Road. Telephoto from Toll Memorial Nature Trail. (Photo by D. Harris)

Trail Ridge Road, one of the most scenic roads in North America, climbs steeply to the remarkable, gently rolling topography of the Summit Erosion Surface and connects the west and east sides of the park. The road crosses Milner Pass (10,758 feet; 3280 m) at the Continental Divide and Fall River Pass (11,796 feet; 3600 m) to the east and reaches elevations over 12,000 feet (3660 m)—making it the highest continuous automobile road in North America. At these elevations, one may find his or her automobile, especially if tuned for much lower elevations, puffing and complaining when climbing hills. Car owners may experience similar difficulties on steep trails! Other roads lead into the Fall River, Moraine Park, and Bear Lake areas (Fig. 11–8), and numerous trails take the more ambitious into the remote sections—the real park.

With elevations ranging from 7840 at Park Headquarters to 14,256 feet (2378 to 4346 m) atop Longs Peak, several life zones are represented in the park. Elk, deer, and marmots (rock chucks) are likely to be seen by sharp-eyed visitors. Of the smaller animals, the cony, or pika, is most unusual. An unlikely looking little guinea pig-rabbit, the cony is more often heard than seen because it blends in with

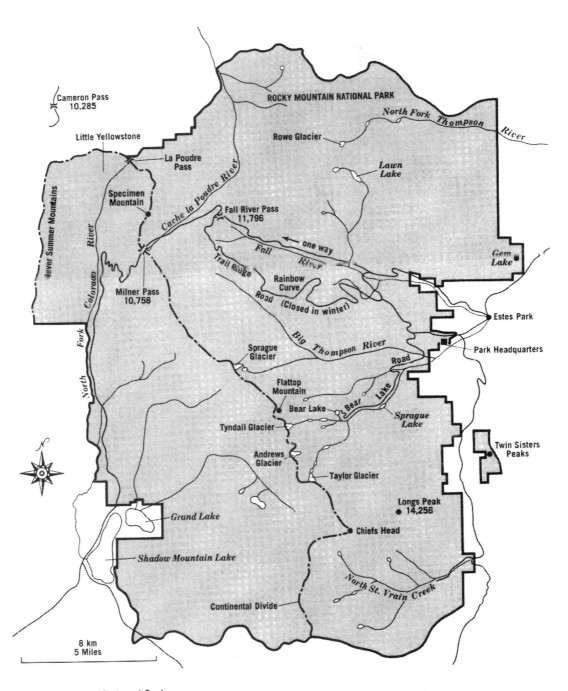

Rocky Mountain National Park.

Figure 11–8   Map of Rocky Mountain National Park.

the gray boulders of its home. Listen for the cony's raucous shriek at Rock Cut on Trail Ridge Road. It does not hibernate but spends its summers storing "hay" for the winter; it is the "haymaker."

Humans and the Park

Evidence of human activity in and near the park dates back 10,000–12,000 years ago when the earliest inhabitants hunted mammoth and the now extinct giant bison in the nearby Great Plains. Clovis and Folsom projectile points found near Trail Ridge Road indicate that these skillful hunters also used the park in their quest for food (Buchholz, 1983). Rock walls and ambush pits constructed in some of the alpine tundra areas were part of game-drive systems where sheep, deer, or elk were driven toward waiting hunters.

Indian use of the park area continued, with the modern Utes arriving perhaps about 1000 years ago and the Arapahoe in about 1790 as they were pushed from their traditional homes in the Great Plains by the advancing Anglo-American settlers. The introduction of the horse greatly changed the Indian life-style although fur traders, mountain men, and explorers had minimal impact on the Indian way of life. However, the 1858 discovery of gold brought perhaps 100,000 fortune seekers, or "fifty-niners," into the Front Range area, ensuring that the Rocky Mountain wilderness would never be the same. Towns and mining camps developed, spelling the end to Indian influence and their eventual relegation to reservations.

Major Stephen Long described the highest peak visible in the northern Front Range in 1820—a peak used by travelers as a landmark and one that would later be named for him. In 1860 Joel Estes discovered, settled, and fell in love with the high, broad valley or park (a meadow rimmed by forest) east of Longs Peak that bears his name. Estes and his family raised cattle and hunted the abundant game in his mountain paradise until 1866.

Interest in the area grew as Coloradoans and vacationers alike found refreshment in this high-elevation wilderness. The lure of the mountains and still unclimbed Longs Peak brought the one-armed Civil War veteran and geology professor, Major John Wesley Powell, to the area in 1868. His group became the first to conquer the "unclimbable" peak. Powell's thoughts on mountain climbing are shared by many: "The trouble with climbing a mountain is that you can't stay on top!" (Buchholz, 1983, p. 50) The ambitious Powell was destined the following year to become the first to float the Colorado River (in rickety wooden boats!) through the Grand Canyon.

The Reverend Elkanah Lamb climbed the 1630-foot-tall (497-m) vertical east face of Longs Peak in 1871. The feat was repeated only once in the next 50 years, when 32 years after the initial climb, Enos Mills, the "father" of the park, again scaled the vertical granite face (Nesbit, 1959). Today, the east face climb still remains one of the most challenging climbs in the Front Range.

Important government surveys were conducted by Clarence King in 1871 and Ferdinand Hayden in 1873. Hayden was famous for his explorations of the Yellowstone region 2 years before—explorations that helped lead to the establishment of the world's first national park. Hayden's group included the famous frontier photographer William Henry Jackson whose breathtaking photos brought further attention to the area. The descriptions of the spectacular scenery published through the 1860s by influential Denver newspaperman William Byers now had photographic support for his claims.

During the 1870s an English nobleman, the Earl of Dunraven, acquired title to much of Estes Park as his own private hunting reserve, following the European model that reserves were for the wealthy and privileged. Many of the claims were acquired by questionable means and he later had to sell the land. Among other activities, Dunraven built a hotel to accommodate the growing number of mountain climbers, hunters, and tourists that came to see "America's Switzerland." Local ranchers were also furnishing accommodations to visitors—an easier and more profitable business than cattle ranching.

A short flurry of mining activity in the 1880s in the Never Summer Mountains (west edge of the park) produced Lulu City and other boom towns along the North Fork of the Colorado River—short-lived mining towns with colorful histories. The ghost town of Lulu City consists of crumbling log-house frames—accessible today only by trail. Commercial meat hunting to supply the miners and townspeople of Colorado began to decimate the "boundless" herds of elk, sheep, and deer in the Rocky Mountains. Under responsible management, many of the animal populations have now partially recovered.

Pioneer, cowboy, miner, mountain-climbing guide, and self-taught naturalist Enos Mills was to Rocky Mountain what John Muir was to Yosemite. He began to develop a new outlook for the mountains that were his home. Rather than using the mountains merely for economic wealth, Mills recognized that the pleasure and adventure that this mountain paradise brought to visitors was its real value. Mills had witnessed the disappearance of the wilderness frontier of the 1890s. He now realized that the effect that the beauty of the mountains had on visitors was the best use of the mountain paradise. After writing 64 newspaper and magazine articles and delivering numerous lectures promoting the cause for a national park (Buchholz, 1983), his efforts succeeded when the area was dedicated as a national park on September 4, 1915. Mill's original proposal for a 1000-square-mile (2600 km^2) park extending from Wyoming south to Pikes Peak was opposed by the mining industry and other political forces. However, today's 414-square-mile park preserves a significant part of the landscape that symbolizes its rugged beauty and the character of the people who have inhabited this land.

Mills was a popular guide who led climbs on Longs Peak from 1888 to 1905. His keen observations enabled him to describe nature to his clients on the summit climb—the beginning of a tradition that evolved into the ranger-led nature walks offered in most parks today.

Geologic History—Rocks and Mountains

About one third of the history of the earth, about 1800 million years, can be read in the rock record at Rocky Mountain National Park (Braddock and Cole, 1990). Thick accumulations of sedimentary and volcanic rock were metamorphosed and twisted into mountains some 1750 Ma during Precambrian (Proterozoic) times. Erosion followed by the intrusion of the Longs Peak and other batholiths between 1450 and 1350 Ma (Braddock and Cole, 1990) again pushed the area upward into mountains—only again to be subjected to the destructive and leveling action of erosion.

By Paleozoic time western North America was the *trailing edge* of the plate, and a few thousand feet of marine sediments accumulated on the beveled surface of the ancient Precambrian mountains. The eastward-moving North American Plate ground to a halt in late Paleozoic time (Pennsylvanian and Permian, about 300–245 Ma) when it collided with the Eurasian Plate. The resulting effects were spectacular—a giant supercontinent (Pangaea) with the massive Appalachian Mountains and their European equivalents squeezed between plates, much like today's Himalayan Range is squeezed between the Indian and Asian plates. Lesser but significant ranges formed along the south edge of the North American Plate (Ouachita Mountains), and the Ancestral Rockies formed in the Colorado and New Mexico area.

The Ancestral Rockies shed coarse sediment into surrounding areas and occupied nearly the same location as today's Front Range—lending credence to the hypothesis that reactivated Precambrian faults localized both ranges. Erosion gradually truncated the Ancestral Rockies, and nonmarine sediments of Mesozoic-age buried the former mountains. Crustal subsidence and deposition of marine sediments over 20,000 feet (6100 m) thick during late Mesozoic (Cretaceous) time formed the short-lived Rocky Mountain Geosyncline. Pangaea had split into smaller plates during the early Mesozoic (Triassic, about 200 Ma) and now the North American Plate was creeping westward—generating the compressive forces that produced the Cordilleran Orogeny. Mountain building slowly migrated eastward, finally reaching the Rocky Mountain area by Late Cretaceous (70 Ma) or early Tertiary (60 Ma) time where it is known as the Laramide Orogeny.

Uplifts during the Laramide Orogeny were modest in the Southern Rocky Mountains and by 50 Ma, or earlier in some parts of the Southern Rockies, the thick cover of Mesozoic strata was stripped away exposing the ancient Precambrian granites, gneisses, and schists in the cores of the anticlinal mountains (Fig. 11–9). By late Eocene time (about 40 Ma) the mountains were a gently rolling plain, less than 3000 feet (915 m) above sea level (Eaton, 1986; Richmond, 1974) with a few monadnocks (residual hills) rising above.

Considerable heat was accumulating beneath the crust in the newly formed mountain range. The heat manifested itself as widespread volcanic eruptions that included an eruptive center in the Never Summer Range along the west edge of the park. Mid- and late Tertiary volcanism, mainly pyroclastics, covered the Eocene

West   East

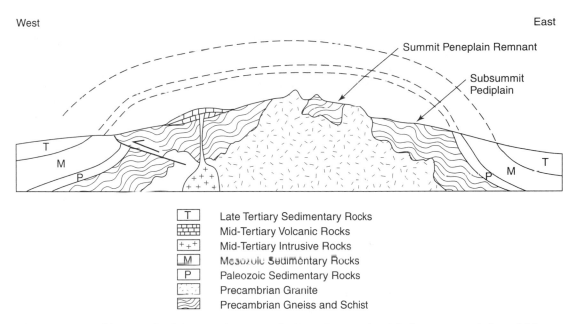

T	Late Tertiary Sedimentary Rocks
	Mid-Tertiary Volcanic Rocks
+ +	Mid-Tertiary Intrusive Rocks
M	Mesozoic Sedimentary Rocks
P	Paleozoic Sedimentary Rocks
	Precambrian Granite
	Precambrian Gneiss and Schist

FIGURE 11–9   Highly generalized cross section across the Front Range as it might have appeared about 8 Ma. Note the cut-and-fill late Tertiary erosion surface (Subsummit Pediplain) in the process of forming.

erosion surface. A remnant of one of the ash flows is readily seen from Trail Ridge Road at Iceberg Lake. The 28 million-year-old ash flow here rests directly on the Eocene erosion surface.

Uplift History—Rampant Inflation!

The relatively low elevation of the Eocene erosion surface was soon to change as the high concentration of heat beneath the Southern Rocky Mountains took hold. The expansion of the warmed mantle rock lowered its density, causing it to rise and form an elongate thermotectonic uplift (Eaton 1987) that would ultimately produce today's towering mountain peaks.

Erosion vigorously attacked the uplifted Eocene surface leaving remnants in the high country along Trail Ridge Road (the Tundra Curves section), Flattop Mountain, and elsewhere (Fig. 11–7). These delightful tundra-covered surfaces at 11,000–12,000 feet (3350–3650 m) elevation are part of the Summit Erosion Surface or Summit Peneplain. Another episode of equilibrium or balance between erosion at higher elevations and burial at lower elevations may have created yet another cut-and-fill surface, the Subsummit Erosion Surface (Fig. 11–9) during mid- to late Tertiary (late Miocene, perhaps 6–8 Ma) time (Mears, 1993).

Examination of the remnants of the Subsummit Surface combined with studies of the paleoclimate record in Cenozoic-age sediments in the Rocky Mountain basins and the Great Plains led to the conclusion that the Subsummit Surface was formed under relatively dry climatic conditions and is therefore, strictly speaking, a pediplain. This lower elevation surface was also largely destroyed when the latest major uplift occurred about 2–5 Ma. Deep canyons were cut below the pediplain surface in the ongoing fluvial cycle, sometimes referred to as the "canyon-cutting cycle." Other interpretations of the surfaces are possible. For example, Richmond (1974) and Scott and Taylor (1986) recognize only one erosion surface and believe that later fault displacements account for surfaces at different elevations.

Glaciers and Glacial Landforms

Uplift of nearly a mile in the last few million years combined with the worldwide cooling trend during Cenozoic time caused a profound transformation of climates—especially in higher latitudes and elevations. Winter snow no longer completely melted during the summer but piled up in favorable locations to form glaciers. Initially those in the park likely formed sheetlike ice caps. However, deepening of valleys, especially during intervals of reduced or nonglaciation, led to the generation of valley glaciers during later glaciations. Evidence of older glaciers has been mostly obscured by weathering and scouring by younger streams of ice. Thus, although the mountain scenery is the work of glaciers, it was primarily the glaciers that advanced several times during the Wisconsin (term used in North America to describe late Pleistocene glaciations from about 60,000–10,000 years ago) that produced the rugged topographic features of today's high country.

Small incipient basins in the high country were enlarged into large cirque basins such as that at the east base of Longs Peak and at hundreds of other locations in the park. Valley glaciers spawned in these basins crept downslope, merged with their neighbors, and flowed as much as 25 miles (40 km) to Grand Lake, Moraine Park, Horseshoe Park, Glacier Basin, and elsewhere. At the lower parts of the glaciers large *lateral moraines,* or debris ridges, formed along the lateral margins of the glacier, and *end,* or *terminal, moraines* formed along the front edge of the ice stream. Large *glacial erratics* strewn about on the moraine surface (Fig. 11–10) are reminders of the power of large glaciers. The lateral moraine of the Fall River glacier is particularly intriguing where it diverted a small stream and formed Hidden Valley. The stream in Hidden Valley now follows a new and longer route in order to join Fall River (Fig. 11–11). At their maximum extent valley glaciers reached 8000 feet (2440 m) elevation on the east side but were melted at a slightly higher elevation on the warmer west side.

Chains of rock-basin lakes (*paternoster lakes*) at different elevations are on the treads of the "Giants Stairways," or Cyclopean Stairs, in certain of the gorges—Fern Canyon for one. Cirques etched their way headward into the mountain core, narrowing the flat peneplain area and sharpening rock divides into *aretes* and

FIGURE 11–10   Huge boulder, or glacial erratic, transported by a large valley glacier and deposited in the moraine at Bear Lake Parking Area, Rocky Mountain National Park, Colorado, (Photo by D. Harris)

forming high mountain passes, or *cols*, where ridge divides were lowered by erosion. *Glacial horns*, or *matterhorn peaks*, are abundant where cirques and glacial valleys outline or enclose residual rock masses.

Topographically subdued moraines of the Bull Lake Glaciation are perhaps 150,000 years old, and the most extensive moraines of the more prominent, younger Pinedale Glaciation are about 15,000 years old. Glaciers retreated between 14,600 and 13,000 years ago and were mostly gone by 12,000 years ago (Kaye, 1987). Climatic cooling during the past few thousand years has permitted small glaciers to reform and produce Neoglacial deposits—small moraines and rock glaciers near to or within a mile downvalley of cirque headwalls.

A few of the cirques are occupied by tiny glaciers—such as Andrews, Tyndall, Taylor, Sprague, and Rowe—which were born during the last few hundred years, the Little Ice Age portion of Neoglacial times (Fig. 1–25). The five small existing glaciers in the park are all located on the east side of the Continental Divide where winter winds, sometimes of hurricane force and over 200 miles (320 km) per hour, blow snow over the divide where it drifts into the Wisconsin-age cirques and nourishes these small glaciers. By the late 1930s, at the end of a warming trend, they had dwindled to the point where they appeared doomed. Slight cooling during the 1940s to the 1970s enabled them to grow and stabilize. However, global warming, aggravated and perhaps caused entirely by human activities, puts the future of these small glaciers in doubt.

(a)

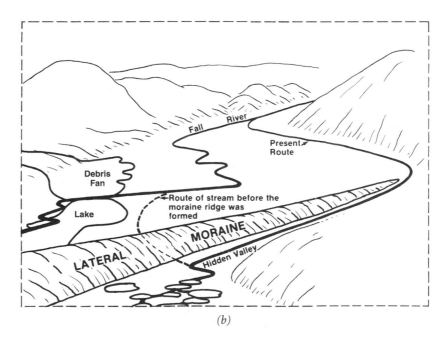

(b)

FIGURE 11–11   (a) Looking down from Rainbow Curve Overlook, over Horseshoe Park (middle ground) and Hidden Valley (lower right). Note the debris fan and lake formed during the Lawn Lake flood in 1982. (Photo by John Buchanan) (b) Sketch showing the origin of Hidden Valley as pictured in (a) (Sketch by Gregory Nelson)

Other Geologic Features

Periglacial features are found high in the tundra areas of the park. The term *periglacial* means "around the glaciers," but it is often used to denote a cold environment in which "frost action"—freezing and thawing—has formed unusual but incompletely understood features. Frost riving of layered rocks in the park—mainly gneisses and schists—produces flat slabs, and by frost heaving some of the slabs are left standing on edge as *tombstone rocks.* Freezing and thawing also produces *patterned ground* in which an inner core of finer sediment is surrounded by a line of large rock fragments.

Where fine-textured material becomes saturated, mass wasting in the form of solifluction takes place, and small *solifluction terraces* are formed, even on very gradual slopes. The mass of saturated soil creeps slowly downslope, wrinkling the mat of vegetation above and forming the wavy slopes so prominent in the alpine zone and readily visible along Trail Ridge Road. In some areas, solifluction may be related to permafrost, which is present at depths of a few feet, especially where willow thickets provide sufficient insulation. A frozen layer below prevents water from infiltrating deeper and encourages the upper soil to liquify, become "squishy," and move slowly downslope.

Tors are isolated rock towers rising prominently above the high slopes in the park. They were formed by differential weathering along rock fractures and other weak areas while the rock was still buried by soil. Removal of the soil exposes these rock cores, some of which resemble sea stacks. Excellent examples can be seen on the nature walk above the Rock Cut Stop on Trail Ridge Road.

Unloading, or removal, of overlying rock permits massive, homogenous rock such as granite to expand and develop a system of parallel fractures. These fractures, if prominent, often result in *sheeting* and sometimes form bizarre landforms (Fig. 11–12).

FIGURE 11–12　Vertical jointing in granite (the Keyhole) along with frost action near Longs Peak produced this unusual landform. Note emergency cabin for climbers. (Photo by E. Kiver)

Most lakes in the park were formed by glaciers, either by erosion or by deposition. However, a few are man-made. Sprague Lake, along the road to Bear Lake, has special interest. The nature trail around the lake is a "Five-Senses-Trail" designed to accommodate handicapped visitors, including those in wheelchairs. Innkeeper Abner Sprague needed a fishing spot for his guests, so he built a low dam across the creek. The Park Service purchased the property years ago, and in 1957 they razed the old hotel.

Lawn Lake (10,987 feet; 3330 m), high in the mountains north of Fall River, is partially artificial. At the end of a 6-mile (9.6-km) trail, it has long been a favorite of ambitious hikers. One of the early-day hikers had an idea; if a dam were built on top of the morainal dam, a very substantial amount of water could be stored for irrigation down on the flatlands. This "improvement" of nature was built in 1902, several years before the park was established. The building of a dam on morareal material of unknown composition has proved to be poor engineering practice, here and elsewhere. Early on July 15, 1982, the aging Lawn Lake Dam failed, sending a torrent down the steep channel of Roaring River, carrying everything with it, including car-size boulders. At the mouth of the canyon, the water spread out and deposited an enormous boulder fan on the floor of Fall River Canyon (Figs. 11–11 and 11–13). The floodwaters raced down Fall River Canyon and through the business district of Estes Park. Property damage was over $31 million, but miraculously only three lives were lost. Fall River Road in the park was buried beneath many gigantic boulders; the road was reopened after being built up and over the boulder fan (Fig. 11–13).

FIGURE 11–13   Road building across boulder field formed by the catastrophic failure of Lawn Lake Dam on July 15, 1982. (Photo by John Buchanan)

Lawn Lake is still there, but it is even smaller than before the dam was built, perhaps 5 or 6 feet (1.5 or 2.0 m) below its 1902 level. The disaster had one beneficial result: The program of inspection of dams in Colorado, including old dams, has high priority.

This was the second flood to disrupt normal visitation in the park in recent years. On July 31, 1976, the worst flood and natural disaster in Colorado's history devastated Big Thompson Canyon between the towns of Estes Park and Loveland. Although the cloudburst storm, and consequently the flood, was downstream from the park, large sections of U.S. Highway 34 were ripped out, 316 homes, 45 mobile homes, and 52 tourist-oriented businesses were destroyed, and a major access to the park was closed for an extended period. More important by far, 146 lives were lost during the flood. Rather than avoiding the next flood disaster, many landowners chose to rebuild in the same flood-prone areas. Will disaster-relief funds from the public treasury help them or their successors rebuild again after the next flood?

Visiting the Park

If possible, visit Rocky Mountain National Park soon after the main tourist season is over. If you should hike up the Bierstadt Trail you can look down on the valley and see an ocean of golden aspen leaves quaking in the breeze. Perhaps you will be high on the trail at one of those one-in-a-million moments when whirling winds updraft showers of gold leaves high into the air, then release them to float gently down.

When you climb Longs Peak, you will be up where the air is pure and the views are superb—to the south, west, and north. But do not look to the east; on many days, far too many, you will see a pall of orange-brown smog sitting on top of Denver and other Front Range cities and towns. On occasion, fortunately not often so far, upslope winds carry the smog high up into the eastern part of the park. It is indeed a sad commentary on modern living when the scramble to develop more land and industry to support a rapidly increasing number of people leads to the deterioration of our home and park environments. Is the quality of life really getting better?

REFERENCES

Braddock, W.A., and Cole, J.C., 1990, Geologic map of Rocky Mountain National Park and vicinity: U.S. Geological Survey Map I-1973.

Buchholz, C.W., 1983, Rocky Mountain National Park, a history: Associated University Press, Boulder, CO, 255 p.

Eaton, G.P., 1986, A tectonic redefinition of the Southern Rocky Mountains: Tectonophysics, v. 132, p. 163–193.

Eaton, G.P., 1987, Topography and origin of the southern Rocky Mountains and Alvarado Ridge: in, Coward, M.P., Dewey, J.F., and Hancock, P.L., eds., Continental extensional tectonics: Geological Society Special Publication No. 28, p. 355–369.

Kaye, G., 1987, Rocky Mountain National Park, the story of its origin: text accompanying U.S. Geological Survey National Park Series topographic map, Rocky Mountain National Park, 1:50,000, U.S. Geological Survey, Reston, VA.

Mears, B., Jr., 1993, Geomorphic history of Wyoming and high-level erosion surfaces, in Snoke, A.W., Steidtmann, J.R., and Roberts, S.M., eds., Geology of Wyoming: Geological Survey of Wyoming Memoir No. 5, p. 608–626.

Nesbit, P.W., 1959, Longs Peak, its story and a climbing guide: Colorado Springs, CO, Paul W. Nesbit, 48 p.

Richmond, G.M., 1974, Raising the roof of the Rockies: Rocky Mountain Nature Association, Estes Park, CO, 81 p.

Scott, G.R., and Taylor, R.B., 1986, Map showing late Eocene erosion surface, Oligocene-Miocene paleovalleys, and Tertiary deposits in the Pueblo, Denver, and Greeley 1° × 2° quadrangles, Colorado: U.S. Geological Survey Map I-1626.

Park Address

Rocky Mountain National Park
Estes Park, CO 80517

GREAT SAND DUNES NATIONAL MONUMENT (COLORADO)

North America's tallest sand dunes lie tucked away along the east edge of the San Luis Valley in southern Colorado (Fig. 11–14). The 750-foot-tall (230-m) dunes are part of a dunefield that covers over 150 square miles (389 km^2)—39 square miles (100 km^2) of which were set aside as a national monument in 1932. These outstanding examples of dunes provide opportunities to experience *eolian* (wind) processes—geology in action. Visitors can become geologic detectives and discover for themselves some of the critical evidence that leads to an understanding of nature's workings in the only true desert in the Southern Rocky Mountains.

Humans and Dunes

Only broken spear points and other artifacts bear mute testimony that the first inhabitants, individuals belonging to the Folsom and Clovis cultures, hunted mammoth, bison, and other game in and near the monument some 10,000–11,000 years ago (National Park Service, 1988). The Utes were the most recent Indians to hunt in the grasslands adjacent to the dunes and were here when the first Europeans arrived. Don Diego de Vargas described the dunes in 1694, and Lieutenant Zebulin Pike vividly recorded them in his journal in 1807. The Utes signed a series of treaties beginning in 1855 that led to their giving up the San Luis Valley but giving them "forever" the west slope of Colorado and their traditional hunting

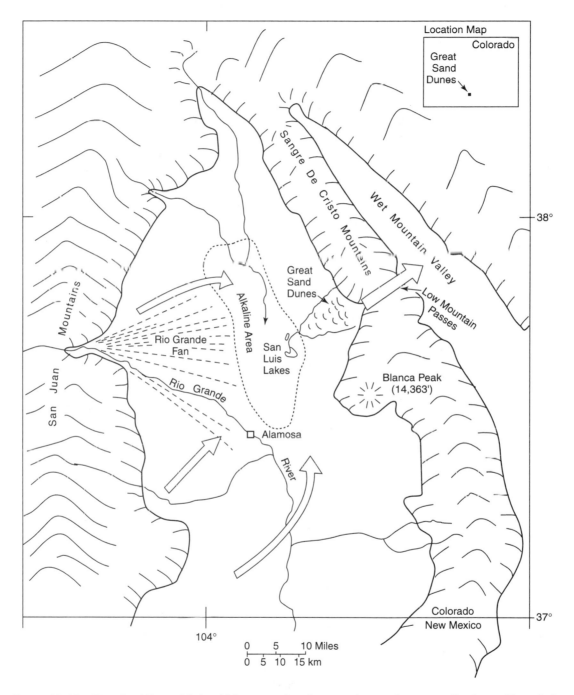

FIGURE 11–14   Great Sand Dunes National Monument location map. Arrows show generalized prevailing wind directions toward low passes in the Sangre de Cristo Mountains. (Modified from Upson, 1939)

grounds. By 1883 and several broken treaties later that were justified by "rifle fire," the Utes were pushed from their mountain homes to a small reservation in southwestern Colorado (Trimble, 1972). Homesteaders and ranchers moved in during the late nineteenth and early twentieth centuries, but the dry climate and harsh winters (occasionally the coldest spot in the United States!) limited development.

However, modern developers are not so easily discouraged and are proposing to tap the rich groundwater supply beneath the monument and construct over 13,000 houses along or near the monument edge. The resulting lowering of the water table would have detrimental effects on the preservation of the dunes as well as park plants and wildlife. Also, the clutter of civilization would greatly diminish the park experience. Thirsty Front Range cities would also like to build an aqueduct and pump the precious groundwater—enabling them to grow larger before becoming thirsty once again!

Geologic and Geographic Setting

Following the Laramide Orogeny, further uplift of the Southern Rocky Mountains was accompanied by the development of a series of down-dropped crustal blocks (*grabens* and *half-grabens*) along their crest (Chapin and Cather, 1994). These collapsed structures include the San Luis Valley and extend from central Wyoming through Big Bend in Texas to northern Mexico. They are part of the Rio Grande Rift, a major crustal pull-apart zone that, at least locally, remains active today. Extension began about 30 Ma and was accompanied by local volcanism, including the massive volcanoes that built the 14,000-foot-tall (4268-m) San Juan Mountains along the west edge of the 50-mile-wide (80-km) San Luis Valley from about 17 to 35 Ma. Like Death Valley and certain other "valleys," the San Luis Valley is not a valley of erosion but one of tectonic origin. The Sangre de Cristo fault-block mountains form the lofty topographic barrier along the east edge of the San Luis Valley. Both the glacially sculpted San Juan and Sangre de Cristo mountains rise over 6000 feet (1830 m) above the 7000- to 8000-foot-high (2135- to 2440-m) valley. The two ranges merge northward into the complex of ranges of the northern part of the Southern Rockies.

Since its initiation, the boundary fault along the Sangre de Cristo Mountains has moved over 30,000 feet (9150 m) vertically. As a result of fault movement, as much as 21,000 feet (6400 m) of Tertiary and Quaternary age sediment underlies the San Luis Valley (Kluth and Schaftenaar, 1994). As basins filled with sediment, the Rio Grande River gradually acquired a through-flowing path to the Gulf of Mexico—probably as recently as early to middle Pleistocene time (Chapin and Cather, 1994).

A large debris fan formed as the Rio Grande flowed from its headwaters in the San Juan Mountains into the San Luis Valley (Fig. 11–14). Debris shed from the mountains was particularly voluminous during episodes of glaciation when large valley glaciers scraped and plucked the volcanic bedrock. A large lake once occupied the valley, as indicated by widespread lake sediments; now, a relict lake, San Luis Lake, occupies the bottom of the enclosed basin in the northern part of the valley (Fig. 11–14). As the Rio Grande gradually shifted its course farther to the southwest, it left a thick cover of sand-rich sediment—sediment that would play a major role during the past 12,000 years in the formation of North America's tallest dunes.

Work of the Wind

The end of the Pleistocene (about 12,000 years ago) brought an end to the large valley glaciers in the mountains and the cooler, wetter climate that had prevailed in the San Luis Valley during the last episode of glaciation. As the once more-abundant vegetation in the San Luis Valley disappeared, the prevailing southwest winds began their relentless task of moving sand along the valley floor to its east edge where it began to accumulate. With about 10 inches (25 cm) of precipitation per year, the resulting sparse vegetation in many areas is unable to hold sediment in place. Winds of 15 miles per hour (24 km/hr) or more cause the dunes and the desert floor to spring to life as sand grains begin to roll and bounce. Nearly 75 percent of the sand movement is by bouncing—a process called *saltation* (Trimble, 1972). Except for the dune crests where sand that moves up the gentler windward face is launched over the steep leeward face, sand grains seldom rise more than a few inches above the surface. Frequent winds of 40 mph (64 km/hr) and seasonally up to 60 mph (97 km/hr) are particularly effective movers of sand. Overgrazing during the early twentieth century further destroyed the hardy native grasses—freeing additional sediment to add to the growing dunefield.

The interaction between moving sand and sparse vegetation along the southwest edge of the dunefield (just outside of the monument) produces small crescent-shaped *parabolic,* or *blowout, dunes* (Johnson, 1967) whose tails point upwind (Fig. 11–15). Other crescent-shaped dunes with their tails facing downwind (*barchan dunes*) locally occur on the east-central edge of the dunefield (Chatman and others, 1997). *Transverse dunes,* with long linear ridges oriented perpendicular to the wind (Fig. 11–16) are common in the center of the dunefield. All three of these dune types have asymmetric shapes with the steep *slip-face* downwind toward the east. Dry sand will maintain a slope of only about 32 degrees—the *angle of repose* and the maximum angle of the slip-face.

Most of the dunes here are *reverse* and *star dunes*—configurations that form with highly variable wind directions (Figs. 11–15 and 11–17). Although the prevailing winds come from the southwest, especially during the spring, variable winds occur throughout the remainder of the year, causing sand to blowback westward and stack itself to higher levels. Thus, part of the answer for the unusually large concentration of thick sand deposits and the stationary appearance of the dunefield is the frequent occurrence of reversing winds. A high level of groundwater beneath the valley surface and creeks along the north, south, and east edges

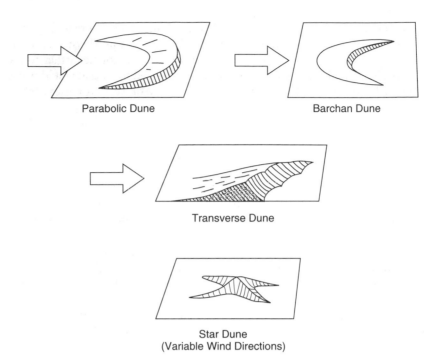

FIGURE 11–15   Common types of sand dunes; arrows indicate wind direction.

FIGURE 11–16   Air view north across transverse dunes at Great Sand Dunes to the Sangre de Cristo Mountains in background. (Photo by R.L. Burroughs)

FIGURE 11–17   Reversing sand dunes, with Sangre de Cristo Mountains in background, Great Sand Dunes National Monument. (Photo by E. Kiver)

of the dunefield also play important roles in maintaining the dunefield boundaries. A critical topographic factor accounting for the sand concentration is the unusual configuration of the mountain barrier on the east side of the San Luis Valley.

The sand dunes are located at the base of the Sangre de Cristos in an area where the range bends and forms a topographic pocket (Fig. 11–14). To the east are Mosca, Medano, and Music passes—low areas in the lofty range that funnel winds out of the San Luis Valley (Fig. 11–18). Medano Pass, at about 9700 feet (2960 m) elevation, is the lowest of the three and plays a significant role in localizing dune formation. At one time, it was assumed that most of the sand was derived from deposits of Medano and Sand Creeks that flow out of the Sangre de Cristos and along the north, south, and east edges of the dune mass. However, mineralogical studies by Johnson (1967) and others indicate that essentially all of the sand was derived from the volcanic rock source in the San Juans about 50 miles (80 km) to the west. Streams carried sediments eastward onto the valley floor where the prevailing southwest winds pick up sand-size material and transport it northeastward across the valley to the dune area. The wind is funneled through the high mountains and forced to rise through the low passes—effectively forming a sand trap. Rising wind drops its sand load near the mountain base. Quartz and feldspar are the main minerals in the dunes, but there is an unusually wide variety of accessory minerals, including garnet and magnetite.

FIGURE 11–18   High-altitude air view of Great Sand Dunes. Note the low passes (arrow) in the Sangre De Cristo Mountains in background and Medano Creek along the south edge of the dunefield in the foreground. (Photo by National Park Service)

Magnetite is not a common mineral in most sand dunes, yet black bands of magnetite sand are common on some of the dunes in the monument. With a specific gravity almost twice that of quartz and feldspar, magnetite is not transported as readily by the wind. The magnetite was not derived from the San Juans but from the Sangre de Cristos on the east. Medano Creek, which flows along the eastern side of the dunes, deposits the heavy magnetite in its broad streambed at the base of the dunes. But how does it get up on the crests of the dunes? Although the westerlies are the prevailing winds, for short periods early in the year violent winds roar down out of the canyons to the east and lift the sand, including the magnetite, high up on the dunes. Concentration of the magnetite is accomplished by gentler winds winnowing out the lighter minerals, and leaving the black bands of magnetite along the dune crests (Wiegand, 1977).

Why are the tallest dunes located a few miles west of the mountain edge rather than up against the mountain slopes? Why is a small valley located between the mountains and the east edge of the dunefield? Sand Creek on the north edge of the dunefield and Medano on the east and south help confine the dunes by trapping and moving sand in upwind (south and west) directions. Eventually the creeks sink into the valley floor and deposit their load of sand. The pervasive prevailing wind eventually blows it back onto the dunes. Thus, the thickest sand deposits are located upwind from Medano Creek (Valdez, 1996). At times, dunes encroach on and even block the intermittent streams and form temporary lakes. The next flood out of the mountains removes the dune front; thus, there is a battle for occupation of this boundary area—one that appears to be a standoff at the present time. Sand that does cross the seasonally dry stream bed is blown eastward into the wind-tunnel-like valley leading toward Medano Pass and creates *longitudinal* and *climbing dune* forms (Johnson, 1967). Shifting sand in this area can invade and suffocate portions of the ponderosa forest. Further sand migration exposes the forest skeletons—forming the ghost forests along the trail to Medano Pass.

Visiting the Dunes

Although there are good views from the valley or—after an arduous climb up the shifting sands—from the top of the 750-foot-high (230 m) dunes, it is also impressive to look out over the top of this mountain of sand from the Montville Trail leading up to Mosca Pass. The darker color of the obsidian-rich sand absorbs heat from the sun and can become surprisingly hot—up to 140°F (60°C) during midday. The cool morning sand feels pleasant on bare feet; however—those who leave their shoes in camp may regret it later in the day!

REFERENCES

Chatman, M., Sharrow, D., and Valdez, A., 1997, Water resources plan Great Sand Dunes National Monument, Colorado: National Park Service, 156 p.

Chapin, C.E., and Cather, S.M., 1994, Tectonic setting of the axial basins of the northern and central Rio Grande rift, in, Keller, G.R., and Cather, S.M., eds., Basins of the Rio Grande Rift: Structure, stratigraphy, and tectonic setting: Geological Society of America Special Paper 291, p. 5–25.

Johnson, R.B., 1967, The Great Sand Dunes of southern Colorado: U.S. Geological Survey Professional Paper 575-C, p. C177–C183.

Kluth, C.F., and Schaftenaar, C.H., 1994, Depth and geometry of the northern Rio Grande rift in the San Luis Basin, south-central Colorado, in Keller, G.R., and Cather, S.M., eds., Basins of the Rio Grande Rift: Structure, stratigraphy, and tectonic setting: Geological Society of America Special Paper 291, p. 27–37.

National Park Service, 1988, Great Sand Dunes National Monument: Statement of Management.

Trimble, S.A., 1972, Great Sand Dunes: Globe, Arizona, Southwest Parks and Monuments Association, 33 p.

Upson, J.E., 1939, Physiographic subdivisions of the San Luis valley, southern Colorado: Journal of Geology, v. 47, no. 7.

Valdez, A.D., 1996, The role of streams in the development of the Great Sand Dunes and their connection with the hydrologic cycle, in, Hydrogeology of the San Luis Valley and environmental issues downstream from the Summitville Mine: Geological Society of America field guide, Geological Society of America, Denver.

Wiegand, J.T., 1977, Dune morphology: Colorado State University, Ph.D. thesis.

Park Address

Great Sand Dunes National Monument
Mosca, CO 81146

BANDELIER NATIONAL MONUMENT (NEW MEXICO)

Adolph F.A. Bandelier was a prominent scholar who studied the ruins and the Pueblo people of New Mexico from 1880 to 1891. His pioneer research on prehistoric and modern Pueblo people provided the basis for later archeological and anthropological studies. In 1880 nearby Cochiti Pueblo Indians guided Bandelier to their ancestral homes in Frijoles Canyon. Bandelier described it as "the grandest thing I ever saw!" Prominent Southwest archeologist Edgar L. Hewitt excavated some of the ruins from 1908 to 1920 and was instrumental in establishing and naming a 32,737-acre (51 square miles, 133 km²) area of the Pajarito Plateau in north-central New Mexico after Bandelier in 1916. Although primarily an archeological park, an unusual set of geologic circumstances played a major role in determining the life-styles of people living in the Bandelier area.

Bandelier National Monument is located near the south end of the Southern Rocky Mountains and on the east flank of the Jemez Mountains (Fig. 11–19)—a major volcanic pile located along the west bank of the Rio Grande about 45 miles (72 km) west of Santa Fe and 48 miles (77 km) north of Albuquerque. The Los Alamos National Laboratory, where J. Robert Oppenheimer and his colleagues developed the first atomic bomb, is located immediately north of the monument. The Tsankawi section is a small detached unit located about 5 miles (8 km) northeast of the park entrance adjacent to the San Ildefonso Indian Reservation—another group of Indians who trace their ancestry to the Bandelier Anasazi.

The Jemez volcanic center has erupted more or less continuously for at least the past 13 million years (Wolff and Gardner, 1995). Thick layers of consolidated volcanic ash (a rock called *tuff*) known as the Bandelier Tuff were deposited 1.6 and 1.22 Ma during some of the earth's largest known volcanic eruptions. The ash filled entire valleys and formed gently sloping plateaus around the volcano flanks, including the Pajarito Plateau on the east flank. Rivers flowing in a *radial drainage pattern* (like spokes on a wheel) from the volcano summit slashed deep canyons in the plateaus and exposed the pink-gray Bandelier Tuff. The pre-European Anasazi,

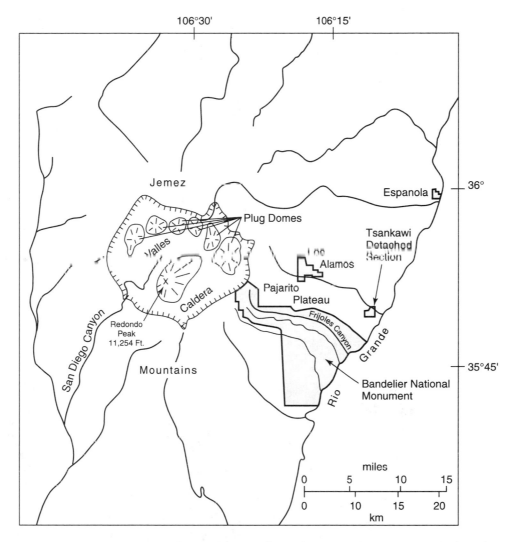

FIGURE 11–19 Jemez Mountains area showing location of Bandelier National Monument and related features. Note the postcaldera plug domes and the Redondo Peak resurgent dome in the Valles Caldera.

or Hisatsinom, people found that the tuff was strong enough locally to hold vertical cliffs but weak enough to be hollowed out with stone, wood, and bone tools—an ideal situation for the construction of cliff dwellings.

Bandelier Anasazi

The Anasazi civilization began in the Four Corners area about 2000 years ago as nomadic hunters began to rely more on an agricultural life-style. As their technologies grew, they progressed from basket-making to pottery and improved methods of agriculture and house construction. Their numbers expanded, and large settlements developed along small but permanent streams and springs where their crops of maize (corn), squash, and beans could be irrigated. The Long Drought from 1276 to 1299 forced abandonment of Anasazi areas such as Mesa Verde and caused settlements to increase along more permanent streams such as the Frijoles River. The "ancient ones" began to settle the Frijoles and other canyons at Bandelier in the late 1100s; their population increased dramatically after 1300 as other Anasazi migrated to the Bandelier area as a result of the Long Drought.

Perhaps small caves weathered into the Bandelier Tuff were utilized initially before *talus houses* were constructed along the base of the cliffs (Fig. 11–20). Long

FIGURE 11–20   North Cliff, Frijoles Canyon. Note cliff-side pueblo, weathering pits, and excavations in welded tuff. (Photo by E. Kiver)

House, an 800-foot-long (244 m) apartment house structure contained over 300 rooms in the largest cave-and-masonry dwelling found on the Pajarito Plateau. Rooms could be hollowed out in the easily excavated tuff and three- or four-story structures were common. *Viga holes* line the rock walls where floor joists and roof beams were inserted. On the flat valley floor the free-standing, three-story-high Tyuonyi Village pueblo with approximately 400 rooms was constructed (Fig. 11–21). The walled village was easily defended as access ladders were hoisted up and the narrow zig-zag entrance to the central plaza placed invaders at a distinct disadvantage.

A number of circular underground chambers called *kivas* were constructed for ceremonial and religious purposes. Several partially reconstructed kivas are present on the Frijoles Canyon floor. Cave kiva is fully reconstructed and well worth visiting. Adorning many of the rock walls in the area are *petroglyphs* (rock art that is pecked into the rocks walls) and *pictographs* (rock paintings) of turkeys, snakes, faces, and birds.

Perhaps even the Frijoles Creek was susceptible to serious drought. During the 1500s the grand villages in the canyons of Bandelier were abandoned, presumably because of drought conditions or perhaps because of exhaustion of the soil and other resources. The last of the cliff-dwelling Anasazi migrated to other areas by 1550. Subsequent village sites were located in broad valleys and on mesa tops rather than in cliff houses. Some Anasazi apparently joined the nearby Cochiti and San Ildefonso pueblos where their descendants reportedly live today. The area was in serious decline as Coronado, the first of the European explorers, arrived in the area. No mention of the Pajarito Plateau villages is made in his chronicles.

Figure 11–21   Excavated ruins of Tyuonyi Village near the visitor center at Bandelier National Monument. (Photo by D. Harris)

Geologic Setting

The Jemez Mountains are an important volcanic center that is larger and more long-lived than the other numerous volcanic areas located along the Rio Grande Rift. The opening of the Rio Grande Rift—a huge north–south trending fracture zone that extends from at least central Colorado to Mexico—promoted melting of the mantle and provided pathways for magma to reach the surface during the past 30 million years. A second rift zone, the northeast–southwest trending Jemez Lineament, is also marked by numerous volcanic centers. The Jemez Lineament extends southwest from Capulin Mountain volcano in northeastern New Mexico to at least the San Carlos volcanic field in southwestern Arizona—crossing the Rio Grande Rift at the site of the Jemez Mountains. The location of the Jemez Mountains at the intersection of these two important rift zones is undoubtedly related to its unusual longevity and large eruptions (Christiansen and Yeats, 1992).

The Jemez Mountain area is one of several in the Rio Grande Rift where magma bodies lie some 9–19 miles (15–30 km) beneath the surface. The ground above at least one of these magma bodies near Socorro (about 75 miles, 120 km, south of Albuquerque) has been slowly rising since at least 1909 when accurate survey lines were first measured (Reilinger and others, 1980). Thus, the Socorro magma chamber is expanding—an ominous sign. Similar studies in the Valles Caldera in the Jemez Mountains have yet to be undertaken, but Wolff and Gardner (1995) suspect that a new batch of magma has been newly generated beneath the volcano.

Geologic Sequence

The current serenity at Bandelier hides a turbulent geologic past where eruptions of almost unimaginable fury swept across the landscape. Destruction of plants and animals in the path of eruptive activity was commonplace during the past 13 million years as low silica basalt and higher silica, more explosive andesite built a large stratocone complex whose eroded remnants form today's Jemez Mountains. Quiet basaltic eruptions were overshadowed by more explosive Mount St. Helens types of andesitic and dacitic eruptions as the restless giant has continuously rumbled into activity during the past 16 million years.

Eruptions changed dramatically about 1.8 Ma as high-silica magma, of rhyolitic composition, erupted violently. The explosive mixture of rhyolite magma and high gas pressure created a huge eruption cloud about 1.61 Ma that sent airfall debris at least as far as Kansas and Oklahoma and produced a turbulent cloud of ash that roared down the mountain flanks as an *ash flow* or *pyroclastic flow*. The pyroclastic debris filled in valleys and other topographic lows and formed the lower part of the Bandelier Tuff. Expulsion of 12 cubic miles (50 km^3) of magma removed support for the mountain top, causing it to collapse and form the Toledo Caldera—the first of two major calderas in the summit area (De Nault, 1987).

About 400,000 years later (about 1.22 Ma), a second eruption of over 12 cubic miles (50 km³) deposited the upper part of the Bandelier Tuff and formed the Pajarito Plateau, destroyed the older Toledo Caldera, and formed the spectacular 18-mile-diameter (29-km) Valles Caldera just west of the monument. The two eruptions together were immense—over 24 cubic miles (100 km³) of ash were ejected—over 80 times larger than the 1980 Mount St. Helens eruption! The ash near the mountain was locally up to 1000 feet (305 m) thick, and its temperature was several hundred degrees—hot enough to remelt fragments of volcanic ash and pumice and form a *welded tuff*. Welding was especially dense in the middle of the flow where heat was trapped longer.

Gases trapped in the pyroclastic flow escaped to the surface through temporary vents and must have appeared much like the Valley of Ten Thousand Smokes in Alaska after its violent eruption in 1910. The fossil vent areas are better cemented and welded than the surrounding tuff and weather into cone-shaped landforms known locally as *tent rocks*.

A caldera lake existed for some time following the eruption but was eventually drained as canyons cut headward into the caldera rim and the caldera floor again began to swell upward. Over 2000 feet (610 m) of sediment accumulated in the caldera before the Jemez River drained the caldera through San Diego Canyon on the southwest side of the caldera. Redondo Peak (11,254 feet; 3431 m), the highest in the caldera, is a *resurgent dome* that began to rise shortly after the last catastrophic eruption. Small rhyolitic plug domes have intermittently formed along ring fractures and other faults in the caldera. Eruptions continued throughout the Pleistocene Epoch with the last one occurring about 50,000–60,000 years ago (Reneau and others, 1996; Wolff and Gardner, 1995).

Home, Home in the Tuff

The Frijoles River cut its canyon into the Bandelier Tuff on the southeast side of the Jemez Mountains. Here, some 10 miles (16 km) from the vent, the tuff is only moderately welded—strong enough to locally form vertical cliffs but weak enough to be excavated by stone, bone, and wood tools. The tuff here is also case hardened by small amounts of dissolved silica that have reprecipitated as a thin crust on the canyon wall. Breaking through the outer hard layer provides access to more easily worked material—an ideal condition for Anasazi cliff dwellers.

The orientation of the valley walls to the rays of the sun significantly influences weathering processes. The observant visitor will note that the north walls (south-facing) often form cliffs and the south walls have much gentler slopes. Effective moisture is the key here: drier south-facing walls receive more sunlight and often form cliffs, while cooler, moister south walls form deeper soils and are more heavily vegetated and less steep. Pockets of weathering formed the "Swiss cheese" effect on the tuff walls, perhaps first attracting the attention of bands of cliff-dwelling Anasazi. The presence of a permanent stream enabled the Anasazi to live in relative comfort in the placid mountain canyon for nearly 300 years before the whims of climate likely caused crop failures and abandonment of their homes and fields.

The Future

Bandelier is an ideal site for the quality experience that most people desire. With only 3 miles (4.83 km) of public road and 70 miles (113 km) of trails, opportunities for a wide range of experiences exists. Short wheel-chair accessible trails from the visitor center to the ruins enable all to enjoy and contemplate the amazing accomplishments of the Anasazi. Longer trails lead downcanyon to magnificent waterfalls that plunge over the older layers of basalt beneath the Bandelier Tuff and to the Rio Grande beyond. Other trails lead into the backcountry wilderness where one can feel closer to land that is little changed since the Anasazi roamed over these same mesas and canyons.

Unfortunately, on an increasing number of days, heavy visitation requires waiting for up to an hour for a parking place! More parking spaces are not the answer—they would only increase crowding on the already crowded trails on such days. More parks and monuments are part of the answer—at least on a temporary basis until the world is willing to seriously address our growing population problem. Increasing world population by 70 million people each year cannot continue indefinitely without causing increased environmental degradation, resource depletion, poverty, food shortages, and ultimately mass starvation!

A visit to the Valles Caldera just west of the monument along the New Mexico 4 highway ties the Bandelier story together. From the overlooks along the highway one can view this large area of private land and observe some of the caldera walls, the postcatastrophic eruption plug domes, and the Redondo Peak resurgent dome. Hot springs and *solfataras* (gas vents emitting sulfur-rich fume) occur in the caldera but most are not accessible to the public. The proposed Valles Grande–Bandelier National Park would set aside this special area for the long-term benefit of future generations. The private owners of this remarkable area are willing to sell to enable this to be accomplished. Such ideas do not have high priority for politicians who wish to cut costs on a short-term basis. However, if nothing is done soon, some of our remaining outstanding natural areas like the Valles Caldera will eventually be seriously degraded or lost to the American public. Others will be much more costly to acquire. Postponing acquisition will cost taxpayers more in the long run.

The long record of volcanic activity, evidence from the behavior of earthquake waves, and the abundance of hot springs and solfataras in the caldera all indicate that an active magma chamber still underlies the Valles Caldera. Future eruptions at the scale of the plug domes and small ash eruptions of the past million years would be bothersome but relatively nondestructive. Eruptions of the magnitude of those 1.61 and 1.22 Ma would be another story.

The Anasazi have provided important lessons from which we all could bene-fit. They were acutely tuned-in to what nature can provide on a sustainable basis. Their world depended mostly on replenishable resources—a life style that could be maintained in perpetuity as long as human populations were maintained below critical levels. Modern life-styles in developed countries depend on using materi-als and energy sources that are mostly nonreplenishable. Will we need to wait until conditions are at a critical level before we relearn what the Anasazi had discovered centuries ago?

REFERENCES

Christiansen, R.L., and Yeats, R.S., 1992, Post-Laramide geology of the U.S. Cordilleran re-gion, in Burchfiel, B.C., Lipman, P.W., and Zoback, M.L., eds., The Cordilleran Oro-gen: Conterminous U.S.: Geological Society of America, Boulder, CO, The Geology of North America, v. G-3, p. 201–406.
De Nault, K.J., 1987, The Valles Caldera, Jemez Mountains, New Mexico, in Beus, S.S., ed., Geological Society of America, Centennial Field Guide, Rocky Mountain Section of the Geological Society of America: Boulder, CO, v. 2, p. 425–429.
Reilinger, R., Oliver, J., Brown, L., Sanford, A., and Balazs, E., 1980, New measurements of crustal doming over the Socorro magma body, New Mexico: Geology, v. 8, p. 291–295.
Reneau, S.L., Gardner, J.N., and Forman, S.L., 1996, New evidence for the age of the youngest eruptions in the Valles caldera, New Mexico: Geology, v. 24, no. 1, p. 7–10.
Wolff, J.A., and Gardner, J.N., 1995, Is the Valles caldera entering a new cycle of activity? Geology, v. 23, p. 411–414.

Park Address

Bandelier National Monument
HCR 11, Box 1
Suite 15
Los Alamos, NM 87544

FLORISSANT FOSSIL BEDS NATIONAL MONUMENT (COLORADO)

Florissant Fossil Beds is located about 35 miles west of Colorado Springs and the eastern edge of the Front Range. The high mountain valley containing the monu-ment is within view of the west side of 14,110-foot-high (4302-m) Pikes Peak. The Florissant Valley is underlain by lake and volcanic sediments of mid-Tertiary (late Eocene–early Oligocene, about 34–35 Ma) age that contain the "most exten-sive fossil record of its type in the world" (Hutchinson and Kolm, 1987). Abundant insect and plant fossils, along with some mammal, bird, and fish fossils give an un-usually complete record of a short moment in the evolution of lifeforms and the geologic history of North America (Meyer and Weber, 1995). In combination with the record of mammals found at John Day National Monument and the Badlands, a particularly good picture of North American life during the Oligocene has emerged. Essentially all of the butterflies and most of the record of Tertiary-age in-sects in the New World comes from Florissant.

To the Ute Indians who still had encampments in the Florissant Valley as re-cently as the early 1900s, this was the "Valley of the Stone Trees." When Judge James Castello founded the small town 2 miles (3.2 km) from what is now the national monument in about 1870, he named it Florissant after the town in Missouri where he grew up. The word is French for "flowering"—an appropriate name for the val-ley and its wildflower displays during the early summer. No doubt many of the an-cient plants now preserved as fossils at Florissant put on a similar show some 35 Ma.

The fossil beds were discovered by Dr. A.C. Peale of the Hayden survey in 1873. A short time later, Harvard paleoentomologist, Dr. Samuel Scudder, exca-vated over 25,000 insect fossils. Scientific excavations and identifications to date have yielded over 60,000 specimens representing 1144 species of insects and over 140 species of plants! To find one soft-tissue organism preserved is a rare occur-rence—to find tens of thousands specimens is truly a major discovery. Some of the better specimens are on display at the visitor center and in museums around the world. The partially excavated stumps of petrified trees are located along the over 10 miles (16 km) of trails through this relatively small (6000 acres, 9.4 square miles, 15 km^2) but interesting park.

Geologic Sequence

Following the Laramide Orogeny and the development of a widespread, low-relief surface of erosion in the Rocky Mountains during late Eocene time (about 40 Ma), widespread volcanism occurred. Andesitic lavas and ash covered the ancient sur-face of erosion and, in the Florissant area, a mudflow dam created a large lake dur-ing the late Eocene–early Oligocene about 35 Ma (MacGinitie, 1953; Meyer and Weber, 1995).

Abundant plants and forest occupied the swampy edges of Lake Florissant and the lower areas of the valley. A type of giant Sequoia grew in the swampy lake edges; hardwood forests covered wet areas along the lake and streams. A scrub forest and grass covered drier areas away from water bodies (Hutchinson and Kolm, 1987). The air and land teemed with millions of insects living in the warm temperate climate that prevailed here some 35 Ma. A series of eruptions over a 700,000-year interval from a large volcano complex located about 16 miles (26 km) to the southwest produced dense clouds of ash that turned the area into an "insect Pompeii," with large num-bers of insect bodies settling to the lake bottom with the ash. Later mudflows and lava flows further shielded the lake shales with a protective, air-tight cover.

By late Oligocene time the area probably had extensive rolling surfaces of moderate relief with elevations of less than 3000 feet (915 m). Present elevations are 8500 feet (2590 m), indicating that many changes were yet to come. Over a mile of vertical uplift during the past 30 million years was accompanied by local faulting that dropped the Florissant beds downward against the underlying Pikes Peak Granite (Precambrian age, about 1 billion years old), thereby delaying the erosion that ultimately cut the present valley and exposed the buried lake shales. Erosion eventually prevailed, and for the first time in 35 million years the sun's rays once again fell on the remains of ancient inhabitants of Florissant Valley.

Snapshot in Time

The incredible fossil record at Florissant is unique in that it enables paleontologists and geologists to reconstruct a relatively brief moment of time 35 Ma using types of organisms that are not ordinarily fossilized. Fossilization is a rare event and is most likely to occur in organisms that possess hard bone or wood tissue. Rapid burial and the presence of mineral-rich groundwater further encourages fossilization. Here, insects and plants were snuffed out rapidly as ash fell in large volumes. Both ash and insects settled to the bottom of Lake Florissant and compression under the weight of additional sediment and rock produced carbon films. The finer than talcum-powder-size ash faithfully preserves minute anatomical details enabling paleoentomologists to study everything except the internal organs and the original color of the insects. Delicate antennae, legs, and hairs on the bodies are often preserved. Butterfly and moth wings often show the patterns of spots and other markings (Fig. 11–22). Some insects were identified by comparison to mod-

FIGURE 11–22   Well-preserved moth, one of more than 1100 species of insects found in Oligocene ash at Florissant Fossil Beds National Monument. (Photo by E. Kiver)

ern forms by using a microscope to help count the number of facets in the eyes! A wide variety of insects are preserved here including mosquitos, spiders (over 50 species!), grasshoppers, earwigs, beetles, lice, ants (over 24 species), aphids, dragonflies, and even two species of tsetse fly! Insects were much like those today, although many no longer occur in Colorado and others are now extinct.

Soft plant fossils include leaves, cones, and even delicate petals of wild rose—also preserved as carbon films. Rain mixing with ash on nearby slopes mobilized as mudflows and swiftly flowed into the lake, depositing a thick mud-rich layer that further prevented oxygen from reaching and destroying the delicate organic remains entombed below. Earlier mudflows had buried the lower parts of trees growing along the streams and lake edge—including a species of giant Sequoia that lived along the lake edge. Only the lower 14 feet (4.3 m) of the stumps are preserved (Fig. 11–23)—the upper parts of what were likely 300-foot-tall (91-m) trees stuck up above the mudflow layer and probably rotted away. Silica from the volcanic ash was mobilized by groundwater and impregnated the woody fabric, thereby converting it to petrified wood. Details of the woody tissue enable paleobotanists to identify tree types. In addition to Sequoia, also present are palm, magnolia, cedar, oak, maple, beech, willow, and elm. The largest of the excavated Sequoia trees is 13

FIGURE 11–23   Giant Sequoia tree stump in Florissant Fossil Beds National Monument. Stumps were exposed by excavation of volcanic mudflow material. Note person for scale and metal band placed around stump to slow the weathering processes. (Photo by E. Kiver)

feet (4 m) in diameter and has a circumference of 42 feet (12.8 m). Fossilized tree rings indicate that this tree was at least 1000 years old! An unknown number of outer rings were lost to erosion since the stump was excavated in 1920.

The combination of botanical evidence indicates that a warm temperate, nearly subtropical climate, similar to that in the southeastern United States or Northeastern Mexico prevailed 35 Ma in Colorado. Further inferences from the fossil record, particularly the physical characteristics of leaves, can be used to infer paleoelevation. Depending on the method used, the results are contradictory. MacGinitie's (1953) method indicates that elevations were less the 3000 feet (915 m); Gregory and Chase (1992) believe that late Eocene elevations were similar to the present (about 8200 ft; 2500 m). The paleoelevation is important to geologists who attempt to reconstruct the erosional, structural, climatic, and uplift history of the Rocky Mountain region (Madole and others, 1987). More research is needed to resolve the conflicting elevation interpretations.

Birth of a National Monument

In spite of the incredible paleontologic discoveries, the Florissant area remained in private ownership through the years. Owners allowed visitors to remove fossils for a fee. In many cases people looted the area, especially to steal some of the abundant fossilized wood specimens. Areas that were once impassable by wagons because of the rocky litter of petrified wood were soon passable as the fossil treasures were hauled off and disappeared into oblivion. How many important specimens were removed as souvenirs and what valuable information they might furnish will never be known. The Henderson and later Pike Petrified Forest companies built a small museum in 1924 and allowed digging and unlimited collecting of fossils into the 1960s. Proposals to place the area under National Park Service protection were made as early as 1911, but private ownership, apathy toward science, concern about cost to the government (taxpayers), and Congress's assumption that the fossils were infinitely abundant and the huge fossil tree stumps were adequately protected from vandals by existing private companies proved a deterrent to establishing a park. Agnes Singer, owner of the Colorado Petrified Forest noted that "tourists began sneaking in under the fence and taking all they could carry." Other owners noted that they wished that their land belonged to the government while there were still fossils left! Private owners and operators were willing and anxious to sell but Congress remained apathetic. Eventually after waiting for 8 years for Congress to act, 1800 acres of the proposed monument land was sold to a real estate company.

Developers then discovered a new way to make money—why not sell parcels of land for building summer homes? This would, of course, sacrifice forever a world-class paleontological area. Researchers such as Estella Leopold (daughter of renowned conservationist Aldo Leopold) and the general public would lose an important site critical to understanding how North America and its life have changed.

An estimated millions of unexcavated fossils still remain in the undisturbed ground beneath this part of the Florissant Valley. Fortunately, the 1960s were a time of environmental awareness and concern, and groups of individuals protested—some willing to lie down in front of the real estate developer's bulldozers (Meyer and Weber, 1995). With the realization that destruction of the area was imminent, politicians responded and President Nixon authorized purchase of private lands for the establishment of Florissant Fossil Beds National Monument in 1968. The area officially became a national monument in 1969. The old Henderson Museum was purchased in 1974 and now houses the museum and park headquarters.

REFERENCES

Gregory, K.M., and Chase, C.G., 1992, Tectonic significance of paleobotanically estimated climate and altitude of the late Eocene erosion surface, Colorado: Geology, v. 20, p. 581–585.

Hutchinson, R.M., and Kolm, K.E., 1987, The Florissant Fossil Beds National Monument, Teller County, Colorado, in Beus, S.S. ed., Centennial Field Guide, Rocky Mountain Section of the Geological Society of America: Geological Society of America, Boulder, CO, v. 2, p. 329–330.

MacGinitie, H.D., 1953, Fossil plants of the Florissant beds, Colorado: Washington, DC, Carnegie Institution of Washington Publication 599, 198 p.

Madole, R.F., Bradley, W.C., Loewenherz, D.S., Ritter, D.F., Rutter, N.W., and Thorn, C.E., 1987, Rocky Mountains, in Graf, W.L., ed., Geomorphic systems of North America: Geological Society of America, Boulder, CO, Centennial Special Volume 2, p. 211–257.

Meyer, H.W., and Weber, L., 1995, Florissant Fossil Beds National Monument: Preservation of an ancient ecosystem: Rocks and Minerals, v. 70, p. 232–239.

Park Address

Florissant Fossil Beds National Monument
Box 185
Florissant, CO 80816

TWELVE

Great Plains Province

The Great Plains Province lies east of the Rocky Mountains and the Basin and Range Province and extends from the Rio Grande in south-central Texas far northward into Canada (Plate 12). It is regarded as part of the stable interior, or *craton,* of North America. Much of the province is characterized by extensive low-relief topography. However, its proximity to the Rocky Mountains and the Laramide forces that built them has produced some localized mountain structures and volcanism near its west edge.

The Black Hills are the easternmost of these isolated Rocky Mountain uplifts in the Great Plains. In its Precambrian core (Fig. 12–1) is found 7242-foot-high (2208-m) Mount Harney (highest point east of the Rocky Mountains) and Mount Rushmore National Memorial. Paleozoic limestone encircling the Precambrian core hosts a number of caves, including the outstanding Wind and Jewel caves systems, two of the longest in the world.

In addition to the more spectacular uplifts and areas of igneous activity such as those at Devils Tower and Capulin Mountain national monuments, running water has produced intricately eroded areas of *badland topography* at the Badlands and Theodore Roosevelt parks in the Dakotas. Exposed in the layers of the Tertiary-age sedimentary debris cover that was shed from the eroding Rocky Mountains is an excellent record of terrestrial life in western North America that is well illustrated by fossils found at Theodore Roosevelt, Agate Fossil Beds, and the Badlands. Thus, local topographically interesting areas and the important geologic and fossil record in the rocks and sediments well qualify the inclusion of a number of Great Plains areas into the National Park System.

Geography

The western boundary is the sharp topographic break where the flat-lying or gently inclined Paleozoic and Mesozoic strata of the Great Plains are bent dramatically

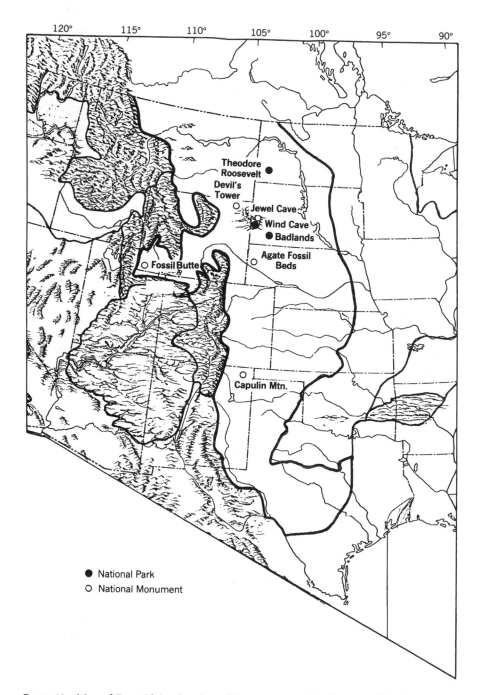

FIGURE 12–1   Precambrian core rock in the Black Hills uplift in the Needles area. (Photo by E. Kiver)

upward into *hogbacks* along the Middle and Southern Rocky Mountains (Fig. 11–1) or where resistant Precambrian rocks in the Northern Rocky Mountains have been thrust over and tower above the Mesozoic and Cenozoic strata of the Great Plains (Fig. 9–7). Elevations along the west edge are between 4000 and 6000 feet (1220 and 1830 m) dropping to about 1500 feet (460 m) along the eastern boundary. The boundary with the Interior Lowlands Province to the east is a scarp or elevation rise in the northern and southern portions of the Great Plains in the United States but is transitional and less distinct in the Nebraska and Kansas area. Within a 100-mile-wide (162-km) zone, the dryness of the plains gives way to a more humid environment; widely spaced intermittent streams are replaced by closely spaced perennial streams. Here the essentially treeless plains become wooded and the underlying Paleozoic rocks are widely exposed. Also, the extensive cover of Tertiary rocks so prominent in many Great Plains areas is nearly absent in the Interior Lowlands.

Sufficient differences in landforms exist in the mostly flat-lying, rolling topographic surfaces of the Great Plains to cause Fenneman (1931) to subdivide the province into 10 sections. Of interest here is the Missouri Plateau Section in North Dakota where Theodore Roosevelt National Park is located; the Black Hills Section in South Dakota that contains Mount Rushmore and Wind and Jewel caves; the High Plains Section where Devils Tower, Agate Fossil Beds, and the Badlands are located, and the Raton Section in New Mexico where Capulin Mountain and other volcanic features are abundant. In addition, we include the Wyoming Basin, an extension of the Great Plains topography into the Middle Rocky Mountains (Plate 12). Thus, Fossil Butte National Monument and its Eocene-age sediments and fossils are also discussed in this chapter.

Geologic History

Precambrian mountains were reduced to low-relief by the time that shallow Cambrian (early Paleozoic) seas began to creep across and cover much of the North American continent. Fluctuating shallow seaways and relatively thin layers of Paleozoic and Mesozoic strata characterize the history of the craton until Late Cretaceous time when the Laramide Orogeny began. Uplift of the Rockies produced a corresponding downwarp in the Great Plains where extensive shallow basins formed. The Williston Basin formed in the northern Great Plains in the Montana–Dakota area, the Denver–Julesburg Basin in eastern Colorado, and the Midland Basin in Texas (Thornbury, 1965), as well as a number of smaller downwarps or basins (Wayne and others, 1991).

Alluvium (stream-deposited sediment) from the eroding mountains filled these shallow basins to overflowing. Local basins that formed in the Fossil Butte area during Eocene time filled with lake sediments [about 45 million years ago (Ma)]. However, shifting stream channels, floodplains, swamps, and occasional air-fall ash (especially in Oligocene and Miocene time) mostly characterize the Tertiary-age Great Plains sediments. The landscape some 6–8 Ma was quite monotonous as the vast fluviatile plain sloped gradually eastward from the mountains—with only the Black Hills and a few other local mountains rising above them. The depositional surface of the Great Plains ramped up to the mountain edges where erosional plains were cut across the hard Precambrian and younger rocks to form the high-level-erosion surfaces found in Rocky Mountain National Park and other Rocky Mountain areas.

Uplift of the Rocky Mountains and eastward tilting of the Great Plains during the Pliocene initiated a new regime of downcutting and removal of sedimentary debris. Tertiary sediments were completely removed from the Colorado Piedmont and certain other areas. However, the High Plains from South Dakota to the Texas border forms an extensive area where the late Miocene-Pliocene surface has been little eroded. A narrow remnant of this old surface extends westward in southeastern Wyoming, north of the Colorado Piedmont; it is called the Gangplank. The Gangplank demonstrates that the formation of the Sherman erosion surface (subsummit erosion surface, see Rocky Mountain National Park discussion in Chapter 11) and the Oligocene-Miocene sedimentary caprock of the High Plains are related, simultaneous events. Modern day Union Pacific tracks and Interstate 80 make use of this natural sedimentary rock ramp to cross the Laramie Mountains.

Pleistocene ice sheets pushed southwestward out of Canada and covered the northern part of the Great Plains, the area north of the Missouri River. Prior to glaciation, the Missouri flowed northeastward into Hudson Bay. Blocked off by ice during the Wisconsin advance, the river followed a new course along the front of the glacier. Once established, the Missouri maintained essentially this course after the glacier retreated.

The contrast in landforms on the two sides of the glacial boundary is striking. The glaciated section is noted for its hummocky topography made up of irregular moraines in which kettles, kettle lakes, and morainal lakes are abundant. The streams wander aimlessly about through the glacial deposits, into and out of lakes and swamps that are seasonally filled with ducks and geese—an extremely critical environment for their survival. There are those who would drain these wetlands to gain added cropland and boost our gross national product! Those who measure the value and quality of life in dollar signs probably have more dollars than the rest of us—but they are indeed poor.

In the area south of the glacial boundary, streams have developed regular drainage patterns and typical fluvial landforms. Excellent summaries of the Quaternary history of the area are available in Wayne and others (1991) and Gustavson and others (1991). Dry climatic episodes during the past 1000 years have generated desert conditions—conditions that would make modern agriculture impossible. Muhs and colleagues (1993) believe that we are close to the threshold where windblown sediment will again move and convert thousands of square miles into dunes and desert—a frightening scenario—especially in light of global warming.

Therefore, although extensive monotonous plains still exist, there are many areas where interesting, even spectacular, geologic features have been developed. A number of such areas are now national parks and monuments. Rather than discuss these areas in the geographic order in which one might visit them, our tour of the Great Plains will follow the geologic events that made today's province. We will begin in the Black Hills at Mount Rushmore National Memorial where Precambrian rocks in the core of the uplift have been sculpted into the faces of four of our most significant presidents. A short distance away are the Paleozoic limestones that contain the remarkable Wind and Jewel cave systems. Magma generated during and after the Laramide uplift of the Black Hills produced a number of localized intrusions in surrounding areas, including the spectacular monolith of Devils Tower. We will then follow the Tertiary strata from Paleocene (very early Tertiary) at Theodore Roosevelt National Park to Eocene lake sediments and the finest fish fossils in the world at Fossil Butte National Monument. Oligocene sediments and a fossilized zoolike collection of early mammals that lived in the vast Serengetti-like plains of western North America are present at Badlands National Park. Fluvial sediments containing fossil mammal bones at Agate Fossil Beds National Monument continue the story of mammal evolution into Miocene time. Finally, the volcanic outpourings during post-Pleistocene time at Capulin Mountain will complete our Great Plains tour.

FIGURE 12–2   Bison in Wind Cave National Park are some of the only remaining remnants of the 60 million that were annihilated by Euro-Americans in the late 1800s. (Photo by E. Kiver)

Excellent museums in Denver, Rapid City, and elsewhere also help tell the story of the area's geology and evolving lifeforms. Missing from the story in the national parks is the latest chapter of life that includes the mammoths, mastodons, sloths, and other large mammals of Pleistocene and Holocene times. These animals suspiciously disappeared from the face of the earth about 8000 years ago, about the same time as the rising importance of another mammal species—*Homo sapiens*. The latest upheaval in animal populations occurred as Euro-Americans exterminated over 60 million bison during the late 1800s and forever changed the Great Plains environment. Small herds of the noble animals can still be seen in the Badlands and at Wind Cave National Park and Custer State Park in the Black Hills (Fig. 12–2).

REFERENCES

Fenneman, N.M., 1931, Physiography of western United States: McGraw Hill, New York.
Gustavson, T.C., Baumgardner, R.W., Jr., Caran, S.C., Holliday, V.T., Mehnert, H.H., O'Neill, J.M., and Reeves, C.C., Jr., 1991, Quaternary geology of the Southern Great Plains and an adjacent segment of the Rolling Plains, in Morrison, R.B., ed., Quaternary nonglacial geology; Conterminous U.S.: Geological Society of America, Boulder, CO, The Geology of North America, v. K-2, p. 477–501.

Muhs, D.R., Millard, H.T., Jr., Madole, R.F., and Schenk, C.J., 1993, History of desertification on the Great Plains: a Holocene history of eolian sand movement: in Kelmelis, J.A., and Snow, K.M., eds., Proceedings of the U.S. Geological Survey global change research forum, Herndon, Virginia, March 18–20, 1991: U.S. Geological Survey, Reston.

Thornbury, W.D., 1965, Regional geomorphology of the United States: Wiley, New York, 609 p.

Wayne, W.J., Aber, J.S., Agard, S.S., Bergantino, R.N., Bluemle, J.P., Coates, D.A., Cooley, M.E., Madole, R.F., Martin, J.E., Mears, B., Jr., Morrison, R.B., and Sutherland, W.M., 1991, Quaternary geology of the Northern Great Plains, in Morrison, R.B., ed., Quaternary nonglacial geology; Conterminous U.S.: Geological Society of America, Boulder, CO, The Geology of North America, v. K-2, p. 441–476.

MOUNT RUSHMORE NATIONAL MEMORIAL (SOUTH DAKOTA)

The four presidential faces carved into Mount Rushmore in the core of the Black Hills are recognized worldwide as unique and significant (Fig. 12–3). The 60-foot-high (18-m) faces of four important U.S. presidents also represent four important phases in the country's history: George Washington as the commander of the revolutionary army and the first U.S. president; Jefferson as author of the Declaration of Independence, our third president, and a proponent of westward expansion; Lincoln as the 16th president who preserved the Union and ended slavery on U.S. soil; and Theodore Roosevelt, the 26th president who helped initiate the movement to protect public lands for the benefit of all. Mount Rushmore is a "must see" for all who visit the Black Hills. The monument also provides a convenient locale to briefly describe the general geology of the Black Hills and to begin our tour of the Great Plains parks.

FIGURE 12–3   The four presidents carved into granite on Mount Rushmore in the Black Hills of South Dakota. Note the contact of the light-colored granite with the darker-toned metamorphic rock below the sculptures about half way down Mount Rushmore. Compare with the "before" photo in Figure 12–6. (Photo by Stanley A. Schumm)

Black Hills—Geologic Setting

The Black Hills are a Laramide uplift—a bit of Rocky Mountain geology that formed in the Great Plains. Because it is topographically detached from the Rocky Mountains some 120 miles (193 km) to the west, it must be included in the Great Plains Province rather than the Rocky Mountain Province (Plate 12). The 125- by 65-mile-elliptical (200–105 km) Black Hills dome is elongated in a north–south direction and has experienced at least 9000 feet (2745 m) of vertical uplift (Thornbury, 1965). Erosion has removed the Paleozoic and Mesozoic strata from its central area, exposing the Precambrian core of the uplift (Figure 12–4) where the geologic story of Mount Rushmore begins. The younger rocks of Paleozoic and Mesozoic age appear as concentric rings of strata around the Precambrian core—thus creating an elongated bullseye pattern on geologic maps.

Elevations in the surrounding Great Plains are about 3000–3500 feet (915–1070 m) compared to 7242 feet (2208 m) at Harney Peak, the highest peak in the Black Hills. Rivers flow radially from the rugged central area across hogbacks (ridges underlain by steeply inclined strata) on the east side, *cuestas* (ridges underlain by gently inclined strata) on the west side, and the Red Valley—a scenic *strike valley* underlain by the red shales and soft red sandstones of the Triassic-age Spearfish Formation that encircles the uplift.

Laramide mountain building (65–45 Ma) also produced deep faults and fractures that formed conduits for magma to rise toward the surface in the northern Black Hills and nearby areas in the Great Plains, including the Devils Tower area. Erosion of the Black Hills and Rocky Mountains sent vast quantities of sediment down the ancient stream systems into the Great Plains, burying the plains and the lower parts of the mountain areas. Much of this debris apron still remains in the Great Plains as we will discover later in this chapter when we visit Theodore Roosevelt National Park, the nearby Badlands, and Agate Fossil Beds.

Although the Black Hills are topographically high, conditions were apparently too dry for glaciers to form during the Pleistocene (Rahn and others, 1985). Thus the rugged, jagged landscape in the core of the range is one produced by differential weathering and erosion, mostly by running water.

Precambrian Geology

The minerals in the Precambrian rocks exposed in the core of the Black Hills require extremely high temperatures and pressures to form—the type of conditions that existed when sedimentary and igneous rocks were buried deeply during an ancient episode of mountain building. Erosion has exposed to view the deep "innards" of this Precambrian-age mountain range for all to see and contemplate. A few small areas contain metamorphic and igneous rocks produced during mountain building at least 2500 Ma, but most of the schists, phyllites, quartzites, and other rocks were intensely metamorphosed about 2100 Ma during a younger episode of mountain building (Redden, 1985).

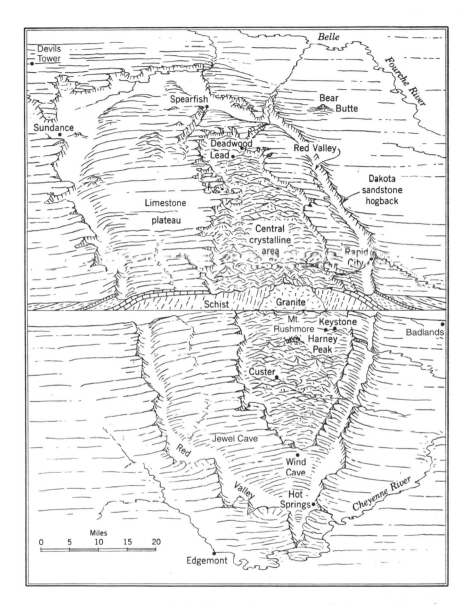

FIGURE 12–4 Diagram of the Black Hills area showing a generalized cross section and the locations of national parks and monuments. (Modified slightly from A.N. Strahler, 1960, *Physical Geography,* John Wiley & Sons)

Crustal melting generated a large magma body that squeezed and melted its way into the surrounding metamorphic rocks, eventually crystallizing into the Harney Peak Granite about 1700 Ma (Shearer and Papike, 1987). An eastward extension (sill?) of the Harney Peak Granite was injected into the schists in the Mount Rushmore area and would much later furnish the media for noted sculptor Gutzon Borglum to apply his creative genius. Remnant magma from the cooling batholith was injected into the crystallized granite and the metamorphosed country rock as numerous dikes and sills. Very late-stage, volatile-rich magmas in turn forced their way through earlier formed dikes and rocks. The volatile-rich fluids formed *pegmatites*—intrusive igneous rocks with unusually large mineral grains that are of interest to mining companies and mineral collectors. The granite dikes and pegmatites show up as white bands extending through the sculpted faces at Mount Rushmore (Fig. 12–5).

Quartz and feldspar are common in the pegmatites, but the unusual chemistry of many of these pegmatites caused some rarer minerals to form, some of which were mined in nearby areas. *Spodumene,* a mineral and an ore of the rare element lithium, often occurs in some of the pegmatites outside of the memorial as single crystals 3–10 feet long (1–3 m), although crystals up to 40 feet long (12 m) are known (Shearer and Papike, 1987)!

Intense heat and pressure accompanying the emplacement of the granite complex bent and folded the surrounding schist. The contorted appearance of the schist in the lower part of Mount Rushmore (particularly well shown below the head of Washington, Fig. 12–3) attests to the intense pressures exerted some 1700 Ma when the Black Hills lay under a cover of perhaps 6–9 miles (10–15 km) of rock (Redden and others, 1985).

The Harney Peak Granite was locally deeply fractured during the Laramide Orogeny. Erosion concentrating in these weakened areas formed the pinnacles and rock fins in the scenically spectacular Needles area (Fig. 12–1). Fortunately, the thick granite sill at Mount Rushmore remained more monolithic, preventing the massive sculptures from collapsing along joints and fractures. Hopefully the sculptures will continue to remain intact for many centuries and perhaps millenia.

Making of a Rock Monument

Although the right geologic conditions are absolutely necessary for Mount Rushmore Memorial to exist, even more important is the imagination, creativity, and willingness of people to invest time and money in such a bold adventure. South Dakota's state historian, Doane Robinson, became intrigued with the idea of a supercolossus rock sculpture. In 1924 he contacted the well-known American artist, Gutzon Borglum. Robinson visualized a monument to western heroes. Borglum believed that it should have a broader significance—one that symbolized some of the more important historic phases of a relatively new republic.

Borglum spent days on horseback investigating the Harney Peak Granite looking for a favorable site. The granite needed to have a relatively uniform size of min-

FIGURE 12–5   The head of Lincoln during construction. Note the light-colored pegmatite dikes extending through the sculpture. (National Park Service photo)

eral grains, it must be free of extensive and deep fractures, and it must be on a south-facing cliff to receive maximum sunlight. The remotely located Mount Rushmore satisfied the geologic requirements admirably (Fig. 12–6).

Borglum was an accomplished sculptor who had worked on the carving of the huge Confederate Memorial at Stone Mountain in Georgia and was one of the only individuals in the world who knew how to produce such large rock sculptures. Art and dynamite do not usually mix, but Borglum was able to combine art, engineering, skilled workmen trained on the job, and dynamite to get the job done! Work began in 1927 and continued intermittently (because of a lack of dependable funding) for 14 years. Work halted in 1941 about 8 months after Borglum's death. His son was then the sculptor-in-charge and used the remaining funds to refine the

FIGURE 12–6   Mount Rushmore as it appeared in 1925 prior to the removal of the outer, weathered rock and the beginning of the sculpture process. Compare with finished sculptures in Figure 12–3. (National Park Service photo)

faces to make them look more lifelike. The original plan to finish the figures to their waists showing their hands and period clothing was abandoned as the country became embroiled in World War II (Borglum, 1996).

The four presidents depicted here, Washington, Jefferson, Lincoln and Theodore Roosevelt, brought the United States through important phases of history including the founding, expansion, preservation of freedom for all, and preservation of some of the natural beauty of a fledgling nation (Borglum, 1996). It is a memorial not just to a few individuals but also to the millions of people who through their hard work and creativity built a nation of opportunity for all.

Visiting Mount Rushmore

The incredible story of a man determined to create a lasting work of art for future generations without concern for monetary gain is inspiring to those who take time to learn the history of the memorial. The scale of the faces is such that as one stands on the viewing terrace 400 feet (122 m) below the top of the heads, the feet of each figure, if completed, would be located another 65 feet (20 m) below the terrace! The fascinating details of the construction process that used a simple but ingenious system of transferring measurements from a model to the rock face by using a horizontal bar and a plumb bob is explained in the visitor center and in Borglum (1996).

The rocks themselves tell their story of the turbulent past when the pressure of some 9 miles (14 km) of rock buried the area, and magma engulfed and injected itself into and domed the metamorphosed Precambrian sediments. The Laramide Orogeny produced the present uplifted mountain range and triggered the erosion that resulted in today's landscape. Mount Rushmore provides an excellent jumping-off place to follow the rest of the geologic story of the Great Plains Province.

REFERENCES

Borglum, L., 1996, Mount Rushmore: The story behind the scenery: KC Publications, Las Vegas, 48 p.

Rahn, P.H., Bump, V.L., and Steece, F.V., 1985, Engineering geology of the central Black Hills, South Dakota, in Rich, F. J., ed., Geology of the Black Hills, South Dakota and Wyoming: Geological Society of America, Boulder, CO, Field Trip Guidebook for the Rocky Mountain Section, 1981 Annual Meeting, p. 135–153.

Redden, J.A., 1985, Summary of the geology of the Nemo area, in Rich, F.J., ed., Geology of the Black Hills, South Dakota and Wyoming: Geological Society of America, Boulder, CO, Field Trip Guidebook for the Rocky Mountain Section, 1981 Annual Meeting, p. 193–209.

Redden, J.A., Norton, J.J., and McLaughlin, R.J., 1985, Geology of the Harney Peak Granite, Black Hills, South Dakota, in Rich, F.J., ed., Geology of the Black Hills, South Dakota and Wyoming: Geological Society of America, Boulder, CO, Field Trip Guidebook for the Rocky Mountain Section, 1981 Annual Meeting, p. 225–240.

Shearer, S.K., and Papike, J.J., 1987, Harney Peak Granite and associated pegmatites, Black Hills, South Dakota, in Beus, S.S., ed., Rocky Mountain Section of the Geological Society of America: Geological Society of America, Boulder, CO, Centennial Field Guide Vol. 2, p. 227–232.

Thornbury, W.D., 1965, Regional Geomorphology of the United States: Wiley, New York, 609 p.

Park Address

Mount Rushmore National Memorial
P.O. Box 268
Keystone, SD 57751

WIND CAVE NATIONAL PARK (SOUTH DAKOTA)

Wind Cave National Park is two parks in one—an underground park and an above-ground prairie park that provides a glimpse of the Great Plains before Euro-Americans changed forever an ecosystem that was over 10,000 years in the making. The 28,295-acre (44 square mile, 115 km²) park on the southeastern flank of the Black Hills uplift (Fig. 12–4) contains Wind Cave, the fifth longest cave in the United States and the sixth longest in the world! It also contains grassy prairies where bison, deer, antelope, prairie dogs, and other animals live in a landscape relatively unchanged from the late 1800s.

The park lies on the southeast flank of the Black Hills dome where rocks of Paleozoic and Mesozoic age are inclined (*dip*) gently to the southeast. The northwestern part of the park includes a small part of the Precambrian core of the Black Hills. Younger layers of Paleozoic and Mesozoic rocks to the south of the core are stacked on top like pages in a book. The cave is developed in the Pahasapa Limestone of Mississippian age (about 340 Ma) and contains excellent examples of relatively rare cave features called *boxwork, frostwork,* and *cave linings.* Wind Cave is well named. It is a *blowing cave,* a cave where strong currents of air move alternately in and out of the cave, sometimes at velocities exceeding 50 miles/hour (80 km/hr) that require closing the cave temporarily!

History of a Park

The Paha Sapa, or "Hills Black," were the sacred, traditional hunting grounds of the Sioux Indians, a tradition that was reinforced by the treaty of 1868 that gave the Sioux the Black Hills "forever." The definition of "forever" in this treaty was apparently 8 years. Lieutenant George Armstrong Custer's 1874 exploratory expedition to the Black Hills violated the treaty of 1868. Custer's report indicated that gold was present in significant quantities in the Precambrian core of the range—a finding that fired-up public interest. However, the Sioux were unwilling to sell their sacred lands. Custer returned in 1876 and he and 260 of his cavalrymen were annihilated in the Battle of the Little Bighorn. The stunning Sioux victory was to no avail as the U.S. Army ultimately prevailed. Large numbers of miners and settlers entered the area in the late 1870s—quickly converting a wilderness into mining camps and ranches.

Wind Cave is likely the Sioux's "Sacred Cave of the Winds" where the spirits of their ancestors and the buffalo emerged long ago. The movement of air through the narrow cave entrance can be quite impressive. Local discovery stories involve a hat being blown off a cowboy's head. Returning the next day with a friend, the cowboy threw a hat at the 1-foot-wide (30-cm) cave entrance. It was reportedly sucked into the cave! Such reversing behavior of air currents does indeed occur in

Wind Cave as a result of changes in outside air pressure. Thus, Wind Cave is a natural barometer with air moving out of the cave when a low-pressure weather system moves into the area and air moving into the cave when high pressure arrives. The narrow natural entrance intensified the wind velocity as a large volume of moving air was forced through a small opening.

The official discoverer is a local settler, Tom Bingham, who in 1881 followed a loud whistling sound to the tiny entrance where air was violently blowing. He dug a larger entrance and became the first to enter the cave. A mining claim was filed in 1890, and Jesse McDonald and his sons became the first caretakers and explorers of the cave. Jesse's son Alvin was a determined explorer who kept a detailed diary of his discoveries. He was one of the first of many who through the years would "follow the wind." Stairways were constructed and the cave was opened to the public in 1892. Occupation of the land and operation of the cave by the McDonald family and later the Stabler family on public land was illegal. The land reverted to the federal government, and the cave became the nation's first cave park and its seventh national park in 1903. The park was enlarged in 1935 when the adjacent Wind Cave Game Reserve was added.

Geological Background

The Precambrian core rock of the Black Hills dome exposed in the northwest part of the park gives way to Paleozoic- and Mesozoic-age sedimentary rocks to the south. Seaways covered and uncovered the Black Hills and Great Plains area many times over the past 500 million years. In addition, nonmarine sediments of Mesozoic age record the presence of a thriving population of dinosaurs. Of particular interest at Wind Cave is the mid-Paleozoic (Mississippian) shallow seaway in which innumerable shells and fragments of shells accumulated to form the Pahasapa Limestone, the cave-forming limestone of the Black Hills. Pockets of gypsum in the limestone record localized areas where intense evaporation and poor water circulation occurred. The domal uplift of the Black Hills began during the Laramide Orogeny (about 60 Ma in the Black Hills), and subsequent erosion exposed the Mississippian and other sedimentary strata. The strata are exposed today in concentric layers that form an oval pattern around the Black Hills uplift. The Pahasapa Limestone forms the Limestone Plateau in the Wind Cave–Jewel Cave area on the south flank of the Black Hills (Fig. 12–4).

The uplift produced radiating fracture systems (joints and faults) that greatly influenced the location of cave passages and the eventual character of the cave. The gentle slope of the limestone layers and more soluble beds within the Pahasapa at Wind Cave also influenced cave development. Three zones in the upper part of the Pahasapa Limestone are particularly soluble, thus accounting for the three prominent levels of cave passages.

Cave Genesis

The process by which water combines with carbon dioxide in the atmosphere and soil occurs whenever nonmarine conditions exist. If a soluble rock such as limestone is within reach of the slightly acidified groundwater, then the process of solution begins. A few million years after deposition of the Pahasapa Limestone, the emergence from the sea resulted in the formation of a *karst* topography, characterized by such solutional features as caves and sinkholes. Thus, at least some of the passages in the upper level of today's cave formed some 320 Ma. The reddish surface soils as well as the overlying marine sands of the Minnelusa Formation filled these ancient sinkholes and are exposed in the upper level passages as a *paleokarst* surface. Cave formation was interrupted as sedimentary rock layers deeply buried the Pahasapa Limestone and the paleokarst surface until early Tertiary time (about 60 Ma) when the Laramide Orogeny uplifted the Black Hills and subjected them to a long interval of erosion that continues today.

The brittle limestone was severely fractured during the uplift. The more continuous fractures are oriented northwest–southeast and have concentrated the movement of acidic groundwater and therefore the location of cave passages (Fig. 12–7). The major fractures and lesser cross fractures enabled solutional openings to develop a complicated three-dimensional maze pattern that has yet to be completely explored. Cave formation may have started shortly after the uplift of the Black Hills, perhaps 50 Ma according to Deal (1962). If so, this is one of the oldest caves in the world. Most caves have a history of development that goes back at most only 1 million years. In spite of intensive exploration beginning in the 1950s by qualified *speleologists* (scientists and others who study caves) under the supervision of the National Park Service, new passageways continue to be discovered. The underground maze is located in an area of less than 2 square miles (5 km^2). However, so far the total combined length of all mapped passages in Wind Cave exceeds 80 miles (130 km)!

The myriad of smaller fractures produced during uplift had gypsum squeezed into the tiny openings. Later the gypsum was chemically converted to dense veins of calcite as hot thermal waters circulated through the limestone (Bakalowicz and others, 1987). As cave enlargement progressed, the dense calcite veins proved more resistant to the solution process than the limestone between the veins. Thus, as the cave enlarged, the complicated arrangement of criss-crossing calcite veins protruded into the cave as boxwork (Fig. 12–8), an uncommon feature in most caves. Wind Cave, especially in its middle and lower level passages, contains the best and most abundant examples of boxwork in any known cave. To the early explorers Jesse McDonald and his sons, the calcite projections from the ceiling and walls resembled post office boxes. Early visitors even left notes in the Post Office Room for subsequent visitors to read. Eventually the name "post office boxes" was shortened to boxwork. Where the fractures are extremely numerous and closely spaced, the delicate pattern of protrusions are called *lacework*. On the opposite end of the spectrum are widely spaced fracture fillings up to 5 feet (1.5 m) apart that are called *cratework!*

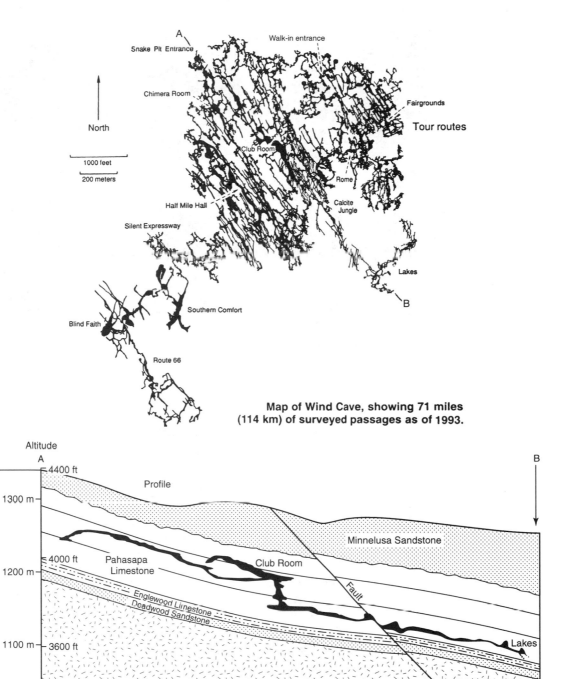

FIGURE 12–7 Generalized cross section and map of Wind Cave showing the northwest–southeast orientation of the main passages. (From Palmer, 1981; courtesy of Black Hills Parks and Forests Association)

FIGURE 12–8   Boxwork, a complex of calcite projections in Wind Cave National Park. (Photo by Tom Miller)

Other Cave Features

In addition to the remarkable boxwork, an intricate maze of random solutional openings called *spongework* forms a Swiss-cheese-like pattern—especially in the upper cave levels. Enlargement of passageways usually destroyed such spongework that had developed earlier in the middle and lower levels of the cave.

The usual *speleothems* (secondary deposits of minerals in a cave) like dripstone and flowstone are not abundant here except in cave passages located directly beneath surface streams where more water reaches the cave below. However, a knobby, grapelike calcite growth on walls and ceilings called *popcorn* (technical name is *globulites*) is relatively abundant. Popcorn can form in a variety of ways including the seepage of mineral-charged water from cave walls, dripping water splashing from the floor or rock ledges, or precipitation of knobby growths when the cave was water-filled.

Delicate needles of calcium carbonate in the mineral form of *aragonite* rather than the more common calcite decorate many surfaces of boxwork, popcorn, and cave walls. Their resemblance to needles of ice that cover surfaces on a frosty morning led the early cave guides to name these features frostwork—a term now used to describe similar features found in other caves around the world. Frostwork seems to form best in areas where air movement is significant, leading some speleologists to suspect that they form where rapid evaporation or loss of CO_2 is important.

Although not abundant along the tour route, linings of calcite crystals called *dogtooth spar* (if crystals are pointy on the ends as in Fig. 12–9) or *nailhead spar* (more blunt crystal ends) cover walls and ceilings in parts of the middle and lower levels of the cave. These crystal linings likely formed during the past few million years when passages were completely flooded and calcite precipitated from mineral-rich groundwater. Cave linings are better developed and displayed along the tourist trail in nearby Jewel Cave.

Interesting complications in the Wind Cave story include those discussed by Bakalowicz and others (1987). Groundwater trapped in the limestone between impermeable layers was hydraulically pushed through the Pahasapa Limestone. Such a groundwater system under hydraulic pressure is known as an *artesian* system and enabled the three passage levels to enlarge simultaneously. In contrast, most caves are nonartesian and each level forms at a different time as the location of the water table changes. The slow movement of water through the Wind Cave system apparently enabled all of the larger joints in the fracture system to enlarge and create the complicated three-dimensional maze. A change in the location of a surface stream about 1–2 million years ago reduced groundwater flow, lowered the water table, and terminated cave enlargement (Palmer, 1981).

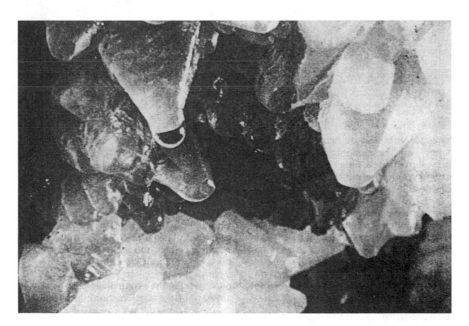

FIGURE 12–9　Dogtooth spar calcite crystal lining present in both Wind and Jewel caves. (National Park Service photo)

Analyzing *isotopes* (varieties of the same element that have different atomic weights) of oxygen and carbon that are sensitive to temperature at the time of formation determined that some of the calcite deposits such as the boxwork and cave linings in the cave were deposited by water at elevated temperatures. Additional support for higher-than-normal rock temperatures comes from cave air and rock temperature studies. In most caves air temperature reflects the average surface temperature of the area. In the Wind Cave area this is 47°F (8.3°C)—yet the upper levels of the cave are always about 53°F (12°C), and deeper areas of the cave are at a year-round temperature of 57°F (13.9°C)! Apparently the unusually warm rock temperatures that influenced cave development millions of years ago still remain today. The presence of thermal springs in the nearby town of Hot Springs (Fig. 12–4) are likely part of this same geothermal system.

Unusual speleothems in the lower part of the cave in an area known as the Calcite Jungle are many hours away from the tourist trail. Here are found a number of underground lakes whose waters are saturated with calcium. Thin crystal plates called *calcite rafts* form and float on the water surface due to surface tension. Some of these amazing rafts can reach lengths of 6–7 inches (15–18 cm) (Tom Miller, personal communication) before becoming too heavy and sinking. Disturbing the water surface also causes the rafts to sink. More rafts then reform at the lake surface. Calcite rafts littering the floor of now-dry sections of cave indicate that these passages were once flooded. Some unusual calcite growths called *helictite bushes* up to 6 feet tall (2 m) also form below the surface of these strange subterranean lakes.

Visiting Wind Cave

Wind Cave is open all year but more tours are available during the summer season. Besides a number of short tours, a special candlelight and a spelunking tour are also conducted. Only a few visitors at a time can be accommodated on these latter two tours—early registration at the visitor center is necessary. If you visit the cave when outside air pressures are changing, you will experience the movement of air, sometimes at very high velocities moving either in or out of the cave entrances. Construction of a walk-in entrance greatly changed air movements and the natural cave environment. Installation of air locks have more closely restored the natural cave meteorology.

Driving along park roads in the grassland prairie provides a taste of what the Great Plains from Texas to well into Canada were like before the western wilderness was eliminated. No longer can we see migrating herds of bison that numbered in the millions. Columns of bison 20–100 miles (32–160 km) wide were common in the 1830s; today only isolated managed herds remain. A population of about 60 million animals roamed through 30 of the contiguous states in the early 1800s. Intensive hunting, encouraged by the U.S. Army, eliminated all wild bison in the United States by 1883. By 1900, fewer than 600 bison remained on the entire North

American continent (National Park Service, 1979)! Present population is about 30,000 animals—all in wildlife refuges, parks, and private herds. About 350 of these magnificent animals inhabit the park today, an impressive sight to see. The bison is North America's largest mammal, standing as high as 6 feet (2 m) at the shoulder. Remember to enjoy these animals from the safety of your automobile. These usually docile animals, like any large wild animal, can be unpredictable. An 1800-pound bull charging at 30 miles/hour (48 km/hr) can be unnerving and could ruin your day!

The pronghorn antelope had a nearly similar history as a result of over-hunting. Their population dropped from 40 million animals in the early 1800s to less than one thousand by 1925. However, with good game management this animal has re-established a sizable population in Wyoming and other parts of the Great Plains, including the Wind Cave area. The pronghorn is North America's fastest animal. It can maintain speeds of 25 miles per hour (40 km/hr) for long distances and can run in spurts up to 70 miles per hour (113 km/hr)!

Do not miss the prairie dog towns in the park. These once incredibly abundant creatures were eliminated by the new settlers except for local coteries (family units) in remote areas or in protected areas such as parks. Their numbers are growing rapidly here and may expand into adjacent private land. It is hoped that one of their important predators, the endangered black-footed ferret, may ultimately be brought back from the brink of extinction and once again play its natural role in the ecosystem. Perhaps by understanding how each element in an ecosystem fits in with each of the others, humans will some day discover how to coexist peacefully with each other and all of nature.

REFERENCES

Bakalowicz, M., Ford, D.C., Miller, T.E., Palmer, A.N., and Palmer, M.V., 1987, Thermal genesis of dissolution caves in the Black Hills, South Dakota: Geological Society of America Bulletin, v. 99, p. 729–738.

Deal, D., 1962, Geology of Jewel Cave National Monument, Custer County, South Dakota, with special reference to cavern formation in the Black Hills: Laramie, M.S. thesis, Univ. of Wyoming.

National Park Service, 1979, Wind Cave National Park, National Park Service handbook no. 104, 145 p.

Palmer, A.N., 1981, Geology of Wind Cave: Wind Cave Natural History Association, 44 p.

Park Address

Wind Cave National Park
Hot Springs, SD 57747

JEWEL CAVE NATIONAL MONUMENT
(SOUTH DAKOTA)

A mere 20 miles (32 km) northwest of Wind Cave in the southern part of the Black Hills lies Jewel Cave (Fig. 12–4), one of the earth's most significant caves from the standpoint of both scientists and visitors. Like nearby Wind Cave, this is one of the world's longest and geologically oldest caves—a cave with a complicated history whose details are still being worked out.

Jewel Cave, like Wind Cave, is also located in the Mississippian age (about 350 Ma) Pahasapa Limestone. Boxwork, frostwork, and cave popcorn are here as well. However, Jewel Cave's distinctive hallmarks are its incredible length [over 110 miles (177 km) of passages] and the presence of "jewels." Calcite crystals with well-developed crystal faces—each a scalene triangle—are the "jewels" (Fig. 12–9). Some are white, others transparent, but most are highly colored and relatively large; all are extremely attractive. In contrast to the thin cave lining formed in Wind Cave (about 2 inches; 5 cm), a crystal lining from 6 inches (15 cm) to as thick as 1.5 feet (50 cm) decorates many of the ceilings and walls in Jewel Cave.

History of Discovery

The cave was discovered in 1900 by Frank and Albert Michaud who felt a strong breeze blowing from a small bedrock hole in Hell Canyon (Fig. 12–10). Two sticks of dynamite later, the brothers became the first people to enter the cave. The Michauds filed a mining claim on the government land but quickly realized that the real potential of the cave was as a tourist attraction. Ladders and ropes were installed and they guided visitors through the less than half mile (0.8 km) of known passages. The cave became a national monument in 1908, and the National Park Service took over its administration in 1933. Park Service employees quickly discovered new passages and extended the known cave to about 1 mile (1.6 km). The early years of cave development before the Park Service took over were hard ones for the cave. Lack of supervision permitted vandalism painful to see. The crystals were broken and carted off—probably to obscure boxes in attics and eventually to trash heaps. New discoveries now permit visitors to experience pristine passages and chambers—this time with adequate protection from despoliation.

Geologist Dwight Deal and a couple from nearby Custer, Herb and Jan Conn, began mapping and studying the three-dimensional-maze cave in 1959. By 1961 mapped passages totaled 5 miles (8 km). The Conns and others "followed the wind" and over the next few decades over 110 miles (160 km) of cave passages were mapped in a 2-square-mile (5 km²) area (Fig. 12–10)—and still more cave remains! For those who enjoy reading about the exciting history of exploration, the *Jewel Cave Adventure* by Herb and Jan Conn (1981) is highly recommended.

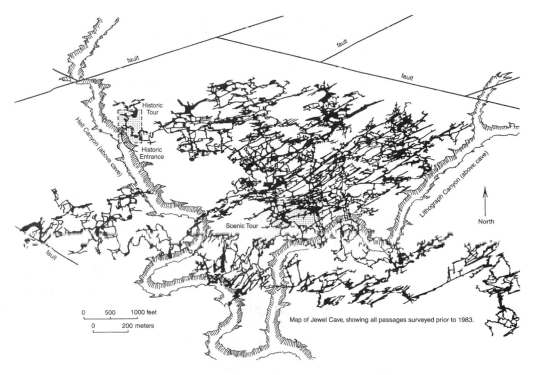

FIGURE 12–10   Generalized map showing cave passages mapped prior to 1983. Note the dominant northeast–southwest orientation of cave passages. (From Palmer, 1984, courtesy of the Black Hills Parks and Forests Association)

Speleogenesis—Cave Formation

The geologic story and cave formation story at Jewel Cave is the same as that at nearby Wind Cave. Some minor differences include the orientation of major fractures (joints and faults) and cave passages. Fractures tend to form radially on a domal uplift. Thus the location of Jewel Cave on the southwest flank and Wind Cave on the southeast flank adequately accounts for the different cave passage orientations (southwest at Jewel Cave and southeast at Wind Cave). Acid waters slowly flowing though the Pahasapa Limestone along these fracture zones during Eocene (about 50 Ma) time account for the initial episode of solutional enlargement.

Cave temperature at Jewel Cave is 47°F (8.3°C) compared to 53°F (11°C) at Wind Cave where an unusual amount of geothermal heat is present. Both caves had thermal waters circulating through them many millions of years ago when the crystal linings and boxwork formed. Some of that heat source apparently still remains at Wind Cave, while Jewel Cave's temperature reflects average yearly surface temperature, a more normal condition in a cave.

The following brief outline summarizes the story described in greater detail in the discussion of Wind Cave. More details are available in Palmer's (1984) excellent summary available in the visitor center.

1. Pahasapa Limestone deposited in a shallow sea during Mississippian time about 340 Ma.

2. Brief emergence forms karst surface as caves and sinkholes form in upper zone of Pahasapa Limestone.

3. Area submerged as Pennsylvanian-age (about 330 Ma) sands of the Minnelusa Formation fill in caves and sinkholes and bury the area, forming a paleokarst surface. Small voids in debris that fills the ancient cave develop dogtooth spar (pointy calcite crystals) linings. Some individual crystals are quite large—about the size of a goose egg!

4. Both marine and nonmarine sediments of Paleozoic and Mesozoic age continue to bury the Black Hills region until about 60 Ma (Paleocene) when the Laramide Orogeny begins.

5. Formation of the Black Hills dome causes fracturing of the brittle limestone and other rock units. Sedimentary cover is removed by erosion, and groundwater circulates through buried limestone and begins to dissolve the voids that will become Jewel Cave.

6. Most cave enlargement occurs during the Eocene Epoch about 50–40 Ma. At that time, streams draining the Black Hills core sink into the exposed limestone and circulate under artesian pressure through the numerous joints—especially those developed in the more soluble layers in the Pahasapa Limestone. The three-dimensional maze of enlarged fractures and bedding planes thus formed the passages of today's cave. Boxwork at least partially developed during the cave enlargement process as the limestone between crack fillings was removed faster than the dense, less-soluble calcite that filled the innumerable small bedrock fractures.

7. The water table lowered and cave walls were deeply weathered (further emphasizing the boxwork) before the area was buried by sediment of the Oligocene-age White River Formation—a rock unit we will meet again at Badlands National Park. Burial of the Jewel Cave and foothills area of the Black Hills by debris eroded from higher parts of the range, as well as climate fluctuations, allowed the water table to rise and fall, producing at least five intervals of reflooding and draining of passages. The slowly circulating, calcium-rich groundwater during intervals of flooding forms the thick crystal lining, or jewels, that suggested the cave name. Most of the lining on the floor and walls is dogtooth spar. Nailhead spar, the blunter form of calcite crystals, occurs most frequently on the ceilings. The groundwater was at an elevated temperature of about 95–104°F (35–40°C) at this time (Hill and Forti, 1997)—perhaps a thermal remnant of the igneous activity that affected the Black Hills during Eocene time.

8. Continued erosion and uplift, especially during the minor uplift of the Black Hills during Pliocene (about 4 Ma) time, drained the cave and permitted lo-

calized dripstone (soda straws, stalactites, stalagmites, and helictites) and flowstone as well as cave popcorn and frostwork (aragonite needles) to form. In less humid sections of the cave, calcium sulfate (the mineral *gypsum*) formed needle- and flowerlike crystal clusters.

9. Pleistocene climates produced alternately drier and wetter intervals during the past 1.6 million years, greatly influencing speleothem growth (Hill and Forti, 1997). Further analysis of dripstone and flowstone speleothems indicates that they began to form at least 500,000 years ago. Streams in Hell and Lithograph canyons deepened their channels. Eventually the valley wall in Hell Canyon intersected part of the cave (Fig. 12–10) and formed the small 1-foot-diameter (0.30 m) hole that permitted air to move in and out of the cave—the feature that led to its discovery.

Visiting Jewel Cave

Jewel Cave is unquestionably one of North America's most remarkable caves—one well deserving national park status. Although many details of its origin are still uncertain, its development, as well as that of nearby Wind Cave and other Black Hills caves, appears to be unique. The cave is truly a "gift from the past." Nearly all caves are, from the geologic standpoint, recent additions to the landscape—mostly formed in the past million years. To have one of the world's longest caves persist for perhaps 50 million years is indeed unique! Erosion will eventually destroy the known cave. As this occurs, the water table will lower—perhaps exposing now flooded caves or caves that are slowly forming beneath our feet.

A number of tours are available at Jewel Cave, especially during the summer months. These guided tours (historic and scenic tours) can only accommodate 25–30 people at a time, so early registration for a tour is recommended. A 4-hour-long *spelunking* (recreational caving) tour is also given during the summer for those who want to experience the rigors and beauty of an undeveloped section of cave. Registration for the spelunking tour (limited to 10 people) should be made up to a month before your visit in order to ensure a spot. Besides registering for the tour, all participants must be able to crawl through an 8.5-inch-high (22-cm), 24-inch-wide (61-cm) concrete-block tunnel—a skill that aspiring spelunkers might want to practice at home! Should you become stuck, calling 911 at home is much simpler than in the depths of Jewel Cave!

REFERENCES

Conn, H., and Conn, J., 1981, The Jewel Cave adventure: Cave Books, St. Louis, MO, 238 p.

Hill, C., and Forti, P., 1997, Cave minerals of the world: National Speleogical Society, Huntsville, Alabama, 463 p.

Palmer, A.N., 1984, A gift from the Past: Jewel Cave: Wind Cave/Jewel Cave Natural History Association, 41 p.

Park Address

Jewel Cave National Monument
RR 1, Box 60AA
Custer, SD 57730

DEVILS TOWER NATIONAL MONUMENT (WYOMING)

Devils Tower in northeastern Wyoming is on the northwestern edge of the Black Hills uplift and is the most "upstanding" feature of the Great Plains (Fig. 12–11). Devils Tower does not "sneak up" on an approaching traveler—it appears unusu-

FIGURE 12–11   Devils Tower, Wyoming. Note the flaring columns near the tower base and the more fractured and weathered appearance of the upper one third of the tower. (Photo by E. Kiver)

ally large even 30 or more miles (50 km) away and grows in significance with a closer approach. To stand at the base of this 867-foot-high (264-m) shaft of nearly vertical igneous rock is a humbling experience. The summit elevation is 5070 feet (1546 m), some 1267 feet (386 m) above the north-flowing Belle Fourche River—the river that excavated the buried igneous feature and continues to cut away at its south flank. Although one of our smaller park areas (1367 acres, 2.1 square miles, 5.5 km²), Devils Tower has the distinction of being our first national monument—established by Theodore Roosevelt in 1906 shortly after Congress passed the Antiquities Act. This important legislation was enacted to speed up protection of areas of national significance and continues to work effectively. Most recently the Grand Staircase–Escalante National Monument in Utah was added to the park system—another gift to ourselves and, more importantly, to future generations.

Humans at Devils Tower

One can sense that this is a special place—spectacularly impressive yet serene with the surrounding ponderosa forest, grasslands, and cottonwood-lined banks of the Belle Fourche River. Not surprisingly, the tower figures strongly in the creation stories of at least 20 Native American tribes, including the Cheyenne and Lakota Sioux who called it Mateo Tepee, or Grizzly Bear Lodge. One story relates that a group of Indians were chased by a giant bear. Standing on a rock, they prayed to the Great Spirit to save them. He did so by having the rock grow in elevation. The angry bear clawed at the sides of the tower forming the long cracks (Fig. 12–12) that geologists call columnar jointing!

The 1875 U.S. Geological Survey expedition to the Black Hills was under military escort by Colonel Richard Dodge. His interpreters misunderstood the Indian's description of the feature and translated it as "Bad God's Tower." Dodge reported it as Devils Tower, a name that is offensive to many Native Americans.

Early settlers were drawn to the area, and it became a special place for social gatherings and celebrations. Local ranchers William Rogers and Willard Ripley amazed a crowd of neighbors at the Fourth of July gathering in 1893 by climbing to the top using 30-inch-long (76-cm) stakes that they had secretly pounded into a continuous crack that led to the summit! Portions of the stake ladder are still visible from the Tower Trail that encircles the base.

The first successful rock climb using mountaineering techniques occurred in 1937 and was followed in 1938 by a group led by Jack Durrance. Durrance was called upon in 1941 to repeat his climb and rescue George Hopkins who was stranded on top of the tower. As a stunt, Hopkins had parachuted on to the top of the tower. His 1000-foot-long (305-m) rope became hopelessly entangled on the rocks below. After considering a rescue by the Goodyear blimp and an experimental helicopter being developed by the U.S. Navy, the Park Service contacted Durrance who, along with a party of rock climbers brought Hopkins down from his lofty perch six days later (Norton, 1991)!

FIGURE 12–12   Origin of Devils Tower, or Tepee Mateo, according to Sioux legends. (Painting by Herbert A. Collins, Used by permission of the National Park Service)

Regional Setting

Laramide uplift was most intense in the northern part of the Black Hills where deep fractures formed and later acted as conduits for magma. The northern Black Hills has 13 major igneous centers where shallow intrusions and volcanic activity occurred during Eocene time about 55–33 Ma (Robinson, 1998). The westernmost of these centers is at Devils Tower and nearby Missouri Buttes, located about 4 miles (6.4 km) northwest of the tower. A zone of unusually high heat flow extends northward from the Rio Grande Rift in New Mexico, through the Southern Rocky Mountains, and into the Black Hills. Some geologists consider this zone to be an aborted rift where the North American plate began to split apart but then slowed or stopped (Karner and Halvorson, 1987).

Unusually high temperature at shallow depths and lessening of pressure along the rift caused partial melting of the mantle and generation of basalt magma that accumulated at or near the base of the crust. The basalt magma in the Devils Tower area in turn melted a considerable amount of the overlying granitic-composition crustal rocks. The composition of the resulting silica-rich magma is mostly that of a relatively uncommon group of igneous rocks that include *phonolite,* the rock found at Devils Tower. Phonolite is a rock whose silica composition is between dacite and rhyolite and whose content of elements such as sodium and potassium is unusually high. The name phonolite comes from the musical ringing sound often emitted when the unweathered rock is struck with a hammer. Because large, easily visible white crystals of feldspar and other minerals are surrounded by finer-grained crystals, the rock texture is called a *porphyry.* Thus, most of the rock at Devils Tower and nearby Missouri Buttes is a *phonolite porphyry.*

The rising magma formed a variety of shallow intrusions in the Black Hills area including sills, laccoliths, volcanic necks, and small stocks (Fig. 1–7) as well as some extrusive igneous features. *Breccia pipes* also occur in the Black Hills area. Violent gas explosions or rising magma carry angular fragments of granitic, metamorphic, and sedimentary rocks ripped from the walls of the feeder vents toward the surface. Outcrops of breccia a few hundred feet northwest and southwest of the tower could have formed by such a violent explosion or perhaps a different process. However, their presence provides an important clue in the mystery of the tower's origin.

Origin of the Tower

The exact history of the magma and the original shape of the igneous feature called Devils Tower has been a source of debate since geologists first examined it in 1875. Devils Tower presents an interesting study in how the scientific method is used to explain geological phenomena. Geologists typically develop multiple explanations for a geological problem and then gather evidence that supports or conflicts with each suggested explanation. This information is shared with the scientific community and, based on careful evaluation, a tentative or perhaps a strong conclusion or explanation may be accepted. New evidence might require an explanation to be modified or perhaps even discarded. Thus science slowly refines its knowledge and proceeds ever closer to understanding natural processes and events. Unfortunately, erosion and later geologic events may remove critical evidence that makes strong conclusions difficult.

The original suggestion by Carpenter in 1888 was that Devils Tower is a volcanic-neck remnant—part of the feeder pipe of an ancient volcano that remained filled with magma that cooled and contracted to form the spectacular pattern of columnar jointing visible today. Many geologists today still favor the volcanic-neck hypothesis (Karner, 1985; Karner and Halvorson, 1987). Others noted that part of the base of the tower rests on top of the surrounding Mesozoic-age sedimentary rocks and that the tower and Missouri Buttes were once much more extensive. Perhaps they were emplaced as a sill or as laccoliths (Fig. 1–7). A large volume of phonolite porphyry rock fragments derived from Devils Tower is strewn about on and in stream terrace deposits along the Belle Fourche River— thus lending support to the suggestion that Devils Tower was once much more extensive.

Another hypothesis suggests that the tower is a small *pluton*—a stocklike intrusion not much larger than the present tower (Robinson, 1956, 1998). This hypothesis proposes that an igneous plug pushed its way close to the surface without actually breaching or erupting at the surface. Both Devils Tower and Missouri Buttes are in shallow structural depressions where the sedimentary rocks are gently bowed downward. Perhaps magma flowing back into the magma chamber caused nearby rocks to collapse (Karner and Halvorson, 1987).

Such depressions can also form when rising magma encounters groundwater near the earth's surface, producing an *explosion crater*. Later filling of the depression with lava or hot pyroclastic material that welded together, cooled, and contracted into columnar-jointed rock might also explain Devils Tower. According to this latter hypothesis, Devils Tower had a very violent past. Later erosion would remove most of the basin fill material. Only the thickest part of the igneous fill remains—the part that we call Devils Tower.

Columnar jointing tends to form perpendicular to the surface of cooling. Fractures or jointing form because igneous rock occupies less space than the magma from which it was derived. The magma or hot welded pyroclastic material in an explosion crater, volcanic neck, or shallow intrusion cools and contracts from the top down and from the vent walls inward. Vertical columns form where heat is lost mostly to the air or rock surface above or a flat surface below—more horizontal or inclined columns would tend to form where heat is mostly lost to the surrounding vent or basin walls. Thus, a volcanic neck, shallow intrusion, or a sloping basin surface could account for the nearly vertical columns in the center (best seen on the southeast side of the tower) and the outward flaring columns elsewhere around the base. Most of the columns have five sides although many are four and six sided.

All of the origin hypotheses explain most of the observed features, but some apparent discrepancies exist. The vertical columns flaring outward at the base are similar to the columnar-jointing patterns that occur in known volcanic necks, yet no evidence of extrusive rock or phonolite debris occurs in the nearby White River sediments (Oligocene) that formed a few million years after the Devils Tower rock crystallized (Robinson, 1998). The size of the mineral grains in the phonolite matrix suggests a slower rate of cooling, one that would occur with slightly deeper burial. Thus a lava- or pyroclastic-debris-filled explosion crater at the earth's surface might seem less likely. Perhaps a better understanding of the few breccia outcrops in the area would provide additional evidence to more firmly establish the geologic history.

Whatever its original form might have been, erosion by the Belle Fourche River and its tributaries, combined with the processes of weathering, account for much of the topographic expression of the tower. A two-stage erosion process likely occurred. The upper 150 feet (45 m) of the tower exhibits more deeply weathered rock than the lower section (Fig. 12–11) suggesting a longer exposure to weathering. Perhaps erosion was slowed as a resistant sandstone in the Fall River Formation (Cretaceous age) was slowly removed. Rapid removal followed as the underlying shales and less resistant strata were removed down to the present level of the Belle Fourche River. The wetter climates during the Pleistocene (Ice Age) likely led to more frequent flood-swollen streams and contributed to intervals of more intense erosion.

Visiting Devils Tower

A short 1.25-mile long (2 km) hike takes visitors completely around the base of Devils Tower and provides spectacular views. The views change hourly, so those who camp overnight and repeat the hike or take other trails in the park at different times of the day will be treated to the changing moods of the tower and surrounding landscape.

Most visitors will be content to watch from a distance as mountain climbers inch their way up the long cracks outlining the columnar jointing. For those who have to "be on top of things," the climb is considered outstanding and relatively safe for those who have the necessary skills. For those who would like to more safely examine the summit and columnar-jointed rock on its nearly vertical flanks, large rock-fall boulders strewn about the base of the tower as *talus* provide convenient opportunities. Many of the four- to six-sided columns are relatively intact, even after what must have been a spectacular descent from the upper slopes. The diameters of the columns decrease from as much as 15 feet (4.6 m) near the tower base to about 3 feet (1 m) near the top.

The Great Plains wildlife is here in abundance—including the white-tailed deer, jack rabbits, and prairie dogs. The prairie dog towns form the basis of a food chain that supports a population of hawks, eagles, prairie falcons, coyotes, and fox. Pack rats or wood rats inhabit the rocky crevices and talus areas and are occasionally seen at the summit where they have been known to practice their habit of collecting shiny souvenirs for their nests—like coins, car keys, and wrist watches from unwary climbers! Chipmunks, mice, and even rattlesnakes are seen on the rounded summit of Devils Tower—apparently using the same cracks (without the benefit of ropes!) that their human counterparts use to gain access!

The interaction and interdependence of the plants and animals here is beautiful to watch and understand—a microcosm of the entire planet. Preservation of each environmental microcosm and of all elements in the larger planet earth system ensures that all things living now will continue to survive, hopefully even those at the very top of the food chain.

REFERENCES

Karner, F.R., 1985, Geologic relationships in the western centers of the northern Black Hills Cenozoic igneous province, in Rich, F.J., ed., Geology of the Black Hills, South Dakota and Wyoming: American Geological Institute, Alexandria, VA, 2nd ed., p. 126–133.

Karner, F.R., and Halvorson, D.L., 1987, The Devils Tower, Bear Lodge Mountains, Cenozoic igneous complex, northeastern Wyoming, in Beus, S.S., ed., Rocky Mountain Section of the Geological Society of America: Boulder, CO, Centennial Field Guide Vol. 2, p. 161–164.

Norton, S., 1991, Devils Tower, the story behind the scenery: KC Publications, Las Vegas, NV, 48 p.

Robinson, C.S., 1956, Geology of the Devils Tower National Monument, Wyoming: U.S. Geological Survey Bulletin 1021-I.

Robinson, C.S., 1998, Geology of Devils Tower National Monument, Wyoming: Devils Tower Natural History Association, 96 p.

Park Address

Devils Tower National Monument
P.O. Box 10
Devils Tower, WY 82714

THEODORE ROOSEVELT NATIONAL PARK
(NORTH DAKOTA)

In the southwestern corner of North Dakota along the Little Missouri River is an area of rugged beauty, the so-called badlands. Here one can look out at a landscape that is little changed from 1883 when 24-year-old Theodore Roosevelt came west to hunt buffalo, North America's largest mammal. Roosevelt was too late to see the vast herds of bison that had roamed the Great Plains a few decades earlier—by 1883 almost all of the 60 million animals had already been slaughtered by the Euro-Americans! However, the ruggedness of the badlands and the man matched one another—Roosevelt purchased an interest in the Maltese Cross Ranch and moved onto it the following year. Here he began to develop his conservation ethic based on sustainable use of public land for the benefit of the general public rather than for a few greedy individuals and corporations—a policy that would later be initiated under his presidential administration. He further realized that "the greatest good for the greatest number" really applies to our obligation to those who have yet to be born. Thus, our natural resources would be evaluated for their scientific, scenic, and historic values for the generations yet to come rather than exclusively for their present transitory economic value.

The scenic badlands (Fig. 12–13) are developed in 60-million-year-old (early Tertiary Period, Paleocene Epoch) sedimentary rocks that began to be deposited

FIGURE 12–13   Little Missouri River badlands in Theodore Roosevelt National Park. (Photo courtesy of Bruce Kaye and National Park Service)

about the time that the Rocky Mountains and Black Hills were uplifted. Debris from these uplifts was carried by streams into the Great Plains and deposited as the sedimentary layers known as the Fort Union Group. Also present in Paleocene strata in the Great Plains are plant and animal fossils that record the subtropical climate of the time, along with evidence of the early development of the mammals—the animal group that filled the void left by the rapid disappearance of the dinosaurs a few million years earlier. From Theodore Roosevelt National Park our time-travel story of the Great Plains will continue later in the chapter as we visit Fossil Butte, Badlands, Agate Fossil Beds, and Capulin Volcano.

Geographic Setting

The Little Missouri River flows northward from the Devils Tower and Black Hills area cutting deeply into the Missouri Plateau Section of the Great Plains. The river flows through the South Unit of the park, past Roosevelt's Elkhorn Ranch Site, and into the North Unit where the river turns abruptly eastward and joins the Missouri River about 50 miles (80 km) away. Seasonally and during flash floods, small tributaries flow down the steep valley sides along the Little Missouri River, cutting deeply and extensively into the Paleocene-age Fort Union sediments. The poorly consolidated sediments are eroded rapidly by these tributaries forming a highly dissected landscape known as badland topography (Fig. 12–13). Badlands characteristically form where the combination of steep slopes, weak rocks, and the lack of abundant vegetation occurs. The park units lie mostly within the confines of the Little Missouri River valley in a relatively remote area of North Dakota.

Human History

The Lakota Indians called it "Mako Shika" or "land bad" because of difficulties in traveling across it—particularly after a rain. The Euro-Americans agreed, especially when they tried to move horse-drawn wagons across the rugged topography.

Roosevelt lost both his mother and wife on the same day in February 1884 and was greatly saddened and demoralized. He moved to the badlands and the Maltese Cross Ranch a few months later to regain his vigor by pursuing outdoor activities and becoming part of the open-range ranching operation. He further enlarged his cattle operation activities by establishing the Elkhorn Ranch about 30 miles (50 km) north of the Maltese Ranch. Cows grazed on the unfenced public domain lands in North Dakota in the late 1800s, often in numbers larger than the range would support.

Roosevelt soon learned that land use has limits and that the winters in the northern Great Plains can be brutal (Schoch and Kaye, 1993). Whereas bison have heavy coats to protect them from the bitter cold winds and could paw through deep snow for grass, cattle survival was more tenuous. The cottonwood thickets along the Little Missouri furnished some protection. However, late frosts and drought conditions in 1886 reduced the grass crop. Lack of feed exacerbated by overgrazing, and the severe winter of 1886–1887 resulted in death for most of Roosevelt's 5000 head of cattle. He left the badlands in 1886 but returned for brief visits through 1896, selling his ranching interests in 1898. Fortunately for the nation, Roosevelt went on to other activities—such as becoming the 26th president of the United States!

Roosevelt was intensely interested in conservation and, as president, he greatly enlarged the fledgling forest service, established over 50 wildlife refuges, designated 15 areas as national monuments, and was instrumental in establishing 5 national parks, thereby doubling the number of national parks. Ironically, he originally came to North Dakota to hunt buffalo; later in life he helped establish refuges and preserves that, among other benefits, helped save the buffalo from extinction!

Roosevelt's ranch land was later divided and homesteaded. However, the Dust Bowl days of the 1930s were devastating for farmers and ranchers and the submarginal land was abandoned—reverting back to the federal government and making the establishment of a park easier to accomplish. The movement for a national park had begun soon after Roosevelt's death in 1919. Eventually, as a tribute to this remarkable man—big game hunter turned conservationist—in 1947 Congress established a national memorial park in the South Unit and the Elkhorn Ranch Site and in 1948 added the North Unit. In 1978 the 70,448-acre (110-square mile, 285-km^2) area, consisting of three units, was upgraded to national park status.

Geologic Development

The Paleozoic and Mesozoic rocks that are still buried deep beneath the Theodore Roosevelt National Park record a long period of crustal stability and the thick accumulation of sediments, mainly marine. Minor warping of the earth's crust produced a number of broad arches and structural basins in the midcontinent area. The park is located along the southern edge of one of these downwarped areas—the huge Williston Basin. The basin covers parts of North Dakota, Montana, Alberta, Saskatchewan, and Manitoba and contains important accumulations of oil and gas.

The inland sea, in which Cretaceous sediments were laid down, retreated from most of the continent by the end of the period; locally, the Cannonball Sea was the last holdout, as recorded in the early Paleocene (about 65 Ma) coastal-plain rocks (lower layers in the Fort Union Group) still buried beneath younger Paleocene rocks that are exposed in the park (Fig. 12–14). Most of the overlying Paleocene rocks exposed in the badlands are stream-deposited sediments about 55–60 million years old that were derived from the rising Rocky Mountains to the west.

Sandwiched between the stream-deposited sediments, pencil-thin to foot-thick (30 cm) beds of *lignite*, a low-grade coal, are locally present. Lush vegetation grew in low-lying floodplain swamps during Fort Union (Paleocene) time. Organic debris accumulating in the swamps was compacted by the weight of overlying sediments into *peat*; additional compression converted the peat to lignite. Had

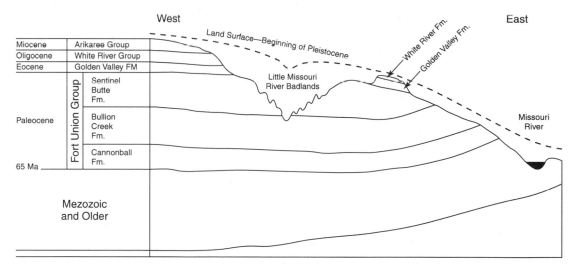

FIGURE 12–14   Generalized cross section in Theodore Roosevelt National Park area. (Modified from Bluemle and Jacob, 1981)

additional pressure been applied, *bituminous* (soft coal) would eventually form. The lack of oxygen in these buried sediments limited bacterial decay permitting a carbon-rich deposit to form—a deposit that contains abundant fragments of plants that have persisted for 60 million years (Bluemle and Jacob, 1981)!

As mountain building continued during Paleocene and later geologic time, Mount St. Helens types of volcanoes in the Rockies to the west belched forth large quantities of ash and dust that blanketed large areas of the Great Plains, including the park area. Silica derived from these ash beds played a major role in the petrification of cypress, sequoia, and other deciduous and cone-bearing trees that grew in this low-lying, subtropical environment during Fort Union time. The petrified wood here is the third most important deposit in North America—surpassed only by those in Petrified Forest and Yellowstone national parks. Weathering of ash beds produced sticky (when wet), purplish-colored clay layers (called *bentonite*) such as the conspicuous layer locally called "Big Blue" in the North Unit. When dry, the clay contracts forming a "popcorn" surface of loose clay aggregates—a difficult surface for vegetation to become established on. Thus, the presence of these swelling and contracting bentonite clays contributes to the bare slopes and rapid erosion that leads to badland development.

During the remainder of the Tertiary, streams flowing eastward and northward deposited their sediments over the Great Plains. Then, only a few million years ago, uplift of the Rockies and Great Plains rejuvenated the streams, causing them to downcut and erode away much of the poorly consolidated upper rock layers. Thus the younger Tertiary rocks that once covered the Theodore Roosevelt National Park area were stripped off. Eocene and thick Oligocene (White River Formation) layers occur as caprock on buttes and plateaus located out of the park and a short distance away from the Little Missouri River valley (Fig. 12–14). The story of these younger Tertiary times in the Great Plains can be read in the rock record at Badlands National Park (South Dakota) and elsewhere.

During the Pliocene (about 2 Ma) the ancestral Little Missouri River and other streams in the northern Great Plains flowed northeasterly, joined, and flowed through Saskatchewan and Manitoba to Hudson Bay. During the Pleistocene when the continental ice sheets moved down out of Canada and covered the area, stream erosion was obviously impossible; also, the north-flowing streams, now blocked off by the ice, were diverted to more southerly courses. The lower sections of the Little Missouri, the Missouri, the Yellowstone, and other streams were relocated to their present southeastward courses about 600,000 years ago (Schoch and Kaye, 1993). The new river paths were shorter and steeper—speeding up erosional processes and causing tributary streams to cut rapidly headward through the relatively soft rocks of the Fort Union Group to form the intricately dissected topography of the Little Missouri badlands.

During the Wisconsin glacial advance, the ice blocked off the Little Missouri River, creating a large lake that covered the North Unit of the park. When the glacier melted away, the lake was drained and badland-forming stream erosion resumed, and it continues today. As the streams downcut into the poorly consolidated sediments, some of the side slopes became oversteepened and unstable, especially those containing clay layers. Therefore, mass movement or landslides are relatively common in the park.

Visiting Theodore Roosevelt

To look out and see the same landscape, relatively unchanged from what Roosevelt first saw in 1883, indeed presents a rare opportunity in an increasingly crowded world. Although all the bison, antelope, elk, bighorn, grizzly, cougars, and wolves were completely eliminated by hunters and settlers, with the exception of the grizzly, wolves, and cougar, all of these animals have been reintroduced and can be seen in the park. Recognizing that prairie dogs ate some of the range grass but not recognizing their importance in maintaining a healthy prairie environment, a nearly successful campaign to eliminate the prairie dog was initiated by ranchers and farmers. Today a number of prairie dog towns are located in the park and even a small herd of wild horses is found in the South Unit of the park.

The North Unit is within view of the former glacier edge—the glacier that barely entered the north end of the park. The presence of a formidable wall of ice effectively diverted the northern Great Plains rivers to more easterly courses. Boulders of granite and metamorphic rocks occur only north of the river where the glaciers had been—none of these rocks derived from Canada are located south of the river. In addition to a visitor center, prairie dog towns, landslides, cannonball-shaped *concretions*,[1] and magnificent views of badlands, a number of short trails and longer wilderness trails criss-cross the area. The Elkhorn Ranch Site (Roosevelt's second ranch) where Roosevelt spent many hours on his veranda reading and putting his thoughts on paper is difficult to reach. The more adventurous visitor should check at the visitor center for directions and road and trail conditions.

The South Unit also contains a visitor center where Roosevelt's Maltese Cross Ranch cabin has been relocated. Along the trails and roads one can enjoy the prairie dog towns and excellent views of the badlands. Trails lead to the Petrified Forest, the Coal Vein area, and Scoria Point where a burnt coal seam occurs. The lignite beds exposed on the side slopes are susceptible to grass fires, lightning strikes, and perhaps spontaneous combustion. Once ignited, the lignite may burn for many years. In places the heat is so intense that the overlying sediment will melt or bake into a hard, red-brick-colored material that, for reasons unknown, has locally been called "scoria." This is definitely a misnomer, as this material is

---

[1] A concretion is a hard deposit of mineral material precipitated by groundwater, oftentimes around a nucleus of a leaf, bone, or other fossil.

in no way related to volcanic scoria. Most of the scoria formed a few thousand years ago; however, the results of a more recent coal vein fire can be observed in the South Unit. Here in 1951 a lightning strike started a prairie fire that in turn ignited a lignite bed that smoldered until 1977. When the lignite burns deeply into the hillside and oxygen can no longer reach the fire front, the coal fire extinguishes.

One might also consider a canoe trip (requires about 3 days) and float the 110 miles (177 km) of river between the South and North units; a backpacking expedition into some of the designated wilderness in the park; or renting a horse to explore some of the backcountry in the same fashion that Roosevelt did in the late 1800s. The Maah-Daah-Hey Trail connects the North and South units for those who want a multiday hike through the badlands. The 17 hours of summer daylight is particularly interesting to those who live in more middle latitudes and the display of northern lights can be awesome.

We have merely glimpsed some of the features in the "fantastically beautiful" badlands that Roosevelt admired so much. Roosevelt once wrote: "I would not have been President, had it not been for my experience in North Dakota" (in Theodore Roosevelt National Park Website, 1998). Although it would involve some zigzagging, you might consider visiting or at least thinking of the Tertiary parks of the Great Plains in geologic sequence in order to better appreciate the physical, biological, and climatic changes that took place there during the past 65 million years. You would start at Theodore Roosevelt (Paleocene), then visit Fossil Butte (Eocene), the Badlands of South Dakota (Oligocene), and finish with Agate Fossil Beds (Miocene). Following Roosevelt's philosophy about nature will also add greatly to one's life experiences:

> It is an incalculable added pleasure to anyone's sum of happiness if he or she grows to know, even slightly and imperfectly, how to read and enjoy the wonderbook of nature. (in Schoch and Kaye, 1993, p. 48)

REFERENCES

Bluemle, J.P., and Jacobs, A.F., 1981, Auto tour guide along the South Loop Road, Theodore Roosevelt National Park: Medora, ND, Theodore Roosevelt Nature and History Association, 14 p.

Schoch, H.A., and Kaye, B.M., 1993, Theodore Roosevelt National Park, the story behind the scenery: KC Publications, Las Vegas, NV, 48 p.

Park Address

Theodore Roosevelt National Park
Box 7
Medora, ND 58645

FOSSIL BUTTE NATIONAL MONUMENT (WYOMING)

The flat, rolling topography characteristic of the Great Plains extends through central Wyoming where it is known as the Wyoming Basin. The basin is surrounded by mountains on three sides and forms a natural divide between the ranges of the Middle and Southern Rocky Mountains. Although the surface topography is that of the Great Plains, the surface beneath the thick cover of Tertiary sediments consists of a series of uplifts and deep basins produced during the Laramide Orogeny.

On the west edge of the Wyoming Basin in southwestern Wyoming lies the Green River Basin. A subsidiary basin along its west edge is Fossil Basin where Fossil Butte National Monument is located, about 11 miles (18 km) west of the coal town of Kemmerer (Fig. 12–15).

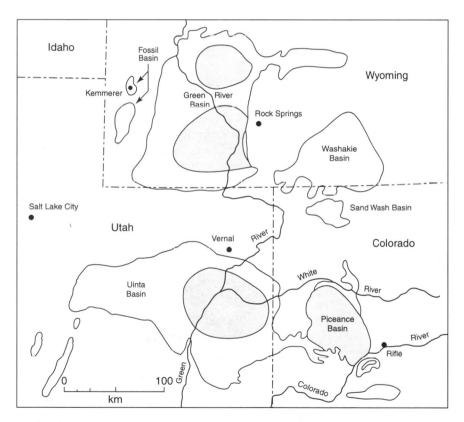

FIGURE 12–15   Laramide basins in the Wyoming Basin and nearby areas containing lake sediments of Green River age (Eocene). Darker-shaded areas are underlain by oil shale containing over 18 gallons per ton of rock. (Modified slightly from Skinner and Porter, 1987)

The monument was established in 1972 to preserve an unusually complete assemblage of plant and animal fossils, especially the world-famous Green River fish fossils that accumulated in a system of great lakes that occupied the Green River Basin, Fossil Basin, and the Uinta Basin (Fig. 12–15) from late Paleocene (about 55 Ma) to late Eocene time (about 40 Ma). The fossilized bones and carbonaceous films preserve amazing details of the fish, including their articulated skeletons, teeth, scales, and even their skin! The dark brown to black fossils make a handsome display when outlined against the buff- to white-colored shaly limestone of the Green River Formation (Fig. 12–16). Specimens are displayed in museums around the world. Its only rival of scientific interest is a fish quarry near Solenhofen, Germany, where Jurassic-age fish fossils and some of the oldest bird fossils in the world are found. For nearly 150 years geologists have pondered why the Green River fish are so well preserved and why some layers contain hundreds of thousands of fish bodies—the apparent result of mass die-offs. The answers, or probable answers, to most of the big questions are known, but many lifetimes of work are needed to uncover additional detail and to solve some of the smaller mysteries in reconstructing the Eocene environment.

History of Discovery

Dr. John Evans discovered fossil fish along the Green River in southwestern Wyoming in 1856. Soon after, the Union Pacific Railroad made a large cut through a hill 2 miles (3 km) west of the town of Green River, exposing large numbers of fish. This exposure, known as the Petrified Fish Cut, was mentioned in Hayden's

FIGURE 12–16 *Diplomystus,* an Eocene relative of the herring, 18.25 inches (46 cm) long. (Colorado State University collection)

1871 report of the first geological survey of the territory. Discovery of fish fossils at Fossil Butte (the Twin Creek locality) occurred during the early 1870s when the Union Pacific tracks were rerouted. The Fossil Butte site was described in the 1877 Hayden survey report. Scientists were not the only "fossil fishermen." Some of the local people, sometimes whole families, were fascinated by the fossils. Some supplemented their income by selling fossils to train passengers; others spent their lives preparing and selling specimens to museums and to private collectors.

More comprehensive studies of the area were made in the early 1900s and especially by William Bradley (1948) of the U.S. Geological Survey from the 1920s through the 1960s. The regional geology of the Green River and Fossil basins was further studied by Roehler (1993) and other researchers. McGrew and Cassilliano (1975), Buchheim (1994), and Grande and Buchheim (1994) and others also contributed greatly to our understanding of the rock and fossil record in the Green River Formation.

National Monument status for such a scientifically important area was inevitable; in 1972 a small (8198 acres, 12.8 square miles, 33 km²) area including some important fossil-fish quarries was set aside at Fossil Butte on the north side of Twin Creek. Commercial collecting continues on Fossil Ridge on the south side of the Twin Creek valley where "good fishing" will continue for generations as long as excessive quarrying is not permitted.

The "Big Picture"

Mountain building along the west edge of North America migrated eastward with time and culminated in the Laramide Orogeny during late Mesozoic through early Tertiary time (about 70–50 Ma). Thrust faulting and crustal warping in western Wyoming produced mountains and other topographic irregularities—including the 170- by 140-mile (275- by 225-km) Green River Basin (Thornbury, 1965) and the smaller, 32- by 120-mile (20- by 75-km) Fossil Basin to the west (Thornbury, 1965). Water trapped in these and other nearby basins in Colorado and Utah (Fig. 12–15) formed a series of lakes that expanded and contracted through mid-Tertiary time—sometimes merging into a huge single lake. Seasonal and climatic changes also brought wide fluctuations in lake extent and produced a dynamic environment in which a wide variety of plants and animals flourished.

The area was close to sea level and warm, humid, subtropical conditions prevailed during the early history of the lake. Streams deposited the red-colored sediments of the Wasatch Formation, which in Fossil Basin both underlies and intertongues with the thinly laminated lake sediments of the Green River Formation. Expansion and contraction of the lake produced a complex shifting of the boundary zone between lake and stream deposition producing the interfingering between the Wasatch and Green River Formations (Fig. 12–17). Because basin slopes were very gentle, a slight change in lake elevation produced a wide variation in lake extent. In response to changes in lake extent local environments often

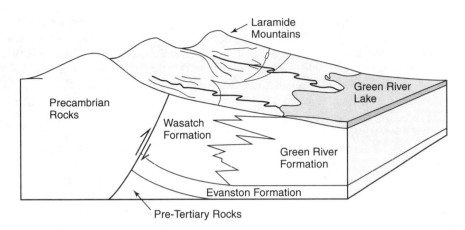

FIGURE 12–17   Generalized block diagram showing the relationship of the Laramide mountains to Eocene paleo-stream and lake sediments near Fossil Butte.

changed from freshwater lake to salty lake, playa lake, mudflat, and back to a freshwater lake. These cycles were repeated over 77 times during deposition of the Green River Formation (Roehler, 1993). Occasional volcanic eruptions in Yellowstone and other areas to the west sent ash clouds into the area and produced ash layers that enable researchers to correlate and radiometrically date the ash and associated lake beds in the various basins.

The lakes disappeared by Oligocene time (38 Ma) as the basins filled with debris. Leveling of the landscape by erosion and deposition continued until about 5 Ma when the rates of erosion quickened and streams began to uncover the buried Laramide-age mountains in the Rockies and Wyoming Basin areas. The Green River established its present-day course as it flowed from the nearby Wind River Mountains across the Green River Basin to Lodore Canyon in the Uinta Mountains (Dinosaur National Monument area) and on into the Colorado River. Twin Creek and other streams in the Fossil Basin area drained west and north, eventually joining the westward-flowing Snake River in Idaho. Downcutting produced a dissected badland topography containing ridges and buttes—including the 1000-foot-high (305-m) Fossil Butte. Along the edges of the valley and butte walls are the exposed edges of strata that contain a record of the past, including the record of the Eocene lakes that formed the Green River Formation.

Green River Fossils

With literally millions of fish and other fossils available for study, this is indeed a paleontological Mecca—one in which the Eocene environment of southwestern Wyoming can be reasonably reconstructed. The opportunity to study the interrelationships of numerous plants, insects, fishes, reptiles, and mammals from an ancient ecosystem is indeed rare.

Fish fossils are rarely found in the rock record. Here they are found in great numbers, with over 20 species represented. A freshwater herring is most abundant, but other species such as paddlefish, gar, bowfins, catfish, and perch are also present. A rare species of freshwater sting ray has also been found. Preservation is amazing—bones remain articulated, tail rays are visible, and even fish skin is often preserved. Thermal stratification of the lake during the summer months produced a warm, less dense oxygen-rich surface zone and a colder, deeper zone where oxygen is absent and toxic gases such as hydrogen sulfide accumulate. Organic materials falling into this lifeless zone were not subjected to destructive oxidizing bacteria or scavengers. Deposition of a limestone mud layer further protected the dead fish or other organism from disturbance and decay.

In addition to the fish, the Green River Formation contains both invertebrates, mainly molluscs and insects, and vertebrates—birds and reptiles including a 13-foot-long (4-m) crocodile and a nearly complete snake skeleton. Bats are rarely fossilized because of their fragile skeletons; nevertheless, the Green River shales yielded a complete skeleton of North America's oldest known flying mammal.

The plant kingdom is well represented—pollen from spruce, fir, and pine, and leaves from oak, elm, maple, beech, and palm. A 10-foot-long (3-m) palm frond, crocodile remains, and the presence of figs, cypress, and other subtropical trees and shrubs support the subtropical climate interpretation for the area immediately around the lake during early Eocene time. Farther up the nearby mountain slopes were broad-leaf forests, deciduous hardwood forests, and finally subalpine forests at elevations of perhaps 6000 feet (1830 m) (Roehler, 1993).

Most of the mammal fossils are found in near-shore deposits or in the Eocene Wasatch Formation. Nearby Pliocene beds also contain bone fossils. The number of different mammals that made Fossil Basin their home is more than a little exciting; the mere listing of the names fills an entire page in McGrew and Casilliano's (1975) report. The dawn horse, *Hyracotherium,* no larger than a fox terrier, pranced about on its four toes (front) and three toes (hind feet) during the early Eocene. Ancestors of rhinoceroses and elephants browsed on the abundant plant life. And during the middle Eocene, the lemurlike and tarsierlike primates put in their appearance.

Green River Lakes

Two questions in particular about the Green River Formation have long fascinated geologists. What environmental conditions caused the alternating dark and light layers that characterize most of the deposit? Second, what environmental condition or catastrophe produced the concentrations of thousands of fish skeletons in certain layers? Apparently there were occasional events that caused huge die-offs or mass mortalities of fish.

As in many lake deposits, the Green River sediments are thinly bedded or laminated as fine particles accumulated quietly on the lake bottom. Most of the lake sediments are best called *calcareous shale*. The term *shale* describes their strong tendency to split along bedding plane weaknesses. They are also calcareous because their chemical composition is mostly that of calcium carbonate (the mineral calcite). Roehler (1993) and other researchers equate the light layers to winter and spring conditions when very little organic material was present to darken the clay-size particles of calcite, dolomite, and quartz that fell to the lake bottom. The dark layers formed "during the summer and fall seasons when the lake waters were warm," algal blooms occurred, and organic productivity was high (Roehler, 1993, p. 10). If this indeed occurred as an annual cycle, then these yearly accumulations are *varves,* analogs to the summer–winter sediment couplets that form in glacial lakes. Each foot of lake sediment contains from 3500 to 5000 (average is 4200) couplets that represent nearly 2 million years in the life history of the lake.

As appealing as the varve idea may be, varvelike layers can be produced when increased inflow events cause more extensive carbonate layers to cover the organic-rich layers—perhaps numerous times during the year. Thus, multiple layers could be produced in a given year (Roehler, 1993), especially near the lake margins (Buchheim, 1994). Laminae farther from shore are more likely to reflect an annual cycle, but even here large inflow events could also produce multiple annual laminae.

The concentration of organic matter in the dark layers produces, over time, a waxy petroleum compound called *kerogen*. Processing of the kerogens in the Green River *oil shales* into petroleum products is currently noncompetitive with conventionally produced oil and gas. However, with the world's largest oil-shale reserve of 80 billion barrels of oil in the Green River Formation and the world's supply of abundant oil and gas to be effectively gone by the year 2040, more interest in oil-shale production can be expected in the next few decades. The consequences of strip mining such a vast area (Fig. 12–15) and the disposal of huge volumes of waste rock has the makings of an environmental and sociological nightmare for the oil-shale states and the nation. About 10–140 gallons of oil can be extracted from each ton of ore-grade oil shale. Each billion barrel of oil will require disposal of about 900 million to 4 billion tons of waste rock. To make matters worse, after being processed this enormous mass of rock increases in volume, and thus the treated rock will not fit back into the hole from which it was excavated!

The occasional huge fish kills evident in the lake history require a special explanation. Although fish fossils occur in many layers, some layers contain an unusually high abundance. A particularly prolific fossil zone called the "18-inch layer" is located about one-third of the way down from the top of Fossil Butte. Possible causes of the mass mortality represented by the 18-inch and similar layers elsewhere in the Green River Shale include (1) expansion of saline lake waters into areas along the lake margin where freshwater fish were living, (2) excessively warm water temperatures during the summer months, (3) an algal bloom that poisoned the water, (4) rapid overturn of oxygen-poor lake bottom waters in the fall that suffocated gill-breathing organisms, (5) some other cause, or (6) perhaps a combination of factors.

Fossil Butte Experience

Do not expect to be permitted to "fish" in the monument; instead, merely cast your eyes on the excellent visitor center exhibits. The fossilized remains of a 13-foot-long (4-m) crocodile, the oldest known North American bat, and an actual slab of Green River Shale with 356 fish fossils exposed are impressive. Short trails lead to the historic quarries, and wayside exhibits interpret the geology and the present high desert environment. A climb to the top of Fossil Butte, 7570 feet (2308 m) above sea level and over 1000 feet (305 m) above the valley below, provides excellent views of an area still relatively unspoiled by humans—an area where not only the "deer and the antelope play," but also moose, elk, coyotes, beaver, and eagles.

For those whose imaginations are well developed, the views across the vast surfaces of the Wyoming Basin are similar to those one would have seen 5–10 Ma elsewhere in the Rocky Mountains—before extensive erosion was reactivated and uncovered the mountains that were awash in their own debris. That slow process continues today where buried uplifts and mountains as well as partially exhumed mountains are found in the broad expanse of the Wyoming Basin. Today is merely one frame in a movie that has run for over 4.5 billion years—the next few frames will bring a very different appearance to the Wyoming Basin as sediments are slowly removed—changes that we as individuals will never see but can only imagine.

REFERENCES

Bradley, W.H., 1948, Limnology and the Eocene lakes of the Rocky Mountain region: Geological Society of America Bulletin, v. 59, no. 7, p. 635–648.

Buchheim, H.P., 1994, Paleoenvironments, lithofacies and varves of the Fossil Butte Member of the Eocene Green River Formation, southwestern Wyoming: Contributions to Geology, University of Wyoming, v. 30, no. 1, p. 3–14.

Grande, L., and Buchheim, H.P., 1994, Paleontological and sedimentological variation in early Eocene Fossil Lake: Contributions to Geology, University of Wyoming, v. 30, no. 1, p. 33–56.

McGrew, P.O., and Cassiliano, M., 1975, The geological history of Fossil Butte National Monument and Fossil Basin: National Park Service Occasional Paper no. 3.

Roehler, H.W., 1993, Eocene climates, depositional environments, and geography, greater Green River Basin, Wyoming, Utah, and Colorado: U.S. Geological Survey Professional P 1506-F, 74 p.

Skinner, B.J., and Porter, S.C., 1987, Physical Geology: Wiley, New York, 750 p.

Thornbury, W.D., 1965, Regional geomorphology of the United States: Wiley, New York, 609 p.

Park Address

Fossil Butte National Monument
P.O. Box 592
Kemmerer, WY 83101

BADLANDS NATIONAL PARK (SOUTH DAKOTA)

Because of the unusual abundance and significance of vertebrate fossils present in the sediments in southwestern South Dakota, the White River Badlands were set aside as a national monument in 1929 and enlarged and upgraded to a national park in 1978. This prehistoric graveyard is considered to be the world's richest Oligocene-age (38–24 Ma) vertebrate fossil deposit and is one of eight areas in the National Park System where fossils are a major reason for the park's existence. Pictures and discussions of these fossil beds are found in nearly every historical geology textbook. The bones of the ancestors of many of today's land animals are there for scientists and others to see. With over 250 species represented, the fossil record here helps to better put the later history of the earth and its inhabitants into perspective.

The fossils just happen to be exposed along the edges of the rock layers that are so vividly exposed in the spectacular, moonscape-like topography known as the White River Badlands—a feature by itself worthy of national park status (Fig. 12–18). Here erosion has cut a myriad of channels into the poorly consolidated mid-Tertiary-age stream deposits to produce the pinnacles and rounded topographic forms known collectively as badlands. The Badlands of South Dakota are the type locality, that is, the area where this special type of topography was first recognized and described. Thus, the area gives its name to other areas that display similar intricately eroded topography.

The 244,000-acre (380-square mile, 988-km²) Badlands National Park is located about 50 miles (80 km) east of the Black Hills (Fig. 12–4) in an area where the Oligocene-age White River sediments are extensively exposed. The White River, about 3–13 miles (4.8–21 km) south of the Badlands escarpment, is so named because of its light-cream-colored water produced by the presence of very tiny submicroscopic particles derived from the weathering and erosion of the White River sediments. Because of electrical charges on the tiny particles, they remain in suspension, thus never allowing the particles to settle and the water to become clear, a difficult problem for thirsty early settlers and travelers. Drinking the water would temporarily quench one's thirst but would produce gastrointestinal discomfort.

History of a Park

The Indians have used the area for over 11,000 years. The most recent occupants were the Oglala Sioux or Lakota who first entered the area during the mid-eighteenth century, pushed westward by disruptions of Indian populations as the Euro-Americans advanced. Their life-style centered around the vast herds of bison that roamed the grassland prairies from Texas to well up into Canada. Even for the resourceful Indians, travel was difficult through the jagged topography between the White River on the south and the Cheyenne River to the north. Their name for the

FIGURE 12-18 Chadron (lower, rounded hills in foreground) overlain by the Brule Formation (upper, steep, angular cliffs) in Badlands National Park. Note the person standing in the mouth of the V-shaped canyon in the foreground. (Photo by E. Kiver)

land, "mako sica" (land bad) reflected the descriptions of the early French trappers. Subsequent explorers and settlers agreed.

Fur trader Jedediah Smith and his fur-trading party were the first Euro-Americans to see the area. Dr. Hiram A. Prout described a jaw from a strange creature called a titanothere, a primitive ancestor of the modern rhinoceros, in 1846. Bones studied by Dr. Joseph Leidy of Philadelphia led to the description of an early primitive camel—an animal along with the horse that first appeared in North America. Leidy went on to become the world authority on the vertebrate fossils of western North America and published a number of papers including a historic paper on the evolution of the horse—one of the best documented evolutionary transitions of an animal through time.

Museums and universities sent expeditions to collect the thousands of fossils that had been exposed by eons of erosion. Private and amateur collectors were also attracted to the area, causing far-sighted individuals in the South Dakota state legislature to petition the federal government to establish a national park as early as 1909. Senator Peter Norbeck actively championed the park concept that bore fruit in 1929 when a national monument was authorized (Hauk, 1969). Norbeck Pass in the park commemorates his efforts.

Rock Record

The layer-cake arrangement of strata is well exposed in the rugged cliffs and gullies of Badlands National Park. The strata appear horizontal but have a slight inclination (dip) to the east—the result of direction of flow of the depositing streams during Oligocene (about 38–24 Ma) times and the regional uplift and eastward tilting of the Great Plains area during late Tertiary (about 5 Ma) times.

Exposed only in the deepest gullies near the foot of the Badlands escarpment (locally called "The Wall") are the oldest rocks, the Late Cretaceous black marine muds of the Pierre (pronounced "peer") Shale. The last of the great inland seas retreated as the land rose upward, initiating a new chapter in the development of the Great Plains.

The area was barely above sea level when the newly exposed strata became deeply weathered during the prevailing subtropical climate. A deep soil zone with brilliant yellows and reds formed—the Interior Paleosol (Fig. 12–19).

The overlying White River Group contains vast quantities of sediment eroded from the Laramide-age Black Hills and Rocky Mountains—sediment that was carried eastward and southward by streams into the Great Plains during late Eocene and Oligocene times, about 40–24 Ma. Because of the long distance of transport

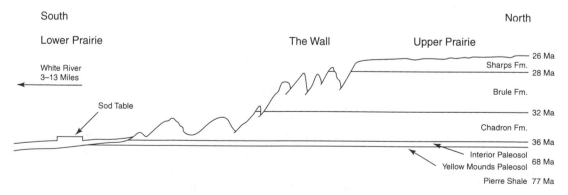

FIGURE 12–19   Generalized cross section from the Lower Prairie across The Wall to the Upper Prairie in Badlands National Park showing important stratigraphic units and their approximate ages.

[50 miles (80 km) or more] and the gentle stream gradients, mostly fine clay-size sediment was deposited to form the gray-colored claystone of the Chadron Formation (Martin, 1987). The overlying claystone and siltstones of the Brule Formation contain significant amounts of volcanic ash derived from erupting volcanoes in the Yellowstone and other Rocky Mountain areas. At least 87 distinct soil zones formed (Rettalack, 1983)—each one subsequently buried by sediment to produce the distinct color bands in the Brule (Fig. 12–18). Zones of rounded, calcium-rich *nodules* and concretions, also produced in soil zones, form resistant ledges in the White River Group and give a whitewashed appearance to many sediments. Also present in both the Chadron and Brule formations are stream channels that were filled with sand (now sandstone). Animal carcasses and bones were buried in these channel sands as well as in the finer sediments deposited in floodplain marshes and ponds.

Volcanism increased during deposition of the Brule Formation and during the deposition of the overlying Miocene age Sharps Formation (Fig. 12–19). A 10- to 23-foot-thick (3- to 7-m) ash layer at the base of the Sharps Formation represents a catastrophic eruption somewhere to the west that must have been deadly to life here and in other areas of thick ashfall.

Additional Tertiary-age deposits were no doubt present but were removed during late Tertiary time when renewed vertical uplift of the Rocky Mountains and Great Plains area rejuvenated stream erosion and lifted the Badlands area up to its present elevation of 2400–3200 feet (732–976 m). Lack of deep burial contributed to the poor consolidation of the White River rocks—rocks that are rapidly eroded into the splendid cliffs and gullies of the scenic Badlands.

Fossil Record

The extinction of the dinosaurs about 65 Ma created a void in the utilization of nature's potential food sources—a void that was soon to be exploited by an opportunistic group of warm-blooded vertebrates known as mammals. Presented with unoccupied environmental niches, organisms rapidly evolve or modify into new forms better adapted to the conditions. The group's diversity initially increases but eventually decreases as better-adjusted forms develop and less-adapted forms become extinct. Oligocene fauna were increasing in diversity toward Miocene times—the Golden Age of Mammals. Later we will visit Agate Fossil Beds National Monument where the Miocene chapter of mammal development is recorded.

By today's standards, the mammals of early and middle Tertiary times appear bizarre and radically different. The basic four-legged form with fur and the ability to nurse its young were there, but its resemblance to modern forms was vague at best.

The White River beds contain over 50 kinds of plant eaters (*herbivores*) and 14 types of meat eaters (*carnivores*)—"the most complete succession of mammal fossils known anywhere in the world" (Wicander and Monroe, 1993, p. 541) and the "birthplace of the science of vertebrate paleontology" (National Park Service Website, 1997). Isolated bones or groups of bones are common, and occasionally almost entire skeletons are found in the channel sandstones and swampy floodplain sediments. After burial, mineral material, mostly calcium or silica, is deposited by groundwater in the porous bony material creating a rocklike mass. Teeth are the hardest body parts and often show amazing detail of their original structure. They are also the most diagnostic part of the skeleton—often a single tooth is adequate to identify a specific genus.

All sizes of animal fossils from tiny rodents to rabbits to elephant-size titanotheres, distant relatives of the modern rhinoceros, occur in the White River beds. The titanotheres made phenomenal progress from the hog-sized beast of the Eocene to the *Brontotherium* of the Oligocene and were the largest land animals in North America at that time. They were much bulkier than the largest living rhinoceros, perhaps too bulky. They are abundant in the Chadron Formation, they are rare in the Brule, and they become extinct before the Oligocene ended. Another group of mammals that were very abundant but became extinct in late Tertiary (Pliocene) time were the piglike oreodonts. These cud-chewing browsers were fox-size and must have been important elements in the food chain. Fortunately, some of the more distant relatives of the oreodonts fared better in Europe and Asia, eventually giving rise to today's domesticated cows and sheep.

Hyracotherium, or dawn horse, made its appearance in North America during Paleocene time, about 60 Ma. This bizarre, house-dog-size, four-toed creature would evolve through Tertiary time into the modern single-toed *Equus* during Pleistocene time (Fig. 12–20). *Mesohippus,* the collie-sized three-toed horse, is common in the White River sediments. Camels also first appeared in North America, and along with the horse evolved to more modern forms before becoming extinct in North America during the Pleistocene. Fortunately, the camel and horse migrated to Asia before the Pleistocene extinctions, otherwise human history would be quite different. The cause of these extinctions is unknown.

Many other animals also wandered about on the floodplains, including saber-tooth or stabbing cats, doglike carnivores, and the early ancestors of pigs. *Proteroceras,* a sheep-size animal with up to five pairs of horns or knobs on its snout was a strange animal indeed. The variety and abundance of mammals was still increasing during Oligocene time—peaking during the Miocene and declining slowly afterward. A major disruption in the usual slow processes of biologic change was initiated about 10,000 years ago as another mammal species, *Homo sapiens,* increased in numbers and developed more efficient technologies. Extinctions of plants and animals has greatly accelerated during the past 200 years and have now reached unprecedented numbers. The current extinction event promises to be the most serious mass extinction of lifeforms that the earth has ever experienced. These present extinctions are more serious than the great extinctions at the end of the Paleozoic Era, when perhaps 90 percent of all life was extinguished, or at the end of the Mesozoic, when the dinosaurs along with perhaps 70 percent of all life was eliminated.

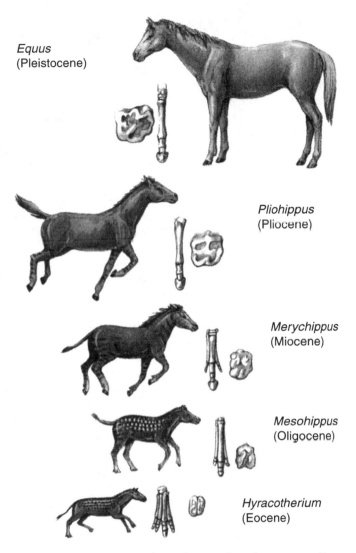

Equus
(Pleistocene)

Pliohippus
(Pliocene)

Merychippus
(Miocene)

Mesohippus
(Oligocene)

Hyracotherium
(Eocene)

FIGURE 12–20   Changes in size, numbers of toes, and tooth structure of horses during the past 50 million years. (Diagram from Wicander and Monroe, 1993)

Erosion at Work

Pliocene (5 Ma) uplift and eastward tilting of the Great Plains rejuvenated the streams draining the Rocky Mountains and Great Plains area enabling them to cut downward and begin the process of sediment removal. The White River, which is south of the Badlands, the South Fork of the Cheyenne River to the northwest, and the Bad River to the north are the master streams of the area. Wetter Pleistocene

(Ice Age) climates also contributed to increased erosion and formation of the modern badlands. Initially, a narrow valley was cut along the course of the White River. Small, closely spaced tributaries flowed southward down the steep north valley wall, causing the valley side to erode and migrate northward as a line of cliffs—locally known as "The Wall." Those badlands began about 500,000 years ago and were a few miles south of the present park—closer to the White River. Another 500,000 years from now they will have migrated north—destroying the Upper Prairie, the drainage divide between the White River and the Cheyenne River (Fig. 12–19). The Lower Prairie is formed as the cliffs migrate. Later downcutting leaves isolated, flat-topped buttes in the Lower Prairie called *sod tables* because of their grassland surfaces.

All of the essential elements are here to produce the classic erosional forms known as badland topography—weakly consolidated, easily eroded fine-grained rock; steep slopes for streams to flow down; the lack of abundant vegetation to stabilize hillslopes; and the semiarid climate that often produces short, intense rains that form "gully washers." Bentonite clays derived from the volcanic ash swell when wet and contract when dry—forming a "popcornlike surface" that makes root establishment difficult—further encouraging high rates of erosion. Loose sediment is also vulnerable to mechanical dislodgement and removal as raindrops beat down upon the surface. Thus, some of the highest rates of erosion known occur in the Badlands. Some surfaces are reduced by an inch (2.54 cm) every year. Other catastrophic changes such as collapsing ridges and pinnacles have literally happened overnight as a result of a single thunderstorm.

Today The Wall separates the flat grassy Upper Prairie to the north from the Lower Prairie or pediment surface to the south. The Wall divides the vigorous streams flowing south to the White River from the slow, sluggish streams flowing north to the Cheyenne and Bad Rivers. The appearance of the resulting landforms is controlled by the characteristics of the rocks themselves. The clay-rich sediment in the Chadron Formation causes hills to develop a rounded, haystacklike form. The silty, ash-rich Brule Formation is more resistant to erosion and forms sharpened pinnacles and knife-edge ridges. Rapid erosion produces the development of interesting short-lived features such as natural arches and bridges.

Vertical cracks caused by sediment shrinkage and compaction were filled with sediment, mostly volcanic ash, forming *clastic dikes*. Groundwater circulating through these filled cracks often deposits silica cement, often in the form of *chalcedony* (quartz with extremely tiny, submicroscopic crystals) that adds great strength and resistance to the clastic dike. Thus these crack fillings often form straight microridges that project above the surrounding sediment. A good place to see these features is at Clastic Dike Overlook along the loop road.

The channel-fill sandstones are also more resistant than the claystones and siltstones and often form nearly vertical cliffs and *balanced rocks*—locally known as *toadstool rocks*. The red and yellow paleosols produce the prominent color banding in the Brule Formation and the colorful views of the Interior Paleosol at the Yellow Mounds Overlook and elsewhere in the park.

Visiting Badlands

Most visitors only spend a few hours in the Park. Leaving Interstate 90 they stop at the visitor center, drive the 30-mile-long (48-km) loop road, and perhaps take one or two hikes along the developed trails. Cliff Shelf Trail leads across a large *slump block*—a section of cliff that has detached from the cliff behind and has slipped downward and rotated backward—much as one might slip downward in a comfortable chair. Water trapped on the back section of the slump block promotes a more luxurious growth of vegetation and occasionally, especially in wetter years, continues to slip—much to the detriment of the loop road. Fossil Exhibit Trail has partially exposed bones for visitors to examine.

The more adventurous visitors who set aside adequate time to see the park may backpack into the Badlands wilderness or may hike some of the longer trails. A walk up one of the numerous undeveloped canyons to observe firsthand the work of erosion and to "get in touch" with the rock can be an interesting experience. Flash floods continually remove sediment and often produce rounded, baseball-size clay masses called *armored mudballs* because their outer surfaces are embedded with rock fragments (Fig. 12–21).

Bone hunting in the Badlands can be frustrating, particularly so if you find one sticking out of a gully bank. This is a park area, and removal of any type of

FIGURE 12–21 Armored mudballs (and shoes) washing out from one of the numerous unnamed canyons that cut back into the line of badland cliffs along The Wall at Badlands National Park. (Photo by E. Kiver)

material is forbidden. Although the bones have been preserved in the rock for over 25 million years, once exposed they will crumble to useless fragments unless they have been stabilized by adding a special cement and are removed by skilled technicians or vertebrate paleontologists. The fossil is most valuable to science when properly excavated and its exact stratigraphic and geographic location is noted. Report all such finds to a naturalist and go on to the Ben Reifel Visitor Center or especially to Rapid City where, at the museum of the South Dakota School of Mines and Technology, you can see the fossil record come "alive" in the reconstructed skeletons of many of the inhabitants who roamed the area 24–40 Ma.

For those with an interest in western history, the story of the numerous fossil digs and the individuals who did the pioneering science in a land unknown to the developed world is fascinating. Also of interest are the proud Oglala Sioux and the unfortunate stories of their people at the hands of the Euro-Americans. Their history can be better understood by visiting the White River Visitor Center in the South Unit of the park. The 1890 massacre at Wounded Knee in the southern end of the park resulted in the death of 220 Indian men, women, and children and was the last major Indian battle fought in the West.

The prairie ecosystem that once covered about one-third of the North American continent was nearly eliminated by Euro-Americans. However, here is the largest protected remnant of the vast grasslands for all to see. The superabundant wildlife present when the Euro-Americans first arrived was decimated until by 1919 a game investigator reported that not even a coyote could be heard in the Badlands area. The bear, gray wolf, and Audubon Bighorn are still gone; but the bison, pronghorn antelope, Rocky Mountain bighorn, deer, and elk are here thanks to natural repopulations and reintroductions. From a nucleus of 40 bison acquired by the Bronx Zoo in 1903, some 450 of their descendents now roam the Badlands prairies. Without the gray wolf, the natural predator of the bison, the herd must be managed by humans.

Prairie dog towns are still here in spite of the intense eradication efforts of the early settlers and ranchers. Reintroduction of the black-footed ferret, North America's most endangered mammal, will hopefully add some balance to the remnants of what was once the world's largest mixed-grass prairie ecosystem.

REFERENCES

Hauk, J.K., 1969, Badlands: Its life and landscape: Badlands Natural History Association, 64 p.

Martin, J.E., 1987, The White River Badlands of South Dakota, in Beus, S.S., ed., Rocky Mountain Section of the Geological Society of America: Geological Society of America, Boulder, CO, Centennial Field Guide, p. 233–236.

Rettalack, G.J., 1983, Late Eocene and Oligocene paleosols from Badlands National Park, South Dakota: Geological Society of America, Special Paper 193, 82 p.

Wicander, R., and Monroe, J.S., 1993, Historical geology: West Publishing, Minneapolis/St. Paul, 640 p.

Park Address

Badlands National Park
Post Office Box 6
Interior, SD 57750

AGATE FOSSIL BEDS NATIONAL MONUMENT
(NEBRASKA)

Agate Fossil Beds National Monument is a small, 1970-acre (3.1-square mile, 8-km^2) area in northwest Nebraska located in the sparsely populated ranchland along the Niobrara River. Abundant agates in the local rocks provided the inspiration to name the nearby small town and later the national monument. The monument is about 130 miles (210 km) southwest of Badlands National Park where the fossil-rich Oligocene-age White River Beds (38–24 Ma) are spectacularly exposed in the badland cliffs. Here at Agate the White River beds remain covered by lower Miocene-age (about 24–13 Ma) Arikaree Group rocks—rocks that contain the next chapter in the history of the Great Plains and the Age of Mammals. The fossil graveyard here was the first major concentrated deposit of Tertiary-age (last 65 million years) mammals discovered in North America. Outstanding specimens, some virtually complete, help science to piece together the fascinating past that eventually leads to the terrestrial animals of today's world.

History of Agate Fossil Beds

James H. Cook bought the Agate Springs Ranch from his father-in-law in 1887. Cook was a man of broad interests and skills. He was a cowboy, big-game hunter, U.S. Army scout, and eventually an author who recorded some of his frontier experiences. He was a close friend of the famous Chief Red Cloud of the Oglala Sioux and often mediated problems between the Sioux and the settlers (National Park Service, 1980). Riding past a pair of conical buttes not far from his ranch house (Fig. 12–22), Cook discovered an animal leg bone protruding from one of the buttes, the first of a series of discoveries that would greatly influence the lives of the Cook family and the science of paleontology.

A keen competition among paleontologists was underway in the American West to find the best fossils to reconstruct the life of the past. In particular, the record of mammal evolution that was poorly preserved in most areas of the world was extremely well-preserved in the West where Cenozoic-age rocks of terrestrial origin that span the entire 65 million years of the Cenozoic are relatively abundant. In 1891 Professor Erwin Barbour of the University of Nebraska explored the Cook Ranch (Agate Springs) area. His fascination with a strange spiral feature in the rock

FIGURE 12–22   View of University and Carnegie hills in Agate Fossil Beds National Monument. (Photo by E. Kiver)

called *Daemonelix* caused him to ignore some of the interesting bone fossils that one of his students discovered. O.A. Peterson from the Carnegie Museum of Pittsburgh began excavations in 1904 in one of the two conical hills (Carnegie Hill) in what was soon recognized as a bone bonanza of international significance (Hunt, 1992). Barbour began excavations in 1905 at University Hill, and numerous museums and universities conducted major fossil digs up until about 1923. Cook graciously hosted these groups and eventually acquired a rare collection of natural history specimens, fossils, and Indian artifacts. Many of the artifacts were gifts from the Oglala Sioux, including Chief Red Cloud. Part of Cook's house turned into a popular museum as tourists visited the area.

Cook's oldest son Harold was brought up among all of this scientific excitement and went on to become a geologist. He married another geologist—Eleanor Barbour—the daughter of the distinguished Nebraska geologist who was the first to study the fossils at the Agate Springs Ranch! Harold went on to his own distinguished career in paleontology and requested that the ranch become part of the National Park System. Through the diligent efforts of the Cook family and Nebraska Senator Roman Hruska and Representative Dave Martin, in 1965, three years after Harold Cook's death, this significant scientific area was authorized as a national monument by Congress.

Geologic Background

The dominating influence on the geologic history of the Great Plains is a series of events occurring in the Rocky Mountains during and after the Laramide Orogeny. Much of the sediment eroded from the rising mountains was initially trapped in structural basins in and near the mountains. During Oligocene time as these basins filled, eastward-flowing rivers deposited vast sheets of the White River sediment in the Great Plains, including the Agate area. These fine-grained, volcanic ash-rich White River beds are splendidly exposed in southwestern South Dakota in Badlands National Park but are buried in the Agate area by early Miocene sediments, including the bone deposits in the Harrison and Marshland formations.

The area was close to sea level and humid during early Tertiary time—thus accounting for the warm subtropical climate recorded in the plant and animal fossils in Theodore Roosevelt National Park and in Fossil Butte and Florissant national monuments, as well as at other sites. By early Miocene time the uplift of the Rocky Mountains was increasing the rain-shadow effect and climates were becoming drier and cooler. Vegetation changed from the subtropical forests of early Tertiary time to a grassland with abundant patches of trees (savannah) during Oligocene and Miocene time. Increasing aridity eventually produced the grassland prairies that persist to the present. The slow transition from savannah to grassland prairie is reflected in the fossil record in the Great Plains.

Shifting streams and episodes of downcutting during early Miocene time produced valleys that were later filled with sediment as the surface of the Great Plains continued to aggrade or build upward. During times of drought, animals congregated around the shallow waterholes in these channels. Thousands of animals perished during some of these droughts forming the two major bone layers at Agate. The older bone layer lies below an ash layer that is radiometrically dated at 21 million years. The younger layer is above the ash layer and is perhaps about 20 million years old.

Continued uplift, particularly during the past 5 million years, raised the area to its present elevation of 4400 feet (1340 m), resulting in erosion and removal of previously deposited sedimentary material. Erosion, particularly during the wetter episodes of the Pleistocene, or Ice Age, allowed the modern Niobrara River to establish its course and erode its valley to its present configuration—including the two conical-shaped hills known as Carnegie and University hills.

Fossil Record

The early Miocene (about 20 Ma) savannah was teeming with life—much like the great animal herds that swarmed across East Africa as recently as the early twentieth century. Miocene life would seem odd to us but would certainly appear more familiar than that of earlier Tertiary time. All of the fossil mammal species at Agate are now extinct although many are distant relatives of modern forms.

The Agate Springs quarry in the Marshland Formation is dominated by the bones of a 3-foot-high (1-m) two-horned rhinoceros called *Menoceras* (Fig. 12–23).

The abundance of mammals reached a peak in the Miocene Epoch. The time was marked by refinements in life forms, and many animals and plants developed features recognizable in some species today. Forests and savannas persisted in some parts of North America; treeless plains expanded where cool, dry conditions prevailed.

Many mammals adapted for life on the prairie by becoming grazers, runners, and burrowers. Large and small carnivores evolved to prey on these plains-dwellers. Great intercontinental migrations occurred throughout the Miocene, with various animals entering and leaving North America.

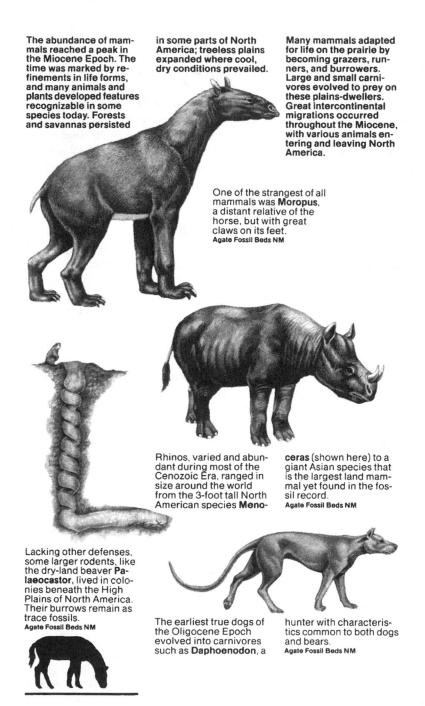

One of the strangest of all mammals was **Moropus**, a distant relative of the horse, but with great claws on its feet.
Agate Fossil Beds NM

Rhinos, varied and abundant during most of the Cenozoic Era, ranged in size around the world from the 3-foot tall North American species **Meno-ceras** (shown here) to a giant Asian species that is the largest land mammal yet found in the fossil record.
Agate Fossil Beds NM

Lacking other defenses, some larger rodents, like the dry-land beaver **Palaeocastor**, lived in colonies beneath the High Plains of North America. Their burrows remain as trace fossils.
Agate Fossil Beds NM

The earliest true dogs of the Oligocene Epoch evolved into carnivores such as **Daphoenodon**, a hunter with characteristics common to both dogs and bears.
Agate Fossil Beds NM

FIGURE 12–23   Reconstructions of some of the early Miocene animals that lived in the Agate area about 20–21 Ma. (Illustrations from National Park Service brochure)

Dinohyus, "the terrible hog," had bone-crushing teeth, which allowed it to scavenge the remains of other grassland animals.
Agate Fossil Beds NM

The tiny gazelle-camel **Stenomylus** probably grazed in herds for protection against predators.
Agate Fossil Beds NM

FIGURE 12–23   (*Continued*)

The rhino bones are mixed in with the bones of pig and a large piglike scavenger called *Dinohyus*, or "terrible pig," and those of camels, oreodonts, sabretooth cats, carnivorous beardogs, and numerous other animals. One species of horse, *Miohippus*, would soon become extinct. However, one evolutionary line led to *Parahippus* (also found in the bone beds) that was to become the "Horse of Destiny" (National Park Service, 1980). Its teeth had an extra wrinkle in the enamel, a specialization that strengthened the tooth and permitted *Parahippus* to eat the abrasive silica-rich grasses that were beginning to become more widespread in the Great Plains. Climates were becoming more arid and colder as the Rockies continued their slow rise. Trees were decreasing in abundance and the browsing animals such as the *Menoceras* rhinoceros and the *Miohippus* horse would be extinct by late Miocene time.

Parahippus went on to develop high-crowned teeth that were even longer-lasting and would, in middle Miocene time, give rise to *Merychippus, Pliohippus* during late Miocene, and the one-toed modern horse *Equus* near the end of the Pliocene (Fig. 12–20). Fortunately, *Equus* migrated to Eurasia over the Bering Land Bridge before becoming extinct in North America at the end of the Pleistocene, about 10,000 years ago. The horse did not return to its land of origin until the Spaniards brought horses back to the New World in 1519.

One of the stranger-appearing animals was *Moropus*—a 7-foot-tall (2-m) sloth-like animal (Fig. 12–23) that was a distant relative of the rhinoceros and horse. *Moropus* looks like it could have been designed by a committee in which compromises in design were made between individual committee members. The head was horselike, the front legs were like those of a rhinoceros, and the hind legs were like those of a bear. Most unusual were the clawed feet, perhaps used for defense and to dig for plant roots. Another strange-looking animal was *Syndyoceras,* a deerlike beast with two pairs of curving horns on its head—a descendant of the strange *Proteroceras* that was present in the White River beds at Badlands National Park.

Of special interest is the strange spiral-shaped feature as tall as a human that mystified early paleontologists who named the feature *Daemonelix*—or Devil's Corkscrew. One early idea was that it was a filled-in cavity left by some unknown tree that grew a corkscrew-shaped tap root. Discovery of the bones of *Paleocastor,* an extinct beaver at the bottom of some of these corkscrews indicated that these were their filled-in burrows (Fig. 12–23). The abundance of *Daemonelix* suggests that these ancient dry-land beavers lived in colonies—much like prairie dogs do today.

About 1.5 miles (2.4 km) east of University and Carnegie hills is a small detached unit of the monument containing the *Stenomylus* quarry. *Stenomylus* was a 2-foot-tall (60-cm), gazellelike camel (Fig. 12–23) that probably traveled in large herds for protection. The bone bed is in the Harrison Formation and must be slightly older than the 21 million-year-old ash bed above. A group of 100 or more mummified, mostly articulated bodies of these ancient camels occur in a group. Their demise remains a mystery—perhaps it was the result of an earlier drought.

Journey to the Past

At first view, the visual impact at Agate Fossil Beds is not as spectacular as in our more scenic parks. However, the open spaces, as far as the eye can see, is impressive and reminds us of the vastness of the continent, the grassland prairie, and the millions of animals that once inhabited the area. For many who live in cities and towns (which is most Americans), looking across vast, sparsely settled landscapes, or knowing that such landscapes still exist, is refreshing and necessary for one's psyche.

Here at Agate our minds and imaginations are called into action to absorb the significance of the information provided in the visitor center and by observing the partly excavated fossils in University and Carnegie hills. An estimated 75 percent of all the fossils in the bone bed still remain buried for future generations to study—perhaps in ways that are not technologically feasible at present. Recommended reading, perhaps while seated on an outcrop on University Hill, is James and Laurie Macdonald's excellent word description of an imaginary field trip to the

Agate area 21 Ma to observe the early Miocene landscape and the animals and their activities. Their realistic descriptions are in the National Park Service (1980) handbook—available in the visitor center or by writing to the park before your visit.

Be sure to take the trail to see the partly excavated *Daemonelix* fossil in the outcrop to appreciate the confusion that this strange fossil evoked in the early paleontologists. A list of museums around the country that display some of the Agate fossils is available in the handbook or at the visitor center.

A continuation of the fascinating history of mammal evolution is found in eastern Nebraska at Ashfall Fossil Beds State Historical Park. Here animals also congregated around a waterhole—this time because 10 feet (3 m) of ash from a huge volcanic eruption in Idaho buried the landscape! Hundreds of animals migrated to the waterhole where they suffered a slow death by suffocation. Unlike the Agate fossil deposits, carnivores and scavengers did not extensively gnaw and churn up the bone deposit—at Ashfall entire articulated animals and their young lie in the exact position in which they died—some of the young trying to nurse from their dead or dying mothers.

REFERENCES

National Park Service, 1980, Agate Fossil Beds: Handbook 107, 95 p.

Hunt, Jr., R., 1992, Death at a 19 million year-old waterhole: The bonebed at Agate Fossil Beds National Monument, western Nebraska: Museum Notes, University of Nebraska State Museum, no. 83, 6 p.

Park Address

Agate Fossil Beds National Monument
301 River Road
Harrison, NE 69346-2734

CAPULIN VOLCANO NATIONAL MONUMENT
(NEW MEXICO)

Far-sighted members of Congress increasingly recognized the scenic, scientific, and historical importance of special areas within our borders and began to withdraw some of the public lands from settlement during the late 1800s. Capulin Mountain was withdrawn in 1891 and given national monument status in 1916. The name of the tiny 793-acre (1.2-square mile, 3.2-km^2) monument was changed to Capulin Volcano National Monument in 1987.

Capulin is a *cinder cone*—a type of volcano constructed of magma blobs and other particles that were blasted into the air as *pyroclastic* material from the vent. Gases trapped in the airborne magma blobs expanded and formed solidified popcornlike masses (*cinder*) that accumulated into a steep-sided, conical-shaped volcano—a cinder cone. Some of the ejected blobs formed more solid lava masses called *volcanic bombs*. Wind direction greatly influenced the trajectory of the cinder and the resulting shape of the cone. The slight northeast elongation of the cone and higher elevation on the east rim indicate that the prevailing westerly and southwesterly winds were active while Capulin was ejecting its immense volume of cinder. Capulin is one of the tallest, most symmetrical, and most readily accessible of North America's most recent volcanoes (Fig. 12–24).

Geographic Location

The boundary between the High Plains Section of the Great Plains and the Southern Rocky Mountains in the northern New Mexico–southern Colorado area is the east base of the Sangre de Cristo Mountains, the southernmost range in the Southern Rocky Mountains. The sawtooth-profile peaks of the range loom above the Great Plains about 80 miles (130 km) west of Capulin Mountain.

Capulin is in about the center of the Raton–Clayton Volcanic Field, an extensive area of lava flows and over 100 volcanic vents in northeastern New Mexico and adjacent Colorado. The volcano's summit is at 8182 feet (2495 m)—an impressive

FIGURE 12–24   View of Capulin cinder cone. Sierra Grande volcano in distance. (National Park Service photo)

1300 feet (396 m) above the surrounding surface of the High Plains. Just west of the Sangre de Cristo Mountains lies the Rio Grande Rift, a major pull-apart structure that extends from at least central Colorado through New Mexico and Texas and into Mexico. Recent crustal stretching along the Rio Grande Rift and southwestern North America has activated other major fault systems and generated local volcanism such as that in the volcanic field associated with Capulin Mountain.

Geologic Setting

The Raton–Clayton Volcanic Field contains the northeasternmost volcanoes along the Jemez Lineament, a 375-mile-long (600-km) fault zone that extends from east-central Arizona to the Capulin Mountain area (Mutschler and others, 1998). Where the lineament crosses the Rio Grande Rift, a major explosive volcanic center occurs in the Jemez Mountains in the Bandelier National Monument area. The Jemez Lineament faults are quite old, perhaps formed in Precambrian time and reactivated during the past 40 million years as new crustal stresses began to pull southwestern North America apart. Extension of the crust in the Rio Grande Rift and along nearby fault zones caused decompression melting of the upper mantle and lower crust and triggered episodes of extensive volcanism—processes that continue today and are expected to continue into the future.

Geologists concerned about understanding these deeper earth processes have much to study here (Stormer, 1987). There is a wide range of volcanic rock types from low-silica basalt to higher-silica andesite and dacite, all geographically close. Although most of the volcanic rocks in the Raton–Clayton Volcanic Field are basalt produced by partially melting rocks in the upper mantle, unusually high amounts of certain chemical constituents such as potassium suggest that contamination by lower crustal rocks occurred. Partial melting of lower crustal rocks or perhaps mixing of magmas of different compositions could account for the high-silica andesitic stratovolcanoes and dacitic plug domes in close proximity to low-silica shield volcanoes and basalt flows.

Fiery Past

Volcanism along the Jemez Lineament is less than 10 million years old with most of the activity occurring during the past 5 million years. Numerous eruptions along the lineament in the Raton–Clayton and other volcanic centers during the past 1.6 million years (Luedke and Smith, 1991) suggests that activity will continue well into the future and should provide more than a little excitement for geologists and others.

Volcanism in the Raton–Clayton Volcanic Field occurred in three separate episodes; the oldest began about 8.2 Ma (Sayre and others, 1995) and produced extensive basalt flows that filled broad valleys (Stormer, 1987). Because the lava is more resistant than the sedimentary rocks in the area, later erosion lowered the sides of the lava-filled valleys, leaving the former valley floors as today's elongated mesa tops—a textbook example of *topographic reversal*. A trip along the length of the mesa surface follows the position of an ancient valley. The second interval of volcanism filled the new valleys and produced lava-capped surfaces at a slightly lower elevation. The most recent eruptive interval ended with flows from Capulin Mountain and Baby Capulin that are "essentially at the present erosional level" (Stormer, 1987, p. 424).

The excellent preservation of the Capulin cinder cone and associated lava surfaces is a result of its recent formation 56,000–62,000 years ago (Sayre and others, 1995; Stroud, 1996) and the dry climate of the High Plains. Recently formed basalt surfaces are highly irregular and often display blocky surfaces, abundant *pressure ridges* where the hot crust was wrinkled as the flow moved, and small *spatter cones* where pasty blobs of magma were ejected above a fissure vent. Such features are absent from the 8-million-year-old lava-capped surfaces, are barely recognizable on the intermediate-age flows, and look as if they formed "yesterday" on the Capulin flows. Older flows are smooth and mostly covered with a thick soil and grassy vegetation; young flows have localized or very thin soils and considerable area of exposed bedrock. The Capulin cone is little modified from its original form—except for a conspicuous notch that spirals up the cone and is paved with asphalt!

Visiting Capulin Volcano

Your experience at Capulin will begin many miles away as the approach roads traverse the lava-capped mesas and the younger lava flows. Volcanic peaks dot the landscape. Sierra Grande (8720 feet; 2660 m) about 10 miles (16 km) southeast of Capulin is one of New Mexico's largest volcanoes (Fig. 12–24). The visitor center exhibits will help "set the scene" for your ride up the road that spirals up to the summit parking area.

Near the base of the mountain on the west flank is the *Boca* (Spanish for mouth) and *lava levees* where magma rising up into the loose pile of cinder that makes up the cone broke through forming four distinct flows. The second flow moved south and then east—terminating in a lobate-shaped front that is easily seen from the crater rim. A 1-mile-long (1.6 km) trail circles the rim and on clear days provides spectacular views when parts of five states (New Mexico, Colorado, Kansas, Oklahoma, and Texas) are visible. The 415-foot-deep (127 m) crater is awesome, and the views of the basalt-capped mesas and scores of surrounding volcanic peaks are superb. The Sierra Grande shield volcano about 10 miles (16 km) to the southeast is the largest volcanic edifice in the region. Conical-shaped Red Mountain and Towndrow Mountain to the northwest are dacitic plug dome volcanoes. The lobate pattern of the Capulin flows and the concentric pattern of pressure ridges on its surface are readily visible as are the Baby Capulin vents that are even younger than Capulin Mountain. Lava from Baby Capulin flowed nearly 20 miles (32 km) to the northeast.

A short trail from the Rim parking area leads into the bottom of the crater—a good place to imagine the events occurring here about 60,000 years ago. The same tectonic forces that were stretching the crust under southwestern North America then are continuing today, thus the release of pressure and other magma-producing processes are still active and make this region a likely site for future eruptions. Cinder cones are usually "one-shot volcanoes" that experience only one episode of eruption—thus additional eruptions from the Capulin vent are unlikely. If they were to occur here again, hopefully they will not occur while you are sitting on the crater floor!

REFERENCES

Luedke, R.G., and Smith, R.L., 1991, Quaternary volcanism in the western conterminous United States, in Morrison, R.B., ed., Quaternary nonglacial geology; conterminous U.S.: Geological Society of America, Boulder, CO, The geology of North America, v. K-2, p. 75–92.

Mutschler, F.E., Larson, E.E., and Gaskill, D.L., 1998, The fate of the Colorado Plateau—A view from the mantle, in Friedman, J.D., and Huffman, A.C., Jr., coordinators, Laccolithic complexes of southeastern Utah: Time of emplacement and tectonic setting—Workshop proceedings: U.S. Geological Survey Bulletin 2158, p. 203–222.

Sayre, W.O., Ort, M.H., and Graham, D., 1995, Capulin Volcano is approximately 59,100 years old: Park Science, Spring, p. 10–11.

Stormer, Jr., J.C., 1987, Capulin Mountain Volcano and the Raton-Clayton Volcanic Field, northeastern New Mexico, in Beus, S.S., ed., Rocky Mountain Section of the Geological Society of America: Geological Society of America, Boulder, CO, Centennial Field Guide Volume 2, p. 421–424.

Stroud, J.R., 1996, The volcanic history and landscape evolution of the Raton-Clayton volcanic field: New Mexico Institute of Mining and Technology, M.S. thesis, 49 p.

Park Address

Capulin Volcano National Monument
P.O. Box 40
Capulin, NM 88414

THIRTEEN

Central Lowlands Province

The Central Lowlands Province is the largest geomorphic province in the United States, covering the north-central portion from just east of the Great Lakes west to the Great Plains in the Dakotas. From the Dakotas the province continues south into the Osage Section, which extends through eastern Kansas, Oklahoma, and ends in north-central Texas (Plate 1). The province continues north into Canada where the Paleozoic sedimentary rocks end and the Precambrian crystalline rocks of the Canadian Shield begin. Elevations are low, ranging from 1500 to 1800 feet (460 to 550 m) in the west to about 400 feet (120 m) in southern Illinois where the upper Mississippi River flows south into the Coastal Plain Province. Elevations on the east along the edge of the Appalachian Plateaus are 1000 feet (305 m) or less. Although mostly lacking the spectacular topography of some of our western parks and parts of the Appalachian Mountains, the province contains areas of unusual geologic, scenic, and ecological interest (Fig. 13–1), many of which form the nucleus of state parks. Discussions of some of these state parks and other areas are found in geological guidebooks. Highly recommended are the *Centennial Field Guides* in which some of the areas of unusual geologic interest are described (Biggs, 1987; Hayward, 1988).

Province Boundaries

The Central Lowlands are mostly surrounded by higher elevation topography including the Appalachian, Ozark, and Interior Low Plateaus on the east and south and an eastward-facing escarpment along most of the Great Plains on the west (Plate 1). The topographic boundary is gradual and less distinct in Nebraska and Kansas. The southern boundary from southern Ohio to central Missouri is essentially the southern limit of Pleistocene glaciation.

FIGURE 13–1　In the Central Lowlands particularly, state parks largely take the place of national parks. Relict plants left over from the Ice Age still thrive in the cool recesses and overhangs of Rocky Hollow located in Turkey Run State Park in western Indiana. Sugar Creek and its tributaries cut deeply into the Mansfield Sandstone, thus forming this unique environment. (Photo by D. Harris)

Structure and Bedrock

The Central Lowlands are part of the stable continental interior, or *craton*—an area where only minor deformation has occurred since Precambrian time. So-called *mobile belts,* or *geosynclines,* have occurred along its southern and eastern margins in more recent geologic time. Like the Great Plains, the Central Lowlands have only locally been subjected to severe tectonic upheavals, and the structures consist largely of broad, gentle upwarps and downwarps such as the Cincinnati Arch and the Illinois and Michigan basins. The Wichita and Arbuckle mountains in Oklahoma where Chickasaw National Recreation Area is located, is one exception; here the rocks are distinctly folded and locally faulted, an anomaly for the midcontinent region.

A large Paleozoic-age rift zone (Southern Oklahoma Rift) intersects the Ouachita Mountains Province along the south edge of the North American continent and extends northwesterly into the Central Lowlands. The Ouachita Geosyncline and the associated Southern Oklahoma Rift received thick accumulations of early Paleozoic sediments (mostly limestone) before being folded and faulted into mountains during late Paleozoic time. The more rugged topography in the Wichita and Arbuckle mountains in the Osage Section in Oklahoma follows along this ancient rift. Recent (Holocene, last 10,000 years) movements along the Meers Fault on the northern edge of the Wichita Mountains in southwestern Oklahoma has produced the best example of an active fault scarp east of the Rocky Mountains (Donovan, 1988)—another anomalous feature for a cratonic area. The Meers Fault last moved about 1300 years ago and is believed capable of generating a 7.5–8.0 earthquake—a large earthquake even by California standards. Platt National Park, now included in Chickasaw National Recreation Area, is located about 80 miles (130 km) east of the Meers Fault in the Arbuckle Mountains.

The "big tectonic picture" for the south and east edges of the craton during Paleozoic time involves the movement and assembling of continental plates into the giant supercontinent Pangaea. The previous tectonically quiet *passive plate margin,* or *trailing edge,* of early Paleozoic time changed to an active *convergent plate margin,* or *leading edge,* during the late Paleozoic, thus enabling a thick section of sediments to accumulate in the Ouachita Geosyncline and the Southern Oklahoma Rift—followed by folding and faulting during late Paleozoic [beginning in Late Mississippian, about 340 million years ago (Ma)] time. The mechanical force driving the mountain building (*orogeny*) during Pennsylvanian time (about 325 Ma) was the collision of the southern assemblage of land masses (Gondwana) with the northern landmass assemblage (Laurentia) to form the Pangaea supercontinent. Coarse Pennsylvanian-age conglomerate shed from the folded and faulted Ouachita–Arbuckle–Wichita mountains covered the eroded edges of upturned strata creating an *angular unconformity.* The use of fossils makes it possible to determine effectively the age of rocks and consequently the age of folding and mountain uplift (Donovan and Heinlen, 1988).

The mostly Paleozoic-age bedrock in the Central Lowlands is largely covered by glacial deposits and outwash materials except in the unglaciated Osage section. However, along the western border extensive areas of younger rocks remind us of the vastness of the Cretaceous Geosyncline. In eastern Kansas and farther north, the Cretaceous-age (about 130 Ma) Dakota Sandstone is locally exposed. When we last saw the Dakota, it was forming the main hogback ridge just east of the Southern Rockies. It lies deep beneath most of the Great Plains in a broad synclinal basin and reappears in places along the west edge of the Central Lowlands.

Igneous rocks are not exposed except locally; in the Arbuckle and Wichita mountains in Oklahoma, Precambrian granites are exposed over significant areas. More recent igneous activity that has occurred in many of the provinces has not affected the Central Lowlands.

Pleistocene Glaciers

During the Pleistocene, great changes were brought about by the huge continental ice sheets (Fig. 13–2) that pushed down out of Canada at least four times. They completely remodeled the drainage systems—resulting in the formation of the Great Lakes and changing the location and even flow directions of many rivers and streams. Most of the low-relief topography of the Central Lowlands is due to the effects of glaciation—mostly deposition and filling of lower areas with glacial sediment rather than significant erosion of topographically high areas (Thornbury, 1965). Glacial landforms such as *eskers* (Fig. 13–3), *drumlins,* and broad looping *end moraines* are locally abundant. Glacial erosion was effective locally, especially in the five Great Lakes where all of the lakes except Lake Erie are below sea level.

The Osage Plains Section is the only large area that remained unglaciated. A relatively small area, mainly in southwestern Wisconsin, was surrounded but bypassed by the glaciers. It is called the Driftless Area.

Most of the deposits, related directly or indirectly to the glaciers, formed excellent parent materials for the soil that developed on them; the Corn Belt of Iowa and Illinois would not have developed otherwise. Distantly related are the *loess* deposits, wind-laid silt derived in part from glacial outwash. Soils formed from loess are highly productive in Iowa, Illinois, and parts of Indiana and help supply the world's food supply.

Parklands

Within the province, there is a wide variety of natural features of special interest, both geologically and ecologically. State parks are abundant in this section of the country, and they serve essentially the same purpose as the national parks in preserving things natural; examples are Turkey Run State Park in Indiana (Fig. 13–1),

FIGURE 13–2   Extent of continental ice sheet in North America approximately 15,000 years ago. (From Huber, 1975)

FIGURE 13–3   View along the top of an esker near St. Paul, Minnesota, before it was converted into a sand and gravel pit. (Photo by D. Harris)

Devils lake in Wisconsin, Taylors Falls in Minnesota, and Turner Falls in Oklahoma. Niagara Falls (Fig. 13–4) in upstate New York is outstanding and also has a spectacular geologic story to tell to those who can comprehend the not-so-distant-past when nearly a mile of ice covered the landscape.

National Park Service areas of geologic interest in the Central Lowlands are Chickasaw National Recreation Area in Oklahoma; Pipestone National Monument in Minnesota; and Pictured Rocks, Indiana Dunes, and Sleeping Bear Dunes National Lakeshores in Michigan. Each has a special story to tell. Pictured Rocks contains Upper Proterozoic and Cambrian (early Paleozoic) sedimentary rocks at the very northern edge of the Central Lowlands. Precambrian rocks of the Canadian Shield lie to the north—hidden beneath the deep waters of Lake Superior. Sleeping Bear Dunes contains beach dunes and dunes perched on rugged bluffs that tower as much as 460 feet (140 m) above Lake Michigan. Glaciers played a major role in developing the Pictured Rocks, Indiana Dunes, and Sleeping Bear landscapes. Pipestone National Monument in the southwestern corner of Minnesota preserves Indian lore. Here the Indians quarried the hard red clay and carved it into tobacco pipes. In this chapter only Chickasaw and Indiana Dunes will be discussed in detail.

FIGURE 13–4   Water from Lake Erie plunges 167 feet (51 m) over Niagara Falls on its way to Lake Ontario. (Photo by Barbara Kiver)

REFERENCES

Biggs, D.L., ed., 1987, North-central section of the Geological Society of America: Geological Society of America, Boulder, CO, Centennial Field Guide, v. 3, 448 p.

Donovan, R.N., 1988, The Meers fault scarp, southwestern Oklahoma, in Hayward, O.T., ed., South-central section of the Geological Society of America: Geological Society of America, Boulder, CO, Centennial Field Guide, v. 4, p. 79–82.

Donovan, R.N., and Heinlen, W.D., 1988, Pennsylvanian conglomerates in the Arbuckle Mountains, southern Oklahoma, in Hayward, O.T., ed., South-central section of the Geological Society of America: Geological Society of America, Boulder, CO, Centennial Field Guide, v. 4, p. 79–82.

Hayward, O.T., ed., 1988, South-central section of the Geological Society of America: Geological Society of America, Boulder, CO, Centennial Field Guide, v. 4, 468 p.

Huber, N.K., 1975, The geologic story of Isle Royale National Park: U.S. Geological Survey Bulletin 1309.

Thornbury, W.D., 1965, Regional geomorphology of the United States: Wiley, New York, 609 p.

INDIANA DUNES NATIONAL LAKESHORE (INDIANA)

Nestled along the southern tip of Lake Michigan, the sixth largest lake in the world, is a fragile area where sand dunes and an amazing variety of plants are found in Indiana Dunes National Lakeshore (Plate 1). The area is within easy driving distance of 10 million people and provides needed rest, reflection, and recreation for over 2 million visitors every year. The area is unique in many ways, one being that this small area of incredibly diverse environments is in the midst of one of the most heavily industrialized areas in the United States. The dune area and the various wetlands and other habitats are the gift of the glaciers that scoured the Lake Michigan Basin as recently as 14,000 years ago and the subsequent work of shore processes—wind, waves, and currents. It is also the story of people who believed that economic development at the cost of destruction of all of our once beautiful landscapes is too high a price to pay.

Battle for the Dunes

In 1916, Stephen T. Mather, the Park Service's first director, held a public hearing and in 1919 recommended establishment of a national park encompassing a 40-mile (65-km) stretch of dunes along the Indiana lakeshore. This effort was thwarted by the concern of Congress about costs to acquire private lands and the desire of industrialists and politicians determined to use every inch of Indiana's limited lakeshore for industrial uses (Franklin and Schaeffer, 1983). The 1916 park hearing mobilized citizen groups who later succeeded in having a small area set aside as Indiana Dunes State Park in 1923. A battle between those who would develop the remaining dunes and those who saw a need for recreational and educational opportunities occurred in the 1960s. Indiana Senator Paul Douglas joined the crusade for park status begun earlier by local citizen Dorothy Buell and others. In 1966, after most of the dunes in the surrounding area had been flattened for housing and commercial developments, Indiana Dunes National Lakeshore was created to preserve the remaining dunes, less than one-third of the original area. The lakeshore, which surrounds the 2182-acre (3.4 square mile, 8.8 km^2) state park, consists of about 13,000 acres (20 square miles, 53 km^2) of dunes and wetlands.

Gift of the Glaciers

The geologic story begins with the Pleistocene glaciers that advanced over and retreated from the area. Late in the Pleistocene, a lobe of the Wisconsin-age ice sheet dug deep in the basin now occupied by Lake Michigan and excavated the lake bottom to over 340 feet (104 m) below sea level. Erosion of the Lake Michigan Basin was particularly effective as it paralleled the dominant southward ice flow from Canada. About 16,000 years ago the retreating ice paused and deposited the rolling ridges of the Valparaiso Moraine around the southern end of Lake Michigan. As climates moderated about 14,000 years ago, the Michigan lobe melted northward, creating a depression between the Valparaiso Moraine and the retreating ice front. Water filling the depression became an early version of the Great Lakes—a lake that earth scientists call glacial Lake Chicago.

Along the edge of the glacial lake, waves reworked the thick glacial sediment. Finer sediment was carried into deeper parts of the lake by suspension, and gravel and sand concentrated in the wave zone, forming a beach along the Glenwood level (about 650 feet elevation; 198 m) of glacial Lake Chicago about 12,000–14,000 years ago. Wind blew some of the sand inland forming a broad belt of sand dunes. The glacial lake lowered from the Glenwood terrace level to the Calumet level (about 630 feet; 192 m) about 11,800–11,200 years ago and to the Tolleston level (about 600 feet; 183 m) about 4000–5000 years ago (Hill and others, 1991). Each terrace level has relic beach and dune features on its lakeside edge. Further ice retreat exposed lower spillways that allowed the lake to eventually drop to near its present elevation of 580 feet (177 m) where the modern beach and associated dunes formed. Thus, at least four distinct levels of beaches and dunes are encountered as one descends northward from the Valparaiso Moraine to the modern-day lakeshore (Fig. 13–5). A stop at the Dorothy Buell Memorial Visitor Center provides an opportunity to examine the Calumet Lake terrace and its relic dune and beach features. A good rule of thumb in the Indiana Dunes area is that geologic features become younger the closer one approaches to Lake Michigan.

Amid the hummocky moraine topography are *kettle holes* where large blocks of ice were partially buried by sediment. Later, when the ice blocks melted, *kettle lakes* formed in the depressions. In time, the lakes and ponds were partially filled with sediment, forming marshes and bogs; Cowles and Pinhook bogs are the two

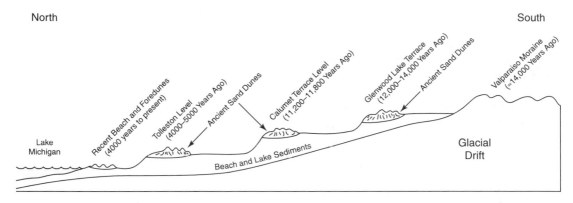

North   South

Lake
Michigan

Recent Beach and Foredunes
(4000 years to present)

Tolleston Level
(4000–5000 Years Ago)

Ancient Sand Dunes

Calumet Terrace Level
(11,200–11,800 Years Ago)

Glenwood Lake Terrace
(12,000–14,000 Years Ago)

Ancient Sand Dunes

Valparaiso Moraine
(≈14,000 Years Ago)

Glacial
Drift

Beach and Lake Sediments

FIGURE 13–5   Generalized topographic profile in the Indiana Dunes area showing levels of glacial Lake Chicago. Drawing is not to scale.

that lie within Indiana Dunes National Lakeshore boundaries. Other wetlands and lakes form in areas between dunes (*interdunal lakes*) and behind the beach dunes where water is trapped. As the lakes fill in with sediment, they shallow and become swamps or marshes, providing yet different habitats for an amazing variety of plants and animals.

Shore Processes

Lakeshores and seashores are some of the earth's most dynamic environments. A single storm, an increase or decrease in sea or lake levels, a change in the supply of sand to the beach and its associated dunes, or destruction of vegetation by natural or human causes can cause significant changes—sometimes overnight or over a mere few months or years.

Waves strike the east edge of Lake Michigan at an angle causing a *longshore current* or *longshore drift* that moves sand sediment toward the south end of the lake. Abundant sand concentrations encourage the wind to move excess sand up the beach where it forms *coastal dunes,* or *foredunes.* Construction of harbors, breakwalls, jetties, and other structures seriously interferes with longshore drift and can cause significant changes in coastal landforms. Such construction to the east now causes sediment to accumulate behind these man-made obstructions. Deprived of their sediment load, currents become more erosive and have seriously eroded public beaches and even parts of Lake Front Drive (Hill, 1987).

Unusually high lake levels and less available sand from longshore drift significantly increased beach erosion below Mt. Baldy on the east side of the Lakeshore —threatening to destroy one of the higher dunes and one of the favorite goals of hikers. Over 250,000 cubic yards (191,000 m³) of sand were placed on the beach by the U.S. Corps of Engineers in 1974, 80,000 cubic yards (61,120 m³) of which were lost to erosion in the first year (Hill, 1987)! Another large volume of sand placed in front of Mt. Baldy in 1981 was gone by 1984 (Hill and others, 1991).

Vegetation-covered dunes that are not excessively disturbed remain stable and relatively unaffected by erosion (Fig. 13–6). However, areas where blowing sand buries and kills plants or where vegetation is destroyed by other natural or human causes allow active moving dunes to form. A variety of dune shapes form depending on the sand volume available at a particular site, on wind patterns, and on how effective vegetation is in stabilizing the dune surface. A dune variety called a *blowout dune* forms where vegetation has been removed from the crest of a fore-dune. A narrow wind channel develops in the foredune allowing wind velocities to increase substantially. A large area is eroded downwind forming a bowl-like depression rimmed by a high dune ridge. Some of the resulting sand mountains are nearly 200 feet (60 m) tall.

Moving sand can overwhelm adjacent forests. A good example is on the north flank of Mt. Baldy where dunes are encroaching and destroying a living forest. Dune migration has increased since 1974 and 1981 when beach nourishment was attempted. Several buildings are now threatened by the advancing dunes (Hill, 1987). Later sand movements can uncover these tree graveyards creating an eerie scene of ghost forests.

FIGURE 13–6　Partially stabilized dunes at Indiana Dunes National Lakeshore. (Photo by Peggy Gilmour)

Park Attractions

In addition to hiking, swimming, and boating, the opportunity to experience one of the most varied populations of birds and plants in the National Park System in the midst of a major urban and industrial area is unique. Ecologists are intrigued with the unusual assemblage of plants, over 1400 species in a relatively small area, some of which are threatened or endangered species (Hill and others, 1991). Plant representatives from different climatic zones are mixed together here as a result of the climatic extremes of the Pleistocene. Some plants are relics of the glacial period—forced to migrate southward as glaciers advanced. Some survived in special environmental niches in the Indiana Dunes National Lakeshore. The Arctic barberry grows here, along with the jack pine, tamarack, and birch, all of which are usually found at least 100 miles (160 km) farther north. Growing nearby are prickly pear cactus and southern dogwoods—remnants of warmer times.

Animals that humans either killed or drove away due to the fragmentation of the ecosystem include the bear, wolf, lynx, bison, elk, deer, and beaver; but the raccoon, opossum, fox, mink, and squirrel have been able to survive. Migratory birds stop by in droves to feed in the marshes, and shorebirds are on the beaches much of the time. Listen as you walk the beaches just above the water's edge. As the shifting sand-size quartz crystals under your feet compact, a distinct musical ringing sound may be heard. The beaches at Indiana Dunes are one of the few known areas where singing sand occurs.

A climb up Mt. Baldy in the east unit of the park provides an excellent overview of the area where one can see the important landform pieces that make up the geologic puzzle of Indiana Dunes. The distant crest of the Valparaiso Moraine is on the skyline to the east, the beach and dune features of glacial Lake Chicago on the different lake-terrace levels are located to the east and south, and today's beach and dunes are all visible—important pieces of the puzzle that tie together the last 16,000 years of earth history along the Lake Michigan shore. The most recent chapter is apparent in the huge steel mill structures, breakwaters, and other creations of human design. What will the area look like in 100 years? Our actions and inactions today will decide its fate.

REFERENCES

Franklin, K., and Schaeffer, N., 1983, Duel for the dunes: Chicago, Univ. of Illinois Press, 278 p.

Hill, J.R., 1987, The Indiana Dunes area, northwestern Indiana, in Biggs, D.L., ed., North-Central Section of the Geological Society of America: Geological Society of America, Boulder, CO, Centennial Field Guide, v. 3, p. 321–324.

Hill, C.L., Ryan, B.J., McGregor, B.A., and Rust, M., 1991, Our changing landscape, Indiana Dunes National Lakeshore: U.S. Geological Survey Circular 1085, 43 p.

Park Address

Indiana Dunes National Lakeshore
1100 North Mineral Springs Road
Porter, IN 46304

CHICKASAW NATIONAL RECREATION AREA (OKLAHOMA)

Humans have always regarded mineral-rich springs and streams as special places, thus the establishment of Platt National Park (now part of Chickasaw National Recreation Area) in the Arbuckle Mountains of southern Oklahoma was enthusiastically supported. Chickasaw is located in the hills near Sulphur and includes the Lake of the Arbuckles reservoir and Travertine and Rock creeks where the springs are located.

The Arbuckle Mountains are part of a linear zone of ancient folded and faulted mountains that extend outward from the late Paleozoic-age Ouachita Mountains into the normally tectonically stable region (*craton*) of the midcontinent. The Ouachita Mountain belt in turn provides a connection between the late Paleozoic Appalachian Mountains on the east and the ancestral Rocky Mountain belt of the same age to the west. Ecologically it is also a transitional area—one where the hardwoods of the east and the short-grass prairies of the west merge. Here the road runner from the drylands associates with the eastern bluebird, the cardinal, and the bluejay, and more than a hundred other species. You may see a bison herd, particularly from the Bison Viewpoint on the Perimeter Drive east of Bromide Hill. Watch carefully for armadillos, foxes, bobcats, raccoons, and opossums.

History

The banks of Rock Creek and its dozens of cold-water springs were some of the favorite campgrounds of the Chickasaw and Choctaw Indians. The Euro-Americans followed their lead, and the resort town of Sulphur developed around some of the mineral-rich waters. Concern about commercial development prompted the Indians to cede the area to the government in 1902 so that all people would forever have access to this special area (Donovan and Heinlen, 1988). The area was redesignated as Platt National Park in 1906 and was combined with the adjacent Arbuckle National Recreation Area in 1976. The 9521-acre (14.9 square mile, 39 km²) park was named Chickasaw National Recreation Area in honor of the Indians who foresaw the need to place the area under government stewardship to protect this important component of our natural heritage. The spring-rich northeastern area of the recreation area, the former Platt National Park, is called the Travertine District and will be emphasized in the following discussions.

Geologic Setting

Chickasaw is located in the Southern Oklahoma Rift—an important fault-bounded, *graben*-like structure of Paleozoic age that extends outward at nearly 90° from the Ouachita Geosyncline and into the stable midcontinent (craton) region (Fig. 13–7). The Ouachita Geosyncline formed along the southeastern edge of what would later become North America. The thick section of early Paleozoic sediments that accumulated in the geosyncline and rift zone were folded and faulted intensely during late Mississippian and Pennsylvanian time (about 340–300 Ma). The origin of the compressive forces was the collision of the huge assemblage of southern plates (Gondwana) with its northern counterpart (Laurentia) along the Ouachita and Appalachian plate edges—forming the late Paleozoic supercontinent of Pangaea. The Wichita–Arbuckle mountain complex in the Southern Oklahoma Rift formed during these massive plate tectonic events.

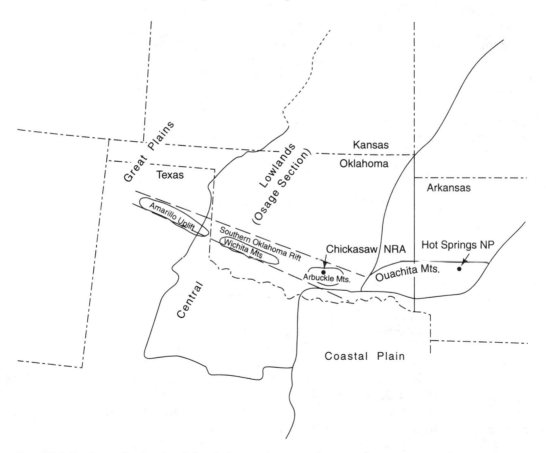

FIGURE 13–7   Generalized regional sketch showing location of geomorphic provinces and major structural features.

Intense erosion accompanying the uplift of the mountains eventually exposed rocks as old as Precambrian in the cores of the Wichita and Arbuckle mountains. The coarse gravels, sands, and muds of the Pennsylvanian-age (about 300 Ma) Vanoss Formation were deposited as large alluvial fan and alluvial plain complexes that covered the eroded edges of the upturned lower Paleozoic sediments. The resulting spectacular angular unconformity between the nearly horizontal Pennsylvanian-age conglomerates and steeply dipping older Paleozoic rocks is well exposed in the Honey Creek area southwest of the park and elsewhere in the Ouachita–Arbuckle mountains area. The Vanoss Formation conglomerate with its abundant fragments of early Paleozoic limestones, and even pebbles of Precambrian basement rocks, can be examined along the trail on Bromide Hill at Chickasaw (Donovan and Heinlen, 1988). Apparently even the Precambrian basement rocks were exposed by erosion during late Paleozoic mountain building.

Mineral Springs and Waterfalls

The mineral springs are most abundant in the Travertine District. Although the Travertine District is small, its size in no way measures its significance in the National Park System. True, its waterfalls are neither deafening nor overpowering, nor are its cliffs breathtaking; nevertheless it is a delightful place, a restful spot in the shade of the sycamores, cottonwoods, and oaks.

Geologically, the Travertine District is unique. Waterfalls are usually the result of erosion where downcutting streams encounter resistant rock materials. Such waterfalls gradually migrate upstream as erosion progresses. Here we have waterfalls formed by deposition by chemical processes—the same processes that build massive flowstone and other speleothems in caves. Waterfall building continues today; slowly the little waterfalls are getting bigger and higher—gradually growing in a downstream direction. Not long ago, geologically speaking, there were no waterfalls; instead there were merely rapids along Travertine Creek. The rapids cause turbulence, allowing carbon dioxide to more readily escape from the calcium bicarbonate in solution; as a result, travertine, or *tufa* (a chemical deposit around a spring or in a stream), a porous form of limestone, has gradually been deposited, forming low dams across the streams. Where most waterfalls erode the resistant strata and migrate upstream, travertine falls enlarge and migrate *downstream*. Evaporation also helps to increase the supersaturation of the water with respect to calcium carbonate. As algae and moss photosynthesize, they further extract CO_2 and cause even more precipitation of calcium carbonate.

The travertine dams are several feet high, as seen in Figure 13–8. Since such travertine dams are not common, Travertine Creek must be carrying an extraordinarily heavy load of calcium bicarbonate in solution. Take the foot path upstream from Travertine Nature Center, and at Antelope Springs and Buffalo Springs you will see the source of the bicarbonate waters.

FIGURE 13–8   Little Niagara, a waterfall built by precipitation of a travertine dam across Travertine Creek near Sulphur, Oklahoma. (Photo by D. Harris)

Except during time of storm runoff, these two springs are the source of the water in Travertine Creek. Normally the flow is several million gallons a day, but has steadily decreased as more water wells are drilled in surrounding areas. During drought periods and increasingly during nondrought periods the springs and the creek are dry.

Where does this large volume of calcium bicarbonate come from? Slightly acidic rainwater descends into the ground through small openings in the underlying 650-foot-thick (198-m) Vanoss Formation. Here the slightly acidic groundwater contacts the abundant calcium-rich pebbles and cobbles of limestone that make up the conglomerate layers, dissolving large amounts of the mineral calcite.

Antelope and Buffalo Springs are fed from a relatively shallow plumbing system and, although they are rich in calcium in solution, they are called "freshwater springs." Most of the other springs in the park are located in the western part and are "mineral springs"—springs producing water containing significant amounts of dissolved mineral materials that impart a definite taste to the water. These waters contain small amounts of rare elements such as bromine, lithium, iodine, and sulfur, and large quantities of sodium and chlorine, evidently a watered-down solution of seawater. Is it conceivable that water from the Paleozoic sea was trapped in some of these porous limestone formations and that it is now being forced out at the surface? It has been so determined; surface water penetrates down into rocks at elevations higher than the springs and flushes out the old seawater from Ordovician rocks. The diluted seawater moves down-dip and is then forced up to the surface by the hydrostatic head provided by the difference in elevation of the intake area and the springs. These mineral springs are therefore *artesian* springs. Due to the structure of the rocks in the artesian system, however, the water does not penetrate to great depths. Consequently, Bromide Spring, Medicine Spring, Black Sulphur Spring, and the others are cold springs.

The National Park Service does not claim medicinal value for the mineral water; furthermore, they warn against taking it in quantity except on the advice of a physician.

Visiting Chickasaw

As always, the visitor center should be the first or an early stop to learn more about the natural setting as well as trails to the springs, waterfalls, and other features in the park. The trail from Bromide Pavillion to the top of Bromide Hill provides glimpses of the gravels and finer layers of the Vanoss Formation. The view from the top is the best in the park. Zones of dense carbonate concentration in the Vanoss called *caliche* are visible from the trail and mark times when soils formed under the semiarid climate that prevailed here in late Paleozoic time. Intervals of deposition separate a number of *paleosols* (fossil soil zones) in the Vanoss (Donovan and Heinlen, 1988).

Rock Creek, which flows westward through the district, was dammed a few miles downstream—to impound water in the Lake of the Arbuckles. The lake, about 2350 acres (3.7 square miles, 9.5 km^2) in extent, became the Arbuckle National Recreation Area—for camping, boating, and fishing. In 1976 the area was enlarged and joined with Platt National Park to form a single unit called the Chicasaw National Recreation Area.

Turner Falls State Park is about 5 miles (8 km) southwest of the Travertine District and is well worth a visit. The park along Honey Creek contains the most spectacular waterfall in Oklahoma—it too is formed by deposition of travertine from water supersaturated with calcium (Donovan and others, 1988). Here one can examine the travertine, or tufa, up close to see its porous nature. The older travertine is extremely porous with some large connected cavities that can be explored for short distances by the more adventurous.

Unfortunately, even though we learn about past mistakes in resource management, we continue to repeat many of those same mistakes. Parks are not "safe" islands that are unaffected by what happens in surrounding areas or in some cases, such as air pollution, elsewhere on other continents on "spaceship Earth." At Chickasaw the presence of springs is dependant on groundwater flow from private lands outside of the artificial boundaries established for the park. Over 1000 water wells have been drilled within 7 miles (11 km) of the park and predictably spring flow has reduced. Over one half of the 30 major springs are now dry, and others are expected to disappear in the next few decades—another result of our inability to manage population and the empty goal of economic development without regard to the limits of resources, space, and preservation of natural beauty.

REFERENCES

Donovan, R.N., and Heinlen, W.D., 1988, Pennsylvanian conglomerates in the Arbuckle Mountains, southern Oklahoma, in Hayward, O.T., ed., South-central section of the Geological Society of America: Geological Society of America, Boulder, CO, Centennial Field Guide, v. 4, p. 159–164.
Donovan, R.N., Ragland, D.A., and Schaefer, D., 1988, Turner Falls Park; Pleistocene tufa and travertine and Ordovician platform carbonates, Arbuckle Mountains, southern Oklahoma, in Hayward, O.T., ed., South-central section of the Geological Society of America: Geological Society of America, Boulder, CO, Centennial Field Guide, v. 4, p. 153–158.

Park Address

National Park Service
Chickasaw National Recreation Area
P.O. Box 201
Sulphur, OK 73086

FOURTEEN

Ouachita Province

The Ouachita Province, in eastern Oklahoma and western Arkansas, lies south of the Ozark Plateaus and north of the Gulf Coastal Plain. On the west are the Arbuckle Mountains in the Osage Section of the Central Lowlands (Plate 1). The province is readily divisible into two sections, the Arkansas Valley and the Ouachita Mountains to the south. It is in the latter section that Hot Springs National Park is located.

Plates in Motion

Here in the Midwest we catch our first glimpse of Appalachian geology. Appalachian structures are continuous westward from Alabama, surfacing only in the Ouachita Mountains, the Llano area in central Texas, and the Marathon Mountains in southwestern Texas. Intervening mountain structures are buried by younger, mostly horizontal strata of Mesozoic and Cenozoic age. The late Paleozoic mountain belt extends northwestward in the subsurface into New Mexico where it connects structurally with the Ancestral Rockies of the same age. Thus, during late Paleozoic time much of the stable interior of what was soon to become the North American Plate was ringed by mountains—the result of the closing of the gaps between the northern and southern assemblages of plates to form the Pangaea supercontinent. Erosion of the tightly folded and thrust-faulted Paleozoic sedimentary rocks produced a valley-and-ridge topography in both the Appalachians and Ouachitas.

Geologic Development

The geologic history began with a geosyncline early in the Paleozoic, with the deposition of shales, sandstones, conglomerates, and cherts. Notable is a variety of chert called novaculite, a dense, even-textured siliceous rock that, when pure, is as white as new snow. It was highly prized by the Indians who used it for arrowheads, spearpoints, and other tools; later it was cut into perhaps the best whetstones in the world. Geologists have long pondered the problem of its origin—how thick layers of pure or nearly pure silica could have developed. The novaculite likely formed from the recrystallization of tiny siliceous skeletons of microscopic organisms that accumulated in a deep marine basin during middle Paleozoic [Devonian-Mississippian, about 350–370 million years ago (Ma)] time. Although several hundred feet of Arkansas Novaculite accumulated in the geosyncline, shale is the most abundant rock, with as much as 8500 feet (2591 m) accumulating in a single formation.

Near the end of the Paleozoic, the geosynclinal rocks were intensely folded and thrust faulted, forming the Ouachita Mountains. Where orogenic movements were most intense, shales were metamorphosed to form slates, and sandstones were changed to quartzites.

The mountains were eroded as they were being uplifted, and this erosion continued after mountain building ceased. Remnants of two erosion surfaces indicate that the area was at least twice worn down and twice uplifted prior to the development of the valley-and-ridge topography that we see today (Fig. 14–1). Igneous activity was not widespread, judging from the general lack of exposed igneous rocks. Intrusive igneous rocks are found in a few areas; all were emplaced during the Cretaceous.

The topography of the Ouachitas is solely the result of weathering, mass-wasting, and stream erosion; continental glaciers did not push southward into any part of the province. The ridges were formed because of the superior resistance of certain formations, particularly the quartzites and novaculite beds, as is readily observed in the Hot Springs National Park area.

HOT SPRINGS NATIONAL PARK (ARKANSAS)

Within the National Park System, parks and monuments have been set aside for several purposes. About 95 were established primarily to preserve classic geologic features, and 140 other units contain significant geologic features (Applegate, 1997). Other areas were established for their archeological and historic values, still others for ecological reasons. Hot Springs National Park is unique in that it was established to preserve hot springs that were believed by some to have extraordinary healing qualities. According to legend, the area was used as a healing ground by the Indians; here, regardless of tribe, their weary and wounded could come in peace to recuperate in this quiet area—the Valley of the Vapors. In addition to the historic

FIGURE 14–1   View north from Hot Springs Mountain out over dissected erosion surface. (Photo by D. Harris)

spring area, some of the surrounding valley-and-ridge topography of the Ouachita Mountains is contained in the park.

Historical Background

To the Caddo Indians, this was "Tah-ne-ca," the place of the hot waters. Spanish conquistadors and later settlers were also intrigued by the area. Water emerging in large quantities from a hillside, especially when the water is too hot to touch, was mysterious to early visitors and was thought to have therapeutic values. Thomas Jefferson, the only scientist-president that the United States has ever had, dispatched two scientists in 1804 to investigate the origin of the springs and the source of the heat—questions that were not adequately resolved for well over a century.

As the fame of the springs spread among the westward-moving immigrants, so did the opportunity for the exploitation of those who came. Therefore, in 1832 Congress established the 4-square-mile-area (10-km^2) around the springs as Hot

Springs Reservation, the first National Reservation—an area to be protected for future generations. When the Park Service was created in 1916, Hot Springs Reservation was transferred to its jurisdiction. The name was changed to Hot Springs National Park in 1921. Originally the city of Hot Springs nearly encircled the park, but, by extensive enlargement, the 5839 acres (9 square miles, 24 km^2) of the park now surrounds the northern part of the city.

With the advent of modern science and medicine, the belief in the curative power of hot, mineralized water declined. However, interest then shifted toward the "scientific, esthetic, and recreational values (that) are more appreciated by today's visitors" (Bedinger and others, 1979). Addition of land to the north preserves some of the valley-and-ridge topography in the tight anticlines and synclines found in the Zigzag Mountains Section—part of the Ouachita Mountain belt. The nearby ridge tops are protected from rapid erosion by the resistant layers of the Arkansas Novaculite—a silica-rich rock composed of submicroscopic (*microcrystalline* or *cryptocrystalline*) quartz. The long anticlinal and synclinal folds dive, or *plunge,* beneath the surface on their ends, producing a zigzag pattern of ridges that provides a challenge (nightmare?) to beginning geology students.

The Waters

While surface waters in the form of streams have played an important role in shaping the valley-and-ridge topography of the area, the subsurface waters are of greater significance in the park. They descend along faults and fractures on the southwest flank of Hot Springs Mountain and surface as springs. Numerous cold springs and warm springs occur in the area, but the hot springs are the main attraction (Bedinger and others, 1979; McFarland, 1988). At present there are 47 hot springs in the park, of which 45 are covered to prevent pollution.

The water emerges from the springs as *artesian* water—water under pressure because it is in a confined system in which its intake area is higher in elevation that its discharge point. Surface water soaks, or *infiltrates,* into chert and novaculite beds in the nearby hills. The water follows the steeply inclined strata and faults to a depth of at least 4500–7500 feet (1370–2285 m) where the water is heated by the warm rock. The water moves rapidly to the surface through zones of highly faulted and fractured rock (Bedinger and others, 1979). Analysis of carbon-14 and tritium (a radioactive form of hydrogen) indicate that most of the spring water is fossil water—water that infiltrated the ground about 4400 years ago and is just now emerging at the surface! Further analysis of the silica concentration in the water (the higher the silica the higher the temperature of the source rock) indicates that the water is only a few degrees less than the source temperature. Thus the water must move relatively quickly from deep areas to the surface.

An early hypothesis for the heat and water source was that a hot igneous body was present in the subsurface. According to this idea, as the magma cooled, water would be expelled and eventually surface in the hot springs. Small dikes, sills, and larger igneous bodies of Late Cretaceous age occur as close as 6 miles (10 km) southeast of the springs. However, to maintain significant heat for nearly 100 million years seems unlikely. Further, chemical analyses indicate that water from Hot Springs does not contain the elements and isotopes, such as those at Yellowstone, that are commonly found in water derived at least partially from cooling magma.

Flow from the springs fluctuates from a maximum during the winter and spring to a minimum in the fall. The maximum flow from the 47 springs is about a million gallons per day, and the temperature ranges from 95°F to more than 147°F (35–65°C) (McFarland, 1988). Gradual changes have been recorded in both temperature and flow; both are declining, but at a rate too slow to threaten the bathing industry.

Visitors are impressed with the bubbles that grow and then break from the surface of the pools, resulting from a concentration of dissolved gases, mainly carbon dioxide, nitrogen, and oxygen. Impressed also by the presence of small amounts of radium and radon gas in the waters, many health seekers are convinced of their curative powers.

The main hot springs are located in downtown Hot Springs where the Park Service maintains the historic bath buildings and a short trail to the two open springs where one can see the natural springs and deposits of *calcareous tufa* (a porous form of limestone). A museum staffed by park rangers and volunteers help to explain the geologic and human history of the area. The Fordyce Bath House, remodeled in 1988 to restore it to its 1915 appearance, is the current visitor center/museum. Of the remaining seven bath houses, the Buckstaff Bath House is the only other one presently open to the public.

The remainder of the park occupies the Hot Springs Anticline and other nearby ridges where scenic drives and 26 miles (42 km) of trails can be explored and where one can observe the plants and wildlife in the park.

The Mountains

The park is located in the Zigzag Mountains Section of the Ouachitas. The ridges reflect the structure and lithology, intensely folded and faulted beds of extremely resistant rocks, mainly novaculite, which are interbedded with poorly resistant valley-forming shales. The mountains rise abruptly above the lowland areas to heights as great as 1410 feet (427 m) on Music Mountain. Although from a distance they appear to have a rounded, rolling topography, there are numerous precipitous cliffs.

The drives and hiking trails take one into country usually considered more typical of some of the better known national parks. One road leads through dense forests of oak, hickory, and short-leaf pines to the top of Hot Springs Mountain which is located immediately east of the springs (Fig. 14–1). You may enjoy seeing the abandoned novaculite quarries where Indians mined the stone to use for the shaping of knives and projectile points. A conducted hike takes one up on Indian

Mountain during the summer months. The West Mountain section is equally attractive, with scenic overlooks of the various ridges of the Ouachitas.

Rock outcrops are abundant along the roads and trails, even in this humid climate. Many are novaculite, which resists weathering, regardless of the high humidity. Be sure to see the excellent exposure of white novaculite at the turnaround at the top of West Mountain.

Along the trails you will not see the bison, elk, or bear that once inhabited the area, but where there are hickory nuts and acorns there are squirrels who will keep track of you, chattering and scolding you for invading their domain. You are not likely to see the resident raccoons or opossums unless you also are nocturnal, and you will be lucky to catch a glimpse of a fox or wild turkey.

The vegetation in the park is one of its finest assets, with wildflowers blooming the year around. The most flamboyant of the many trees is the southern magnolia, which has been introduced in the lower areas of the park.

There is much of interest in the country surrounding the park. When you have exhausted the immediate area, there are other scenic drives, notably Highway 7 north from Hot Springs. You will wind through the Ouachitas in Ouachita National Forest; perhaps you will stop at the crystal mine where fine rock crystal quartz and smoky quartz are on display. Also, for a small sum you can dig for real diamonds at the Crater of Diamonds near Murfreesboro, about 60 miles (100 km) southwest of Hot Springs.

REFERENCES

Applegate, D., 1997, National parks: Geology matters: Geotimes, v. 42, no. 8, p. 15.

Bedinger, M.S., Pearson, Jr., F.J., Reed, J.E., Sniegocki, R.T., and Stone, C.G., 1979, The waters of Hot Springs National Park, Arkansas—Their nature and origin: U.S. Geological Survey Professional Paper 1044-C, 33 p.

McFarland, J.D., III, 1988, Geological features at Hot Springs, Arkansas, *in* Hayward, O.T., ed., South-Central Section of the Geological Society of America: Geological Society of America, Boulder: CO, Centennial Field Guide Volume 4, p. 263–264.

Park Address

Hot Springs National Park
P.O. Box 1860
Hot Springs, AK 71902-1860

FIFTEEN

Superior Upland Province

The Superior Upland is a small part of the 2-million-square-mile (5.2 million km²) Canadian Shield, probably the largest area of exposed Precambrian rocks in the world. Although Thornbury (1965) views the Superior Upland as a section of the Laurentian Upland Province, it is here regarded as a separate geomorphic province. In the United States, the Superior Upland occupies the Precambrian outcrop areas of Minnesota, Michigan, and Wisconsin, although its boundary with the Central Lowland Province is locally indefinite because of the cover of glacial sediment.

The Precambrian rocks of the Canadian Shield continue southward and westward into the Central Lowlands where these long-time stable basement rocks are buried by horizontal or nearly horizontal sedimentary rocks of Paleozoic age. Farther west in the Great Plains the Precambrian basement is buried by rocks as young as Cenozoic. The Precambrian rocks reappear at the surface in the twisted wreckage of the Rocky Mountains, which marks the edge of the *craton* or region of the continent where significant deformation has not occurred in a billion or more years. The New England, Adirondack, Appalachian, and Ouachita mountains mark the craton edge on the east and south edges of the craton (Plate 1). Thus the Superior Upland and Canadian Shield provide a window to view some of the oldest rocks and events recorded on the North American continent—rocks that are mostly obscured by younger rocks elsewhere in North America. Also recorded in the Superior Upland Province are some of the most recent geologic events to significantly affect the continent—the Pleistocene glaciers (Fig. 13–2) that formed the huge system of lakes known as the Great Lakes and left glacial landforms and deposits that are as young as 10,000 years in the Lake Superior area.

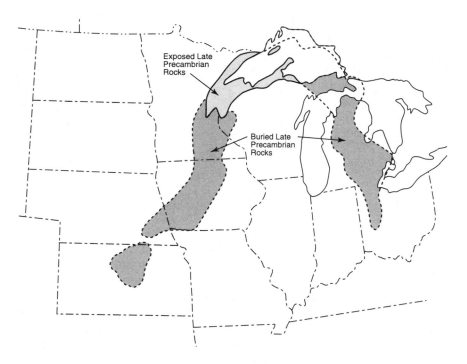

FIGURE 15–1   Location of late Precambrian-age rocks associated with the Midcontinent Rift—dashed lines where covered by younger rocks.

Early Craton Evolution—Archean

During the Archean or earliest Eon of Precambrian time [2500–4500 millions of years ago (Ma)], a number of isolated areas of continental crust formed by plate-tectonic processes on the primordial surface of the newly formed earth. Radiometric dates of over 3 billion years ago occur in rocks from these ancient continental nuclei, including the Wyoming area, Minnesota, parts of Canada, and Greenland. Some of the oldest known rocks come from Minnesota (3800 Ma) in the Superior Upland Province.

A series of these microplates collided and formed the 2.5- to 2.7-billion-year-old granite and gneiss bedrock found at Voyageurs National Park. Each collision produced a rift basin that filled with basalt and lesser amounts of sediment that were later compressed, folded and faulted, and intruded by granite. Metamorphism of the basalt produced the mineral chlorite that gives the rock a green color—hence the name *greenstone*—a common rock in the Superior Upland and southern Canada. Thus, at the end of Archean time a series of isolated regions composed of continental crust occupied the area that would become connected during Proterozoic time to form the North American Shield. The Ely Greenstone in northern Minnesota formed at this time and has the distinction of being the oldest known rock in Minnesota.

Craton Evolution—Proterozoic

The Proterozoic Eon is an immense episode of geologic time—from 2500 to 545 Ma—about 40 percent of all geologic time. The isolated Archean cratons experienced the formation of geosynclines and mountain building along their edges. By 1800 Ma these areas were amalgamated into a single large craton that contained Greenland, central Canada, and the north-central United States. Additional mountain building added more continental material along the south margin of the growing land mass from 1800 to 1600 Ma (Yavapai-Mazatzal terrane) and along the east edge from 1300 to 1000 Ma (Grenville Mountains). During the latter mountain-building event a rift system, the Midcontinent Rift, opened up in the Lake Superior region (Fig. 15–1)—an event that had a profound influence on the local geology, especially at Isle Royale National Park (Fig. 15–2).

The Midcontinent Rift is likely the surface expression of a rising plume of hot mantle material (Hauser, 1996). Upward expansion produced splits or rifts whose arms are approximately 120° apart. The northwest arm is marked by a number of faults and igneous dikes along this trend but is subdued in its topographic expression.

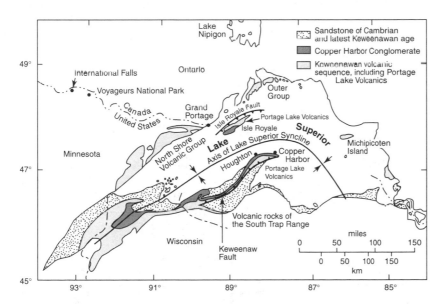

FIGURE 15–2   Map of Lake Superior region showing the location of the Lake Superior Syncline, late Precambrian rock outcrops, and Voyageurs and Isle Royale National Parks. (Modified slightly from Huber, 1975)

The southeast arm extends from northern Lake Superior through Michigan and into Ohio, and the southwest arm extends through Minnesota and Iowa and into Kansas (Fig. 15–1). These latter two rifts developed large synclines that filled with several miles of lava and sediment before the rift began to close and further deform the rocks. The distinct arcuate shape of Lake Superior with its southwest and southeast arms is controlled by the location of these ancient rifts and associated syncline (Fig. 15–2).

The first lava flows poured out of long fissures that are parallel to the axes of the now-existing synclines. The lavas contain ellipsoidal masses called *pillow* structures and thus must have been extruded into a body of water. Evidently the water was shallow because later lava lacks pillows and must have flowed across dry land. Gravels were deposited along the edges of the rifts and some were carried out and deposited on the lava flows.

At or near the end of the eruption period, a huge mass of gabbro was emplaced within the basalt flows. The Duluth gabbro is exposed in a number of places north and east of Duluth. Associated with the intrusion are valuable copper–nickel deposits.

During the long period of igneous activity, as more and more magma was removed from below the surface and piled up as lava flows, the Lake Superior area gradually subsided, particularly along the axis where the flows were thickest. Thus, the Lake Superior Syncline was formed. Downfolding was followed by compression as the rift closed, producing large thrust faults parallel to the fold axis. One fault of large displacement lies between Isle Royale and the north shore of the lake (Figs. 15–2 and 15–3). Cross fractures and small faults also developed as the basin subsided; many are parallel and cut across the linear ridges of basalt.

After the eruptions ceased, streams carried large quantities of coarse debris southward across the basin and deposited them on the youngest lava. These conglomerates are now exposed extensively along the south side of the west end of Isle Royale.

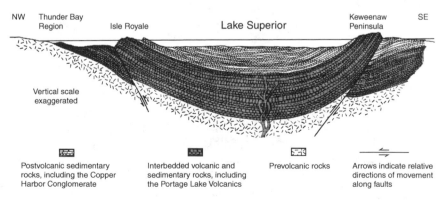

FIGURE 15–3   Cross section of Lake Superior Syncline in the Isle Royale area. (From Huber, 1975)

Early Paleozoic times brought extensive seaways that covered much of and at times the entire craton area. Small remnants of Paleozoic rocks are found on and at the edges of the shield, and a good Paleozoic record is found in the adjacent Central Lowlands where thick sedimentary sections are preserved. The sedimentary cover buries the Precambrian rocks that contain the earlier history of the continent—including the continuation of the Midcontinent Rift that can only be followed by using geophysical techniques and findings from the occasional deep wells that penetrate through the Paleozoic rock cover. During the late Paleozoic, the Mesozoic, and the Cenozoic, the province stood above sea level and suffered severe erosion that helped lower the land surface and reduce the topography to a flat rolling plain, even before the extensive attack of the Pleistocene glaciers.

Enter the Glaciers

During the Pleistocene, glaciers pushed down out of Canada and relandscaped the area (Fig. 13–2). Lake Superior's basin was scooped out by glaciers along the trend of the Midcontinent Rift. At one point on its east end the lake bottom is 1302 feet (397 m) below the surface at an elevation of 700 feet (213 m) below sea level! Glaciers left large deposits of morainal debris around the southwestern end of the lake. Areas where the rocks were unusually resistant to glacial abrasion were left as prominent hills, some of which, including Isle Royale, rise above the present lake level.

The surface of Lake Superior is 602 feet (184 m) above sea level, but at an earlier time the depth of water was even greater. When the lake level remained constant for a significant period, wave erosion formed a terrace and beach. In order to distinguish the stillstands of the past, separate names are used. Glacial Lake Minong is an example: The results of Lake Minong wave erosion are clearly marked on the north side of Lake Superior and on Isle Royale.

When the ice sheet covered the Great Lakes area, the tremendous weight of the ice caused significant depression of the crust; when the ice melted, uplift to its original position began. The uplift is continuing today at about 1 foot every century at Isle Royale.

REFERENCES

Hauser, E.C., 1996, Midcontinent rifting in a Grenville embrace, in van der Pluijm, B.A., and Catacosinos, P.A., eds., Basement and basins of eastern North America: Geological Society of America Special Paper 308, p. 67–75.

Huber, N.K., 1975, The geologic story of Isle Royale National Park: U.S. Geological Survey Bulletin 1309, 66 p.

Thornbury, W.D., 1965, Regional geomorphology of the United States: Wiley, New York, 609 p.

VOYAGEURS NATIONAL PARK (MINNESOTA)

About 219,000 acres (342 square miles, 886 km²) of northern Minnesota's wilderness was authorized as Voyageurs National Park in 1971. The park is in the lake country about 15 miles (24 km) east of International Falls and follows the international boundary for 55 miles (89 km) where it connects with the Boundary Waters Canoe Area in Superior National Forest (Fig. 15–2). The roadless Kabetogama Peninsula, south of Rainy Lake and north of Kabetogama Lake, is the main land area—covering about 75,000 acres (117 square miles, 304 km²) of the park. Parts of Rainy, Kabetogama, Nanakan, and Sand Point lakes, together with more than 30 smaller lakes, make up almost 40 percent of the park area. Water flows northwest to Lake-of-the-Woods and Athabasca Lake in Canada and eventually to Hudson Bay—the only United States park whose waters discharge into this subarctic water body. This is a water park—a fitting tribute to the French-Canadian voyageurs who annually made a 3000-mile (4830 km) water passage from Montreal to Lake Athabasca from the 1600s to about 1832. Trading materials were annually transported westward and many tons of furs were shipped eastward—all in birch-bark canoes powered by men with incredible stamina.

Here at Voyageurs National Park are some of the oldest Precambrian rocks on the North American continent. These ancient (Archean, older Precambrian) rocks formed one of the nuclei around which a landmass would grow to continental-size proportions. Locked up in the minerals and the metamorphic and igneous rocks that they form is the story of moving plates, ancient subduction zones, mountain building, volcanoes, and extensive lava flows. However, the rocks do not give up their dark mysterious past easily—as several generations of geologists have discovered.

Glaciers were the most recent geologic process to significantly affect the area. Ice left its characteristic calling cards—the dozens of scooped-out lake basins, the scratched and polished rock surfaces where over a mile (1.6 km) of ice dragged many tons of loose rocks over the bedrock, and the lake, outwash, and till deposits left mostly during ice retreat.

History

Native Americans have used the area for over 8000 years. The Chippewa, the most recent group to occupy the area, guided the early Voyageurs and showed them the best water routes to follow from Grand Portage, east of the park on the northwest shores of Lake Superior (Fig. 15–2), on their 2000-mile (3220-km) journey to Lake Athabasca in northeastern Alberta. Here they exchanged trade goods for valuable furs. The Voyageurs used 36-foot (11-m) Montreal canoes between Montreal and Grand Portage and lighter 25-foot (7.6-m) birch-bark canoes powered by five to six men for the remainder of the journey to Lake Athabasca. This colorful, boisterous breed of men faded from the scene sometime during the 1830s, leaving behind their story written in song.

With the declining fur trade in the 1830s came a 50-year episode in which practically all of the virgin north woods forest was logged. Gold mining created a minor boom in the 1890s but was relatively short-lived and unprofitable. Local interest in establishing a national park was officially noted in 1891 by the Minnesota legislature. Ontario felt likewise and succeeded in establishing Quetico Provincial Park to the east on the Canadian side. Boundary Waters Canoe Area was established on the United States side of the border and Voyageurs was authorized as a national park in 1971.

Bedrock Geology—Archean Times

Although some isolated areas of extremely ancient rocks over 3 billion years old occur in Minnesota, Canada, Greenland, and elsewhere in the world, these were apparently localized, early-formed areas of continental crust. More extensive areas of mountain building and larger areas of continental crust developed during Late Archean time, about 2800–2500 Ma. This event, the Algoman Orogeny in North America, formed the gneisses, schists, and granites exposed today in the glacially scoured landscape at Voyageurs National Park and elsewhere (Miller and others, 1987). Subduction of the oceanic plate produced a series of volcanic islands whose rocks were intensely metamorphosed. Squeezed between these areas of lighter continental crust were zones of thick, basaltic lava that accumulated in basins behind the volcanic islands (so-called *back-arc basins*). These linear, lava-filled basins form today's *greenstone belts* (Southwick, 1987). A number of these strips of continental crust and greenstone belts were added, or *accreted,* to form the Superior Craton of south-central Canada and the northern Great Lakes—including the strip that occupies the Voyageurs area. Intrusion of granitic plutons later in the accretion process completes the assemblage of rocks found at Voyageurs. Precambrian rocks younger than Archean (2500 Ma) are called Proterozoic (see time chart, Fig. 1–23) and occur on the outer edges of the Archean cratons and in areas such as Isle Royale National Park in the Lake Superior area.

Mineral-rich fluids from the plutons produced dikes and veins of quartz, mostly in the greenstone belts. The gold-mining boom from 1893 to 1898 saw numerous prospects and small mines opened, including the Little American Mine and Big American Mine on islands now within the park boundaries. None of the mines produced significant amounts of ore nor were they economically rewarding to their owners.

Although the topography is low and lacks high, bold cliffs, scouring by Pleistocene glaciers as recently as 10,000 years ago has left many rocks exposed or covered only by thin soils (Mikelson and others, 1983). Additional exposures of bedrock are common around the edges of the abundant lakes where waves have cleaned off soils and the thin spotty deposits left by the massive Pleistocene ice sheets.

Pleistocene Glaciers

Glaciers have covered and uncovered the land in the North Temperate zones numerous times in the last 2 million years. The most recent glaciers (Fig. 13–2), those of late Wisconsinan age, covered the Voyageurs area 10,000 years ago or less (Mikelson and others, 1983). The massive, bulky end moraines of Wisconsinan age are mostly in the southern Great Lakes area. In the more northern areas the ice was thicker and more erosive—carving out lake basins, thousands of them, linking the west to the east in southern Canada and the northern United States. The resulting watery highway provided passage for the Indians, Voyageurs, explorers, and missionaries. The lake basins are carved from the hard Precambrian rocks of the Canadian Shield and have remained essentially unchanged except for the recent environmental disruptions wrought by humans.

The scratch marks (*striations*) and rock polish produced by the passage of debris-laden ice is abundant and relatively untouched by weathering. Also abundant are ice-deposited rocks, or *glacial erratics,* that were carried from sometimes great distances. Some are the size of automobiles or trucks, and some were deposited on other rocks and left in precarious positions as the great ice sheet thinned, stagnated, and melted—thereby lowering the ice-enclosed blocks randomly onto the land surface beneath.

Visiting Voyageurs

A trip to the North Woods is a pleasant step back in time for those who wish to sample the relatively unchanged landscape of the northern Indian tribes and the Voyageurs (Fig. 15–4). There are no roads in the park except for those that lead to the visitor centers at Kabetogama and Rainy lakes where boat launching and parking areas are located.

A few trails lead from the visitor centers and provide an opportunity to experience the North Woods on foot, but most of the park must be seen from a canoe or boat in the style of the French-Canadian Voyageurs. Park visitors will relive the life of the Voyageurs, canoeing and portaging from one lake to the next, and sleeping under the stars. The master plan calls for a wilderness park. Thus, the Park Service salutes the deeds of those men who contributed much in the development of this continent. Another reason for the park is to preserve a small patch of the vast area of virgin timber—fir, spruce, pine, aspen, and birch—and its moose, deer, wolves, and beaver. Perhaps in time the caribou will return. The forest is a transitional one—a subarctic forest that has characteristics of the arctic boreal forests to the north and the northern temperate forests to the south.

The summer season is a short one from the ice melt sometime in May through August when temperatures begin to decline. Fall colors are spectacular—especially enjoyable because the numerous biting insects are gone. Winter brings snowmobilers and cross-country skiers to the frozen lakes and snow-bound forests.

FIGURE 15–4   Arm of Rainy Lake from Bear Pass, Voyageurs National Park (National Park Service photo)

In time, when all of the surrounding area has been conquered and converted to the alleged needs of humans, Voyageurs National Park with all of its natural beauty and historical significance will hopefully be one of our truly outstanding national treasures. Today's decisions of elected officials will determine the type of world that succeeding generations will inherit. Will it be a crowded world of large metropolises devoid of natural beauty or will it contain significant areas where one can experience and feel what is really important?

REFERENCES

Mikelson, D.M., Clayton, L., Fullerton, D.S., and Borns, H.W., Jr., 1983, The late Wisconsin glacial record of the Laurentide Ice Sheet in the United States, in Porter, S.C., ed., Late-Quaternary environments of the United States: Volume 1, The Late Pleistocene: Minneapolis, Univ. of Minnesota Press, p. 3–37.

Miller, J.D., Jr., Morey, G.B., and Weiblen, P.W., 1987, Seagull Lake–Gunflint lake area: A classical Precambrian stratigraphic sequence in northeastern Minnesota, in Biggs, D.L., ed., North-Central Section of the Geological Society of America: Geological Society of America, Boulder, CO, Centennial Field Guide, v. 3, p. 47–51.

Southwick, D.L., 1987, Geologic highlights of an Archean greenstone belt, western Vermilion district, northeastern Minnesota, in Biggs, D.L., ed., North-Central Section of the Geological Society of America: Geological Society of America, Boulder, CO, Centennial Field Guide, v. 3, p. 53–58.

Park Address

Voyageurs National Park
3141 Highway 53
International Falls, MN

ISLE ROYALE NATIONAL PARK (MINNESOTA)

Isle Royale National Park in Michigan is composed mainly of water. The park consists of about 684 square miles (1772 km²) of Lake Superior and about 210 square miles (544 km²) of land of which there are over 400 tiny islands along with the 45-mile-long (72 km) by 9-mile-wide (14 km) Isle Royale (Fig. 15–5). Its isolation and northern location combined with the fiordlike edges of the island and its tree-covered valley-and-ridge topography give this island a unique character. Isle Royale can be reached by boat from Grand Portage, Minnesota, 22 miles (35 km) to the west, or by 4½- to 6½-hour-long ferry trips from Copper Harbor and Houghton, Michigan, 56 and 73 miles (90 and 118 km), respectively, to the south on the Keewenaw Peninsula (Fig. 15–2). Seaplanes also bring visitors into the park. Isle Royale is a hiker's park, for there are no roads and no automobiles. The moose, fox, and smaller animals coexist, largely free of human interference (Fig. 15–6). Two trails extend from Rock Harbor, near the eastern end of the main island, to Windigo, 40 miles (64 km) to the southwest. These trails are for those who are physically fit and well equipped with rain gear, warm clothes, and mosquito repellent! But there are also many shorter trails leading into the back country—some to high areas where the island and Lake Superior are sprawled out magnificently below.

Here at Isle Royale is the next chapter—the Proterozoic (late Precambrian) story of the Superior Uplands Province and the Canadian Shield to which it belongs. The ancient Archean rocks at Voyageurs and elsewhere in the upper Great Lakes region are not exposed here. Rather, over a billion years ago huge outpourings of lava, much like those of the much younger Columbia River Basalt of Washington and Oregon, filled a huge rift valley that formed on the craton or relatively stable interior of the continent.

No clear record of the last billion years of mostly erosion remains except for the most recent chapter when the giant glaciers of continental proportions smothered

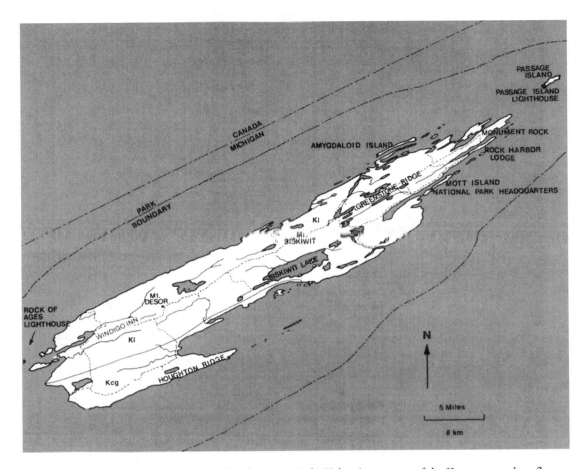

FIGURE 15–5   Map of Isle Royale National Park. Portage Lake Volcanics are part of the Keweenawan lava flows (Kl) and cover most of the island. The younger Copper Harbor Conglomerate (Kcg) is restricted to the southern tip of the island.

the Great Lakes area. Ice over a mile thick put the finishing touches on Isle Royale, producing most of today's topography. Added during the last 10,000 years were whatever vegetation and animals could find their way to this island wilderness.

History

Isle Royale was explored by Pre-Columbian Indians at least 8000 years ago. No habitation sites have been found, according to Rakestraw (1965), and except for one thing we probably would have no record of their being on the island. Native copper occurs in the rocks of Isle Royale, and the Indians mined it from more than a

FIGURE 15–6   Moose grazing in one of many small lakes on Isle Royale. (National Park Service photo)

thousand hand-dug pits. The copper occurs as *amygdule* fillings of the *vesicles* (gas-bubble holes) in some of the lavas and as fillings around pebbles in the conglomer-ates. By using hard beach cobbles as hammerstones, the Indians freed the malleable copper for ornaments and knives. Charred wood from their fires in the mining pits provides the radiocarbon dates of their earliest known visits. Isle Royale is likely the legendary "Island of Minong" where the Chippewas gathered loose copper frag-ments. Their mysterious predecessors were the miners who had worked the out-crops. Copper weapons, tools, and ornaments from Isle Royale and the copper-rich Keewenaw Peninsula of northern Michigan are found in widespread archeological sites in eastern and central North America (Huber, 1983).

French exploration of the Lake Superior area began in 1618, and early on they rediscovered copper on the island. In 1669 they named the island Isle Royale after King Louis XIV. The British acquired Canada, including Isle Royale, from the French in 1763 and in turn ceded the island to the United States in the Treaty of Paris that formally ended the Revolutionary War.

A flood of prospectors descended on the island after the Chippewas relin-quished their claims in 1842. A flurry of activity from 1843 to 1855 and 1873 to 1881 deepened the pits originally excavated by the Indians and removed much of the accessible copper deposits—all at a financial loss to the investors!

Because of the high mineral potential of the Precambrian rocks of the Cana-dian Shield and upper Great Lakes, geologic studies were done quite early by the Euro-Americans. Important iron, copper, and nickel deposits were discovered that fueled the growing industrial centers in the Great Lakes area. Early geologic stud-ies by John Foster and Josiah D. Whitney in the 1850s were surprisingly accurate—especially considering the remoteness and difficult terrain of Isle Royale. Whitney went on to a distinguished career as the first state geologist of California—eventu-ally the highest mountain in the conterminous United States was named after him.

Interest in using the island as a resort during the short summer season in this northern climate began in the early 1900s. Newspaperman Albert Stoll began efforts to establish Isle Royale National Park in 1921, but it was not until 1931 that Congress passed the necessary legislation and 1940 before the park was officially accepted into the National Park System.

Geographic Setting

Isle Royale is located on the north limb of the Lake Superior Syncline, and the lava flows and sedimentary rocks—sandstones and conglomerates—dip beneath the water along its south shore (Fig. 15–3). Between Isle Royale and the Minnesota shore the water is in places more than 900 feet (275 m) deep, thus extending down to about 300 feet (91 m) below sea level. With the water at its present level of 602 feet (184 m), there are more than 400 islands of the Isle Royale Archipelago. Almost all of the islands are distinctly elongate, reflecting the structure of the rocks. Amygdaloid Island on the north side and Mott Island, where the park headquarters are located, are typical. It is more than 50 miles (80 km) from the easternmost lighthouse on Passage Island to the one on the Rock of Ages west of the main island.

Large peninsulas extend out at each end of the 45-mile-long (72-km) island with fiordlike harbors between them. Mt. Desor, the highest point, reaches 1394 feet (425 m) above sea level and 792 feet (241 m) above Lake Superior. Elongate lakes, some of them large, are prominent features on the main island (Fig. 15–5).

The corrugated topography of the park is distinctive (Figs. 15–7 and 15–8). In a distant view from the air, the ridges appear to continue unbroken from one end of the main island to the other. A closer view reveals many breaks and offsets

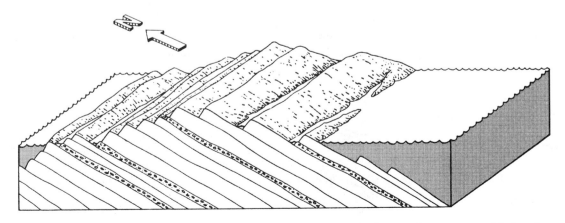

FIGURE 15–7 Block diagram showing relation of ridge-and-valley topography to rock structure on Isle Royale. (Diagram from Huber, 1983)

FIGURE 15–8   Corrugated topography on Isle Royale. Tilted layers of resistant basalt form the ridges; poorly resistant layers of amygdaloidal basalt were more deeply eroded, forming the valleys. (Photo by National Park Service)

related to cross fractures and faults. The valley-and-ridge topography is the work of glaciers that overrode the area at least four times. Greenstone Ridge, composed of massive basalt flows, was better able to withstand the onslaught of the glaciers and forms the high backbone of the island. Views from more open areas on the high ridges are superb.

Geologic Setting

As discussed in the chapter introduction and elsewhere in the book, continents grow from their centers outward as *mobile belts,* or *geosynclines,* add, or accrete, new material. The interior areas become stable cratons that experience minimal tectonic activity over hundreds of millions of years unless interior splits, or *rifts,* form. Such was the case in the Lake Superior area about 1100 Ma when the Midcontinent Rift formed (Green, 1987). The eastern rift extends southeast along Lake Superior and through Michigan, and the western arm extends southwest to

Kansas. Except in the Lake Superior area, the rift is buried by younger rocks (Fig. 15–1). A massive rise of hot mantle material, or a *mantle plume,* may be the immediate cause of rift formation. However, simultaneous mountain building along the east edge of the craton in the Grenville Geosyncline (1300–1000 Ma) is no doubt involved. The details and causal connections are still uncertain.

What is certain is that huge fissures opened up along the axis of the rift and massive lava flows (the Keewenawan Basalts), each flow containing tens or hundreds of cubic miles of lava, welled up to the surface to fill the growing gap. Such *flood basalts* are relatively rare in the geologic record. After the Keewenawan flows ended about 1000 Ma, large flood basalt eruptions did not occur in North America until 15 Ma when the 2-mile-thick (3.2-km) section of Columbia River Basalt erupted in the eastern Washington–Oregon area. Early flows on the Keewenawan Peninsula in upper Michigan have rounded forms called pillows that indicate underwater eruptions occurred initially as lava filled a shallow water body. Later flows moved quietly across a landscape that was then above sea level. Between some of the eruptions the thickening pile of lava subsided and slopes reversed—where lava had flowed from the center of the rift to the edges, gravel and sand sediment from the rift basin edges now washed toward the rift center. The next flows buried the sediment creating an *interbed* and additional flows again built the rift center higher. On Isle Royale, the Portage Lake Volcanics contain over 100 individual lava flows, many of which are separated by interbeds (Huber, 1983).

Subsidence continued after the eruptions ceased, and large alluvial fans deposited the thick Copper Harbor Conglomerate. The subsidence process also produced rock layers that slope, or *dip,* toward the basin center—a structure formally called the Lake Superior Syncline (Fig. 15–2). An episode of compression followed—squeezing the crust and further steepening the limbs of the Lake Superior Syncline (Fig. 15–3). Thus the lava and sedimentary layers on Isle Royale are on the north flank of the syncline and dip about 15°–20° to the southeast beneath the waters of Lake Superior. The layers reappear on the south side of Lake Superior on the Keewenaw Peninsula where they dip northward toward the lake basin and the axis of the syncline.

Because the lava layers are inclined like leaning books on a bookshelf, more layers are exposed at the surface and permit a thicker section of rock to be measured (Fig. 15–7)—an estimated 10,000 feet (3050 m) or more of lava and interbeds on Isle Royale alone (Huber, 1983). Additional flows are hidden beneath the waters of Lake Superior (Fig. 15–3) and about 5.4 miles (8.7 km) of basalt are exposed on the Keewenaw Peninsula (Green, 1987)!

Bedrock Details

With the big picture in mind of rock layers of the Portage Lake Volcanics dipping southeasterly beneath the waters of Lake Superior, an examination of the eroded edges of the beds on Isle Royale shows numerous interesting characteristics.

Subaerial lava flows typically have a dense interior and a basal and top zone where numerous spherical or elongated gas bubble holes, or vesicles, are formed. Subsequent filling by secondary mineral material produces an amygdaloidal texture. These amygdaloidal, or filled vesicle, zones are less resistant to weathering and erosion than the denser flow interiors and have been carved out as valleys. Thus exposures of amygdaloidal rocks are not common. Amygdaloid Island on the northeast end of Isle Royale is an exception. Interestingly, details of the present topography reflect what happened a billion years ago.

On Isle Royale, many of the amygdules are composed of an unusual mineral, *prehnite,* a semiprecious stone. Originally it was believed to be thomsonite, a more highly prized gemstone. Even on Thomsonite Beach, on the north side of the island, most of the amygdules are prehnite.

Chlorastrolite, the state gem of Michigan, is sometimes called "Isle Royale Greenstone" and also occurs as amygdules. They occur as pea-size pebbles on certain beaches where one can examine them and toss them back onto the beach for the next modern-day explorer to discover. Collecting in any national park or monument is illegal—and more importantly, it shows a lack of concern for most of the people who will visit the island—those who have yet to be born! Chlorastrolite has a mosaic pattern and when polished is usually *chatoyent,* reflecting light like the eye of a cat. Much more common, however, are the agates that are found on most of the pebble beaches. They are composed of concentric bands of chalcedony, a translucent variety of quartz. Mainly they formed as amygdules in the lavas; later, some were incorporated into the conglomerates, others in the glacial drift. Still later they were freed by wave action and deposited on the beaches.

Native copper in abundance is rare in nature and occurs in abundance only in a few areas in North America. In the Lake Superior region it is associated with the Keewenawan Volcanics. On Isle Royale it occurs as amygdules, but mainly it either fills fractures or acts as the matrix in conglomerate. One copper "nugget" weighing 5720 pounds was unearthed by prehistoric Indians; frustrated by their inability to cut it into workable pieces, they abandoned it. Legal battles later fought over ownership of the giant nugget resulted in federal ownership and eventually adding it to the Smithsonian collection in Washington.

The dense interior of the flows is more resistant to weathering and erosion than the vesicular outer layers, and supports the series of parallel rock ridges in the park (Figs. 15–7 and 15–8). Some of these flow interiors are fine grained and break with arcuate fractures—similar to fractures in massive glass or obsidian. Other interiors are *porphyritic* and contain large feldspar or pyroxene crystals surrounded by darker fine-grained minerals.

The Copper Harbor Conglomerate that overlies the lavas records an ancient environment far different from that which exists today. At the western tip of the island, boulders as large as 2 feet (0.6 m) in diameter are found at the base of the Copper Harbor Conglomerate. Because the size of the materials decreases toward the east, Huber (1983) concludes that the source area lay to the west, the north shore of Lake Superior. In order to transport boulders, the streams must have had considerable velocity resulting from substantial relief along the edges of the rift.

Perhaps flash floods were at work in the Lake Superior area almost a billion years ago as these huge alluvial fans spread over the Keewenawan lava flows.

Gap in the Record

No rocks of Paleozoic, Mesozoic, or pre-Pleistocene Cenozoic age are found in the area. The evidence in adjacent areas, however, suggests that the early Paleozoic sediments were deposited here but were later eroded away. From the late Paleozoic to the present, the Lake Superior area was apparently stable and has remained at essentially its present position above sea level except for elevation changes resulting from loading of the crust by a mile-thick glacier during the Pleistocene.

Work of the Ice Sheets

During the Pleistocene Epoch, huge ice sheets (continental glaciers) developed in Canada, expanded southward, and covered nearly all of the United States north of the Missouri and Ohio rivers (Fig. 13–2). The glaciers followed and deepened an ancestral river valley system that ultimately became the Great Lakes. There were at least four major advances separated by long periods of warm climate, called *interglacials*. As the glaciers advanced, they essentially obliterated the evidence of former advances—especially in more northern locations where erosion was more effective. Therefore, the records of all but the last advance, the Wisconsin, were completely destroyed in the more northern areas such as that at Isle Royale.

During the Wisconsin advance, the Lake Superior Basin was scoured deeply and depressed by the weight of the ice. In places, the present lake depth reaches about 300 feet (91 m) below sea level with lakewaters between Isle Royale and the Canadian shore over 1000 feet (305 m) deep! The glaciers dug deep into some of the rocks on Isle Royale and formed many rock-basin lakes. Lake Desor and Siskiwit Lake (Fig. 15–5) are two of the larger of this type. Some of the shallow lakes have since been converted to swamps.

The main island is a huge *roche moutonnee, whaleback,* or glacially elongated bedrock ridge; if Lake Superior were emptied, we would become aware of its true size. Superimposed on the big one are smaller roches moutonnees that are hosts to still smaller ones. Although glacial erosion dominated, some deposition of till occurred, especially on the southwest end of the island west of Siskiwit Lake. Glacial erratics are numerous, and in places the ice deposited elongate, streamlined ridges of till behind bedrock knobs to form a type of *drumlin*. Most of them are small, but one is almost 2 miles (3 km) long (Huber, 1983).

About 11,000 years ago the climate became warmer and the ice sheet began to retreat. However, at the beginning of the recession, the ice in the Lake Superior Basin was thousands of feet thick and the ice front was many miles to the southwest. Therefore, it was not until much later that any appreciable part of the area was freed from ice.

The retreat was not without interruption. The climate, although warming generally, was fluctuating, and at times melting merely equaled the forward movement of the ice; at such times the ice front remained almost stationary for extended periods. When the western end of Lake Superior was ice free, glacial retreat paused and meltwater formed Lake Duluth at the glacier front. Glacial Lake Duluth was much smaller than today's Lake Superior, but the water surface was much higher, more than 1100 feet (335 m) above sea level compared to today's 602 feet (184 m). At this time—about 11,000 years ago—Isle Royale was still buried beneath the glacier.

Glacial Lake Duluth remained at this level for a period sufficient for wave action to develop a wave-cut terrace along the southern margins of Lake Superior—more than 500 feet (152 m) above the present lake level. In like manner, stage by stage as the glacier withdrew, eventually the present Lake Superior level was reached. As the ice melted off of Isle Royale, morainal deposits—relatively thin ground moraines and irregular recessional moraines—were laid down on the southwest end of the island. The Glacial Lake Minong developed a shoreline about 10,500 years ago, forming isolated *stacks* (Fig. 15–9) and *wave-cut platforms* on which beaches were deposited—all at elevations on Isle Royale that are now 200 feet (61 m) above Lake Superior. Subsequent lake levels left similar features preserved at lower elevations on Isle Royale and elsewhere.

Your Visit to Isle Royale

Isle Royale provides an opportunity to contemplate the processes that operated in Precambrian time to build the craton—the stable nucleus that forms the core of the North American continent. The immense Pleistocene ice sheets relandscaped the islands into a series of long parallel ridges and valleys. Rocky fingers of land sep-

FIGURE 15–9   Monument Rock, a lake stack formed when lake level was much higher than now. (National Park Service photo)

arated by drowned fiordlike bays project into Lake Superior, especially on the northeast end of Isle Royale. The island is covered by a verdant growth of vegetation as the forest reestablishes itself after the significant deforestation experienced during the early days of mining and logging.

Another scientific problem to contemplate is how revegetation and repopulation of the isolated landscape of Isle Royale occurred. No living thing could exist beneath the mile-thick ice sheet—now a jungle of plants inhabits the island along with moose, fox, timber wolves, and numerous small mammals. Seed dispersal by wind and floating debris and transport by birds are all possible. Moose are strong swimmers or they might have walked across the frozen lake surface of Lake Superior during the early 1900s when their presence was first noted. The lake surface does occasionally freeze over during unusually cold winters and the Canadian shore is only 15 miles (24 km) from Ontario and 22 miles (35 km) from Grand Portage, Minnesota. The caribou, lynx, and pine martens that were present when the miners first arrived could have reached the island in a similar manner. Deforestation and perhaps hunting eliminated these species.

A pack of eastern gray wolves walked from Canada during the extremely cold winter of 1948–1949 and helped keep the moose population in balance for a number of years. Overpopulation of moose had caused serious mass starvation problems prior to the arrival of the wolf. Inexplicably the wolf population plummeted in the early 1980s. A slight rise in numbers during the 1990s is encouraging. Inbreeding of a small population of animals and the appearance of canine parvovirus in the region are suspected to be at least part of the problem.

Another biologic mystery involves the thorny, people-unfriendly devils club found in a small area along the Blake Point Trail. Devils club is a common plant in the temperate rainforests of the Pacific Northwest but is unknown east of the Rocky Mountains—except for this small area on Isle Royale! Perhaps a visitor who had previously been to a western park had a seed stuck on a boot or in camping gear and unknowingly released it on Isle Royale.

Your visit must be carefully planned, including advance reservations. A two-day tour on a boat such as the *Ranger* or perhaps the *Voyageur,* with an overnight at Rock Harbor or Windigo, will take you around the main island. With the stops at various landings, you get a reasonably good view of the shore area, the historic lighthouses, and some of the numerous shipwreck sites. But the quality park experience comes as a result of some foot work on the trails, for at least a day or two on some of the 164 miles (266 km) of trails. To walk alone, deep into the woods, is to know Isle Royale.

REFERENCES

Green, J.C., 1987, Plateau basalts of the Keweenawan north shore volcanic group, in Biggs, D.L., ed., North-Central Section of the Geological Society of America: Geological Society of America, Boulder, CO, Centennial Field Guide, v. 3, p. 59–62.

Huber, N.K., 1983, The geologic story of Isle Royale National Park (U.S. Geological Survey Bulletin 1309): Isle Royale Natural History Association, 66 p.

Rakestraw, L., 1965, Historic mining on Isle Royale: Isle Royale Natural History Association.

Park Address

Isle Royale National Park
800 East Lakeshore Drive
Houghton, Michigan 49931

SIXTEEN

Interior Low Plateaus Province

The Interior Low Plateaus Province is at the southeastern edge of the *craton,* or stable continental interior. It lies between the Central Lowlands on the northwest, the Mississippi embayment part of the Coastal Plain Province on the southwest, and the Cumberland Escarpment at the edge of the Appalachian Plateaus Province to the east (Plate 1). The Cumberland Escarpment is where the craton ends and the Appalachian Mountains begin. In elevation, the Interior Low Plateaus Province ranges from about 500 feet (150 m) to 1100 feet (335 m) above sea level. Lying mostly in Kentucky and Tennessee, it extends into the southern parts of Illinois, Indiana, and Ohio, and southward into Alabama. The world-renowned Mammoth Cave National Park is located in this province in west-central Kentucky and will be emphasized in this chapter.

The geologic structures, generally similar to those in the Central Lowlands, are broad uplifts and basins with low dips on their flanks. The Cincinnati Arch extends northeast–southwest through Kentucky and Tennessee with a distinct structural bulge or uplift along its axis called the Nashville Dome. The Nashville Dome is interesting because, although it is structurally a dome, it is topographically a basin— a classic case of *topographic reversal.* Downcutting streams encountered weak rocks in the center of the dome and more resistant strata along its flanks, resulting in the present anomaly where the lowest part of the basin coincides with the highest part of the dome.

Another important structure is the Illinois Basin that extends through part of western Kentucky and Tennessee. The basin played a major role in the development of the extensive cave systems found at Mammoth Cave and other areas in this part of the Midwest. The gentle northward-dipping Paleozoic strata in the Mammoth Cave area form the Chester Cuesta, or Chester Plateau—a broad sandstone-capped plateau underlain by cavernous limestone. The distinct east–west trending escarpment along the Chester Cuesta in western Kentucky is called the Dripping Springs Escarpment. South of the escarpment where the limestone is at the surface in the Pennyroyal Plateau is a vast *sinkhole plain,* a virtual "Land of 10,000 sinks."

Rivers such as the Green River at Mammoth Cave, the Cumberland, the Tennessee, and others drain westward and northward into the Ohio River. The Green River has cut deeply into the Chester Plateau and acts as a *local base level* for the subterranean drainage in southwestern Kentucky. As will be discussed in the following section on Mammoth Cave, important connections exist between the geologic history of Mammoth Cave and the downcutting history of the Green River.

MAMMOTH CAVE NATIONAL PARK (KENTUCKY)

Mammoth Cave is perhaps the most famous cave in the world and one of the oldest tourist caves in the United States. The name came from early explorers and visitors who noted the extremely large passages present near the cave entrance. The name became particularly appropriate in September of 1972 when a group of Cave Research Foundation *speleologists* (people who study caves) who were exploring the lower section of the nearby Flint Ridge (Fig. 16–1) cave system (at that time the largest cave in the world) crawled and dragged their tired bodies through miles of a long wet passage and found themselves in a larger cave passage—a little while later they were standing on an asphalt walkway in Mammoth Cave! They had established that the world's largest and third largest caves were in reality one cave—a cave that contained more than 300 miles (480 km) of passage—three times longer than the world's second largest cave located in the Ukraine! By going on all four of the regularly scheduled summer tours, the public can experience about 10 miles (16 km) of the over 330 miles (530 km) of explored cave and many of its significant geological and historical features.

The cave has a particularly long geologic, archeologic, and historical past that adds greatly to the significance of this national treasure. The history of westward expansion, the struggle of early settlers, the strategic value of the cave during the War of 1812, and the development of a society with environmental concern and curiosity are all part of the Mammoth Cave story.

Human History at Mammoth Cave

Who the first humans to enter the cave were will never be known. Prehistoric Indians camped in the natural entrance for thousands of years and began to enter the cave at least 4120 (plus or minus 70) years ago (Kennedy and Watson, 1997). They used cane and reed torches to light their way on dangerous expeditions many miles back into the cave to remove wall linings of gypsum and other minerals to use for some unknown purpose. Remnants of torches, woven baskets, fiber slippers, footprints in the cave sediment, and a number of mummified bodies have been found deep in the cave.

One unfortunate soul was discovered in 1935, crushed by a 6.5-ton boulder that had apparently rolled on him about 2000 years ago while he was removing

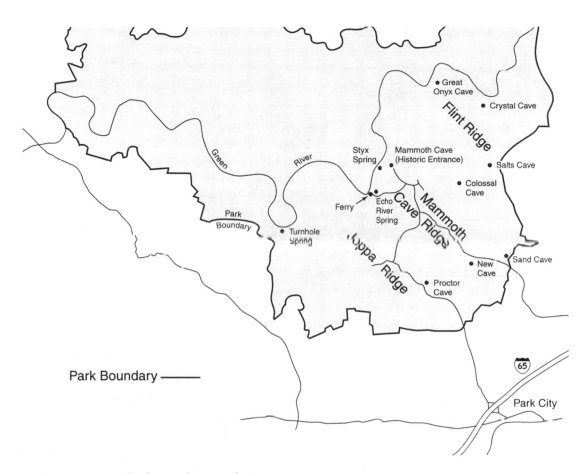

FIGURE 16–1   Generalized map of Mammoth Cave area.

gypsum from the cave wall. The mummified remains of "Lost John" were on display in the cave for many years. Other mummies were discovered during the 1800s but were destroyed before they could be adequately described and preserved.

The modern Cherokee, Iroquois, and Chickasaw tribes used the Kentucky area as a hunting and fighting ground but did not establish permanent settlements (Bridwell, 1971). The area became a "hunting paradise" for the Euro-Americans in the mid-1700s as they depleted game in the east. Settlers soon reached the new frontier and began cutting the forest to sell timber and establish small farms where they eked out a meagre living. The legendary story of the rediscovery of the cave is that in 1799 a hunter named Houchins pursued a wounded black bear into the cave entrance. Sadly, black bear, bison, and the passenger pigeon that once inhabited the western Kentucky area are now gone—the passenger pigeon forever—a result of "progress." Beaver, whitetail deer, and turkey have been restocked and are flourishing (Bridwell, 1971).

Sediment in the cave was found to be rich in nitrates, a material called *saltpeter* that was used for preservation of meat and the manufacture of gunpowder. Mammoth Cave and other caves acquired new significance during the War of 1812 when the shipping embargo by England deprived the fledgling United States of their needed supply of gunpowder. A major mining and processing operation at Mammoth Cave used about 80 slaves to produce many tons of saltpeter for the war effort. The nitrates were shipped by ox cart and barge to Philadelphia where the gunpowder was manufactured. The hand-hewn pipes and leaching vats are amazingly well preserved and are one of the highlights of the historic tour (Fig. 16–2).

After the War of 1812 visitors began to arrive in great numbers at the now famous cave that helped save the country. Visitation was exceeded only by that at Niagara Falls during the early 1800s. Exploration begun by the saltpeter miners was continued by the early guides—some of whom were miners who had worked the cave during the war years. The guiding tradition continued in many families for over 100 years as generations of cave guides devoted their working lives to the cave. Many are buried in the historic Guides Cemetery in the park. One of the first guides and the nation's first truly outstanding cave explorer and speleologist was Stephen Bishop, a self-educated black slave. In 1838 Bishop was the first to cross the Bottomless Pit on a slender cedar pole placed precariously across the yawning

FIGURE 16–2   Hand-hewn log water pipes used in Mammoth Cave during the War of 1812. (Photo by E. Kiver)

abyss. Extensive discoveries followed—including the first discovery of the blind cave fish in the Echo River in the lowest accessible depths of the cave (Bridwell, 1971).

Concern over private ownership of the cave and commercial logging in the area was expressed by the public as early as 1870. Efforts to establish a park began in 1905 when the state delegation recommended park status to a Congress that was skeptical of the cave's significance. Stephen Mather, the first director of the National Park Service, indicated in his yearly reports to Congress beginning in 1918 that more parks were needed in the east and that Mammoth Cave would be an excellent addition.

During the early 1900s the "Kentucky Cave War" began as people searched for new entrances to Mammoth Cave, and others used questionable tactics to lure tourists to other nearby commercial caves. A local man, Floyd Collins, discovered and opened to tourists the Great Crystal Cave on nearby Flint Ridge, the next ridge east of Mammoth Cave Ridge (Fig. 16–1). In an effort to find a cave closer to the main tourist roads, Floyd entered Sand Cave on January 30, 1925, where he was trapped in a tight crawlway when a basketball-size rock fell and pinned his leg. In spite of national attention and concern, Floyd was unable to be freed and died 18 days later. The nationwide publicity again focused attention on the Mammoth Cave area. The Mammoth Cave National Park Association had formed the year before Floyd's death and was, this time, well received in Washington where park legislation was passed in 1926. Money then had to be raised and land purchased before the park was officially established in 1941.

Additional exploration by cave guides and later by dedicated speleologists found connections to previously discovered caves as well as discovering extensive new passageways, thereby extending the length of the cave to 330 miles (530 km)! The 52,714 acres (82 square miles, 213 km²) in today's park include only a fraction of the total subterranean cave system in this part of Kentucky.

Geologic and Hydrologic Setting

Because of its location at the southeastern edge of the broad Illinois structural basin, the rock layers in western Kentucky incline, or dip, gently to the northwest. Inclinations are so gentle that they are measured in feet per mile (30 feet/mile at Mammoth Cave; 5.69 m/km) rather than degrees (less than 0.5°). This gentle northwest dip exposes the edges of the rock layers, which are progressively younger toward the center of the Illinois Basin. Where resistant rock layers such as the Big Clifty Sandstone occur, a *cuesta* landform with a prominent updip escarpment forms (Fig. 16–3).

In the Mammoth Cave area the rocks are Mississippian in age, about 350–325 millions of years old, whereas in Illinois, in the center of the basin, coal-bearing rocks of Pennsylvanian age (about 300 millions of years old) occur. The limestones are composed of carbonate particles and fragments of shells that accumulated in a

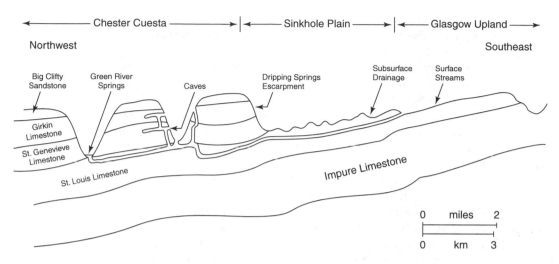

FIGURE 16–3   Generalized cross section of the Mammoth Cave area.

shallow, widespread sea that covered much of the southern part of the continent during Mississippian time (Palmer, 1981). The limestones are exposed in the Sinkhole Plain Section of the Pennyroyal Plateau to the south of the park (Fig. 16–3). Overlying the limestone is the 50- to 60-foot-thick (15- to 18-m) Big Clifty Sandstone that was deposited by a large south-flowing river and delta complex along the edge of the Mississippian seaway. The sandstone holds up the prominent, 400-foot-high (122-m) Chester Cuesta or escarpment and overlies and protects the same limestone units exposed in the Sinkhole Plain (Fig. 16–3).

The St. Louis, Saint Genevieve, and Girkin limestones contain an extensive system of caves (human-size openings) and smaller solutional openings that allow surface waters to enter the limestone and flow downdip in underground conduits. As the slightly acidic water flows initially through tiny openings along the bedding planes of the limestone, the openings are enlarged as the mineral calcite that makes up the limestone bedrock is dissolved. Water flows many miles northward from the Glasgow Upland to where the subterranean waters discharge from a number of springs into the Green River (Fig. 16–3), the major surface stream and the only permanent surface stream in the park. From there water flows into the Ohio River, a tributary to the Mississippi.

Where limestone is exposed at the surface in the Pennyroyal Plateau, solution produces a unique assemblage of landforms collectively known as *karst*—a name derived from a limestone plateau in Jugoslavia where such features are widely developed. Acidic water descending through fractures or joints in a rock widens the joints and can form *solutional sinkholes* (Fig. 16–4); collapsing cave roofs form *collapse sinkholes*. Some of the collapse sinkholes are very large, and it might be inferred that a huge mass of rock crashed down at one time. However, this seldom

FIGURE 16–4   Typical karst topography with sinkholes and sinkhole ponds near Mammoth Cave. (Photo by National Park Service)

occurs; instead, collapse begins early when the cavern is small. Later as more of the limestone support is removed, the roof rocks fall in, block by block, eventually enlarging the sinkholes to their present size. Hundreds of thousands of sinkholes dot the surface of the Pennyroyal Plateau. Sinkholes are also abundant in the limestone valleys separating the sandstone-capped Mammoth Cave, Joppa, and Flint ridges on the Chester Plateau (Fig. 16–1).

Surface streams are short and quickly disappear into sinkholes or *swallow holes* where they flow subterraneously to the Green River. Where the limestone is protected by the sandstone and shale of the Big Clifty Sandstone, water cannot penetrate easily into the limestone below. However, the lateral flow of water from the Pennyroyal Plateau accounts for solutional openings here. Further, caves beneath the protective caprock survive much longer than their counterparts on the Pennyroyal Plateau.

The size of a cave passage depends on a number of factors. A major factor at Mammoth Cave is believed to be the length of time that groundwater flows at a particular level, a concept that helps explain not only passage size but also the six distinct levels of passageways recognized at Mammoth Cave (Palmer, 1981; Miotke and Palmer, 1972).

Cave Formation—Speleogenesis

A long interval of weathering and erosion during late Tertiary times removed the Big Clifty Sandstone and developed the relatively flat, low relief surface of the Pennyroyal Plateau. The vertical control here is the Green River, which acts as a base level—the elevation to which streams and groundwater adjust. Water passing through the atmosphere and especially the soil zone combines with carbon dioxide (CO_2) to form carbonic acid (H_2CO_3)—a weak but effective acid when it reacts with limestone over long periods of time. Because the Green River and therefore the groundwater levels remained unchanged for a considerable length of time during the late Tertiary, an inordinately large system of cave passages formed in what is now the uppermost and largest of the cave passages, including Broadway, Kentucky, and Audubon Avenues—all of which are visited on the historic tour (Fig. 16–5). These are the passages that originally inspired Mammoth Cave's name.

FIGURE 16–5   Broadway Avenue, one of the large rectangular passages found in the upper level of Mammoth Cave. (National Park Service photo)

The stable condition of late Tertiary times changed drastically during the Pleistocene, or Ice Age. The advancing ice sheets permanently deflected north-flowing rivers in Ohio, Indiana, and Illinois southward where they combined with other streams to form the Ohio River. The greatly enlarged Ohio River was now more erosive and deepened its channel, causing its tributaries, including the Green River, to do likewise. Thus new lower base levels formed, particularly during intervals between glaciations (*interglacials*). Six important cave levels are recognized in the Mammoth Cave system (Miotke and Palmer, 1972). During interglacials and since the last ice age (Holocene time beginning about 10,000 years ago), sediment accumulated in the stream valleys, causing groundwater levels to rise and the lowermost cave passages to flood. Such is the case along the Echo River in the lowest accessible part of Mammoth Cave and in unnamed flooded passages below the level of the Green River.

Cave passage shapes also give clues to the cave history. Below the water table where water completely fills passages in the *zone of saturation, tubular passages* (or *tubes*) with round or horizontally elliptical cross sections form. Above the water table in the *zone of aeration* underground streams follow the gently inclined limestone layers and form *canyon* passages (Fig. 16–6). Initially these canyon passages are narrow. When the streams are given a long period of flow at a particular level, the passages can widen greatly. Some of those in the upper cave level are gigantic— over 50 feet wide (15 m) and 100 feet high (30 m) (Fig. 16–5). Where vertical cracks or joints intersect, groundwater may flow downward and create a vertical shaft or *domepit*. Domepits deepen with time and are up to 200 feet (61 m) from top to bottom in the Mammoth Cave area (Palmer, 1981) and often connect long-abandoned passages with lower passages that are actively forming. Vertical grooves or channels are dissolved as water descends along the domepit walls. Water escapes or drains the domepit through a lower passage. The Ruins of Karnak, Bottomless Pit, Cathedral Dome, and Mammoth Dome are all visited on one or more of the cave tours.

The distribution of passage types in the cave indicates that groundwater moved as underground streams toward the base level, the Green River. Canyon

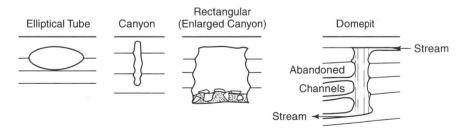

FIGURE 16–6 Cross sections of some passage types found in Mammoth Cave. Based on Palmer (1981) and White and Deike (1989)

passages formed in limestone above the level of the Green River in the zone of aeration, and tubular passages formed simultaneously near the Green River at or below the water table. Changes in base level produced the six distinct levels of development. Deposition of sediment in the Green River channel since the last ice age has raised base level and flooded or partly flooded the lower cave levels where the famous eyeless fish live. Where domepits form, connections are made between cave levels—accounting for the vertically connected stacked network of caves called Mammoth Cave.

A commonly asked question is "How old is Mammoth Cave?" The previous discussion indicates that evolution of surface topography is genetically related to cave development and that the development of the Pennyroyal Plateau and the upper-level cave passages formed simultaneously during late Tertiary time. Thus the beginning of the present caves goes back at least a few million years, perhaps 5 or 10. Travertine deposits in the middle levels of the cave have been radiometrically dated and are as much as 800,000 years old. The earth's magnetic record is recorded in the cave sediment and suggests that some of these sediments are older than 2 million years (Schmidt, 1982). Thus because the upper cave passages must predate the sediments contained in them, cave development must be at least late Tertiary in age.

The older passages have remained intact only where the Big Clifty Sandstone remains as a protective cap. In the Pennyroyal Plateau and in the limestone-floored valleys separating Mammoth Cave Ridge from Joppa and Flint ridges (Fig. 16–1), cave destruction occurs about as rapidly as new caves form. As these limestone valleys deepen and widen, the intervening flat-topped ridges collapse along their edges into the older cave passages. Eventually the Chester Cuesta will take on the same appearance as the Pennyroyal Plateau to the south—a large sinkhole plain lacking today's complex multilevel cave system. Fast forward even further into the future and the soluble limestone will be gone and surface streams will once again appear and begin their particular brand of landscape change.

Cave Minerals—Speleothems

Secondary deposits of minerals in a cave are called *speleothems*. Most caves contain speleothems composed of the mineral calcite (calcium carbonate; $CaCO_3$) derived from descending acidic water containing dissolved calcium. The loss of carbon dioxide (CO_2) from water droplets that enter the cave atmosphere produces deposits of *travertine,* a type of limestone that takes many forms. *Dripstone* varieties include slender hollow tubes called *soda-straw stalactites;* larger icicle-shaped *stalactites* hang from the ceiling; *stalagmites* project upward from the floor; *columns* connect the floor to the ceiling; and *helictites* twist and turn and grow in any direction. *Draperies* form along sloping cave ceilings, and *flowstone* that coats walls and floors resemble frozen rivers and waterfalls. An abundance of soda-straw stalactites ornament the Onyx Chamber, which has been dubbed "The Macaroni Factory." Nearby is the Onyx Colonnade, a series of stalactites, stalagmites, and columns, many highly colored.

Surprisingly, calcite speleothems do not occur everywhere in this extensive cave system. Where present, such as in the Frozen Niagara section, they are spectacular (Fig. 16–7). How is it that speleothem development is limited to certain areas and is not widespread throughout the cave? The key here is the Big Clifty Sandstone, the relatively unfractured, impermeable caprock on the Chester Cuesta. Only near limestone valleys along the edges of the Mammoth Cave Ridge and other ridges is the caprock fractured sufficiently to allow water to enter the underlying cave. Where entering water is charged with calcium, speleothems form in the cave below. If the water entering the ground is acidic, fractures are enlarged or domepits may form.

The most abundant speleothems in Mammoth Cave are composed of *gypsum* (calcium sulfate; $CaSO_4$). Gypsum is more soluble than calcite and forms only in drier cave areas where small amounts of water penetrate the sandstone caprock and emerge slowly from the porous cave walls. Slow growth by evaporation here enables crystals to form snowball-blister forms, gypsum crusts on walls, and graceful curving flowerlike petals. The source of sulfur is believed to be from weathering of pyrite (iron sulfide) contained in nearby sedimentary rocks. Other sulfate minerals present include *mirabolite* ($Na_2SO_4 \cdot 10H_2O$) and *epsomite* ($MgSO_4 \cdot 7H_2O$). The sulfate minerals were the ones sought by the Paleo-Indian miners on their expeditions into Mammoth Cave.

FIGURE 16–7   Frozen Niagara, the largest travertine flowstone deposit in Mammoth Cave. (National Park Service photo)

Fragile Cave Environment

The Mammoth Cave system contains over 336 miles (540 km) of mapped cave and is unquestionably a national treasure. Caves by definition are only human-size openings in rock. Thus smaller but significant openings that are here by the thousands are excluded! Two hundred years of exploration have enabled humans to connect what were originally thought to be separate cave systems beneath Joppa, Mammoth Cave, and Flint ridges. Other passages, either undiscovered, presently lacking entrances that humans can enter or submerged beneath the water table unquestionably exist—a thousand or more miles in total length is likely! Scuba divers have entered some of these drowned passages from springs along the Green River and have reappeared in the Echo River passage in the lower part of Mammoth Cave! Other underwater passages go back 1000 or more feet (305 m) and have yet to be completely explored (Hess and others, 1989). Cave diving is extremely dangerous and should be attempted only by experienced people whose plans are approved by the Park Service.

Mammoth Cave's location south of the boundary of the great ice sheets has minimized disruption in the cave environment and has permitted over 200 species of animals to adapt to this special environment. Some, such as the bats and cave rats, are temporary visitors. Cave crickets are the most abundant and important life forms in the cave. They make forages to the surface for food and help supply the food chain that permits other cave species to exist. Species such as the blind cave fish, crayfish, cave shrimp, and certain millipedes would not survive in the sunlit realm of the surface world—they depend for food on other species or on its introduction by floodwaters that back up into the cave. Forty-two species of *troglobites* (species dependent on caves for survival) inhabit Mammoth Cave, making this one of the most diverse cave environments known.

This underground world is extremely fragile and can easily be destroyed by careless visitors or improper land use in and outside of the park. Breaking speleothems (accidentally or willfully) that are thousands of years old is an obvious problem. However, merely touching a speleothem deposits an oil that inhibits further speleothem growth. Lint brought in by tourists covers and dulls speleothems and cave walls. Electric lighting enables algae to grow and discolor walls and speleothems. Of even greater threat are the dangers imposed beyond the park boundaries. Acid rain generated by burning fossil fuels affects the chemical balance in karst areas. Of more immediate concern is that most of the water beneath the park comes through the karst aquifer from areas many miles away (Fig. 16–3)—leading the Karst Waters Institute, an international nonprofit organization, to identify Mammoth Cave area as one of the top 10 threatened or endangered karst ecosystems in the world (Mylorie and Tronvig, 1998). Some of the water flows beneath Park City and other areas where sewage and other pollutants constantly enter the aquifer and move quickly into the park. The unique fauna including the blind cave fish could be threatened or destroyed by such pollution as it already has in several caves near Mammoth Cave (Quinlan and others, 1986).

Hidden River Cave, not far from Mammoth Cave, was shown commercially until 1943 when, because of the concentration of sewage and also nauseous waste from a nearby creamery, it was forced to close down. A major effort has been launched to prevent Mammoth Cave from meeting a similar fate—a project that deserves vigorous support.

About 10 miles (16 km) of cave are open for public tours—each tour showing some new aspect of the cave environment and history. Surface trails lead through the woodlands and to the Green River where dozens of springs discharge water that may have entered the subterranean drainage system tens of miles away. Just as the caves and their inhabitants are closely connected to and dependent on the surface environment, so too are we dependent on an environment that cannot be modified or changed beyond certain limits. The number of species and the populations of most organisms (except humans) are decreasing worldwide and in some cases plummeting. Have we already exceeded those limits?

REFERENCES

Bridwell, M.M., 1971, The story of Mammoth Cave National Park, Kentucky: Mammoth Cave National History Association, 64 p.

Hess, J.W., and White, W.B., 1989, Water budget and physical hydrology, in White, W.B., and White, E.L., editors, Karst hydrology: Concepts from the Mammoth Cave area: Van Nostrand Reinhold, New York, p. 105–174.

Kennedy, M.C., and Watson, P.J., 1997, The chronology of early agriculture and intensive mineral mining in the Salts Cave and Mammoth Cave region, Mammoth Cave National Park, Kentucky: Journal of Cave and Karst Studies, v. 59, no. 1, p. 5–9.

Miotke, F.-D., and Palmer, A.N., 1972, Genetic relationship between caves and landforms in the Mammoth Cave National Park area: Published by authors, 69 p.

Mylroie, J., and Tronvig, K., 1998, Karst Waters Institute creates top ten list of endangered karst ecosystems: National Speleological Society News, v. 56, no. 2, p. 441–443.

Palmer, A.N., 1981, A geological guide to Mammoth Cave National Park: Zephyrus Press, Teaneck, NJ, 196 p.

Quinlan, J.F., Ewers, R.O., and Palmer, A.N., 1986, Hydrogeology of Turnhole Spring groundwater basin, Kentucky, in Neathery, T.L., ed., Southeastern Section of the Geological Society of America: Geological Society of America, Boulder: CO, Centennial Field Guide Volume 6, p. 7–12.

Schmidt, V.A., 1982, Magnetostratigraphy of sediments in Mammoth Cave, Kentucky: Science, v. 217, p. 827–829.

White, W.B., and Deike, G.H., III, 1989, Hydraulic geometry of cave passages, in White, W.B., and White, E.L., editors, Karst hydrology: Concepts from the Mammoth Cave area: Van Nostrand Reinhold, New York, p. 223–258.

Park Address

Mammoth Cave National Park
Mammoth Cave, KY 42259

SEVENTEEN

The Appalachian Provinces

The closeness of the Appalachian Mountains to population centers, and especially to the early concentration of major universities in the eastern United States, resulted in many geologists being first exposed to the mysteries of mountain building and other geologic problems in an Appalachian setting. This giant outdoor laboratory has not only been the training ground for generations of geologists but has also acted as a proving ground in developing and testing important geologic concepts and techniques. The geosynclinal theory that associates mountains with unusually thick accumulations of sediments and the recognition that landscapes change in regular ways through time were first developed in the Appalachians.

With more studies additional problems surfaced. How could rocks in one area have a completely different age, deformation, and metamorphic history from those immediately adjacent? Why do the Paleozoic sedimentary rocks thicken and become coarser to the east where there is no apparent major landmass to furnish such debris? The answers to these and other questions had to wait for the development of another concept—the concept of plate tectonics.

Provinces and Geology

The term Appalachians, as used here, includes the northern Appalachians of New England and Canada and the four geomorphic provinces of the central and southern Appalachians that extend southwest from the New England Province. Together, these areas cover about 10 percent of the conterminous United States. From west to east the southern provinces are: (1) Appalachian Plateaus, (2) Valley and Ridge (Folded Appalachians), (3) Blue Ridge, and (4) Piedmont provinces (Plate 17, Fig. 17–1). Because the sequence of geologic events is generally similar throughout, the following discussion includes New England.

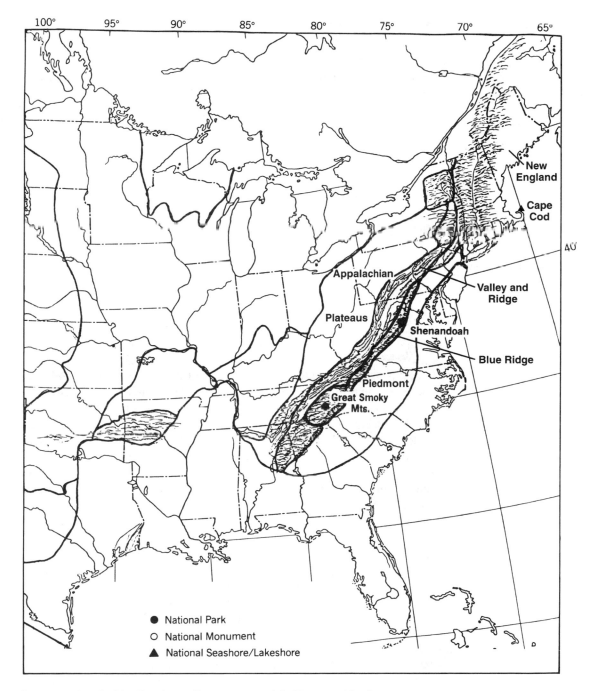

PLATE 17   Appalachian Provinces. (Base map copyright Hammond Inc.)

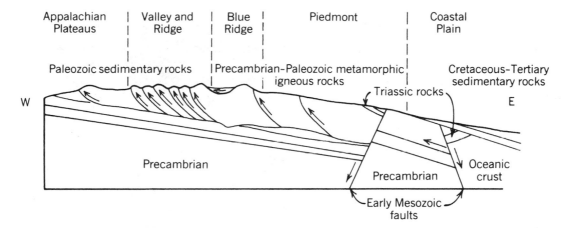

FIGURE 17–1   Generalized west-to-east cross section, showing major structures and rock types in the Appalachian Mountains.

The Fall Line—an important boundary in the eastern United States—is the boundary between the hard Precambrian-Paleozoic metamorphic and igneous rocks of the Piedmont and the weaker, easily eroded Mesozoic and Cenozoic sedimentary rocks of the Coastal Plain. On a number of streams, notably the Delaware and the Potomac, the Fall Line is marked by falls and rapids (Fig. 17–2). In some places, the Fall Line is less distinct, where rapids are found over reaches of several miles. In essentially all cases, whether falls or rapids, boat transportation was interrupted at the Fall Line; consequently, towns—which grew into cities such as Baltimore, Richmond, and Raleigh—were located at the "end of the run." This is an excellent example of effect of bedrock geology on the topography, which in turn affected human use and development of the region.

The low, rolling Piedmont, like the topographically rugged New England Province to the north, is composed of similar Precambrian and Paleozoic metamorphic and plutonic rocks. In addition to the schists, gneisses, slates, and granite plutons, elongate grabens contain Triassic-Jurassic sandstones, shales, and basalt. Although the Piedmont is generally a lowland area that rises to 980–1480 feet (300–450 m) above sea level at the foot of the Blue Ridge Province to the west, it reaches as much as 1800 feet (552 m) in the Dahlonega Plateau of Georgia.

The Blue Ridge rises abruptly above the Piedmont and forms the backbone of the Appalachians. Although somewhat subdued as compared to the mountains of the West and Alaska, the Blue Ridge contains some exciting topography—precipitous cliffs, spires, and waterfalls, particularly in the southern part. Because the only national parks in the Appalachians are in the Blue Ridge Province, a separate section is devoted to it.

FIGURE 17–2　Falls of the Potomac, at the boundary of the Piedmont and Coastal Plain provinces, near Washington, D.C. (Photo by B. Kiver)

West of the Blue Ridge lies a narrow system of limestone-floored valleys collectively called the Great Valley, or Appalachian Valley, that belongs to the Valley and Ridge Province. The Shenandoah, Cumberland, Lebanon, and other valleys are part of a connected system of valleys that extends from Alabama into Canada. Both the Valley and Ridge and the Appalachian Plateaus to the west contain folded, unmetamorphosed sedimentary rocks and are underlain by large thrust faults produced during a collision of continents during late Paleozoic time. The intensity of deformation decreases westward away from the Blue Ridge and Piedmont. Rocks thrust westward onto the more stable interior of the continent are tightly folded in the Valley and Ridge Province and have broad, open folds in the Appalachian Plateaus (Fig. 17–1). Most of the rocks in the Valley and Ridge are early Paleozoic sandstones, limestones, and shales, although Pennsylvanian sandstones and coal beds are common at the north end.

The Appalachian Plateaus on the western side are the most extensive of the Appalachian provinces and occupy a broad synclinal basin that is up to 220 miles (350 km) wide. The plateaus consist mainly of flat-lying sandstones, conglomerates, shales, and coal beds of middle and late Paleozoic age. Several broad folds and faults are present, but this part of the geosyncline was least affected by the mountain building and metamorphism that so profoundly changed the rocks in most of the geosynclinal area. The Devonian- to Pennsylvanian-age rocks thicken substantially from west to east—a reflection of sediments shed from mid-Paleozoic-age mountains in the Piedmont and Blue Ridge areas. A dissected mountain front separates the Plateau from the eastern provinces and a lower out-facing escarpment marks the province boundaries in most areas along the west edge (Fig. 17–3) as described by Thornbury (1965).

A well-developed *dendritic* drainage pattern characterizes the area and easily distinguishes it from the linear, *trellised* pattern of the Valley and Ridge Province to the east. Along the eastern border, dissection is sufficiently complete to form "mountains of erosion" such as the Alleghenies and the Catskills.

Small but interesting areas containing unusual plutonic rocks called *kimberlites* occur in the plateau as well as in other parts of the Appalachians. These dark-colored dense rocks formed when mantle-derived magma moved rapidly upward through deep fractures and cooled. The fractures and release of pressure that triggered magma production resulted from crustal tension that followed the last of the

FIGURE 17–3   Devonian shales exposed near the west edge of the Appalachian Plateaus Province near Cleveland, Ohio. Outfacing or west-facing scarp of province is nearby. (Photo by E. Kiver)

Appalachian orogenies (Figs. 17–4*h* and 17–4*i*). Dennison (1983) believes the tensional fractures formed because of tectonic uplift or isostatic uplift after some of the overlying sedimentary rocks were eroded away, whereas Parrish and Lavin (1982, 1983) suggest that separation of North America, Europe, and Africa by plate movements in late Paleozoic and Mesozoic times created the necessary tension. Kimberlites—such as those in South Africa, in the Murfreesboro, Arkansas area in the Ouachita Province, and in northern Colorado—often have diamonds associated with them. Many gem-quality stones have been found in stream deposits in widely separated areas of the Appalachians; however, their source localities remain a mystery (Sinkankas, 1959).

Birth of a Mountain Range

The Appalachian Geosyncline, or *geocline,* played a key role in the formation of the Appalachian Mountains. Although the concept of geosynclines was first recognized by James Hall, who worked in the Appalachian region in the 1850s, a more thorough understanding of the mountains and the forces creating them had to await many more detailed investigations and the acceptance of plate tectonics concepts. Hall realized that the accumulation of sedimentary rocks thickened markedly as one approached the Appalachian Mountains from the west, thereby recognizing that areas with thick sedimentary sequences later become mountains. He did not specify adequately the mechanics of how these extremely thick accumulations of sediments become high mountains. This led some geologists to remark that his was a theory of mountain building with the mountains left out! Others proposed that the earth was cooling and shrinking, thus accounting for the depression, or geosyncline, and for the horizontal forces so evident in most structural features in mountainous areas. With the realization that the earth is not losing appreciable heat and that plates of the earth's crust move laterally along its surface, a new interpretation of the geologic development of the Appalachian area became necessary.

The broad picture of plate movements now recognized in the Appalachians is both exciting and startling. We know that certain continents and plates are today moving toward each other, while others are splitting and moving apart in other parts of the world. It seems likely that many collisions and much breaking apart of plates must have occurred in the past. Indeed, it is a complex sequence that was involved in the location, timing, and deformation recorded in the rocks and structures in the Appalachians. The following summary is greatly simplified, but more detail can be found in numerous journal articles. Papers in the Appalachian–Ouachita Orogen volume edited by Hatcher and others (1989) are particularly useful. A study of the diagrams in Figure 17–4 will help one visualize the general sequence of plate tectonic events that produced the Appalachian Mountains.

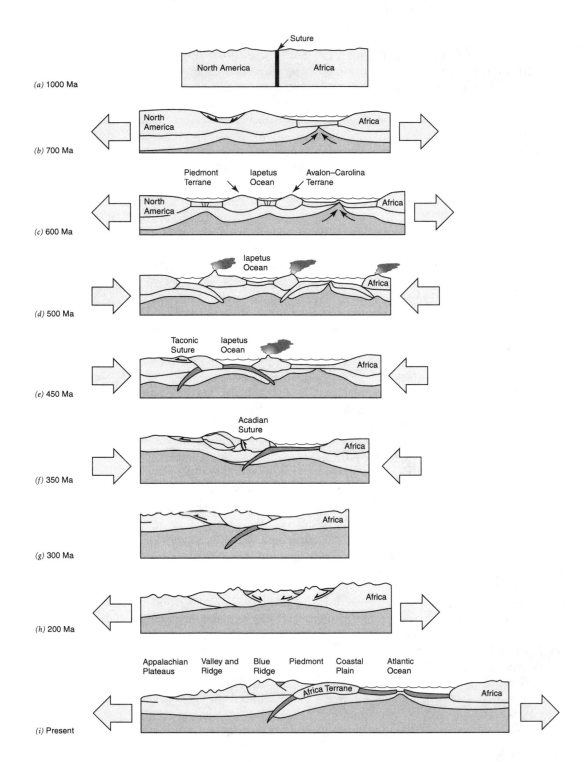

A Late Precambrian orogeny (the Grenville Orogeny) deformed and metamorphosed the rocks along the eastern edge of the plate from about 1300–1000 Ma (Fig. 17–4*a*), perhaps resulting in one or more supercontinents containing most of the continental crustal areas then in existence. Rising heat trapped beneath the extensive crustal blanket eventually overcame the mechanical strength of the crust and produced a rift along what was soon to become the Appalachian Geosyncline. Pull-apart basins in the Shenandoah area filled with flood basalts from about 800 to 1000 Ma as the continent continued to stretch and fragment. Other areas farther south in the Smoky Mountain area received tens of thousands of feet of sediment as the newly formed Iapetus Ocean developed. One crustal section would eventually become the North American plate, another would become Africa (Fig. 17–4*b*) with one or more seaways between. Other smaller crustal fragments (*microplates*) and volcanic arcs (Fig. 17–4*c*) would play important roles in the mountain building that was soon to follow.

During early Paleozoic time (Ordovician, about 500 Ma) plates began to move toward one another—changing the *passive plate margins* to *active margins* (Fig. 17–4*d*) and beginning the elimination of the ancient seaways. Although the Iapetus Ocean lay off the east coast of North America, it was not the present Atlantic Ocean—much was yet to happen before the present Atlantic would come into being. With unimaginable slowness by human standards, huge plates containing North America, Africa, Europe, and South America crept ever closer to each other. The closing of the giant vise occurred sooner in the northern Appalachians as large microplates episodically docked and produced orogenic events at different times (Figs. 17–4*e* and 17–4*f*). The Ordovician–Silurian event is part of the Taconic Orogeny (about 420 Ma) and the Devonian event is called the Acadian Orogeny (about 350 Ma). The timing and significance of these events is different between the Northern and Southern Appalachians. The final event began in the Southern Appalachians about 270 Ma during late Paleozoic time (Pennsylvanian-Permian) when the African, European, North American, and South American plates collided

FIGURE 17–4 (left)   Simplified time sequence diagrams of the Southern Appalachian mountain-building history. Differences between the Northern and Southern Appalachians include the timing of events. Acadian was most important in the Northern Appalachians and Taconic and Alleghanian were most important in the Southern Appalachians. (After Hatcher, 1989) (*a*) Continents welded together during Grenville Orogen 1300–1000 Ma. (*b*) Continents separate, proto-Atlantic Ocean (Iapetus Ocean) forms. (*c*) Microplates form between continents, quiet trailing-edge conditions prevail. (*d*) Island arcs form, plates move toward each other. (*e*) Piedmont microplate or Carolina microplate collides with North America; Taconic Orogeny results. (*f*) Avalon–Carolina microplate collides, Acadian Orogeny (Devonian-Mississippian) most important in Northern Appalachians. (*g*) Continent–continent collision, Alleghanian Orogeny (Pennsylvanian-Permian), Pangaea supercontinent forms. (*h*) Pangaea breakup begins; fault basins form and are filled with sediment and volcanics. (*i*) Continents separate and form Atlantic Ocean producing the trailing-edge conditions that continue to the present.

and sutured together to form the supercontinent of Pangaea (Fig. 17–4g). This final orogeny is known as the Alleghanian Orogeny in the Appalachians.

The collisions left their marks on the plate edges in the form of faults and folds. Rocks subjected to intense pressure and heat changed to the gneisses, schists, and other metamorphic rocks of the New England, Blue Ridge, and Piedmont provinces. The collision also produced huge thrust faults when large blocks of crust were shoved onto the continent edge like so many slabs of ice driven ashore and stacked against each other by the wind. The sedimentary rocks away from the plate edges in the Valley and Ridge Province and Appalachian Plateaus were less affected, but they too were changed—deformed by folds and faults.

Although structures at the surface were mapped many years ago, controversy arose concerning what happened to these structures at depth. All agreed that a relatively thin slice of crust (thin-skinned tectonics) was thrust over more rigid basement rock in the Appalachian Plateaus and Valley and Ridge Province. Most workers a few years ago believed that these thrusts were analogous to sheets of toothpaste and that the tube or root of the mountains underlies the Blue Ridge and Piedmont. Information from recent drilling and geophysical studies strongly indicates that thin-skinned deformation also extends under the entire Blue Ridge and at least part of the Piedmont (Fig. 17–1). Whether shallow thrust faults end under the eastern part of the Piedmont or continue under the Coastal Plain cannot be determined yet with certainty. How far the thrusting continues could have important economic consequences. If some of the oil-rich rocks in the Valley and Ridge extend eastward beneath the thrust slices, then the potential for oil in what were assumed to be barren areas could be high.

At the end of Paleozoic and during early Mesozoic time, what is now the east edge of the North American Plate was an impressive range of mountains in the interior of Pangaea. Again the thermal energy that drives the plate tectonics machine began to pull the plates apart. Large fault-block valleys, or grabens, extend from Canada to Florida in the Piedmont, in New England, and beneath the Coastal Plain Province. Late Triassic and Early Jurassic rivers washed debris into the troughs, and large numbers of dinosaurs roamed the Connecticut Valley, New Jersey, and other areas; they left their tracks in the muddy sediments, particularly on the edges of freshwater lakes. Deep-reaching fractures channeled basaltic magma toward the surface where it formed dikes, sills, and lava flows in the grabens. The Palisade Sill along the Hudson River west of New York City is one of the most striking and famous of these intrusions. The splitting apart of the continents continued, and by mid-Mesozoic time marine sediments were being deposited on the coastal plains along the newly opened ocean that would later be named the Atlantic.

But where is the rest of the mountain range that formed during the late Paleozoic? The intensity of metamorphism and crustal thickness is greatest in the Blue Ridge Province area. There should be more mountains to the east—counterparts to the Valley and Ridge and Appalachian Plateaus. The northeast-trending Appalachian Mountains abruptly end in Newfoundland at the edge of the Atlantic, and the metamorphic roots of the Appalachians disappear eastward beneath the

Coastal Plain. The missing parts of the mountains are there—however, one has to visit the British Isles, Norway, and Africa to find the continuation of the mountains and their missing components.

Thus, during Mesozoic time eastern North America became the trailing edge of the westward-moving continent, thereby ushering in the tectonic quiet that persists there today. However, the relentless force of erosion will continue to tear away at the mountains, and the stresses that move the plates will in the very distant future produce another geosyncline and ultimately another generation of mountains.

Appalachian Geomorphology

Not only did the Appalachians inspire the geosynclinal idea that mountains form in areas of unstable crust where thick sediment accumulations form, but many ideas relating to the erosional history of mountains were developed and tested in the Appalachians. A classic study published by the noted geomorphologist William Morris Davis in 1889 suggested that low relief surfaces in the Piedmont and Plateaus provinces resulted from long intervals in which sea level, or *base level,* had remained stable and streams had erased many of the topographic irregularities in the landscape and lowered the surface close to sea level. Such a surface formed under humid climate conditions is called a *peneplain.* Considerable debate followed over the origin of such surfaces—a debate that is not yet completely resolved. Thornbury (1965) provides an excellent historical review of ideas.

Whether or not base leveling produced these low-relief surfaces and accordant elevations of ridges in an area, topographic elevations are closely tied to the ability of the underlying rock to resist weathering and erosion (Hack, 1989). Thus, higher peaks and surfaces are usually underlain by sandstone, quartzite, or slightly metamorphosed basalt. Shale, slate, and in this humid climate, limestone underlies most of the valleys. These observations support the hypothesis that rock characteristics can adequately account for low-relief surfaces and accordant elevations of ridges.

For example, in the Valley and Ridge the topography is clearly related directly to rock type and structure. The anticlines and synclines are truncated by erosion—valleys are carved into the shales and limestones; the resistant sandstones, quartzites, and dolomites stand high as ridges. Thus, mountain ridges are on the resistant rocks on the flanks of an anticline or syncline. Seldom are ridge tops on the crest of an anticline and a valley bottom along the axis of a syncline. In fact, more major valley bottoms are along the axes of anticlines than along the axes of synclines.

When the Piedmont is seen from the Blue Ridge, with few exceptions it appears as a monotonous plain (Fig. 17–5). Actually, the topography holds interest both esthetically and geologically. Much of the area is a gently rolling plain but numerous ridges and ravines occur on the deeply weathered bedrock (Hack, 1989).

FIGURE 17–5   View of Piedmont from Blue Ridge Parkway in Virginia. (Photo by E. Kiver)

Unusually resistant rocks such as quartzite or quartz-rich igneous rocks underlie higher areas. One well-known prominent hill is the site of Monticello, near Charlottesville, Virginia; another is the granite knob, Stone Mountain, in Georgia. These and similar hills are interpreted by some as *monadnocks* or hilly remnants on a peneplain that have not yet been dissected by streams. John Hack (1989) points out that all of the higher hills that he examined owe their elevation to the more resistant rocks that underlie them and are therefore the result of a continuously eroding topography. According to this interpretation, weathering and erosion are continuous, and differences in rock characteristics account for local differences in elevation. Whatever the exact history of erosion may be, only by being spared rapid erosion for a long period could the Piedmont have developed a 45- to 60-foot-thick (15- to 20-m) mantle of weathered soil. Soil up to 330 feet (100 m) thick occurs in some areas (Hack, 1989).

Rock structures and their consequent topography predetermined early transportation routes, making travel in the Valley and Ridge Province both tedious and tortuous. The roads follow the valleys between the linear ridges; roads connecting those valleys are widely spaced and confined mainly to the water gaps where streams have cut through the ridges. Now, however, a few major highways such as the Pennsylvania Turnpike cut across the ridges. Some of the rock structures, as shown in the much-simplified cross section of Figure 17–1, can be seen in the huge road cuts.

Other major questions arose concerning the drainage history in the Appalachians. The original drainage was westward. However, in the northern Appalachian

Plateaus the drainage divide is well to the west—suggesting a major reversal in drainage direction. Rivers such as the Susquehanna, Potomac, and Delaware flow eastward *across* the resistant ridges of the Appalachians through a series of *water gaps* such as that found at Delaware Water Gap National Recreation Area (Fig. 17–6). Understanding the history of these water gaps is intricately related to the evolution of the drainage system.

Again, numerous hypotheses (summarized in Thornbury, 1965, and Hack, 1989) attempt to explain the present-day drainage direction. Because of hundreds of millions of years of erosion, evidence of older landscapes disappears as new landscapes are carved from the bedrock. Rivers change location as erosion continues, making a detailed reconstruction of the drainage history, especially for older events, difficult or impossible. However, one process that explains at least some of the drainage changes is *stream piracy*.

Streams flowing eastward have a shorter route to the sea than those flowing westward. On average, eastward-flowing streams drop faster (have a higher gradient) and erode headward into the drainage divides more rapidly than westward-flowing

FIGURE 17–6   View downstream to Delaware Water Gap, with New Jersey's Mt. Tammany beyond (left), Delaware Water Gap National Recreation Area, near Stroudsburg, Pennsylvania. The Delaware River has cut a gap through the steeply dipping, highly resistant Tuscarora Quartzite ridge, forming one of the numerous enigmatic examples of transverse drainage in the Appalachians. (Photo by National Park Service)

streams. Thus a series of stream piracies or drainage diversions have occurred where east-flowing streams have become larger by capturing the headwaters of west-flowing streams. Further piracies can in turn divert streams from the watergaps. As the rivers continue to deepen their valleys, these former watergaps are left high and dry as *windgaps*—areas where streams no longer flow completely through the gap. Numerous examples of geologically soon-to-happen piracies exist throughout the Appalachians. An excellent example of an abandoned watergap, a wind gap, is found at Cumberland Gap National Historical Park on the Kentucky–Virginia–Tennessee border.

Where limestone (calcium carbonate) or *dolomite* (a magnesium-rich carbonate rock) is at or near the surface, extensive karst areas and caverns form. As a reminder, although limestones generally form ridges in the arid West, here they weather and erode rapidly, and thus form valleys. Here also, the abundance of subsurface water in this humid climate is actively dissolving out the calcium carbonate, forming large cavern systems in the limestones. Sinkholes, disappearing streams, and extensive cave systems occur in many areas—especially in areas where Mississippian-age limestones occur. Thousands of caves occur in the Appalachian Plateaus and Valley and Ridge provinces. A number of commercial caves are in operation, but only Russell Cave National Monument in the extreme northeast corner of Alabama in the Valley and Ridge Province is operated by the National Park Service.

Also widely known are Natural Bridge (Fig. 17–7), which supports highway traffic near Lexington, Virginia, and Natural Tunnel, which is the route of the Southern Railway near Clinchport, Virginia. Both were developed by subsurface water. Natural Bridge represents a more advanced stage in the collapse of cavern roofs than Natural Tunnel.

Glaciers significantly altered the entire New England Province and modified the northern Appalachian Plateaus and northern Piedmont provinces in New York, New Jersey, Pennsylvania, and Ohio (Fig. 13–2). Erosion was severe in New England and locally severe in areas such as the Finger Lakes of upstate New York. Large, extensive moraine systems such as those in the Great Lakes area are generally absent, but some *drumlins* (ridges of till and other material elongated in the direction of ice movement), *eskers* (sinuous ridges deposited by streams flowing beneath a stagnant sheet of ice), and some ice-marginal moraines, deltas, and glacial-lake deposits occur. The crust subsided beneath the weight of the ice, enabling salt water to encroach into areas such as the Saint Lawrence Valley during ice retreat. Crustal rebound eventually destroyed this arm of the sea as the land again rose above sea level. When ice retreated north of the Saint Lawrence Valley a new outlet to the sea for the Great Lakes was established.

Although glaciers never moved very far south of the northeastern part of Pennsylvania and northern New Jersey, climates to the south, especially those closest to the ice margin were substantially modified during glacial episodes. In these *periglacial* climates, frost action was severe, and jumbled areas of boulders called *block fields* formed. Those on steeper slopes moved slowly downslope during colder times to form *block streams* and *stone stripes*. Severe frost action can also form *stone polygons* where lines of rocks surrounding finer sediment form five- or six-

FIGURE 17–7   Natural Bridge in the Valley and Ridge Province near Lexington, Virginia. Because of its historic and geologic interest, Natural Bridge compares favorably with many of our national monuments. (Photo by Barbara Kiver)

sided polygonal arrangements. The polygons occur in groups and are up to 30 feet (10 m) in diameter (Denny, 1956). Such features are actively forming today in parts of Alaska and other areas where glacial or near-glacial conditions exist. As would be expected of periglacial features, their abundance decreases to the south where climatic changes were less severe.

REFERENCES

Davis, W.M., 1889, The rivers and valleys of Pennsylvania: National Geographic Magazine, v. 1, p. 183–253.

Dennison, J.M., 1983, Comment on "Tectonic model for kimberlite emplacement in the Appalachian Plateau of Pennsylvania": Geology, v. 11, p. 252–253.

Denny, C.S., 1956, Surficial geology and geomorphology of Potter County, Pennsylvania: U.S. Geological Survey Professional Paper 288, 72 p.

Hack, J.T., 1989, Geomorphology of the Appalachian Highlands, in Hatcher, R.D., Jr., Thomas, W.A., and Viele, G.W., eds., The Appalachian-Ouachita Orogen in the United States: Geological Society of America, Boulder, CO, The Geology of North America, v. F-2, p. 459–470.

Hatcher, R.D., Jr., 1989, Tectonic synthesis of the U.S. Appalachians, in Hatcher, R.D., Jr., Thomas, W.A., and Viele, G.W., eds., The Appalachian-Ouachita Orogen in the United States: Geological Society of America, Boulder, CO, The Geology of North America, v. F-2, p. 511–535.

Parrish, J.B., and Lavin, P.M., 1982, Tectonic model for kimberlite emplacement in the Appalachian Plateau of Pennsylvania: Geology, v. 10, p. 344–347.

Parrish, J.B., and Lavin, P.M., 1983, Reply to comment on "Tectonic model for kimberlite emplacement in the Appalachian Plateau of Pennsylvania": Geology, v. 11, p. 254–256.

Sinkankas, J., 1959, Gemstones of North America: Van Nostrand Reinhold, New York.

Thornbury, W.D., 1965, Regional geomorphology of the United States: Wiley, New York, 609 p.

BLUE RIDGE PROVINCE
(VIRGINIA AND NORTH CAROLINA)

The Blue Ridge is the backbone of the Appalachians, rising distinctly above the adjacent provinces and extending southwestward from southern Pennsylvania into northern Georgia. The term *Blue Ridge* was first used in the northern section where it is a single prominent ridge, in places flanked by lower ridges, and is 250 miles (400 km) long and about 9 miles (15 km) wide. South of the Roanoke River in Virginia, the Blue Ridge is 350 miles (560 km) long and up to 75 miles (120 km) wide and consists of several irregular ridges, of which the Great Smoky Mountains are outstanding (Hack, 1989).

For the early settlers who were intent on pushing westward, the Blue Ridge was the "front range," the first real physical barrier encountered on their westward trek. It rises abruptly above the gently rolling hills of the Piedmont, in places as much as half a mile. With an essentially impenetrable forest, water gaps such as those along the Shenandoah and the Potomac rivers were all-important gates to exploration. Therefore, the geologic development of the water gaps had a profound effect on the exploration and settlement of this area, which at that time was "The West." But there are no water gaps through the Blue Ridge south of Roanoke, Virginia. Consequently, penetration into the southern Blue Ridge was slow; and once there, having given so much of themselves in establishing their homes, the settlers and their offspring remained, forming an island of tranquility surrounded by the swirling turmoil known by some as progress. It was an exercise in survival not just for a weekend but for a lifetime, and these isolated mountain people were extremely ingenious in making their own tools and in developing their own culture. The life of the past continues today in certain areas, and park visitors should not fail to take advantage of the opportunity to see it in Cades Cove in Great Smoky Mountains National Park.

Geologic Story of the Blue Ridge

Early interpretations of Blue Ridge geology were controversial, chiefly because of the lack of data on the ages of the older rocks. Today we have radiometric dates on many of the rocks, but even so, the precise sequence of development of the older rocks—those more than a billion years old—has yet to be completely established. Therefore, the metamorphic and granitic rocks are together referred to as the Basement Complex (King and others, 1968).

The Basement Complex formed as part of the most recent episode of Precambrian mountain building (Grenville Orogeny) along the eastern edge of the Laurentian Plate (Fig. 17–4*a*)—a plate that contained North America and other land masses. Sedimentary rocks, some of which were derived from rocks as old as 1870 Ma, as determined from minerals whose "radiometric clocks" were not reset, were metamorphosed and intruded by plutonic rocks prior to and during the intense Grenville metamorphism that produced the high-grade gneisses of the Blue Ridge. Radiometric ages indicate that Precambrian metamorphism, and therefore the Grenville Orogeny, occurred between 1000 and 1200 Ma (Rankin and others, 1989).

Granitic magmas intruded into the Basement Complex in several places in the Appalachian area; one example is the Old Rag Granite in Shenandoah National Park, now known to be about 1.1 billion years old. This determination was made using uranium–lead analyses of zircon, one of the accessory minerals in the granite (Lukert, 1982). Later, after prolonged erosion, sedimentary rocks were laid down in many places; elsewhere, basaltic lavas and volcanic ash covered wide areas. The sedimentary and volcanic rocks formed the floor of the Appalachian Geosyncline, which developed at or near the end of the Precambrian, when the continental landmass rifted and a narrow ocean, the Iapetus Sea, formed along what is now the eastern edge of North America (Fig. 17–4*c*).

As the Cambrian seas advanced onto what was now the *trailing edge* of the North American Plate, coarse clastics were laid down near shore while the silts and clays were carried farther out and deposited in deeper water. Later in the Cambrian and during the Ordovician, several thousand feet of limestone and dolomite were laid down. Conditions in the Appalachian Geosyncline changed dramatically during early Paleozoic time when the direction of plate movements reversed, causing the plates to move toward each other. The Iapetus seaway would eventually be eliminated as a series of mountain-building events occurred episodically through the remainder of the Paleozoic.

Initially geologists believed that deposition continued without interruption until near the end of the Paleozoic, when the Appalachian Orogeny formed the Appalachian Mountains. Later, it became evident that there was not one single orogeny. Rather, significant tectonic disturbances occurred periodically, beginning with the Taconic Orogeny in Late Ordovician time. The docking of a volcanic-island chain as a *microplate,* or *exotic terrane,* a detached islandlike mass off of the North American coast, provided the energy to thrust masses of basement rock and the overlying later Precambrian basalts and early Paleozoic sedimentary rocks westward onto the continent (Fig. 17–4e). Thus the major framework for what was to become the Blue Ridge was established following the Taconic Orogeny.

Docking of a major microplate, the Avalon-Carolina exotic terrane, occurred during the Acadian Orogeny, mostly in Devonian time and mostly affecting the Northern Appalachians. The timing of this event varies in different parts of the Appalachians and is not well constrained in the Southern Appalachians. Dates of metamorphism and intrusion of granitic rocks range from 315 to 430 Ma (Osberg and others, 1989). The final Appalachian orogeny, the Alleghanian, produced large thrust sheets and folds in the Valley and Ridge Province and Appalachian Plateaus to the east as the North American, African, and European continents collided during late Paleozoic time. During this latter orogeny, the Blue Ridge and Piedmont provinces were thrust westward as an intact crystalline body—transferring stresses to the thick pile of Paleozoic sedimentary rocks to the west and producing the large thrust faults and anticlines and synclines in the Valley and Ridge and Appalachian Plateaus (Hatcher and others, 1989).

So intense were the pressures during the orogenies that most of the Blue Ridge rocks were highly metamorphosed, thus forming schists, phyllites, quartzites, greenstones, and gneisses. In places, however, the original shales, sandstones, lava flows, and granites were less altered, even essentially unaltered.

The Alleghanian Orogeny was particularly severe because it involved collision of two major crustal plates: the North American and the Gondwanan (Africa, South America, Australia) plates. The ocean between the plates was eliminated, and large-scale thrusting of geosynclinal rocks occurred toward the interior of the continent. Both the Blue Ridge and at least some of the Piedmont are part of a giant thrust sheet that is believed to have been shoved at least 163 miles (260 km) to the west (Cook and others, 1979, Harris and Bayer, 1979). The greater elevation of the Blue Ridge as compared to other Appalachian areas may be due in part to subsidiary faults that branch steeply from the main overthrust and bring more resistant crystalline rocks to the surface (Fig. 17–1).

Geomorphology

As the mountains were being thrust up, erosion was at work tearing them down. The topography of the Blue Ridge is the result of prolonged stream erosion. Flattish summit and subsummit areas have been interpreted by most workers to represent the final products of stream erosion cycles. However, the number of cycles has long been a matter of controversy. Those interested in the full account are referred to Fenneman (1938) and Thornbury (1965). The classical interpretation involves three erosion cycles, the last of which is still in progress. Remnants of two peneplains are recognized by many workers, but there is diversity as to the degree of perfection and the age of each of the two. More recently, even the cyclic development concept was challenged. Hack (1960, 1989) proposed instead a "dynamic equilibrium" theory that emphasizes the significance of rock resistance to weathering and erosion; in this way, he attempts to explain the flattish areas at different elevations.

In a recent summary of his observations and ideas concerning the Appalachian areas, Hack (1989) reaffirms the correlation of rock resistance and elevation but also stresses that recent, probably continuous, uplift is necessary to maintain significant elevation of a landmass. Although many geologists do not question the importance of rock resistance, there remains the question as to whether this idea replaces or merely supplements the cyclic development concept.

There is almost a complete absence of sharp peaks in the Blue Ridge, even though in many places the side slopes are very steep. Instead, the tops are characteristically rounded, and the term *dome* is widely used. In this warm, humid climate, chemical weathering is effective in rounding off the sharp corners and edges.

Glaciation, which profoundly changed the mountains of New England, did not affect the Blue Ridge. The colder climate that prevailed at times during the Pleistocene, however, left its mark on the Blue Ridge in the form of frost-heaved boulders and other features.

Within the Blue Ridge, there are two national parks—Shenandoah in the northern section and Great Smoky Mountains in the southern part. Another beautiful unit of the National Park System is the Blue Ridge Parkway, a 469-mile-long (756 km) scenic highway that connects the two parks.

REFERENCES

Cook, F.A., Albaugh, D.S., Brown, L.D., Kaufman, S., Oliver, J.E., and Hatcher, R.D., Jr., 1979, Thin-skinned tectonics in the crystalline southern Appalachians: COCORP seismic-reflection profiling of the Blue Ridge and Piedmont: Geology, v. 7, p. 563–568.

Fenneman, N.M., 1938, Physiography of eastern United States: McGraw-Hill, New York.

Hack, J.T., 1960, Interpretation of erosional topography in humid temperate regions: American Journal of Science, v. 258-A, p. 80–97.

Hack, J.T., 1989, Geomorphology of the Appalachian Highlands, in Hatcher, R.D., Jr., Thomas, W.A., and Viele, G.W., eds., The Appalachian-Ouachita Orogen in the United States: Geological Society of America, Boulder, CO, The Geology of North America, v. F-2, p. 459–470.

Harris, L.D., and Bayer, K.C., 1979, Sequential development of the Appalachian Orogen above a master decollement—a hypothesis: Geology, v. 7, p. 568–572.

Hatcher, R.D., Thomas, W.A., Geiser, P.A., Snoke, A.W., Mosher, S., and Wiltschko, D.V., 1989, Alleghanian Orogen, in Hatcher, R.D., Jr., Thomas, W.A., and Viele, G.W., eds., The Appalachian-Ouachita Orogen in the United States: Geological Society of America, Boulder, CO, The Geology of North America, v. F-2, p. 233–318.

King, P.B., Neuman, R.B., and Hadley, J.B., 1968, Geology of the Great Smoky Mountains National Park, Tennessee and North Carolina: U.S. Geological Survey Professional Paper 587, 23 p.

Lukert, M.T., 1982, Uranium-lead isotope age of the Old Rag Granite, northern Virginia: American Journal of Science, v. 282, p. 391–398.

Osberg, P.H., Tull, J.F., Robinson, P., Hon, R., and Butler, J.R., 1989, The Acadian Orogen, in Hatcher, R.D., Jr., Thomas, W.A., and Viele, G.W., eds., The Appalachian-Ouachita Orogen in the United States: Geological Society of America, Boulder, CO, The Geology of North America, v. F-2, p. 179–232.

Rankin, D.W., Drake, A.A., Jr., Glover, L., III, Goldsmith, R., Hall, L.M., Murray, D.P., Ratcliffe, N.M., Read, J.F., Secor, D.T., Jr., and Stanley, R.S., 1989, Pre-orogenic features, in Hatcher, R.D., Jr., Thomas, W.A., and Viele, G.W., eds., The Appalachian-Ouachita Orogen in the United States: Geological Society of America, Boulder, CO, The Geology of North America, v. F-2, p. 7–100.

Thornbury, W.D., 1965, Regional geomorphology of the United States: Wiley, New York.

SHENANDOAH NATIONAL PARK (VIRGINIA)

Shenandoah National Park in north-central Virginia, along with its southern continuation, the Blue Ridge Parkway, which in turn connects with Great Smoky Mountains National Park in North Carolina, provides a convenient and unique opportunity for those living in the large population centers of the East to drive and hike through the delightful forests along the backbone of the Appalachians—the Blue Ridge Mountains (Plate 17). The Blue Ridge holds key parts of the geologic story of the Appalachian Highlands and also interesting elements in the nation's history.

Park History

Projectile points and other archeologic evidence indicate that Native Americans used the area for hunting-and-gathering activities for at least 10,000 years (Crandall, 1992). The abundant game and the nuts from the chestnut and hickory trees were major food sources for the Indians as well as the early European settlers. The Monacan and Manahoac Piedmont tribes were the most recent Indians to use the area. Their way of life left no lasting effect on the land or the environment and could have been perpetuated indefinitely. However, the exploding Euro-American population and their technological developments would soon end their life-style.

The establishment of the Jamestown settlement in 1607 had little effect on the Appalachian Highlands for over 100 years. Explorers had reached the crest of the Blue Ridge by 1669, but settlers did not begin to cross over into the Shenandoah Valley just west of the Blue Ridge until about 1725. By 1830 large game animals such as the woodland bison, elk, and deer had been exterminated and the poor mountain soils were becoming even less productive. Farms in the lower elevation Shenandoah Valley and Piedmont prospered while the Blue Ridge mountain people lived a marginal existence. During the Civil War the Shenandoah Valley acted as the "breadbasket of the Confederacy" by producing significant amounts of food for the army and people of the South. No such surpluses could be produced from the small mountain farms. However, the mountain people were a determined lot who were drawn to their mountains in spite of economic difficulties.

By 1910 lumber companies had stripped away much of the hardwood forest, thereby increasing erosion of the already poor soils. Importation of chestnut trees from China also brought a blight to which the American chestnut had no resistance. The loss of all of the mature chestnut trees between 1910 and 1934 (about 20 percent of the forest) removed yet another supply of food for the hardy mountain people.

Resorts began to appear in the early 1900s, and visitors were exposed to the charm and beauty of the area—the true value of the area and the one that could furnish a sustainable way of life for many people. President Herbert Hoover loved the area and used it as an escape from the hot summer weather in Washington D.C. and as a fishing camp where he entertained important guests at what is known today as Camp Hoover.

A fortuitous coincidence of events led to the establishment of a national park. Most of the mountain people who had eked out a bare living on the steep Blue Ridge slopes had finally moved out, abandoning their worn-out, badly eroded farmlands. The lumber, mineral, and most of the game resources were gone. Stephen Mather, the first director of the National Park Service, had early on recommended that national parks needed to be established in the eastern United States. A committee was formed in 1924 to address this need, and their recommendation to Congress was that parks be established in the Blue Ridge and Mammoth Cave areas—a proposal that was approved in 1926.

Parks are generally created to preserve the pristine environment; at Shenandoah the immediate objective was the acquisition and restoration of a badly misused area. Eastern parks presented a different challenge to a government experimenting with a new concept—how to pass on to future generations a sample of the relatively unspoiled landscape and the nation's special heritage areas. Establishing western parks was relatively easy and low cost because the federal government already owned the land. In the East the land was mostly privately owned, and there was no precedent for the federal government to buy back lands to establish a park. Thus the park was authorized—but only on land donated to the government.

The exact sites of the new parks were not determined under the authorizing legislation. Resort owner George Freeman Pollock became a spokesperson for the park, and his enthusiasm and showmanship helped convince private contributors and the State of Virginia to establish the park in its present location. Even before the park was formally established, President Hoover authorized funds to construct a roadway along the crest of the northern Blue Ridge—a road that would eventually be known as Skyline Drive. The project was undertaken partly to provide jobs to counteract the economic turmoil of the Great Depression. Work on the road, overlooks, campgrounds, and other facilities was mostly accomplished by the newly created Civilian Conservation Corps (CCC)—an organization that employed thousands of people who were anxious and willing to work. The CCC gave the National Park Service and other public works projects a jump start that prepared the nation for the economic miracles that followed.

In 1935 the State of Virginia presented over 250 square miles (648 km²) of land to the federal government and in 1936 President Franklin Delano Roosevelt dedicated the park at the Big Meadows site in the center of the park. The park was eventually enlarged to 196,466 acres (307 square miles, 795 km²) in the northern Blue Ridge from Front Royal in the north to Rockfish Gap near Waynesboro in the south where it connects to the Blue Ridge Parkway. Only 4 percent of the park has roads or other developments; the rest remains in forest that over the next few generations will once again closely resemble what the first Euro-American explorers saw when they reached the crest of the Blue Ridge in 1669—an amazing accomplishment considering the previous condition of the land.

In the protective hands of the Park Service, the area is now in amazingly good condition. Essentially all of the old wounds have healed to the extent that many visitors find it difficult to accept the descriptions of conditions that obtained in the early 1900s. High precipitation and a long growing season are favorable for rapid revegetation, and in a short time locusts, pines, and other trees began to reclaim the cutover forests and abandoned fields. The trees, together with vines and shrubs, held the soil and rocks in place, greatly reducing the erosion that had been going on unchecked. Many of the animals have returned or were reintroduced to the area so that black bear, white-tailed deer, raccoons, opossums, skunks, squirrels, wild turkey, and others have returned along with a host of songbirds, to restore in large part the original ecosystem.

Geographic Setting

The Blue Ridge in north-central Virginia rises prominently to an elevation of 4050 feet (1235 m) at Hawksbill Mountain, some 3500 feet (1070 m) above the low-lying Piedmont Province to the east and 2000 feet (610 m) above the Shenandoah Valley (part of the Valley and Ridge) to the west. The Shenandoah Valley is one of a number of prominent limestone valleys that extend nearly continuously from Alabama northward into Canada. The Blue Ridge is narrow here, about 9 miles (15 km) wide with locally a single- or double-ridge crest.

The Potomac River crosses the Blue Ridge through a water gap at Harpers Ferry to the north, and the James and Roanoke rivers cross to the south—the only rivers to cut through the mighty Blue Ridge. Although no rivers cut across the Blue Ridge in the park, a number of wind gaps (low areas or passes in a ridge that lack through-flowing streams) such as Thornton Gap, Powell Gap, and others mark ancient stream courses. During the erosional lowering of the landscape, streams that cut downward faster than neighboring streams that cut across resistant ridges would divert their headwaters and leave a wind gap along the former course. Such a process is called *stream capture,* or stream piracy.

Access to the park from the north is at Front Royal, a mere 72 miles (116 km) west of Washington D.C. The south entrance is 75 air miles (120 km) to the southwest near Waynesboro in central Virginia. Skyline Drive, the only major road in the park, winds its way some 105 miles (169 km) near the Blue Ridge crest and connects Front Royal to the Blue Ridge Parkway. On a clear day, one can look eastward across the Piedmont to the Coastal Plain beyond, and westward across the Shenandoah Valley to the Allegheny Front—an impressive escarpment that marks the east edge of the mountains in the Valley and Ridge Province. Beyond the blue haze in the distance are the Appalachian Plateaus. The Appalachian Trail from Maine to Georgia also follows the ridge, crisscrossing the highway in many places (Fig. 17–8). This trail affords an opportunity to turn the clock back and to see this part of the Blue Ridge as people saw it long before Skyline Drive was there. Even an hour or two on this well-beaten path through the forest will be long remembered. For long hikes involving the use of shelter cabins, advance reservations are necessary.

Geologic Story of Shenandoah

The general sequence for the Blue Ridge applies in the Shenandoah National Park area. However, certain events were more significant in the northern section than elsewhere. Fortunately, a good source of general geology of Shenandoah National

FIGURE 17–8   Trail marker on the famous Appalachian Trail in Shenandoah National Park. (Photo by D. Harris)

Park is available in Gathright's (1976) publication available at the visitor centers. Frye's *Roadside Geology of Virginia* (1986) also provides a general introduction for the southern part of the park. More recent detailed summary articles about the geology of the entire Appalachian Highlands of the United States are available in the Appalachian–Ouachita volume edited by Hatcher and others (1989). The general stratigraphy of the Shenandoah area is shown in Figure 17–9 and will help in understanding the following discussion.

Beginning slightly more than a billion years ago, granitic magma intruded into older rocks—gneisses of the Pedlar Formation (Figs. 17–9 and 17–10) that formed during the Middle Proterozoic (Precambrian) Grenville Orogeny (Fig. 17–4a). The granite, which Lukert (1982) determined to be about 1.1 billion years old, was named the Old Rag Granite, after Old Rag Mountain ("Old Raggedy") where the granite is well exposed (Fig. 17–11). Likely, a supercontinent consisting of most of the world's then existing continental crust was amalgamated into one landmass during Middle Proterozoic time.

Prolonged erosion finally stripped off the older rocks into which the magma was intruded, exposing the Old Rag and other granitics. Breaking apart of the Late Proterozoic-age supercontinent (Fig. 17–4b) produced deep rifts or fractures that triggered the eruption of flood basalts. Lavas of the Catoctin Formation poured out of the fissures and buried the late Precambrian erosion surface, perhaps about 600 Ma (Lukert and Mitra, 1986). Magma cooled in the fissures, forming the dikes that cut the Pedlar Formation and Old Rag Granite in many areas of the park (see generalized sketch in Fig. 17–9). Along with interbeds of volcanic ash and clastic sediments, the Catoctin Formation reached a thickness of as much as 2000 feet (610 m) and covered all but the higher hills. The basalt covered the exposed Pedlar and Old Rag Granite and was later metamorphosed to a more resistant rock called greenstone. The Catoctin Formation greenstones (Fig. 17–12) are of particular interest. Because of their resistance to erosion and weathering, they cap many of the higher mountains in the park (Fig. 17–13). Most of the lava flowed out onto the land surface and locally formed *columnar jointing* during cooling. A good example of columnar jointing is well displayed at Indian Run Overlook. Elsewhere, the presence of *pillow lava* indicates that some lava flowed into lakes or other bodies of water (Lukert and Mitra, 1986).

Plate separation continued at the seemingly slow rate of a few inches each year. However, after a few million years saltwater from the ocean flowed into the widening gap between plates—the beginning of the Iapetus Ocean (Fig. 17–4b)—the predecessor of the Atlantic Ocean. Erosion of the lava surface stopped as the Cambrian seas advanced over the plate edge and deposited the gravels, sands, and muds of the Chilhowee Group (Fig. 17–9). These rocks were later metamorphosed to metaconglomerates, quartzites, and phyllites but are usually not as resistant to erosion as the Precambrian granite and greenstones that underlie some of the higher ridges and peaks in the park. Deposits of younger Paleozoic carbonate and other rocks were present but are mostly eroded away from the Blue Ridge.

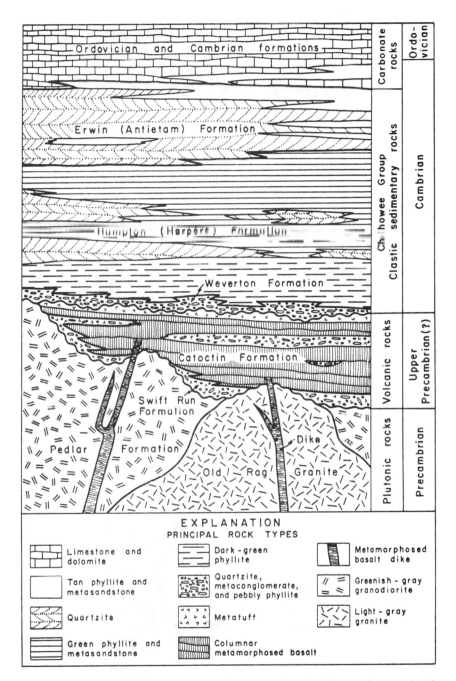

FIGURE 17–9   Rock formations in Shenandoah National Park. (Diagram from Gathright, 1976, Virginia Division of Mineral Resources)

FIGURE 17–10   Gneiss outcrop at Hazel Mountain Overlook on Skyline Drive. (Photo by D. Harris)

FIGURE 17–11   Residual boulders of Old Rag Granite on top of Old Rag Mountain ("Ole Raggedy"), with the Piedmont far below. (Photo by M.T. Lukert)

FIGURE 17–12   Catoctin Greenstone along roadcut near Skyline Drive. (Photo by E. Kiver)

FIGURE 17–13   Recumbent profile of Stony Man, at the end of Stony Man Trail. (National Park Service photo)

Plate tectonics processes are driven by heat trapped and generated within the earth. Changes in plate locations, collisions with other plates, and other factors can change the distribution of heat energy and the resulting plate motions. Such was the case in the Iapetus Ocean as plates began once again to move toward each other. The collision of a volcanic terrane with North America during Middle Ordovician (about 465 Ma) to Early Silurian (about 425 Ma) time produced the Taconic Orogeny and associated metamorphism and thrust faulting. The exotic terrane today makes up part of the inner Piedmont to the east. Other major collision events occurred during the Acadian (mostly in the Northern Appalachians) and Alleghanian orogenies (Fig. 17–4e, 17–4f, and 17–4g) during middle to late Paleozoic time. The final orogeny, the Alleghanian, marked the joining of the major continental crust areas of the earth to once again unite into another supercontinent—this one named Pangaea. Crustal squeezing and shortening amounted to perhaps 75 miles (125 km) in the Shenandoah area producing an area characterized by a greatly thickened crust, intense metamorphism, folding and thrust faulting (Fig. 17–1), and igneous intrusions. The Blue Ridge was also folded into a broad anticlinal uplift containing smaller folds and faults—a feature called an *anticlinorium*.

During and after mountain building, erosion stripped off most of the Paleozoic rocks from the Blue Ridge. There are, however, a number of Cambrian quartzite ridges along the western border; associated shales are generally covered by alluvium. Also, Cambrian Chilhowee beds outcrop in a few places along Skyline Drive, notably near Jeremys Run Overlook in the northern part and for some distance north of the south entrance. The anticlinorium was breached by erosion forming a prominent west-flank ridge—the Blue Ridge—and a lesser range of hills and ridges considered as part of the Piedmont Province.

Triassic events are sometimes overlooked because much of the evidence is found beyond the Blue Ridge in the Piedmont Province and buried beneath the Coastal Plain. Separation of plates along newly formed rifts produced fault basins, or *grabens*, beginning about 200 Ma (Fig. 17–4h). Streams carried boulders and gravel down the steep east slopes of the Blue Ridge and deposited them in various fault basins. Igneous activity, which was extensive in the Triassic fault basins, continued into the Jurassic (Manspeizer and others, 1989). Diabase dikes, sills, and lava flows are common in the grabens, and dikes also occur in the fractures in Blue Ridge rocks, although few have been found within the park.

Since Triassic time, erosion and isostatic—and even some tectonic (Manspeizer and others, 1989)—uplift have been the dominant processes. As indicated earlier, there are alternative interpretations of the erosional sequence. Several high-level flattish areas, notably at Big Meadows, have through the years been regarded as remnants of a widespread erosion surface (peneplain), one that was uplifted and largely destroyed during later cycles of erosion. The evidence, chiefly in the Valley and Ridge and in the Piedmont, appears to support this time-honored interpretation—that the Appalachians have been eroded down and then uplifted and eroded, at least twice. Blue Ridge rocks are considerably more resistant to erosion than the adjacent rocks; therefore, the Blue Ridge stands high above the Piedmont and the Valley and Ridge.

Pleistocene glaciers did not reach the Blue Ridge Province; however, the cold climate is clearly recorded in Shenandoah. Large inactive *boulder fields* and *rock streams* on the slopes and huge *talus* cones and aprons at the base of cliffs are the products of frost wedging produced when colder climates prevailed. Spruce and fir forests replaced the oak–maple–hickory hardwoods during colder climatic episodes when continental glaciers covered New York, northeastern Pennsylvania, and northern New Jersey. The northern or boreal forests retreated northward with the receding glaciers and upward to higher, cooler elevations where today remnants of the boreal forest occur as scattered spruce and fir trees in the park.

Visiting Shenandoah

Visitors may elect to see the park from Skyline Drive, stopping at the many overlooks. Those interested in the geology should obtain Gathright's (1976) "Geology of Shenandoah National Park," which contains a detailed road log beginning at the north entrance and ending where Skyline Drive changes to the Blue Ridge Parkway, 105.2 miles (169.3 km) to the south. At Signal Knob Overlook (mileage 5.7; 9.2 km), red sandstone overlies vesicular basalt in which the pyroxenes have been partially altered to pistachio-green epidote. Also, good columnar jointing in a younger basalt flow is well exposed here. At Hogback Overlook, you will see typical Precambrian granodiorite, in which ancient fractures are outlined with iron-oxide stains. From Thorofare Mountain Overlook, you will have one of the best views of Old Rag Mountain (3291 feet; 10,003 m), which stands alone east of the crest of the Blue Ridge. These few stops provide merely an introduction to the many points of geologic interest along Skyline Drive.

Although some visitors are content with the views from overlooks, others wish to really become a part of Shenandoah by hiking the trails that lead off from Skyline Drive. In addition to the Appalachian Trail, there are many others that test, to varying degrees, such devotion. The trails up to the top of Old Rag Mountain are perhaps the most strenuous, and advice from a park ranger will be helpful in selecting the best route. If you are on the Ridge Trail, you will climb a natural stairway. The steps are horizontally oriented columnar-joint blocks in a dike. The sheer walls that rise above you are of Old Rag Granite, much more resistant than the diabase dike.

Of the others, the Stony Man Self-Guiding Nature Trail is especially interesting, as the many geological and botanical features are identified by numbered markers. At one point, a prospect pit marks the spot where an early optimist had vain hopes of mining copper. From the cliffs at the end of the trail you have an excellent view of the Catoctin Greenstone cliffs of Stony Man (Fig. 17–13) and sweeping views of the Shenandoah Valley far below.

Shenandoah is now a beautifully forested area, consisting mainly of hardwoods of many species. When you are there in the forest, perhaps on the Appalachian Trail, observing the spectacular mountain laurel in bloom or the animals and the birds, recall that this is a "restored" area, one that was allowed to regenerate from its former "disaster area" status. Without park status, this area would continue to be blighted by development and by those who would continue to manipulate the environment for financial gain while providing a handful of jobs for local people. The success of this park project required a relatively short period of time for regeneration and environmental healing to occur. Its success makes one ask, "what other areas can be brought back from the brink of ruin and provide this many people as much pleasure?"

REFERENCES

Crandall, H., 1992, Shenandoah: The story behind the scenery: KC Publications, Las Vegas, NV, 48 p.

Frye, K., 1986, Roadside geology of Virginia: Mountain Press, Missoula, MT, 278 p.

Gathright, T.M., II, 1976, Geology of the Shenandoah National Park, Virginia: Virginia Division of Mineral Resources Bulletin 86.

Hatcher, R.D., Jr., Thomas, W.A., and Viele, G.W., eds., 1989, The Appalachian–Ouachita Orogen in the United States: Geological Society of America, Boulder, CO, The Geology of North America, v. F-2.

Lukert, M.T., 1982, Uranium-lead isotope age of the Old Rag Granite, northern Virginia: American Journal of Science, v. 282, p. 391–398.

Lukert, M.T., and Mitra, G., 1986, Extrusional environments of part of the Catoctin Formation, in Neathery, T.L., ed., Southeastern Section of the Geological Society of America: Geological Society of America, Boulder, CO, Centennial Field Guide, v. 6, p. 207–208.

Manspeizer, W., DeBoer, J., Costain, J.K., Froelich, A.J., Coruh, C., Olsen, P.E., McHone, G.J., Puffer, J.H., and Prowell, D.C., 1989, Post-Paleozoic activity, in Hatcher, R.D., Jr., Thomas, W.A., and Viele, G.W., eds., The Appalachian-Ouachita Orogen in the United States: Geological Society of America, Boulder, CO, The Geology of North America, v. F-2, p. 319–374.

Park Address

Shenandoah National Park
3655 U.S. Highway 211 E
Luray, VA 22835-9036

GREAT SMOKY MOUNTAINS NATIONAL PARK (TENNESSEE AND NORTH CAROLINA)

The Smoky Mountains sit astride the Tennessee–North Carolina border (Fig. 17–14)—the most rugged and spectacular range of the Blue Ridge. The geologic sequence is similar to that at Shenandoah National Park and elsewhere in the Blue Ridge, but the intensity here was greater—more metamorphism and greater

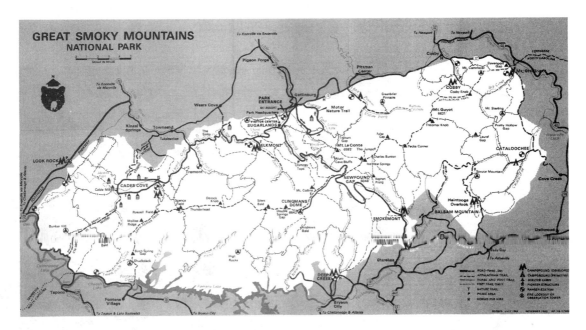

FIGURE 17–14   Map of Great Smoky Mountains National Park.

amounts of crustal shortening as manifested in greater displacements along the numerous thrust faults. That humans are part of nature and dependent on geology is often overlooked in our modern society. Here, however, both the Native Americans and early pioneers quickly found those areas where geology provides more hospitable conditions for survival. Their story of what geology has provided and how they adjusted their life-styles to it is an important element in the heritage preserved in our most visited national park.

Geographic Setting

The Blue Ridge Province in its northern section in southern Pennsylvania and northern Virginia is a narrow single or double ridge. In southwestern Virginia, northwestern North Carolina, and eastern Tennessee it widens significantly where it forms row after row of forested ridges (Fig. 17–15). The east edge of the Blue Ridge Province here begins abruptly where the Brevard Fault zone separates it from the Piedmont Province to the east. Elsewhere the Blue Ridge Escarpment separating the Blue Ridge and Piedmont may be controlled by a monoclinal flexure (Hack, 1989).

The Great Smoky thrust fault on the west separates the Blue Ridge from the Appalachian Valley—the easternmost section of the Valley and Ridge Province.

FIGURE 17–15   Several ridges of the Blue Ridge accentuated by "scenic haze," Great Smoky Mountains National Park. (Photo by National Park Service)

Water vapor and gaseous vapors called *terpenes* emitted from the forest form a blue haze during the summer months giving a "smoky" appearance to the views. The Cherokee name is Shaconage—"the place of blue smoke"—an appropriate name for the entire province and especially for their mountain homeland in North Carolina.

The range forms the divide between the Atlantic Ocean and the Gulf of Mexico drainages and contains 16 peaks over 6000 feet (1830 m) in elevation, making this area the highest mountainous area in the United States east of the Black Hills of South Dakota. Northeast of the park near the Blue Ridge Parkway is Mt. Mitchell—at 6684 feet (2040 m) the highest peak in the East. Clingmans Dome (6642 feet; 2025 m) and Mt. Guyot (6621 feet; 2020 m), the highest peaks in the Great Smoky Mountains National Park, are not far behind in elevation. The peaks were named for geologists T.L. Clingman and Arnold Guyot who conducted the first reconnaissance of the range in the 1860s.

Air masses moving inland from the sea cool as they rise over the range and release large amounts of precipitation; thus, this area has the highest rainfall in the eastern United States. Precipitation ranges from 65 inches (165 cm) in sheltered valleys to over 88 inches (225 cm) on Clingmans Dome and other areas (National Park Service, 1981). Soils are thick and productive at lower elevations where slopes are flat or gentle but much thinner and less productive at higher elevations. A thick forest covers 95 percent of the park—a reminder of what the first settlers faced when they arrived on the Atlantic coast—an unbroken canopy of trees that stretched from the ocean to the Mississippi River.

Humans and the Mountains

Humans left the first traces of their presence in the Smoky Mountain area about 12,000 years ago, perhaps only a few thousand years after they had crossed the Bering Land Bridge and entered the New World. Little is known about these early hunter-and-gatherer people—much more is known about their Cherokee successors who have lived in the Smoky Mountain area for about 1000 years. When the first Euro-American pioneers arrived in the mid-1700s, the Cherokee were living much like the pioneers. They depended on growing crops of corn, squash, beans, melons, and tobacco and supplemented their food supply by hunting and gathering stores of nutritious chestnuts and acorns to supplement their winter food supply. They lived in log cabins with thatched roofs rather than tepees and attempted to coexist with the new settlers.

The Cherokee developed a republican form of government and had a written constitution. An amazing accomplishment occurred in the early 1800s when Chief Sequoyah created a Cherokee alphabet for his people. Two years later nearly every Cherokee could read and write! They even acquired a printing press and published their own newspaper.

Land-hungry Euro-Americans continued pressure as treaty after treaty was broken and the Cherokee lands were reduced. An unfortunate discovery was made in the 1820s that spelled the temporary end to Cherokee influence—gold was found on Cherokee land! The U.S. government decided to relocate the Cherokee to Oklahoma onto reservation land. The "Trail of Tears" was the forced march of 13,000–20,000 Cherokee in 1838–1839 that resulted in the death of an estimated one fourth of the marchers—a dark page in American history. Another 1000 hid in the mountains, their spiritual homeland, where the army was unable to find them. The gold deposits were not significant and in 1843 the government allowed the Cherokee to return. The descendants of the Cherokee who remained in Oklahoma were pleased in the early 1900s when large reservoirs of oil were discovered beneath their new reservation—this time tribal members benefited! Today, a small reservation is located immediately southeast of the park just outside of the Oconaluftee entrance.

When the Euro-Americans arrived in the New World they found a people who had adjusted to the land in a style of living that could be perpetuated indefinitely. The new settlers, like the Cherokee, were also farmers who needed open land in a land of trees. It was said of the eastern forests in those days that a squirrel could travel from the Atlantic to the Mississippi River without leaving the tree tops. Thus, many of the early efforts of the settlers were spent in clearing land for agricultural purposes.

The pioneers settled the productive coastal areas first before westward expansion began. The Piedmont Province was next and then the fertile limestone-floored Appalachian valleys immediately west of the Blue Ridge. Isolated limestone valleys like Cades Cove in the north end of the park were discovered about 1818. Here the interaction of geology and humans is striking. Limestone soils are more productive than soils derived from quartzites and other metamorphic rocks—a feature quickly recognized by the early pioneers. Other settlers in search of a way-of-life migrated farther and higher into the mountains. These nearly self-sufficient people living in the mountains and coves led an isolated life-style for nearly a century and developed their own particular culture that involved handicrafts, carpentry, and music. Glimpses of the past are possible in the park (Fig. 17–16) but especially in Cades Cove where some of the original log houses, an overshot water wheel that

FIGURE 17–16   Typical Blue Ridge log house, restored at Ocanaluftee Visitor Center, Great Smoky Mountains National Park. (Photo by D. Harris)

even today grinds corn at Cable Mill, and a general store open to visitors is available as part of the "living history" demonstrations. At the right time of the year you will have an opportunity, rare in these times, to see a walking plow in operation, pulled by a mule.

Gradually the mountain soils became less productive and life became more difficult. A temporary boost to the economy occurred as lumber companies, after exhausting the timber farther east, bought huge tracts of land in the Smokies. Large-scale logging operations began about 1900 to systematically remove the virgin forest. Between logging and the unintended introduction of the chestnut blight from China in 1904 that killed all mature American chestnut trees (about 20 percent of the forest) by 1938, about 65 percent of the forest was destroyed. Destructive logging practices along with unintentional fires started by logging equipment also contributed to tree cover destruction—soil erosion increased by many magnitudes. With a greatly reduced tree cover, increased runoff generated large floods whose turbulent waters carried thousands of tons of topsoil from the mountains. Slopes were destabilized and large mudslides roared down the mountain side for many years after trees were cut. Stream valleys were also ruined and the once plentiful fish population crashed (Cantu, 1989).

Making of a Park

Citizens in Tennessee and North Carolina were irate with the wanton long-term destruction for short-term economic gain by logging companies. The eloquent writings of Horace Kephart in the early 1900s alerted people to the destruction—Kephart became the first significant advocate for the establishment of a national park to protect the damaged Smoky Mountain landscape. A Knoxville couple, Mr. and Mrs. Willas P. Davis visited some of the western national parks and were so taken with the park concept that they, along with Colonel David Chapman, formed the Great Smoky Mountains Conservation Association in 1923. Congress authorized the park in 1926 but was in a quandary concerning how to acquire privately owned land for the park. The federal government had been in the business of giving or selling land—not purchasing it for park purposes. Local groups began to raise money, and the state legislatures of Tennessee and North Carolina donated $2 million each for the cause. The Great Depression made conditions appear bleak for raising sufficient funds. The day was saved when John D. Rockefeller donated $5 million to the cause and Congress allocated the remainder of the funds to finish the acquisition of lands for the park—a new direction in the establishment of national parks.

Fortunately, the logging companies sold out (under the threat of condemnation proceedings) before they could remove all of the virgin forest. About 35 percent of the Smoky Mountain forest has never been cut—the largest tract of virgin forest in the eastern United States (Cantu, 1989). Authorization for full development of the park was given to the National Park Service in 1934 and the official dedication occurred in 1940. Today, of the 520,269 acres (800 square miles, 2072 km²) of the park, 95 percent of the park is forested and about 90 percent is managed as wilderness. Pioneer clearings are being reclaimed by the forest, reverting them back to wilderness. Thus the heart of the park is kept intact and most visitors experience it on the edges—mostly along the roadways.

Record in the Rocks

The Smoky Mountains do not give up their geological secrets easily. Deep soils and thick forests combined with a complex assemblage of rocks and geologic structures make geologic interpretations difficult. Although pioneer investigations by Keith and others in the 1890s laid the groundwork, it was not until the 1960s that King and others (1968) of the U.S. Geological Survey published the first comprehensive geologic history of the region. Moore's (1988) roadside guide is recommended for those interested in more details while touring in the park.

Important summary papers integrating plate tectonic concepts with geologic data accumulated from the Appalachian Mountains over the previous 200 years are found in the Appalachian–Ouachita volume edited by Hatcher and others (1989b). In spite of considerable agreement about the big picture in the Appalachians, some discrepancies and different interpretations exist, leading Hatcher and others (1989a, p. 3) to note that "many controversies remain, and new ones arise with new data," and that "much remains to be learned."

The broad picture of geologic events parallels that of Shenandoah Park about 300 miles to the north and is highly generalized in Figure 17–4. The timing and type of deformation to affect both areas are similar except that much broader and more intense deformation occurred in the Great Smoky Mountains. The major plates along the Atlantic margin closed like a giant vise about 1100 Ma (Fig. 17–4a), opened during Late Proterozoic–early Paleozoic time (Figs. 17–4b, 17–4c, and 17–4d), and closed again during middle Paleozoic to late Paleozoic time (Figs. 17–4e, 17–4f, and 17–4g). A second cycle of opening began about 200 Ma in early Mesozoic time (Fig. 17–4h), the one that continues today (Fig. 17–4i).

The oldest rocks, designated by King and others (1968) as the earlier Precambrian Basement Complex, consist of gneisses, schists, and granitic rocks formed during the Middle Proterozoic Grenville Orogeny and exposed along the southeastern boundary of the park. These rocks have experienced multiple episodes of metamorphism, making the original relationships difficult to determine. The "radiometric clocks" in minerals have also been reset and mostly reflect the younger episodes of deformation; some minerals that were not recrystallized yield dates of the Grenville Orogeny about 1100 Ma. Metamorphism of rocks decreases in the younger rocks exposed toward the Valley and Ridge Province—rocks that experienced fewer episodes of deformation.

Erosion of basement rock was followed by rifting (Fig. 17–4b) and deposition of the Late Proterozoic Ocoee sediments in the developing Iapetus Seaway. Perhaps

50,000 feet (15,200 m) of sediment accumulated in the Ocoee Basin in Tennessee, western North Carolina, and northern Georgia. The Ocoee Supergroup sediments and their metamorphic equivalents are the most abundant rocks in the park. Sandstones were metamorphosed to quartzites (Fig. 17–17), and depending on the degree of metamorphic intensity, the mudstones were changed to slate, argillite, phyllite (Fig. 17–18), or even mica shist. Virtually all of the summits and waterfalls in the park are on this extensive group of rocks.

Deposition was continuous in the expanding rift ocean, and the Ocoee rocks are conformably overlain by the Early Cambrian Chilhowee clastic sediments that are exposed in the northwest part of the park on Chilhowee Mountain. These same rocks can be followed northward to Shenandoah National Park, indicating that similar conditions existed in many areas of the geosyncline. The oldest known fossils from the Great Smoky Mountains—trilobites and other invertebrates—have been collected from the Chilhowee Group. Fossils, however, need not be the actual remains or impressions of the once-living organisms but can include any physical record left by an organism. Thus, the remains of vertical tubes infilled by sediment in the sandstones and quartzites are evidence that a primitive Cambrian-age sea worm (*Scolithus*) burrowed through these sediments.

Deposition on the trailing edge of the continent occurred in later Cambrian and Ordovician time as limestones record an increasingly deep and extensive sea. These rocks are exposed in the Appalachian Valley Section of the Valley and Ridge Province immediately west of the park and also in Cades Cove within the park. Some younger Paleozoic rocks of at least Silurian age (Unrug and Unrug, 1990) occur as an isolated mass in the park, suggesting that younger rocks were once present but have been removed by erosion.

FIGURE 17–17 Thunderhead quartzite, part of the Ocoee Supergroup, at Clingman Dome Parking Area on Forney Ridge. (Photo by D. Harris)

FIGURE 17–18   Phyllite, a low grade metamorphic rock, in road cut near Look Rock, Foothills Parkway, west of Cades Cove. (Photo by D. Harris)

Plates, Microplates, and Thrust Faults

The dominant structures in the Great Smoky Mountains are thrust and overthrust faults—the result of collisions of westward-moving plates and microplates with North America. A series of large thrust faults underlies the mountains generally paralleling the crest of the range (Fig. 17–1). The Brevard Thrust fault marks the boundary between the Blue Ridge and Piedmont provinces through part of its length, and the Great Smoky–Guess Creek faults mark the Blue Ridge–Valley and Ridge boundary. All of the thrust faults are believed to flatten with depth and are considered splays from a relatively flat-lying *detachment fault* or faults at depth (Fig. 17–1). The mapping of the faults was a formidable task because almost everywhere they are buried beneath alluvium or slope wash. However, it is clear that the Precambrian rocks were thrust north-northwestward up and over the Paleozoic rocks that, in the adjacent Valley and Ridge Province, have been reexposed by erosion.

Faulting occurred here at different times, mostly during the Ordovician-age Taconic Orogeny about 450 Ma when the Piedmont or Carolina microplate docked, and during the Alleghanian Orogeny about 300 Ma when North America collided with the African and European plates. Some of the faults, such as the Greenbrier Thrust (Fig. 17–19), affect only rocks of Ocoee age and are believed to be of early Paleozoic, Taconic-Orogeny age. The Greenbrier Fault was later folded and eroded, thus locally exposing basement rock at the surface (King and others, 1968). Some of the movement along the Great Smoky Fault may also date from the Taconic Orogeny that affected the Blue Ridge and Piedmont provinces, but most of the movement occurred in late Paleozoic time when the last of the Iapetus Ocean was eliminated and the plates were sutured together to form the supercontinent of Pangaea. The mid-Paleozoic Acadian Orogeny had little effect in the Southern Appalachians except for an episode of granitic intrusions during Devonian (about 390 Ma) time.

The Brevard Fault zone and the Great Smoky fault system are unquestionably the big structures in or near the park and in the Blue Ridge Province. The Brevard

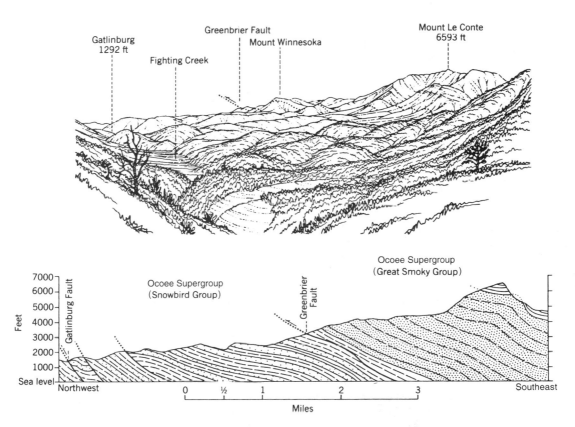

FIGURE 17–19 View from Maloney Point on Tennessee Highway 73, east of Fighting Creek Gap. The cross section shows a number of typical eastward-dipping thrust faults, including the Greenbrier Fault that formed during the Taconic Orogeny and was folded during the later Alleghanian Orogeny. (Sketch by Philip B. King, in King and others, 1968)

extends from the Coastal Plain of Alabama some 370 miles (600 km) to near the North Carolina–Virginia border where other faults continue the northeasterly trend through Virginia (Horton and Butler, 1986). The Great Smoky fault system can also be traced for hundreds of miles, from Virginia to Alabama (Rodgers, 1953). Cook and others (1979) indicate that at least 74 miles (120 km) of late Paleozoic displacement must have occurred. Others suggest that crustal shortening was over 125 miles (200 km) in the Great Smoky Mountain area during late Paleozoic time (Horton and Zullo, 1991)! The incredible amount of mechanical work accomplished during the Alleghany Orogeny would seem to be an appropriate consequence of the collision of major plates such as those containing the continents of North America and Africa (Fig. 17–4g).

Although the Great Smoky Fault is exposed just west of the park between the Appalachian Valley and the Blue Ridge, the fault is also exposed in Cades Cove (Fig. 17–20a) within the park and in Tuckaleechee and Wear coves to the north. A cove in this part of the Appalachians is a valley that is walled-in by high ridges. Here erosion has stripped off part of the thrust sheet creating a *fenster*, or *tectonic window*, where the Ordovician carbonate rocks beneath the thrust sheet are exposed (Fig. 17–20b). The overlying rocks in the surrounding mountains are the Late Proterozoic Ocoee rocks, thus older rocks sit atop of younger here. Where freshly exposed, the contact between these rocks displays finely ground and crushed rock and linear scratches oriented in the direction of movement called *slickensides* (Fig. 17–21); all are products of fault movement.

Folding both preceded and followed the thrust faulting (Fig. 17–22), in places warping the thrust planes (note warped or folded surface of Greenbrier Fault in Fig. 17–19). Further complicating the pattern, there was additional faulting, this time by steep-angled faults that offset the older structures.

Geomorphology

Erosion has been continuous for 250 or more million years since sometime in the Paleozoic Era, yet the area remains relatively high and rugged. How can mountains persist for so long without being worn to sea level? Erosion without additional uplift should reduce the landscape to 10 percent of its former height in 18–20 million years (Ahnert, 1970; Hack, 1982). Thus, the Appalachians would be a rolling plain at or close to sea level if vertical adjustments, either tectonic or isostatic, had not occurred. Paleozoic mountain building had greatly thickened the continental crust, enabling isostatic uplift to occur for hundreds of millions of years. Hack (1982) believes that these adjustments must be going on now or must have occurred very recently in order to account for the high elevations in the Blue Ridge.

Many of the streams developed their valleys in the fractured rocks of the fault zones; others eroded their valleys in the less resistant formations. A number of high ridges and peaks are held up by quartzites, the most resistant of rocks. Resistant

(a)

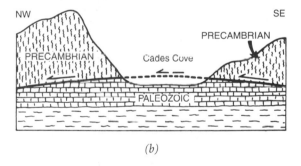

(b)

FIGURE 17–20   (*a*) Cades Cove on the west side of the Smokies is a tectonic window where erosion has cut through a thrust sheet and exposed the younger rocks below (Photo by E. Kiver). (*b*) Generalized cross section across Cades Cove, showing thrust fault. (Adapted from King and others, 1968)

rocks form sharp peaks and ridges (Fig. 17–23), while those capped by nonresistant rocks erode rapidly, forming rounded mountain tops.

Geomorphic features of special interest include the Sinks (Fig. 17–24) on the Little River west of Gatlinburg. Ordinarily the term *sink* refers to solution sinkholes, generally in limestone; here, the Sinks are plunge pits formed when the river shortened its course by cutting through a meander neck.

FIGURE 17–21   Smoky Mountain Thrust Fault (arrow) beneath overhang; Precambrian resting on Ordovician rocks. (Photo by R.J. Larson)

Alum Cave Bluffs south of Mt. Le Conte is one of the many points of interest available to the hiker. It is not a true cave formed by solution but is merely a large overhang of highly resistant metamorphic rocks. The trail to Alum Cave Bluffs also leads to one of the *balds,* conspicuous but incompletely understood features in the Great Smokies.

From afar, balds appear to be grassy openings (about 50–1000 feet wide; 15–305 m) in the forested ridges. Some indeed are grassy and seem to have a degree of permanency about them, perhaps maintained by exposure to prevailing winds and perhaps by stock grazing prior to the establishment of the park. Other "heath balds," "laurel slicks," or "hells" are covered by an impenetrable tangle of shrubs, mainly mountain laurel and rhododendron. As aptly described by Cantu (1989, p. 12), these balds can be examined only up close "on hands and knees, the only possible way to get through them." But why are they treeless in this land of trees? Although some are on the tops of mountains, others are lower down. The highest mountain in the park, Clingmans Dome, is heavily forested. Therefore, a climatic cause of the balds, involving a colder, more rigorous climate during the Ice Age, appears untenable. Perhaps they are related to landslides, blowdowns, or fires, either lightning fires or fires set by Indians for lookouts or to improve hunt-

FIGURE 17–22   Tight folding in Late Proterozoic Ocoee Supergroup rocks within one of the numerous thrust sheets in the Smokies. (Photo by R.J. Larson)

ing by attracting wildlife. Whatever their origin, bald is a complimentary term here in the Smokies where magnificent displays of azaleas and rhododendron in bloom are the outstanding attraction of the park during June and July.

Although glaciers did not affect the Smokies, the climate of that time was sig-nificantly colder than now. Snowline never lowered far enough to intercept the high peaks in the Smoky Mountains, but colder temperatures lowered timberline (the elevation above which the climate is too rigorous for trees to survive). Rem-nants of the spruce–fir forest that covered much of the mountains during the Ice Age still cling to the higher peaks and ridges. The entire area, and especially the re-gions of exposed rock above timberline, were subjected to intense frost wedging that loosened blocks of bedrock and concentrated them in boulder fields and talus slopes—none of which are still forming under today's milder climatic regime.

FIGURE 17–23   Chimney Tops Mountain capped by a resistant phyllite of Late Proterozoic age. (Photo by D. Harris)

FIGURE 17–24   The Sinks on Little River near Gatlinburg, Tennessee. (Photo by National Park Service)

Visiting the Great Smoky Mountains

The park is open throughout the year and is not far removed from heavily populated areas; consequently, more people visit Great Smokies than any other national park. If you drive the Blue Ridge Parkway you enter the park on the south side, at the Oconaluftee entrance and visitor center, in North Carolina. On the north side of the park, near Gatlinburg, Tennessee, is the Sugarland Visitor Center and park headquarters, on the West Prong (fork) of the Little Pigeon River. U.S. Highway 441, which climbs up and over Newfound Gap, is the connecting link between these two centers (Fig. 17–14). The Oconaluftee Visitor Center emphasizes the Cherokee history and traditions, the Sugarland Center concentrates on natural history, and the Cades Cove Center (open summer only) describes the mountain people and their life-style.

Automobile roads offer only an introduction to the park or access to the more than 800 miles (1290 km) of trails to high peaks, waterfalls, and other points of interest. The famed Appalachian Trail—the Maine to Georgia trail along the eastern backbone of the United States—traverses the park through country with elevations over 6000 feet (1830 m) above sea level. For over 70 miles (113 km) one can walk with one foot in Tennessee and the other in North Carolina! Exploring forests that contain an abundance of plants (over 1400 species!) and animals, seemingly endless views of forested ridges, and the opportunity to share the important things in life with family or friends—or to be alone—are all possibilities in the Smokies.

Forney Ridge Road winds southwestward along the divide from Newfound Gap to Clingmans Dome (6643 feet; 2026 m), the high point in the park (Fig. 17–14). This side trip is a "must" because of the superb view from the tower, where you look out over the treetops. Most of the park's forests are hardwoods, but on the mountaintop the climate is similar to that of central Canada, with similar balsam forests. On this Clingmans Dome side trip, allow an extra half-hour for a "bear-jam," where the little black moochers hold up traffic. Only a small percentage of the estimated 400–600 bears in the park have acquired panhandling habits, but feeding these wild bears is illegal and is ultimately harmful to them and potentially to visitors.

Other wildlife is abundant and enjoyed by many. In addition to deer and small mammals, over 200 species of birds use the park with about 80 species making this their year-round home. The bison were hunted out by the late 1700s, the elk by 1840, and the gray wolf and eastern mountain lion by about 1900. Unfortunately, an unwanted species, the European wild boar, escaped from a nearby game farm in the early 1920s and entered the park in the 1940s. These three-foot-high digging machines weigh about 220 pounds and uproot acres of wildflowers each year in search of grubs and roots, prey on small animals, and compete with the black bear and other species for acorns and other important food sources.

The Smokies are not without other problems. The large number of visitors and developments in the surrounding areas all detract from the escape that so many seek in our national parks. Air pollution knows no political boundaries and reflects the influence of surrounding industrial and urban activities. Visibility has reduced 30 percent since the 1950s as the "smoke" in the Smokies turns more to the white haze that city dwellers know too well. Acid rain and nitrous oxides are often as high in the Smokies as in many major cities—a sad commentary on what a lack of a national population policy and greed for more and more material possessions can do to the basic quality experiences of life. Humans and the earth are connected. If that connection is not made periodically, living becomes an artificial space-ship-like existence.

Fortunately for the Midwest and East, a remarkable system of Appalachian parks is located within a day's drive of one-half of the nation's population—an amazing gift from our predecessors, those with great vision. A society that acts to protect wildlands is encouraging—to wait until the last minute to snatch such an area from inevitable destruction is discouraging and often unsuccessful. Perhaps the people of today will look closely at what is left of our natural world and act sooner rather than later to protect it, and along with it, our own species. For what quality-of-life gifts will our generation be remembered?

REFERENCES

Ahnert, F., 1970, Functional relationships between denudation, relief, and uplift in large, mid-latitude drainage basins: American Journal of Science, v. 268, no. 3.

Cantu, R., 1989, Great Smoky Mountains: The story behind the scenery: KC Publications, Las Vegas, 48 p.

Cook, F.A., Albaugh, D., Brown L., Kaufman, S., Oliver, J., and Hatcher, R., Jr., 1979, Thin skinned tectonics in the crystalline southern Appalachians: COCORP seismic reflection profiling of the Blue Ridge and Piedmont: Geology, v. 7, p. 563–567.

Hack, J.A., 1982, Physiographic divisions and differential uplift in the Piedmont and Blue Ridge: U.S. Geological Survey Professional Paper 1265.

Hack, J.A., 1989a, Geomorphology of the Appalachian Highlands, in Hatcher, R.D., Jr., Thomas, W.A., and Viele, G.W., eds., The Appalachian-Ouachita Orogen in the United States: Geological Society of America, Boulder, CO, The Geology of North America, v. F-2, p. 459–470.

Hatcher, R.D., Jr., Thomas, W.A., and Viele, G.W., 1989b, The Geology of North America, v. F-2, The Appalachian–Ouachita Orogen in the United States: Geological Society of America, Boulder, CO.

Hatcher, R.D., Jr., Thomas, W.A., and Viele, G.W., 1989, Appalachians introduction, in Hatcher, R.D., Jr., Thomas, W.A., and Viele, G.W., eds., The Appalachian–Ouachita Orogen in the United States: Geological Society of America, Boulder, CO, The Geology of North America, v. F-2, p. 1–6.

Horton, J.W., Jr., and Butler, J.R., 1986, The Brevard fault zone at Rosman, Transylvania County, North Carolina, in Neathery, T.L., ed., Southeastern Section of the Geological Society of America: Geological Society of America, Boulder, CO, Centennial Field Guide, v. 6, p. 251–256.

Horton, J.W., Jr., and Zullo, V.A., 1991, An introduction to the geology of the Carolinas, in Horton, J.W., Jr., and Zullo, V.A., eds., The Geology of the Carolinas: University of Tennessee Press, Knoxville, p. 1–10.

King, P.B., Neuman, R.B., and Hadley, J.B., 1968, Geology of the Great Smoky Mountains National Park, Tennessee and North Carolina: U.S. Geological Survey Professional Paper 587.

Moore, H.L., 1988, A roadside guide to the geology of the Great Smoky Mountains National Park: University of Tennessee Press, Knoxville, 192 p.

National Park Service, 1981, Great Smoky Mountains: National Park Service Handbook 112, Washington, DC.

Rodgers, J., 1953, Geologic map of east Tennessee with explanatory text: Tennessee Division of Geology Bulletin 58.

Unrug, R., and Unrug, S., 1990, Paleontological evidence of paleozoic age for the Walden Creek Group, Ocoee Supergroup, Tennessee: Geology, v. 18, p. 1041–1045.

Park Address

Great Smoky Mountains National Park
107 Park Headquarters Road
Gatlinburg, TN 37738

EIGHTEEN

New England Province

The Northern Appalachians include New England and Canada up to New-foundland where the Appalachian structures end in North America. The sequence of geologic events affecting the New England Province are generally the same as those described in Chapter 17. Because of more frequent and more extensive igneous activity, the landforms are different. The province boundaries correspond to the northeastern states with a narrow extension of Paleozoic metamorphic rocks into the area east of the Hudson River and Lake Champlain valleys and a belt of late Precambrian crystalline (metamorphic and igneous) rocks called the Reading Prong, which extends across northern New Jersey into eastern Pennsylvania, ending at Reading (Fig. 18–1). The Adirondack Mountains of upstate New York are geologically related to the Canadian Shield and are not considered here.

The province is mainly an upland area, a low plateau that has been dissected to varying degrees. The area is rich in history and in scenery (Fig. 18–2) but surprisingly poor in national parks. From west to east across New England, as shown in Figure 18–1, are the Appalachian Valley Section, which separates the Adirondacks from the Taconic Highlands Section, the Vermont Valley Section, which is underlain by carbonate rocks, the Green Mountains–Reading Prong composed of late Precambrian and Cambrian crystalline rocks, the Connecticut Valley fault basin containing Triassic-Jurassic volcanic and sedimentary rocks, the Central Highlands, which include the White Mountains and 6186-foot-high (1886-m) Mt. Washington (Fig. 18–3), the highest peak in New England, and the Coastal Lowlands Section containing Cape Cod. In eastern Maine is the New Brunswick Highlands Section, which includes low hills, coastal lowlands, estuaries, peninsulas, islands, and the fiordlike sounds in Acadia National Park and other areas. On Mt. Desert Island in Acadia resistant rocks rise to 1530 feet (466 m) above the Atlantic on Cadillac Mountain, the highest point on the eastern coast.

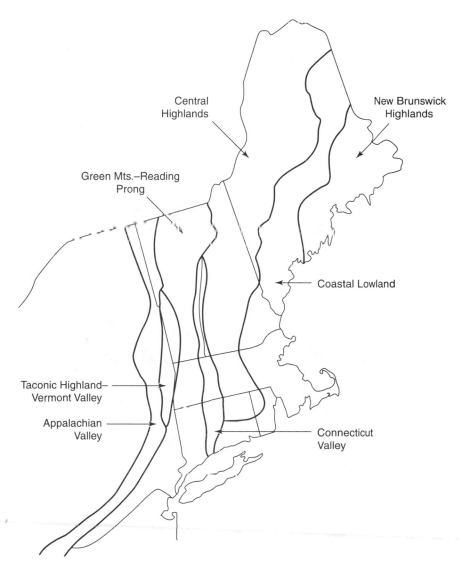

FIGURE 18–1   Geomorphic sections in the New England Province. (Section boundaries after Hack, 1989)

FIGURE 18–2   Beginning about 2 Ma, huge glaciers moved down out of Canada and relandscaped the New England area. With each advance more and more Canadian rocks and boulders were left strewn over the land. In clearing their fields, the early settlers piled the stones along the borders; thus, the stone fences are of geologic and historic, as well as scenic, interest. And, as Robert Frost's neighbor said, "Good fences make good neighbors." (Photo by D. Harris)

Geologic History

Following the Middle Proterozoic Grenville Orogeny [about 1000–1200 million years ago (Ma)] that added more area to the growing North American continental nucleus, a Late Proterozoic rift broke the plate edge into smaller microplates and created a new seaway, the Iapetus Ocean (Fig. 17–4*b*). Fragments of these Grenville rocks and early Paleozoic sedimentary rocks were thrust back onto the continent edge and folded (Fig. 18–4) during the Taconic Orogeny (beginning in Middle Ordovician, about 470 Ma) when a volcanic island system collided with the continental plate. Other masses of Grenville rocks rose through the upper crust as *gneiss domes* during the orogeny. The Grenville rocks along with granitic plutons (Dallmeyer, 1989) of late Precambrian and Devonian age (from a later orogeny) form the crystalline core of the Taconic Highlands, Central Highlands, and Green Mountain–Berkshire–Reading Prong rocks in the western part of the New England

FIGURE 18–3   Mt. Washington, New Hampshire's highest mountain, is only 6288 feet (1917 m) high, but its weather may be the worst on the earth—it has winds up to 231 miles (372 km) per hour and temperatures as low as –60°F (–51°C). Not clearly shown here are the effects of glaciation; it was overridden by continental ice sheets and later modified locally by cirque glaciers. (Photo by Fairchild Aerial Survey, Inc.; reprinted from Thornbury, 1965, by permission of John Wiley & Sons, Inc.)

Province (Rankin and others, 1989). The Taconic Orogeny was the first of several tectonic events that marked the closing of the Iapetus Ocean.

A Devonian-Mississippian mountain building–metamorphic–igneous intrusive event called the Acadian Orogeny was triggered in the New England Province when a large microplate or a number of smaller microplates called the Avalon terrane collided with North America (Fig. 17–4*f*). This exotic terrane contains Upper Proterozoic rocks and Cambrian volcanic and sedimentary rocks overlain by sedimentary rocks of early Paleozoic age. Fossil invertebrates in the Paleozoic rocks are more closely related to similar-age fossils found in Europe and northwestern Africa than to those found elsewhere in North America, strengthening the conclusion that this terrane is more closely connected to the European and African plates and is therefore exotic to North America. The rocks are similar to those found on the Avalon Peninsula in Newfoundland, hence the name Avalon microplate or terrane. This exotic terrane and its equivalents extend the length of the Appalachians from

FIGURE 18–4   Fold in Paleozoic sedimentary rocks caught up in mountain-building processes in New Hampshire. (Photo by Jennifer Thomson)

Newfoundland to Alabama. The southern part (Carolina terrane) makes up much of the Piedmont Province in the Southern Appalachians. The northern part is exposed in the Coastal Lowlands and New Brunswick Highlands sections along the New England and Canadian coasts—including northern Maine at Acadia National Park (Rankin and others, 1989).

Near the end of the Acadian Orogeny in Mississippian time (about 330 Ma) northern North America and northern Europe were connected as one landmass. The Iapetus Ocean was still open in the southern Appalachians, and a number of small nonmarine basins were present in the Coastal Lowlands Section of New England (Dallmeyer, 1989). These small basins and the oceanic areas offshore of the Southern Appalachians would be eliminated during the late Paleozoic Alleghanian Orogeny as described in Chapter 17. The Alleghanian had little effect on the Northern Appalachians except for some local metamorphism. The continent-to-continent collision between the northern and southern landmasses formed the

Pangaea supercontinent during late Paleozoic time and affected mostly the Southern Appalachians. Thus the late Paleozoic geography was strange indeed compared to our present arrangement of land areas. The huge Pangaea supercontinent was an amalgamation of most of the world's land areas—with a massive range of interior mountains rising majestically above the seemingly endless land mass.

Such an arrangement is not stable. Heat rising from deeper in the earth is trapped beneath the crustal blanket where it eventually manifests itself as rising mantle rock and forms a rift system. Such a rift system cut indiscriminately through the Appalachian Highlands east of the Piedmont and eastern New England provinces and initiated the development of a new ocean—one that 200 million years later would be called the Atlantic, named for the Greek god Atlas (Figs. 17–4*h* and 17–4*i*). Atlas's father's name, Iapetus, is then an appropriate name to use for the sea that preceded the Atlantic. A number of fault basins developed in the Piedmont and eastern New England sections as Pangaea was pulled apart. Huge alluvial fans and lakes formed as sediment and lava (Fig. 18–5) filled these fault troughs during early Mesozoic (Triassic and Jurassic) time.

FIGURE 18–5 Pillow lava, exposed near Amherst, Massachusetts, was formed when lava flowed into a lake about 190 Ma. (Photo by E. Erslev)

Later Mesozoic and Cenozoic crustal movements involved both uplift and subsidence with some movement along faults (Manspeizer and others, 1989). Without crustal uplift a landmass would be eroded to sea level in a few tens of millions of years. Hack (1979, 1989) estimates that the Appalachians must have been uplifted by both tectonic and *isostatic* (crust tends to float at an equilibrium level on denser mantle and rises as surface layers are removed) processes at a rate of about 130 feet (40 m) every million years in order to keep up with downwasting of the land surface. Downwasting slightly exceeds isostatic uplift. Thus the land surface is likely to be eroded to near sea level one or more times during the extremely long period of time since the Appalachian orogenies ended. Such low-lying, low-relief areas in humid-climate areas are called *peneplains*.

The province is mainly an upland area, a low plateau that has been dissected to varying degrees. This uplifted peneplain has several *monadnocks* or hills rising above the general level of the surface, including Mt. Monadnock (Fig. 18–6) and Old Man Mountain (Fig. 18–7) in southwestern New Hampshire. Monadnock in the Algonquin Indian language means "a mountain that stands alone," an appropriate name for a rock island surrounded by low country.

FIGURE 18–6   Mt. Monadnock, southwestern New Hampshire. (Photo by J.S. Shelton; reprinted from Thornbury, 1965, by permission of John Wiley & Sons, Inc.)

FIGURE 18–7   Old Man Mountain in New Hampshire, a roches moutonee composed of the 190-million-year-old Conway Granite. (Photo by Wallace Bothner)

Mt. Monadnock is not only the classic area selected by the noted geomorphologist William Morris Davis to represent all mountains that stand above old erosion surfaces, but it is also of historical and literary interest. According to newspaper accounts, more than 400 people were tramping up the mountain on a single day in August of 1860. According to Carter (1831) people hike to the top "for the sake of the wide prospect which they can there behold." Ralph Waldo Emerson paid tribute to the mountain in his 1847 poem "Monadnoc." Henry David Thoreau hiked to the top on four occasions and kept a fascinating "natural history journal." It was Thoreau who, pointing out that the Algonquins regarded the mountain as a sacred place, suggested that it be set aside for the benefit and enjoyment of everyone—a philosophy that would soon take hold and lead to the world's first and finest systems of national parks. It would also instill in thoughtful people an attitude of responsibility toward nature—rather than one of unlimited use and abuse.

Many monadnocks are composed of rock that is harder and more resistant to erosion than the surrounding rocks. Interestingly, the prototype is different; the top of Mt. Monadnock is composed of schist that is *less* durable than the granite on the lower slopes, suggesting that this and other mountains of the area owe their existence to their position (location) and not to the superior resistance of the rocks topping the mountains. The area that lies at the headwaters of several streams is the area least eroded and therefore remains high. Consequently, although the tops of some monadnocks are made up of highly resistant rocks, this is by no means always the case.

Erosion of the New England landscape by streams and other surface processes has persisted for over 350 million years since at least middle Paleozoic time. Streams had no doubt adjusted their courses to areas where structural weaknesses such as folds and faults or nonresistant rocks were located. That pattern would soon be more strongly emphasized as the first of at least three major episodes of glaciation began during the Pleistocene (Hack, 1989).

Finishing Touches—Glaciers

Glacial flow was strongly influenced by preglacial topography, which in turn further emphasized the differences in rock resistance. Valleys were deepened and widened, soil and loose debris was moved, and eventually even the highest mountains in the Central Highlands Section, such as Mt. Washington, were overtopped by the continental ice sheet advancing from Canada. It is unclear whether the glacial cirques on higher peaks formed before or after the last invasion of continental ice in late Wisconsin time.

The Wisconsin ice advance destroyed much of the evidence of older glaciations. The terminal moraine lies south of New England in northeastern Pennsylvania and New Jersey (Fig. 13–2). The moraine follows Long Island and continues onto the continental shelf where it was deposited subaerially. Glaciers had locked up such vast amounts of water during times of maximum glaciation that sea level was reduced by as much as 330 feet (100 m), exposing large areas of the continental shelf. A younger moraine lies to the north and extends through Rhode Island and inner Cape Cod. Glacial retreat left deposits of outwash and glacial lake sediments in the lowlands. Deposits of till and isolated erratics (Fig. 18–8) are common and esker and drumlin landforms are found locally in larger valleys.

As climates warmed and ice melted, water again returned to the sea, causing sea level to rise. Waves and currents attacked the rocks and glacial debris along the coast forming the special coastal landscapes found at Acadia National Park, Cape Cod National Seashore, and other areas along the New England coast.

FIGURE 18–8   Glacial erratic, so delicately balanced on smooth bedrock that it can easily be rocked back and forth. One of New Hampshire's famous "rocking stones." (Photo by D. Harris)

REFERENCES

Carter, J.G., 1831, A geography of New Hampshire for families and schools: Portsmouth, N.H., March, Boston, Hillard, Gray, Little, and Wilkins.

Dallmeyer, R.D., 1989, Late Paleozoic thermal evolution of crystalline terranes within portions of the U.S. Appalachian Orogen, in Hatcher, R.R., Jr., Thomas, W.A., and Viele, G.W., eds., The Geology of North America, v. F-2, The Appalachian-Ouachita Orogen in the United States: Geological Society of America, Boulder, CO, p. 417–444.

Hack, J.T., 1979, Rock control and tectonism: Their importance in shaping the Appalachian Highlands: U.S. Geological Survey Professional Paper 1126-B, p. 1–17.

Hack, J.T., 1989, Geomorphology of the Appalachian Highlands, in Hatcher, R.D., Jr., Thomas, W.A., and Viele, G.W., eds., The Geology of North America, v. F-2, The Appalachian-Ouachita Orogen in the United States: Geological Society of America, Boulder, CO, p. 459–470.

Manspeizer, W., DeBoer, J., Costain, J.K., Froelich, A.J., Coruh, C., Olsen, P.E., McHone, G.J., Puffer, J.H., and Prowell, D.C., 1989, Post-Paleozoic activity, in Hatcher, R.D., Jr.,

Thomas, W.A., and Viele, G.W., eds., The Appalachian-Ouachita Orogen in the United States: Geological Society of America, Boulder, CO, The Geology of North America, v. F-2, p. 319–374.

Rankin, D.W., Drake, A.A., Jr., Glover, L., III, Goldsmith, R., Hall, L.M., Murray, D.P., Ratcliffe, N.M., Read, J.F., Secor, D.T., Jr., and Stanley, R.S., 1989, Pre-orogenic terranes, in Hatcher, R.D., Jr., Thomas, W.A., and Viele, G.W., eds., The Appalachian-Ouachita Orogen in the United States: Geological Society of America, Boulder, CO, The Geology of North America, v. F-2, p. 7–100.

Thornbury, W.D., 1965, Regional Geomorphology of the United States: Wiley, New York.

ACADIA NATIONAL PARK (MAINE)

Most of Acadia National Park is on Mt. Desert Island just off the Maine coast, southeast of Bangor. The attraction of a rugged, rocky seacoast among the evergreen forest of the north lands holds a fascination for all who experience this landscape, luring many people back for yet another visit. What Acadia lacks by being one of our smaller national parks, it makes up for by being one of the most scenic areas on the Atlantic coast. The influence of the sea is strong with lighthouses, small fishing and crabbing boats, foghorns, harbor seals, and an abundance of seabirds.

The area has a rich history that includes conflicts between the French, British, and later an upstart group of colonists to gain control of the Mt. Desert area and part of the New World. Creation of a national park here was possible only because of a unique set of circumstances where a group of wealthy individuals organized to preserve the inspiring scenery for the enjoyment of all. To those who take time to listen, the rocks, mountains, and landforms tell a remarkable story of moving plates, incredible pressure, the work of massive glaciers, and the unrelenting attack of the sea on the land's edge.

Geographic Setting

Acadia National Park lies in the New Brunswick Highlands Section of the New England Province (Fig. 18–1) about 70 miles (113 km) southwest of the U.S.–Canadian border. This is an unusual section of the Atlantic Coast in the United States where mountains come down to meet the sea. The rugged, deeply indented, fiordlike coast and glacially carved mountain valleys form an appealing landscape that first began to attract summer visitors in the mid-1800s. Mt. Desert Island forms the core of the 41,888-acre (65-square mile, 170-km^2) park and is reached by crossing the Mt. Desert narrows on the Trenton Bridge on the north end of the park (Fig. 18–9). All or parts of the offshore islands of Baker, Great and Little Cranberry, and others, as well as Isle au Haut and the Schoodic Peninsula about 10 miles (16 km) to the east are also under park ownership. About half of Mt. Desert Island is park land—the rest is private property.

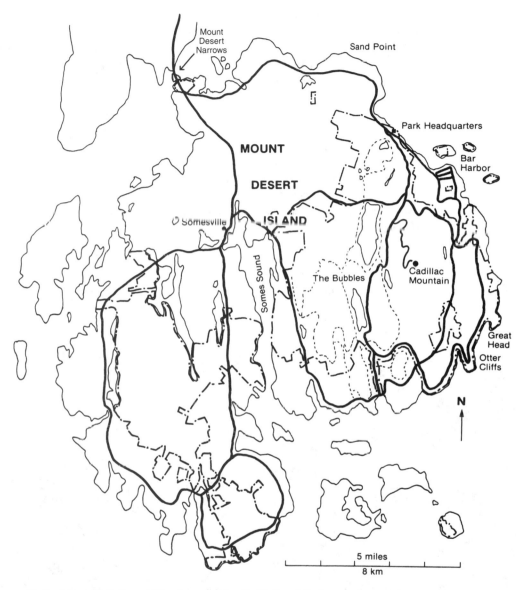

FIGURE 18–9   Map of Mt. Desert Island area of Acadia National Park.

The topography on Mt. Desert Island contains north–south oriented ridges and valleys initially excavated by streams and most recently by the huge continental ice sheets during the late Wisconsin. Arms of the sea extend into the island, especially in Somes Sound along the south coast, and a number of lake basins are scoured in the bottoms of the half dozen or so glacial troughs on the island. Mountain peaks rise up to 1530 feet (466 m) on Cadillac Mountain, the tallest peak on the Atlantic Seaboard (Fig. 18–10). Summer temperatures are a cool 69°F (20°C), providing even more incentive for visitors to leave the hot and humid summer conditions farther south along the Atlantic Coast. The cooler climate promotes the growth of the balsam fir and spruce forests—forests that are more typical of Canada.

Human History

Native Americans have used the area for at least 7000 years. The land called Pemetic was the homeland of the Abnaki Indians—an Algonquin-speaking group that inhabited Mt. Desert Island and surrounding areas when the first Europeans arrived. The name Arcadia given by early explorers to the North Carolina area was extended northward by map makers and eventually shortened to Acadia (Rothe, 1995).

Samuel de Champlain landed in Pemetic in 1604 and named it Isle des Monts Deserts—"Isle of Bare Mountains." Champlain went on to found Quebec and is

FIGURE 18–10   View from Cadillac Mountain across Frenchman Bay to the Schoodic Peninsula section of Acadia National Park. Note the glacially scoured bedrock surfaces. (Photo by D. Harris)

considered the founder of "New France"—16 years before the Pilgrims arrived at Plymouth Rock to found "New England." The French adventurer, Lord Cadillac, lived on Mt. Desert Island for a while and hoped to establish a feudal state. Cadillac was restless and soon abandoned that enterprise and moved on to establish another settlement—that one would later be named Detroit. His name lives on in the highest peak on the island and as a prestigious automobile!

After the British defeated the French in 1759, land speculators became interested in the Maine coast. Farming, logging, fishing, and shipbuilding industries were the economic base in the mid-1800s. Paintings of the seascapes and other scenery by Thomas Cole beginning in 1844, by Frederick Church, and by other well-known artists of the day inspired visitors to see the area for themselves. Mostly people of modest means, the early visitors stayed with local families, spawning a welcome boost to the local economy. By 1860 resorts and hotels were built and tourism became the major industry of the area.

The extremely affluent discovered the area in the late 1800s and early 1900s—a discovery that protected much of the area from logging and willy-nilly development and would eventually lead to the establishment of the national park. Large estates were established as summer retreats for such notables as the Rockefellers, Morgans, Vanderbilts, Carnegies, Fords, and Kennedys. One of the so-called "cottages" built by E.T. Stonesbury in 1925 had 80 rooms with 28 bathrooms (Rothe, 1995)! Many of the wealthy landowners were sensitive to things natural. Of particular concern was the overdevelopment in the Bar Harbor area and signs that logging companies, which had devastated forests elsewhere, were preparing to move into the area.

One person in particular, George B. Dorr, became extremely concerned, devoting much of the next 43 years of his life to preserving as much of the Mt. Desert area landscape as possible. Dorr formed a nonprofit corporation that included people such as John D. Rockefeller, Jr., Charles Eliot (president of Harvard), John S. Kennedy, and others. The corporation solicited and received land donations and purchased points of interest such as Cadillac Mountain to preserve them for public use. This amazing person was able to place thousands of acres under corporation control and began to campaign for national park status to assure the perpetual use of the area by the public.

The area was designated Sieur de Monts National Monument in 1916 and became Lafayette National Park in 1919. The area was further enlarged and changed to Acadia National Park in 1929. Rothe (1995) describes Dorr's efforts as the "greatest one-man shows in the history of land conservation."

John D. Rockefeller, Jr., donated over 10,000 acres (15 square miles, 40 km^2) to the cause and constructed 57 miles (92 km) of narrow carriage roads to provide access to the public as an alternative to automobile travel. Expert stone masons constructed 16 artfully designed stone bridges along the carriage roads—a significant attraction in its own right. The land and all of this construction was given to the American public at no cost to the taxpayers—an amazing gift to be enjoyed by countless millions.

Geologic Story

The big picture of Appalachian mountain building generalized in Figure 17–4 and described in the chapter introduction was pieced together by combining information from many areas of the Appalachians—the equivalent of many lifetimes of work by hundreds of geologists. The piece of the puzzle from the New Brunswick Highlands and Acadia National Park is unique. The older sedimentary and volcanic rocks here, the Ellsworth Schist of Early Cambrian age (about 540 Ma), formed somewhere to the east on the Avalon microplate. Metamorphism converted the pre-existing rocks to schist and gneiss, and erosion exposed these deeply buried rocks—well before the area was added to the North American continent (technically called Laurentia at this stage).

Deposits of sandstone, siltstone, and volcanic rocks of the Bar Harbor Formation and the Cranberry Island series unconformably covered the Ellsworth Schist during Silurian-Devonian time about 380–400 Ma (Gilman and others, 1988) prior to the docking of the Avalon microplate.

The collision between Avalonia and North America beginning in mid-Devonian time (about 380 Ma) produced the Acadian Orogeny and also eliminated the Iapetus Ocean, the large rift ocean that had separated the Atlantic plates during Late Proterozoic and early Paleozoic time. Thus the rift between continents in the north Atlantic area was closed at the end of the Acadian Orogeny, although the seaway in southern New England and the Southern Appalachians persisted until late Paleozoic time (Hatcher, 1989). The continent-to-continent collision in the Southern Appalachians produced the Alleghanian Orogeny, completely eliminated the Iapetus Ocean, and resulted in the formation of the super-continent of Pangaea.

Near the end of the Acadian Orogeny and beginning about 360 Ma, melting deep in the crust and upper mantle caused magma to form and rise toward the surface. The invasion of magma was one of the highlights of Paleozoic activity in the Northern Appalachians, sending first dark-colored basaltic magma (called gabbro or diorite because of the large crystals formed during cooling) as dikes and sills that cut into the Ellsworth and Bar Harbor rocks in the Mt. Desert Island area (thus the gabbro and diorite are younger); followed by light-colored, fine-grained granite that cuts all of the older rocks; and followed in turn by the massive, coarse-grained granite of Cadillac Mountain and other areas about 360 Ma (Osberg and others, 1989). Igneous activity ended with intrusion of more diabase dikes.

The Cadillac Mountain magma apparently so weakened the roof rocks that they began to sag and then sink, en masse, into the magma. As the foundering of the crustal block occurred, magma surged upwards to fill the spaces created. Around the borders of the intrusion the rocks were badly shattered and brecciated. The extensive shatter zone occurs along the island edge on the north, east, and south and contains blocks of gabbro and Ellsworth Schist fragments that are up to hundreds of feet across—blocks that were engulfed by the granite magma that intruded into the brecciated shatter zone and into fractures as dikes. Where the fractures were curved, so-called *ring dikes* formed; several are now exposed on both the west and east sides of Mt. Desert Island. The shatter zone is exposed in road cuts and along trails, and in places is as much as a mile wide. The final episode of igneous intrusion occurred near the end of the Acadian Orogeny when basaltic magma formed diabase (medium-size crystals formed during cooling) dikes that cut the granites as well as older rocks. The dark-colored dikes cutting the light-colored granites make for striking contrasts at Schoodic Point and elsewhere in the park.

The following 300 or so million years of erosion prior to the Pleistocene left very little of a record to decipher. Plate separation again occurred about 200 Ma as rift basins opened between the separating plates. Sediments and volcanic rocks poured into these basins elsewhere in the Appalachians. Uplift and erosion likely produced a number of erosional surfaces with resistant, deep-seated rocks such as those of the Mt. Desert Range standing high as monadnock-like features. By the end of Tertiary time (about 2 Ma) the Mt. Desert area was probably a peninsula attached to the mainland. During the Tertiary, uplift rejuvenated the streams and they deepened their valleys. Most of the valleys were aligned in a north–south direction, generally parallel to the main fractures. Additional fractures formed where the massive granites expanded slightly as erosion of overlying rock reduced the confining pressure. Such an *unloading,* or *exfoliation* process, forms fractures that parallel the rock surface. Thus the stage was set for the coming of the glaciers that put the final touches on today's landscapes.

Glacial History

At times during the past 2 million years, worldwide climates cooled from 5 to 10°F (3 to 6°C). Continental glaciers developed around three main centers in Canada and pushed southward into the northern United States. The Labrador center was the source of the ice that covered the New England Province (Fig. 13–2). Pleistocene glaciers rode over the Mt. Desert Island area and rounded off the high ridges. Each episode of glaciation removed much of the evidence of the previous glaciation so that only evidence of the Wisconsin Glaciation is found at Acadia.

As the late Wisconsin ice advanced into Maine, ice piled up on the north side of the Mt. Desert Range. As ice topped the range, tongues of ice followed the valleys southward creating much of today's erosional topography—the deep, steep-sided valleys and the elongate rock-basin or trough lakes. These linear, *dorr valleys* (Gilman and others, 1988) are named in honor of George Dorr, without whose enthusiasm and perseverance there would not be a national park here. One lobe scoured particularly deep into the rocks, thus forming a fiordlike valley known as Somes Sound (Fig. 18–11). At one point, the wall of the fiord rises more than 1000 feet (303 m) above the floor. Other ice lobes excavated the elongate trough lakes, such as Eagle Lake and Jordan Pond east of Somes Sound, and Echo Lake, Long Pond, and Seal Cove Pond to the west (Fig. 18–9). If the excavations had been

FIGURE 18–11   Somes Sound, believed to be the only true fiord along the New England coast. It was excavated by a lobe of the Wisconsin glacier. (Photo by D. Harris)

deeper and longer by a few miles, they too would have been converted into fiords. Ice continued to thicken and by 21,000 years ago the entire island and all of New England were covered by ice—including Cadillac Mountain (Fig. 18–10) and other high mountains on Mt. Desert Island (Gilman and others, 1988).

About 18,000 years ago glaciers were in general retreat—thinning and melting back at their edges occurred more rapidly than ice was moving forward to replace the melting ice. Ice had thinned and retreated sufficiently by 13,000–14,000 years ago that the rocks of Mt. Desert Island once again began to be exposed to the light of day. As the land was uncovered by the glacier, the ice lobes in the dorr valleys behaved like valley glaciers and built small end moraines and outwash deltas as the sea began to once again lap up against Mt. Desert Island. The end moraine holding in Jordan Pond is readily accessible and is an excellent place to examine glacial features.

As the glacier ground its way across the mountains, it not only rounded off the tops but also sheared off the sides, forming elongate hills called *whalebacks,* or *roches mountonees.* Typically they are asymmetrical in long dimension, with a gradual slope formed by abrasion as the ice moved up and over the top; the down-glacier end is steep and jagged where, by quarrying, the glacier lifted out and removed blocks of rock. North and South Bubbles north of Jordan Pond are outstanding examples (Fig. 18–12).

Glacial *striations* (scratch marks), *grooves* (deeper furrows), and *polish* occur in a number of places, particularly on the bare whalebacks (Fig. 18–10). Glacial *erratics,* boulders carried far from where they were originally quarried out of the bedrock, are abundant. Look for the one perched precariously on the slope of South Bubble, plainly visible from the Loop Road (Fig. 18–12).

Battle with the Sea

As glaciers became larger, they locked up additional water in the form of ice, and sea levels dropped accordingly. A sea level drop of as much as 330 feet (100 m) exposed vast areas of the continental shelf. At the same time, the glaciers caused the land beneath to bend under the load of ice—about 100 feet (30 m) for every 300 feet (91 m) of ice. Ice thickness was likely about 1 mile (1.6 km). Each acre of ice 1-mile thick weighs about 7 million tons—a significant load on the crust!

During deglaciation interesting interactions occurred between rising sea level and the rebounding crust. Depending on how fast each occurred, sea level relative to Mt. Desert Island and the entire New England coast varied. When the Jordan Pond moraine and the outwash delta were forming about 12,500 years ago, sea level was at the same relative elevation as the top of the delta deposits (Gilman and others, 1988). Elsewhere on the island, former shoreline deposits occur as much as 210 feet (64 m) above present sea level. Later, as melting of the continental ice sheet continued, the land emerged above the sea and subaerial valleys were cut by streams. However, as rebound slowed and stopped, the sea continued its rise as glaciers continued to melt. Eventually sea level rose to its present level, drowning the lower ends of valleys and producing the irregular shoreline so typical of Acadia and New

FIGURE 18–12   South Bubble, a roche moutonee, or whaleback, reshaped by ice that rode over its top, from right to left. Note large erratic on skyline, which was carried almost to the top of South Bubble. (Photo by D. Harris)

FIGURE 18–13   A "Northeaster" hits the shores of Acadia National Park. (Photo by Art Hathaway)

England. Such a coast is known as a *submerged,* or *drowned,* coastline. Today, sea level continues to rise slowly along the Atlantic Coast at a rate of about 2 inches (0.8 cm) each century. This slow rise of sea level should continue, particularly in light of the reality that worldwide global warming due to human-caused factors will continue into the foreseeable future.

The sea continues to relentlessly batter the Acadian shoreline, particularly during the famous "Northeasters" (Fig. 18–13). By hydraulic action, *sea cliffs* are formed (Fig. 18–14) and blocks of granite are pulled from the cliffs and used as abrasive tools to pulverize other rocks. Weak fracture zones are converted into sea caves or narrow

FIGURE 18–14   Bald Porcupine, like other islands in Acadia National Park, has a prominent sea cliff facing the direction from which waves attack the island. (Photo by National Park Service)

chasms where less resistant rocks are removed. Vertical-walled slots sometimes form where vertical diabase dikes are attacked and removed by the sea.

The more resistant rocks are left extending out into the water as *headlands, heads,* or *points*. Great Head and Schooner Head on Mt. Desert island are typical. A large sea cave that extends back into the east face of Great Head can be readily seen from a boat. A short distance south of Great Head is Thunder Hole (Fig. 18–15), which can be reached by a short hike from the parking area. When the seas are heavy and a giant wave suddenly compresses the air and then displaces it from the cave, the thunderlike sound can be felt as well as heard.

In places, sea caves form along the sides of headlands and when enlarged sufficiently penetrate entirely through the rock wall, forming a *sea arch*. At Sand Point

FIGURE 18–15　Thunder Hole, a sea cave formed by the battering of waves, Acadia National Park. (Photo by D. Harris)

on the north shore, the sea cave was extended back along a prominent vertical fracture, forming the high narrow arch (Fig. 18–16).

As the waves continue to undercut the cliffs, the wave-cut platform at the base is enlarged; as the cliffs retreat with each rockfall, they become more prominent. At Otter Cliffs south of Thunder Hole the 107-foot-high (33-m) cliffs rise vertically above a well-developed wave-cut bench or platform.

Visiting Acadia

Access to the island is across the Mt. Desert narrows on the Trenton Bridge. Loop roads lead around the island to Bar Harbor and other towns and provide access to trail heads and points of interest along the craggy coast and inland areas, includ-

FIGURE 18–16   Arch formed near Sand Point on the northeast part of Mt. Desert Island at Acadia National Park. (Photo by D. Harris)

ing a spur road to the high point, the top of 1530-foot (466-m) Cadillac Mountain (Fig. 18–10, the highest point on the Atlantic Coast. From here, you can see the southern tip of Schoodic Peninsula, about 5 miles (8 km) to the east, the only part of the park on the mainland. About 25 miles (40 km) southwest of Mt. Desert Island is Isle au Haut, which can be reached by boat from Stonington. For those seeking solitude, this section of the park is a wilderness area. Four other islands near Mt. Desert Island, including Bald Porcupine (Fig. 18–14) are also parts of the Acadia National Park complex.

For those who wish to explore Acadia's landscapes in detail, a large topographic map, trail maps, and a visitors guide to the geology by Gilman and others (1988) can be purchased at the visitor center near Eagle Lake in the north-central part of the island (Fig. 18–9). Maps in Gilman and others' visitor's guide contain directions to sites of geologic interest where many of the special bedrock and geomorphic features described in this chapter can be best seen.

In 1986 legislation was passed to fix permanently the boundaries of Acadia National Park. Apparently the occasional expansion of the park was interfering with economic development and "progress" on the remaining private lands. Perhaps this is a microcosm of the remainder of the country as we continue down the path of eliminating open spaces and sacrificing quality of life to accommodate increasing numbers of people. Is this the direction we really want to go?

REFERENCES

Gilman, R.A., Chapman, C.A., Lowell, T.V., and Borns, H.W., Jr., 1988, The geology of Mount Desert Island: A visitor's guide to the geology of Acadia National Park: Maine Geological Survey, 50 p.

Hatcher, R.D., Jr., 1989, Tectonic synthesis of the U.S. Appalachians, in Hatcher, R.D., Jr., Thomas, W.A., and Viele, G.W., eds., The Appalachian-Ouachita Orogen in the United States: Geological Society of America, Boulder, CO, The Geology of North America, v. F-2, p. 511–535.

Osberg, P.H., Tull, J.F., Robinson, P., Hon, R., Butler, J.R., 1989, The Acadian orogen, in Hatcher, R.D., Jr., Thomas, W.A., and Viele, G.W., eds., The Appalachian-Ouachita Orogen in the United States: Geological Society of America, Boulder, CO, The Geology of North America, v. F-2, p. 179–232.

Rothe, R., 1995, Acadia: The story behind the scenery: KC Publications, Las Vegas, NV, 48 p.

Park Address

Acadia National Park
P.O. Box 177
Bar Harbor, ME 04609

CAPE COD NATIONAL SEASHORE (MASSACHUSETTS)

Cape Cod National Seashore is located along the eastern end of Massachusetts' prominent, oddly shaped peninsula, about 65 air miles (110 km) southeast of Boston. It is within easy access of a large population, as the 5-million yearly visitation suggests. Quaint and picturesque villages—some very old—shipwrecks, lighthouses, and sailboats add both flavor and variety to the beaches, dunes, and nearby cranberry bogs.

The attraction of Cape Cod was so strong that private and commercial development was proceeding at an alarming rate, and even many of the fiercely independent residents whose ancestors had lived there for hundreds of years realized that some form of protection was needed. In 1959 a bill was proposed by Massachusetts Senators John F. Kennedy and Leverett Saltonstall which would protect a 40-mile (64-km) stretch of coast as a national seashore, but would also allow recreational activities to continue and homeowners to remain within its boundaries. On August 7, 1961, President Kennedy signed into law the park he had proposed 3 years earlier!

Cape Cod is a uniquely shaped appendage of land that, as we shall see, also has a unique geologic history. On maps, this land area resembles an upturned right arm with a flexed muscle. The "bicep" connected to the mainland is upper Cape Cod, and the "forearm" and "fist" are lower Cape Cod, which is the location of the national seashore. The fist is a sandy extension of land called a *spit* that curves around to create the ideal harbor near the town of Provincetown. Here the *Mayflower* stopped in 1620 before proceeding to nearby Plymouth on the mainland, and here also was anchorage for some of the Yankee whaling and fishing fleets. Other pieces of the geologic puzzle of this area are nearby Nantucket and Martha's Vineyard, islands lying south of Cape Cod (Fig. 18–17).[1]

Before the Glaciers

The basement rocks, a few hundred feet below sea level, are granites, schists, and gneisses. Like the rocks of parts of the northern Appalachians, they are related to, and were once connected with, the rocks of Morocco in northern Africa (Rodgers, 1972). According to Schenk (1971), these rock masses—along with parts of Nova Scotia—are "microcontinental fragments broken off Africa." When the huge landmass split apart, perhaps 200 Ma, seawater filled the gaps.

By Late Cretaceous time, deltaic, swamp, and marine clay deposits, such as those exposed on Martha's Vineyard, were being deposited on the newly formed coastal plain. Similar Cretaceous deposits probably underlie Nantucket and Cape

---

[1]Strahler's (1966) *A Geologist's View of Cape Cod* is a widely used reference. See also Melham (1975), Kaye (1987), and Oldale (1980).

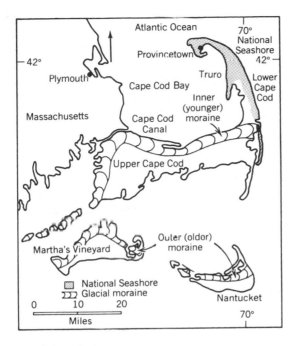

FIGURE 18–17   Map of Cape Cod area.

Cod. Tertiary coal deposits exposed beneath the sands near Provincetown, in combination with other deposits from nearby areas, indicate that alternate marine and nonmarine conditions persisted through the Tertiary Period. At the beginning of the Pleistocene there probably were low hills parallel to the present coastline. These hills would significantly influence the glaciers that were soon to cover the Cape Cod area.

Arrival of the Glaciers

Although vast continental glaciers originating in Labrador moved southward across New England and retreated a number of times, the Wisconsin glaciers left the best record and created most of the major geologic features of this area. As the glaciers accumulated additional ice, sea level was lowered, gradually exposing much of the continental shelf. Between 25,000 and 16,000 years ago, the ice sheet moved across Cape Cod and temporarily stabilized on what is now Nantucket and Martha's Vineyard (Oldale, 1980). Topographic irregularities produced different rates of ice movement and resulted in the extension of some sections of the glaciers—called lobes—farther than others, creating a scalloped ice front. Thus, some of the arcuate features here were glacier-caused but others were reshaped by the sea.

Moraines, some of them large, were formed at the front of the glacier as the frontal section melted and the rock material was freed from the ice. The prominent morainal ridges across Martha's Vineyard and Nantucket indicate that the ice front was essentially stationary for a long period. Meltwater carried materials out away from the ice front and deposited them on the *outwash plain*. Here, isolated blocks of ice that were partially buried in the outwash deposits eventually melted, leaving depressions called *kettles*. Those that were later filled with water are now *kettle lakes*.

The glacier began to retreat, and about 15,500 years ago tundra plants grew on Martha's Vineyard (Oldale, 1980). Later, the ice front was located on Upper Cape Cod where large morainal ridges were built up. Much of lower Cape Cod (the forearm part) was located between the two ice lobes and received mostly sandy outwash debris. Additional retreat left a depression between Cape Cod and the ice front (now Cape Cod Bay) that filled with meltwater and formed a glacial lake that spilled through what is now Cape Cod Canal.

When the glacial ice retreated farther, it left a vast, rolling landscape on the continental shelf. Tundra vegetation grew initially, later followed by pine and spruce, and finally by the maples and black oak hardwoods of today.

Return of the Sea

Shrinkage of glaciers freed vast quantities of water, and once again the level of the sea began to rise; the outer shelf was inundated first. Still the water levels rose, rapidly at first and then at gradually reducing rates. Lower hills became islands and then *shoals* as rising water levels and erosion by waves overwhelmed these areas. Whole forests were submerged, as drowned stumps off Provincetown and other areas indicate. Mammoth and mastodon teeth dredged from the continental shelf indicate that these giant animals roamed these woodlands several thousand years ago. About 6000 years ago, a morainal area southeast of Cape Cod was inundated and became the important fishing area known as Georges Bank. The wave action then concentrated on the next land area to the west, Cape Cod. At this time Cape Cod must have extended at least 2 miles (3.2 km) farther into the Atlantic (Strahler, 1966). Presently, Great Beach, the continuous beach facing the sea on the east side of lower Cape Cod, is retreating at a rate of about 3 feet (0.9 m) each year.

The wave-cut cliff behind Great Beach is also retreating at about 3 feet (0.9 m) each year—a victim of increasing sea level and large waves that crash incessantly against the cliffs during the famous "Northeasters" and other storms (Fig. 18–18). Highland or Cape Cod Lighthouse near Truro in the north part of Cape Cod is the oldest lighthouse on the peninsula. The lighthouse was a comfortable 527 feet (160 m) from the cliff edge when it was built in 1857. By 1996 it was only 110 feet (34 m) from the brink and oblivion. The entire structure was moved another 450 feet (137 m) inland (Berinato, 1997)—postponing its inevitable destruction by another 150 years.

FIGURE 18–18 Wave-cut cliff in glacial deposit, illustrating two of the geological processes involved in Cape Cod National Seashore. (Photo by M.W. Williams, National Park Service)

Storm waves that crash onto Cape Cod's beach wash away the loose glacial sediment that makes up the sea cliffs. Some of the sediment is carried offshore out of reach of the waves, but much is moved laterally by the longshore drift generated when waves impinge at an angle rather than parallel to the seashore. Sand washed up onto the beach is carried laterally a short distance along the beach before backwashing into the ocean again. Each wave repeats the process, and the net effect is that large volumes of sand are moved each year and, where the water deepens, longshore drift builds extensions of the beach as shoals and spits. Longshore transport near Truro on the north end of Cape Cod carries sand northward, where it is added onto the impressive spit that helps form the harbor at Provincetown. South of Truro, sediment moves southward, building Nauset and other spits.

Unfortunately for the longevity of Cape Cod, for each acre of land added by the sea, 2 acres are destroyed each year. At the present rate of erosion, the northern part of Cape Cod will become an island in 4000–6000 years when the narrow section south of Truro is severed by the sea (Kaye, 1987). More islands will form, and each will in time be reduced to shoals unless glaciers reverse the process by causing sea level to lower. Fortunately, for many generations we will be able to enjoy the rich cultural and natural features of this changing landscape.

REFERENCES

Berinato, S., 1997, The Fate of Cape Cod: Earth, v. 6, no. 1, p. 58–59.
Kaye, G., 1987, Cape Cod, the story behind the scenery: KC Publications, Las Vegas, 48 p.
Melham, T., 1975, Cape Cod's circle of seasons: National Geographic, v. 148, no. 1, p 40–65.
Oldale, R.N., 1980, Geologic history of Cape Cod, Massachusetts: U.S. Geological Survey, Popular Publications Series, 23 p.
Rodgers, J., 1972, Latest Precambrian rocks of the Appalachian region: American Journal of Science, v. 272, p. 507–520.
Schenk, P.E., 1971, Southeastern Atlantic Canada, northwestern Africa, and continental drift: Canadian Journal of Earth Science, v. 8, p. 1218–1251.
Strahler, A.N., 1966, A geologist's view of Cape Cod: Natural History Press, Garden City, NY.

Park Address

Cape Cod National Seashore
99 Marconi Site Road
Wellfleet, MA 02667

NINETEEN

Coastal Plain Province

The Coastal Plain Province consists of the seaward-sloping lowlands along the Atlantic Ocean and Gulf of Mexico and the submerged section, the continental shelf. Fluctuations in sea level and uplifts of the land have both affected the area of the two sections.

The majority of Congress was far sighted and believed that the general public and not just the wealthy should have access to the beaches and the sea. Public lands were needed along the seashores so that all who wish to can "feel the wind in their faces and the surf on their feet." Thus Cape Hatteras was established in 1937, and eight more seashores and a recreation area were added from 1961 to 1975. In this province, the Park Service is given stewardship for a number of areas including those discussed in this chapter—Cape Hatteras, Cape Lookout, and Padre Island National Seashores and Biscayne and Everglades national parks (Plate 19).

At the beginning of the Cenozoic Era, the shoreline of the continent was inland from its present position, particularly in the Mississippi Embayment, which extended northward to the southern tip of Illinois. Part of the present coastline is extremely irregular, with deep indentations where submergence has caused the drowning of river mouths; Chesapeake Bay is a good example.

Essentially all of the rocks of the Coastal Plain Province fall into one of three groups. Discontinuously around the inner border of the province are marine sedimentary rocks deposited when the Cretaceous sea invaded this part of the continent. They rest on Paleozoics in most places and Precambrian metamorphics elsewhere—the products of accretion of microplates and mountain building during the Appalachian Orogenies. In the middle section, marine Tertiary-age rocks rest on the Cretaceous and dip gently toward the sea. Along the coast sediments of Quaternary age form a more or less continuous band of varying width from southern Texas to Long Island.

Salt domes, structures unknown in most areas, are abundant in the coastal section of Texas, Louisiana, and Mississippi, and offshore. (The principles involved in their formation were discussed in connection with the salt anticlines in the

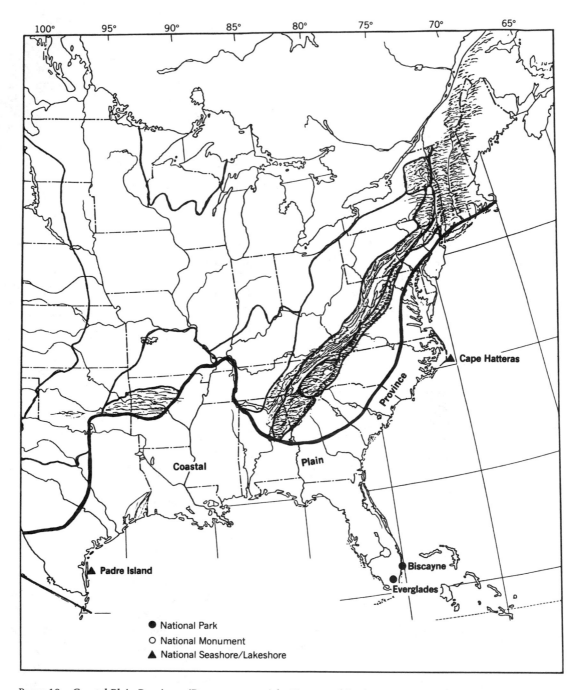

PLATE 19   Coastal Plain Province. (Base map copyright Hammond Inc.)

Paradox Basin in the Colorado Plateaus Province.) Salt domes are fascinating to geologists, but they are even more intriguing to oil companies that have obtained millions of barrels of oil from these structures.

The development of the vast Mississippi delta is an interesting story summarized by Thornbury (1965). By delta building, this great river has been extending the shoreline out into the Gulf of Mexico, and it is continuing its work today. Here is a unique environment, one of the very few not now included in the National Park System.

Glaciers directly affected only the northern area; many of the glacial deposits are beneath the sea, but large moraines cover extensive sections of Cape Cod and Long Island. Farther south, terraces generally believed to be of marine origin are prominent features of the subdued landscape.

Karst features, essentially absent elsewhere in the province, are abundant in central and northern Florida. Sinkholes and the caverns beneath them are widespread features north of Lake Okeechobee. Only in the higher areas are the caverns open; most of the vast cavern system is below sea level.

Coral reefs are attractive features around the southern coast of Florida, and offshore. Extending for about 150 miles (242 km) south and then west, the Florida Keys, an arcuate chain of islands, end at Key West.

Nearly 300 long, narrow barrier islands, formerly called offshore bars, extend discontinuously for 2000 miles (3220 km) around the Gulf Coastal Plain and from Cape Cod to Georgia (Dolan and Lins, 1985). The islands are "barriers" that experience the full fury of the sea—thereby sparing the mainland the full impact of waves.

Regarding the origin of these offshore islands, the only consensus at present is that there is no consensus, except that the classical interpretation of Douglas Johnson (1919) of emergent shorelines is untenable. A gently sloping coastal plain and an abundance of sand seem to be the keys to their development.

Possibilities that have been considered include the formation of coastal dune ridges that are later submerged, the development of complex *spits* (linear sand bars created by longshore currents) on a shoreline of submergence, the emergence of offshore bars, and the shoreward migration of offshore bars. Whatever the theory on the mechanism or combination of processes, it must take into account the rise in sea level that resulted from the melting of Wisconsin-age glaciers, beginning about 11,000 years ago. The recent sea level rise of about 6 inches (15 cm) every century contributes strongly to the instability and landward migration of the loose piles of sand known as barrier islands. The expected increase in the rate of sea level rise to 18 inches (46 cm) per century due to global warming will greatly increase the destabilization of the barrier islands and flooding of coastal areas—a frightful prospect as an exploding human population moves seaward as the shoreline moves landward.

Clearly these island-forming processes are the results of the work of the waves and the wind. All coasts are lashed periodically by storms, sometimes with hurricane force. The dune ridge superimposed on all barrier islands may be locally shifted landward; both the seaward and the landward beaches are likely to be eroded; in some cases the waves break through, forming new inlets or passes. Thus island migration occurs mostly by overwash and the opening and closing of inlets.

An understanding of shore dynamics adds greatly to one's appreciation of the barrier islands and the beach environment. Wind-generated water disturbances called *swells* in the open ocean organize into distinct waves as the sea shallows. The waves grow higher and topple over in the zone of breakers where their energy moves large amounts of sand and other sediment. Gentle waves bring sediment onto the beach as the *swash* runs up onto the edge of the land. Storm waves erode beaches and carry the sediment into deeper water where it is stored in offshore bars. Because waves strike the shore at an angle, sediment also moves in a lateral direction under the influence of *longshore drift*. Details of shore processes and shore morphology are found in Bascom (1964) and summaries are found in most physical geology textbooks. Other excellent summaries of shore processes and landforms are available in Leatherman (1982) and Toops (1993).

The Park Service had two objectives in mind when they established national seashores such as Cape Cod, Cape Hatteras, and Padre Island. One purpose was to provide the public with additional recreation; the other was to preserve the beach areas in their natural state. It soon became apparent that their newly adopted areas were changing—in some cases drastically—with each violent storm. Their response was to initiate a program of stabilizing the dunes by building sand fences, and the shorelines by installing elaborate erosion control structures. After spending large amounts of money, in most cases without overwhelming success, the policy was reexamined. The Park Service had not attempted to thwart nature by trying to stop the erosion in Grand Canyon or in the Badlands. The decision was made in 1973 to let nature take its course, even with this highly dynamic natural system (Dolan and Hayden, 1974). One bonus result from the new policy—it provides the opportunity for scientists to study natural processes at work, without any modifying influences by humans.

REFERENCES

Bascom, W., 1964, Waves and beaches: The dynamics of the ocean surface: Anchor/Doubleday, Garden City, NY.

Dolan, R., and Hayden, B., 1974, Adjusting to nature in our national seashores: National Parks and Conservation Magazine, v. 48, no. 6, p. 9–14.

Dolan, R., and Lins, H., 1985, The Outer Banks of North Carolina: U.S. Geological Survey Professional Paper 1177-B, 103 p.

Johnson, D.W., 1919, Shore processes and shoreline development: Wiley, New York.

Leatherman, S.P., 1982, Barrier island handbook: Coastal Publications, Charlotte, NC.

Otvos, E.G., Jr., 1970, Development and migration of barrier islands, northern Gulf of Mexico: Geological Society of America Bulletin, v. 81, p. 241–246.

Thornbury, W.D., 1965, Regional geomorphology of the United States: Wiley, New York.

Toops, C., 1993, National seashores: The story behind the scenery: KC Publications, Las Vegas, NV, 48 p.

PADRE ISLAND NATIONAL SEASHORE (TEXAS)

The National Park Service broadened its coverage of natural environments by creating national seashores, beginning with Cape Hatteras in North Carolina in 1937. They now dot our shores, from Point Reyes near San Francisco to Cape Cod on the East Coast.

Padre Island National Seashore, established in 1962, occupies the 80-mile-long (130-km) midsection of the 113-mile-long (180-km) Padre Island near Corpus Christi, Texas. First called the White Islands because of its almost-white sands, this seashore was renamed around 1800 when Padre Nicolas Balli received a large Spanish land grant that included the island. Actually, Padre Island is a chain of elongate islands separated by narrow "passes" that were broken through during violent storms, only to be sealed shut by sand deposited by longshore currents. But new passes were broken through to once again connect the Gulf of Mexico with Laguna Madre, the long lagoon that lies between Padre Island and the Texas mainland.

The Intracoastal Waterway extends through the length of Laguna Madre; when the channel was dredged in 1949, the dredged material was piled up alongside. These spoil piles became island nesting sites for such waterbirds as the laughing gull, foresters tern, great blue heron, roseate spoonbill, and the white pelican. Because Laguna Madre is shallow and essentially cut off from the gulf, its waters have become increasingly saltier; according to Chief Naturalist Robert G. Whistler (personal communication, 1981) the salinity is at times almost twice that of gulf waters. Therefore, only salt-tolerant plants such as shoalgrass can survive in the marshlands near the lagoon.

Padre Island ranges in width from a few hundred yards to about 3 miles. The northern end of the seashore is sufficiently wide for the various geomorphic features to be clearly represented. High sand dunes, the foredunes, form a ridge that rises abruptly above the beach, as shown in Figure 19–1a. In the area beyond the foredunes there are many clusters of dunes, some temporarily stabilized by low-growing vegetation and others that reform on an annual basis (Kocurek and others, 1992). Many of the interdunal areas are marshlands, but small lakes occupy some of the depressions. Farther west are the back-island dunes, with the tidal flats and Laguna Madre beyond. The best way to see a representative section of the western part of the seashore is to take the Bird Island Basin Road, which leads west a short distance north of the ranger station.

Your first stop should be the visitor center where you can obtain valuable information, including warnings of hazards such as the Portuguese man-of-war jellyfish, stingrays, rattlesnakes, and unexploded ammunition (relics of the days when the U.S. Navy used parts of the island as target sites). Next, get an overview

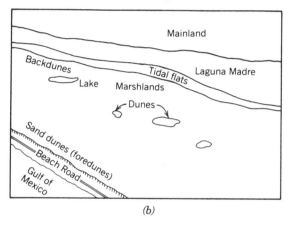

FIGURE 19–1   (*a*) Padre Island National Seashore: air view southwest across Padre Island (Photo by R. Whistler, National Park Service). (*b*) Padre Island National Seashore: explanatory sketch showing zones and features.

of the island from the observation tower on Malaquite Beach. With binoculars, you can see all of the features mentioned above.

Origin and Development of Padre Island

Padre Island is young, less than 5000 years old (Fisk, 1959). Several mechanisms have been proposed for the origin of barrier islands, but no one explanation appears to be satisfactory for all of the world's offshore islands.

For Padre Island, Weise and White (1980) favor the offshore bar origin. They picture a sandbar, a ridge of sand piled up by the waves, in shallow water near shore. With continued wave action the bar is built up above sea level, thus forming the embryonic island that will be widened and heightened by subsequent wave and wind action. But problems arise when we examine this sequence carefully. Anything exposed to the lashing of the waves will be destroyed in time; a low ridge of loose sand would not withstand the onslaught of even one of the hurricanes that occur here every few years. Also, it is significant to note that, although there are submerged bars offshore now, at no point have they been built up to the surface.

Other workers visualize an entirely different sequence. Frank Ethridge (personal communication, 1981) begins the sequence with a long, high ridge of sand bordering the mainland coast—coastal dunes like those found in many places today. Sea level has been rising since the great glaciers began to melt back some 11,000 years ago; eventually the waters of the gulf invaded the area back of the dune ridge, thus forming Laguna Madre and isolating Padre Island from the mainland. Hoyt (1967) and others agree that this explanation is the logical one for the origin of Padre Island, at the same time recognizing that other barrier islands may have been formed in other ways.

As is true of all barrier islands, Padre Island is being reshaped. During heavy storms, the waves erode the sand away from the beach side of the foredunes as much as 9 feet (3 m) per year, thus shifting the beach–dune boundary westward (Mathewson, 1975). From time to time, hurricane-driven water overwashes the dunes in places and forms new passes through the ridge. The floodwaters rush across the island and deposit sand as washover fans where the water spreads out over the tidal flats, even out in the lagoon. Thus, the island is being extended westward at the expense of Laguna Madre.

On the gulf side of the island, the wind picks up the sand from the beach and carries it up onto the ridge, where much of it is deposited. Some, however, is carried far beyond the ridge and is deposited on the flats, where it is shaped into dunes. Until the dunes are stabilized, they are moved about by the wind, as "wandering" dunes. The back-island dunes form a low, irregular ridge along the east side of Laguna Madre. In many places the dunes are at least partially stabilized by grasses, low-growing shrubs, and by the long railroad vine. In the interdunal areas, grasses thrive in most places, providing good grazing for cattle prior to the time that the national seashore was established. Live oak trees, once fairly widespread, are almost entirely gone, having succumbed to droughts and to burning by the ranchers—to improve the grass.

When fierce winds blow, the least stabilized sand is carried away, forming "blowouts," bowl-shaped "wounds" on the windward side of the dunes. Once the deflation process starts, it is likely to continue for some considerable period, sometimes enlarging the blowout to enormous size. A blowout that will probably be there for years is along the Grassland Nature Trail in the northern section of the seashore. An hour or two on this nature trail is time well spent, assuming that you keep a sharp eye out for rattlesnakes.

Visiting Padre Island

The *Padre Island Field Guide* by Hunter and others (1981) contains a road log and other information that will help you make the best possible use of your time. The

Malaquite Beach section is off-limits for all vehicles. Most of the roads are travelable by road car, but a 4-wheel-drive vehicle or a good pair of walking shoes and a backpack is a must if you want to explore the 55-mile (89-km) primitive south end of the island. You will be well rewarded if you spend 2 or 3 hours on the beach near the northern end of the seashore, at the end of the North Beach Access Road. To the south of the parking area, the beach is for walking only; as a result, burrowing animals such as the ghost crabs, mole crabs, and ghost shrimps have the opportunity to live their lives without being disturbed or prematurely ended by cars, trucks, and motorcycles. If you are there at the right time, you will see the beautiful blue Portuguese man-of-war jellyfish. *Do not touch;* if you get stung, *do not rub!* Go immediately to the nearest first-aid station for treatment.

Padre Island attracts birds of many kinds, especially during the winter. One big bird that probably will not be there is the whooping crane. To see it, you will have to go to Aransas National Wildlife Refuge about 50 miles (83 km) north of Corpus Christi. To see these big fellows "gal-loping" low over the water is an experience to long remember.

The animals on and around Padre Island include the coyote, jackrabbit, the Mexican freetail bat, several kinds of rodents, a few Ridley sea turtles, lizards, and snakes. All sea turtles are threatened or endangered—partially the result of humans continuing to develop barrier islands and eliminate nesting sites. Padre Island has one colony of Kemp's Ridley, the world's most endangered sea turtle species. Their activities can be observed from mid-April through August.

There are also fish in the gulf and in Laguna Madre. However, with an abundance of fascinating things to see at Padre Island, surely some of them will be of sufficient interest to distract you from swimming and fishing, at least for part of the time. Padre Island is open throughout the year; winters are delightful except when a "norther" moves in to chill the air.

Most barrier islands are "graveyards for ships." Since hurricanes are a common occurrence, Padre Island has had its share of shipwrecks. Long ago, when the island was the home of cannibalistic Indians, the hazard was even more frightening. Of the many who were "fortunate" in surviving the storm in 1533, when a Spanish fleet ran aground, only two escaped from the Indians.

REFERENCES

Fisk, H.N., 1959, Padre Island and the Laguna Madre Flats, coastal south Texas, in Russell, R.J., chair, 2nd Coastal Geography Conference (April), p. 103–151.

Hoyt, J.H., 1967, Barrier island formation: Geological Society of America Bulletin, v. 78, no. 9, p. 1125–1135.

Hunter, R.E., Watson, R.L., Dickinson, K.A., and Hill, E.W., 1981, Padre Island National Seashore Field Guide: Corpus Christi Geological Society, Corpus Christi.

Kocurek, G., Townsley, M., Yeh, E., Havholm, K.G., and Sweet, M.L., 1992, Dune and dune-field development on Padre Island, Texas, with implications for interdune deposition and water-table-controlled accumulation, Journal of Sed. Pet., v. 62, no. 4, p. 622–635.

Mathewson, C.C., and Piper, D.P., 1975, Mapping the physical environment in economic terms, Geology, v. 3, no. 11, p. 627–629.

Weise, B.R., and White, W.A., 1980, Padre Island National Seashore: A guide to the geology, natural environment and history of a Texas barrier island: Bureau of Economic Geology, Austin, University of Texas.

Park Address

Padre Island National Seashore
9405 S. Padre Island Drive
Corpus Christi, TX 78418-5597

EVERGLADES NATIONAL PARK (FLORIDA)

At the southern tip of Florida, where the waters of the Gulf of Mexico meet those of the Atlantic, is a vast, flat lowland known as the Everglades. Large areas are covered by shallow water—in lakes, freshwater sloughs, estuaries, and bays. Plant material accumulates faster than the decay process here, thus producing layers of a low-grade coal called peat and providing geologists with a modern analog to understand how extensive deposits of coal formed in the geologic past (Bond, 1986).

This aquatic, subtropical environment of sawgrass (Fig. 19–2) and mangrove trees is alive with crocodiles, alligators, West Indian manatees, roseate spoonbills, anhingas (Fig. 19–3), wood storks, and a host of more common animals and birds (de Golia, 1993). It was for the purpose of preserving this unique environment that about 2190 square miles (5670 km²) of land and water were established as Everglades National Park in 1947. In spite of the seeming abundance of wildlife, their numbers have drastically declined since the park's inception as serious environmental modifications occurred from outside of the park boundaries. Water diversion and pollutants have led scientists to grim predictions about the area's future. Their concerns are reflected in Vice President Gore's summary that "we are dealing with an extremely fragile ecosystem that is on the verge of collapse" (in O'Connell, 1996, p. 13).

The park extends westward from Homestead, where the park headquarters are located, to Everglades on the Gulf of Mexico (Fig. 19–4). On the west side the park extends out into the gulf and includes the Ten Thousand Islands; on the south it extends across Florida Bay almost to the Florida Keys. Highway 27 leads west from Homestead and then south to Flamingo at the southern tip, with stub roads to Royal Palm, Pa-hay-okee Overlook, Mahogany Hammock, and other points of interest.

The land that is now the Everglades was beneath the sea until recent geologic time. Now, after being above the sea for a short period, the Everglades are gradually losing ground (Gleason and others, 1974). Although the boundaries of the park remain constant, the land area is gradually shrinking.

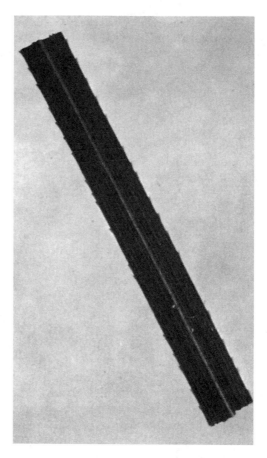

FIGURE 19–2   Sawgrass, a sedge, has teeth of opaline silica capable of inflicting severe flesh wounds.

Geology

The geology of Everglades National Park is relatively simple. Along the western side, limestones of the Miocene Tamiami Formation lie beneath the swamps and bogs (Gleason, 1972). Apparently no Pliocene sediments were laid down in this area. Most of the park is underlain by the Pleistocene Miami Oolite, or oolitic limestone, composed largely of tiny, spherical concretions that resemble fish roe.

The oolite is porous and permeable, and the surface layers are unusually susceptible to chemical weathering. Consequently, the surface is in many places pitted and irregular. On this surface, peat accumulation has been rapid because of the luxuriant vegetation—mangrove trees, water-lilies, and sawgrass. Perhaps this is a present-day coal-forming swamp.

FIGURE 19–3 The Anhinga, or snake-bird or water turkey; also called the "drip-dry" bird. Because of lack of oil in its feathers, it must sit with wings spread in order to dry out. (Photo by National Park Service)

FIGURE 19–4 Everglades National Park headquarters, from along Anhinga Trail.

If we probe deeply into the past, we find that beneath the surface rocks are Paleozoic metamorphic rocks similar to those exposed north of the Florida Peninsula. This information, obtained from wells, leads to the realization that long before the Miami Oolite was laid down, a projection of the continent occupied the same general area as the present Florida Peninsula. That landmass was submerged and deposited upon, most recently by the oolite, and then became emergent again in relatively recent times.

Everglades Unique Environment

Three words—low, limy, and flat—describe the Everglades of today. The only features that break that level line as you look out across the sawgrass flats are the *hammocks,* which rise several feet above the swamps (Fig. 19–5). They are elongate, tree-covered mounds aligned in a north–south direction. Although their origin is puzzling, it is probable that the trees play an important role in hammock development. Once established, possibly along fracture zones in the limestone, the trees protected these "islands" from devastating floods which, until recently, periodically rushed southward from Lake Okeechobee. Then, by solution and reprecipitation, "rock reefs" developed into elevated mounds. Additional investigations will probably reveal the solution to this intriguing problem. Regardless, visitors will not want to miss Mahogany Hammock, which is along the road to Flamingo.

FIGURE 19–5   Sawgrass flats with tree-covered hammocks rising above the vast plains. (Photo by National Park Service)

The Everglades are one part of a large system of rivers, lakes, and swamps that drain the Florida Peninsula from just south of Orlando near the center of the state to Florida Bay on the south tip. The Kissimmee River, Lake Okeechobee, and the adjacent Big Cypress National Preserve along the north edge of the Everglades are all part of the south Florida aquifer. The environment as nature had designed it for the past 5000 years has been drastically changed by humans—so much so that Interior Secretary Bruce Babbitt noted in 1993 that the Everglades is the most imperiled of our national parks and the one that is in most danger of extinction. Unfortunately, conditions have not improved significantly since then. In the pre-1940s the Everglades was in reality a huge river—a 50-mile-wide (80-km) river that was 6 inches (15 cm) deep—a virtual river of grass. Today it is half as large as a result of drainage modifications promoted by industry and land developers.

Originally, Lake Okeechobee, a large lake about 75 miles (125 km) north of the Everglades, overflowed periodically, and its floodwaters swept southward across Big Cypress Swamp and the Everglades. Thus the water in the lakes and swamps was replenished, and the habitats for fish, alligators, and water birds were secure. "Reclamation" projects included the construction of canals that carry Lake Okeechobee's excess water southeastward to the Atlantic—thereby bypassing the Everglades. Consequently, the Everglades water is manipulated by humans, depriving the area of floodwaters from the north, and during prolonged droughts many of the lakes and swamps dry up and large numbers of animals and birds die (de Golia, 1993). Nearly 90 percent of what had seemed in the 1930s like a limitless population of wading birds are now gone.

Residential development and agriculture moved into the former wetlands and large quantities of fertilizer from sugar cane fields poured into the Everglades. Cattails replaced the sawgrass and further choked off water flow. The decrease of freshwater flowing from the Everglades into Florida Bay to the south has caused the bay to become saltier, warmer, and to contain less oxygen—causing die-offs of shrimp and fish—important components of Florida's fishing industry.

Visiting the Everglades

Along the coastal area, beach deposits and low, wave-cut cliffs are the main features. In places, the results of recent hurricanes that sweep across the Everglades will remind the park visitor of this unique aspect of the environment.

Many of the interesting features of America's largest swamp can be seen along the highway to Flamingo. Others may be observed along the Tamiami Trail from Miami west to the town of Everglades, and on Western Water Gateway to the Ten Thousand Islands. Parts of the park must be seen from a boat, preferably with an experienced guide. If you are manning your own boat, navigation charts are essential. And be sure to file a "float plan" so that the park ranger will know which swamp to search in case you do not report in as scheduled. Needless to say, the Everglades may be visited at any time of the year.

Finally, a new chapter is being written in the Everglades. The century-long program of channelization of the Kissimmee and other rivers has seriously damaged the environment of the Everglades. The lengthy and expensive process of removing levees and canals in order to restore the original meandering character of the streams was begun in 1984. Litigation requires that the sugar industry reduce its pollution to lower levels. Will these changes be enough and soon enough to save this unique ecosystem or will this be another black mark on our ability to be good stewards of our planet?

REFERENCES

Bond, P., 1986, The Everglades National Park, in Neathery, T.L., ed., Southeastern Section of the Geological Society of America: Geological Society of America, Boulder, CO, Centennial Field Guide Volume 6, p. 343–344.

de Golia, J., 1993, Everglades, the story behind the scenery: KC Publications, Las Vegas, NV, 64 p.

Gleason, P.J., Cohen, A.D., Brooks, H.K., Stone, P., Goodrick, R., Smith, W.G., and Spackman, W., 1974, The environmental significance of Holocene sediments from the Everglades and saline tidal plain: Miami Geological Society, Miami.

O'Connell, K.A., 1996, Gore unveils Everglades plan. National Parks, v. 39, no. 5-6, p. 13–14.

Park Address

Everglades National Park
40001 State Road No. 9336
Homestead, FL 33034

BISCAYNE NATIONAL PARK (FLORIDA)

Biscayne National Park—289 square miles (750 km^2) of blue water, mangrove shoreline, a linear group of limestone islands called keys, and coral reefs—is located about 20 miles (33 km) south of Key Biscayne and immediately north of Key Largo, Florida, within view of downtown Miami. Biscayne is mostly a water park—one where the fascinating biologic processes that produce limestone reefs can be observed in action beneath the warm waters of the Atlantic Ocean, and the end result can be viewed in the limestone bedrock exposed on the keys. Thus the reef and other carbonate (mostly limestone)-forming environments of southern Florida, along with those in the Bahama Banks some 40 miles (km) east across the Straits of Florida, are recognized worldwide as modern analogs that help us understand the huge accumulations of limestone found in the rock record (Bond, 1986).

Native Americans used the area for centuries. Most recently in the 1830s the Seminole Indians occupied the area as they were driven southward from their previous homes by the advancing Euro-American settlers. During the seventeenth and eighteenth centuries pirates had "easy pickings" as numerous storm-crippled sailing ships limped into the area or wrecked on the treacherous shoals in southern Florida. Planned development of the area in the 1960s would have forever changed the environment and locked up this tropical paradise for the benefit of a few wealthy individuals. Fortunately the area was declared a national monument in 1968 and upgraded to park status in 1980 (Fig. 19–6).

Geographic Setting

Biscayne National Park is located in southeast Florida with headquarters at Convoy Point, about 9 miles (15 km) east of the city of Homestead. Nearby areas of interest are Everglades National Park several miles west and John Pennekamp Coral Reef State Park immediately to the south.

From west to east the park consists of the mainland mangrove shoreline, Biscayne Bay (part of the Intracoastal Waterway route), with several elongate islands or keys along its east side, then Hawk Creek Channel beyond which North America's northernmost living coral reefs rise above a shallow platform (Fig. 19–7). Beyond the reefs in the Straits of Florida lies the eastern boundary of the park.

FIGURE 19–6   Headquarters, Biscayne National Park.

The keys extend along southeast Florida from Ragged Keys in Biscayne National Park southwest to Key West off the southwestern coast of Florida. Elliot Key, the largest of the islands, stretches for 8 miles (12.9 km) along the east side of Biscayne Bay; much smaller, Sands Key extends 2 miles (3.2 km) to the north of Elliot Key. South of Elliot Key, across Caesar Creek (a channel or passageway between the islands), Old Rhodes and Totten Keys are the main islands; farther west, beyond the Intracoastal Waterway, are the Arsenicker Keys, which rise from the shallows of Biscayne Bay (Fig. 19–7).

The days of the pirates are long gone, but Caesar Creek is a reminder that Black Caesar and his rogues once struck fear into the hearts of the unfortunates who were shipwrecked in these dangerous waters. On Turkey Point, just south of park headquarters on the mainland, the cooling towers of a nuclear power plant rise 415 feet (126 m) into the sky, lending an ominous air to the otherwise tranquil natural scene.

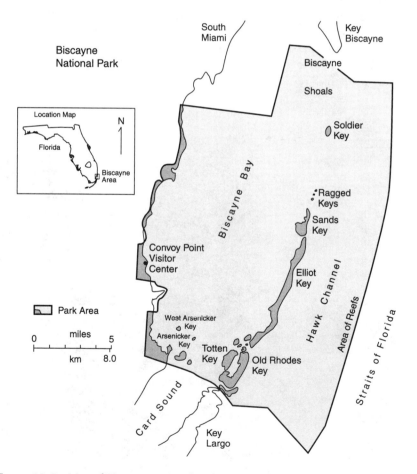

FIGURE 19–7   Map of Biscayne National Park.

Biology

The lifeforms in the park are highly varied and seemingly endless, both beneath its waters and on land. Colorful flower (Fig. 19–8), star, brain, staghorn, elkhorn, and finger corals on the reefs resemble plants in that they are "fixed." Free to move are fish of many colors, including the four-eyed butterfly and the French angelfish. Biscayne Bay is a few feet (1 m) to perhaps 15 feet (5 m) deep and contains vast seagrass beds utilized by fish, spring lobsters, and loggerhead turtles. The bottle-nosed dolphin also uses the underwater park and on land, the raccoon, the marsh hare, and the exotic Mexican red-bellied squirrel scamper about. Mangrove swamps contain numerous rookeries, as well as alligators, snakes, and other animals.

Unlike at Padre Island, where trees are essentially absent, the Biscayne keys support a dense and luxuriant forest. Tropical hardwoods of many types, including the mahoganies, dominate the interior of the islands, which are ringed by the spraddle-legged mangroves that defend the shoreline. A quiet few minutes in the Hurricane Creek passageway, with the mangroves hovering over you, is a delightful, relaxing interlude—when the winds are gentle breezes. In many places stately palms sway gently when the wind is light and bend well when gales are blowing.

The salt-tolerant mangroves with their sturdy system of prop roots serve a geologic purpose by defending the keys and mainland coasts during storms and hurricanes. Removal of some of the protecting mangroves by developers in the 1950s resulted in drastic shoreline changes, destruction, and loss of lives during hurricane Donna in 1960 and Betsy in 1965. State law now requires construction setbacks to help protect the delicate Florida shoreline and ecosystems.

Park visitors find it amazing that the dense forest on Elliott Key is not virgin and that most of the island was under cultivation for a time—until the late 1930s.

FIGURE 19–8   Flower coral (*Eusmilia fastigiata*), one of many fascinating corals on the reef at Biscayne National Park. (Photo by National Park Service)

Surprisingly, the existing forest is believed to be similar in type and diversity to the one that was destroyed. What mars the forest of today, when viewed from the air, is the gash that extends from one end almost to the other. It was the work of the developers' bulldozers, immediately before Biscayne National Monument was established. But even this ghastly wound will heal; and even the Sargent's palm, known also as the buccaneer palm, rare and almost wiped out, will recover under the protection of the National Park Service.

The bird life of the park is abundant, changing with the seasons. Herons and cormorants are year-round residents. There are many nesting areas in Biscayne; particularly popular are the Arsenicker Keys in the southwestern part of the park.

Biscayne's Past

The geologic story involves only the last part of the last chapter in the geologic history book—the latter part of the Pleistocene Epoch. The Key Largo Limestone, with type-section on Key Largo, outcrops on the chain of keys and lies beneath the waters of the eastern part of the park. It was a living coral reef over 100,000 years ago (Hoffmeister, 1974). Although corals were probably dominant, many other marine invertebrate animals and plants (mainly algae) contributed their energies and their skeletons in the reef-building process.

Beneath Biscayne Bay and exposed along its western shores is the Miami Oolite, a limestone only slightly younger than the Key Largo. Although other materials, fossils and fragments of fossils ("hash"), are present in varying amounts, the Miami is largely made up of tiny spherical concretions (oolites). The oolites, resembling fish roe, are formed by accretion of one thin layer upon another, generally around a tiny nucleus, as the limy ooze on the seafloor is agitated. The snow-white sand on the bottom of Biscayne Bay is derived in large part from the Miami Formation.

Reef building was interrupted when sea level was lowered significantly during the Wisconsin glacial advance; left high and dry, the corals and other fixed life-forms died, leaving the reef exposed to the elements. The top part of the reef was destroyed by weathering and wave action, leaving an eroded reef platform.

Later, about 11,000 years ago, when the glaciers began to melt, the seas rose again. Soon, a new generation of corals began to build new reefs, particularly in the eastern section of the park. As before, essentially the same assemblage of animals and plants assisted in the reef-building process. *Symbiosis* was again a way of life on the reef: One animal or plant paired off with some other species, for their mutual benefit. For example, corals provided protection for the algae (plants) and the algae provide life-sustaining oxygen to the corals. It is a beautiful system, one that is difficult to improve upon. It will continue to work well until some major change takes place, either as a result of natural causes or by human ignorance, perhaps plain carelessness.

At Biscayne we have the dead reefs—the keys—and living reefs abuilding on top of the eroded reef platform. In time they will be as large as those of the past, unless sea level is lowered again—or instead if all of the water tied up in existing glaciers melts and returns to the sea.

If global warming and the resultant rise in sea level currently underway proceeds too rapidly, delicate natural systems such as that in Florida will be greatly stressed. "High" elevations on the keys are 18 feet (5.5 m), and much of the area is barely above high tide level! Elevations in all of south Florida are less than 60 feet (18 m). Thus even small changes in sea level will have devastating effects on coastal environments both here and worldwide. That some or most of the current climatic warming is of our own making does not speak well of our stewardship of the planet.

Of more immediate concern is the slow die-off of coral reefs that began in the 1970s in the Caribbean and especially in the Florida Keys. Die-offs of the prolific staghorn corals and sea urchins allowed algae to rapidly expand. Again, humans have upset the environmental equation. Warmer sea temperatures, decreased freshwater flow through the Everglades system into Florida Bay, and nutrients from agricultural sources and human sewerage are all likely related to the diminishing coral reefs in the Florida area (Shinn, 1996).

Visiting Biscayne

Hurricane Andrew scored a "direct hit" on Biscayne in 1992. Its arrival at high tide with sustained winds of 150 miles (240 km) per hour devastated man-made structures including the now rebuilt Convoy Point Visitor Center. The visitor center is accessible by automobile from the north. Concessioner-run tour boats with glass bottoms enable visitors to see the keys and glimpse the reef environment. Snorkel and dive tours (divers must be SCUBA certified) allow an up-close look at the underwater part of the park. Touching coral makes them more vulnerable to disease and should be avoided. Those with their own boats should use the detailed navigational charts that are available and have a good radio should storm warnings be issued.

REFERENCES

Bond, P.A., 1986, Carbonate rock environments of south Florida, in Neathery, T.L., ed., Southeastern Section of the Geological Society of America: Geological Society of America, Boulder, CO, Centennial Field Guide Volume 6, p. 345–349.

Hoffmeister, J.E., 1974, The geologic story of south Florida: Univ. of Miami Press, Coral Gables, FL.

Shinn, E.A., 1996, No rocks, no water, no ecosystem: Geotimes, v. 41, no. 4, p. 16–19.

Park Address

Biscayne National Park
P.O. Box 1369
Homestead, FL 33090-1369

CAPE HATTERAS AND CAPE LOOKOUT NATIONAL SEASHORES (NORTH CAROLINA)

Cape Hatteras, established in 1937, is the oldest of our national seashores. Cape Lookout, which extends southwestward from Cape Hatteras, was established in 1966. Together these elongate barrier islands bracket about 125 miles (210 km) of North Carolina's coastal waters (Fig. 19–9)—part of a 2000-mile-long (3220-km) chain of islands that extends discontinuously from Cape Cod, Massachusetts, southward to Florida and westward along the Gulf Coast to the southern tip of Texas at Padre Island National Seashore (Plate 19).

The two North Carolina seashores form the Outer Banks, where the struggle between sand, sea, and wind produces "one of the most dynamic areas in the National Park System" (Dolan and others, 1973). The shifting, sandy shoals and the frequent severe storms and hurricanes combine to make this the "Graveyard of the Atlantic" where over 600 ships and hundreds of mariners have met their demise. The history of these shipwrecks, pirates, the lighthouses, and the daring deeds of the U.S. Lifesaving Service are all part of the total experience in the Outer Banks. A short distance north of Cape Hatteras are other elements of American history at the Wright Brothers National Memorial and nearby Fort Raleigh (1580), first but unsuccessful English settlement in the New World.

Origin of the Barrier Islands

The geologic story begins only 10,000—11,000 years ago when the huge continental ice sheets melted, returning vast quantities of water to the sea. As sea level rose, the waters encroached over the land, inundating extensive areas of the continental shelf. As at Padre Island in Texas, the sequence begins about 5000 years ago with a coastal dune ridge that became an elongate island when the shoreward area—now Pamlico Sound—was covered with seawater. Pamlico Sound now separates the barrier islands from the mainland by as much as 25 miles (40 km). In places, the islands are as much as 2.5 miles (4.0 km) wide, but there are long stretches less than a quarter of a mile in width. Some sections of the dune ridges are low, but others rise as high as 40 feet (12 m) above the sea.

Storms have broken through the ridges in places, and these inlets permit tidal currents to carry sediment out into the ocean, where it is deposited on the shoals that extend as much as 10 miles (16 km) out from the barrier islands. Hundreds of ships have foundered on the shoals, and some of their masts are visible, reminding today's mariner to beware.

Barrier Islands in Dynamic Balance

On a quiet day, when the breeze is light and the sea is calm, here is a truly tranquil scene. It is the time to see what a barrier island is made of. As shown in

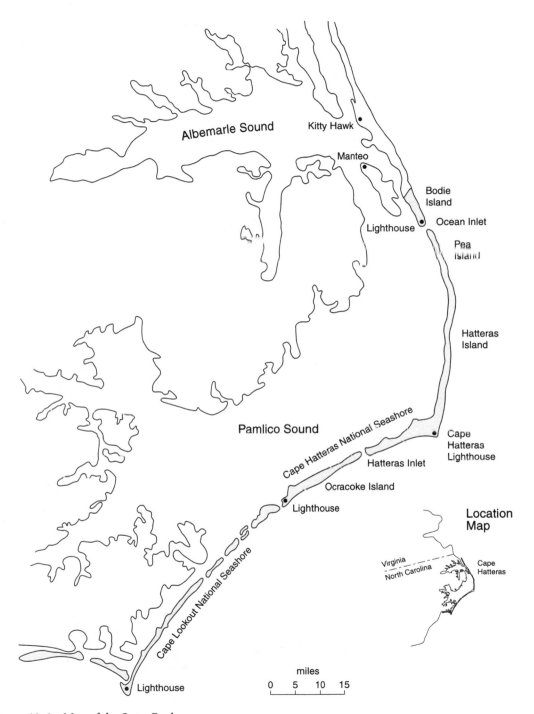

FIGURE 19–9 Map of the Outer Banks.

Figure 19–10*a,* back of the beach area is the dune ridge, partially stabilized by a scattering of hardy grasses. Considerably more vegetation grows on the overwash terrace that slopes westward toward the salt marshes along Pamlico Sound. On such a day it seems that nothing is changing. However, even on such days the relentless surf lazily moves sand along the beach.

But even one storm, particularly one of hurricane force, can profoundly alter the landscape. Gigantic waves override the ridge at low points and cut new inlets across the island; terrific winds move the sand about, depositing it on the lee side or in Pamlico Sound. The static situation cannot weather the storm.

On unmodified barrier islands such as Cape Lookout, the wave energy that is not utilized by washing up the wide beaches is spent washing over the dune ridge and down into the salt marsh behind. Fine sands brought into the marsh by overwash slowly build the area upward and toward the mainland. The net effect is that the barrier island will slowly migrate and thousands of years hence will become a mainland beach, unless more sand is brought to the beaches by longshore currents than is removed by overwash and other processes. This is not presently the case (Dolan and Lins, 1985).

Also contributing to barrier island migration is global warming that warms and expands seawater, reduces storage of water in the world's glaciers, and thus causes sea level to rise. However, as rising sea level speeds up barrier island migra-

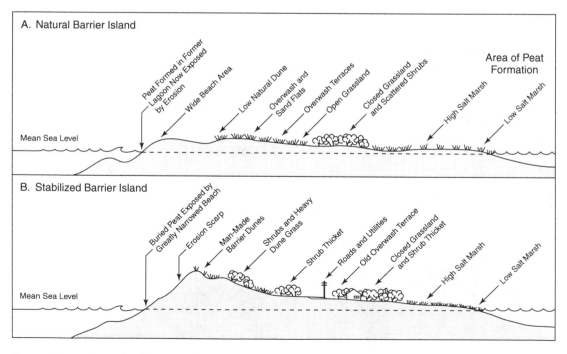

FIGURE 19–10   Generalized topographic profiles showing environments across (*a*) an undeveloped barrier island and (*b*) a developed island. (Modified slightly from Dolan and Lins, 1985)

tion, the mainland shore is also inundated, and it too migrates inland, albeit at a slower rate—especially in those areas where the shoreline topography is steep.

Measurement of the migration of the seaward edge of the Outer Banks indicates shoreline retreat of about 16 inches (40 cm) per year for the past 134 years (Fenster and Dolan, 1993). Island migration is a natural, expectable process that can only be temporarily slowed. When humans seek to control these natural processes, the costs in dollars, resources, and lives are high.

People as Geologic Agents

Early settlers used the Outer Banks mostly for grazing and built few permanent structures. Their occupations included whaling and the salvaging of the numerous shipwrecks. Development was heavier in the northern section, although permanent roads were impossible because of the occasional overwash during storms.

In the 1930s, the decision was made to establish a boundary along the northern islands across which the sea was not to pass (Dolan and others, 1973). Sand fences were constructed on the dunes along the northern islands from 1936 to 1940, creating what was then thought to be an inexpensive, artificial, protective dune system. Special grasses were planted, and by the late 1950s a much higher dune system and a continuous mass of vegetation covered the northern Outer Banks. A new but unjustifiable feeling of security led to construction of permanent roads, interisland bridges, and buildings on privately owned land behind the barrier dunes.

The artificial barrier dune no longer permitted the big waves to expend their excess energy by overwashing into the sound; instead, all the energy was concentrated between the dune and the ocean. The wide beaches that were one of the major attractions at Cape Hatteras began to disappear (Fig. 19–10*b*). In contrast to Cape Lookout, where no artificial dune system exists and where the beaches are over 410 feet (125 m) wide, many of Hatteras's beaches are now only 100 feet (30 m) or less in width. The big waves now crash into the barrier dune, which is slowly being destroyed. Dune breaches and overwashes are becoming more common each year (Dolan and others, 1980). The destruction of the structures behind the dune is inevitable, even if many hundreds of millions of dollars are spent on temporary stabilization measures.

A similar unstable and dangerous situation exists on most other barrier islands where development has occurred. How is it that someone could build a million-dollar "cottage" and afford to rebuild it after storms and shifting inlets destroy it? As pointed out by Ackerman (1997), it is well known that even the Bible reports that only a fool would build his house on sand. However, in Biblical days National Flood Insurance and federal disaster relief was not available so that taxpayers could subsidize private property by helping to rebuild wrecked "cottages"!

One area where wave erosion is particularly well documented, and an area whose historic values have been deemed important enough to preserve, is the candy-striped Cape Hatteras Lighthouse, the symbol of Cape Hatteras. The first lighthouse was built in 1802 but was razed after it was severely damaged by cannon fire during the Civil War. Although the original lighthouse stood about a mile from the water's edge, because of the westward migration of the barrier island, the 1802 foundation was washed into the sea by the big storm of 1980 (Fig. 19–11). This same storm washed around the steel groin protecting the present lighthouse and allowed the ocean to reach within 50 feet (15 m) of its foundation (Lisle and others, 1983). This lighthouse, built in 1870, was originally about 1500 feet (457 m) from the ocean! By 1988 Hatteras Lighthouse was 160 feet (49 m) from the sea, and by 1998 it was 120 feet (37 m). Current estimates are that sea level in the year 2018 will be 2.4 inches (6.0 cm) to 6.1 inches (15.5 cm) higher than 1988 levels and that shoreline retreat is expected to be 157–407 feet (48–124 m). Thus unless something is done in the very near future, this historic treasure is only a few storms away from destruction. Careful unbiased studies concluded that the only effective long-term (100 years or more) protection is to physically move this 2800-ton historical monument about 1600 feet (488 m) inland (Clarke, 1998).

The lessons are clear: what humans regard as a disruptive change in nature is often essential to the maintenance of a particular ecosystem, and interference with these natural processes can be very costly. Meanwhile, we can enjoy the delightful seascapes and moods of the ocean, knowing that each process and plant community has a place in the continually changing picture at Cape Hatteras.

FIGURE 19–11   Historic Cape Hatteras Lighthouse, at 208-feet high (63-m) it is the tallest on the continent. Rocks in the foreground are part of the foundation for the lighthouse built in 1802. The foundation was washed away since this photo was taken. (Photo by E.H. Wrenn)

REFERENCES

Ackerman, J., 1997, Islands at the edge: National Geographic Society, v. 192, no. 2, p. 2–31.

Clarke, W.M., 1998, Moving Cape Hatteras Lighthouse: National Parks, v. 72, no. 5–6, p. 20–23.

Dolan, R., Godfrey, P.J., and Odum, W.E., 1973, Man's impact on the barrier islands of North Carolina: American Scientist, v. 61, no. 2, p. 152–162.

Dolan, R., Hayden, B., and Lins, H., 1980, Barrier islands: American Scientist, v. 68, no. 1, p. 16–25.

Dolan, R., and Lins, H., 1985, The Outer Banks of North Carolina: U.S. Geological Survey Professional Paper 1177-B.

Fenster, M.S., and Dolan, R., 1993, Historical shoreline trends along the Outer Banks, North Carolina; processes and responses: Journal Coast. Res., v. 9, no. 1, p. 172–188.

Lisle, L.D., and Dolan, R., 1983, Coastal erosion at Cape Hatteras: A lighthouse in danger: National Park Service Research/Resource Management Report SER-65.

Park Address

Cape Hatteras National Seashore
Route 1, Box 675
Manteo, NC 27954

Index

Terms defined or explained in the text are shown in boldface page numbers; photographs and other illustrations are indicated by italics.